Computational Biology for Stem Cell Research

T0276178

Computational Biology for Stem Cell Research

Edited by

Pawan Kumar Raghav
BioExIn, Delhi, India

Rajesh Kumar
BioExIn, Delhi, India

Anjali Lathwal
Department of Computational Biology, Indraprastha Institute of Information Technology (IIIT), Delhi, India

Navneet Sharma
Amity University, Uttar Pradesh Sector-125, Noida, Uttar Pradesh, India

ACADEMIC PRESS
An imprint of Elsevier

Academic Press is an imprint of Elsevier
125 London Wall, London EC2Y 5AS, United Kingdom
525 B Street, Suite 1650, San Diego, CA 92101, United States
50 Hampshire Street, 5th Floor, Cambridge, MA 02139, United States

Copyright © 2024 Elsevier Inc. All rights are reserved, including those for text and data mining, AI training, and similar technologies.

No part of this publication may be reproduced or transmitted in any form or by any means, electronic or mechanical, including photocopying, recording, or any information storage and retrieval system, without permission in writing from the publisher. Details on how to seek permission, further information about the Publisher's permissions policies and our arrangements with organizations such as the Copyright Clearance Center and the Copyright Licensing Agency, can be found at our website: www.elsevier.com/permissions.

This book and the individual contributions contained in it are protected under copyright by the Publisher (other than as may be noted herein).

Notices

Knowledge and best practice in this field are constantly changing. As new research and experience broaden our understanding, changes in research methods, professional practices, or medical treatment may become necessary.

Practitioners and researchers must always rely on their own experience and knowledge in evaluating and using any information, methods, compounds, or experiments described herein. In using such information or methods they should be mindful of their own safety and the safety of others, including parties for whom they have a professional responsibility.

To the fullest extent of the law, neither the Publisher nor the authors, contributors, or editors, assume any liability for any injury and/or damage to persons or property as a matter of products liability, negligence or otherwise, or from any use or operation of any methods, products, instructions, or ideas contained in the material herein.

ISBN: 978-0-443-13222-3

For information on all Academic Press publications visit our website at
https://www.elsevier.com/books-and-journals

Publisher: Stacy Masucci
Acquisitions Editor: Elizabeth Brown
Editorial Project Manager: Kathy Padilla
Production Project Manager: Fahmida Sultana
Cover Designer: Mark Rogers

Typeset by TNQ Technologies

Working together
to grow libraries in
developing countries

www.elsevier.com • www.bookaid.org

Dedication

This book is dedicated to my cherished parents, my incredible wife Rajni, and my amazing son Ranbir for their unwavering support and affection, which brightens each day.

I further thank coeditors and authors, without whom this book would not have been possible. To the readers, for lending their time and curiosity, may this book offer you insights and enrichment.

Above all, to Almighty God, who always gives me strength, knowledge, and wisdom.

Dr. Pawan Kumar Raghav (PhD)
Editor-in-Chief,
Computational Biology for Stem Cell Research
Elsevier

Contents

Section I
In silico tools and approaches in stem cell biology

1. Advancement of in silico tools for stem cell research

*Ambuj Kumar, Keerthana Vinod Kumar,
Kunjulakshmi R., Kavita Kundal, Avik Sengupta
and Rahul Kumar*

2. Paradigm shift in stem cell research with computational tools, techniques, and databases

*Arnab Raha, Prateek Paul, Samriddhi Gupta,
Shruti Kaushal and Jaspreet Kaur Dhanjal*

3. Stem cell Informatics: Web resources aiding in stem cell research

*Rabiya Ahsan, Lubna Maryam and
Salman Sadullah Usmani*

Section IV
Computational approaches for stem cell tissue engineering

26. Tissue engineering in chondral defect

Madhan Jeyaraman,
Arulkumar Nallakumarasamy,
Naveen Jeyaraman and
Swaminathan Ramasubramanian

27. Recent advances in computational modeling: An appraisal of stem cell and tissue engineering research

Pinky, Neha and Suhel Parvez

32. Computational analysis in epithelial tissue regeneration

Priyanka Chhabra and Khushi Gandhi

Abbreviations

2-DE	Two-dimensional gel electrophoresis
2D-DIGE	Two-dimensional differential gel electrophoresis
2DLL	Low-glucose, low-density two-dimensional
3DP	Three-dimensional printing
ACI	Autologous chondrocyte implantation
ACPCs	Articular cartilage resident chondroprogenitor cells
ACR	Acute cell-mediated rejection
ADAP	Degranulation-promoting adapter protein
ADHD	Attention deficit hyperexcitability disorder
adMSC-Exos	Adipose-derived mesenchymal stem cells-sourced exosomes
ADSCs	Adipose tissue-derived stem cells
AFC	Alveolar fluid clearance
AGM	Aorta-gonad mesonephros
AI	Artificial intelligence
AIH	Autoimmune hepatitis
aiNSC	Artificially induced neural stem cells
aiPSCs	Artificially induced pluripotent stem cells
AIS	Adolescent idiopathic scoliosis
ALK	Anterior lamellar keratoplasty
ALL	Acute lymphoblastic leukemia
ALS	Amyotrophic lateral sclerosis
AM	Amniotic membrane
AMD	Age-related macular degeneration
AML	Acute myeloid leukemia
AMR	Antibody-mediated rejection
ANK	Ankyrin repeat domain
ANN	Artificial neural network
APC	Adenomatous polyposis coli
APCs	Antigen-presenting cells
ASCs	Adipose-derived stem cells
ASCs	Adult stem cells
ASCT	Autologous stem cell transplantation
ASD	Autism spectrum disorder
AT-MSCs	Adipose tissue–derived MSCs
ATAC-seq	Assay for transposase-accessible chromatin using sequencing
BBB	Blood–brain barrier
BBB	Basso, Beattie, Bresnahan
BCECs	Bovine corneal epithelial cells
BCSCdb	Biomarkers of cancer stem cell database
BCSCs	Breast cancer stem cells
BDNF	Brain-derived neurotrophic factor
bFGF	Basic fibroblast growth factor
BioGRID	Biological General Repository for Interaction Datasets
BLI	Bioluminescence imaging
BM-MSCs	Bone marrow mesenchymal stem cells
BMP	Bone morphogenetic protein
BMSCs	Bone marrow–derived stem cells

BN	Bayesian network
BP	Bipolar
CAD	Computer-aided design
CancerSCEM	Cancer Single-Cell Expression Map
CARs	Chimeric antigen receptors
CARS	Coherent anti-Stokes Raman scattering
CCD	Central composite design
CCL5	Chemotactic cytokine ligand
CFU-F	Colony-forming units-fibroblasts
CGH	Comparative genomic hybridization
CHARM	Comprehensive high-throughput arrays for relative methylation
ChIP	Chromatin immunoprecipitation
ChIP-seq	Chromatin immunoprecipitation followed by sequencing
CLOUD	Continuum of Low-primed UnDifferentiated
CM	Conditioned medium
CM-hUESCs	Conditioned media from human uterine cervical stem cells
CML	Chronic myeloid leukemia
CMs	Cardiomyocytes
CNNs	Convolutional neural networks
CNS	Central nervous system
CNVs	Copy number variations
CoIP-MS	Coimmunoprecipitation-mass spectrometry
COMP	Cartilage oligomeric matrix protein
CPM	Counts per million
CRISPR/Cas9	Clustered regularly interspaced short palindromic repeats
CSCB	Computational stem cell biology
CSCdb	Cancer stem cell database
CSCs	Cancer stem cells
CSCTT	Cancer stem cells therapeutic target database
CT	Computed tomography
CTA	Computed tomography arthrography
CTP	CT perfusion
DBNs	Dynamic Bayesian networks
DDBJ	DNA Data Bank Japan
DECT	Dual-energy CT
DEGs	Differentially expressed genes
DEMs	Differentially expressed miRNAs
DenseNet	Densely connected convolutional network
DEPs	Differentially expressed proteins
dGEMRIC	Delayed gadolinium-enhanced magnetic resonance imaging of cartilage
DNN	Deep neural network
DoE	Design of experiment
DRA	DDBJ Sequence Read Archive
DRS	Nanopore direct RNA sequencing
EB	Embryoid body
EBiSC	European Bank of induced Pluripotent Stem Cells
EBs	Embryoid bodies
EC	Embryonic carcinoma
ECACC	European Collection of Authenticated Cell Cultures
ECM	Extracellular matrix
EGF	Epidermal growth factor
EK	Endothelial keratoplasty
ELISA	Enzyme-linked immunosorbent assay
EMBL	European Molecular Biology Laboratory
EMBL-EBI	European Molecular Biology Laboratory—European Bioinformatics Institute
EMT	Epithelial—mesenchymal transition
ENA	European Nucleotide Archive
Epo	Erythropoietin

ER	Endoplasmic reticulum
ERCC	External RNA control consortium
ERK	Extracellular signal−regulated kinase
ES	Embryonic stem
ESCAPE	Embryonic Stem Cell Atlas from Pluripotency Evidence
ESCD	Embryonic stem cell database
ESCs	Embryonic stem cells
EVs	Extracellular vesicles
FAC	Fibronectin−aggrecan complex
FACS	Fluorescence-activated cell sorting
FAIR	Findability, accessibility, interoperability, and reuse
FC	Flow cytometry
FDA	Food and Drug Administration
FESP	18F-fluoroethylspiperone
FGF	Fibroblast growth factor
FISH	Fluorescence in situ hybridization
FISSEQ	Fluorescent in situ RNA sequencing
FPKM	Fragments per kilobase of exon per million mapped fragments
FVS	Feedback vertex sets
GAGs	Glycosaminoglycans
GAL-1	Galectin-1
GBM	Glioblastoma multiforme
GC	Gas chromatography
GCNs	Graph convolutional networks
GDNFs	Glial-derived neurotrophic factors
GEO	Gene Expression Omnibus
GEPs	Gene expression profiles
GFP	Green fluorescence protein
GPUs	Graphics processing units
GRN	Gene regulatory network
GSC	Glioma stem-like cell
GSCA	Gene set control analysis
GSEA	Gene set enrichment analysis
Gsk-3	Glycogen synthase kinase-3
GvHD	Graft vs. host disease
GWAS	Genome-wide association studies
HA	Hydroxyapatite
HCC	Hepatocellular carcinoma
HCL	Hierarchical clustering
HCT	Hematopoietic cell transplantation
HD	Huntington's disease
hECCs	Human embryonal carcinoma cells
hESC	Human embryonic stem cell
HFSCs	Hair follicle stem cells
HGF	Hepatocyte growth factor
HIFs	Hypoxia-inducible factors
hIPFP	Human infranatellar fat pad
hiPSC	Human-induced pluripotent stem cell
Hiniei	Human induced pluripotent stem cell initiation
HLA	Human leukocyte antigen
HMGB1	High mobility group protein B1
HPA	Human Protein Atlas
HPC	High-performance computing
HPLC	High-performance LC
hPSC	Human pluripotent stem cell
HSC	Hematopoietic stem cell
HSCI	Harvard Stem Cell Institute
HSCT	Hematopoietic stem cell transplantation

HSPC	Hematopoietic cells at the stem-progenitor stage
HTMs	High-throughput markers
hUCB-MSCs	Human umbilical cord blood–derived MSCs
ICAT	Isotope-coded affinity tag
ICM	Inner cell mass
ICMR	Indian Council of Medical Research
IFs	Instructive factors
IGF	Insulin-like growth factor
IGF-1	Insulin like growth factor 1
IHC	Immunohistochemistry
IHH	Indian hedgehog protein
IL	Interleukin
IL-1	Interleukin-1
IL-3	Interleukin-3
INSDC	International Nucleotide Sequence Database Collaboration
iPSCs	Induced pluripotent stem cells
IRENE	Integrative gene REgulatory NEtwork model
ISCBI	International Stem Cell Banking Initiative
ISCs	Intestinal stem cells
ISCT	International Society for Cell Therapy
ISSCR	International Society for Stem Cell Research
iTRAQ	Isobaric tag for relative and absolute quantitation
IVF	In vitro fertility
IVM	Intravital microscopy
JK	Janus kinase
k-NN	k-nearest neighbors
KD	Knockdown
KEGG	Kyoto Encyclopedia of Genes and Genomes
KFDA	Korean Food and Drug Administration
KGF	Keratinocyte growth factor
LAA	L-ascorbic acid
LC	Liquid chromatography
LCK	Lymphocyte-specific protein tyrosine kinase
LICs	Leukemia-initiating cells
LIF	Leukemia inhibitor factor
LOH	Loss of heterozygosity
LSCD	Limbal stem cell deficiency
LSCs	Leukemic stem cells
LTMs	Low-throughput markers
LVNC	Left ventricular noncompaction
M-FISH	Multiplex fluorescence in situ hybridization
MACI	Matrix-autologous chondrocyte implantation
MALDI-TOF	Matrix-assisted laser desorption/ionization–time of flight
MAPCs	Multipotent adult progenitor cells
MC	Monte Carlo
MD	Molecular dynamics
MECs	Microvessel endothelial cells
MEFs	Mouse embryo fibroblasts
MEP	Magnetoelectroporation
MeRIP-Seq	Methylated RNA immunoprecipitation sequencing
mESC	Mouse embryonic stem cell
MET	Mesenchymal to epithelial transition
MFAT	Microfragmented adipose tissue
MIACARM	Minimum Information About a Cellular Assay for Regenerative Medicine
MIB1	Mindbomb homolog 1
micro-CT	Microcomputed tomography
miRNA	Micro-RNA
ML	Machine learning

MM	Multiple myeloma
MMP	Matrix metalloproteinase
MNPs	Magnetic nanoparticles
MOP	Multiomics platform
MPPs	Multipotent progenitors
MRA	Magnetic resonance arthrography
MRI	Magnetic resonance imaging
MRS	Magnetic resonance spectroscopy
MS	Mass spectrometry
MSC-Exos	MSC-derived exosomes
MSCs	Mesenchymal stem cells
MSL	Material script language
MTFs	Master transcription factors
MVs	Microvesicles
NCBI	National Center for Biotechnology Information
NEI-RECS	National Eye Institute Refractive Error Correction Study
NGS	Next-generation sequencing
NIS	Sodium/iodide symporter
NK	Natural killer
NLP	Natural language processing
NMF	Negative matrix factorization
NMR	Nuclear magnetic resonance
NPCs	Neural progenitor cells
NRR	Negative regulatory region
NSCs	Neural stem cells
OA	Osteoarthritis
OATS	Osteochondral autograft transfer system
OCT	Optical coherence tomography
Oct4	Octamer-binding transcription factor 4
ODEs	Ordinary differential equations
OMIM	Online Mendelian Inheritance in Man database
ONT	Oxford nanopore technology
LSM	Laser-scanning microscope
PAN	Polyacrylonitrile
PB-MNCs	Peripheral blood mononuclear cells
PB-MSCs	Peripheral blood-derived MSCs
PBN	Probabilistic Boolean network
PCA	Principal component analysis
PCL	Polycaprolactone
PCM	Progressive cardiomyopathy
PD	Parkinson's disease
PDEs	Partial differential equations
PDGF	Platelet-derived growth factor
PECAM-1	Platelet endothelial cell adhesion molecule
PEDF	Pigment epithelium-derived Factor
PEEK	Polyether ether ketone
PEG	Polyethylene glycol
PEGDA	Poly ethylene glycol diacrylate
PET	Positron emission tomography
PF4	Platelet factor 4
PGA	Polyglycolic acid
PHEMA	Poly 2-hydroxyethyl methacrylate
PINA	Protein interaction network analysis
PK	Penetrating keratoplasty
PLA	Polylactic acid
PLC/CS	Polycaprolactone/chitosan
PM	plasma membrane
PP1	Protein phosphatase 1

PPIs	Protein–protein interactions
PRF	Platelet-rich fibrin
PRP	Platelet-rich plasma
PRS	Polygenic risk score
PSC	Pluripotent stem cell
PTA	Posttraumatic arthritis
PVA	Poly vinyl alcohol
qPCR	Quantitative polymerase chain reaction
QSAR	Quantitative structure–activity relationship
RAM	RBPJ-associated module
RBCs	Red blood cells
RDA	Representational difference analysis
RF	Random forest
RFS	Relapse-free survival
RGCs	Retinal ganglion cells
RNA-Seq	RNA sequencing
RNN	Recurrent neural network
ROIs	Reactive oxygen intermediates
ROS	Reactive oxygen species
RPKM	Reads per kilobase million
RT	Reverse transcription
S-GAGs	Sulfated glycosaminoglycans
SAGE	Serial analysis of gene expression
scBS-seq	Single-cell bisulfite sequencing
SCDE	Stem Cell Discovery Engine
SCENT	Single-cell entropy
SCF	Stem cell factor
SCID	Severe combined immunodeficiency disease
SCIs	Spinal cord injuries
SCNT	Somatic cell nuclear transfer
scRNA-seq	Single-cell RNA sequencing
SCs	Stem cells
SCT	Stem cell transplantation
SDF-1	Stromal cell–derived factor-1
SF	Synovial fluid
SFA	Signal flow analysis
SGZ	Subgranular zone
SHG/THG	Second- and third-harmonic generation
SILAC	Stable isotope labeling with amino acids in cell culture
SKA3	Spindle and kinetochore-associated complex subunit 3
SKIP	Stem Cell Knowledge and Information Portal
SKY	Spectral karyotyping
SLE	Systemic lupus erythematosus
SM	Synovial membrane
SMDS	Single-molecule DNA sequencing
SMRT	Single-molecule real-time
SNP	Single-nucleotide polymorphism
SNVs	Single-nucleotide variants
SOX2	Sex-determining region Y-box 2
SOXs	Sex-determining box
SPECT	Single-photon emission computed tomography
SRA	Sequence Read Archive
SRY	Sex-determining region Y
SSCs	Adult/somatic stem cells
STAT3	Signal transducer and activator of transcription 3
STFs	Signaling transcription factors
STR	Short tandem repeat
STRING	Search Tool for the Retrieval of Interacting Genes/Proteins

SVF	Stromal vascular fraction
SVM	Support vector machine
SVZ	Subventricular zone
SWATH	Sequential window acquisition of all theoretical mass spectra
T-ALL	T cell acute lymphoblastic leukemia
t-SNE	t-distributed stochastic neighbor embedding
TAD	Transcriptional activation domain
TCF3	Transcription factor 3
TDA	Topological data analysis
TE	Tissue engineering
TFs	Transcription factors
TGF	Transforming growth factor
TGF-β	Transforming growth factor beta
TGP	Thermoreversible gelation polymer
TGS	Third-generation sequencing
TGSTs	Third-generation sequencing technologies
TI	Trajectory inference
TNBC	Triple-negative breast cancer
TNF	Tumor necrosis factor
TNF-α	Tumor necrosis factor
TOF	Time-of-flight
TRAIL	TNF-α-related apoptosis-inducing ligand
TSB	Translational systems biology
TSCs	Transcriptional signaling centers
TSO	Template-switching oligo
TSP1	Thrombospondin 1
TSS	Transcription start sites
TWAS	Transcriptome-wide association studies
UGIC	Upper gastrointestinal cancer
UMAP	Uniform manifold approximation and projection
UMI	Unique molecular identifier
VEGF	Vascular endothelial growth factor
VPA	Valproic acid
WB	Western blot
WGCNA	Weighted gene co-expression network analysis
WGS	Whole-genome sequencing
WHA	Wound healing assay
WHO	World Health Organization
wwPDB	Worldwide Protein Data Bank
ZFP217	Zinc finger protein 217

List of figures

List of tables

Contributors

Achala Anand, Department of Biotechnology, Faculty of Life and Allied Health Sciences, M. S. Ramaiah University of Applied Sciences, Bangalore, Karnataka, India

Aditya Arya, Vector Biology Laboratory, National Institute of Malaria Research, New Delhi, Delhi, India

Aditya Raghav, BioExIn, Delhi, India

Agnieszka Maria Jastrzębska, Warsaw University of Technology, Faculty of Materials Science and Engineering, Warsaw, Poland

Ahalya N., Department of Biotechnology, Ramaiah Institute of Technology, Bangalore, Karnataka, India

Aiindrila Dhara, Molecular Oncology Laboratory, Department of Biological Sciences, Bose Institute, Kolkata, West Bengal, India

Akib Mohi Ud Din Khanday, Department of Computer Science and Software Engineering-CIT, United Arab Emirates University, Al Ain, United Arab Emirates

Ambuj Kumar, Department of Biotechnology, Indian Institute of Technology Hyderabad, Kandi, Telangana, India

Amisha Singh, Department of Biotechnology, JAIN University, Bangalore, Karnataka, India

Amrita Kumari Panda, Department of Biotechnology, Sant Gahira Guru University, Ambikapur, Chhattisgarh, India

Anamta Ali, Toxicoinformatics and Industrial Research, CSIR-Indian Institute of Toxicology Research, Vishvigyan Bhawan, Lucknow, Uttar Pradesh, India

Anil Vishnu G.K., Centre for BioSystems Science and Engineering, Indian Institute of Science, Bangalore, Karnataka, India

Anjali Lathwal, Department of Computational Biology, Indraprastha Institute of Information Technology (IIIT), Delhi, India

Ankit Choudhary, Department of Pharmacology, All India Institute of Medical Sciences (AIIMS), New Delhi, Delhi, India

Annapoorni Rangarajan, Department of Developmental Biology and Genetics, Indian Institute of Science, Bangalore, Karnataka, India

Anshuman Chandra, School of Physical Sciences, Jawaharlal Nehru University, New Delhi, Delhi, India

Anugya Sengar, Army Hospital Research and Referral (R&R), Delhi, India

Arnab Raha, Department of Computational Biology, Indraprastha Institute of Information Technology Delhi, New Delhi, Delhi, India

Arulkumar Nallakumarasamy, Department of Orthopaedics, ACS Medical College and Hospital, Dr. MGR Educational and Research Institute, Chennai, Tamil Nadu, India

Asif Adil, Department of Computer Sciences, Baba Ghulam Shah Badshah University, Rajouri, Jammu and Kashmir, India

Avik Choudhuri, Stem Cell Program and Division of Hematology/Oncology, Boston Children's Hospital and Harvard Medical School, Boston, MA, United States; Harvard Department of Stem Cell and Regenerative Biology, Harvard University, Cambridge, MA, United States

Avik Sengupta, Department of Biotechnology, Indian Institute of Technology Hyderabad, Kandi, Telangana, India

Ayushi Gupta, Department of Bioinformatics & Applied Sciences, Indian Institute of Information Technology-Allahabad, Allahabad, Uttar Pradesh, India

B. Suresh Pakala, Department of Biochemistry, School of Life Sciences, University of Hyderabad, Hyderabad, Telangana, India

B.S. Dwarakanath, Indian Academy Degree College-Autonomous, Bangalore, Karnataka, India

Basudha Banerjee, BioExIn, Delhi, India

Bhawna Sharma, Department of Biosciences, Jamia Millia Islamia, New Delhi, India

D. Thirumal Kumar, Faculty of Allied Health Sciences, Meenakshi Academy of Higher Education and Research, Chennai, Tamil Nadu, India

Daizy Kalpdev, Department of Pharmacology, All India Institute of Medical Sciences (AIIMS), New Delhi, Delhi, India

Daniel C. Hoessli, Dr. Panjwani Center for Molecular Medicine and Drug Research, International Center for Chemical and Biological Sciences, University of Karachi, Karachi, Sindh, Pakistan

Debabrat Baishya, Department of Bioengineering and Technology, Gauhati University, Guwahati, Assam, India

Dibyabhaba Pradhan, Indian Biological Data Centre, Regional Centre for Biotechnology, Faridabad, Haryana, India; Department of Bioinformatics, Centralized Core Research Facility, All India Institute of Medical Sciences, New Delhi, Delhi, India

Emmet Flynn, Stem Cell Program and Division of Hematology/Oncology, Boston Children's Hospital and Harvard Medical School, Boston, MA, United States; Pediatric Hematology/Oncology Program of Boston Children's Hospital and Dana Farber Cancer Institute, Harvard Medical School, Boston, MA, United States

Gerardo Caruso, Department of Biomedical and Dental Sciences and Morphofunctional Imaging, Unit of Neurosurgery, University of Messina, Messina, Italy

Himanshu Sehrawat, Department of Biosciences, Jamia Millia Islamia, New Delhi, India

Jaiganesh Inbanathan, Department of Pediatric and Preventive Dentistry, Meenakshi Ammal Dental College and Hospital, Meenakshi Academy of Higher Education and Research, Chennai, Tamil Nadu, India

Jaspreet Kaur Dhanjal, Department of Computational Biology, Indraprastha Institute of Information Technology Delhi, New Delhi, Delhi, India

Jayasha Shandilya, Amity Institute of Molecular Medicine and Stem Cell Research, Amity University, Noida, Uttar Pradesh, India

Johannes Pernaa, The Unit of Chemistry Teacher Education, Department of Chemistry, Faculty of Science, University of Helsinki, Helsinki, Finland

Joyeeta Talukdar, Department of Biochemistry, All India Institute of Medical Sciences (AIIMS), New Delhi, Delhi, India

K. Ponnazhagan, Department of Biochemistry, Meenakshi Medical College Hospital & Research Institute, Meenakshi Academy of Higher Education and Research, Kanchipuram, Tamil Nadu, India

Kaushala Prasad Mishra, Ex Radiation Biology and Health Sciences Division, Bhabha Atomic Research Center, Mumbai, Maharashtra, India

Kaushik Kumar Bharadwaj, Department of Bioengineering and Technology, Gauhati University, Guwahati, Assam, India

Kavita Kundal, Department of Biotechnology, Indian Institute of Technology Hyderabad, Kandi, Telangana, India

Keerthana Vinod Kumar, Department of Biotechnology, Indian Institute of Technology Hyderabad, Kandi, Telangana, India

Khalid Raza, Department of Computer Science, Jamia Millia Islamia, New Delhi, Delhi, India

Khushboo Bhutani, Department of Biomedical Engineering, SRM University Delhi-NCR, Sonepat, Haryana, India

Khushi Gandhi, Amity Institute of Biotechnology, Amity University, Noida, Uttar Pradesh, India

Kunjulakshmi R., Department of Biological Sciences, Indian Institute of Science Education and Research, Berhampur, Odisha, India

Leonard I. Zon, Stem Cell Program and Division of Hematology/Oncology, Boston Children's Hospital and Harvard Medical School, Boston, MA, United States; Pediatric Hematology/Oncology Program of Boston Children's Hospital and Dana Farber Cancer Institute, Harvard Medical School, Boston, MA, United States; Harvard Department of Stem Cell and Regenerative Biology, Harvard University, Cambridge, MA, United States; Howard Hughes Medical Institute, Boston, MA, United States

Lubna Maryam, Independent Researcher, Bronx, NY, United States

M.R. Sanjana, Department of Oral Medicine & Radiology, Meenakshi Ammal Dental College and Hospital, Meenakshi Academy of Higher Education and Research, Chennai, Tamil Nadu, India

Madhan Jeyaraman, Department of Orthopaedics, ACS Medical College and Hospital, Dr. MGR Educational and Research Institute, Chennai, Tamil Nadu, India

Magali Cucchiarin, Center of Experimental Orthopaedics, Saarland University Medical Center and Saarland University, Homburg, Germany

Mahesh Mahadeo Mathe, Kusuma School of Biological Sciences, Indian Institute of Technology Delhi, New Delhi, Delhi, India

Manisha Sengar, Department of Zoology, Deshbandhu College, University of Delhi, Delhi, India

Manoj Kumar Yadav, Department of Biomedical Engineering, SRM University Delhi-NCR, Sonepat, Haryana, India

Manvee Chauhan, Amity Institute of Molecular Medicine and Stem Cell Research, Amity University, Noida, Uttar Pradesh, India

Meera Prasad, Stem Cell Program and Division of Hematology/Oncology, Boston Children's Hospital and Harvard Medical School, Boston, MA, United States; Pediatric Hematology/Oncology Program of Boston Children's Hospital and Dana Farber Cancer Institute, Harvard Medical School, Boston, MA, United States

Mitrabasu Chhillar, Naval Dockyard, Mumbai, Maharashtra, India

Mohammed Asger, Department of Computer Sciences, Baba Ghulam Shah Badshah University, Rajouri, Jammu and Kashmir, India

Musharaf Gul, Departmetn of Zoology, Hemvati Nandan Bahuguna Gharwal University (A Central University), Srinagar, Uttarakhand, India

N.S. Amanda Thilakarathna, Department of Biotechnology, Faculty of Life and Allied Health Sciences, M. S. Ramaiah University of Applied Sciences, Bangalore, Karnataka, India

Nagendra Singh, School of Biotechnology, Gautam Buddha University, Greater Noida, Uttar Pradesh, India

Nainee Goyal, School of Biotechnology, Gautam Buddha University, Greater Noida, Uttar Pradesh, India

Naveen Jeyaraman, Department of Orthopaedics, ACS Medical College and Hospital, Dr. MGR Educational and Research Institute, Chennai, Tamil Nadu, India

Neelam Chhillar, Department of Biochemistry, School of Medicine, DY Patil University, Navi Mumbai, Maharashtra, India

Neha Deshpande, Department of Developmental Biology and Genetics, Indian Institute of Science, Bangalore, Karnataka, India

Neha, Department of Medical Elementology and Toxicology, School of Chemical and Life Sciences, Jamia Hamdard, New Delhi, India

Netra Pal Sharma, Department of Zoology, Kumaun University, Nainital, Uttarakhand, India

Nirbhay Raghav, Department of Interdisciplinary Mathematical Sciences, Indian Institute of Science, Bangalore, Karnataka, India

Nirmalya Sen, Molecular Oncology Laboratory, Department of Biological Sciences, Bose Institute, Kolkata, West Bengal, India

Nishant Tyagi, Stem Cell and Tissue Engineering Research Group, INMAS, DRDO, Delhi, India

Pawan Kumar Raghav, BioExIn, Delhi, India

Pinky, Department of Medical Elementology and Toxicology, School of Chemical and Life Sciences, Jamia Hamdard, New Delhi, India

Prashanthi Karyala, Department of Biotechnology, Faculty of Life and Allied Health Sciences, M. S. Ramaiah University of Applied Sciences, Bangalore, Karnataka, India

Prateek Paul, Department of Computational Biology, Indraprastha Institute of Information Technology Delhi, New Delhi, Delhi, India

Princy Choudhary, Department of Bioinformatics & Applied Sciences, Indian Institute of Information Technology-Allahabad, Allahabad, Uttar Pradesh, India

Pritish Kumar Varadwaj, Department of Bioinformatics & Applied Sciences, Indian Institute of Information Technology-Allahabad, Allahabad, Uttar Pradesh, India

Priyanka Chhabra, Center for Medical Biotechnology, Amity Institute of Biotechnology, Amity University, Noida, Uttar Pradesh, India

Priyanka Narad, Amity Institute of Biotechnology, Amity University, Noida, Uttar Pradesh, India

Protyusha Guha Biswas, Department of Oral Pathology and Microbiology, Meenakshi Ammal Dental College and Hospital, Meenakshi Academy of Higher Education and Research, Chennai, Tamil Nadu, India

R. Anitha, Department of Oral Medicine & Radiology, Meenakshi Ammal Dental College and Hospital, Meenakshi Academy of Higher Education and Research, Chennai, Tamil Nadu, India

Rabiya Ahsan, Department of Pharmacology, Faculty of Pharmacy, Integral University, Lucknow, Uttar Pradesh, India

Rahul Kumar, Department of Biotechnology, Indian Institute of Technology Hyderabad, Kandi, Telangana, India

Rajesh Kumar, BioExIn, Delhi, India

Rajiv Kumar, University of Delhi, New Delhi, Delhi, India

Rajni Chadha, BioExIn, Delhi, India

Ramakrishnan Parthasarathi, Toxicoinformatics and Industrial Research, CSIR-Indian Institute of Toxicology Research, Vishvigyan Bhawan, Lucknow, Uttar Pradesh, India; Academy of Scientific and Innovative Research (AcSIR), Ghaziabad, Uttar Pradesh, India

Ratnesh Singh Kanwar, Division of Clinical Research and Medical Management, Institute of Nuclear Medicine and Allied Sciences (INMAS), DRDO, Delhi, India

Rayees Ahmad Magray, Department of Environmental Studies, Government Degree College, Kupwara, Jammu and Kashmir, India

Reena Wilfred, Division of Clinical Research and Medical Management, Institute of Nuclear Medicine and Allied Sciences (INMAS), DRDO, Delhi, India

Saifullah Afridi, Department of Chemistry, Kohat University of Science and Technology, Kohat, Khyber Pakhtunkhwa, Pakistan; Department of Allied Health Sciences, Faculty of Life Sciences, Sarhad University of Science & Information Technology (SUIT), Mardan Campus, Mardan, Khyber Pakhtunkhwa, Pakistan

Salman Sadullah Usmani, Department of Molecular Pharmacology, Albert Einstein College of Medicine, Bronx, NY, United States

Samriddhi Gupta, Department of Computational Biology, Indraprastha Institute of Information Technology Delhi, New Delhi, Delhi, India

Sangeeta Singh, Department of Bioinformatics & Applied Sciences, Indian Institute of Information Technology-Allahabad, Allahabad, Uttar Pradesh, India

Sanghamitra Pati, ICMR-Regional Medical Research Centre, Bhubaneswar, Odisha, India

Sangramjit Mondal, Molecular Oncology Laboratory, Department of Biological Sciences, Bose Institute, Kolkata, West Bengal, India

Santosh Kumar Behera, Department of Biotechnology, NIPER, Ahmedabad, Gujrat, India

Satpal Singh Bisht, Department of Zoology, Kumaun University, Nainital, Uttarakhand, India

Seeta Dewali, Department of Zoology, Kumaun University, Nainital, Uttarakhand, India

Shaban Ahmad, Department of Computer Science, Jamia Millia Islamia, New Delhi, Delhi, India

Shilpi Agarwal, School of Physical Sciences, Jawaharlal Nehru University, New Delhi, Delhi, India

Shivalika Pathania, Department of Biotechnology, Panjab University, Chandigarh, Punjab, India

Shivi Uppal, Division of Clinical Research and Medical Management, Institute of Nuclear Medicine and Allied Sciences (INMAS), DRDO, Delhi, India

Shraddha Pandit, Toxicoinformatics and Industrial Research, CSIR-Indian Institute of Toxicology Research, Vishvigyan Bhawan, Lucknow, Uttar Pradesh, India

Shruti Kaushal, Department of Computational Biology, Indraprastha Institute of Information Technology Delhi, New Delhi, Delhi, India

Simran Tandon, Amity School of Health Sciences, Amity University, Mohali, Punjab, India

Sonali Rawat, Stem Cell Facility, DBT-Centre of Excellence for Stem Cell Research, All India Institute of Medical Sciences, New Delhi, Delhi, India

Song Yang, Stem Cell Program and Division of Hematology/Oncology, Boston Children's Hospital and Harvard Medical School, Boston, MA, United States; Pediatric Hematology/Oncology Program of Boston Children's Hospital and Dana Farber Cancer Institute, Harvard Medical School, Boston, MA, United States; Harvard Department of Stem Cell and Regenerative Biology, Harvard University, Cambridge, MA, United States

Sorra Sandhya, Department of Bioengineering and Technology, Gauhati University, Guwahati, Assam, India

Subodh Kumar, Stem Cell and Tissue Engineering Research Group, INMAS, DRDO, Delhi, India

Suhel Parvez, Department of Medical Elementology and Toxicology, School of Chemical and Life Sciences, Jamia Hamdard, New Delhi, India

Sujata Mohanty, Stem Cell Facility, DBT-Centre of Excellence for Stem Cell Research, All India Institute of Medical Sciences, New Delhi, Delhi, India

Sunil Kumar, ICAR-Indian Agricultural Statistical Research Institute, New Delhi, Delhi, India

Sushanth Adusumilli, Kusuma School of Biological Sciences, Indian Institute of Technology Delhi, New Delhi, Delhi, India

Swaminathan Ramasubramanian, Department of Orthopaedics, Government Medical College, Omandurar Government Estate, Chennai, Tamil Nadu, India

Swati Sharma, Department of Pharmacology, All India Institute of Medical Sciences (AIIMS), New Delhi, Delhi, India

Sweta Singh, Stem Cell and Tissue Engineering Research Group, INMAS, DRDO, Delhi, India

Tabassum Zahra, Department of Chemistry, Kohat University of Science and Technology, Kohat, Khyber Pakhtunkhwa, Pakistan

Tanya Jamal, Toxicoinformatics and Industrial Research, CSIR-Indian Institute of Toxicology Research, Vishvigyan Bhawan, Lucknow, Uttar Pradesh, India

Tapan Kumar Nayak, Kusuma School of Biological Sciences, Indian Institute of Technology Delhi, New Delhi, Delhi, India

Umar Nishan, Department of Chemistry, Kohat University of Science and Technology, Kohat, Khyber Pakhtunkhwa, Pakistan

Usha Agrawal, ICMR-National Institute of Pathology, Safdarjung Hospital Campus, New Delhi, Delhi, India

V.S. Vipin, Division of Clinical Research and Medical Management, Institute of Nuclear Medicine and Allied Sciences (INMAS), DRDO, Delhi, India

Vandana Gupta, Department of Microbiology, Ram Lal Anand College, University of Delhi, New Delhi, India

Vijay Kumar Goel, School of Physical Sciences, Jawaharlal Nehru University, New Delhi, Delhi, India

Vinay Bhatt, Amity Institute of Biotechnology, Amity University, Noida, Uttar Pradesh, India

Vinay Randhawa, Division of Cardiovascular Medicine, Brigham and Women's Hospital, Harvard Medical School, Boston, MA, United States

Vivek Kumar, R&D Dept, Shanghai Proton and Heavy Ion Centre (SPHIC), Shanghai Key Laboratory of Radiation Oncology and Shanghai Engineering Research Centre of Proton and Heavy Ion Radiation Therapy, Shanghai, China; Department of Radiation Oncology, Albert Einstein College of Medicine, Bronx, New York, United States

Yashvi Sharma, Stem Cell Facility, DBT-Centre of Excellence for Stem Cell Research, All India Institute of Medical Sciences, New Delhi, Delhi, India

Yi Zhou, Stem Cell Program and Division of Hematology/Oncology, Boston Children's Hospital and Harvard Medical School, Boston, MA, United States; Pediatric Hematology/Oncology Program of Boston Children's Hospital and Dana Farber Cancer Institute, Harvard Medical School, Boston, MA, United States; Harvard Department of Stem Cell and Regenerative Biology, Harvard University, Cambridge, MA, United States

Yogesh Kumar Verma, Stem Cell and Tissue Engineering Research Group, INMAS, DRDO, Delhi, India

Zarrin Minuchehr, National Institute of Genetic Engineering and Biotechnology (NIGEB), Tehran, Iran

Biographies

Editor-in-chief: Dr. Pawan Kumar Raghav

Email: pwnrghv@gmail.com

Dr. Pawan Kumar Raghav, PhD, PGDCI, MPhil, MSc, BSc, is a distinguished Scientist with over a decade of experience who has significantly contributed to research in stem cells and computational biology. Dr. Raghav received his PhD in Life Sciences from Bharathiar University, Coimbatore, India. During his PhD at the Institute of Nuclear Medicine and Allied Sciences (INMAS), Defense Research Development Organization (DRDO), Delhi, India, he delved into designing and evaluating molecules with applications in stem cell regulation, focusing on proliferation, differentiation, and apoptosis. As a research associate at INMAS, DRDO, he contributed to hematopoietic stem cell research, formulated peptides-PLGA nanoparticles, and validated their efficacy in *in vitro* and *in vivo* studies. Dr. Pawan Kumar Raghav is the recipient of a National Postdoctoral Fellowship from the Department of Science and Technology (DST), SERB, Government of India. As a principal investigator of the project at the Department of Computational Biology, IIIT Delhi, India, he developed tools using machine learning algorithms and databases. As a Scientist 'D' in the Stem Cell Facility, AIIMS, Delhi, India, Dr. Pawan Kumar Raghav carried out stem cell informatics and tissue engineering experimental studies to generate clinical-grade mesenchymal stem cells (MSCs). Dr. Raghav's career has centered on experimental and computational biology, with research spanning stem cells, immunology, cheminformatics, structural biology, tissue engineering, and molecular biology. Dr. Raghav's contributions are evident through numerous publications in prestigious scientific journals, book chapters, books, and patents that provide insight into stem cells and immunology. He also serves as an editor and reviewer for various international journals and books. Dr. Pawan Kumar Raghav earned several international recognitions including travel, research excellence, and best research awards. He is a member of the American Society for Histocompatibility and Immunogenetics (ASHI) and the International Society for Computational Biology (ISCB).

Editor: Dr. Rajesh Kumar

Email: b.rajesh1130@gmail.com

Dr. Rajesh Kumar is a Clinical Bioinformatician with 7+ years of experience intrinsically rooted in the scientific exploration of immunotherapy, multiomics data analytics, genomic structural aberrations in tumors, peptide therapeutics, epitope-focused vaccine design, and the discerning application of machine learning for precision medicine. During his doctoral tenure, Dr. Kumar made significant contributions, unraveling the intricacies of fragile sites in the human genome, elucidating the role of enhancer elements in gene expression modulation, and forging prognostic biomarkers for pancreatic adenocarcinoma. Dr. Kumar has a vast experience in analyzing and interpreting the large scale clinical trial multiomics data and designing and developing custom bioinformatics algorithms/data analysis pipelines. Dr. Kumar's profound intellectual endeavors have manifested in the form of over 35 peer-reviewed publications, coauthored books, and his standing as a distinguished member of the Asia-Pacific Bioinformatics Network.

Editor: Dr. Anjali Lathwal

Email: anjali06lathwal@gmail.com

Dr. Anjali Lathwal (B.Tech, M.Tech, Ph.D.) is a computational biologist and computer scientist, with strong research acumen cultivated through diverse experiences in academia and industry. Her academic journey was ignited by a deep-seated passion for biological science during her Master's thesis project. She holds a Ph.D. in Computational Biology from IIIT-Delhi, where she led groundbreaking research in biomarker discovery, cancer immunotherapy, and computational modeling. Her expertise spans Applied Machine Learning, Mathematical Modeling, Computer-aided Vaccine Design, Biomarkers Discovery, and Bio-resource Development. During her postdoctoral work at NIHNational Institute on Aging, USA, Dr. Lathwal pioneered drug repurposing for Alzheimer's and age-related neurological disorders. She employed mechanism-based systems biology, network science principles, and extensive multi-omics data analysis, all underpinning her exceptional contributions. Her remarkable career encompasses 10+ peer-reviewed publications, collaborative book chapters, and influential consultancy projects. Dr. Lathwal is a member of the Asia-Pacific Bioinformatics Network society and a recipient of the esteemed Research Excellence Award from the Institute of Scholars (InSc). Her research embodies excellence in viral therapeutics, biomarker discovery, and computational biology, reflecting her profound dedication to scientific advancement.

Editor: Dr. Navneet Sharma

Email: navneetrssharma@gmail.com

Navneet Sharma's (MPharm, PhD, PGDRA) work is inclined more toward applied R&D, especially needs-based product development. As an expert in biomaterials, he had developed dermal decontamination formulation for the radiological decontamination of the skin. He had devised and applied pharmacoscintigraphic procedures for its preclinical evaluation of the drug delivery systems in various experimental models of disease. Further, he had developed the dermal formulations for the broad spectrum decontamination and healing of radiation-induced wounds. For the past 5 years, his work has been exclusively focused on the medical management of chemical, biological, radiological, and nuclear emergencies. Currently, he is working in the field of material sciences as a Scientist at IIT Delhi. In his new research career of 5 years, he had acclaimed eight national and international awards. The most prominent among them are the SCO and Ministry of External Affairs, Government of India, COVID-19 Best Innovation Award 2020, and the Department of Science and Technology, Young Scientist Award for the years 2018 and 2022. Currently, he is an investigator in the three projects from DST-India, filed and granted 10 patents, 4 technologies successfully transferred to the industry. He had 45 publications, including 5 book chapters, authored and edited 4 books.

Preface

Stem cells have taken the center stage as invaluable tools in unraveling the intricate mechanisms governing development, tissue regeneration, and the complexities of diseases. Nevertheless, the multifaceted landscape of stem cell biology, coupled with the staggering volumes of data generated through diverse experimental techniques, presents formidable challenges for conventional research methodologies. In the recent years, computational biology has emerged as an indispensable instrument for deciphering, analyzing, and harnessing the potential of these remarkable cells.

This book, *Computational Biology for Stem Cell Research*, stands as a comprehensive guide, offering an in-depth exploration of the computational methodologies and approaches that have been instrumental in unraveling the mysteries of stem cell biology. This book comprises 32 chapters, divided into four distinct sections, each covering a wide range of computational tools and techniques tailored to meet stem cell research's unique requirements. This book covers many topics, ranging from stem cell differentiation and epigenetic control to gene expression analysis, network analysis, and modeling strategies. This book contains numerous examples and case studies to illustrate the practical application of computational approaches in stem cell research. Every topic of the computational stem cell biology domain is comprehensively addressed, encompassing crucial areas such as bioinformatics tools, genomics, proteomics, single-cell and bulk RNA-seq, next-generation sequencing (NGS), and machine learning.

The authors have adopted a pragmatic approach, offering clear and concrete explanations of complex concepts. They demonstrate the active use of computational methodologies in stem cell research through numerous case studies. This book is designed for diverse readers, including researchers and students interested in stem cells and computational biology. Moreover, it is of great value to bioinformaticians, computer scientists, data scientists, and systems biologists engaged in stem cell research efforts.

This book's organization offers a comprehensive survey of the computational tools and techniques instrumental in advancing our understanding of stem cell biology. Its accessible style ensures it is equally suitable for newcomers and experienced researchers in the field. In *Computational Biology for Stem Cell Research*, an indexed section features keywords and abbreviations, serving as a valuable resource for individuals who may need to be better acquainted with the content and scientific terminology. It represents the combined effort of editors and more than 100 scholars and scientists whose pioneering work has defined our understanding of computational stem cells. We hope the new knowledge and research outlined in this book will help contribute to new therapies for stem cell transplantation and various other diseases. We aspire that this book will inspire and empower researchers to seamlessly integrate computational methods into their stem cell investigations, leading to groundbreaking insights and discoveries in this dynamic and exhilarating domain.

Dr. Pawan Kumar Raghav (PhD)
Editor-in-Chief,
Computational Biology for Stem Cell Research
Elsevier

Rajesh Kumar (PhD)
Editor,
Computational Biology for Stem Cell Research
Elsevier

Anjali Lathwal (PhD)
Editor,
Computational Biology for Stem Cell Research
Elsevier

Navneet Sharma (PhD)
Editor,
Computational Biology for Stem Cell Research
Elsevier

Acknowledgments

We would like to express our deep appreciation to all the individuals and organizations who contributed to the creation of this book, *Computational Biology for Stem Cell Research*. Their support, expertise, and dedication were invaluable in bringing this project to fruition.

We extend our gratitude to the esteemed authors of this book, whose insightful contributions have made it a comprehensive resource in stem cell research. Their knowledge and commitment to advancing *Computational Biology for Stem Cell Research* are evident throughout the chapters.

We would like to thank our anonymous reviewers, whose meticulous feedback and constructive suggestions enhanced the quality and relevance of the content. Your expertise and dedication to ensuring the accuracy and clarity of the book are greatly appreciated.

Our heartfelt thanks go to the Elsevier publishing team, Elizabeth Brown (Elsevier, Senior Acquisitions Editor), Tim Eslava (Editorial Project Manager), Kathy Padilla (Content Team Manager), Fahmida Sultana (Project Manager), and production team who worked tirelessly to turn this vision into a reality. Your professionalism and commitment to excellence have been vital to the success of this project.

We also acknowledge the support of our family, colleagues, mentors, and institutions, whose encouragement and resources have been instrumental in our journey to create this book.

Lastly, we extend our gratitude to the readers and researchers who will find value in this book. Your curiosity and pursuit of knowledge drive progress in the field of stem cell research, and we hope this book contributes to your endeavors. Thank you all for your unwavering support and dedication.

Dr. Pawan Kumar Raghav (PhD)
Editor-in-Chief,
Computational Biology for Stem Cell Research
Elsevier

Rajesh Kumar (PhD)
Editor,
Computational Biology for Stem Cell Research
Elsevier

Anjali Lathwal (PhD)
Editor,
Computational Biology for Stem Cell Research
Elsevier

Navneet Sharma (PhD)
Editor,
Computational Biology for Stem Cell Research
Elsevier

Introduction

Stem cell research has been one of the most exciting and rapidly growing areas of biomedical research over the past few decades. Stem cells have the potential to revolutionize medicine by providing new treatments for diseases and injuries that were previously untreatable. They can be used to replace or repair damaged tissues and organs and to develop new drugs and therapics. However, the complexity of stem cell biology and the vast amount of data generated by various techniques pose significant challenges for traditional experimental methods. In recent years, computational biology has emerged as a powerful tool to analyze and integrate large datasets and to develop models that can predict the behavior of stem cells.

This book, *Computational Biology for Stem Cell Research*, comprehensively explores how computational tools and approaches are reshaping our understanding and application of stem cell research. Each chapter in this book is written by experts who have used a practical approach to explain the concepts, providing step-by-step instructions and code examples to understand the methods. This book is divided into four sections, each covering a specific aspect of the topic.

Section I, *In silico Tools and Approaches in Stem Cell Biology*, commences the exploration by delving into the advancements in computational tools for stem cell research. From comparative studies of bioinformatics approaches to web resources aiding stem cell research, this section provides a strong foundation in the computational landscape of stem cell biology. Integrating artificial intelligence and machine learning is also explored, underscoring their potential to revolutionize the field.

Section II, *Application of Genomic and Proteomic Approaches in Stem Cell Research*, includes chapters from authors focusing on genomics and proteomics in stem cell studies. Single-cell transcriptome profiling, systematic genomics and proteomics, and advanced omics approaches are discussed in detail. These chapters unravel the molecular intricacies of stemness and provide insights into therapeutic applications.

Section III, *Stem Cell Network Modeling and Systems Biology*, explores the power of network modeling and systems biology in understanding stem cell fate and regulatory networks. From integrating multiomics data to deciphering the complexities of stem cells through network biology, this section showcases how computational approaches illuminate the inner workings of these remarkable cells. The chapters include case studies, discuss the challenges and limitations of computational modeling in stem cell research, and provide insights into future directions in the field.

Section IV, *Computational Approaches for Stem Cell Tissue Engineering*, ventures into the exciting intersection of stem cells and tissue engineering. Computational tissue engineering for various tissues, from chondral defects to corneas, is explored. Additionally, the chapters in this section discuss the transformative potential of combining cheminformatics, metabolomics, and computational tools in tissue engineering and highlight their relevance in cancer stem cell targeting and radiotherapy. This section also demonstrates how computational modeling can assist in designing and optimizing biomaterials, ensuring compatibility with stem cell growth and differentiation. Machine learning, multiscale computational models, and computational analysis of tissue regeneration are also dissected, showcasing the pivotal role of computational approaches in designing regenerative medicine therapies.

As editors, we are thrilled to present this comprehensive compilation of chapters contributed by experts in the field. For researchers, students, or practitioners in stem cell biology or computational biology, this book provides a rich source of knowledge and insights into the dynamic and fascinating realm of computational approaches in stem cell research. We hope this book is a valuable resource, inspiring further innovation and collaboration in this dynamic field.

Dr. Pawan Kumar Raghav (PhD)
Editor-in-Chief,
Computational Biology for Stem Cell Research
Elsevier

Rajesh Kumar (PhD)
Editor,
Computational Biology for Stem Cell Research
Elsevier

Anjali Lathwal (PhD)
Editor,
Computational Biology for Stem Cell Research
Elsevier

Navneet Sharma (PhD)
Editor,
Computational Biology for Stem Cell Research
Elsevier

In silico tools and approaches in stem cell biology

Chapter 1

Advancement of in silico tools for stem cell research

Ambuj Kumar[1,a], Keerthana Vinod Kumar[1,a], Kunjulakshmi R.[2,a], Kavita Kundal[1], Avik Sengupta[1] and Rahul Kumar[1]

[1]Department of Biotechnology, Indian Institute of Technology Hyderabad, Kandi, Telangana, India; [2]Department of Biological Sciences, Indian Institute of Science Education and Research, Berhampur, Odisha, India

1. Introduction

Undifferentiated cells in the body with the potential to differentiate into various other cell types are called stem cells (Ilic & Polak, 2011; National Institutes of Health, 2006). Stem cells are categorized as embryonic stem cells (ESCs), induced pluripotent stem cells (iPSCs), and adult stem cells. Adult stem cells can be obtained from both human and animal sources and are further divided into different types, such as mesenchymal stem cells (MSCs), hematopoietic stem cells (HSCs), etc. (Ullah et al., 2015). They either remain as stem cells or divide to produce daughter cells with specialized activities such as cardiac cells, muscle cells, bone cells, etc., under appropriate conditions (Ilic & Polak, 2011; Mathur & Martin, 2004). Key features of stem cells include their ability to self-renew, asymmetric replication, potency, and ease of genetic manipulation using techniques such as clustered regularly interspaced short palindromic repeats (CRISPR/Cas9) (Zhang et al., 2017). The potency of stem cells varies between unipotent (restricted to a single cell type), multipotent (all cell types of a particular lineage), pluripotent (can give rise to all types of cells), and totipotent (can produce all adult cell types and can enter the germ line) (Alison et al., 2002). The zygote is the only totipotent stem cell in the human body, with the exceptional ability to differentiate into all cell types, and gives rise to an entire organism through transdifferentiation (Fortier, 2005). ESCs, derived from the inner cell mass of preimplantation blastocysts, are remarkable for their pluripotency within the three germ layers (Evans & Kaufman, 1981). These cells have unique features such as self-renewal capacity, genomic stability (Yoon et al., 2014), and the potential to generate almost all lineages. Their incredible potential has made them a promising candidate for cellular therapy, immunotherapy, drug discovery, and regenerative medicine (Wei et al., 2013). The ability of stem cells for indefinite cell division and transdifferentiation into different cell types made them a leading source of regenerative medicine in recent years, with the potential to repair tissues and organs affected by congenital disabilities, diseases, and age-related effects (Mason & Dunnill, 2008).

Stem cell research helped in understanding of tissue development and disease conditions (Harschnitz & Lorenz, 2021). To study the efficacy of neural stem cells (NSCs) therapy in 3D models of brain tumors, engineered NSCs have recently been developed to mimic tumor growth and malignancy (Carey-Ewend et al., 2021). Stem cell therapy has become a promising immune reconstitution method after cancer treatment with chemotherapy and radiation therapy to restore blood and immunological function (Sagar et al., 2007). Chimeric antigen receptors (CARs) genetically engineered natural killer (NK) cells isolated from HSCs and iPSCs have demonstrated their antitumor activity against malignancies (Daher et al., 2021). Autologous stem cell transplantation (ASCT) has significant immunogenic effects that provide antitumor efficacy in patients with lymphoma (Merryman et al., 2022). It is a potential therapy for other nonmalignant diseases such as severe autoimmune disorders. A retrospective study of systemic lupus erythematosus (SLE) patients revealed significant remission of disease activity within 6 months (Jayne et al., 2004). Allogeneic

---

[a]Equal contribution.

Computational Biology for Stem Cell Research. https://doi.org/10.1016/B978-0-443-13222-3.00018-6
Copyright © 2024 Elsevier Inc. All rights are reserved, including those for text and data mining, AI training, and similar technologies.

hematopoietic cell transplantation (HCT) provides effective therapy for various malignant and nonmalignant diseases. A metaanalysis and systematic review of prospective trials revealed that allogeneic stem cell transplantation (SCT) provides significant relapse-free survival (RFS) and overall survival benefit for intermediate- and poor-risk acute myeloid leukemia (AML) (Koreth et al., 2009). HSC transplantation is another standard procedure for treating multiple myeloma and leukemia, in which cells are harvested from bone marrow or peripheral blood and transplanted intravenously to the recipient (Copelan, 2006). Cell-based therapies have proven effective in cancer treatment but have also been used to prevent or reverse myocardial ischemia and promote heart tissue regeneration (Schulman & Hare, 2012). Also, stem cells have played a significant role in dentistry and ophthalmology. Regenerative endodontics can replace diseased or damaged pulp tissue with healthy tissue to restore teeth to their natural function (Murray et al., 2007). The prevalence of tooth decay and loss has necessitated the development of innovative dental tissue replacement therapies. Recent advances in dental stem cells and tissue engineering strategies suggest that the use of bioengineering techniques may potentially lead to the successful regeneration of whole teeth and dental tissues (Chen et al., 2009). The discovery of iPSCs offers a promising source for corneal cell regeneration and treatment of corneal maladies such as limbal stem cell deficiency (LSCD) and bullous keratopathy (Mahmood et al., 2022). Recent studies report the beneficial effects of mesenchymal stromal cells central nervous system (CNS) diseases (Wang et al., 2022). Combining nanotechnology with nanomaterials such as quantum dots, polymeric nanoparticles, etc. has opened new avenues for studying neurological illnesses and anticancer and gene delivery (Abdollahiyan et al., 2021; Dong et al., 2021). Integration of experimental and computational methods has dramatically advanced the field of stem cell biology. Recent years have seen the development of several in silico techniques for evaluating stem cells' differentiation potential and clinical relevance (Mottini et al., 2021). Improvements in fidelity and efficiency in engineering cell destiny resulted from in silico techniques in stem cell biology. They have also been used to better comprehend lineage commitment, maturation, and disease modeling (Yoshida & Yamanaka, 2010), regenerative medicine (Migliorini et al., 2021), drug discovery (Esmail & Danter, 2021), and toxicology studies (Davila et al., 2004). This chapter focuses on recently developed in silico tools that have not been covered before. We anticipate that this chapter will enhance understanding of these tools among stem cell researchers with no computational background.

2. In silico tools and approaches

We have summarized some of the recently developed stem cell tools in stem cell biology. Fig. 1.1 and Table 1.1 show the recent in silico tools compiled in this chapter. We have collated these tools by considering their accessibility and recentness, while excluding outdated ones. Fig. 1.1 is a flowchart showing the classification of all the approaches (note that the classification used here is subjective due to unclear boundaries between the techniques). Fig. 1.2 depicts the timeline of the tools discussed in this chapter, and Table 1.2 depicts the tools discussed in previous review studies.

2.1 SCENT: Single-Cell Entropy

Single-cell entropy (SCENT) (Teschendorff & Enver, 2017) is a novel algorithm, which can accurately identify the varying potencies of different cell subpopulations allowing cell-lineage trajectory reformation. Signaling entropy can quantify different levels of unpredictability of gene expression levels in a cell, similar to how it works in cellular interaction networks. Overall, this tool helps to understand single-cell populations in terms of their differentiation potency and transcriptomic heterogeneity. In other words, normal or malignant single cells with differential potency and phenotypic plasticity can be predicted by in silico methods using their signaling entropy from an RNA-seq profile. Some important results that supported the validity of SCENT are (1) the correct prediction of the human embryonic stem cell (hESC) population, which contains a small percentage of cells with lesser potency that are ready for differentiation; (2) SCENT correctly predicted the two subsets in neural progenitor cells (NPCs) and also found that subset with lower potency represented actual NPCs; (3) in a human myoblast time course differentiation experiment, SCENT was able to correctly identify a cell population of contaminating interstitial mesenchymal, the potency of which was not able to reasonably alter during the differentiation process. Using signaling entropy, the authors validated SCENT by analyzing over 7000 single-cell RNA-sequencing (scRNA-seq) profiles. They demonstrated all 24 main differentiation stages, including time-course data. To summarize, researchers can use SCENT directly to identify single-cell differentiation potency and plasticity with unbiased quantification of intercellular heterogeneity. Thus, it can provide numerous ways to recognize normal and cancer stem cell phenotypes. Installation details with the package's user manual are accessible on GitHub (https://github.com/aet21/SCENT).

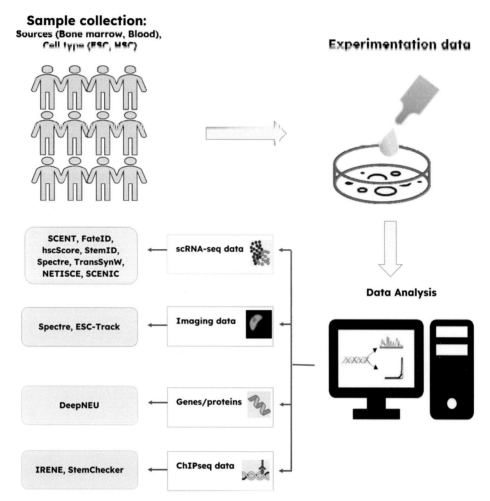

FIGURE 1.1 Schematic representation of the flow of information of primary data collected from wet lab experiments for in silico analysis and the classification of the tools based on their input derived from primary data.

Limitations: The accuracy of SCENT depends on the quality and coverage of the scRNA-seq data. Low coverage or poor-quality data may lead to inaccurate estimations of differentiation potency. SCENT may not be effective for comparing cells at different stages of differentiation, as it is designed to estimate the differentiation potential of cells based on their current transcriptomes.

2.2 FateID

FateID (Herman et al., 2018) unveils the heterogeneity of multipotent progenitors (MPPs) along with transcriptome modulations to lineage biases. Prognostication of this tool elucidates early regulators' cell fate, allowing the isolation of de novo identified marker genes and functional analysis of populations. This tool uses an iterative method to move retrogressively along the differentiation trajectories from dedicated cell populations, aiding in quantifying individual progenitor cells. Cells from the locale of the target cluster can be identified by FateID using random forests (RFs). Those that show a notable bias toward a specific lineage are included in the corresponding target cluster and used for categorization in the following repetition. The process continues until it deduces the fate biases of all cells. FateID is designed with the intent that it will be a beneficial tool, which will identify lineage choices in multilineage differentiation systems. This tool is available as an R package from CRAN. The link and manual are available at https://github.com/dgrun/FateID.

Limitations: FateID was developed and validated using data from mouse hematopoietic stem and progenitor cells. While it may be applicable to other types of MPP cells, further validation is needed to determine its generalizability to other cell types.

TABLE 1.1 Summary of the latest in silico tools for stem cell research. The table provides concise descriptions of the tool, specifies the required input data, outlines the resulting output, and includes the respective URL for tool access. A detailed explanation of each tool can be found in Section 2.

Method	Description	Input	Output	URL	References
SCENT	Identify distinct subpopulation of cells with differing potency, which can be utilized to reconstruct the trajectories of cell lineages.	Single-cell RNA sequencing data	Estimates differentiation potency of cell	https://github.com/aet21/SCENT	(Teschendorff & Enver, 2017)
FateID	To quantify fate biases in individual progenitor cells.	Single-cell RNA sequencing data	Detects fate bias in various multi-lineage differentiation systems and elucidates lineage choice in multipotent progenitors	https://github.com/dgrun/FateID	(Herman et al., 2018)
hscScore	It robustly identifies hematopoietic stem cells within a single-cell RNA sequencing data.	Gene count matrix for mouse single-cell RNA sequencing data with dimensions cells × genes	Identifies bone marrow hematopoietic stem cells	https://github.com/fionahamey/hscScore	(Hamey & Gottgens, 2019)
StemID	It can identify stem cells among all detectable cell types within a population.	Single-cell RNA sequencing data	Predicts multipotent cell populations	https://github.com/dgrun/StemID	(Grün et al., 2016)
Spectre	For analysis of high-dimensional cytometry data from different batches or experiments. It can also analyze data generated by single-cell RNA sequencing.	Datasets generated by both flow (including spectral) and mass cytometry, single-cell RNA sequencing data, high-dimensional imaging data: such as that generated by imaging mass cytometry	Gives summary tables and quantitative plots from datasets providing an overview of relative changes between samples	https://github.com/immunedynamics/spectre	(Ashhurst et al., 2022)
TransSynW	Guides cell conversion experiments by predicting transcription factors required for conversion using single-cell RNA sequencing data.	Single-cell RNA sequencing data	Predicts specific transcription factors and nonspecific pioneer factors and ranks them by expression fold change between starting and targetted cell type and also predicts the marker genes of targetted cell type	https://transsynw.lcsb.uni.lu/	(Ribeiro et al., 2021)
NETISCE	Cell fate reprogramming model.	Single-cell RNA sequencing data	Identifies cell fate reprogramming targets	https://github.com/VeraLiconaResearchGroup/Netisce	(Marazzi et al., 2022)
SCENIC	A computational method for identifying cell fate and reconstruction of gene regulatory network.	Single-cell RNA sequencing data	Identification of groups of genes that are coexpressed across cells, to determine the different cell states	https://github.com/aertslab/SCENIC	(Aibar et al., 2017)

Tool	Description	Input	Function	URL	Reference
ESC-Track	Computer workflow for four-dimensional segmentation, tracking, lineage tracing, and dynamic context analysis of embryonic stem cells.	Live four-dimensional confocal image datasets of embryonic stem cells	Tracks cell division, lineage tree, fluorescence dynamics, cell morphology, and motility	https://repisalud.isciii.es/handle/20.500.12105/9895	(Fernández-de-Manuel et al., 2017)
DeepIU	Machine learning platform to simulate artificially induced pluripotent stem cells using transcription factors.	Genes/proteins involved in human stem cell (embryonic stem cells, induced pluripotent stem cells) signaling pathways	Expression profiles (expression/repression) of genes or proteins in the simulated artificially induced pluripotent stem cells	https://123genetix.com	(Canter 2019)
IRENE	Computer-guided design tool to improving cell conversion efficiency.	Cell type–specific transcription factor chromatin immunoprecipitation-sequencing data	Predicts the most efficient instructive factors necessary for cell conversions	https://github.com/saschajung/IRENE	(Jung et al., 2021)
StemChecker	Web server for analysis of stress signatures in gene set	ChIP-ChIP and ChIP-seq data for genes annotated as being related to stem cells and pluripotency	Overview/Stats, Stemness Signature Match, and Transcription Factor Match	http://stemchecker.sysbiolab.eu/	(Pinto et al., 2015)

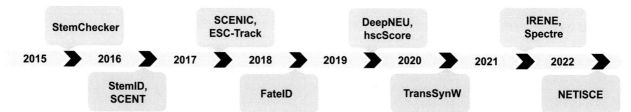

FIGURE 1.2 Timeline illustrating the in silico tools reported for stem cell research from 2015 to 2022. The figure highlights the specific tools discussed in each year: 2015 - StemChecker; 2016 - StemID, SCENT; 2017 - SCENIC, ESC-Track; 2018 - FateID; 2019 - DeepNEU, hscScore; 2020 - TransSynW; 2021 - IRENE, Spectre; 2022 - NETISCE. This chronological depiction shows the evolving landscape of in silico methodologies in stem cell research.

TABLE 1.2 List of in silico tools already covered in previous review studies.

Method	Description	Input	References
CellNet	A platform to guide cellular engineering by quantifying how similar the engineered cells are to their target type	Gene expression (microarray-based) data from 56 published reports and predefined gene regulatory networks	(Cahan et al., 2014)
Mogrify	A prediction tool for identifying the cell conversion reprogramming factors	Gene expression data of 173 human cell types and 134 tissues	(Rackham et al., 2016)
SLICE	An algorithm centered on single-cell entropy directed quantification of differentiation states of cells and their lineage prediction	Single-cell RNA sequencing data	(Guo et al., 2017)
SeesawPred	A web tool for predicting cell fate determinants in cell differentiation	Gene expression data and predefined gene regulatory networks	(Hartmann et al., 2018)
TransSyn	A computational platform for the determination of cellular identities by their synergistic transcriptional cores	Single-cell RNA sequencing data	(Okawa et al., 2018)
SigHotSpotter	It predicts key signaling molecules (hotspots) accountable for controlling cell phenotype	Single-cell RNA sequencing data and signaling interactome network	(Ravichandran et al., 2020)
ScoreCard	It predicts the differentiation propensity of pluripotent stem cell lines	Gene expression data (nanostring and quantitative PCR), DNA methylation data	(Liu & Zheng, 2019)
Pluritest	It evaluates the pluripotency of cells based on their gene expression profiles	Gene expression data from Illumina HT12v3 and HT12v4 microarrays (.idat)	(Müller et al., 2011)
Teratoscore	Evaluates the differentiation potential of human pluripotent stem cells in teratomas, which involves analyzing their gene expression patterns	Gene expression data from Affymetrix Human Genome U133 Plus 2.0 Array (.CEL file)	(Avior et al., 2015)
KeyGene	It calculates the tissue differentiation efficiency from data generated by RNA-sequencing or based microarray based on gene signatures of 21 different human fetal tissues at several developmental stages	Gene expression profile (Affymetrix, Illumina), next-generation sequencing (NGS) data	(Roost et al., 2015)
Capybara	To dissect cell identity and fate transitions	Cells by gene matrix	(Kong et al., 2022)

2.3 hscScore

Identifying potential rare populations is difficult in single-cell data of HSCs, and validation of the function of cell population remains a challenge for single-cell analysis. To address these problems, researchers developed hscScore (Hamey & Gottgens, 2019) that could easily be adapted to scRNA-seq data to detect transcriptional profiles of HSCs and track down likely HSCs. This tool is tested with machine learning methods to assess the similarity between the gene expression patterns of validated HSCs and single-cell transcriptomes obtained from murine bone marrow. It enables rapid and accurate detection of HSCs in a dataset. After identifying the occurrence of known stem cells in a sample, researchers can select a threshold to classify cells based on their hscScore. Even though these details may not be available for all datasets, this tool can still unveil most likely stem cells. One can also use this approach to justify stem cell sorting procedures through the identification of genes encoding the surface marker proteins whose expression level corresponds to hscScore. hscScore robustly identifies HSCs within a scRNA-seq data and provides novel understanding of disrupted hematopoiesis in situations like blood disorders. This tool helps in rapid recognition of HSCs (mouse bone marrow) by analyzing gene expression data from scRNA-seq data. The code for hscScore is freely available online on GitHub (https://github.com/fionahamey/hscScore).

Limitations: The authors noted that the method may not be suitable for identifying rare or quiescent HSCs, as these cells may exhibit lower levels of gene expression. The method is designed to predict putative HSCs based on gene expression data, but it does not provide information on other important features of HSCs, such as their functional properties or epigenetic signatures.

2.4 StemID

Stem cell identification is important to understand tissue homeostasis and its oddity upon disease. StemID (Grün et al., 2016) is a computational tool that helps in predicting multipotent stem cells by inferring genealogical tree from sc-RNA data. StemID can differentiate stem cells from any other identifiable cell types in a given population. Determination of identity of stem cells in all rare cell types discovered also need deviation of a lineage tree. StemID used RaceID2, an enhanced version of RaceID (Grün et al., 2016), to address this problem since it has a more robust initial clustering step. The algorithm for obtaining transcript counts with unique molecular identifiers forms the basis of this approach. StemID uses RaceID2 results and predicts multipotent cell identities. This approach recovered the stem cell's identification from a population of two well-defined systems, intestine and bone marrow. The position of a cell type within the genealogical tree can be revealed by correlating number of branches and transcriptome homogeneity with degree of pluripotency. The number of branches and the transcriptome homogeneity of a cell type have been used to correlate with the degree of pluripotency, thus revealing the position of a cell type within the lineage tree. StemID predicted well-defined ductal cells (subpopulations) with different differentiation potentials. Additionally, it was noted that human and mouse may differ in their maintenance of B cell differentiation in the adult pancreas. Code for StemID algorithm (R package) is present online at https://github.com/dgrun/StemID.

Limitations: The study used a limited set of genes as features for the de novo prediction method, which may not be comprehensive enough to capture the full range of stem cell identities. The inclusion of additional genes or features could improve the accuracy of the method.

2.5 Spectre

Spectre (Ashhurst et al., 2022) is an R package that provides seamless integration of allogeneic stem cells and also has the ability to analyze data generated from experiments such as high-dimensional cytometry. Spectre has well-organized stages of data integration, preprocessing, dimensionality reduction, batch alignment, clustering, visualization, and statistical analysis. It uses basic data structures and implements machine learning classifiers, which allows it to scale up the HD datasets generated by flow, mass, or spectral cytometry. It can also analyze scRNA-seq data. The presence of certain populations within individual samples such as HSCs in the bone marrow is at very low frequency. Therefore, representing these rare populations is quite difficult to sparse on plots such as dimensionality reduction (DR), which are more relative for large populations. To cluster and analyze such low-frequency populations is always desired. Thus, to address this, Spectre was developed to access the data of these populations at multiple levels. After finishing clustering of these datasets, researchers can isolate the clusters of low-frequency populations. Similarly, multiple lineages can be isolated and profiled independently. One can merge these clusters to maintain the cluster annotations that facilitate combined plotting and quantitative analysis if needed. The researchers used Spectre to examine the mass cytometry dataset of bone marrow HSCs, where they extracted clusters with HSCs and progenitors from a complete dataset and then used for independent clustering and DR. Therefore, it allowed independent analysis and more detailed assessment of rare populations, which in full datasets cannot be assessed easily. Spectre (R code) is accessible on GitHub (https://github.com/immunedynamics/spectre).

Limitations: Spectre requires a certain level of expertize in preprocessing single-cell cytometry data as the quality of the data, and its preprocessing will significantly impact the downstream analysis and high computational resources, and the analysis can take significant time to run especially for large datasets, limited ability to handle time-series data, which is an important feature of stem cell research.

2.6 TransSynW

Cell conversion can potentially revolutionize regenerative medicine by generating desired cell types, but it presents several challenges that researchers must overcome. These challenges include identifying cell type—specific markers, heterogeneity among starting cell populations, optimization of conversion protocols, etc.

To overcome these challenges using single-cell technologies, a web application, TransSynW (Ribeiro et al., 2021), was developed that predicts transcription factors (TFs) required for cell conversion using scRNA-seq data for user-specified cell populations. This algorithm works by predicting TFs and prioritizing only the pioneers among the nonspecific ones necessary to promote chromatin opening, an important event in cellular conversion. It anticipates those genes that are markers of the target cell type, which also help the researchers to evaluate their cell conversion experiment performance. One can apply this approach to various levels of cell conversion specificity and recapitulate the familiar conversion factors at each level. This algorithm can make novel prognosis inclusive of the cellular conversion of phenotype in organoids and their in vivo counterparts. Hence, it is a valuable tool for developing cell conversion experiment protocols in stem cell and regenerative medicine research. When applied to different cell systems, it captures the biologically relevant synergistic interactions between predicted TFs that modulate each other and the marker genes. The process takes place in three steps: (1) identification of both specific and nonspecific TFs that are expressed (only previously reported pioneers), (2) selecting those combinations with the highest synergistic interactions, and (c) ranking the predicted TFs based on their change in expression fold between the initial cell population and target, thus helping in effectively determining which TFs should be given priority for further experimental investigation. This tool can be accessed using the link: https://transsynw.lcsb.uni.lu/.

Limitations: TransSynW is currently limited to only a few model organisms, which may not be applicable to all research contexts. Further experimental validation is required to confirm the accuracy of the predicted conversion conditions.

2.7 NETISCE: NETwork-drIven analysiS of CEllular reprogramming

A new computational method called NETwork-drIven analysiS of CEllular reprogramming (NETISCE) (Marazzi et al., 2022) uses Nextflow pipeline (Cruz, 2017) and Galaxy Project workflow (Afgan et al., 2018) to determine the targets of cell fate reprogramming via dynamical systems and control theory methods. The method successfully identified the experimentally determined perturbations leading to cell fate reprogramming and converted primed stem cells to naive stem cells. Thus, becoming a promising tool in recognizing new series of perturbations that can activate required pluripotent genes to get desired phenotype in cell fate reprogramming. This recognition can be helpful in disease modeling, tissue regeneration, or reversing the phenotype of targets with drug resistance. The first step involves collecting data from cells that display both the desired and undesired phenotypes. After collecting the data, researchers use it to build a network of signals and regulations that displays various cell reprogramming mechanisms. This is achieved using signal flow analysis (SFA), which undergoes a network-based analysis. SFA builds an attractor landscape containing a network from the topological information conveyed by the gene regulatory network (GRN) reactions. The normalized gene expression level is the basic network node (Lee & Cho, 2018). Hence, the attractor landscape with experimentally observed states undergoes k-means clustering, depicting clusters of desired and undesired phenotype. The next step involves identifying the feedback vertex sets (FVS) control nodes using simulated annealing algorithms. A series of perturbations are then performed on these nodes. The perturbations are then virtually screened using SFA (Mochizuki et al., 2013). Out of the series of perturbations on the FVS control nodes, the perturbations that lead to cell fate reprogramming are narrowed down using two filtering criteria. The first criteria is using supervised machine learning algorithms such as RF, naive Bayes, and support vector machine (SVM) to categorize the perturbation cluster(s) that form undesired/desired phenotypes. Two out of three algorithms must have the same classification. The second criteria for filtration use the expression value of perturbations passed from first criteria to determine if they have shifted to expected phenotypical range for users-defined internal marker nodes. The majority (90%) of the same are inside the expression value range of the desired phenotype. Therefore, the final set is the one that can produce perturbations that can cause cell fate reprogramming by changing the undesired cell phenotype to the desired one. The following link directs to NETISCE repository https://github.com/VeraLiconaResearchGroup/Netisce.

Limitations: Before using this tool to treat any illnesses, it needs to undergo experimental validation. Only empirically verified perturbations may be performed using the SFA algorithm employed in NETISCE with an accuracy of 85%; time-delayed perturbations cannot be performed by SFA. The tool demands the necessity to have epigenetic data for both the desired and undesirable states to add epigenetic profiles to the network that has been constructed.

2.8 SCENIC: single-cell regulatory network inference and clustering

SCENIC (Aibar et al., 2017) identifies specific GRNs that regulate stable cell states. For demonstrating cis-regulatory analysis, it uses brain and tumor scRNA-seq data. As a result, users can gain valuable biological insights into the mechanisms via TF identification that drive cellular heterogeneity. SCENIC comprises three steps: (1) identifying coexpressed genes with TFs, (2) analyzing their cis-regulatory motifs, and (3) identifying cells with significantly higher subnetwork activity, the activity of each regulon in each cell is scored, utilizing the R packages GENIE3, RcisTarget, and AUCell. This method is robust against dropouts and can be used to identify important master regulators for each cell type. The authors have applied SCENIC to different scRNA-seq data sets, including tumors, the brain, and oligodendrocytes, demonstrating its robustness and ability to reduce batch or technical effects. Using TFs and cis regulatory sequences in SCENIC enables the identification of cell states using scRNA-seq data. The findings demonstrate that GRNs can effectively trace the underlying gene regulatory programs that drive cell type—specific transcriptomes in scRNA-seq data. SCENIC can be accessible from the link: https://github.com/aertslab/SCENIC.

Limitations: SCENIC relies on prior knowledge of TF target genes. If the TF—target gene interactions are not well characterized or documented in the literature, then SCENIC may not accurately infer regulatory networks. SCENIC is primarily designed to analyze gene expression data from homogenous cell populations. It may not be as effective for analyzing data from heterogenous cell populations or samples with low cell numbers.

2.9 ESC-T: ESC-Track

Visualizing the dynamics of stem cell differentiation using fluorophores is essential for exploring methods for tissue repair. To overcome the problem of low spatiotemporal resolution pertaining to 3D imaging, a 4D (3D + time) tracking tool, ESC-T (Fernández-de-Manúel et al., 2017), was designed for automated segmentation and tracking of cell motility and morphology. Identifying cellular divisions and membrane contacts permits single mouse embryonic stem cell (mESC) lineage and localization remodeling. It takes 4D confocal image generated datasets of live cells as input, which also allows analysis of fluorescence signal dynamics. ESC-T has been used to examine dynamics of Myc in mESC cultures and evaluate microenvironmental and genealogical signals responsible for maintaining the fitness of ESCs. This tool helps to understand stem cell fate with a wide range of applications, from exploring the etiology of diseases to its use in regenerative medicine. It analyzes behavior of cells, lineage, and neighborhood relationship, which is of great importance to stem cell biology. The pipeline involves (1) processing and analysis of image, (2) automated segmentation of cell and nuclei with cell tracking, (3) manual edition and feature extraction, and (4) lineage tracing with context analysis. After tracking completion, the aggregated measurements are calculated for total cell lifetimes, which allows the visualization of single-cell tracks and depiction of any specification listed in the feature table. The outputs produced could be (1) trajectories and color-coded values with point referring to temporal cell division, (2) Z-projection 2D videos of the traced cells with color-coded values, and (3) a merged output incorporating the preceding two outputs, depicting a singular video illustrating feature values and cell positions over time along with trajectories from the start of the experiment to the current time point.

Limitations: The accuracy of this tool is dependent on the quality and resolution of the microscopy images used as input. If the images are of low quality or resolution, the tool may not perform well. The tracking algorithm may not always accurately track cells and lineages. The accuracy of the tracking algorithm may depend on factors such as cell density, cell movement, and cell division.

2.10 DeepNEU

Inducing pluripotency in the somatic cells, the formation of iPSCs is problematic due to the likelihood of iPSCs reverting to the initial state. Thus, a computational model for determining genes/proteins essential for producing and sustaining iPSCs was developed utilizing human PSC relevant genes. DeepNEU (Danter, 2019), an unsupervised deep machine learning technology, utilizes a specific set of reprogramming TFs, to simulate artificially induced pluripotent stem cells (aiPSCs) systems. The aiPSC model contains all upregulated biomarkers responsible for the pluripotency of human pluripotent stem cells (hPSCs) thus validating this model for representing hPSC. The authors have programmed DeepNEU to generate disease-specific models of aiPSCs, whose simulation results of gene expression profiles are consistent with that of iPSCs. This helps in interpreting disease pathologies and to come up with new targeted therapies. It uses recurrent neural network (RNN) architecture to accomplish deep learning and analyze the regulation of various genes and pathways in embryonic and programmed somatic cells. This database accommodates 3589 genes/proteins (human) and 27,566 nonzero relationships accounting to copious flow of data in and out of every node in the network, where each node has more than seven inputs and seven outputs. Every input has a connection to its output nodes (feedforward neurons) and vice versa (feedback neurons). RNN uses feedback neurons for storing information in the form of "memory" and can hold up sequential data of arbitrary length. The RNN architecture can mimic the relation between a specific gene to another gene (one-one), specific gene to many genes (one to many), many genes to a specific gene (many to one), and many to many genes. Various other models including artificially induced neural stem cells (aiNSCs) have been developed using DeepNEU to study rare neurological disorders' pathology. Hence, it is an efficient, programmable, and cost-effective tool for modeling disease, prototyping results of wet lab experiments, identifying biomarkers, etc.

Limitations: The platform requires a high level of expertize in machine learning and genetics to use it effectively, which may limit the number of researchers who can benefit from the platform. It provides predictions without providing any insight into the biological mechanisms underlying the predictions, which may limit the platform's utility in some applications, particularly those where understanding the underlying biology is important.

2.11 IRENE: Integrative gene REgulatory NEtwork

Cell conversion technology, a significant tool in stem cell research, exhibits great potential in producing different cell types in vitro but with low conversion efficiency. An Integrative gene REgulatory NEtwork model (IRENE) (Jung et al., 2021) was developed to guide computer-based design. It helps in the effective amalgamations of instructive factors (IFs) crucial for cell conversions and aims to improve cell conversion efficiency. Researchers systematically integrate chromatin accessibility, gene expression, TF ChIP-seq, protein—protein interaction (PPI), and histone modification data to reconstruct the core GRN for a specific cell type. This approach allows them to achieve their goal. The transposon-based genomic integration system forms the basis of the model, which considers the possible combinations of IFs and addresses all main factors governing cell conversion efficiency. It utilizes a stochastic Markov Chain approach, identifies those overexpressed IFs of the initial cell type, and reinforces it at the epigenetic and transcriptional levels between the target and converted cells, thus increasing the percentage of successfully converted cells. Researchers use piggyBac-integrable TF-overexpression cassettes to upregulate the predicted combinations of IFs and overcome the constraint of genetic silencing while avoiding concerns related to viral vector gene delivery. IRENE will notably expand its contribution of cell sources for gene therapies and cell transplantations as it provides the most reliable and efficient approach for using TFs in direct conversions of cells. The source code for IRENE is available online at https://github.com/saschajung/IRENE.

Limitations: IRENE is designed to optimize conversions that only rely on the expression of transcription factors. While IRENE can identify the optimal combination of transcription factors, experimental validation is still required to confirm the effectiveness of the conversion process, which can be time-consuming and resource-intensive. It requires a certain level of prior knowledge about the biology of the starting and target cell types, including the key transcription factors involved in the conversion process.

2.12 StemChecker

StemChecker (Pinto et al., 2015) is a freely accessible web-based stemness analysis tool with the curation of RNA interference screens, TF-binding sites, gene expression, literature review, and in silico techniques so that it could be easily available to the research community. It integrates a huge comprehensive collection of data containing target genes of cell identity and pluripotency and reviews of already published papers reporting human or mouse stem cell signatures. StemChecker thus curates 132 signatures of stemness and TF target genes. MySQL database stores all the curated datasets. The Java and MySQL components use the Hibernate library to communicate.

StemChecker serves as a primary resource where users can quickly search for genes associated with stemness, transcription factors that regulate multipotency or pluripotency genes, and the types of stem cells that express them. This process eliminates the requirement to conduct long curation of literature. StemChecker categorizes stemness signatures as follows based on their source: (1) target genes of the transcription factor, (2) 34 sets of upregulated genes in expression profiles in nine types of stem cells, (3) RNAi screen containing five sets of genome-wide experiments for genes based on RNAi, which are crucial for self-renewal, (4) curation of genes from publicly available literature resources, and (5) computationally derived gene sets. The StemChecker workflow comprises four steps:

1. User input: The user provides input (gene(s) of interest) as gene symbols, entrez IDs, or both, which StemChecker accepts.

2. Selecting data: Based on the accepted input, datasets of stemness signatures for the gene(s) of interest are selected. Although the default species is human, StemChecker searches for all existing datasets, including mice.

3. Searching data: It searches for the required data once the user inputs it

4. Analysis: It displays the results in three pages within the same tab.

The first page of StemChecker displays statistics for the input dataset and their significance. The second page shows a table with a checkerboard pattern that displays the match of each query gene in various stemness signatures. The third page shows the match in the query genes that are targets of TFs. StemChecker can be accessed online at the following link http://stemchecker.sysbiolab.eu/.

Limitations. StemChecker relies on a limited number of gene sets to perform its analysis. It provides a list of differentially expressed genes associated with stemness signatures but does not provide functional analysis of these genes. This lack of functional analysis could limit the tool's ability to identify potential stem cell regulators or pathways.

3. Conclusion and future prospects

"Computational stem cell biology" is similar to the transition of cancer biology to cancer genomics. With the increasing amount of single-cell research data, there is still a need to find more subtle ways to deal with large data, especially scRNA-seq data. We must develop tools/models for different levels of complexity in organisms, such as cell—cell interaction and multiscale modeling. These tools/models can enable the design of stem cell therapies. We can achieve this by bringing experimental and computational research collaborators closer together. This chapter discusses the available computational tools and algorithms in stem cell research ranging from single-cell analysis to cell identity to fate characterization. Each of the tools has shown some added value to stem cell research. We summarized tools or algorithms recently developed but not covered in any previous review study on stem cell tools. We also listed tools already included in the previous review study.

Even though the leap from observing and predicting stem cell differentiation in a T-flask to a laptop or computer screen is remarkable, the area is also open to many challenges. The methods will be much more fruitful if they provide a clear-cut division in identifying coding and noncoding genetic variants leading to specific lineages. There is a need to polish epigenomics, multiomics, and proteomics at the single-cell level to understand stem cell differentiation pathways. Besides GRN, intercellular signaling network prediction can give information on tissue homeostasis and regeneration. The place where stem cells adhere is important for stem cell maintenance. Hence, niche-dependent markers are an emerging tool in computational methods. Researchers should conduct proper validation studies in parallel with experimental setups to ensure the reliability of emerging rules.

Acknowledgments

AK, KVK, and KK acknowledge the fellowship from Ministry of Education, Government of India. AS acknowledges the fellowship from Council of Scientific and Industrial Research (CSIR), Government of India. We would like to acknowledge the research infrastructure provided by the Indian Institute of Technology Hyderabad.

Author contribution

AK, KVK, KR, and RK conceptualized the manuscript framework. AK, KVK, KR wrote the manuscript under the supervision of RK. KK and AS helped in collecting the literature. All authors read the manuscript and approved it for submission.

References

Abdollahiyan, P., Oroojalian, F., & Mokhtarzadeh, A. (2021). The triad of nanotechnology, cell signalling, and scaffold implantation for the successful repair of damaged organs: An overview on soft-tissue engineering. *Journal of Controlled Release, 10*. Retrieved October 21, 2022, from https://www.sciencedirect.com/science/article/pii/S0168365921001115. Retrieved October 21, 2022, from.

Afgan, E., Baker, D., Batut, B., Van Den Beek, M., Bouvier, D., Čech, M., Chilton, J., Clements, D., Coraor, N., Grüning, B. A., Guerler, A., Hillman-Jackson, J., Hiltemann, S., Jalili, V., Rasche, H., Soranzo, N., Goecks, J., Taylor, J., Nekrutenko, A., & Blankenberg, D. (2018). The Galaxy platform for accessible, reproducible and collaborative biomedical analyses; 2018 update. *Nucleic Acids Research, 46*(W1), W537—W544. https://doi.org/10.1093/NAR/GKY379

Aibar, S., González-Blas, C. B., Moerman, T., Huynh-Thu, V. A., Imrichova, H., Hulselmans, G., Rambow, F., Marine, J. Č., Geurts, P., Aerts, J., Van Den Oord, J., Atak, Z. K., Wouters, J., & Aerts, S. (2017). SCENIC: Single-cell regulatory network inference and clustering. *Nature Methods, 14*(11), 1083—1086. https://doi.org/10.1038/nmeth.4463

Alison, M. R., Poulsom, R., Forbes, S., Wright, N. A., & Alison, M. (2002). An Introduction to stem cells. *Wiley Online Library, 197*(4), 419—423. https://doi.org/10.1002/path.1187

Ashhurst, T. M., Marsh-Wakefield, F., Putri, G. H., Spiteri, A. G., Shinko, D., Read, M. N., Smith, A. L., & King, N. J. C. (2022). Integration, exploration, and analysis of high-dimensional single-cell cytometry data using spectre. *Cytometry, Part A, 101*(3), 237−253. https://doi.org/10.1002/CYTO.A.24350

Avior, Y., Biancotti, J., & Benvenisty, N. (2015). TeratoScore: assessing the differentiation potential of human pluripotent stem cells by quantitative expression analysis of teratomas. *Stem Cell Reports, 4*. Retrieved October 21, 2022, from: https://www.sciencedirect.com/science/article/pii/S2213671115001496. Retrieved October 21, 2022, from:.

Cahan, P., Li, H., Morris, S., Rocha, E. Da, Daley, G. Q., & Collins, J. J. (2014). CellNet: network biology applied to stem cell engineering. *Cell, 158*. Retrieved October 21, 2022, from: https://www.sciencedirect.com/science/article/pii/S0092867414009349.

Carey-Ewend, A. G., Hagler, S. B., Bomba, H. N., Goetz, M. J., Bago, J. R., & Hingtgen, S. D. (2021). Developing bioinspired three-dimensional models of brain cancer to evaluate tumor-homing neural stem cell therapy. *Tissue Engineering Part A, 27*(13−14), 857−866. https://doi.org/10.1089/TEN.TEA.2020.0113

Copelan, E. A. (2006). Hematopoietic stem-cell transplantation. *New England Journal of Medicine, 354*(17), 1813−1826. https://doi.org/10.1056/NEJMRA052638

Cruz, S. (2017). Nextflow enables reproducible computational workflows. *Nature, 35*. https://doi.org/10.1093/database/bar026

Daher, M., Garcia, L. M., Li, Ye, & Rezvani, K. (2021). CAR-NK cells: The next wave of cellular therapy for cancer. *Wiley Online Library, 10*(4). https://doi.org/10.1002/cti2.1274

Danter, W. R. (2019). DeepNEU: Cellular reprogramming comes of age - a machine learning platform with application to rare diseases research. *Orphanet Journal of Rare Diseases, 14*(1). https://doi.org/10.1186/S13023-018-0983-3

Davila, J. C., Cezar, G. G., Thiede, M., Strom, S., Miki, T., & Trosko, J. (2004). Use and application of stem cells in toxicology. *Toxicological Sciences, 79*. https://doi.org/10.1093/toxsci/kfh100

Dong, Y., Wu, X., Chen, X., Zhou, P., Xu, F., & Liang, W. (2021). Nanotechnology shaping stem cell therapy: Recent advances, application, challenges, and future outlook. *Biomedicine & Pharmacotherapy, 137*. Retrieved October 21, 2022, from https://www.sciencedirect.com/science/article/pii/S0753332221000214. Retrieved October 21, 2022, from.

Esmail, S., & Danter, W. (2021). Viral pandemic preparedness: A pluripotent stem cell-based machine-learning platform for simulating SARS-CoV-2 infection to enable drug discovery and repurposing. *Stem Cells Translative Medicine, 10*. Retrieved October 21, 2022, from https://academic.oup.com/stcltm/article-abstract/10/2/239/6404028.

Evans, M. J., & Kaufman, M. H. (1981). Establishment in culture of pluripotential cells from mouse embryos. *Nature, 292*(5819), 154−156. https://doi.org/10.1038/292154A0

Fernández-de-Manúel, L., Diaz-Diaz, C., Jimenez-Carretero, D., Torres, M., & Montoya, M. C. (2017). ESC-track: A computer workflow for 4-D segmentation, tracking, lineage tracing and dynamic context analysis of ESCs. *Future Science, 62*(5), 215−222. https://doi.org/10.2144/000114545

Fortier, L. A. (2005). Stem cells: Classifications, controversies, and clinical applications. *Veterinary Surgery : Vysokomolekulyarnykh Soedinenii, 34*(5), 415−423. https://doi.org/10.1111/J.1532-950X.2005.00063.X

Grün, D., Muraro, M., Boisset, J., et al. (2016). De novo prediction of stem cell identity using single-cell transcriptome data. *Cell Stem Cell, 19*. Elsevier. Retrieved October 21, 2022, from https://www.sciencedirect.com/science/article/pii/S1934590916300947.

Guo, M., Bao, E., Wagner, M., Whitsett, J. A., & Xu, Y. (2017). SLICE: determining cell differentiation and lineage based on single cell entropy. *Nucleic Acid Research, 45*. Retrieved October 21, 2022, from https://academic.oup.com/nar/article-abstract/45/7/e54/2725372.

Hamey, F., & Gottgens, B. (2019). Machine learning predicts putative hematopoietic stem cells within large single-cell transcriptomics data sets. *Experimental Hematology, 78*. Retrieved October 21, 2022, from https://www.sciencedirect.com/science/article/pii/S0301472X19309956.

Harschnitz, O., & Lorenz, S. (2021). Human stem cell models to study host−virus interactions in the central nervous system. *Nature Reviews Immunology, 21*. Retrieved October 21, 2022, from https://www.nature.com/articles/s41577-020-00474-y.

Hartmann, A., Okawa, S., Zaffaroni, G., & Antonio, del. Sol (2018). SeesawPred: a web application for predicting cell-fate determinants in cell differentiation. *Scientific Reports, 8*. Retrieved October 21, 2022, from: https://www.nature.com/articles/s41598-018-31688-9. Retrieved October 21, 2022, from:.

Herman, J., Sagar, & Grun, D. (2018). FateID infers cell fate bias in multipotent progenitors from single-cell RNA-seq data. *Nature Methods, 15*. https://doi.org/10.1101/218115

Ilic, D., & Polak, J. (2011). Stem cells in regenerative medicine: Introduction. *British Medical Bulletin, 98*. https://doi.org/10.1093/bmb/ldr012

Jayne, D., Passweg, J., Marmont, A., Farge, D., Zhao, X., Arnold, R., Hiepe, F., Lisukov, I., Musso, M., Ou-Yang, J., Marsh, J., Wulffraat, N., Besalduch, J., Bingham, S. J., Emery, P., Brune, M., Fassas, A., Faulkner, L., Ferster, A., … Tyndall, A. (2004). Autologous stem cell transplantation for systemic lupus erythematosus. *Lupus, 13*(3), 168−176. https://doi.org/10.1191/0961203304LU525OA

Jung, S., Appleton, E., Ali, M., Church, G., & Antonio, Del Sol (2021). A computer-guided design tool to increase the efficiency of cellular conversions. *Nature Communications, 12*. Retrieved October 21, 2022, from: https://www.nature.com/articles/s41467-021-21801-4.

Kong, W., Fu, Y., Holloway, E., Garipler, G., Yang, X., Mazzoni, E. O., & Morris, S. A. (2022). Capybara: A computational tool to measure cell identity and fate transitions. *Cell Stem Cell, 29*. https://doi.org/10.1101/2020.02.17.947390

Koreth, J., Schlenk, R., Kopecky, K. J., Honda, S., Sierra, J., Djulbegovic, B. J., Wadleigh, M., DeAngelo, D. J., Stone, R. M., Sakamaki, H., Appelbaum, F. R., Döhner, H., Antin, J. H., Soiffer, R. J., & Cutler, C. (2009). Allogeneic stem cell transplantation for acute myeloid leukemia in first complete remission: Systematic review and meta-analysis of prospective clinical trials. *JAMA, 301*(22), 2349−2361. https://doi.org/10.1001/JAMA.2009.813

Lee, D., & Cho, K. H. (2018). Topological estimation of signal flow in complex signaling networks. *Scientific Reports, 8*. Retrieved October 21, 2022, from; https://www.nature.com/articles/s41598-018-23643-5.

Liu, L., & Zheng, Y. W. (2019). Predicting differentiation potential of human pluripotent stem cells: Possibilities and challenges. *World Journal of Stem Cells, 11*. Retrieved October 21, 2022, from https://www.ncbi.nlm.nih.gov/pmc/articles/PMC6682503/.

Mahmood, N., Suh, T. C., Ali, K. M., Sefat, E., Jahan, U. M., Huang, Y., Gilger, B. C., & Gluck, J. M. (2022). Induced pluripotent stem cell-derived corneal cells: Current status and application. *Stem Cell Reviews and Reports, 18*. https://doi.org/10.1007/S12015-022-10435-8

Marazzi, L., Shah, M., Balakrishnan, S., Patil, A., & Vera-Licona, P. (2022). NETISCE: A Network-Based Tool for Cell Fate Reprogramming. *NPJ Systems Biology and Applications, 8*. https://doi.org/10.1101/2021.12.30.474582

Mason, C., & Dunnill, P. (2008). A brief definition of regenerative medicine. *Regenerative Medicine, 3*(1), 1–5. https://doi.org/10.2217/17460751.3.1.1

Mathur, A., & Martin, J. (2004). *Stem cells and repair of the heart.* Elsevier. Retrieved October 21, 2022, from: https://www.sciencedirect.com/science/article/pii/S0140673604166324.

Merryman, R., Redd, R., Meier, E., Therapy, J. W., et al. (2022). Immune Reconstitution following High-Dose Chemotherapy and Autologous Stem Cell Transplantation with or without Pembrolizumab Maintenance Therapy in patients with lymphoma. *Transplantation and Cellular Therapy, 28*. Retrieved October 21, 2022, from: https://www.sciencedirect.com/science/article/pii/S2666636721013014. Retrieved October 21, 2022, from:.

Migliorini, A., Nostro, M., & Sneddon, J. B. (2021). Human pluripotent stem cell-derived insulin-producing cells: A regenerative medicine perspective. *Cell Metabolism, 33*. Retrieved October 21, 2022, from: https://www.sciencedirect.com/science/article/pii/S1550413121001285.

Mochizuki, A., Fiedler, B., Kurosawa, G., & Saito, D. (2013). Dynamics and control at feedback vertex sets. II: A faithful monitor to determine the diversity of molecular activities in regulatory networks. *Journal of Theoretical Biology, 335*. Retrieved October 21, 2022, from: https://www.sciencedirect.com/science/article/pii/S0022519313002816.

Mottini, C., Napolitano, F., Li, Z., Gao, X., & Cardone, L. (2021). Computer-aided drug repurposing for cancer therapy: approaches and opportunities to challenge anticancer targets. *Seminars in Cancer Biology, 68*. Retrieved October 21, 2022, from: https://www.sciencedirect.com/science/article/pii/S1044579X19301397.

Müller, F., Schuldt, B., Williams, R., Mason, D., et al. (2011). A bioinformatic assay for pluripotency in human cells. *Nature Methods, 8*. Retrieved October 21, 2022, from https://www.nature.com/articles/nmeth.1580.

Murray, P., Garcia-Godoy, F., & Hargreaves, K. M. (2007). Regenerative endodontics: a review of current status and a call for action. *Journal of Endontics, 33*. Retrieved October 21, 2022, from: https://www.sciencedirect.com/science/article/pii/S0099239906008843.

National Institutes of Health. (2006). *Stem cell.* Retrieved October 21, 2022, from: https://scholar.google.com/scholar?hl=en&as_sdt=0%2C5&q=National+Institutes+of+Health.+%282006%29.+Stem+cell+basics+%5BStem+cell+information%5D.&btnG=.

Okawa, S., Saltó, C., Ravichandran, S., et al. (2018). Transcriptional synergy as an emergent property defining cell subpopulation identity enables population shift. *Nature Communications, 9*. Retrieved October 21, 2022, from: https://www.nature.com/articles/s41467-018-05016-8.

Pinto, J., Kalathur, R., Oliveira, D., et al. (2015). StemChecker: a web-based tool to discover and explore stemness signatures in gene sets. *Nucleic Acids Research, 43*. Oup.Com. Retrieved October 21, 2022, from: https://academic.oup.com/nar/article-abstract/43/W1/W72/2467972. Oup.Com. Retrieved October 21, 2022, from:.

Rackham, O., Firas, J., Fang, H., Oates, M., et al. (2016). A predictive computational framework for direct reprogramming between human cell types. *Nature Genetics, 48*. Retrieved October 21, 2022, from: https://www.nature.com/articles/ng.3487.

Ravichandran, S., Hartmann, A., & Sol, A. Del (2020). SigHotSpotter: scRNA-seq-based computational tool to control cell subpopulation phenotypes for cellular rejuvenation strategies. *Bioinformatics, 36*. https://academic.oup.com/bioinformatics/article-abstract/36/6/1963/5614427.

Chen, Y., Huang, H., Li, G., Yu, J., Fang, F., & Qiu, W. (2009). Dental-derived stem cells and whole tooth regeneration: an overview. *Journal of Clinical Medicine, 13*. Retrieved October 21, 2022, from: https://www.ncbi.nlm.nih.gov/pmc/articles/PMC3318856/. Retrieved October 21, 2022, from:.

Ribeiro, M., Okawa, S., & Antonio, Del Sol. (2021). TransSynW: A single-cell RNA-sequencing based web application to guide cell conversion experiments. *Stem Cells Translational Medicine, 10*. Retrieved October 21, 2022, from: https://academic.oup.com/stcltm/article-abstract/10/2/230/6404021. Retrieved October 21, 2022, from:.

Roost, M., Iperen, L. Van, Ariyurek, Y., et al. (2015). KeyGenes, a tool to probe tissue differentiation using a human fetal transcriptional atlas. *Stem Cell Reports, 4*. Retrieved October 21, 2022, from: https://www.sciencedirect.com/science/article/pii/S2213671115001319.

Sagar, J., Chaib, B., Sales, K., Winslet, M., & Seifalian, A. (2007). Role of stem cells in cancer therapy and cancer stem cells: A review. *Cancer Cell International, 7*. https://doi.org/10.1186/1475-2867-7-9

Schulman, I. H., & Hare, J. M. (2012). Key developments in stem cell therapy in cardiology. *Regenerative Medicine, 7*(6), 17–24. https://doi.org/10.2217/RME.12.80

Teschendorff, A., & Enver, T. (2017). Single-cell entropy for accurate estimation of differentiation potency from a cell's transcriptome. *Nature Communications, 8*. Retrieved October 21, 2022, from: https://www.nature.com/articles/ncomms15599.

Ullah, I., Subbarao, R. B., & Rho, G. J. (2015). Human mesenchymal stem cells - current trends and future prospective. *Bioscience Reports, 35*(2), 191. https://doi.org/10.1042/BSR20150025

Wang, D., Zhang, S., Wu, X., Yin, Z., Li, M., Cao, M., Hu, T., Han, Z., Kong, X., Li, D., Zhao, J., Wang, L., Liu, Q., Chen, F., & Lei, P. (2022). Mesenchymal stromal cell treatment attenuates repetitive mild traumatic brain injury induced persistent cognitive deficits via suppressing ferroptosis. *Journal of Neuroinflammation, 19*(1). https://doi.org/10.1186/S12974-022-02550-7

Wei, X., Yang, X., Han, Z. P., Qu, F. F., Shao, L., & Shi, Y. F. (2013). Mesenchymal stem cells: A new trend for cell therapy. *Acta Pharmacologica Sinica, 34*(6), 747–754. https://doi.org/10.1038/aps.2013.50

Yoon, S. W., Kim, D. K., Kim, K. P., & Park, K. S. (2014). Rad51 regulates cell cycle progression by preserving G2/M transition in mouse embryonic stem cells. *Stem Cells and Development, 23*(22), 2700. https://doi.org/10.1089/SCD.2014.0129

Yoshida, Y., & Yamanaka, S. (2010). Recent stem cell advances: Induced pluripotent stem cells for disease modeling and stem cell-based regeneration. *Circulation, 122*(1), 80−87. https://doi.org/10.1161/CIRCULATIONAHA.109.881433

Zhang, Z., Zhang, Y., Gao, F., Han, S., Cheah, K. S., Tse, H. F., & Lian, Q. (2017). CRISPR/Cas9 genome-editing system in human stem cells: current status and future prospects. *Molecular Therapy Nucleic Acids, 9*. Retrieved October 21, 2022, from https://www.sciencedirect.com/science/article/pii/S2162253117302597.

Chapter 2

Paradigm shift in stem cell research with computational tools, techniques, and databases

Arnab Raha, Prateek Paul, Samriddhi Gupta, Shruti Kaushal and Jaspreet Kaur Dhanjal
Department of Computational Biology, Indraprastha Institute of Information Technology Delhi, New Delhi, Delhi, India

1. Introduction

1.1 Early embryonic development and potency

Development of a multicellular organism from a single-cell zygote with a diverse range of cells with varying functions and phenotype is a complex and intriguing process. Highly regulated genetic and molecular events along with surrounding environmental conditions play a massive role in this functional specificity. During the development stage, after fertilization of the oocyte by the sperm, the unicellular zygote gets formed. This zygote actively multiplies to give rise to a 14-cell stage called morula where all the cells can differentiate into any cell type. This ability of the cells to be able to differentiate into any possible cell type and give rise to a healthy organism is called totipotency/omnipotency (Mitalipov & Wolf, 2009). The morula then further divides into a ball-shaped 32–64 cell stage called blastula with a fluid-filled cavity called blastocoel. This cavity also houses an inner cell mass. The outer cellular margin of the blastula is called trophoblast and is special since these cells are not as omnipotent as that of the cells in the inner cavity. Thus, the cells of the blastula are pluripotent in nature. It is during this blastula stage that it gets implanted in the uterus and is then called an embryo. Further, multipotency is when a given cell can differentiate only into a closely related family of cells (Schöler, 2016).

1.2 Characteristics of stem cells

For a cell to be termed a "stem cell," it is quintessential for it to have the following two properties (Fig. 2.1A):

1. *Self-renewal*: A stem cell undergoes "asymmetric cell division," which means that a stem cell divides giving rise to one daughter cell, which is a stem cell (to preserve its population), and the other daughter cell is committed to lineage-specific differentiation (He et al., 2009).
2. *Potency*: It is the capability of a cell type to give rise to different cell types as discussed before (Schöler, 2016).

Due to the aforementioned properties of stem cell, engineering stem cells is lucrative. Thus, it is an active and dynamic field of ongoing research (Fig. 2.1B). Stem cell engineering majorly includes the following:

1. *Somatic cell reprogramming*: Here, differentiated somatic cells are reconverted or dedifferentiated into pluripotent stem cells. For example, the dedifferentiation of neural progenitors (Kim et al., 2009) and fibroblasts (Takahashi & Yamanaka, 2006) to pluripotent stem cells.
2. *Directed differentiation*: Pluripotent cells are directed to specific lineages by providing specific growth factors and inducing ectopic gene expression (Murry & Keller, 2008).

Computational Biology for Stem Cell Research. https://doi.org/10.1016/B978-0-443-13222-3.00019-8
Copyright © 2024 Elsevier Inc. All rights reserved, including those for text and data mining, AI training, and similar technologies.

A. Characteristics of Stem Cells

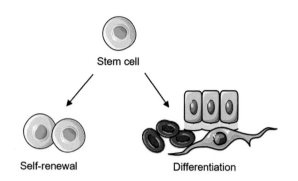

B. Stem Cell Engineering

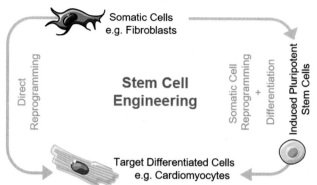

C. Applications of Stem Cell Engineering

Regenerative Medicine	Toxicology Screening & Drug Discovery	Research & Development
• Cell Therapy • Tissue/Organ Engineering • Drug Delivery (exosomes)	• Toxicology testing • Drug Target validation	• Developmental Biology • Disease Modelling • Pathogenesis Research

D. Big Data in Stem Cell Biology

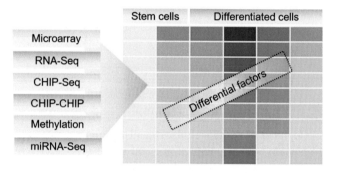

E. Computational approaches for big data analysis

a. Clustering

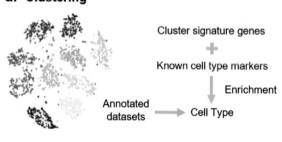

b. Network-based approach

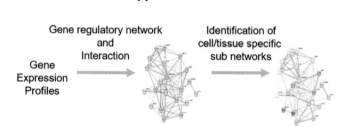

c. Machine learning based approach

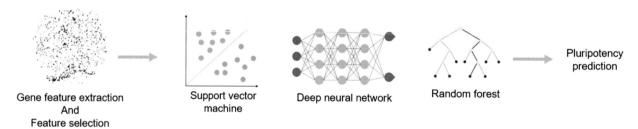

FIGURE 2.1 (A) The properties of stem cells exhibiting differentiation and self-renewal. (B) Stem cell engineering, comprising of reprogramming a terminally differentiated cell to another (direct reprogramming). For instance, here, fibroblasts are directly reprogrammed to cardiomyocytes. However, fibroblasts can be converted to cardiomyocytes by first dedifferentiating fibroblasts to stem cells, which can then be differentiated to cardiomyocytes (somatic cell reprogramming + differentiation). (C) Prospects of stem cell engineering in toxicology screening, regenerative medicine, drug discovery, and further research advancement. (D) Various types of cellular big data and their utilization in stem cell biology. These cellular data generally enable us to infer the characteristic differences between stem cells and the terminally differentiated cells. (E) Different computational approaches for big data analyses (a) clustering novel dataset based on signature genes and known cell type markers to determine cell type, (b) network-based approach wherein gene expression data are viewed as gene regulatory network (GRN) and interaction information for identification of cell/tissue-specific subnetworks is derived, (c) machine learning—based approach to extract prominent gene features or signatures for a given lineage, which can be leveraged further to classify given cell lines into their respective cellular fates or potency.

3. *Direct reprogramming/transdifferentiation*: Instead of entering a proliferative pluripotent stem cell stage, somatic cells are directly transformed into a different lineage. There are multiple literature studies showing transdifferentiation, for instance, production of insulin-generating beta cells in exocrine pancreatic cells (Zhou et al., 2008), and conversion of fibroblasts to hepatocytes (Huang et al., 2011).

1.3 Need to study stem cells

Stem cell research provides incredible promise for new medical understanding and treatments (Fig. 2.1C). Broadly, stem cells are currently being used in treatment and research in the following ways:

1. To understand developmental systems, including the inaccessible cell types.
2. To understand the occurrence of a disease phenotype on a macroscale and bridge it with their underlying cellular and molecular biology.
3. In regenerative medicine, human organs, tissues, or cells are replaced with engineered, programmed stem cells to restore healthy functioning.
4. New drugs are tested on engineered stem cells to check for safety and effectiveness.

In this chapter, we have discussed the generated high-throughput data and various bioinformatics and computational tools that have been used to gain insights into the understanding of stem cell biology.

2. Big data in stem cell research

Regulation and expression of key genes hold paramount importance in defining the functional state of stem cells, the attributes being identity, genomic integrity, sterility, and purity. However, effective analytics to decide stem cells' potency holds a major challenge (Sebastião et al., 2021). Nevertheless, with the advent of detection pipelines, collecting large-scale high-throughput biological data has become easier and simpler. However, owing to the complexity of the nature of biology, integration, and management of the vast and diverse data types, the different hierarchies or dimensions of cellular processes are pivotal to understand the regulation and characterization of cellular process as a whole (Wooley et al., 2005). Since, in stem cells, tight regulation of gene expression holds a significant role, integrating the cross-talk and information flow between different hierarchy/dimensions of cellular processes is important; thus, usefulness of omics data comes into picture (Fig. 2.1D).

2.1 Genomics

It is a multidisciplinary domain comprising the evolution, structure, function, and editing of genomes. The complete genomic sequence of an organism can be obtained by sequencing its cellular DNA, which can be further analyzed to derive meaningful information as follows:

1. *Single-nucleotide polymorphisms (SNPs)*: SNP characterization can identify a stem cell's origin with specificity and surveil its genomic integrity (Gunderson et al., 2005).
2. *Copy number alterations (CNAs)*: It has been observed that genes involved in cell proliferation and pluripotency in human embryonic stem cells tend to show an increase in CNA during somatic reprogramming (Laurent et al., 2011).

2.2 Transcriptomics

Transcriptomic data are the assemblage of all expressed RNA transcripts in a given cellular condition, including protein coding and nonprotein coding, in an individual (single cell) or a population of cells (bulk). The data encompass all information related to RNAs, including their expression concentration, biological or molecular functions, cellular locations, trafficking and degradation. It also includes the parent genes, transcript, splicing patterns, and post-transcriptional modifications (Wang et al., 2009). It covers all kinds of transcripts, including mRNAs (messenger RNAs) encoding protein, miRNAs (microRNAs), and lncRNAs (long noncoding RNAs). miRNAs and lncRNAs have a regulatory role in controlling the expression of protein-coding mRNAs. miRNAs are regulatory RNAs that can bind to an mRNA forming a dsRNA (double-stranded RNA), which then further gets degraded by cellular RNases, thus controlling the availability of the mRNA for its translation. lncRNAs also have regulatory roles wherein they act as molecular decoys, sequestering miRNAs, and thus, evading their interaction with their target mRNA (Mercer et al.,

2009). Thus, miRNAs and lncRNAs regulate a plethora of biological functionalities via their cross-talk with each other and, in turn, regulate protein-coding mRNAs. Since, in stem cells, regulation of differentiation and commitment to different cell stages involve a lot of regulatory affairs, miRNA and lncRNA data hold great promise in understanding stem cells better (Dhanoa et al., 2018). Apart from whole transcriptome analyses, a defined set of RNAs can also be studied using microarray profiling.

The molecular information that can be derived from the transcriptome of the stem cell as observed in the following studies are as follows:

1. *mRNA transcripts that get expressed and their concentration:* In a study, microarray-based gene expression from six human embryonic stem cell (hESC) lines was compared with the mRNA pooled from multiple tissues to investigate for the genes that are enriched in hESCs. Detailed characterization and identification of unique markers of the undifferentiated hESCs were possible with these data (Bhattacharya et al., 2004).

2. *Regulatory RNAs specific to a given cellular condition:* miRNA data compilation and comparison between the F-class (fuzzy colony-forming cell lines) state and iPSCs (induced pluripotent stem cells) suggested that there is a novel set of miRNAs that are responsible for and support pluripotency in the F-class state (Clancy et al., 2014). miRNA expression data collected from and compared among hESCs, multiple stem cells, and differentiated cells, in another study, lead to the finding of a set of novel miRNAs that are differentially regulated in hESCs. These miRNAs were further validated to have a key role in regulating differentiation and pluripotency (Lienert et al., 2011). lncRNAs have also been shown to exhibit multiple roles in processes such as embryonic stem cell pluripotency to cell proliferation (Guttman et al., 2009).

2.3 Proteomics

A proteome refers to the complete set of proteins produced in an organism, cell, tissue, or any system in a biological context. The structure and physiological functions of various proteins expressed in a cell define its biological behavior. The proteins that are present in a given cell in any given condition can be measured using techniques such as matrix-assisted laser desorption/ionization (MALDI) or ribosome profiling (Ribo-Seq).

Studying the proteome of stem cells can be helpful in understanding various aspects, some of the examples for the same are as follows:

1. *Set of protein present in a given cellular condition:* The factors that are responsible for the differentiation of human mesenchymal stem cell into osteocyte were found by comparing differential protein abundance. This transition was found to be induced by epidermal growth factor (Smith et al., 2009).

2. *Post-translational modifications:* Translation in mESCs (murine embryonic stem cells) was analyzed to reveal multiple unannotated translational products such as upstream open reading frames, truncations, and amino-terminal extensions. Such modifications have been suggested to have a strong regulatory role in the maintenance of stem cell state (Ingolia et al., 2011).

2.4 Epigenomics

Epigenetic changes are stable modifications on DNA that do not change the DNA sequence but chemically modify the histones or nucleotide bases of the DNA (Dupont et al., 2009). Epigenomics is thus the study of the whole collection of epigenetic modifications on the DNA/genome of the cell.

Epigenetic changes turn the genes on and off, thus controlling their expression as per the cellular and environmental requirements. The importance of studying the epigenome has been highlighted in the following points:

1. *Protein−DNA interactions:* Most transcription factors are generally proteins. These transcription factors play a huge role in differentiating a stem cell into different specialized cell states. OCT4 is one such transcription factor that was found to be essential for reprogramming differentiated somatic cells into iPSCs (Jerabek et al., 2014). Large-scale protein−DNA interactions can be studied using ChIP-Seq (chromatin immunoprecipitation sequencing) and ChIP-chip methods.

2. *RNA−DNA interactions:* Regulatory RNAs such as miRNAs and lncRNAs can also bind to DNA and regulate the transcription of a given gene. lncRNA033862 binds to GDNF receptor alpha1 (*Gfra1*) chromatin and regulates its expression levels. This binding was observed in spermatogonial stem cells (SSCs), where the interaction was found to be important for the survival of SSCs (Li et al., 2016). High-throughput RNA−DNA interactions can also be derived from the ChIP-chip method.

3. *Histone modifications*: Histone proteins that are present inside nucleosomes are used to pack negatively charged DNA effectively and are often modified by adding chemical moieties. These moieties, when added to the histone proteins, affect the folding of DNA onto itself, thus regulating the accessibility of the DNA sequence by RNA polymerase. These modifications are majorly acetylation, phosphorylation, ubiquitinylation, and methylation (Kim & Kim, 2012). mESCs and hESCs require histone methylation at sites H3K4 to maintain pluripotency (Shiraki et al., 2014). Histone modification at genome level can also be studied using the ChIP-chip and ChIP-seq method.

4. *DNA methylation*: DNA methylation information across the whole genome at single-base resolution was also collected in multiple human iPSC lines to account for significant variability seen in somatic reprogramming and somatic memory (Lister et al., 2011). DNA methylation data can be derived from bisulfite sequencing and DNA hypersensitivity mapping methods

3. Computational methods for effective analyses of high-throughput stem cells data

The distinction between differentiated cells (such as fibroblast and hepatocyte cells) and pluripotent stem cells has attracted a lot of attention recently and is at the central hub of basic research in developmental biology. In addition, there are numerous additional possible uses in fields outside of regenerative medicine. Exciting advancements have also been made in the in vitro modulation of fundamental cellular phenotypes. Comprehensive investigations on pluripotency and differentiation have been effectively conducted recently using computer techniques such as principal component analysis (PCA) and hierarchical clustering, frequently using data on differential gene expression (Lenz et al., 2016).

In the study of stem cell and developmental biology, high-throughput data are routinely being used to supplement research on the principles of pluripotency. Computational methods are thus gaining importance for handling and effectively analyzing this massive amount of data. For instance, data mining using microarrays (gene expression patterns) identifies gene-based biomarkers that could be effective targets for regenerative medicines and offers priceless knowledge about how cells function. Some of the commonly used computational approaches to analyze the high-throughput stem cell data and gain insights are discussed in the following sections.

3.1 Clustering methods

Clustering is an unsupervised method for data analysis that divides the data such as transcriptomics into groups/clusters based on gene expression patterns. A cluster is made up of all the genes that have similar expression patterns and depict similar functions. Clustering is a very useful approach to classify heterogenous data into novel subclasses and recognize the genes that distinguish these subclasses. The two most common clustering approaches are hierarchical and k-mean clustering. The agglomerative and divisive methods are used in the hierarchical clustering algorithm as described:

1. *Agglomerative clustering approach:* Every single data input is primarily accepted as a cluster, and the closest cluster is repeatedly combined into even larger clusters that make up a tree. This procedure includes the following steps:
   a. A distance matrix is computed for each clustered gene.
   b. The score of similarity between the respective clusters is determined by considering two clusters with the maximum similarity. The Pearson correlation coefficient measures the similarity of the two clusters.
   c. The distance between this newly discovered cluster and all the others is calculated. This procedure is continuously repeated for a certain number of iterations until all the clusters get merged into a single cluster. Methods for comparing two clusters include single, complete, average, and centroid linkage.
2. *Divisive clustering approach:* An entire data input is taken as a single cluster in the divisive clustering approach and is further divided into more than one cluster, which shows indistinguishable patterns of gene expression. This procedure is reiterated until all of the data have been separated out. This algorithm does not necessitate the number of clusters to be described in advance. Top-down clustering necessitates a method for splitting a cluster that contains all of the data and then splitting clusters recursively until all of the data have been split into singletons (Everitt et al., 2011).

FateID is one such tool that incorporates the principle of clustering to classify or predict the type of terminal differentiation of a given cell. The tool has been built by training around 2000 murine hematopoietic progenitors that are predominantly enriched for lymphocytic lineages. The tool is built on RaceID, an existing supervised learning algorithm, which takes the query cell's RNA-seq data to predict its cellular fate as a probability (Herman et al., 2018).

3.2 Machine learning approach

Classification of highly expressed genes using high-throughput data has been an elemental step in bioinformatics, and analysis of the connate set of crucial features/descriptors enhances the understanding of various rudimentary mechanisms governing a biological functionality, providing fundamental insights into the biology causing a specific state, phenotype, or condition. For example, characteristics such as 'pluripotency,' as well as its heterogenous/specialized variant, 'differentiated' cells. Various computational approaches characterize and identify genes responsible for stem cell pluripotency by assorting and analyzing gene expression data from pluripotent and differentiated cells. Recent advances in the development of high-throughput data analyses pipelines have resulted in a massive growth in data collection to determine the source of disease, including gene−environment, gene−gene interaction, and cell classification at both the phenotypic and molecular levels. Machine learning (ML) classification algorithms such as naive Bayes classifiers (NBCs), support vector machine (SVM), and random forest (RF), all have a significant impact on biological processes. SVM is one of the data-efficient ML methods used to inspect large amounts of data. SVM has been extensively used as a classification ML model in biological functions such as biomarker discovery, prediction of gene function, disease diagnosis, and identification of protein homolog. NBCs are among the most straightforward and efficient classification techniques. Bayesian networks are the foundation of this methodology to assess the chance that a sample belongs to a class with a known probability distribution. Random forests are created by sampling with replacing each event in the training set, and additionally selecting a sample of potential descriptor variables (for example, genes), then subsequently constructing a number of decision trees that are used together to categorize new data.

Using an ML approach, different gene regulatory network (GRN) configurations respective to various stages of pluripotency, including primed and naive states, have been identified in a study (Assenov et al., 2008). An analysis of chromatin accessibility and gene expression data employing ML and gene regulatory networks revealed that lineages that failed to develop into induced endoderm progenitors (iEPs) were due to the reprogramming transcription factors were incapable of correctly regulating genes crucial to iEP function and distinctiveness. Using a linear regression model, CellOracle can only simulate single transcription factors and cannot anticipate nonlinear, combinatorial effects. However, the implemented simple linear regularized ML model does provide a number of advantages, which include lowering overfitting and providing outcomes that are comprehensible. This advantage is made possible by the initial breakdown of population heterogeneity, which ordinarily throws off the linear modeling of mixed complicated data with many regulatory states. These benefits are apparent in the CellOracle simulations accuracy as well as the biological insights they offer (Kamimoto et al., 2023). HCS (high content screening) with ML support was employed to study in vitro embryogenesis based on stem cells. Credible, nonbiased, and automated ML-based protocols were created to analyze the ability of various iPSC and ESC lines to form embryos, recognize favorable conditions for embryo generation, and bring out cytokine and small-molecule screening to find factors that can encourage embryo generation. This method was adopted to analyze the cellular structures formed from various mouse iPSC lines and ESCs using high-content analysis supported by ML. The study provided a brand-new approach for improving the efficacy and impartiality of multidimensional analyses of stem cell−based embryos (Walcher et al., 2020). Using a live cell imaging device, the reprogramming procedure was documented 48 h after infection. Within 3−5 days of infection, feeder fibroblasts and iPSCs were designated by retroactively tracing the time-lapse microscope image. The XGBoost method was finally utilized to construct a prediction model using six features and the optimal time windows (Zhang, Shao, et al., 2019).

3.3 Network-based approach

The molecular mechanism underlying pluripotency is a multiplex interplay of proteins/genes involving epigenetic regulators, transcriptional factors, and signaling pathways. Protein−protein interactions (PPIs) are involved in these mechanisms, which are involved in primary biological processes in cells. Moreover, today's high-throughput genome sequencing technology generates massive amounts of high-dimensional data. Integrative bioinformatics face the challenge of capturing, integrating, and analyzing these data in a uniform manner to understand novel and deeper insights into the system level (Warsow et al., 2010).

To reconstitute a pluripotent cell's specificities from single-cell data as a weighted sum of regulatory network archetypes, a three-part technique has been developed (Stumpf & Macarthur, 2019). Single-cell expression data were initially projected onto an ensemble model of pluripotency regulatory circuitry to get snapshots of regulatory activity within individual cells. Second, PCA was used to extract all of the regulatory network archetypes and associated weights for individual cells from these snapshots. Third, to improve the classification of observable cell identities resulting from ensemble network dynamics, the weights were assigned to low-dimensional representations of regulatory network activity.

These representations were then fitted with a Gaussian mixture model. A schematic representation showing the use of these computational methods for stem cell research is shown in Fig. 2.1E.

4. Computational resources for stem cell data analysis

With advancements in the field of stem cell biology and availability of massive raw and analytical data, many computational resources have been built to properly archive these data for easy availability and further use. Also, a lot of computational tools have been developed to accelerate discoveries by data mining and innovative algorithms. Table 2.1 lists a few of these computational resources being used for stem cell research. A brief description of these tools has also been included further.

- **BloodExpress**: A database that houses expression profiles of mouse blood cell types, including differentiated cells and progenitors. During development of different blood cell types to their terminally differentiated type, the database enables the discovery of continuous changes in gene expression profile. A gene-centric interface provides a list of all the other genes with comparable gene expression profile patterns and functional annotation, hematopoietic expression patterns, and other information. With the extra and helpful option to filter by certain gene functional categories, a cell type—focused interface enables the discovery of genes that are expressed at particular phases of blood formation. Thus, BloodExpress is a platform for investigating new gene roles throughout the hematopoietic system (Miranda-Saavedra et al., 2009).
- **CellAssign**: To determine the likelihood that each cell belongs to the currently known cell types, CellAssign utilizes a hierarchical statistical framework, while employing a combined expectation—maximization inference approach to estimate all model parameters. CellAssign is designed to identify cell types in situations when widely studied marker genes are present, which may make poorly described or unidentified cell types undetectable. In the scRNA-seq analytic tools, CellAssign plays a crucial role by producing interpretable results from previous information that is biologically informed. As a result, it naturally reduces problems related to current unsupervised clustering algorithms, such as batch effects on clustering and the requirement for post hoc understanding of clusters in the context of existing cell states (Zhang, O'Flanagan, et al., 2019).
- **CellBLAST**: To optimally train a nonlinear projection from a high-dimensional transcriptome space to a low-dimensional cell embedding space in an unsupervised way, CellBLAST is based on a neural network—based generative model. This is done with the help of reference single-cell transcriptomes and adversarial alignment to correct the intra-reference batch effect. A generative model based on neural networks and a special cell-to-cell similarity metric are the foundation of CellBLAST, a dependable and accurate cell-querying method. CellBLAST reliably detects and exposes the existence of undetected query cells rather than making false-positive predictions to find novel cell kinds or aberrant cell states (Cao et al., 2020).
- **CellNet**: It is a computational tool based on graph algorithms that analyze the fidelity of cellular engineering with precision. It also suggests hypotheses for improving cell derivations. It is based on tissue/cell-specific GRNs. It has been constructed to analyze gene expression profiles of human and mouse cell populations derived from engineered stem cells: somatic cells reprogrammed to pluripotency, pluripotent stem cell differentiation, or cells produced through trans-differentiation (Cahan et al., 2014). It can be accessed as a web application wherein users can investigate the data and download the software so as to run it locally, or expression data can be uploaded to be further processed in their servers.
- **CODEX**: Nearly 1000 samples, 221 distinct transcription factors, and 93 different cell types make up CODEX. Thus, CODEX offers one of the largest sets of publicly available next-generation sequencing data, for the direct study of transcriptional programs that regulate cell identity and fate during mammalian development, homeostasis, and disease for the direct study of transcriptional programs that regulate cell identity and fate during mammalian development, homeostasis, and disease (Sánchez-Castillo et al., 2015).
- **CytoTRACE**: Utilizing data from single-cell RNA sequencing, the computational method CytoTRACE (Cellular (Cyto) Trajectory Reconstruction Analysis using gene Counts and Expression) predicts the differentiation status of the cells. It is possible to learn more about immature luminal progenitor cells and the genes associated with them in breast cancer by undertaking scRNA-seq profiling of breast tumor epithelial cells and the surrounding normal epithelial cells (Gulati et al., 2020).
- **ESCAPE**: The online interface allows users to browse, search for, and download various types of cellular information related to embryonic stem cells, including information from gene expression microarrays, ChIP-chip/seq, RNA-seq, genome-wide RNAi screens, phosphoproteomics immunoprecipitation, and mass spectrometry proteomics. For the purpose of interacting with the data in the database, a number of web-based tools were also developed. When given a set of

TABLE 2.1 Record of significant bioinformatics web resources for assembling, displaying, and analyzing data relating to stem cells.

Name	Features	Platform types/Datasets	Species	Link	Reference
BloodExpress[a]	Covers the majority of murine blood cell types, including both progenitors and terminally differentiated cells	Microarray	Mouse	http://hscl.cimr.cam.ac.uk/bloodexpress/	Miranda-Saavedra et al. (2009)
CellAssign	Automated, probabilistic assignment of cell types in scRNA-seq data	scRNA-seq	–	https://github.com/Irrationone/cellassign	Zhang, O'Flanagan, et al. (2019)
CellBLAST	Prediction of cell type based on input single-cell transcriptomics data	scRNA-seq	–	https://github.com/gao-lab/Cell_BLAST	Cao et al. (2020)
CellNet	GRN-based assessment of the accuracy of stem cell engineering	Microarray/ RNA-seq/ scRNA-seq	Mouse Human	http://ec2-3-89-75-200.compute-1.amazonaws.com/cl_apps/agnosticCellNet_web/	Cahan et al. (2014)
CODEX	Next-generation sequencing data concerning embryonic stem cells, hematopoietic cells, and all other varieties of cells	DNase-seq RNA-seq Transcription-factor and histone modification CHIP-seq	Mouse Human	https://codex.stemcells.cam.ac.uk/	Sánchez-Castillo et al. (2015)
CytoTRACE	Reconstruct cellular differentiation trajectories, prediction of differentiation states of cells	scRNA-seq	–	https://cytotrace.stanford.edu/	Gulati et al. (2020)
ESCAPE	Various cellular data for ESCs from mice and humans, network building, lineage prediction, and enrichment analysis	ChIP-CHIP RNA-seq ChIP-seq IP-MS Genome-wide inhibitory RNA (RNAi) screens	Mouse Human	http://www.maayanlab.net/	Xu et al. (2013)
FunGenES	Mouse ESC differentiation	si-RNA	Mouse	https://bio.tools/fungenes	Schulz et al. (2009)
HSC-explorer	Early steps of hematopoiesis with an interactive graphical display Includes qPCR data	Microarray	Mouse Human	https://github.com/deeptools/HiCExplorer	Montrone et al. (2013)
KeyGenes	Predicts the identity of stem cell derivatives, tissues of origin, and developmental stage	Microarray	Human	http://www.keygenes.nl/	Roost et al. (2015)
LifeMap Discovery	Embryonic development, regenerative medicine, and stem cell database	Microarray	Human	http://discovery.lifemapsc.com/	Edgar et al. (2013)
Moana	Construction of robust cell-type classifiers based on a hierarchical machine learning framework	scRNA-seq	–	https://github.com/yanailab/moana	Wagner and Yanai (2018)

TABLE 2.1 Record of significant bioinformatics web resources for assembling, displaying, and analyzing data relating to stem cells. cont'd

Name	Features	Platform types/Datasets	Species	Link	Reference
Mogrify	Prediction of transcription factors for cell transdifferentiation/conversion	—	Human	http://www.mogrify.net	Rackham et al. (2016)
Monocle	Prediction of pseudotime differentiation stage and trajectory validation	scRNA-seq	—	https://github.com/cole-trapnell-lab/monocle-release	Trapnell et al. (2014)
ORIGINS	Network-based approach to quantify pluripotency using input single-cell transcriptomics	scRNA-seq	—	https://github.com/danielasenraoka/ORIGINS	Senra et al. (2022)
PluriNetWork[a]	Network visualization of pluripotent genes with interactions.	Microarray	Mouse	http://www.ibima.med.uni-rostock.de/IBIMA/PluriNetWork/	Som et al. (2010)
Scmap	Characterization of cell types based on unsupervised clustering of the transcriptome	scRNA-seq	—	https://github.com/hemberg-lab/scmap	Kiselev et al. (2018)
ScoreCard	Cell-line specific differentiation	Microarray DNA methylation	Human	http://scorecard.computational-epigenetics.org/	Tsankov et al. (2015)
scPred	Classification of cells based on a low-dimensional representation of gene expression	scRNA-seq	—	https://github.com/powellgenomicslab/scPred	Alquicira-Hernandez et al. (2019)
StemBase[a]	Analysis of gene expression data of stem cell Correlation and mutual information of gene expression	Microarray SAGE	Mouse Human Rat	http://www.stembase.ca/	Porter et al. (2007)
StemCellDB	Human pluripotent stem cells, gene expression level analyses Incorporates data from SNP genotyping and methylation analysis	Microarray	Human	https://stemcelldb.nih.gov/	Mallon et al. (2013)
StemCellNet	Querying, visualization, examining molecular network, physical and regulatory interactions of stem cell data	Microarray CHIP-seq	Mouse Human	http://stemcellnet.sysbiolab.eu/	Pinto et al. (2014)
Stemformatics	Downloading and visualization of curated stem cell data	Microarray RNA-seq scRNA-seq	Human Mouse	https://www.stemformatics.org/	Choi et al. (2019)
StemMapper	Manually curated stem cells database, interface for visual comparison of gene expression across different stem cell types	Microarray	Mouse Human	http://stemmapper.sysbiolab.eu/	Pinto et al. (2018)
SyStemCell[a]	Distinctive evidence of down and upregulation; examination of colocalization to discover novel correlations	RNA-seq ChIP-chip ChIP-seq Microarray	Mouse Human Rat Rhesus macaque	http://lifecenter.sgst.cn/SyStemCell/	Yu et al. (2012)
TeratoScore	Quantitative estimation of pluripotency	Microarray	Human	priweb.cc.huji.ac.il/stemcell/index.php	Avior et al. (2015)
Waddington OT	Spot developmental trajectories in reprogramming	RNA-seq	—	https://broadinstitute.github.io/wot/	Schiebinger et al. (2019)

[a]These tools were not accessible at the time of review.

seed genes, researchers can do activities such as network extension, analysis of the enrichment of upstream regulatory factors and downstream targets, combinatorial lineage predictions, and more. The experimental data may be organized into a logical and interactive framework using system-level analysis and the development of dynamical models (Xu et al., 2013).

- **FunGenES:** The "Functional Genomics in Embryonic Stem Cells" has explored the transcriptome of mouse ES cells in 11 habitats, corresponding to 67 experimental conditions, to investigate the gene networks that are active in pluripotent ES cells and their progeny. The KEGG signaling and metabolic pathways have been animated, and search engines can show the transcript expression profile (Schulz et al., 2009).

- **HSC-explorer:** Over 7000 experimentally verified interactions between chemicals, bioprocesses, and environmental variables are included in organised information. Manual information curation involves a rigorous reading of the scientific literature by knowledgeable annotators. For the purpose of evaluating the validity and applicability of experimental findings, context information on model organisms and experimental techniques is provided with interactions that are important to hematopoiesis. The conversion of the data into complicated networks and downstream bioinformatics applications is made easier by using recognized vocabulary. Insights into stem cell activity, the stem cell niche, and signaling pathways supporting hematopoietic stem cell maintenance are provided through a number of preconfigured datasets. For researchers exploring the area of hematopoiesis, HSC-Explorer offers a flexible web-based resource that enables users to evaluate the related biological processes via an interactive graphical display (Montrone et al., 2013).

- **KeyGenes:** The differentiated cells from human pluripotent stem cells, when cultured in vitro, are phenotypically underdeveloped in contrast to their adult counterparts. Thus, identifying them in an in vitro culture is often difficult to decide with certainty since human fetal equivalents are less known. To resolve this, KeyGenes can be used to evaluate and monitor the differentiation efficiency in vitro (Roost et al., 2015).

- **LifeMap Discovery:** Future fundamental and applied research, including regenerative medical applications, will benefit from the use of LifeMap Discovery to bring significant and useful insights from in vivo development to stem cell studies and clinical applications. It is a database that curates extensive information on embryonic development. Understanding the intricate developmental pathways, patterns of gene expression in developing cells, and signaling cascade that promotes cellular differentiation is crucial for the identification and classification of differentiated stem cell and progenitor cells as well as the creation of more precise protocols for the differentiation of stem cells into desired target cells (Edgar et al., 2013).

- **Moana:** It is based on a hierarchical ML foundation enabling researchers to contrive robust cell-type classifiers/identifiers from a given pool of scRNA-seq datasets. This tool can be scaled to datasets having more than 10000 cells' RNA-seq. Thus, Moana allows researchers to derive tissue-specific/different cell type markers/identifiers or atlas from novel scRNA-seq data (Wagner & Yanai, 2018).

- **Mogrify:** Transdifferentiation is a process via which differentiated cells can be converted from onecell type to another without having to convert them to a pluripotent cell state. However, the identification of responsible transcription factors for direct conversion is currently restricted by expensive and exhaustive experimental testing of possible factors/groups of factors. Mogrify is a predictive computational system that integrates regulatory network information with gene expression data to predict the direct conversion transcription factors essential to bring about cell conversion (Rackham et al., 2016).

- **Monocle:** Monocle groups the cells in a computationally efficient way depending on their progression through differentiation as opposed to the time they were collected to maximize the transcriptional similarity between subsequent pairs of cells. The method first depicts the expression profile of each cell as a point in a highly dimensional Euclidean space, with the dimension of each gene being indicated by a distinct color. Second, it uses independent component analysis to reduce this space's dimensions. Dimensionality reduction reduces the high-dimensional space in which the cell data were originally stored to a low-dimensional space, retaining the vital connections between the populations of cells while making the data easier to view and comprehend. Third, Monocle builds a minimal spanning tree on the cells, an earlier method that is now often employed in other single-cell situations, such as flow or mass cytometry. Fourth, the method determines the longest sequence of transcriptionally comparable cells, which corresponds to the longest path through the MST. This process is then used by Monocle to create a trajectory of how a particular cell differentiates over time (Trapnell et al., 2014).

- **ORIGINS:** Using the PPI network associated with differentiation and the dataset expression matrix, ORIGINS determines a score (differentiation activity) that describes the pluripotency of each cell. The normalized expression profile is used to determine the signal entropy rate for each individual cell. By following the mass action principle, it determines the edges of protein—protein network, which are translated into interaction probabilities (Senra et al., 2022).

- **PluriNetWork:** Based on data from 177 articles published up to June 2010 that involved 274 mouse genes or proteins, a network of 574 molecular interactions, stimulations, and inhibitions was built. Since the data were all in the same electronic format, free programs such as Cytoscape and its plugins could be used to analyze them. The network consists of the Oct4 (Pou5f1), Sox2, and Nanog core circuit, as well as its periphery (Stat3, Klf4, Esrrb, and c-Myc), connections to upstream signaling pathways (such as activin, WNT, FGF, BMP, insulin, and LIF), epigenetic regulators, and a few other relevant genes and proteins, such as nuclear import/export proteins. With the use of a digital data repository, this resource offers information on understanding pluripotency and doing high-throughput data analysis (Som et al., 2010).

- **scmap:** The tool allows researchers to extrapolate a given cellular data derived from scRNA-seq data onto different cellular phenotypes obtained from other experiments. The given cell whose extrapolation is being drawn is used as a reference to identify/group similar cells from a given population of cells. It is available as a web application and also as an R package (Kiselev et al., 2018).

- **scPred:** scPred is a novel generalized approach that combines ML probability-based prediction with unbiased feature selection from a reduced-dimension space to enable extremely accurate categorization of single cells. scPred is a technique that uses orthogonalized gene expression levels and dimensionality reduction to efficiently predict certain cell types or states of single cells from their transcriptional data. The effectiveness of scPred was evaluated by calculating predictions of tumor and nontumor cells based on their transcriptomes and confirming the classification using a cell-specific independent immunohistochemistry investigation focusing on the expression of MLH1 and PMS2 proteins (Alquicira-Hernandez et al., 2019).

- **ScoreCard:** Expression markers and differentiation into each of the three germ layers that determine functional pluripotency are both evaluated using the ScoreCard technique. However, the original ScoreCard had a number of drawbacks, including the inability to differentiate early germ layers, the use of NanoString technology (which is not widely used in laboratories), and the requirement for specialized downstream analysis. To overcome these limitations, a new version of ScoreCard delivers enhanced statistical analysis, accuracy, and utility for a wider range of applications using qPCR measurements of a modified set of genes (Tsankov et al., 2015).

- **StemCellDB:** Several molecular profiles have been combined to create a complete image of undifferentiated human embryonic stem cells and a widespread indication of their propensity to differentiate. Examining the gene expression data is made easier by the search engine it offers (Mallon et al., 2013).

- **StemBase:** To help in the identification of gene activities crucial for stem cell regulation and differentiation, StemBase was first developed as a simple online interface to DNA microarray data produced by the Canadian Stem Cell Network. One of the largest online databases for information on the gene expression of human and mouse stem cells is currently available there (Porter et al., 2007).

- **Stemformatics:** An online gene expression data portal that offers a wide range of user-friendly tools to make it easier for the stem cell community to explore high-quality and pertinent data sets. All processed data can be easily downloaded. A flexible workflow is produced by the option to host the findings of custom analysis for each dataset as stand-alone reports. Additionally, it is a platform for collaboration that has helped a number of initiatives with data processing, analysis, and hosting (Choi et al., 2019).

- **StemMapper:** It is a comprehensive resource for stem cell research built on integrated data for many lineages of human and mouse stem cells, and a carefully maintained gene expression database. It presently comprises over 960 transcriptomes encompassing a wide variety of stem cell types and is based on meticulous selection, uniform processing, and thorough quality checking of pertinent transcriptomics datasets to reduce artifacts. Each integrated dataset underwent a thorough review and human curation. With its simple interface, StemMapper's quality-controlled stem cell gene expression data can be quickly queried, compared, and interactively shown (Pinto et al., 2018).

- **StemCellNet:** It provides the user with a collection of molecular profiles that collectively offer a thorough picture of human embryonic stem cells in their undifferentiated condition, including a broad indication of their capacity for differentiation. The data were used to create a list of 169 gene probes that serve as a pluripotency fingerprint and show how two distinct differentiation conditions can upregulate genes associated with different lineages (Pinto et al., 2014).

- **SyStemCell:** The database includes experimental data from stem cell studies at various levels. The information currently available covers seven levels of stem cell differentiation-related regulatory mechanisms, including DNA CpG 5-hydroxymethylcytosine/methylation, transcript products, histone modification, microRNA-based regulation, phosphorylation proteins, protein products, and transcription factor regulation. These mechanisms were curated from several peer-reviewed publications that were chosen from PubMed (Yu et al., 2012).

- **TeratoScore:** Formation of teratoma is the gold standard assay to test the potential of human pluripotent stem cells to differentiate into embryonic germ layers. TeratoScore has converted this subjective assay into a quantitative measure to assess the potential of engineered pluripotent stem cells to give rise to differentiated cell types. It distinguishes teratomas derived from pluripotent stem cells from that of malignant tumor cells, thus translating the stem cell potency into a quantitative measure (Avior et al., 2015).

- **Waddington-OT:** Effectively reprogrammed cells exhibit early fibroblast identity loss, high levels of proliferative activity, and a mesenchymal-to-epithelial transition before transforming to an iPSC-like state, among other biological characteristics of reprogramming, which is readily uncovered using Waddington-OT. The importance of recognizing alternative cell fates, such as senescence, apoptosis, neural identity, and placental identity, as well as quantifying the percentage of cells in each state at each time point using single-cell resolution and the new model cannot be overstated (Schiebinger et al., 2019).

5. Cancer stem cells

Apart from the conventional stem cells, there are special kind of stem cells associated with cancer. Cancer stem cells (CSCs) are a tiny subset of tumor cells that can self-renew, differentiate, and become tumor-causing when transplanted into an animal host (Walcher et al., 2020). Because of the symmetry of their cell division and changes in their gene expression, CSCs can be identified from other tumor cells (Rosen & Jordan, 2009). Similar to the case studies mentioned in the previous sections, attempts are being made to understand the biology of these cancer stem cell to exploit their potential to relapse cancer and design interventions. For example, in one of the reported studies, mRNA data from two phenotypically similar leukemia were derived via next-generation RNA sequencing. In spite of them being phenotypically similar, their transcriptomic analyses exhibited distinct characteristics and multiple differentially expressed genes. Owing to this difference in gene expression, cells representing the two different leukemia were shown to have varying self-renewal capacities (Wilhelm et al., 2011).

Table 2.2 lists a few databases that have the potential to assist in cancer stem cell research. A brief description of these databases has also been included in the following section.

- **BCSCdb:** This database holds data on three datasets, including correlations between CSC biomarkers, CSC therapeutic target genes, and biomarkers of CSC that have undergone experimental validation. There are three distinct sorts of markers addressed: low-throughput marker (LTM-525), high-throughput marker validated by low-throughput method (283), high-throughput marker (HTM-8307). Additionally, it contains information on clinical trial drugs targeting 10 biomarkers in CSCs, 445 target genes for CSC therapies, 5 distinct kinds of CSC biomarker interaction information, and more (Firdous et al., 2022).
- **CSCdb:** The CSCdb is based on literature on the study of CSCs that includes marker genes, functional annotations, and genes associated with CSCs. It might be a useful tool for locating possible treatment targets and discovering new CSCs. It may also be useful in creating bioinformatics tools to discover novel genes associated with CSCs. The entire data are available for free download and additional investigation (Shen et al., 2016).
- **ESCD:** The datasets on protein overexpression studies, RNAi knockdown, and essential transcription factor—binding locations are all amassed in Embryonic Stem Cell Database (ESCD). This database also contains information related to

TABLE 2.2 Databases associated with cancer stem cells.

Database	Feature	Species	Link	Reference
BCSCdb	Biomarkers of cancer stem cells	Human	http://dibresources.jcbose.ac.in/ssaha4/bcscdb	Firdous et al. (2022)
CSCdb	Biomarkers of cancer stem cells	Human	http://bioinformatics.ustc.edu.cn/cscdb	Shen et al. (2016)
ESCD	Embryonic stem cells	Mouse	https://biit.cs.ut.ee/escd	Jung et al. (2010)
	Embryonic carcinoma cells	Human		
SCDE	CSC data comparison at pathway and gene level	Mouse	http://discovery.hsci.harvard.edu/	Firdous et al. (2022)
		Human		
		Rat		

embryonic carcinoma cells. Gene IDs and GO words are searchable in ESCD. The limited number of data types and datasets offered by ESCD is one of its main flaws (Jung et al., 2010).

- **SCDE:** The Stem Cell Discovery Engine database (SCDE) acts as a repository for CSC data that has been carefully curated and can be used as a foundation for building techniques to compare molecular data for populations relevant to stem cells. Users can reliably characterize, distribute, and evaluate CSC data at the pathway and gene levels using the SCDE (Ho Sui et al., 2012).

6. Conclusion

In stem cell and development biology research, computational methods are crucial. Characterization of the gene expression profile of the embryonic stem cells and foundation for pluripotency has been made easier by high-throughput data. High-throughput approaches have made it feasible to produce a significant amount of diverse data regarding the various aspects of stem cells, and therefore the need to gather, organize, merge, create an analytical platform, and interpret data becomes critical. Numerous computational approaches have been used in recent years, including clustering, ML, and network-based approach, etc. The creation of effective tools to aid in the comprehension of transdifferentiation, differentiation, and reprogramming is a continuous endeavor. Certain web resources are regularly modified or updated. Undoubtedly, new tools will develop in the future. Enhancing our knowledge of the mechanisms governing the preservation of pluripotency and achieving precise control over differentiation, reprogramming, and direct conversion should be a significant area of study.

Author contribution
All the authors have made significant contribution in compilation of the data, writing, and proof reading the manuscript.

References

Alquicira-Hernandez, J., Sathe, A., Ji, H. P., Nguyen, Q., & Powell, J. E. (2019). scPred: accurate supervised method for cell-type classification from single-cell RNA-seq data. *Genome Biology, 20*(1), 264. https://doi.org/10.1186/s13059-019-1862-5

Assenov, Y., Ramírez, F., Schelhorn, S.-E., Lengauer, T., & Albrecht, M. (2008). Computing topological parameters of biological networks. *Bioinformatics, 24*(2), 282−284. https://doi.org/10.1093/bioinformatics/btm554

Avior, Y., Biancotti, J. C., & Benvenisty, N. (2015). TeratoScore: Assessing the differentiation potential of human pluripotent stem cells by quantitative expression analysis of teratomas. *Stem Cell Reports, 4*(6), 967−974. https://doi.org/10.1016/J.STEMCR.2015.05.006

Bhattacharya, B., Miura, T., Brandenberger, R., Mejido, J., Luo, Y., Yang, A. X., Joshi, B. H., Ginis, I., Thies, R. S., Amit, M., Lyons, I., Condie, B. G., Itskovitz-Eldor, J., Rao, M. S., & Puri, R. K. (2004). Gene expression in human embryonic stem cell lines: Unique molecular signature. *Blood, 103*(8), 2956−2964. https://doi.org/10.1182/blood-2003-09-3314

Cahan, P., Li, H., Morris, S. A., Lummertz Da Rocha, E., Daley, G. Q., & Collins, J. J. (2014). CellNet: Network biology applied to stem cell engineering. *Cell, 158*(4), 903−915. https://doi.org/10.1016/J.CELL.2014.07.020

Cao, Z.-J., Wei, L., Lu, S., Yang, D.-C., & Gao, G. (2020). Searching large-scale scRNA-seq databases via unbiased cell embedding with Cell BLAST. *Nature Communications, 11*(1), 3458. https://doi.org/10.1038/s41467-020-17281-7

Choi, J., Pacheco, C. M., Mosbergen, R., Korn, O., Chen, T., Nagpal, I., Englart, S., Angel, P. W., & Wells, C. A. (2019). Stemformatics: Visualize and download curated stem cell data. *Nucleic Acids Research, 47*(D1), D841−D846. https://doi.org/10.1093/nar/gky1064

Clancy, J. L., Patel, H. R., Hussein, S. M. I., Tonge, P. D., Cloonan, N., Corso, A. J., Li, M., Lee, D.-S., Shin, J.-Y., Wong, J. J. L., Bailey, C. G., Benevento, M., Munoz, J., Chuah, A., Wood, D., Rasko, J. E. J., Heck, A. J. R., Grimmond, S. M., Rogers, I. M., ... Preiss, T. (2014). Small RNA changes en route to distinct cellular states of induced pluripotency. *Nature Communications, 5*, 5522. https://doi.org/10.1038/ncomms6522

Dhanoa, J. K., Sethi, R. S., Verma, R., Arora, J. S., & Mukhopadhyay, C. S. (2018). Long non-coding RNA: Its evolutionary relics and biological implications in mammals: A review. *Journal of Animal Science and Technology, 60*, 25. https://doi.org/10.1186/s40781-018-0183-7

Dupont, C., Armant, D. R., & Brenner, C. A. (2009). Epigenetics: Definition, mechanisms and clinical perspective. *Seminars in Reproductive Medicine, 27*(5), 351−357. https://doi.org/10.1055/s-0029-1237423

Edgar, R., Mazor, Y., Rinon, A., Blumenthal, J., Golan, Y., Buzhor, E., Livnat, I., Ben-Ari, S., Lieder, I., Shitrit, A., Gilboa, Y., Ben-Yehudah, A., Edri, O., Shraga, N., Bogoch, Y., Leshansky, L., Aharoni, S., West, M. D., Warshawsky, D., & Shtrichman, R. (2013). LifeMap Discovery™. The embryonic development, stem cells, and regenerative medicine research portal. *PLoS One, 8*(7), e66629. https://doi.org/10.1371/journal.pone.0066629

Everitt, B. S., Landau, S., Leese, M., & Stahl, D. (2011). *Cluster analysis* (5th ed.) Wiley Series in Probability and Statistics.

Firdous, S., Ghosh, A., & Saha, S. (2022). DCSCdb. A database of biomarkers of cancer stem cells. *Database: The journal of biological databases and curation.* https://doi.org/10.1093/DATABASE/BAAC082

Gulati, G. S., Sikandar, S. S., Wesche, D. J., Manjunath, A., Bharadwaj, A., Berger, M. J., Ilagan, F., Kuo, A. H., Hsieh, R. W., Cai, S., Zabala, M., Scheeren, F. A., Lobo, N. A., Qian, D., Yu, F. B., Dirbas, F. M., Clarke, M. F., & Newman, A. M. (2020). Single-cell transcriptional diversity is a hallmark of developmental potential. *Science (New York, N.Y.), 367*(6476), 405−411. https://doi.org/10.1126/science.aax0249

Gunderson, K. L., Steemers, F. J., Lee, G., Mendoza, L. G., & Chee, M. S. (2005). A genome-wide scalable SNP genotyping assay using microarray technology. *Nature Genetics, 37*(5), 549−554. https://doi.org/10.1038/NG1547

Guttman, M., Amit, I., Garber, M., French, C., Lin, M. F., Feldser, D., Huarte, M., Zuk, O., Carey, B. W., Cassady, J. P., Cabili, M. N., Jaenisch, R., Mikkelsen, T. S., Jacks, T., Hacohen, N., Bernstein, B. E., Kellis, M., Regev, A., Rinn, J. L., & Lander, E. S. (2009). Chromatin signature reveals over a thousand highly conserved large non-coding RNAs in mammals. *Nature, 458*(7235), 223−227. https://doi.org/10.1038/nature07672

He, S., Nakada, D., & Morrison, S. J. (2009). Mechanisms of stem cell self-renewal. *Annual Review of Cell and Developmental Biology, 25*, 377−406. https://doi.org/10.1146/ANNUREV.CELLBIO.042308.113248

Herman, J. S., Sagar, & Grün, D. (2018). FateID infers cell fate bias in multipotent progenitors from single-cell RNA-seq data. *Nature Methods, 15*(5), 379−386. https://doi.org/10.1038/nmeth.4662

Ho Sui, S. J., Begley, K., Reilly, D., Chapman, B., McGovern, R., Rocca-Sera, P., Maguire, E., Altschuler, G. M., Hansen, T. A. A., Sompallae, R., Krivtsov, A., Shivdasani, R. A., Armstrong, S. A., Culhane, A. C., Correll, M., Sansone, S.-A., Hofmann, O., & Hide, W. (2012). The stem cell discovery engine: An integrated repository and analysis system for cancer stem cell comparisons. *Nucleic Acids Research, 40*(Database issue), D984−D991. https://doi.org/10.1093/nar/gkr1051

Huang, P., He, Z., Ji, S., Sun, H., Xiang, D., Liu, C., Hu, Y., Wang, X., & Hui, L. (2011). Induction of functional hepatocyte-like cells from mouse fibroblasts by defined factors. *Nature, 475*(7356), 386−391. https://doi.org/10.1038/NATURE10116

Ingolia, N. T., Lareau, L. F., & Weissman, J. S. (2011). Ribosome profiling of mouse embryonic stem cells reveals the complexity and dynamics of mammalian proteomes. *Cell, 147*(4), 789−802. https://doi.org/10.1016/J.CELL.2011.10.002

Jerabek, S., Merino, F., Schöler, H. R., & Cojocaru, V. (2014). OCT4: Dynamic DNA binding pioneers stem cell pluripotency. *Biochimica et Biophysica Acta, 1839*(3), 138−154. https://doi.org/10.1016/j.bbagrm.2013.10.001

Jung, M., Peterson, H., Chavez, L., Kahlem, P., Lehrach, H., Vilo, J., & Adjaye, J. (2010). A data integration approach to mapping OCT4 gene regulatory networks operative in embryonic stem cells and embryonal carcinoma cells. *PLoS One, 5*(5), e10709. https://doi.org/10.1371/journal.pone.0010709

Kamimoto, K., Stringa, B., Hoffman, C. M., Jindal, K., Solnica-Krezel, L., & Morris1, S. A. (2023). Dissecting cell identity via network inference and in silico gene perturbation. *Nature, 614*, 742−751. https://doi.org/10.1101/2020.02.17.947416

Kim, J., & Kim, H. (2012). Recruitment and biological consequences of histone modification of H3K27me3 and H3K9me3. *ILAR Journal, 53*(3−4), 232−239. https://doi.org/10.1093/ilar.53.3-4.232

Kim, J. B., Sebastiano, V., Wu, G., Araúzo-Bravo, M. J., Sasse, P., Gentile, L., Ko, K., Ruau, D., Ehrich, M., van den Boom, D., Meyer, J., Hübner, K., Bernemann, C., Ortmeier, C., Zenke, M., Fleischmann, B. K., Zaehres, H., & Schöler, H. R. (2009). Oct4-induced pluripotency in adult neural stem cells. *Cell, 136*(3), 411−419. https://doi.org/10.1016/J.CELL.2009.01.023

Kiselev, V. Y., Yiu, A., & Hemberg, M. (2018). scmap: projection of single-cell RNA-seq data across data sets. *Nature Methods, 15*(5), 359−362. https://doi.org/10.1038/nmeth.4644

Laurent, L. C., Ulitsky, I., Slavin, I., Tran, H., Schork, A., Morey, R., Lynch, C., Harness, J. V., Lee, S., Barrero, M. J., Ku, S., Martynova, M., Semechkin, R., Galat, V., Gottesfeld, J., Izpisua Belmonte, J. C., Murry, C., Keirstead, H. S., Park, H.-S., … Loring, J. F. (2011). Dynamic changes in the copy number of pluripotency and cell proliferation genes in human ESCs and iPSCs during reprogramming and time in culture. *Cell Stem Cell, 8*(1), 106−118. https://doi.org/10.1016/j.stem.2010.12.003

Lenz, M., Muller, F. J., Zenke, M., & Schuppert, A. (2016). Principal components analysis and the reported low intrinsic dimensionality of gene expression microarray data. *Scientific Reports, 6*, 25696. https://doi.org/10.1038/srep25696

Lienert, F., Mohn, F., Tiwari, V. K., Baubec, T., & Roloff, T. C. (2011). Genomic prevalence of heterochromatic H3K9me2 and transcription do not discriminate pluripotent from terminally differentiated cells. *PLoS Genetics, 7*(6), 1002090. https://doi.org/10.1371/journal.pgen.1002090

Lister, R., Pelizzola, M., Kida, Y. S., Hawkins, R. D., Nery, J. R., Hon, G., Antosiewicz-Bourget, J., O'Malley, R., Castanon, R., Klugman, S., Downes, M., Yu, R., Stewart, R., Ren, B., Thomson, J. A., Evans, R. M., & Ecker, J. R. (2011). Hotspots of aberrant epigenomic reprogramming in human induced pluripotent stem cells. *Nature, 471*(7336), 68−73. https://doi.org/10.1038/nature09798

Li, L., Wang, M., Wang, M., Wu, X., Geng, L., Xue, Y., Wei, X., Jia, Y., & Wu, X. (2016). A long non-coding RNA interacts with Gfra1 and maintains survival of mouse spermatogonial stem cells. *Cell Death & Disease, 7*(3), e2140. https://doi.org/10.1038/cddis.2016.24

Mallon, B. S., Chenoweth, J. G., Johnson, K. R., Hamilton, R. S., Tesar, P. J., Yavatkar, A. S., Tyson, L. J., Park, K., Chen, K. G., Fann, Y. C., & McKay, R. D. G. (2013). StemCellDB: The human pluripotent stem cell database at the National Institutes of health. *Stem Cell Research, 10*(1), 57−66. https://doi.org/10.1016/j.scr.2012.09.002

Mercer, T. R., Dinger, M. E., & Mattick, J. S. (2009). Long non-coding RNAs: Insights into functions. *Nature Reviews Genetics, 10*(3), 155−159. https://doi.org/10.1038/nrg2521

Miranda-Saavedra, D., De, S., Trotter, M. W., Teichmann, S. A., & Göttgens, B. (2009). BloodExpress: A database of gene expression in mouse haematopoiesis. *Nucleic Acids Research, 37*(Suppl. 1_1), D873−D879. https://doi.org/10.1093/nar/gkn854

Mitalipov, S., & Wolf, D. (2009). Totipotency, pluripotency and nuclear reprogramming. *Advances in Biochemical Engineering, 114*, 185−199. https://doi.org/10.1007/10_2008_45

Montrone, C., Kokkaliaris, K. D., Loeffler, D., Lechner, M., Kastenmüller, G., Schroeder, T., & Ruepp, A. (2013). HSC-Explorer: A curated database for hematopoietic stem cells. *PLoS One, 8*(7), e70348. https://doi.org/10.1371/journal.pone.0070348

Murry, C. E., & Keller, G. (2008). Differentiation of embryonic stem cells to clinically relevant populations: Lessons from embryonic development. *Cell, 132*(4), 661−680. https://doi.org/10.1016/J.CELL.2008.02.008

Pinto, J. P., Machado, R. S. R., Magno, R., Oliveira, D. V., Machado, S., Andrade, R. P., Bragança, J., Duarte, I., & Futschik, M. E. (2018). BioiniMapper: A curated gene expression database for stem cell lineage analysis. *Nucleic Acids Research, 46*(D1), D788−D793. https://doi.org/10.1093/nar/gkx921

Pinto, J. P., Reddy Kalathur, R. K., Machado, R. S. R., Xavier, J. M., Bragança, J., & Futschik, M. E. (2014). StemCellNet: An interactive platform for network-oriented investigations in stem cell biology. *Nucleic Acids Research, 42*, W154−W160. https://doi.org/10.1093/nar/gku455. Web Server issue.

Porter, C. J., Palidwor, G. A., Sandie, R., Krzyzanowski, P. M., Muro, E. M., Perez-Iratxeta, C., & Andrade-Navarro, M. A. (2007). StemBase: A resource for the analysis of stem cell gene expression data. *Methods in Molecular Biology, 407*, 137−148. https://doi.org/10.1007/978-1-59745-536-7_11

Rackham, O. J. L., Firas, J., Fang, H., Oates, M. E., Holmes, M. L., Knaupp, A. S., FANTOM Consortium, Suzuki, H., Nefzger, C. M., Daub, C. O., Shin, J. W., Petretto, E., Forrest, A. R. R., Hayashizaki, Y., Polo, J. M., & Gough, J. (2016). A predictive computational framework for direct reprogramming between human cell types. *Nature Genetics, 48*(3), 331−335. https://doi.org/10.1038/ng.3487

Roost, M. S., Van Iperen, L., Ariyurek, Y., Buermans, H. P., Arindrarto, W., Devalla, H. D., Passier, R., Mummery, C. L., Carlotti, F., De Koning, E. J. P., Van Zwet, E. W., Goeman, J. J., & Chuva De Sousa Lopes, S. M. (2015). KeyGenes, a tool to probe tissue differentiation using a human fetal transcriptional atlas. *Stem Cell Reports, 4*(6), 1112−1124. https://doi.org/10.1016/J.STEMCR.2015.05.002

Rosen, J. M., & Jordan, C. T. (2009). The increasing complexity of the cancer stem cell paradigm. *Science (New York, N.Y.), 324*(5935), 1670−1673. https://doi.org/10.1126/science.1171837

Sánchez-Castillo, M., Ruau, D., Wilkinson, A. C., Ng, F. S. L., Hannah, R., Diamanti, E., Lombard, P., Wilson, N. K., & Gottgens, B. (2015). Codex: A next-generation sequencing experiment database for the haematopoietic and embryonic stem cell communities. *Nucleic Acids Research, 43*(D1), D1117−D1123. https://doi.org/10.1093/nar/gku895

Schiebinger, G., Shu, J., Tabaka, M., Cleary, B., Subramanian, V., Solomon, A., Gould, J., Liu, S., Lin, S., Berube, P., Lee, L., Chen, J., Brumbaugh, J., Rigollet, P., Hochedlinger, K., Jaenisch, R., Regev, A., & Lander, E. S. (2019). Optimal-transport analysis of single-cell gene expression identifies developmental trajectories in reprogramming. *Cell, 176*(4), 928−943.e22. https://doi.org/10.1016/j.cell.2019.01.006

Schöler, Hans R. (2016). The potential of stem cells: An inventory. *Humanbiotechnology as Social Challenge*, 45−72. https://doi.org/10.4324/9781315252933-11

Schulz, H., Kolde, R., Adler, P., Aksoy, I., Anastassiadis, K., Bader, M., Billon, N., Boeuf, H., Bourillot, P. Y., Buchholz, F., Dani, C., Doss, M. X., Forrester, L., Gitton, M., Henrique, D., Hescheler, J., Himmelbauer, H., Hübner, N., Karantzali, E., … Hatzopoulos, A. K. (2009). The FunGenES database: A genomics resource for mouse embryonic stem cell differentiation. *PLoS One, 4*(9), e6804. https://doi.org/10.1371/JOURNAL.PONE.0006804

Sebastião, M. J., Serra, M., Gomes-Alves, P., & Alves, P. M. (2021). Stem cells characterization: OMICS reinforcing analytics. *Current Opinion in Biotechnology, 71*, 175−181. https://doi.org/10.1016/j.copbio.2021.07.021

Senra, D., Guisoni, N., & Diambra, L. (2022). Origins: A protein network-based approach to quantify cell pluripotency from scRNA-seq data. *MethodsX, 9*, 101778. https://doi.org/10.1016/j.mex.2022.101778

Shen, Y., Yao, H., Li, A., & Wang, M. (2016). CSCdb: A cancer stem cells portal for markers, related genes and functional information. *Database: The Journal of Biological Databases and Curation*, baw023. https://doi.org/10.1093/database/baw023

Shiraki, N., Shiraki, Y., Tsuyama, T., Obata, F., Miura, M., Nagac, G., Aburatani, H., Kume, K., Endo, F., & Kume, S. (2014). Methionine metabolism regulates maintenance and differentiation of human pluripotent stem cells. *Cell Metabolism, 19*(5), 780−794. https://doi.org/10.1016/J.CMET.2014.03.017

Smith, K. P., Luong, M. X., & Stein, G. S. (2009). Pluripotency: Toward a gold standard for human ES and iPS cells. *Journal of Cellular Physiology, 220*(1), 21−29. https://doi.org/10.1002/JCP.21681

Som, A., Harder, C., Greber, B., Siatkowski, M., Paudel, Y., Warsow, G., Cap, C., Schöler, H., & Fuellen, G. (2010). The PluriNetWork: An electronic representation of the network underlying pluripotency in mouse, and its applications. *PLoS One, 5*(12), e15165. https://doi.org/10.1371/journal.pone.0015165

Stumpf, P. S., & Macarthur, B. D. (2019). *Machine learning of stem cell identities from single-cell expression data via regulatory network archetypes* (Vol 10, p. 2). https://doi.org/10.1101/208470

Takahashi, K., & Yamanaka, S. (2006). Induction of pluripotent stem cells from mouse embryonic and adult fibroblast cultures by defined factors. *Cell, 126*(4), 663−676. https://doi.org/10.1016/J.CELL.2006.07.024

Trapnell, C., Cacchiarelli, D., Grimsby, J., Pokharel, P., Li, S., Morse, M., Lennon, N. J., Livak, K. J., Mikkelsen, T. S., & Rinn, J. L. (2014). The dynamics and regulators of cell fate decisions are revealed by pseudotemporal ordering of single cells. *Nature Biotechnology, 32*(4), 381−386. https://doi.org/10.1038/nbt.2859

Tsankov, A. M., Akopian, V., Pop, R., Chetty, S., Gifford, C. A., Daheron, L., Tsankova, N. M., & Meissner, A. (2015). A qPCR ScoreCard quantifies the differentiation potential of human pluripotent stem cells. *Nature Biotechnology, 33*(11), 1182−1192. https://doi.org/10.1038/nbt.3387

Wagner, F., & Yanai, I. (2018). *Moana: A robust and scalable cell type classification framework for single-cell RNA-seq data*. Preprint. https://doi.org/10.1101/456129

Walcher, L., Kistenmacher, A.-K., Suo, H., Kitte, R., Dluczek, S., Strauß, A., Blaudszun, A.-R., Yevsa, T., Fricke, S., & Kossatz-Boehlert, U. (2020). Cancer stem cells—origins and biomarkers: Perspectives for targeted personalized therapies. *Frontiers in Immunology, 11*. https://doi.org/10.3389/ fimmu.2020.01280

Wang, Z., Gerstein, M., & Snyder, M. (2009). RNA-seq: A revolutionary tool for transcriptomics. *Nature Reviews Genetics, 10*(1), 57—63. https://doi.org/ 10.1038/nrg2484

Warsow, G., Greber, B., Falk, S. S. I., Harder, C., Siatkowski, M., Schordan, S., Som, A., Endlich, N., Schöler, H., Repsilber, D., Endlich, K., & Fuellen, G. (2010). ExprEssence - revealing the essence of differential experimental data in the context of an interaction/regulation net-work. *BMC Systems Biology, 4*(1), 1—18. https://doi.org/10.1186/1752-0509-4-164/FIGURES/14

Wilhelm, B. T., Briau, M., Austin, P., Lie Faubert, A., Ve Boucher, G., Chagnon, P., Hope, K., Girard, S., Mayotte, N., Landry, J.-R., Hé Bert, J., & Sauvageau, G. (2011). RNA-seq analysis of 2 closely related leukemia clones that differ in their self-renewal capacity. *Blood, 117*(2), e27—e38. https://doi.org/10.1182/blood-2010-07-293332

Wooley, J. C., Lin, H. S., Biology, N. R. C., & (US) C. on F. at the I. of C. and. (2005). *On the nature of biological data.* https://www.ncbi.nlm.nih.gov/ books/NBK25464/.

Xu, H., Baroukh, C., Dannenfelser, R., Chen, E. Y., Tan, C. M., Kou, Y., Kim, Y. E., Lemischka, I. R., & Ma'ayan, A. (2013). Escape: Database for integrating high-content published data collected from human and mouse embryonic stem cells. *Database: The Journal of Biological Databases and Curation*, bat045. https://doi.org/10.1093/database/bat045

Yu, J., Xing, X., Zeng, L., Sun, J., Li, W., Sun, H., He, Y., Li, J., Zhang, G., Wang, C., Li, Y., & Xie, L. (2012). SyStemCell: A database populated with multiple levels of experimental data from stem cell differentiation research. *PLoS One, 7*(7), e35230. https://doi.org/10.1371/journal.pone.0035230

Zhang, A. W., O'Flanagan, C., Chavez, E. A., Lim, J. L. P., Ceglia, N., McPherson, A., Wiens, M., Walters, P., Chan, T., Hewitson, B., Lai, D., Mottok, A., Sarkozy, C., Chong, L., Aoki, T., Wang, X., Weng, A. P., McAlpine, J. N., Aparicio, S., … Shah, S. P. (2019). Probabilistic cell-type assignment of single-cell RNA-seq for tumor microenvironment profiling. *Nature Methods, 16*(10), 1007—1015. https://doi.org/10.1038/s41592-019-0529-1

Zhang, H., Shao, X., Peng, Y., Teng, Y., Saravanan, K. M., Zhang, H., Li, H., & Wei, Y. (2019). A novel machine learning based approach for iPS progenitor cell identification. *PLoS Computational Biology, 15*(12). https://doi.org/10.1371/journal.pcbi.1007351

Zhou, Q., Brown, J., Kanarek, A., Rajagopal, J., & Melton, D. A. (2008). In vivo reprogramming of adult pancreatic exocrine cells to beta-cells. *Nature, 455*(7213), 627—632. https://doi.org/10.1038/NATURE07314

Chapter 3

Stem cell informatics: Web resources aiding In stem cell research

Rabiya Ahsan[1,a], Lubna Maryam[2,a] and Salman Sadullah Usmani[3]

[1]Department of Pharmacology, Faculty of Pharmacy, Integral University, Lucknow, Uttar Pradesh, India; [2]Independent Researcher, Bronx, NY, United States; [3]Department of Molecular Pharmacology, Albert Einstein College of Medicine, Bronx, NY, United States

1. Introduction

Since the beginning of life, organisms have been trying to evade and combat the possibilities and reasons behind morbidity and mortality. Malfunctioning of the vital organs remains one of the prime reasons, and the inadequacy of transplantable organs always remains a challenge. In the past century, two breakthroughs, i.e., James A. Thomson et al. derived the human pluripotent embryonic stem cells (hPESCs) (Thomson et al., 1998) and animal cloning by Ian Wilmut et al. (Wilmut et al., 1997) revolutionized the tissue engineering field, where soft tissue replacement is always a challenge. Although several techniques such as alloplastic implants, autologous fat transplantation, and autologous tissue flap are being used in soft tissue reconstruction, but face severe challenges such as implant migration, donor-site morbidity, and foreign body rejection (Li et al., 2023; Zeiderman & Pu, 2021). The role of autologous stem cells (ASCs) prolonged in vitro as well as collectively with unique biomaterials for organ reconstruction provides a potential solution for organ as well as tissue replacement (Foo et al., 2021; Hong, 2022). Stem cells are unspecialized, undifferentiated cells, majorly existing in bone marrow, adipose tissue, cord blood, etc. and have the ability to differentiate into various specialized cell types as well as self-renewal potential by dividing and creating more stem cells. Due to its vast utility in drug development, disease modeling, and tissue replacement or regeneration, stem cell research is continuously proving as a frontier of regenerative medicine (Mahla, 2016; Rajabzadeh et al., 2019).

Embryonic stem cells (ESCs) have shown great promise in the recovery of spinal cord and liver injuries as well as in cartilage repair (Shroff & Gupta, 2015; Tolosa et al., 2015). Tissue-specific progenitor stem cells (TSPSCs) show remarkable progress in regenerating goblet mucosa, corneal tissue, muscle fibrils, ischemic myocardium, etc. (Schmidt et al., 2019; Wang et al., 2022). Mesenchymal stem cells (MSCs) have been explored for their benefits in several diseases such as bladder deformities, dental problems, bone and muscle degeneration, and alopecia. It has been used in regenerating the hair follicle to treat alopecia (Kim et al., 2020; Krefft-Trzciniecka et al., 2023; Ong et al., 2021). Similarly, umbilical cord stem cells (UCSCs) have been studied for their application in restoring tissue repairs in heart diseases, cartilage, and tendon injuries (Abbaszadeh et al., 2020; Alatyyat et al., 2020). Bone marrow stem cells (BMSCs) have broad applications in treating aplastic anemia, hematological malignancies, blood clotting disorders, orodental deformities, and diaphragm abnormalities (Chu et al., 2020; Comazzetto et al., 2021). In the same way, induced pluripotent stem cells (iPSCs) have shown remarkable utility as a therapeutics in eye defects, neurodegenerative disorders, liver and lung disease, as well as in the treatment of acquired immune deficiency syndrome (AIDS) (Calvert & Ryan Firth, 2020; Chitena et al., 2022; Okano & Morimoto, 2022). Besides, stem cell research is a useful tool to investigate gene expression and physiological pathways during the developmental stage and identify new prognostic and diagnostic biomarkers of human genetic diseases (Ben-Yosef et al., 2008; Nelson et al., 2010). Even using stem cells to treat heart disease is approved by the US Food and Drug Administration (FDA) (Sharma et al., 2021). Even ambitious attempts are being made to conserve wildlife by preserving endangered animals and reanimating extinct species; by the vast utility of stem cells, especially iPSCs, led to the concept of

---

[a]Contributed equally as first author.

Computational Biology for Stem Cell Research. https://doi.org/10.1016/B978-0-443-13222-3.00023-X
Copyright © 2024 Elsevier Inc. All rights reserved, including those for text and data mining, AI training, and similar technologies.

creating a frozen zoo. This concept involves the preservation of the gene pool and germplasm from endangered species, and in future, these preserved tissues can be reprogrammed into different cells and tissues (Korody et al., 2021; Ryder & Onuma, 2018). However, stem cell research faces many challenges, especially ethical concerns related to the use of ESCs, as well as difficulties in directing stem cells to differentiate into other specific cell types. In addition, there are always a few risks associated with stem cell therapies, such as tumor development and immune rejection (Andrews & Kriegstein, 2022; Malchenko et al., 2014).

Advancements in high-throughput sequencing techniques augmented stem cell research, too, by producing a plethora of data. This opens tremendous opportunities for the bioinformatician and computational biologists similar to other relevant biological fields such as antibiotic resistance, immunoinformatics, therapeutic peptides, etc. (Maryam et al., 2021; Usmani et al., 2019; Usmani et al., 2017; Usmani et al., 2018). They have contributed significantly by developing computer models and stem cell simulations to understand their behavior better. These models can help researchers predict how stem cells respond to different stimuli, such as environmental cues or growth factors. They can also aid in identifying new targets for therapeutic interventions. There is a continuous growth in the number of publications related to stem cell−based research deposited in PubMed; even in 2022, ∼15,000 articles are being shortlisted with the keyword "stem cell" used either in the title or abstract of the article, and the cumulative count is 20,076 on May 12, 2023. This number itself speaks for the enormous data available in the field of stem cell research and the need for compilation and curation in a comprehensive manner. Not to be surprised, several bioinformatic and computational researchers noted this opportunity and are trying to maintain this data integrity for the betterment and aid in stem cell−based research. In addition, computational biologists have developed tools for analyzing large-scale datasets generated from stem cell experiments, which can help identify genes and pathways involved in stem cell differentiation and development. Overall, the contributions of computational biologists have helped to accelerate progress in stem cell research, bringing us closer to realizing the potential of stem cells for regenerative medicine. Therefore, in this article, we have tried to provide a comprehensive list of web-based resources, which are freely available to the research community and are continuously aiding the growth of stem cell research (Table 3.1).

Hematopoietic stem cell-Explorer (HSC-Explorer): HSC-Explorer is an integrative database containing full information about the hematopoiesis initial steps, while differentiating from most primitive HSCs to multipotent progenitor cells (MPP) in adult mice. It is a freely available web-based resource and provides fast and easy information related to interacting molecules, cell types, and bioprocess by visualization interfaces. The database contains manually derived structured information from scientific literature for more than 7000 experimentally verified interactions between biological processes, molecules, and environmental components. It also contains several predefined datasets regarding stem cell niche, activity, behavior, and signaling pathways associated with stem cell maintenance. An insight into stem cell behaviors is offered through predefined published datasets. HSC-Explorer is a flexible, useful web-based resource with multiple search options for researchers working in the field of hematopoiesis as it provides an interactive graphical presentation to investigate the related biological processes.

Induced pluripotent stem cells (iPSCs): iPSCs possess the remarkable ability to both self-renew and differentiate into all types of cells within tissues. Most of the advances and discoveries have been made by utilizing pluripotent stem cells (PSCs). In the cellular therapeutic domain, clinically applied human iPSC lines act as a starting material; therefore, demonstration of comparability of cell lines derived across diverse facilities is very crucial. This required the agreement protocol for the iPSCs used of lines and the assays. The Global Alliance for iPSC Therapies (GAiT) is currently building a specified repository for clinical-grade iPSCs, and they recommend that information about candidate lines should be registered on the human pluripotent stem cell registry (hPSCReg, https://hpscreg.eu/), which currently has information about 864 hESC lines, 3498 hiPSC lines, and 110 clinical studies (Kurtz et al., 2022; Sullivan et al., 2020).

CellNet: This computational platform was designed in 2014 to evaluate the cellular population and assess the accuracy of cellular engineering and create accurate hypothesis for the improvement of cell derivation. Patrick Cahan et al. utilized the expression data from 56 reports to conclude that directly differentiated cells bear a resemblance to their in vivo counterparts than the final product via direct conversion (Cahan et al., 2014). Besides, cells undergoing into the direct conversion could not succeed into silencing the expression programs. In the same study, by using the CellNet, authors confirm that iPSCs are identical to ESCs while establishing the gene regulatory network. They have also demonstrated that when iPSCs differentiate directly into the neurons and cardiomyocytes, they establish the target tissue as well as the gene regulatory network more efficiently as compared with the fibroblast-derived neurons and cardiomyocytes. Overall, CellNet is a network biology platform, which evaluates the extent of similarity and compatibility between the engineered cells and their target cell type, and therefore can aid in designing strategies for better cellular engineering.

CODEX: It is a platform for multicenter genomic datasets containing open-source next-generation sequencing (NGS) data and is curated in a manner that is easily accessible and friendly to experimental biologists. The ultimate aim of

TABLE 3.1 Web resources—related to stem cells.

Name	Link	Description	References
HSC-Explorer	https://mips.helmholtz-muenchen.de/HSC	Manually curated database containing information about the hematopoiesis	(Montrone et al., 2013)
iPSCs	www.gait.global/the-ipsc-database/	Containing the information about the quality control instruction for clinical-grade iPSCs lines	(Sullivan et al., 2018)
CellNet	https://github.com/pcahan1/CellNet	Assesses the accuracy of cellular engineering and creating accurate hypothesis for enriching cell fate	(Cahan et al., 2014)
CODEX	https://codex.stemcells.cam.ac.uk	Resource for the HSCs and ESCs related all kinds of NGS data	(Sánchez-Castillo et al., 2015)
UESC	http://scgap.systemsbiology.net/	Powerful web tool cell-type gene expression patterns and bladder and prostate's immunohistochemistry images	(Pascal et al., 2007)
CORTECON	https://cortecon.neuralsci.org/	Database of temporal gene expression dataset covering cerebral cortical development from hESCs	(van de Leemput et al., 2014)
SCDE	https://dataverse.harvard.edu/dataverse/stemcellcommons	Web tool for analyzing CSCs	(Ho Sui et al., 2012)
StemChecker	http://stemchecker.sysbiolab.eu/	Identifying the genes associated with stemness	(Pinto et al., 2015)
StemCellDB	https://stemcelldb.nih.gov/	Database of hiPSCs and hESCs, maintained by National Institute of Health	(Mallon et al., 2013)
StemFormatics	https://www.stemformatics.org/	Contains stem cell—related gene expression data	(Choi et al., 2019)
StemMapper	http://stemmapper.sysbiolab.eu/	Human and mouse stem cell expression database	(Pinto et al., 2018)
ESCdb	http://dbresources.jcbose.ac.in/ssaha4/bcscdb	Database of CSCs biomarkers	(Firdous et al., 2022)
CSCdb	(http://bioinformatics.ustc.edu.cn/cscdb)	Database of CSCs for biomarkers and related gene expression information	(Shen et al., 2016)
GeneSigDB	http://compbio.dfci.harvard.edu/genesigdb/http://www.genesigdb.org	Resource for gene expression signature analysis	(Culhane et al., 2012)

CODEX is to unify hematopoietic and ESC-related NGS data; hence, it contains two specialized components: HAEM-CODE for blood cells and ESCODE for ESCs-related data. DNase-seq, RNA-seq, histone modification, and transcription factor ChIP-Seq (chromatin immunoprecipitation coupled with sequencing) data generated from over 1000 samples, which includes 147 specific cell types and 292 special TFs, have been stored in CODEX. All these data, which are related to cell specification and providence in the framework of mammalian development, homeostasis mechanism, and diseases, are easily accessible for visualization as well as the processed dataset for download.

Urologic epithelial stem cell (UESCs): UESC provides a gene expression and protein informatory database of the prostate cancer cell line, all major prostate, and urinary bladder cell types. It contains two kinds of data, i.e., protein abundance localization data, which were derived from immunohistochemistry images, and transcript abundance data originating mostly from DNA microarray analysis. It basically provides the dataset of cell information and powerful tools to evaluate differential gene expression between normal versus prostate cancer. This web resource is a part of the prostate and bladder component of the NIH/NIDDK Stem Cell Genome Anatomy Projects Consortium.

CORTECON: This online platform provides a complete dataset of temporal gene expression comprehending cerebral cortical development from hESCs. Various neurological diseases affect the cerebral cortex, so this database provides information about gene code and changes in RNAs coding during neuronal-associated diseases. Leemput et al. utilized a modified method to develop cortical neurons from hESCs over a period of 77 days, and RNA sequencing was carried out throughout this process to establish an extensive transcriptome database covering human corticogenesis in vitro (van de Leemput et al., 2014). In addition, various analyses were conducted, such as the Kyoto Encyclopedia of Genes and Genomes (KEGG) pathway analysis, gene ontology (GO) categorization, disease association investigation, and alternative splicing examination. In last, all the data were made publicly available, and the web resource offers a way to search for information on how RNA expression and alternative splicing of transcripts change during the development of the human cortex. Its web portal provides multiple options to search the dataset, such as view by gene, disease, KEGG pathway, and GO categorization.

Stem cell discovery engine (SCDE): The SCDE is a cutting-edge platform designed to easily explore curated cancer stem cells (CSCs) data, thus enabling users to enhance their understanding of the unique and intricate molecular processes associated with CSCs. This platform enables users to access a wealth of CSC-related data, which allows them to both discover and share valuable insights. By utilizing the gene-level analysis, researchers can compare their own data with existing datasets, in that way identifying key pathways associated with CSCs in a straightforward manner. SCDE provides high-quality resources containing 1098 assays and 53 public studies. SCDE is also coupled with Galaxy analytical framework, thus providing the feature of comparing gene lists against molecular signatures available in the Gene Signatures Database (GeneSigDB) and Molecular Signatures Database (MSigDB), as well as comparing with pathways curated in WikiPathways (Liberzon et al., 2011, 2015). SCDE database provides a comprehensive collection of well-organized experimental information encompassing assays, derived gene lists, and pathway profiles. The database offered by SCDE presents a comprehensive collection of well-organized experimental information encompassing assays, derived gene lists, and pathway profiles. What sets SCDE apart is its distinct focus on fostering a community-oriented approach, which involves identifying relevant experiments, integrating advanced analysis capabilities compared with previous resources, and diligently capturing and curating essential study information. To tackle the challenge of heterogeneity, SCDE employs rigorous manual curation to ensure the accuracy and reliability of experimental models, disease states, cell and tissue types, surface markers, and other pertinent data.

StemChecker: It is a novel stemness analysis tool developed by utilizing nearly 50 published stemness signatures defined by transcription factor—binding sites, gene expression, and RNAi screens. They also surveyed various studies related to murine or a set of human genes associated with maintenance and stem cell identity, as well as retrieved stem cell—related transcription factors from published data on ChIP-Seq and ChIP-chip. Thus aggregating 132 stemness signature and transcription factor target gene sets. StemChecker proves to be an invaluable resource for individuals seeking information on the association of their genes of interest with stem cell properties. This powerful tool provides not only detailed statistical results in a user-friendly table format but also presents intuitive graphical displays. By utilizing StemChecker, users can uncover transcriptional regulatory programs, shedding light on the potential involvement of stemness-related processes in diseases such as cancer. With its comprehensive approach, StemChecker significantly enhances the existing collection of online tools dedicated to assisting the research communities of stem cell biology, regenerative medicine, and human disease investigation.

Stem Cell Database (StemCellDB):. StemCellDB contains a wealth of molecular profiling assays for 21 hESC lines and 8 hiPSC lines. It grants users access to comprehensive data pertaining to gene expression, miRNA array analysis, array-based comparative genomic hybridization, and DNA methylation analysis. Additionally, the data can be freely accessed through the NCBI GEO public database. Distinguished as a single-gene platform, this resource empowers users to explore the expression patterns of individual genes across diverse cultural conditions. Furthermore, it provides curated gene lists that highlight enrichments observed during differentiation, indicating potential preferences toward distinct cell fates.

Stemformatics: It encompasses a vast collection of 420 public gene expression datasets obtained from diverse sources such as microarray, single-cell profiling, and RNA sequencing. Developed with a strong focus on meeting the specific needs of the stem cell research community, this platform incorporates numerous user friendly features that simplify data exploration and custom analysis. Notably, Stemformatics offers tools such as Yugene Graph, gene expression profile plots, and other intuitive graph plotting options to enhance data visualization. To ensure data reliability and usefulness, Stemformatics applies stringent quality control metrics when selecting datasets from raw data, resulting in a more compact and valuable resource. One of the key strengths of Stemformatics lies in its ability to facilitate the creation, sharing, and analysis of gene sets. By providing a repository of community-annotated stem cell gene lists, Stemformatics equips researchers with informative insights into common technical artifacts, lineage commitment, and pathways. Overcoming a common challenge in stem cell signature analysis, Stemformatics addresses the scarcity of publicly available stem cell or developmental biology pathways/gene lists. This database empowers biologists with convenient and intuitive tools, allowing them to visually explore the data through interactive gene expression profiles, principal component analysis plots, hierarchical clusters, and more. Additionally, the inclusion of tools for examining snapshots of gene expression in multiple cells and tissues assists in identifying cell-type restricted genes or potential housekeeping genes.

StemMapper: Recognized as the premier freely available and extensive resource for stem cell research, this manually curated gene expression database stands out in its excellence. It encompasses a remarkable collection of 798 mice and 166 human transcriptomes, providing comprehensive coverage of expression profiles for 51 distinct types of murine stem cells, progenitor cells, and their respective lineages, as well as 19 types of human stem cells, progenitor cells, and their progeny. The database boasts a user-friendly query interface, enabling researchers to effortlessly navigate and retrieve desired information. Furthermore, a suite of interactive visualization and analysis tools are seamlessly integrated, empowering the stem cell research community with invaluable resources for in-depth exploration and analysis.

Biomarkers of the Cancer Stem Cells database (BCSCdb): BCSCdb is a meticulously curated database compiled from PubMed literature, offering a comprehensive collection of both high-throughput and low-throughput markers specifically associated with CSCs. This database currently encompasses CSC biomarkers derived from cancer cell lines and primary tissues, originating from 10 distinct types of CSCs. Moreover, BCSCdb provides additional information, such as therapeutic target genes and interaction data, whenever available.

Each biomarker within BCSCdb is assigned two distinct scores: the confidence score and the global score. The confidence scoring system evaluates the experimental validation of the biomarker, thereby aiding in assessing its reliability within the CSC population. On the other hand, the global scoring system analyzes the frequency of CSC biomarkers across the 10 different cancer types documented in BCSCdb. This scoring mechanism aids in the identification of both local CSC biomarkers specific to particular cancer types and global CSC biomarkers that exhibit broader relevance across various cancer types.

Cancer Stem Cells database (CSCdb): CSCdb is an extensive database that encompasses CSCs-related genes/microRNAs, marker genes, and comprehensive functional annotations. It offers in-depth information on 1769 genes known to play a significant role in CSCs functional regulation, as well as 74 marker genes that can aid in the identification or isolation of CSCs. The data within CSCdb have been meticulously collected from approximately 13,000 articles through manual curation. Additionally, CSCdb provides 9475 annotations related to 13 distinct functions associated with CSCs, including radioresistance, differentiation, oncogenesis, tumorigenesis, and more. These annotations offer detailed insights into the identified genes, encompassing protein function descriptions, GO information, protein–protein interaction details, posttranscription modification information, regulatory relationships, and relevant literature. The integration of these annotations within CSCdb facilitates easy access to comprehensive information for users. Overall, CSCdb serves as a comprehensive and invaluable resource for researchers engaged in CSCs research, providing a wealth of information related to various aspects of CSC biology.

Gene Signatures Database (GeneSigDB): It is a precisely curated gene signature database carefully compiled and organized from published literature. It serves as a standardized and valuable resource for the scientific community, providing diagnostic and prognostic gene signatures for cancer and related diseases. Researchers can utilize this resource to compare the predictive power of gene signatures or employ them in gene set enrichment analysis (GSEA). Since its initial release as GeneSigDB 1.0, the database has undergone significant expansion, now comprising 3515 gene signatures derived from 1604 published articles. The focus areas of the collected signatures encompass stem cells, immune cells, gene expression in cancer and development, as well as lung disease. To enhance accessibility and user experience, substantial upgrades have been implemented on the GeneSigDB website. Noteworthy additions include a tag cloud browse function, a "basket" feature for storing genes or gene signatures of interest, and a faceted navigation system.

GeneSigDB provides multiple download options, allowing users to obtain data in various formats, such as the.gmt file format for GSEA or as R/Bioconductor data files. Users can explore and analyze gene signatures from GeneSigDB and even upload their own gene lists to identify significant gene overlaps. The output of the analysis can be visualized through an editable heatmap, which can be downloaded as high-quality images suitable for publication. To ensure transparency and traceability, GeneSigDB provides a normalized gene list, primary gene signature, and a comprehensive gene mapping history for each gene, from its transcription in the original data table to its inclusion in the normalized gene list. In summary, GeneSigDB serves as a user-friendly and extensive resource, enabling researchers to access, analyze, and compare gene signatures while providing valuable tools for GSEA and diagnostic applications.

2. Discussion

Stem cell research is continuously proving to be a very advanced field; therefore, biomedical data integration and analysis gained tremendous importance. Over the decades, we have witnessed the emergence of the bulk of high-throughput sequencing data, which embarked on the need to establish methods to cumulate and share the data among researchers. Therefore, many databases and computational resources have been developed over the years to store the data from several experiments and stem cell projects. We have comprehensively examined the existing stem cell repositories to identify their key characteristic specification and observed a remarkable trend; stem cell therapy is increasingly emerging as a remarkable medical solution. Malfunctioning vital organs and scarcity of transplantable organs are persistent issues in combating the underlying cause of morbidity and mortality. Stem cell therapy offers immense potential for addressing these challenges and revolutionizing healthcare.

However, there is considerable variation in the implementation and structure of the stored data. Efficient and accessible data aggregation for stem cell research presents several challenges. The challenges faced in aggregating data for stem cell research are closely parallel to the challenges encountered in high-throughput sequencing data aggregation. Sequencing technologies generate vast amounts of data but often need more standardization and curation. To overcome these challenges, data aggregates integrating "omics" datasets must employ standardized data formats and establish robust methodologies for data curation. Furthermore, relating raw data to a biological context pose an additional significant hurdle in integrating omics data into research. The evolution and proliferation of sequencing technologies have led to increasingly granular and platform-specific data, making integration across different datasets challenging. Additionally, interpreting these data within a biological framework can be complex. Consequently, utilizing ontologies for biological data aggregation has gained significant importance recently.

By leveraging ontologies, researchers can address the challenges of integrating diverse and platform-specific data in a unified manner. Ontologies provide a structured framework for organizing and harmonizing data, enabling meaningful comparisons and analysis across multiple datasets. Incorporating ontologies in data aggregation efforts enhances the ability to interpret and derive valuable insights from the complex and heterogenous omics data generated in stem cell research. Embracing ontologies in the aggregation of biological data empowers researchers to unlock the full potential of omics technologies, facilitating comprehensive analyses and advancing our understanding of stem cell biology. Furthermore, it enables researchers to leverage the knowledge and annotations already available within these frameworks. It allows for identifying relevant terms, concepts, and relationships associated with stem cell biology, thus promoting a functional understanding of the data generated. In summary, integrating stem cell research into existing ontologies addresses the challenge of standardizing datasets and facilitates the interpretation of data from a functional perspective. This advancement holds immense potential for advancing stem cell database development and driving further discoveries.

3. Conclusion

This chapter presents a comprehensive compilation of diverse web resources focused on stem cell research. These resources provide valuable information on stem cells and offer various types of gene signature data. We have meticulously gathered these resources from esteemed journals, ensuring their credibility and reliability. Our analysis of these stem cell—related computational resources has revealed the inclusion of extensive information on gene expression and gene signatures tailored to specific stem cell types. For instance, CSCdb contains vital information related to cancer cell signatures, while other databases cater to different stem cell types. This chapter is a valuable resource for biologists and researchers keen on exploring gene expression identification within the context of stem cells. Furthermore, it aims to contribute to the growth and advancement of the stem cell community.

Acknowledgements

Conceptualization, S.S.U., writing and original draft preparation, R.A., L.M., S.S.U.; writing review and editing, S.S.U., supervision, S.S.U. All authors have read and agreed to this published version of the manuscript

References

Abbaszadeh, H., Ghorbani, F., Derakhshani, M., Movassaghpour, A., & Yousefi, M. (2020). Human umbilical cord mesenchymal stem cell-derived extracellular vesicles. A novel therapeutic paradigm. *Journal of Cellular Physiology, 235*(2), 706−717. https://doi.org/10.1002/jcp.29004

Alatyyat, S. M., Alasmari, H. M., Aleid, O. A., Abdel-Maksoud, M, S., & Elsherbiny, N. (2020). Umbilical cord stem cells: Background, processing and applications *Tissue and Cell, 65*, 101351. https://doi.org/10.1016/j.tice.2020.101351

Andrews, M. G., & Kriegstein, A. R. (2022). Challenges of organoid research. *Annual Review of Neuroscience, 45*, 23−39. https://doi.org/10.1146/annurev-neuro-111020-090812

Ben-Yosef, D., Malcov, M., & Eiges, R. (2008). PGD-derived human embryonic stem cell lines as a powerful tool for the study of human genetic disorders. *Molecular and Cellular Endocrinology, 282*(1−2), 153−158. https://doi.org/10.1016/j.mce.2007.11.010

Cahan, P., Li, H., Morris, S. A., Lummertz da Rocha, E., Daley, G. Q., & Collins, J. J. (2014). CellNet: Network biology applied to stem cell engineering. *Cell, 158*(4), 903−915. https://doi.org/10.1016/j.cell.2014.07.020

Calvert, B. A., & Ryan Firth, A. L. (2020). Application of iPSC to modelling of respiratory diseases. *Advances in Experimental Medicine and Biology, 1237*, 1−16. https://doi.org/10.1007/5584_2019_430

Chitena, L., Masisi, K., Masisi, K., Kwape, T. E., & Gaobotse, G. (2022). Application of stem cell therapy during the treatment of HIV/AIDS and duchenne muscular dystrophy. *Current Stem Cell Research and Therapy, 17*(7), 633−647. https://doi.org/10.2174/1574888x16666210810104445

Choi, J., Pacheco, C. M., Mosbergen, R., Korn, O., Chen, T., Nagpal, I., … Wells, C. A. (2019). Stemformatics: Visualize and download curated stem cell data. *Nucleic Acids Research, 47*(D1), D841−d846. https://doi.org/10.1093/nar/gky1064

Chu, D. T., Phuong, T. N. T., Tien, N. L. B., Tran, D. K., Thanh, V. V., Quang, T. L., … Kushekhar, K. (2020). An update on the progress of isolation, culture, storage, and clinical application of human bone marrow mesenchymal stem/stromal cells. *International Journal of Molecular Sciences, 21*(3). https://doi.org/10.3390/ijms21030708

Comazzetto, S., Shen, B., & Morrison, S. J. (2021). Niches that regulate stem cells and hematopoiesis in adult bone marrow. *Developmental Cell, 56*(13), 1848−1860. https://doi.org/10.1016/j.devcel.2021.05.018

Culhane, A. C., Schroder, M. S., Sultana, R., Picard, S. C., Martinelli, E. N., Kelly, C., … Quackenbush, J. (2012). GeneSigDB: A manually curated database and resource for analysis of gene expression signatures. *Nucleic Acids Research, 40*, D1060−D1066. https://doi.org/10.1093/nar/gkr901. Database issue.

Firdous, S., Ghosh, A., & Saha, S. (2022). *BCSCdb: A database of biomarkers of cancer stem cells*. Database (Oxford). https://doi.org/10.1093/database/baac082

Foo, J. B., Looi, Q. H., Chong, P. P., Hassan, N. H., Yeo, G. E. C., Ng, C. Y., … Law, J. X. (2021). Comparing the therapeutic potential of stem cells and their secretory products in regenerative medicine. *Stem Cells International, 2021*, 2616807. https://doi.org/10.1155/2021/2616807

Ho Sui, S. J., Begley, K., Reilly, D., Chapman, B., McGovern, R., Rocca-Sera, P., … Hide, W. (2012). The stem cell discovery engine: An integrated repository and analysis system for cancer stem cell comparisons. *Nucleic Acids Research, 40*, D984−D991. https://doi.org/10.1093/nar/gkr1051. Database issue.

Hong, I. S. (2022). Enhancing stem cell-based therapeutic potential by combining various bioengineering technologies. *Frontiers in Cell and Developmental Biology, 10*, 901661. https://doi.org/10.3389/fcell.2022.901661

Kim, G. B., Seo, M. S., Park, W. T., & Lee, G. W. (2020). Bone marrow aspirate concentrate: Its uses in osteoarthritis. *International Journal of Molecular Sciences, 21*(9). https://doi.org/10.3390/ijms21093224

Korody, M. L., Ford, S. M., Nguyen, T. D., Pivaroff, C. G., Valiente-Alandi, I., Peterson, S. E., … Loring, J. F. (2021). Rewinding extinction in the northern white rhinoceros: Genetically diverse induced pluripotent stem cell bank for genetic rescue. *Stem Cells and Development, 30*(4), 177−189. https://doi.org/10.1089/scd.2021.0001

Krefft-Trzciniecka, K., Pietowska, Z., Nowicka, D., & Szepietowski, J. C. (2023). Human stem cell use in androgenetic alopecia: A systematic review. *Cells, 12*(6). https://doi.org/10.3390/cells12060951

Kurtz, A., Mah, N., Chen, Y., Fuhr, A., Kobold, S., Seltmann, S., & Muller, S. C. (2022). Human pluripotent stem cell registry: Operations, role and current directions. *Cell Proliferation, 55*(8), e13238. https://doi.org/10.1111/cpr.13238

Liberzon, A., Birger, C., Thorvaldsdottir, H., Ghandi, M., Mesirov, J. P., & Tamayo, P. (2015). The Molecular Signatures Database (MSigDB) hallmark gene set collection. *Cell System, 1*(6), 417−425. https://doi.org/10.1016/j.cels.2015.12.004

Liberzon, A., Subramanian, A., Pinchback, R., Thorvaldsdottir, H., Tamayo, P., & Mesirov, J. P. (2011). Molecular signatures database (MSigDB) 3.0. *Bioinformatics, 27*(12), 1739−1740. https://doi.org/10.1093/bioinformatics/btr260

Li, G., Chen, T., Dahlman, J., Eniola-Adefeso, L., Ghiran, I. C., Kurre, P., … Sundd, P. (2023). Current challenges and future directions for engineering extracellular vesicles for heart, lung, blood and sleep diseases. *Journal of Extracellular Vesicles, 12*(2), e12305. https://doi.org/10.1002/jev2.12305

Mahla, R. S. (2016). Stem cells applications in regenerative medicine and disease therapeutics. *International Journal of Cell Biology*, 6940283. https://doi.org/10.1155/2016/6940283

Malchenko, S., Xie, J., de Fatima Bonaldo, M., Vanin, E. F., Bhattacharyya, B. J., Belmadani, A., … Bonas, M. B. (2014). Onset of rosette formation during spontaneous neural differentiation of hESC and hiPSC colonies. *Gene, 534*(2), 400−407. https://doi.org/10.1016/j.gene.2013.07.101

Mallon, B. S., Chenoweth, J. G., Johnson, K. R., Hamilton, R. S., Tesar, P. J., Yavatkar, A. S., ... McKay, R. D. (2013). StemCellDB: The human pluripotent stem cell database at the national institutes of health. *Stem Cell Research, 10*(1), 57−66. https://doi.org/10.1016/j.scr.2012.09.002

Maryam, L., Usmani, S. S., & Raghava, G. P. S. (2021). Computational resources in the management of antibiotic resistance: Speeding up drug discovery. *Drug Discovery Today, 26*(9), 2138−2151. https://doi.org/10.1016/j.drudis.2021.04.016

Montrone, C., Kokkaliaris, K. D., Loeffler, D., Lechner, M., Kastenmüller, G., Schroeder, T., & Ruepp, A. (2013). HSC-Explorer: A curated database for hematopoietic stem cells. *PLoS One, 8*(7), e70348. https://doi.org/10.1371/journal.pone.0070348

Nelson, T. J., Martinez-Fernandez, A., & Terzic, A. (2010). Induced pluripotent stem cells: Developmental biology to regenerative medicine. *Nature Reviews Cardiology, 7*(12), 700−710. https://doi.org/10.1038/nrcardio.2010.159

Okano, H., & Morimoto, S. (2022). iPSC-based disease modeling and drug discovery in cardinal neurodegenerative disorders. *Cell Stem Cell, 29*(2), 189−208. https://doi.org/10.1016/j.stem.2022.01.007

Ong, W. K., Chakraborty, S., & Sugii, S. (2021). Adipose tissue: Understanding the heterogeneity of stem cells for regenerative medicine. *Biomolecules, 11*(7). https://doi.org/10.3390/biom11070918

Pascal, L. E., Deutsch, E. W., Campbell, D. S., Korb, M., True, L. D., & Liu, A. Y. (2007). The urologic epithelial stem cell database (UESC) - a web tool for cell type-specific gene expression and immunohistochemistry images of the prostate and bladder. *BMC Urology, 7*, 19. https://doi.org/10.1186/1471-2490-7-19

Pinto, J. P., Kalathur, R. K., Oliveira, D. V., Barata, T., Machado, R. S., Machado, S., ... Futschik, M. E. (2015). StemChecker: A web-based tool to discover and explore stemness signatures in gene sets. *Nucleic Acids Research, 43*(W1), W72−W77. https://doi.org/10.1093/nar/gkv529

Pinto, J. P., Machado, R. S. R., Magno, R., Oliveira, D. V., Machado, S., Andrade, R. P., ... Futschik, M. E. (2018). StemMapper: A curated gene expression database for stem cell lineage analysis. *Nucleic Acids Research, 46*(D1), D788−d793. https://doi.org/10.1093/nar/gkx921

Rajabzadeh, N., Fathi, E., & Farahzadi, R. (2019). Stem cell-based regenerative medicine. *Stem Cell Investigation, 6*, 19. https://doi.org/10.21037/sci.2019.06.04

Ryder, O. A., & Onuma, M. (2018). Viable cell culture banking for biodiversity characterization and conservation. *Annual Review of Animal Bioscience, 6*, 83−98. https://doi.org/10.1146/annurev-animal-030117-014556

Sánchez-Castillo, M., Ruau, D., Wilkinson, A. C., Ng, F. S., Hannah, R., Diamanti, E., ... Gottgens, B. (2015). Codex: A next-generation sequencing experiment database for the haematopoietic and embryonic stem cell communities. *Nucleic Acids Research, 43*, Database issue. D1117−D1123. https://doi.org/10.1093/nar/gku895

Schmidt, M., Schüler, S. C., Hüttner, S. S., von Eyss, B., & von Maltzahn, J. (2019). Adult stem cells at work: Regenerating skeletal muscle. *Cellular and Molecular Life Sciences, 76*(13), 2559−2570. https://doi.org/10.1007/s00018-019-03093-6

Sharma, V., Dash, S. K., Govarthanan, K., Gahtori, R., Negi, N., Barani, M., ... Ojha, S. (2021). Recent advances in cardiac tissue engineering for the management of myocardium infarction. *Cells, 10*(10). https://doi.org/10.3390/cells10102538

Shen, Y., Yao, H., Li, A., & Wang, M. (2016). *CSCdb: A cancer stem cells portal for markers, related genes and functional information.* Database (Oxford). https://doi.org/10.1093/database/baw023

Shroff, G., & Gupta, R. (2015). Human embryonic stem cells in the treatment of patients with spinal cord injury. *Annals of Neurosciences, 22*(4), 208−216. https://doi.org/10.5214/ans.0972.7531.220404

Sullivan, S., Ginty, P., McMahon, S., May, M., Solomon, S. L., Kurtz, A., ... Turner, M. L. (2020). The global alliance for iPSC therapies (GAiT). *Stem Cell Research, 49*, 102036. https://doi.org/10.1016/j.scr.2020.102036

Sullivan, S., Stacey, G. N., Akazawa, C., Aoyama, N., Baptista, R., Bedford, P., ... Song, J. (2018). Quality control guidelines for clinical-grade human induced pluripotent stem cell lines. *Regenerative Medicine, 13*(7), 859−866. https://doi.org/10.2217/rme-2018-0095

Thomson, J. A., Itskovitz-Eldor, J., Shapiro, S. S., Waknitz, M. A., Swiergiel, J. J., Marshall, V. S., & Jones, J. M. (1998). Embryonic stem cell lines derived from human blastocysts. *Science, 282*(5391), 1145−1147. https://doi.org/10.1126/science.282.5391.1145

Tolosa, L., Caron, J., Hannoun, Z., Antoni, M., López, S., Burks, D., ... Dubart-Kupperschmitt, A. (2015). Transplantation of hESC-derived hepatocytes protects mice from liver injury. *Stem Cell Research and Therapy, 6*, 246. https://doi.org/10.1186/s13287-015-0227-6

Usmani, S. S., Agrawal, P., Sehgal, M., Patel, P. K., & Raghava, G. P. S. (2019). *ImmunoSPdb: An archive of immunosuppressive peptides.* Database (Oxford). https://doi.org/10.1093/database/baz012

Usmani, S. S., Bedi, G., Samuel, J. S., Singh, S., Kalra, S., Kumar, P., & Raghava, G. P. S. (2017). THPdb: Database of FDA-approved peptide and protein therapeutics. *PLoS One, 12*(7), e0181748. https://doi.org/10.1371/journal.pone.0181748

Usmani, S. S., Kumar, R., Bhalla, S., Kumar, V., & Raghava, G. P. S. (2018). In silico tools and databases for designing peptide-based vaccine and drugs. *Advanced Protein Chemical Structure and Biology, 112*, 221−263. https://doi.org/10.1016/bs.apcsb.2018.01.006

van de Leemput, J., Boles, N. C., Kiehl, T. R., Corneo, B., Lederman, P., Menon, V., ... Fasano, C. A. (2014). Cortecon: A temporal transcriptome analysis of in vitro human cerebral cortex development from human embryonic stem cells. *Neuron, 83*(1), 51−68. https://doi.org/10.1016/j.neuron.2014.05.013

Wang, X., Wang, R., Jiang, L., Xu, Q., & Guo, X. (2022). Endothelial repair by stem and progenitor cells. *Journal of Molecular and Cellular Cardiology, 163*, 133−146. https://doi.org/10.1016/j.yjmcc.2021.10.009

Wilmut, I., Schnieke, A. E., McWhir, J., Kind, A. J., & Campbell, K. H. (1997). Viable offspring derived from fetal and adult mammalian cells. *Nature, 385*(6619), 810−813. https://doi.org/10.1038/385810a0

Zeiderman, M. R., & Pu, L. L. Q. (2021). Contemporary approach to soft-tissue reconstruction of the lower extremity after trauma. *Burns Trauma, 9*, tkab024. https://doi.org/10.1093/burnst/tkab024

Chapter 4

Stem cell based informatics development and approaches

Anshuman Chandra[1], Nainee Goyal[2], Nagendra Singh[2], Vijay Kumar Goel[1], Shilpi Agarwal[1] and Aditya Arya[3]

[1]School of Physical Sciences, Jawaharlal Nehru University, New Delhi, Delhi, India; [2]School of Biotechnology, Gautam Buddha University, Greater Noida, Uttar Pradesh, India; [3]Vector Biology Laboratory, National Institute of Malaria Research, New Delhi, Delhi, India

1. Introduction

All multicellular living organisms carry out their essential functions by orchestrating the functioning and cross-talk of various cells, most of which are specialized in their operations and commonly referred to as differentiated cells. However, every cell has a finite life span and needs to be replaced periodically. Unlike plants, animals do not have an extensive repertoire of those cells, which are capable to differentiate into any cell, yet, a very specialized set of cells that can perform self-renewal and differentiate into a variety of cells to replenish the ever-needed repertoire are stem cells. Historically, the notion of stem cells originated in the late 19th century. It was reportedly adopted from a German term "Stammzelle" by Ernst Haeckel and Theodor Boveri, describing these specialized cells powered with self-renewal and differentiation abilities (Maehle, 2011; Ramalho-Santos & Willenbring, 2007). Haekel (1868) stated in their lecture that stem cells have likely emerged from some of the primitive forms of life, possibly bacteria and associated them analogous to embryological development. Among the other proponents of the idea of stem cells were hematologists Artur Pappenheim and Ernst Neumann, who did not wholly agree with the dualist view of stem cells proposed by Haekel and kept their focus on developmental possibilities of cells or pluripotency. During this period, the exact meaning of "stem cells" kept on fluctuating across different proposals. Stem cells are also known by alternate names such as primordial germ cells or mother cells. These terms represent relatively broader aspects of stem cells, including their differentiation and functional abilities (Maehle, 2011).

Moreover, aspects such as clonality and potency are also included in the definition and further classification of stem cells. Stem cells are present in both embryos and adult tissues. It is now known that the stem cells are derived from early embryos postblastocyst stage or from fetal, postnatal, or adult sources. As an embryo matures into an adult, the potency for development and differentiation is reduced with progressing age (Zakrzewski et al., 2019). The initial part of the 21st century witnessed a rapid expansion of stem cell research and opened new paradigms owing to their enormous potential from regeneration to therapeutic interventions. With the expansion of stem cell research and the identification of various stem cells, it became imperative to classify them for better comprehension.

The initial types of stem cells discovered were a small cluster of cells within murine blastocysts (Evans & Kaufman, 1981), which were also identified later in human embryos as blastocysts (Thomson et al., 1998), most of such stem cells were categorized as embryonic stem cells or ESCs. As ESCs are pluripotent in nature, and they are capable of differentiating into all somatic cells, it is possible to inject early-stage blastocyst into the embryo and form progeny of all three embryonic germ layers in vitro, a technique that led to the emergence of ESC transfer—based transgenic formation (Smith, 2001). Soon after the embryo develops, it turns into a fetus, and a large proportion of cells get differentiated. However, some cells still maintain their stemness and are categorized as fetal stem cells (O'Donoghue & Fisk, 2004). The fetal stem cells, depending on where they are housed in fetus, are named neural crest stem cells, hematopoietic stem cells, and fetal mesenchymal cells (Gulliot et al., 2006). It is interesting to note that some parts of the extra embryonic tissues, especially umbilical cord, placenta, and amniotic fluid, are also known to harbor their independent stem cells, often called perinatal stem cells (Witkowska-Zimny & Wrobel, 2011).

Computational Biology for Stem Cell Research. https://doi.org/10.1016/B978-0-443-13222-5.00026-5
Copyright © 2024 Elsevier Inc. All rights reserved, including those for text and data mining, AI training, and similar technologies.

Perinatal stem cells have recently gained the attention of regenerative clinicians and are foreseen as a promising tool for regeneration and tissue repair (Si et al., 2015). Not all stem cells are diminished after the embryo is matured into an adult, but they are retained in adults and categorized into a separate category called adult stem cells (ASCs) or tissue-specific stem cells (TSSCs) (Moroni & Fornasari, 2013). TSSCs are also multipotent and self-renewing cells, and they have abilities to perform specialized endogenous functions for tissue renewal and repair (van der Kooy & Weiss, 2000). Adult stem cells are ubiquitously distributed throughout a variety of tissues, yet, it is evident from the literature that not all the cells of the repertoire have equal potentials to differentiate and proliferate or are affected by external factors; rather, ASCs are controlled by the inherent properties for undergoing successful single lineage-specific differentiation (Pizzute et al., 2015). There is significant heterogeneity in the ASCs, and several tissues are now known to have them including lung epithelium, gastrointestinal tract, mammary gland, prostate gland, liver, ovaries, heart, muscles and mesenchymal stroma. Sometimes, mesenchymal stem cells (MSCs) or stromal cells are treated as a separate category of stem cells. It was initially thought that stem cells are always of natural origin, and genetic factors entirely influence their stemness. However, the myth was shattered by the enormous efforts of Shina Yamanaka, who showed the possibilities of transforming somatic cells into stem cells. This modulated category of stem cells got popularized as induced pluripotent stem cells (iPSCs). Yamanaka et al. showed that iPSCs could be generated using adult mouse tail-tip fibroblasts and embryonic fibroblasts as starting cell type. They used retrovirus-mediated transfection of various transcription factors and observed that mouse iPSCs were indistinguishable from ESCs in various molecular and physiological aspects. Some of the transcription factors involved in this process were Klf4, c-Myc, and Oct3/4. Also, when transplanted into blastocysts, mouse iPSCs could form adult chimeras, just like conventional stem cells (Takahashi & Yamanaka, 2006). It is also noteworthy that the potency of all types of stem cells is not alike, some have the ability to differentiate in a complete organism and are called totipotent. ESCs have totipotent nature and can be considered analogous to the meristematic tissue of plants. The stem cells, which can divide into most or all cell types in an organism but cannot develop into an entire organism on their own, are classified as pluripotent stem cells. Some other cells, which can differentiate into many or a few types, are categorized as multipotent and omnipotent, respectively. An outline classification of stem cells based on their potential and origin has been depicted in Fig. 4.1.

Stem cell—based informatics development and approaches have emerged as crucial tools in the field of stem cell research and regenerative medicine. Stem cells possess remarkable potential for tissue regeneration and disease modeling, making them a valuable resource for biomedical applications. Informatics development in this context involves the integration of computational methods, data analysis techniques, and bioinformatics tools to gain a deeper understanding of stem cell biology, lineage differentiation, and therapeutic applications. These approaches enable researchers to analyze large-scale omics data, including genomics, transcriptomics, proteomics, and epigenomics, to unravel the complex molecular mechanisms underlying stem cell behavior. By utilizing informatics tools, scientists can identify key regulatory networks, signaling pathways, and genetic factors involved in stem cell maintenance, self-renewal, and differentiation. Furthermore, stem cell informatics aids in the development of predictive models, computational simulations, and data-driven hypotheses, facilitating the design of targeted therapies and personalized medicine approaches. Through the integration of informatics and stem cell research, novel insights into stem cell biology are being uncovered, paving the way for advancements in regenerative medicine and the potential for tailored therapeutic interventions.

2. Applications of stem cells: a promise to behold

Stems cells hold a promising future, especially in regenerative medicine. Therefore, most research for the past two decades has centered on their regenerative potential and possible use in clinics. Although ESCs have more potency than adult stem cells, yet their use is limited due to ethical concerns (Lo & Parham, 2009). However, ASCs have the potential in future to augment the regenerative potential of the human body and therefore support the development of novel therapies (Fig. 4.2). The use of numerous ASCs (such as bone marrow, skin, and MSCs) as well as umbilical cord stem cells has a promising future for translational regenerative medicine (Bajada et al., 2008). Bone marrow—derived stem cells have been successfully used for the regeneration of various organs, including the skeletal system, heart, and bladder (Mashim et al., 2019). Recent reports have evidently shown that the transfer of stem cell progenitors to the heart results in a favorable impact on tissue perfusion and contractile performance of the injured heart. Various applications of stem cells in regenerative heart therapies have been reviewed recently by Wollert and Drexler (2005).

Moreover, the application of stem cells in regenerative medicine has also been tried in curing type I diabetes (Fiorina et al., 2011), interdisciplinary dentistry (Raj et al., 2013), and orthopedics and musculoskeletal medicine (Khan et al., 2012). There are currently two proposed categories of stem cell application in regenerative medicine. First is autologous transplantation, where stem cells are isolated from the patient, and the second is allogenic transplantation, where stem cells

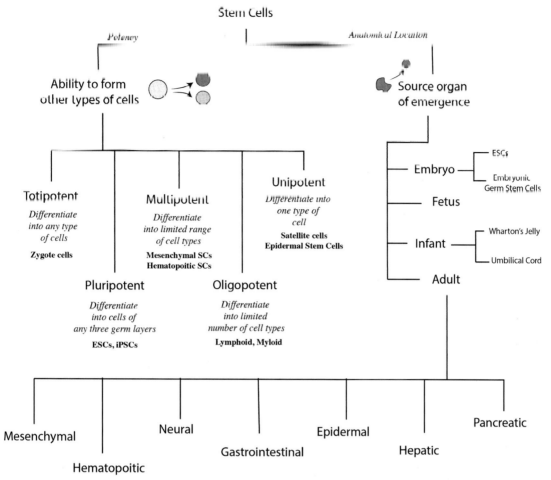

FIGURE 4.1 Various types of stem cells classified on the basis of their potency or origin. (Note: the exact classification of stem cells may still be debatable and subjective in nature and hence could vary across texts).

are obtained from other donors (Jassak & Riley, 1994). Later, the isolated cells are proliferated in vitro and either injected directly into the patient to substitute lost cells or seeded using a three-dimensional scaffold. These cells then differentiate into the required cell type. The former is popular as stem cell therapy, while the latter is known as tissue engineering. Recent studies have established that TSSCs can regenerate cells of tissues from dissimilar organs, which further opens the doors for allogenic transplants. Proof-of-concept studies have already been demonstrated in transmural myocardial infraction (MI) for possible prevention of cardiac failure (Liu et al., 2004). In another study, scleroderma patients were treated with myeloablative CD34+ selected autologous HSCs transplantation along with immunosuppression by means of monthly infusions of cyclophosphamide for a year (Sullivan et al., 2018). Also, Attal et al. (2003) have shown a comparison of single autologous stem-cell transplantation after high-dose chemotherapy and double transplantation, where the latter was found to be better in overall survival among patients with myeloma.

Allogeneic stem cell transplantation is an intensive curative treatment preferred for malignancies with high risk. However, it has been observed that many patients have high early mortality from relapse while others respond. A minor proportion has also shown a second chance at curing (Barret & Battiwalla, 2010). Research also demonstrates the protection and practicability of MSC transfusion in young children, who have already undergone allogeneic stem cell transplantation. Moreover, it is also speculated that patients with posttransplant complications may have advantage from MSC owing to its immunomodulatory effect (Muller et al., 2008). A few studies have identified circumstances where allogeneic stem cell transplantation is possibly considered a preferred treatment in patients with chronic B cell lymphocytic leukemia. Some indicators for the transplant have been defined on the basis of consensus (Dreger et al., 2007). It has been demonstrated that allogeneic stem cell transplant has been able to cure thousands of patients suffering from lethal diseases. Despite this being an excellent promise for the future, it is also crucial that the incidences of post-SCT complications

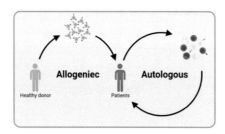

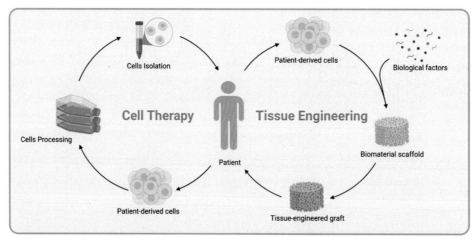

FIGURE 4.2 Approaches in stem cell therapies. (A) Autologous versus autologous transplant, (B) cell therapy versus tissue engineering.

should also be minimized. The two core risk factors, chronic graft versus host disease and pretransplantation conditioning, have been poorly acquiescent to the modifications. However, the arrival of reduced intensity conditioning for stem cell therapy and new immunosuppressive drugs is likely to open the gates for a curative procedure with relatively few ill effects (Socié et al., 2003).

3. Expanding horizons of omics and advent of informatics

Since the advent of DNA sequencing, genomics has escalated rapidly. The availability of next-generation sequencing methods at much lower cost and minimal time consumption has tremendously boosted fundamental biology research. Platforms such as Illumina and oxford nanopore have conferred extensive coverage and portability, respectively, in genome sequencing. Nevertheless, stem cell research has also benefitted to a great extent from genomic advancements. Whole-genome sequencing of hematopoietic stem cells has proven to improve our understanding of hematopoiesis. Lee-Six and David Kent have shown that sequencing of progenitors deciphered mutations that occurred in stem cells. These accumulated mutations may affect the phylogeny and therefore impact the entire process of the hematopoiesis. As they developed a statical model for hematopoiesis based on the genome sequencing data, the mapping of mutations has been beneficial in predicting the odds of development of leukemia in future and therefore may emerge as a valuable diagnostic tool in near future (Lee-Six & Kent, 2020). Besides complete genome sequencing, genomic profiling has proven revolutionary in characterizing stem cells. Documentation of putative biomarkers and identification of key molecules that regulate fundamental biological processes engaged in the survival, growth, and development of stem cells have begun to unravel. Numerous studies have scrutinized the genomic outlines of stem cell progenies isolated from different tissues and shown differential levels of proliferation and diversification abilities (Menicanin et al., 2009). Gallego Romero et al. (2015) have explored distinct applications of iPSC genomics by developing a panel of fully characterized iPSC lines derived from healthy chimpanzee donors and observed slight variation in iPSCs in contrast to somatic cells within the species. This has provided a hint that the stem cell reprogramming process erases several interindividual differences. Another side of the molecular studies on stem cells involves a downstream approach to analyze proteomes. Besides genomics, proteomic strategies have also been used to characterize human MSCs and cells derived from them. Although proteomics outcomes have provided valuable data, the complete inference is still restricted by the number of proteins that have been characterized in these cells, thus demanding a stretchy proteome dataset of MSCs (Park et al., 2007). Furthermore, the proteomic analysis of MSCs has unraveled a distinct set of proteins representing the basic molecular battery, which includes cell surface markers, growth factor responses, signaling cascades, their cross-talk with the extracellular matrix, and combat against cellular stress, suggesting entirely new aspects of stem cell functioning and potency (Maurer, 2011). Proteomics studies of neuronal stem cells are yet another instance of the expansion of proteomic data in stem cell research. It has been shown how, during neuronal repair and stress, various signatures change in proteomic compartments, which provides valuable insight into neuronal cell differentiation (Shoemaker & Kornblum, 2016). Furthermore, concurrent research on metabolomics has also shown that the cells show a marked difference in the metabolome owing to the tissue of origin, tier potency, and the mechanism of differentiation.

The discovery of metabolic markers in rheumatoid arthritis (RA), osteoarthritis (OA), and several other diseases seems to be an emerging aspect of omics oriented stem cell research (Kim et al., 2019). While genomic data of every cell are several gigabases large, proteomic data, on the other hand, are much more diverse, and metabolic data are even further diverse in both temporal and spatial scales. These ever-increasing data and experimental outcomes warrant the need for highly organized and channelized informatics workflow. The use of information technology has continuously been part of genomics, proteomics, and metabolomics studies. Yet, the gap between three popular omics technologies and the unavailability of an integrated omics workflow limits most of the interomics inference derivation in stem cell research. The knowledge gained from interomics studies and the incorporation of computational power aided with retrospective result analysis shall improve our knowledge of the molecular signaling involved in stem cell growth, differentiation, and survival. Also, it will augment the development of novel strategies for the improvement of the tissue regeneration capacities, which is certainly a boon for the alleviation of clinical manifestations. Moreover, the use of bioinformatics tools to compile broader protein datasets and the development of one-stop databases may help us achieve a reference map of their genome, proteome, and metabolome.

4. Network biology and stem cell informatics

Network biology and stem cell informatics have already begun to play a crucial role in disease diagnosis and therapy by providing valuable insights into the complex molecular mechanisms underlying diseases and identifying potential therapeutic targets. Some of the common approaches with which diagnosis and therapies can be explored in stem cell research include analysis of disease networks, biomarker discovery, drug repurposing, drug response prediction, and precision medicine. Network biology allows researchers to construct comprehensive molecular networks that represent the interactions between genes, proteins, and other molecules involved in disease processes. By integrating data from various omics sources and analyzing these networks, researchers can identify key disease-associated genes, pathways, and signaling cascades. This information helps in understanding the underlying molecular mechanisms of diseases and can guide the development of diagnostic biomarkers.

Stem cell informatics facilitates the analysis of large-scale omics data from patient-derived stem cells, enabling the identification of disease-specific molecular signatures. By comparing the gene expression, epigenetic patterns, and protein profiles of healthy and diseased stem cells, researchers can identify potential biomarkers for early disease detection, prognosis, and monitoring of therapeutic responses. Network biology and stem cell informatics can be used to identify potential drug targets and repurpose existing drugs for specific diseases. By integrating molecular networks with drug databases and computational modeling, researchers can predict the effects of drugs on disease-associated pathways and prioritize potential therapeutic candidates. Stem cell informatics can also aid in identifying drugs that can modulate specific stem cell behaviors, such as differentiation or self-renewal, to promote tissue regeneration and repair. Stem cell informatics can help predict drug responses and identify potential adverse effects by analyzing the molecular signatures of patient-derived stem cells. By comparing the gene expression profiles of stem cells from responders and nonresponders to specific drugs, researchers can identify predictive biomarkers for treatment outcomes, allowing for more effective and targeted therapies. Finally, the combination of network biology and stem cell informatics holds great promise for personalized medicine approaches. Patient-specific stem cells, derived through reprogramming or other techniques, can be used to model diseases in the laboratory. By analyzing the omics data from these patient-derived stem cells and integrating them with molecular networks, researchers can identify individual-specific disease mechanisms and develop personalized treatment strategies tailored to a patient's unique molecular profile.

There are several systems biology—based tools, which have emerged in recent times and have proven to be highly valuable for various detailed analysis of high-throughput data generated from scientific experiments on stem cells. Some of the highly useful tools have been tabulated in Table 4.1.

TABLE 4.1 Some commonly used tools in systems biology analysis in stem cell research.

Sr. no.	Tool name	Useful for	Link
1	MINDy	Identification of putative regulatory activity of a gene	http://wiki.c2b2.columbia.edu/workbench/index.php/MINDy
2	SCENIC	Single-cell data–based cell analysis	http://scenic.aertslab.org
3	VIPER	Integrative Genomics viewer with a web application	https://github.com/MarWoes/viper
4	ARACNe	Network reconstruction	https://bio.tools/aracne
5	Pluritest	Prediction of pluripotency	http://www.pluritest.org
6	Teratoscore	Quantitative estimation of pluripotency	http://benvenisty.huji.ac.il/teratoscore.php

The MINDy algorithm, known as Modulator Inference by Network Dynamics, utilizes gene expression data to assess if a potential modulator gene (Mj) affects the regulatory activity of a transcription factor gene (TF) on a group of target genes (Ti). This influence is quantified by examining whether there is a modification in the correlation (measured as mutual information) of expression between the TF and its targets Ti when there is a change in the expression of Mj. The alteration in correlation is determined by calculating the disparity in mutual information (referred to as delta (MI)) for each TF–Ti pair under two conditions: high or low modulator expression. The mutual information values employed in MINDy are computed using the ARACNe algorithm, which is also an integral component of geWorkbench (Margolin et al., 2006).

Aibar et al. have recently developed a computation method SCENIC that helps us understand how genes control each other and identify different types of cells using single-cell RNA-seq data (a technique to study individual cells). By analyzing data from tumors and the brain, we show that studying how genes interact can help us find important factors that control gene activity and identify different cell types. SCENIC gives us valuable information about why cells are different from each other and how they work (Aibar et al., 2017). In yet another computational innovation, Martin Woste et al. have developed VIPER, the Variant InsPector and Expert Rating tool that combines the Integrative Genomics Viewer with a web application (Wöste & Dugas, 2018). This integration enables analysts to efficiently examine and evaluate variants, apply filters, and make decisions using the visual representations and accompanying variant information. VIPER has proven to be effective in analyzing over 10,000 calls through manual inspection. ARACNe, yet another algorithm, demonstrates an accurate reconstruction of networks when the impact of loops in the network structure is minimal, as supported by asymptotic analysis. The algorithm is effective in practical scenarios, even when confronted with intricate topologies and a multitude of loops. It has been tested for transcriptional regulatory network reconstruction, using both synthetic datasets that closely resemble real-world scenarios and an actual microarray dataset obtained from human B cells (Muller et al., 2011). Teratoscore was developed to provide a quantitative evaluation of the differentiation capability of human pluripotent stem cells (hPSCs) based on the gene expression patterns observed in teratomas. This algorithm relies on the fundamental understanding that teratoma formation is widely regarded as a reliable benchmark for assessing the potency of hPSCs, which refers to their capacity to differentiate into cell types derived from all three germ layers. Additionally, Teratoscore has the ability to classify whether a tumor originates from a specific tissue or from pluripotent cells, further enhancing its analytical capabilities (Avior et al., 2015).

5. Recent developments in stem cell informatics

Informatics has now emerged as an important tool to augment stem cell research, and several subdomains of bioinformatics, in particular, docking, molecular dynamics, simulations, genome, and proteome analysis, are playing a vital role in deciphering new molecular mechanisms and reinventing novel cascades of stem cell differentiation. Moreover, this is paving the way for realizing stem cell clinical applications. Babak Arjmand recently reviewed the applications of the computational drug design method and highlighted that a cost-effective and time-saving approach could be developed for cancer treatment. The docking and informatics approaches can increase efficiency and value in pharmaceutical research and therefore provide an added advantage (Arjmand et al., 2022). The in silico experimental data from the studies of Mayank and Jaitak have shown the effect of two drugs namely, emetine and cortistat, can modulate the hedgehog pathway by binding to the sonic hedgehog, Gli, and smoothen proteins. As these proteins are involved in maintenance of cancer stem cells, the outcomes could be a strong motivation for future research. Furthermore, some other drugs targeting cancer stem cells screened by in silico screening methods include solamargine, cyclopamine and sonasomine (Mayank & Jaitak, 2018). These computational models for stem cells are being developed to decipher mutations caused during the DNA replication in stem cells and might answer questions related to the combination of these mutations in the rapid cancer progression. Moreover, it may also shed some light on the impact of more multifaceted regulatory networks on the robustness of the system against malfunctions (Khorasani et al., 2020). Several molecular dynamics studies have also been conducted using computational tools to unravel the molecular mechanisms perturbing or regulating the basic molecular processes of stem cells. In one study, accelerated cell differentiation toward adipose phenotype was observed by the application of tetrahertz radiations. The molecular mechanism revealed by molecular dynamics indicated the activation of transcription factor peroxisome proliferator—activated receptor gamma (*PPARG*) and revealed that the local breathing dynamics of the PPARG promoter DNA coincides with the gene-specific response to the THz radiation (Bock, 2010). Identification of the immunological biomarkers of cholangiocarcinoma has been performed using informatic tools for transcriptomic data analysis with considerable success. The utilization of molecular docking tools, in conjunction with transcriptional biomarkers, has established the potential of novel benzamide-linked small molecules. Furthermore, the use of molecular docking of ligand—receptor interactions has suggested the involvement of CHOL-hub genes (Lawal et al., 2021). Glioblastoma multiform (GBM), a type of tumor formation, is associated with glioma stem cells (GSCs), leading to resistance to conventional therapies. However, random computer-aided drug design (CADD) and target identification were performed, where researchers identified signal transducer and activator of transcription 3 (STAT3), glycogen synthase kinase 3β, and the cluster of differentiation 44 (CD44), and β-catenin, as potential druggable candidates of a novel molecule NSC765689 (Mokagautsi et al., 2021). In yet another study, molecular modeling of c-KIT/PDGFRα dual inhibitors was explored using computational tools for exploring the potential treatment of gastrointestinal stromal tumors. They used a three-dimensional quantitative structure—activity relationship (3D-QSAR) and contour map analysis to identify the best possible orientations of inhibitor binding that could increase the repressive activity against PDGFRα as well as c-KIT (Keretsu et al., 2020). High-throughput data from iPSCs are also useful for providing insight into biomarkers discovery of stem cell behavior. Derivation of transcriptomics signatures of diseases from the transcriptomic profiles of iPSCs of humans has made it possible to perform drug repositioning for various diseases, e.g., a study-integrated transcriptomics with a structure-based approach to discover potential drugs for treating Noonan syndrome (NS) and Leopard syndromes (LS) using informatics approach (Zhu et al., 2020). It is also proposed that the signaling mediated by non-proteinogenic small organic molecules can act as a conduit among complex systems and remains one of the most thrilling challenges for stem cell research. Informatics can provide excellent opportunities for understanding the aforementioned hypothesis. In a recent study, the differentiation of stem cells of the immune system was observed to be affected by the low doses of gamma amino butyric acid (GABA), a potential neurotransmitter. This was concurrent with the vanishing of c-kit+ and Sca-1+ cells in the thymus, pronephros, peripheral blood mononuclear cells, and spleen. Further exploration using the molecular docking approach and multiparametric analysis inferred that stem cell differentiation depends on a subtle balance of adverse and constructive interactions of GABA, and hence, interaction within the niche of HSCs facilitates their differentiation (Vega-López et al., 2019). Metastasis results in nearly 90% of cancer-related deaths and remains one of the major challenges to cancer researchers. Involvement of cancer stem cells has been demonstrated in metastasis; therefore, they remain a potential target (Shiozawa et al., 2013). The use of molecular association, docking, and simulation studies has aided in the refinement of molecular ligands, which can behave as potential inhibitors or drugs for metastasis mediated by stem cells. Mustofa et al. explored the antimigration, antiproliferative, and cancer stemness inhibition activity of derivatives of a drug candidate *N*-phenyl pyrazoline using in vitro culture of cervical cancer. They also suggested the potential protein target using computer-aided predictions in Swiss Target Prediction and AutoDock Vina (Mustofa Satriyo et al., 2022). Those aforementioned and a lot of ongoing research suggest that the role of informatics in stem cell research is making a huge difference. However, it may not always be conclusive, and inference must also be drawn by augmenting in vitro or in vivo research.

6. Future of the stem cell informatics

Systems biology will likely be the future of informatics-related developments in stem cell research. It is now increasingly becoming possible that the integration of omics approaches and availability of high-throughput data of various omics experiments can provide a better bird's eye view and thus help delineate the mechanisms in a realistic sense. Knowing a global regulation at the systems level can facilitate refining the errors made during restricted studies limited to one of two molecular events. It is evident from some of the studies that systems biology has already been steeped in and set the threshold of the modern informatics era in stem cell research. Shreds of evidence show the development of artificial three-dimensional bone marrow–ike scaffolds that model the natural HSC niche in vitro by using polydimethylsiloxane (PDMS). Further analysis making use of network tools and systems biology thereby demonstrated that the expansion of hematopoietic stem cells is involved in the maintenance of their pluripotency (Marx-Blümel et al., 2021). Network pharmacology is yet another emerging aspect of stem cell informatics. A recent report showed that network pharmacology–based identification could provide insight into the mechanisms of drugs during the treatment regimen of cancer. The author demonstrated in a traditional formulation of herbs that the drug–herb-active compounds target gene–disease network and enrich with 190 overlapping candidate targets. Further use of molecular docking led to the identification of NR3C2, and its active compound, naringenin, was a potential drug against triple-negative breast cancer stem cells (Zhang et al., 2021). The involvement of network biology, computational methods, and informatics shall bring new laurels in stem cell research. The realization of stem cell therapies for life-threatening/incurable diseases and tissue regeneration will soon become a reality in clinics.

Acknowledgments
Authors would like to acknowledge ICMR, DHR, and respective funding agencies for supporting the research.

References

Aibar, S., González-Blas, C., Moerman, T., et al. (2017). Scenic: Single-cell regulatory network inference and clustering. *Nature Methods, 14,* 1083−1086.

Arjmand, B., Hamidpour, S. K., Alavi-Moghadam, S., Yavari, H., Shahbazbadr, A., Tavirani, M. R., Gilany, K., & Larijani, B. (February 21, 2022). Molecular docking as a therapeutic approach for targeting cancer stem cell metabolic processes. *Frontiers in Pharmacology, 13.* https://doi.org/10.3389/fphar.2022.768556

Attal, M., Harousseau, J. L., Facon, T., Guilhot, F., Doyen, C., Fuzibet, J. G., et al. (2003). Single versus double autologous stem-cell transplantation for multiple myeloma. *New England Journal of Medicine, 349*(26), 2495−2502.

Avior, Y., et al. (2015). TeratoScore: Assessing the differentiation potential of human pluripotent stem cells by quantitative expression analysis of teratomas. *Stem Cell Reports, 4,* 967−974.

Bajada, S., Mazakova, I., Richardson, J. B., & Ashammakhi, N. (2008). Updates on stem cells and their applications in regenerative medicine. *Journal of Tissue Engineering and Regenerative Medicine, 2*(4), 169−183.

Barrett, A. J., & Battiwalla, M. (2010). Relapse after allogeneic stem cell transplantation. *Expert Review of Hematology, 3*(4), 429−441.

Bock, J., Fukuyo, Y., Kang, S., Phipps, M. L., Alexandrov, L. B., Rasmussen, K.Ø., et al. (2010). Mammalian stem cells reprogramming in response to terahertz radiation. *PLoS One, 5*(12), e15806.

Dreger, P., Corradini, P., Kimby, E., Michallet, M., Milligan, D., Schetelig, J., et al. (2007). Indications for allogeneic stem cell transplantation in chronic lymphocytic leukemia: The EBMT transplant consensus. *Leukemia, 21*(1), 12−17.

Evans, M. J., & Kaufman, M. H. (1981). Establishment in culture of pluripotential cells from mouse embryos. *Nature, 292*(5819), 154−156.

Fiorina, P., Voltarelli, J., & Zavazava, N. (2011). Immunological applications of stem cells in type 1 diabetes. *Endocrine Reviews, 32*(6), 725−754.

Gallego Romero, I., Pavlovic, B. J., Hernando-Herraez, I., Zhou, X., Ward, M. C., Banovich, N. E., et al. (2015). A panel of induced pluripotent stem cells from chimpanzees: A resource for comparative functional genomics. *Elife, 4,* e07103.

Guillot, P. V., O'Donoghue, K., Kurata, H., & Fisk, N. M. (2006). Fetal stem cells: Betwixt and between. *Seminars in Reproductive Medicine, 24*(5), 340−347.

Haeckel, E. (1868). *Natürliche schöpfungsgeschichte (natural history of creation) (15th lecture).* Georg Reimer.

Jassak, P. F., & Riley, M. B. (1994). Autologous stem cell transplant: An overview. *Cancer Practice, 2*(2), 141−145.

Keretsu, S., Ghosh, S., & Cho, S. J. (2020). Molecular modeling study of c-KIT/PDGFRα dual inhibitors for the treatment of gastrointestinal stromal tumors. *International Journal of Molecular Sciences, 21*(21), 8232.

Khan, W. S., Longo, U. G., Adesida, A., & Denaro, V. (2012). Stem cell and tissue engineering applications in orthopaedics and musculoskeletal medicine. *Stem Cells International, 2012,* 403170.

Khorasani, N., Sadeghi, M., & Nowzari-Dalini, A. (2020). A computational model of stem cell molecular mechanism to maintain tissue homeostasis. *PLoS One, 15*(7), e0236519.

Kim, J., Kang, S. C., Yoon, N. E., Kim, Y., Choi, J., Park, N., Jung, H., Jung, B. H., & Ju, J. H. (2019). Metabolomic profiles of induced pluripotent stem cells derived from patients with rheumatoid arthritis and osteoarthritis. *Stem Cell Research and Therapy, 10*(1), 319.

Lawal, B., Kuo, Y. C., Tang, S. L., Liu, F. C., Wu, A. T. H., Lin, H. Y., & Huang, H. S. (2021). Transcriptomic-based identification of the immuno-oncogenic signature of cholangiocarcinoma for HLC-018 multi-target therapy exploration. *Cells, 10*(11), 2873.

Lee-Six, H., & Kent, D. G. (2020). Tracking hematopoietic stem cells and their progeny using whole-genome sequencing. *Experimental Hematology, 83,* 12−24.

Liu, J., Hu, Q., Wang, Z., Xu, C., Wang, X., Gong, G., Mansoor, A., Lee, J., Hou, M., Zeng, L., Zhang, J. R., Jerosch-Herold, M., Guo, T., Bache, R. J., & Zhang, J. (2004). Autologous stem cell transplantation for myocardial repair. *American Journal of Physiology - Heart and Circulatory Physiology*, 287(2), H501—H511.

Lu, H., & Deshmukh, L. (2002). Ethical issues in stem cell research. *Endocrine Reviews, 30*(3), 204—213.

Maehle, A. H. (2011). Ambiguous cells: The emergence of the stem cell concept in the nineteenth and twentieth centuries. *Notes and Records of the Royal Society of London, 65*(4), 359—378.

Margolin, A., Wang, K., Lim, W. K., Kustagi, M., Nemenman, I., & Califano, A. (2006). Reverse engineering cellular networks. *Nature Protocols, 1*(2), 662—671.

Marx-Blümel, L., Marx, C., Sonnemann, J., Weise, F., Hampl, J., Frey, J., Rothenburger, L., Cirri, E., Rahnis, N., Koch, P., Groth, M., Schober, A., Wang, Z. Q., & Beck, J. F. (2021). Molecular characterization of hematopoietic stem cells after in vitro amplification on biomimetic 3D PDMS cell culture scaffolds. *Scientific Reports, 11*(1), 21163.

Mashimo, T., Sato, Y., Akita, D., Toriumi, T., Namaki, S., Matsuzaki, Y., Yonehara, Y., & Honda, M. (2019). Bone marrow-derived mesenchymal stem cells enhance bone marrow regeneration in dental extraction sockets. *Journal of Oral Science, 61*(2), 284—293.

Maurer, M. H. (2011). Proteomic definitions of mesenchymal stem cells. *Stem Cells International, 2011*, 704256.

Mayank, & Jaitak, V. (2016). Molecular docking study of natural alkaloids as multi-targeted hedgehog pathway inhibitors in cancer stem cell therapy. *Computational Biology and Chemistry, 62*, 145—154.

Menicanin, D., Bartold, P. M., Zannettino, A. C., & Gronthos, S. (2009). Genomic profiling of mesenchymal stem cells. *Stem Cell Reviews and Reports, 5*(1), 36—50.

Mokgautsi, N., Wen, Y. T., Lawal, B., Khedkar, H., Sumitra, M. R., Wu, A. T. H., & Huang, H. S. (2021). An integrated bioinformatics study of a novel Niclosamide derivative, NSC765689, a potential GSK3β/β-Catenin/STAT3/CD44 suppressor with anti-glioblastoma properties. *International Journal of Molecular Sciences, 22*(5), 2464.

Moroni, L., & Fornasari, P. M. (2013). Human mesenchymal stem cells: A bank perspective on the isolation, characterization and potential of alternative sources for the regeneration of musculoskeletal tissues. *Journal of Cellular Physiology, 228*(4), 680—687.

Muller, F. J., et al. (2011). A bioinformatic assay for pluripotency in human cells. *Nature Methods, 8*, 315—317.

Müller, I., Kordowich, S., Holzwarth, C., Isensee, G., Lang, P., Neunhoeffer, F., Dominici, M., Greil, J., & Handgretinger, R. (2008). Application of multipotent mesenchymal stromal cells in pediatric patients following allogeneic stem cell transplantation. *Blood Cells, Molecules, and Diseases, 40*(1), 25—32.

Mustofa Satriyo, P. B., Suma, A. A. T., Waskitha, S. S. W., Wahyuningsih, T. D., & Sholikhah, E. N. (2022). A potent EGFR inhibitor, N-Phenyl pyrazoline derivative suppresses aggressiveness and cancer stem cell-like phenotype of cervical cancer cells. *Drug Design, Development and Therapy, 16*, 2325—2339.

O'Donoghue, K., & Fisk, N. M. (2004). Fetal stem cells. *Best Practice and Research Clinical Obstetrics and Gynaecology, 18*(6), 853—875.

Park, H. W., Shin, J. S., & Kim, C. W. (2007). Proteome of mesenchymal stem cells. *Proteomics, 7*(16), 2881—2894.

Pizzute, T., Lynch, K., & Pei, M. (2015). Impact of tissue-specific stem cells on lineage-specific differentiation: A focus on the musculoskeletal system. *Stem Cell Reviews and Reports, 11*(1), 119—132.

Rai, S., Kaur, M., & Kaur, S. (2013). Applications of stem cells in interdisciplinary dentistry and beyond: An overview. *Annals of Medical and Health Sciences Research, 3*(2), 245—254.

Ramalho-Santos, M., & Willenbring, H. (2007). On the origin of the term stem cell. *Cell Stem Cell, 1*, 35—38.

Shiozawa, Y., Nie, B., Pienta, K. J., Morgan, T. M., & Taichman, R. S. (2013). Cancer stem cells and their role in metastasis. *Pharmacology and Therapeutics, 138*(2), 285—293.

Shoemaker, L. D., & Kornblum, H. I. (2016). Neural stem cells (NSCs) and proteomics. *Molecular and Cellular Proteomics, 15*(2), 344—354.

Si, J. W., Wang, X. D., & Shen, S. G. (2015). Perinatal stem cells: A promising cell resource for tissue engineering of craniofacial bone. *World Journal of Stem Cells, 7*(1), 149—159.

Smith, A. G. (2001). Embryo-derived stem cells: Of mice and men. *Annual Review of Cell and Developmental Biology, 17*, 435—462.

Socié, G., Salooja, N., Cohen, A., Rovelli, A., Carreras, E., Locasciulli, A., Korthof, E., Weis, J., Levy, V., Tichelli, A., & Late Effects Working Party of the European Study Group for Blood and Marrow Transplantation. (2003). Nonmalignant late effects after allogeneic stem cell transplantation. *Blood, 101*(9), 3373—3385.

Sullivan, K. M., Goldmuntz, E. A., Keyes-Elstein, L., McSweeney, P. A., Pinckney, A., Welch, B., et al., SCOT Study Investigators. (2018). Myeloablative autologous stem-cell transplantation for severe scleroderma. *New England Journal of Medicine, 378*(1), 35—47.

Takahashi, K., & Yamanaka, S. (2006). Induction of pluripotent stem cells from mouse embryonic and adult fibroblast cultures by defined factors. *Cell, 126*, 663—676.

Thomson, J. A., Itskovitz-Eldor, J., Shapiro, S. S., Waknitz, M. A., Swiergiel, J. J., Marshall, V. S., & Jones, J. M. (1998). Embryonic stem cell lines derived from human blastocysts. *Science, 282*, 1145—1147.

van der Kooy, D., & Weiss, S. (2000). Why stem cells? *Science, 287*(5457), 1439—1441.

Vega-López, A., Pagadala, N. S., López-Tapia, B. P., Madera-Sandoval, R. L., Rosales-Cruz, E., Nájera-Martínez, M., & Reyes-Maldonado, E. (2019). Is related the hematopoietic stem cells differentiation in the Nile tilapia with GABA exposure? *Fish and Shellfish Immunology, 93*, 801—814.

Witkowska-Zimny, M., & Wrobel, E. (2011). Perinatal sources of mesenchymal stem cells: Wharton's jelly, amnion and chorion. *Cellular and Molecular Biology Letters, 16*, 493−514.

Wollert, K. C., & Drexler, H. (2005). Clinical applications of stem cells for the heart. *Circulation Research, 96*(2), 151−163.

Wöste, M., & Dugas, M. (2018). Viper: A web application for rapid expert review of variant calls. *Bioinformatics, 34*(11), 1928−1929.

Zakrzewski, W., Dobrzyński, M., Szymonowicz, M., & Rybak, Z. (2019). Stem cells: Past, present, and future. *Stem Cell Research and Therapy, 10*(1), 68.

Zhang, Y. Z., Yang, J. Y., Wu, R. X., Fang, C., Lu, H., Li, H. C., et al. (2021). Network pharmacology-based identification of key mechanisms of *Xihuang pill* in the treatment of triple-negative breast cancer stem cells. *Frontiers in Pharmacology, 12*, 714628.

Zhu, L., Roberts, R., Huang, R., Zhao, J., Xia, M., Delavan, B., et al. (2020). Drug repositioning for Noonan and LEOPARD syndromes by integrating transcriptomics with a structure-based approach. *Frontiers in Pharmacology, 11*, 927.

Chapter 5

Exploring imaging technologies and computational resources in stem cell research for regenerative medicine: A comprehensive review

Jaiganesh Inbanathan[1], R. Anitha[2], Protyusha Guha Biswas[3], M.R. Sanjana[2], K. Ponnazhagan[4] and D. Thirumal Kumar[5]

[1]Department of Pediatric and Preventive Dentistry, Meenakshi Ammal Dental College and Hospital, Meenakshi Academy of Higher Education and Research, Chennai, Tamil Nadu, India; [2]Department of Oral Medicine & Radiology, Meenakshi Ammal Dental College and Hospital, Meenakshi Academy of Higher Education and Research, Chennai, Tamil Nadu, India; [3]Department of Oral Pathology and Microbiology, Meenakshi Ammal Dental College and Hospital, Meenakshi Academy of Higher Education and Research, Chennai, Tamil Nadu, India; [4]Department of Biochemistry, Meenakshi Medical College Hospital & Research Institute, Meenakshi Academy of Higher Education and Research, Kanchipuram, Tamil Nadu, India; [5]Faculty of Allied Health Sciences, Meenakshi Academy of Higher Education and Research, Chennai, Tamil Nadu, India

1. Introduction

The first concept of stem cells originated in the early 1960s, following which it has undergone multiple evolutions and has garnered considerable attention worldwide. Stem cells now stand at the forefront of research and cell-based therapy in regenerative medicine and have proved to be a potent cure for a myriad of diseases. Stem cells may be categorized as adult stem cells, induced pluripotent stem cells (iPSCs), and embryonic stem cells (ESCs) (Zakrzewski et al., 2019). Initially, several kinds of research were conducted to elicit the regenerative function of blood, eventually leading to the isolation and differentiation of hematopoietic stem cells from mice by Ernest McCulloch and James from the University of Toronto (McCulloch & Till, 2005). Owing to their properties of self-renewing and multipotent differentiation, they were engaged in various therapeutic implications and treatments for myocardial infarction, neurodegenerative diseases, diabetes, and other irreversible diseases attributing to cell death (Sivandzade & Cucullo, 2021). In stem cell research, there was some concrete evidence and success in some studies, which stated that dopaminergic neurons derived from iPSCs and ESCs could improve clinical symptoms of Parkinson's disease in the mouse model (Wernig et al., 2008; Yang et al., 2008). Stem cells have also been differentiated into chondrocytes to repair osteoarthritis, cardiomyocytes to mitigate ischemic heart disease, and insulin-producing cells to potentially treat diabetes (Cao et al., 2008; Djouad et al., 2009; Kroon et al., 2008; Nelson et al., 2009; Zhang et al., 2009).

Stem cell biology is a rapidly evolving field that involves the study of the unique properties of stem cells and their potential clinical and therapeutic applications in regenerative medicine. However, in recent years, regenerative medicine has made an acute drift from laboratory animal studies to human trials, marking a significant advancement in stem cell research (Karathanasis, 2014). To ensure perpetual progress in this field at the same rate, developing a noninvasive and reproducible method to monitor and assess the integration and survival of stem cells was extremely crucial. Therefore, the advent of imaging technologies in stem cell research facilitated the same in studying stem cell characteristics and behavior both in vitro and in vivo. Eventually, the shift from 2D imaging strategies to more advanced 3D and combination imaging technologies has seen a paramount revolution in regenerative medicine and stem cell research. The main difference between 2D and 3D imaging technologies in stem cell research lies in the level of spatial and temporal information they provide. 3D imaging technologies provide a more comprehensive understanding of stem cell behavior in the entire

Copyright © 2024 Elsevier Inc. All rights reserved, including those for text and data mining, AI training, and similar technologies.

structure. They can be used for live tracking of several biological processes, including the distribution, survival, differentiation, and migration of stem cells in vivo, which remains a limitation of 2D imaging (Leahy et al., 2016). These technologies, therefore, allow scientists and clinicians to noninvasively visualize and monitor the behavior of stem cells and regenerative tissues, which is essential for understanding their mechanisms of action and optimizing their therapeutic potential. Some of the imaging modalities employed include magnetic resonance imaging (MRI), positron emission tomography (PET), fluorescence imaging, computed tomography (CT), optical coherence tomography (OCT), ultrasound, live cell imaging, and microcirculation imaging. Each technique has its strengths and limitations, and the choice of imaging modality depends on the specific clinical application.

Another rapidly emerging aspect of stem cell research and regenerative medicine is artificial intelligence (AI) integration. This has the potential to revolutionize stem cell research and finds its application in varied arenas such as differentiation, drug discovery, tissue engineering, establishing the correct diagnosis, selecting suitable medicine, and formulating appropriate therapies. The various image processing software with integrated AI provides advanced tools for automated image analysis, such as segmentation, volume rendering, fusion, image processing, registration, and analysis (Issa et al., 2022). Some commonly used software programs in stem cell imaging include ImageJ, FIJI, Imaris, Velocity, and Zen. These help to analyze large datasets and improve the accuracy and speed of data analysis using superior algorithms to help researchers better understand stem cell behavior and optimize stem cell–based therapies.

However, stem cells still pose many shortcomings in research and stem cell–based therapies due to the lack of gold standard protocols in isolation and differentiation, host response issues, quantification of changes after inducing, cross-contamination, delivery to the host site, etc. (Coelho et al., 2012). So, since the 2020s, stem cell research has taken another leap in regenerative research to understand the behavioral pattern of stem cells. As stem cell research continues to evolve, new software programs will likely be developed to meet the needs of researchers and clinicians in this relatively unexplored and exciting field. Their biodistribution identifies the best source, immunogenicity, and tumorigenicity (Bai, 2020). To facilitate research in stem cell biology, several computer software programs have been developed that aid in the analysis and interpretation of stem cell data. These software programs are designed to help researchers analyze large amounts of data, predict stem cell fate decisions, and identify potential stem cell donors for patients needing stem cell transplants. These programs help researchers analyze and interpret large amounts of data, predict stem cell fate decisions, and identify potential stem cell donors for patients needing transplants (Cahan et al., 2021).

2. Ideal requisite for molecular imaging

Over the past decade, stem cell research has been more targeted to the molecular imaging of stem cells to better understand molecular behavior and in vivo characteristics. There are many molecular imaging strategies employed in stem cell research. The most accepted and standard concepts practiced recently include direct iron particle labeling, direct radionuclide imaging, optical reporter gene imaging, radionuclide reporter gene imaging, magnetic resonance gene imaging, monitoring survival, and biodistribution, monitoring tumorigenicity, and monitoring immunogenicity (Abbas et al., 2019). The ideal requisite of the molecular imaging strategy of stem cells should have the following: (1) optimal visualization of molecular details of stem cells at the time of clinical delivery, (2) continuous real-time assessment of stem cells in the delivery site, (3) ability to reassess and identify the number of viable transplanted stem cells, (4) to aid in the delivery of stem cells to the host, and (5) long-term quantification of transplanted stem cell survival.

3. Imaging strategies employed in regenerative medicine

3.1 Direct iron particle labeling

This is one of the earliest molecular imaging strategies employed for in vivo stem cell research in regenerative medicine, which uses MRI to identify pluripotency and functional information of the stem cells with the aid of superparamagnetic iron oxide nanoparticles (SPIOs), which is composed of the oxidized form of magnetite. These SPIOs are contrast agents that get engulfed by the stem cells and cause T2 relaxation at nanomolar concentration collectively due to endocytosis (Jasmin et al., 2017). This direct iron particle method is of two types.

3.1.1 Magnetoporation

Magnetoporation is a technique used to introduce molecules or particles into cells using magnetic fields. It is a noninvasive method that utilizes the mechanical forces generated by a magnetic field to create temporary pores on the cell membrane, allowing for the introduction of exogenous materials. The process of magnetoporation involves applying a magnetic field to cells suspended in a solution containing the desired material to be introduced. The magnetic field generates mechanical forces acting on the cell membrane, causing it to deform and create pores temporarily. These pores allow for the passage of molecules or particles into the cell. Magnetoporation has been used in various research fields, including gene delivery, drug delivery, and cell engineering. It has several advantages over other methods of introducing materials into cells, such as high efficiency, low toxicity, and ease of use. However, it is still a relatively new technique and requires further optimization and validation for its use in clinical applications (Liu et al., 2012; Yousefian et al., 2022).

3.1.2 Magnetoelectroporation

Magnetoelectroporation (MEP) is a technique that combines the use of magnetic fields and electric fields to introduce molecules or particles into cells. MEP uses magnetic nanoparticles (MNPs) coated with the desired material to be introduced into the cells. These MNPs are then introduced into the cells and exposed to magnetic and electric fields. The magnetic field is used to manipulate the MNPs and bring them into proximity to the cell membrane. The electric field is then applied, which causes the cell membrane to temporarily destabilize and create pores. The MNPs, now close to the cell membrane, can enter the cell through these pores (Liu et al., 2012; Yousefian et al., 2022).

3.2 Clinical implication of direct iron particle labeling

Direct iron particle labeling is a noninvasive and effective molecular imaging method successfully used in clinical scenarios to identify and track mesenchymal stem cells (MSCs) in various therapeutic medical management. Direct iron particle imaging is principally used to track transplanted stem cells or progenitor cells, providing direct real-time information and aiding in the assessment of cell migration, homing, differentiation, and survival of transplanted stem cells. This method does not cause any impact on cellular functions such as morphology, viability, oxidative stress, mitochondrial member energization, etc. (Lu et al., 2017).

4. Imaging methods using nuclear medicine

Direct imaging and indirect imaging are the two primary approaches that are utilized for the identification of stem cells in vivo through the use of nuclear imaging. The two approaches are the direct prelabeling method and the other one being reporter gene approach.

4.1 Direct imaging

Imaging methods derived from nuclear medicine, such as single-photon emission computed tomography (SPECT) and PET, provide potential imaging modalities for surveilling stem cells. To visualize the transport of the administered cells easily, a direct prelabeling approach of the cells before administration is a primary method that may be used. A wide variety of radionuclides, including those that are commercially accessible, have been subjected to testing. Radiotracers are used in nuclear medicine imaging, which are molecules capable of attaching to various ligands. Nuclear medicine imaging methods have an extremely high sensitivity, allowing them to identify molecules at the nanomolar level, which is one of the primary benefits of PET and SPECT. It is also necessary to be equipped with a gamma camera system while working with the SPECT mode. Although SPECT's better adaptability and convenience are a benefit of the technology, PET possesses several inherent benefits that are superior to those of SPECT. These benefits include improved spatial resolution, significantly higher radiotracer sensitivity, established attenuation correction algorithms, and absolute distribution quantification. SPECT also has the advantage of being more accessible. Directly labeling cells is not only easy but also does not require any genetic change in the cells. For example, cells can be actively tagged in vitro using the proper technique and then given to a recipient (Volpe et al., 2021). When cells divide, the label becomes diluted, resulting in the less labeled agent being present in each cell. The loss of radioactivity over time and the escape of radiotracer from cells are two possible drawbacks of the direct labeling method of cell imaging. This is one of the drawbacks of using direct cell labeling techniques. In addition, the label may be passed on to the offspring cells in an asymmetrical manner following the process of cell division, or it may be lost by the cells.

For many years, direct imaging, also referred to as direct labeling of cells using a radioactive probe, has been utilized to monitor cells in vivo. In this technique, cells are incubated with a radiotracer, which creates conditions that enable lipophilic molecules to permeate into the cells and get "trapped" there (Kim et al., 2016). The two isotopes of PET utilized most frequently are fluorine-18, which presents with a half life of 110 min, and copper-64, which presents with a half-life of more than 12.7 h. When using isotopes such as 18F FDG, it may be possible to follow cells for 6−8 h after

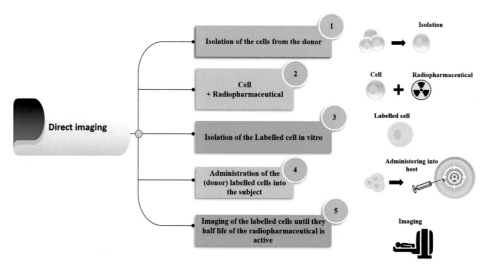

FIGURE 5.1 Diagrammatic representation of processes involved in direct molecular imaging.

transplantation (after accounting for the decay of the isotope due to its physical properties), and it is possible to evaluate the biodistribution of transplanted cells immediately after they have been injected. Contrary to PET tracers that transmit gamma rays in opposing angles, SPECT tracers produce a gamma ray in one direction. Thus, SPECT requires radioisotopes that have a longer half-life and are easier to procure. This reduces overall costs (Kim et al., 2016) (Fig. 5.1).

4.2 Indirect imaging

Indirect imaging is a method that may be used instead of direct imaging to tackle specific challenges. Reporter gene imaging is one type of indirect imaging currently being widely applied. Regarding long-term cell tracking, the reporter gene has an advantage over simple cell labeling since the imaging gene is passed on to the cell offspring. The imaging signal intensity is not reduced owing to diffusion by the entry and exit of the label from the cell. The reporter gene's hereditary nature explains this. This is explained by the possibility of reporter genes being handed down through cell division (Abbas et al., 2019). On the occasion of damage or phagocytosis of a gene-marked cell, the intensity of the signal that images the gene may alter. This is in contrast to simple cell labeling, where the signal used for imaging does not depend on the viability of the cell. In addition, the imaging signal intensity may originate from within immune scavenger cells. This is in contrast to the situation in which simple cell labeling occurs. In general, various types of reporter genes, including enzyme-based, receptor-based, and transporter-based, have been produced for radionuclide imaging. The various imaging strategies or methods employed in nuclear medicine include the following.

4.2.1 Receptor-based gene imaging

A reporter protein usually participates in transcription and synthesis inside the cell, establishing a cell's viability and thus providing the most effective alternatives for measuring cell survival. In addition, since the reporter genes are integrated into the genome of the cell, imaging cannot be hindered by cell changes, which occur when the reporter gene is distributed to the subsequent daughter cells (Ashmore-Harris et al., 2020). Given this, indirect imaging via reporter genes seems to be a potential option. The expression of a receptor that a particular reporter ligand can strike is one alternative way that may be used to image gene expression (probe). One of the receptor systems that has been researched the most is the dopamine D2 receptor (D2R). Few well-known tracers, such as 18F-fluoroethylspiperone (FESP), have been used in these studies. The FESP has a strong affinity for binding to the D2R. After receiving a systemic injection of 18F-FESP, it is possible to identify ectopic expression of the D2R gene while functioning as a PET reporter gene because the reporter probes for PET and SPECT may easily pass through an undamaged blood−brain barrier (BBB), one of the few methods capable of cell tracking in the brain is the D2R reporter. This allows the dopamine receptor to be one of the many units that can be employed. In addition, the D2R is not associated with an immune response since it is encoded by genes already present in the body. However, endogenous D2R in the striatum provides a significant signal, making it impossible to determine whether or not the D2R reporter system is present (Gu et al., 2012).

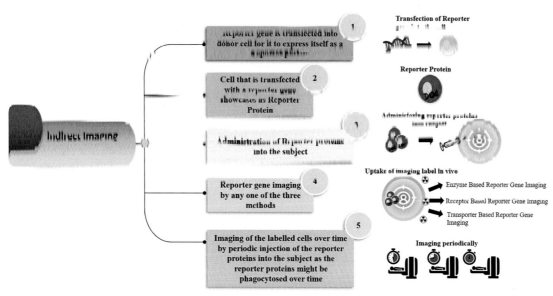

FIGURE 5.2 Diagrammatic representation of processes involved in indirect molecular imaging.

4.2.2 Transporter-based reporter gene imaging system

Large transmembrane proteins called drug and xenobiotic transporters promote the flow of many substances, including pharmaceuticals and xeno- and endobiotics, through cellular membranes. Drug transporters are frequently controlled at the transcriptional and translational levels to protect cells from harmful environmental circumstances. Drug transporters often function well only when localized to the cell surface, while they briefly reside in intracellular membrane compartments during synthesis or membrane reorganization (Mruk et al., 2011). Transport proteins have a high degree of selectivity for radiotracers and can be produced in the membranes of cells. For instance, the sodium/iodide symporter (NIS) is imaged with radioactive iodine or other tracers, which bind to the existing site. As the radioactive iodide is released quickly, cells infused with the NIS gene do not maintain the radio probe after the transfection (Ravera et al., 2017).

Transport proteins have a high degree of selectivity for radiotracers and can be produced in the membranes of cells. For instance, radioactive iodine that binds to the exact location as other tracers, such as 123I for SPECT or 124I for PET, can be used to image the NIS that has been employed as a reporter gene. Due to the quick release of radioactive iodide, cells transfected with the NIS gene do not usually sufficiently maintain the radio probe after transfection (Ravera et al., 2017) (Fig. 5.2).

4.2.3 Optical imaging

Fluorescence and bioluminescence imaging (BLI) are examples of cell imaging techniques that may be accomplished using optical imaging. Both these techniques require the identification of photons released into the atmosphere due to either oxidative chemical reactions or the external activation of a fluorophore. Despite their high levels of sensitivity and improved signal-to-background ratios, these optical imaging techniques exhibit poor depth penetration due to their high levels of absorption and scatter (Chao et al., 2013).

(a) *Fluorescence imaging*: Fluorescence imaging is used in biological and medical research to visualize and study cells and tissues. It involves using fluorescent molecules, such as dyes or proteins, which emit light of a specific wavelength when excited by a particular light source, typically a laser or a special type of light called "excitation light." The fluorescent molecules are injected or genetically introduced into the cells or tissues of interest. The sample is then illuminated with the excitation light, and the emitted fluorescent light is captured by a specialized microscope and analyzed. Fluorescence imaging allows researchers to visualize the location and behavior of specific molecules or structures within cells or tissues in real-time (Ettinger & Wittmann, 2014). It is frequently employed in numerous research fields, including cell biology, genetics, neuroscience, and cancer. In addition to traditional fluorescence imaging, more advanced techniques such as confocal microscopy, multiphoton microscopy, and superresolution microscopy have been developed to enhance the resolution and sensitivity of fluorescent imaging, enabling researchers to observe finer details and structures within cells and tissues (Lidke & Lidke, 2012).

(b) *Bioluminescence imaging*: The process of living things emitting light as a response to a biological reaction is known as bioluminescence. BLI is a technique used in biological and medical research to visualize and study cells and tissues that express bioluminescent molecules, such as luciferase. In BLI, the cells or tissues of interest are genetically engineered to express the luciferase gene, which encodes for the luciferase enzyme. When the luciferase substrate is added to the cells or tissues, it reacts with the luciferase enzyme, producing light that a specialized camera or microscope can detect. BLI allows researchers to visualize the location and behavior of specific cells or tissues in living organisms in real-time. It is instrumental in studying gene expression, cell signaling, and the progression of diseases such as cancer. One advantage of BLI is that it does not need an external light source such as fluorescence, making it the ideal choice for imaging deep tissues within living organisms. However, it has lower sensitivity and requires specialized equipment and reagents (Badr & Tannous, 2011).

4.2.4 Applications for optical imaging methods

Applications for optical imaging methods are numerous. Using these methods, scientists may examine various in vivo stem cell behaviors and outcomes, such as stem cell migration, differentiation, integration, and therapeutic effectiveness. Tracking stem cell migration to specific tissues or organs is one use of optical imaging in stem cell research. Researchers may watch stem cell mobility in real-time and keep track of their distribution inside the body by using fluorescence imaging and BLI to see how stem cells migrate in living things. Evaluation of stem cell development into certain cell types is another use of optical imaging. To detect differentiation processes in vivo, fluorescence imaging can be utilized to track the expression of cell type—specific markers by fluorescently labeled stem cells.

Additionally, reporter genes linked with particular cell types may be identified using BLI, offering a noninvasive way to monitor stem cell development. Investigating the incorporation of stem cells into host tissue may also be done using optical imaging methods. OCT can produce high-resolution pictures of tissue organization, making it possible to see stem cells and how they interact with the tissue around them. Additionally, stem cell integration into host tissue may be detected by BLI, offering a noninvasive way to track the long-term behavior of transplanted cells.

4.3 Other imaging methods

4.3.1 Ultrasound

Ultrasound is an imaging modality based on biological tissues' attenuation of sound waves. This technique allows the capturing of real-time images without the use of ionizing radiation. Ultrasound typically has vibratory frequencies ranging from 1 to 20 MHz for clinical applications. The electrical impulses generated by the scanners are converted into high-frequency sound waves using a transducer. The transducer then emits ultrasonic beams through the tissues being studied and are attenuated by the tissues based on their level of acoustic impedance. The echo reflected by the transducer is then converted and displayed as images on the monitor (Reda et al., 2021). In cellular imaging, however, ultrasound is limited due to poor contrast. To overcome this, some ultrasound contrast agents have been developed to increase the intensity of the ultrasound of the stem cells. Some such agents include microbubbles, silica, and gold nanoparticles. Microbubbles, a few micrometers in diameter, can label and highlight cells. However, due to their size, they may dissociate from the cells and cause aberrant signals. On the other hand, nanoparticles serve as more reliable contrast agents in size, cellular uptake, and acoustic impedance. It can also be employed in the in vivo analysis of injected stem cells (Fu et al., 2019).

4.3.2 Intravital microscopy for live cell imaging

A deep understanding of the biology and dynamics of stem cells could significantly contribute to treating several diseases. For example, in certain degenerative conditions such as Parkinson's disease, leukemia, etc., for which no therapy is available presently, the treatment approach may be cell therapy or gene therapy. In such cases, imaging modalities that track the fate of stem cells in vivo would be hugely beneficial. One such recently developed imaging modality is live cell imaging and intravital microscopy (IVM).

Over the years, live cell imaging has been widely applied to understand stem cell maintenance and differentiation (Chamma et al., 2014). Live cell imaging follows the behavior of individual stem cells over time during the tissue turnover phase. These techniques allow the visualization of cells in their native environment instead of the static images of other imaging techniques that often fail to capture the dynamics of a complex tissue structure. IVM uses optical imaging techniques with cellular and subcellular resolution on living animals. It can be used to study stem cell development, homeostasis, regeneration, and distribution (Fumagalli et al., 2020). Several imaging contrast methods are used to perform IVM, such as follows:

- Fluorescence
- Reflectance
- Second- and third-harmonic generation (SHG/THG)
- Coherent anti-Stokes Raman scattering (CARS)

IVM technique has several advantages such as high spatial resolution, live imaging of an entire system with a total evaluation of tissue behavior and interactions, and dynamic and longitudinal imaging. The contrast method is selected based on the type of tissue that needs to be highlighted. The contrast methods may be of two types, i.e., endogenous and exogenous fluorescence. A laser-scanning microscope (or LSM) configuration is then used for the study.

4.3.3 Microcomputed tomography

CT describes a clinical imaging technique for generating three-dimensional images from multiple image slices. It is a widely used imaging technique in the field of medicine. In stem cell research, CT technologies such as microcomputed tomography (micro-CT or μCT), dual-energy CT (DECT), and CT perfusion (CTP) are more frequently used. Micro-CT or μCT is used in tissue engineering. This imaging technique produces high-resolution three-dimensional (3D) images of a target specimen from two-dimensional (2D) transaxial projections or slices. An x-ray tube, radiation filter, collimator, specimen stand, and phosphor-detector/charge-coupled device camera are the parts of a micro-CT device. Back projection, a digital technique that provides a series of 2D projections integrated to create a 3D image by rotating either the sample (for desktop systems) or the emitter and detector (for live animal imaging), is used.

Some recent studies have thus used CT contrast agents such as gold nanoparticles to image and trace stem cells in vivo and in vitro to evaluate their migration, biodistribution, and final fate (Betzer et al., 2014). However, due to its low sensitivity, elevated cell count is required for CT-based functional imaging. DECT is usually used to gather structural and functional tissue information. The DECT approach uses CT scans at two photon energies for tissue characterization. At various energy levels, different tissues show distinctive attenuation characteristics. However, DECT has limited stem cell applications (Klontzas et al., 2021). Wan et al. conducted a study in which bone marrow MSCs were implanted to promote fracture healing in a rabbit bone defect model. The cells were tagged with gold nanoparticles, covered in silica, and transfected with DNA. DECT successfully revealed it tagged MSCs' migration to the cortex, allowing for the precise calculation of the volume of bone covered by the cells over 14 days following implantation (Wan et al., 2016). CT perfusion (CTP) permits dynamic evaluation of intravascular contrast attenuation. It is used to measure blood volume and blood flow and assessment of tissue vascularity. However, this imaging modality also has limited use in stem cell−based therapies. Although DECT and CTP have recently found some applications in stem cell therapy, tissue engineering, and regenerative medicine, DECT is beneficial for its ability to evaluate material composition. At the same time, CTP shows potential in the study of tissue vascularization. However, both methods have significant potential in studying tissue vasculature and biomaterials, which are frequently coupled with stem cells to develop effective regenerative therapies (Klontzas et al., 2021).

4.3.4 Computed tomography

CT scanning has become an essential tool in stem cell imaging because it provides high-resolution, three-dimensional images of the body. CT imaging can be used in various ways to track and study stem cells. In vivo tracking is one such method that involves labeling transplanted stem cells with a contrast agent visible on CT scans. This allows researchers to noninvasively monitor the distribution, migration, and survival of stem cells in real-time. In vivo, tracking can be particularly useful for assessing the effectiveness of stem cell therapies and optimizing their delivery and dosage. Ex vivo imaging is another method that can be used to study the structure and morphology of stem cells. This involves imaging stem cells removed from the body, such as those grown in a lab. CT imaging can provide detailed information on stem cells' shape, size, and structure and their interactions with other cells and tissues. This can be particularly useful for studying stem cell differentiation and regeneration mechanisms. DECT scanning is a specialized form of CT imaging that uses two different X-ray energies to distinguish between different materials in the body. This can provide more accurate and enhanced contrast imaging of stem cells and other tissues, making it particularly useful for detecting small tumors or other abnormalities. DECT scanning can also track the distribution of contrast agents in the body, providing insights into the pharmacokinetics of stem cell therapies. Micro-CT scanning is a high-resolution form of CT imaging used to study small objects, such as individual stem cells or small tissue samples. Micro-CT imaging can provide detailed information about the structure and morphology of stem cells and their interactions with other cells and tissues at a cellular level. This can be particularly useful for studying the early stages of stem cell differentiation and developing new therapies.

4.3.5 Optical coherence tomography

OCT is a powerful imaging technique that has gained popularity in stem cell imaging due to its noninvasive and high-resolution imaging capabilities. OCT uses light waves to produce cross-sectional images of biological tissues, making it ideal for studying the structure and behavior of stem cells. In stem cell imaging, OCT is primarily used for in vivo tracking and ex vivo imaging of stem cells. In vivo tracking involves labeling transplanted stem cells with contrast agents that can be detected by OCT, allowing researchers to monitor the distribution, migration, and survival of the stem cells in real-time. Ex vivo imaging, on the other hand, involves imaging stem cells removed from the body, such as those grown in a lab, and can provide detailed information on their structure and morphology. OCT can also be used to study the dynamics of stem cell differentiation and proliferation in real-time, providing insights into the mechanisms underlying these processes.

4.3.6 Microcirculation imaging

Microcirculation imaging is a noninvasive imaging technique used to study blood flow and microvascular networks in biological tissues. In stem cell imaging, microcirculation imaging can be used to study the distribution and behavior of stem cells after transplantation. Microcirculation imaging can provide insights into the vascularization of stem cell grafts, the formation of new blood vessels, and the interactions between stem cells and the host tissue. Microcirculation imaging techniques include laser Doppler flowmetry, IVM, and photoacoustic imaging. Laser Doppler flowmetry measures blood flow in small vessels using the Doppler effect, while IVM allows for high-resolution imaging of blood vessels in vivo. Photoacoustic imaging is a noninvasive imaging technique that uses laser light to generate sound waves, which are then used to create high-resolution images of blood vessels and tissues. In stem cell imaging, microcirculation imaging can be used to optimize stem cell transplantation strategies and improve the outcomes of stem cell therapies. Microcirculation imaging can help to identify the most appropriate transplantation site and monitor the distribution and engraftment of stem cells after transplantation. Microcirculation imaging can also provide insights into the mechanisms underlying stem cell—mediated tissue repair and regeneration (Table 5.1).

5. Software programs used in stem cell research

Various software programs are used in stem cell imaging, depending on the imaging modality and the specific analysis required. Some examples of commonly used software programs in stem cell imaging are as follows:

(i) **ImageJ:** ImageJ is a free, open-source image analysis software program that can be used for various imaging techniques, including fluorescence microscopy, confocal microscopy, and time-lapse imaging. It is widely used for image processing, segmentation, and quantification in stem cell research (Schneider et al., 2012).

(ii) **FIJI:** FIJI (Fiji is just ImageJ) is an image processing software based on ImageJ, which provides additional features and plugins. It is widely used in stem cell research for image analysis, segmentation, and 3D rendering (Schindelin et al., 2012).

(iii) **CellProfiler:** CellProfiler is a free, open-source software for high-throughput image analysis. It can be used for automated segmentation and analysis of large-scale images, making it useful for stem cell research applications, such as tracking cell populations and studying differentiation (Stirling et al., 2021).

(iv) **Imaris:** Imaris is a commercial software package designed for 3D and 4D image analysis and visualization. It can be used for stem cell research to visualize and analyze complex 3D structures and to quantify cellular processes, such as cell division and migration.

(v) **MATLAB:** MATLAB is software for numerical computation, data analysis, and visualization. It is widely used in stem cell research for developing custom image processing and analysis algorithms (MathWorks, 2020).

(vi) **Huygens:** Huygens is a commercial software package designed for the deconvolution of fluorescence microscopy images. It can improve the resolution and signal-to-noise ratio of stem cell imaging data.

(vii) **Volocity:** Volocity is a commercial software package for 3D and 4D image analysis and visualization. It can be used for stem cell research to analyze cell behavior in complex 3D environments, such as organoids.

(viii) **Zen:** Zen is a commercial software package for confocal and multiphoton microscopy. It can be used for stem cell research to acquire and analyze high-resolution images of cells and tissues.

TABLE 5.1 Salient features of various imaging modalities in stem cell research.

S.No	Imaging technique	Acquisition time	Imaging depth	Spatial resolution	Advantages	Disadvantages
1.	Magnetic resonance imaging	Within minutes	Limitless	50–200 μm (preclinical), 0.5–1 mm (clinical)	1. Noninvasive 2. Provides better physiological and anatomic imaging details at the molecular level 3. High resolution	1. Time-consuming 2. Less economical
2.	PET	Within seconds to minutes	Limitless	1–2 mm (preclinical), 5–7 mm (clinical)	1. High rate of tissue penetration 2. Very sensitive	1. Mild to moderately invasive 2. Less economical 3. Requires radioactive tracer
3.	SPECT	Within minutes	Limitless	1–2 mm (preclinical), 8–10 mm (clinical)	1. Multiple imaging parameters can be checked at the same time 2. High rate of penetration 3. Good tissue penetration 4. Suitable for clinical translation	1. Have biomolecular interference 2. Low spatial resolution
4.	Optical imaging (BLI)	Within seconds to minutes	1–2 cm	3–5 mm	1. Economical 2. Less technique sensitive 3. High sensitivity	1. Low resolution 2. Nonquantifiable
5.	Optical imaging (fluorescence)	Within seconds to minutes	<1 cm	2–3 mm	1. Economical 2. Good temporal resolution	1. Low resolution 2. Minimal penetration depth into tissues
6.	Ultrasound	Within seconds to minutes	∼1 mm–1cm	∼1–2 mm	1. Noninvasive 2. Economical 3. Good temporal resolution	1. Low tissue penetration depth 2. Less sensitive
7.	Microcomputed tomography	Within seconds	Limitless	50–100 μm	1. Rapid technique 2. Good-resolution images at the tissue level 3. Ease in reconstruction and analysis of images Easy of interpretation 4. Relatively inexpensive	Good-resolution images at the tissue level but not at the cellular level
8.	Computed tomography	Within second	1–10 mm	0.5–0.625 mm	1. Good imaging quality with adequate spatial resolution	1. Limited biocompatibility 2. High radiation exposure
9.	Optical coherence tomography	Within seconds	5–20 μm	2–3 mm	1. Good imaging resolution 2. It can be used as a component in multimodal imaging 3. Suitable for clinical translation	1. Limited penetration depth 2. Limited spatial resolution 3. Limited molecular sensitivity of tissue

(ix) **MetaMorph:** MetaMorph is a commercial software package for automated image acquisition and analysis. It can be used for stem cell research to automate the acquisition of large-scale imaging data and to analyze the data using a variety of algorithms.

(x) **CellSens:** CellSens is a commercial software package for imaging and analyzing fluorescence microscopy data. It can be used for stem cell research to analyze cell behavior responding to stimuli, such as drugs or growth factors.

(xi) **Cytoscape:** Cytoscape is a free, open-source software package for network visualization and analysis. It can be used for stem cell research to analyze the interactions between cells, molecules, and pathways (Shannon et al., 2003).

(xii) **QuPath:** QuPath is a free, open-source software package for digital pathology analysis. It can be used for stem cell research to analyze tissue samples and identify stem cells using automated algorithms (Bankhead et al., 2017).

(xiii) **Arivis:** Arivis is a commercial software package designed to visualize, analyze, and manage large-scale image data. It can be used for stem cell research to analyze complex 3D images, such as brain organoids or tumor spheres.

(xiv) **MetaXpress:** MetaXpress is a commercial software package for high-content imaging and analysis. It can be used for stem cell research to automate image acquisition and analysis of large-scale screening assays, such as drug discovery or toxicity testing.

(xv) **Visiopharm:** Visiopharm is a commercial software package designed for digital pathology analysis. It can be used for stem cell research to analyze tissue samples and identify stem cells using automated algorithms.

6. Current challenges in regenerative medicine imaging

While imaging is essential in regenerative medicine, several challenges must be addressed to improve its effectiveness. Here are some current challenges in regenerative medicine imaging:

(a) **Image resolution:** The resolution of imaging technologies can limit the ability to visualize and analyze the complex structures of tissues and organs. This can make identifying and tracking changes in cell behavior, tissue architecture, and disease progression challenging (Nowzari et al., 2021).

(b) **Imaging artifacts:** Imaging artifacts can occur due to various factors, including motion, image distortion, and variations in image intensity. These artifacts can affect the accuracy and reliability of imaging data, making it difficult to interpret and analyze (Somasundaram & Kalavathi, 2012).

(c) **Limited contrast:** It can be challenging to differentiate between different types of cells and tissues based on imaging data alone. This can limit the ability to identify and track specific cell populations and determine their behavior and function (Malliaras et al., 2013).

(d) **Image processing:** The processing of large datasets of imaging data can be time-consuming and computationally intensive. This can limit the ability to analyze and interpret imaging data promptly and efficiently (Malliaras et al., 2013).

(e) **Standardization:** There is currently a lack of standardization in imaging protocols and analysis methods across different research groups and clinical centers. This can lead to inconsistencies in imaging data and make it difficult to compare results across studies.

Addressing these challenges will require the development of new imaging technologies and analysis methods, as well as the establishment of standard protocols for imaging data acquisition and analysis. This will require collaboration between researchers and clinicians across different disciplines, as well as the use of advanced computational and machine learning techniques to analyze and interpret imaging data.

7. Artificial intelligence in stem cell imaging

The structural behavior of tissues and organs at different phases of growth and regeneration can be better understood through imaging techniques, which are essential in regenerative medicine. With the development of AI, these imaging approaches are expected to become much more potent, allowing researchers and doctors to better understand regeneration processes and provide new treatments for various illnesses. One of the most hopeful applications of AI in regenerative medicine is the analysis of large and complex datasets generated by imaging techniques such as MRI, CT, and PET scans. AI systems can be skilled to identify patterns and identify subtle changes in images that may be missed by human observers, allowing researchers to gain a complete understanding of the biology underlying regeneration. Creating individualized treatment regimens for individuals receiving regenerative treatments is another good use of AI in regenerative medicine. AI algorithms can determine the most efficient treatment plans and anticipate results with more accuracy by examining a patient's imaging data and other clinical data, ensuring that each patient receives the best care possible. In addition to these uses, AI can also be utilized to create new imaging procedures that are less intrusive, quicker, and more accurate than current ones. Researchers are investigating the use of AI-powered microscopes, for instance, which can photograph live tissues at the molecular level and offer hitherto unobtainable insights into the principles of regeneration.

Identifying intricate patterns and structures in stem cell imaging data has shown significant promise for deep learning algorithms. Artificial neural networks are trained as part of deep learning, a type of machine learning, to find patterns and connections in data. Deep learning algorithms have been employed in stem cell biology research to examine huge and complicated imaging datasets, such as confocal microscopy pictures of living cells or 3D reconstructions of tissues. Researchers can learn more about stem cell differentiation and development principles by utilizing deep learning algorithms to recognize and categorize various cell types and architectures. For instance, in a recent work reported in Nature Communications, scientists utilized a deep learning system to examine time-lapse microscopy pictures of human pluripotent stem cells as they differentiated into diverse cell types. The program recognized diverse patterns of cell movement, gene expression, and morphology associated with various differentiation phases, offering fresh information on the molecular processes that control stem cell destiny (Zhu et al., 2021).

8. Future of stem cell imaging with artificial intelligence

Several AI techniques are being used and developed for imaging in stem cell biology. Here are a few examples:

Machine learning: This technique involves training algorithms to recognize patterns in large datasets of images. Machine learning can identify subtle changes in tissues or organs over time or classify images according to specific features.

Deep learning: Deep learning is a branch of machine learning that processes massive volumes of data using neural networks. This technique has been used to analyze medical images and identify specific structures or anomalies that may be difficult to detect by human observers.

Computer vision: Computer vision refers to the ability of machines to interpret and analyze images. In regenerative medicine, computer vision can identify specific structures or features in medical images, such as blood vessels or cellular structures.

Natural language processing (NLP): NLP analyzes and interprets human language. In regenerative medicine, NLP can extract information from medical records or other textual data, which can then be used to inform imaging analysis and treatment planning.

Image segmentation: Image segmentation separates an image into distinct regions or structures. AI algorithms can automatically segment medical images, making it easier for researchers and clinicians to analyze and interpret the data.

These AI techniques are being used to develop new imaging technologies, improve the accuracy and speed of image analysis, and ultimately enable more effective regenerative medicine therapies.

8.1 Computational requirements and artificial intelligence software programs in stem cell biology

Stem cell research involves complex computational tasks, such as data analysis, simulation, and modeling, which require specialized computer hardware. High-performance computing (HPC) systems and graphics processing units (GPUs) are needed to handle large-scale data processing and mathematical computations. Workstations provide a robust computing environment for scientific research, enabling complex data analysis, visualization, and simulation. High-capacity and high-performance storage systems are essential for managing large datasets generated by stem cell research. Finally, 3D visualization systems are essential for visualizing complex stem cell structures and creating simulations and models. The use of specialized computer hardware in stem cell research provides researchers with the computational power, storage capacity, and visualization capabilities necessary to analyze and understand the complex behavior of stem cells.

Several software programs are commonly used in regenerative medicine with AI, including the following:

TensorFlow: It is an open-source software library established by Google for machine learning and deep learning. It is widely used for developing AI applications in regenerative medicine, including image analysis and segmentation.

PyTorch: It is another widespread open-source machine learning library used to develop AI applications in regenerative medicine. It is particularly well-suited for research applications, allowing rapid prototyping and experimentation.

MATLAB: MATLAB is a numerical computing environment widely used in scientific and engineering applications, including regenerative medicine. It includes several toolboxes for image analysis and machine learning, making it a popular choice for researchers and clinicians.

ImageJ: ImageJ is a Java-based image analysis software widely used in biological and biomedical research. It includes several plugins for image segmentation and analysis, making it a helpful tool for regenerative medicine applications.

3D Slicer: 3D Slicer is an open-source software platform for medical image analysis and visualization. It includes several tools for 3D image segmentation, registration, and visualization, making it a popular choice for researchers and clinicians working with medical imaging data.

CDeep3M: The National Institutes of Health (NIH) created CDeep3M, an AI-based program that automates the examination of cell pictures obtained from electron microscopy. In stem cell biology studies, it has been used to detect and examine numerous subcellular structures, including mitochondria and the Golgi apparatus.

DeepCell: DeepCell is another piece of AI-based software that can recognize and categorize various cell types from microscope pictures created by researchers at the University of California, Berkeley. It has been applied to studying and characterizing many stem cell types, including iPSCs and MSCs.

CellCNN: CellCNN is a piece of AI-based software that MIT researchers created to analyze microscope pictures of cells and classify various cell types based on their shape and other characteristics. It has been utilized in stem cell research to distinguish between various stem cell types and to track their differentiation and growth.

These software programs, among others, enable researchers and clinicians to develop and deploy AI applications for a wide range of regenerative medicine applications, from basic research to clinical trials and beyond. AI's use in regenerative medicine imaging methods is generally quite promising. Researchers and doctors will be able to learn more about the biology of regeneration, create better medicines, and eventually enhance the lives of patients with various illnesses by utilizing AI's power.

9. Conclusion

Imaging technology plays a pivotal role in regenerative medicine and stem cell research by providing real-time, noninvasive visualization and tissue growth and regeneration assessment. Imaging technologies enable precise and accurate monitoring of tissue regeneration progress and regenerative therapies' effectiveness. It can also aid in identifying the potential adverse effects of regenerative treatment modalities, facilitating early intervention and management. Integrating imaging technologies into stem cell research has seen significant advancements in personalized therapies, where clinicians provide tailor-made treatment to patients based on their specific needs.

Stem cell research involves complex genomics, epigenomics, transcriptomics, proteomics, and metabolomics data analysis. Developing these advanced and complex regenerative therapeutic modalities has been possible using sophisticated computational resources, including superior software and algorithms, powerful computers, and databases. The high-end computational resources enable researchers to handle and analyze large amounts of data generated by high-throughput sequencing, microarray analysis, and other omics technologies to gain a comprehensive knowledge of the cellular processes involved in stem cell differentiation, proliferation, and function. AI is a supremely advanced computational resource that has revolutionized regenerative medicine and stem cell research. Researchers, therefore, use AI algorithms and machine learning techniques to identify novel targets for stem cell therapies, predict stem cell differentiation outcomes, and optimize stem cell culture conditions. AI has also shown promise in stem cell research for drug discovery, disease modeling, and curating personalized medicine.

As AI advances, its integration into regenerative medicine and stem cell research will only increase, offering newer opportunities to accelerate the development of regenerative therapies and improve patient outcomes. However, it is important to ensure that AI-based models are validated and clinically relevant before they are applied in a clinical setting. Additionally, ethical considerations must be considered to ensure that AI-based models are used responsibly and ethically.

Acknowledgments

The authors express deep gratitude to the management of Meenakshi Academy of Higher Education and Research for all the support, assistance, and constant encouragement to carry out this work.

The authors originally illustrated the images depicted in this chapter using Microsoft Powerpoint.

Author contribution

All the authors were involved in collecting the literature, drafting, and proofreading the manuscript.

References

Abbas, F., Wu, J. C., Gambhir, S. S., & Rodriguez-Porcel, M. (2019). Molecular imaging of stem cells. *StemJournal, 1.* https://doi.org/10.3233/stj-190003

Ashmore-Harris, C., Iafrate, M., Saleem, A., & Fruhwirth, G. O. (2020). Non-invasive reporter gene imaging of cell therapies, including T cells and stem cells. *Molecular Therapy.* https://doi.org/10.1016/j.ymthe.2020.03.016

Badr, C. E., & Tannous, B. A. (2011). Bioluminescence imaging: Progress and applications. *Trends in Biotechnology*. https://doi.org/10.1016/j.tibtech.2011.06.010

Bai, X. (2020). Stem cell-based disease modeling and cell therapy. *Cells, 9*(10). https://doi.org/10.3390/cells9102193

Bankhead, P., Loughrey, M. B., Fernández, J. A., Dombrowski, Y., McArt, D. G., Dunne, P. D., McQuaid, S., Gray, R. T., Murray, L. J., Coleman, H. G., James, J. A., Salto-Tellez, M., & Hamilton, P. W. (2017). QuPath: Open source software for digital pathology image analysis. *Scientific Reports, 7*. https://doi.org/10.1038/s41598-017-17204-5

Betzer, O., Shwartz, A., Motiei, M., Kazimirsky, G., Gispan, I., Damti, E., Brodie, C., Yadid, G., & Popovtzer, R. (2014). Nanoparticle-based CT imaging technique for longitudinal and quantitative stem cell tracking within the brain. Application in neuropsychiatric disorders. *ACS Nano, 8*. https://doi.org/10.1021/nn5051218

Cahan, P., Cacchiarelli, D., Dunn, S. J., Hemberg, M., de Sousa Lopes, S. M. C., Morris, S. A., Rackham, O. J. L., del Sol, A., & Wells, C. A. (2021). Computational stem cell biology: Open questions and guiding principles. *Cell Stem Cell*. https://doi.org/10.1016/j.stem.2020.12.012

Cao, F., Wagner, R. A., Wilson, K. D., Xie, X., Fu, J. D., Drukker, M., Lee, A., Li, R. A., Gambhir, S. S., Weissman, I. L., Robbins, R. C., & Wu, J. C. (2008). Transcriptional and functional profiling of human embryonic stem cell-derived cardiomyocytes. *PLoS One, 3*. https://doi.org/10.1371/journal.pone.0003474

Chamma, E., Daradich, A., Côté, D., & Yun, S. H. (2014). *Intravital microscopy. Pathobiology of human disease: A dynamic encyclopedia of disease mechanisms* (pp. 3959−3972). https://doi.org/10.1016/B978-0-12-386456-7.07607-3

Chao, F., Shen, Y., Zhang, H., & Tian, M. (2013). Multimodality molecular imaging of stem cells therapy for stroke. *BioMed Research International*. https://doi.org/10.1155/2013/849819

Coelho, M. B., Cabral, J. M. S., & Karp, J. M. (2012). Intraoperative stem cell therapy. *Annual Review of Biomedical Engineering, 14*. https://doi.org/10.1146/annurev-bioeng-071811-150041

Djouad, F., Bouffi, C., Ghannam, S., Noel, D., & Jorgensen, C. (2009). Mesenchymal stem cells: Innovative therapeutic tools for rheumatic diseases. *Nature Reviews Rheumatology*. https://doi.org/10.1038/nrrheum.2009.104

Ettinger, A., & Wittmann, T. (2014). Fluorescence live cell imaging. In *Methods in cell biology*. https://doi.org/10.1016/B978-0-12-420138-5.00005-7

Fu, J. W., Lin, Y. S., Gan, S. L., Li, Y. R., Wang, Y., Feng, S. T., Li, H., & Zhou, G. F. (2019). Multifunctionalized microscale ultrasound contrast agents for precise theranostics of malignant tumors. *Contrast Media Molecular Imaging*. https://doi.org/10.1155/2019/3145647

Fumagalli, A., Bruens, L., Scheele, C. L. G. J., & van Rheenen, J. (2020). Capturing stem cell behavior using intravital and live cell microscopy. *Cold Spring Harbor Perspectives in Biology, 12*. https://doi.org/10.1101/cshperspect.a035949

Gu, E., Chen, W. Y., Gu, J., Burridge, P., & Wu, J. C. (2012). Molecular imaging of stem cells: Tracking survival, biodistribution, tu-morigenicity, and immunogenicity. *Theranostics*. https://doi.org/10.7150/thno.3666

Issa, J., Abou Chaar, M., Kempisty, B., Gasiorowski, L., Olszewski, R., Mozdziak, P., & Dyszkiewicz-Konwińska, M. (September 28, 2022). Artificial-intelligence-based imaging analysis of stem cells: A systematic scoping review. *Biology, 11*(10), 1412. https://doi.org/10.3390/biology11101412. PMID: 36290317; PMCID: PMC9598508.

Jasmin, de Souza, G. T., Louzada, R. A., Rosado-de-Castro, P. H., Mendez-Otero, R., & de Carvalho, A. C. C. (2017). Tracking stem cells with superparamagnetic iron oxide nanoparticles: Perspectives and considerations. *International Journal of Nanomedicine, 12*. https://doi.org/10.2147/IJN.S126530

Karathanasis, S. K. (2014). Regenerative medicine: Transforming the drug discovery and development paradigm. *Cold Spring Harbor Perspectives in Medicine, 4*(8). https://doi.org/10.1101/cshperspect.a014084

Kim, M. H., Lee, Y. J., & Kang, J. H. (2016). Stem cell monitoring with a direct or indirect labeling method. *Nuclear Medicine and Molecular Imaging*. https://doi.org/10.1007/s13139-015-0380-y

Klontzas, M. E., Kakkos, G. A., Papadakis, G. Z., Marias, K., & Karantanas, A. H. (2021). Advanced clinical imaging for the evaluation of stem cell based therapies. *Expert Opinion on Biological Therapy*. https://doi.org/10.1080/14712598.2021.1890711

Kroon, E., Martinson, L. A., Kadoya, K., Bang, A. G., Kelly, O. G., Eliazer, S., Young, H., Richardson, M., Smart, N. G., Cunningham, J., Agulnick, A. D., D'Amour, K. A., Carpenter, M. K., & Baetge, E. E. (2008). Pancreatic endoderm derived from human embryonic stem cells generates glucose-responsive insulin-secreting cells in vivo. *Nature Biotechnology, 26*. https://doi.org/10.1038/nbt1393

Leahy, M., Thompson, K., Zafar, H., Alexandrov, S., Foley, M., O'Flatharta, C., & Dockery, P. (2016). Functional imaging for regenerative medicine Functional imaging in regenerative medicine. *Stem Cell Research and Therapy, 7*(1). https://doi.org/10.1186/s13287-016-0315-2

Lidke, D. S., & Lidke, K. A. (2012). Advances in high-resolution imaging - techniques for three-dimensional imaging of cellular structures. *Journal of Cell Science, 125*. https://doi.org/10.1242/jcs.090027

Liu, D., Wang, L., Wang, Z., & Cuschieri, A. (2012). Magnetoporation and magnetolysis of cancer cells via carbon nanotubes induced by rotating magnetic fields. *Nano Letters, 12*, 5117−5121. https://doi.org/10.1021/NL301928Z/SUPPL_FILE/NL301928Z_SI_003.AVI

Lu, C. W., Hsiao, J. K., Liu, H. M., & Wu, C. H. (2017). Characterization of an iron oxide nanoparticle labelling and MRI-based protocol for inducing human mesenchymal stem cells into neural-like cells. *Scientific Reports, 7*(1). https://doi.org/10.1038/s41598-017-03863-x

Malliaras, K., Smith, R. R., Kanazawa, H., Yee, K., Seinfeld, J., Tseliou, E., Dawkins, J. F., Kreke, M., Cheng, K., Luthringer, D., Ho, C. S., Blusztajn, A., Valle, I., Chowdhury, S., Makkar, R. R., Dharmakumar, R., Li, D., Marbán, L., & Marbán, E. (2013). Validation of contrast-enhanced magnetic resonance imaging to monitor regenerative efficacy after cell therapy in a porcine model of convalescent myocardial infarction. *Circulation, 128*. https://doi.org/10.1161/CIRCULATIONAHA.113.002863

MathWorks, T. (2020). *MATLAB (R2020b)*. The MathWorks Inc.

McCulloch, E. A., & Till, J. E. (2005). Perspectives on the properties of stem cells. *Nature Medicine, 11*, 1026–1028. https://doi.org/10.1038/NM1005-1026

Mruk, D. D., Su, L., & Cheng, C. Y. (2011). Emerging role for drug transporters at the blood-testis barrier. *Trends in Pharmacological Sciences.* https://doi.org/10.1016/j.tips.2010.11.007

Nelson, T. J., Martinez-Fernandez, A., Yamada, S., Perez-Terzic, C., Ikeda, Y., & Terzic, A. (2009). Repair of acute myocardial infarction with induced pluripotent stem cells induced by human stemness factors. *Circulation, 120.* https://doi.org/10.1161/CIRCULATIONAHA.109.865154

Nowzari, F., Wang, H., Khoradmehr, A., Baghban, M., Baghban, N., Arandian, A., Muhaddesi, M., Nabipour, I., Zibaii, M. I., Najarasl, M., Taheri, P., Latifi, H., & Tamadon, A. (2021). Three-dimensional imaging in stem cell-based researches. *Frontiers in Veterinary Science, 8*, 352. https://doi.org/10.3389/FVETS.2021.657525/XML/NLM

Ravera, S., Reyna-Neyra, A., Ferrandino, G., Amzel, L. M., & Carrasco, N. (2017). The sodium/iodide symporter (NIS): Molecular physiology and pre-clinical and clinical applications. *Annual Review of Physiology.* https://doi.org/10.1146/annurev-physiol-022516-034125

Reda, R., Zanza, A., Cicconetti, A., Bhandi, S., Miccoli, G., Gambarini, G., & di Nardo, D. (2021). Ultrasound imaging in dentistry: A literature overview. *Journal of Imaging.* https://doi.org/10.3390/jimaging7110238

Schindelin, J., Arganda-Carreras, I., Frise, E., Kaynig, V., Longair, M., Pietzsch, T., Preibisch, S., Rueden, C., Saalfeld, S., Schmid, B., Tinevez, J. Y., White, D. J., Hartenstein, V., Eliceiri, K., Tomancak, P., & Cardona, A. (2012). Fiji: An open-source platform for biological-image analysis. *Nature Methods.* https://doi.org/10.1038/nmeth.2019

Schneider, C. A., Rasband, W. S., & Eliceiri, K. W. (2012). NIH image to ImageJ: 25 years of image analysis. *Nature Methods.* https://doi.org/10.1038/nmeth.2089

Shannon, P., Markiel, A., Ozier, O., Baliga, N. S., Wang, J. T., Ramage, D., Amin, N., Schwikowski, B., & Ideker, T. (2003). Cytoscape: A software environment for integrated models of biomolecular interaction networks. *Genome Research, 13*, 2498–2504. https://doi.org/10.1101/gr.1239303

Sivandzade, F., & Cucullo, L. (2021). Regenerative stem cell therapy for neurodegenerative diseases: An overview. *International Journal of Molecular Sciences, 22*, 1–21. https://doi.org/10.3390/IJMS22042153

Somasundaram, K., & Kalavathi, P. (2012). Analysis of imaging artifacts in MR brain images. *Oriental Journal of Computer Science & Technology, 5*(1).

Stirling, D. R., Swain-Bowden, M. J., Lucas, A. M., Carpenter, A. E., Cimini, B. A., & Goodman, A. (2021). CellProfiler 4: Improvements in speed, utility and usability. *BMC Bioinformatics, 22.* https://doi.org/10.1186/s12859-021-04344-9

Volpe, A., Pillarsetty, N. V. K., Lewis, J. S., & Ponomarev, V. (2021). Applications of nuclear-based imaging in gene and cell therapy: Probe considerations. *Molecular Therapy Oncolytics, 20.* https://doi.org/10.1016/j.omto.2021.01.017

Wan, D., Chen, D., Li, K., Qu, Y., Sun, K., Tao, K., Dai, K., & Ai, S. (2016). Gold nanoparticles as a potential cellular probe for tracking of stem cells in bone regeneration using dual-energy computed tomography. *ACS Applied Materials & Interfaces, 8.* https://doi.org/10.1021/acsami.6b11856

Wernig, M., Zhao, J. P., Pruszak, J., Hedlund, E., Fu, D., Soldner, F., Broccoli, V., Constantine-Paton, M., Isacson, O., & Jaenisch, R. (2008). Neurons derived from reprogrammed fibroblasts functionally integrate into the fetal brain and improve symptoms of rats with Parkinson's disease. *Proceedings of the National Academy of Sciences of the U S A, 105.* https://doi.org/10.1073/pnas.0801677105

Yang, D., Zhang, Z.-J., Oldenburg, M., Ayala, M., & Zhang, S.-C. (2008). Human embryonic stem cell-derived dopaminergic neurons reverse functional deficit in parkinsonian rats. *Stem Cells, 26.* https://doi.org/10.1634/stemcells.2007-0494

Yousefian, B., Firoozabadi, S. M., & Mokhtari-Dizaji, M. (2022). Magnetoporation: New method for permeabilization of cancerous cells to hydrophilic drugs. *Journal of Biomedical Physics & Engineering, 12*, 205. https://doi.org/10.31661/JBPE.V0I0.1256

Zakrzewski, W., Dobrzyński, M., Szymonowicz, M., & Rybak, Z. (2019). Stem cells: Past, present, and future. *Stem Cell Research & Therapy, 10*, 1–22. https://doi.org/10.1186/s13287-019-1165-5

Zhang, J., Wilson, G. F., Soerens, A. G., Koonce, C. H., Yu, J., Palecek, S. P., Thomson, J. A., & Kamp, T. J. (2009). Functional cardiomyocytes derived from human induced pluripotent stem cells. *Circulation Research, 104.* https://doi.org/10.1161/CIRCRESAHA.108.192237

Zhu, Y., Huang, R., Wu, Z., Song, S., Cheng, L., & Zhu, R. (2021). Deep learning-based predictive identification of neural stem cell differentiation. *Nature Communications, 12*, 2614. https://doi.org/10.1038/s41467-021-22758-0

Further reading

Coste, A., Oktay, M. H., Condeelis, J. S., & Entenberg, D. (2020). Intravital imaging techniques for biomedical and clinical research. *Cytometry Part A.* https://doi.org/10.1002/cyto.a.23963

Chapter 6

Application of machine learning–based approaches in stem cell research

Manoj Kumar Yadav[1], Khushboo Bhutani[1], Shaban Ahmad[2], Khalid Raza[2], Amisha Singh[3] and Sunil Kumar[4]

[1]Department of Biomedical Engineering, SRM University Delhi-NCR, Sonepat, Haryana, India; [2]Department of Computer Science, Jamia Millia Islamia, New Delhi, Delhi, India; [3]Department of Biotechnology, JAIN University, Bangalore, Karnataka, India; [4]ICAR-Indian Agricultural Statistical Research Institute, New Delhi, Delhi, India

1. Introduction

As regenerative medicine research progresses, the concept of treating disease has changed a lot. Previously, medical professionals used to develop therapies to combat the disease conditions, but today the tendency is toward using the regeneration concept for treating various illnesses. When it comes to regeneration and replacing damaged cells with new and more effective cells, stem cells are involved. Stem cells are specialized cells having self-renewal capabilities. Stem cells have the potential to differentiate into any type of cell. Stem cells can be further divided into totipotent, pluripotent, multipotent, or unipotent cells based on their differentiation potential. Even though stem cell treatment has not yet received FDA approval, a lot of research is still being done to fully understand its benefits and limitations. Nowadays, stem cell therapy has wide applications in the medical field starting from the treatment of genetic diseases to regeneration of insulin-secreting beta cells.

One of the cutting-edge technologies that might influence the precise use of stem cells in regenerative medicine is artificial intelligence (AI) (Srinivasan et al., 2021). The main goal of AI is to free humans from complex tasks and enable them to create solutions with high accuracy and precision while using fewer resources (Hole & Ahmad, 2019). Machine learning (ML) and deep learning (DL) are two important domains of AI and have a wide range of applications in the prophylactic as well as therapeutic treatments of diseases. Recent advancements in algorithms have enhanced its capability of disease diagnoses by studying complex healthcare data. These algorithms can also forecast the role of different factors in causing a particular disease, and thus, demonstrate the importance of AI in the healthcare industry (Ahuja, 2019). When a study was carried out by Schaub et al. in collaboration with the National Eye Institute, a new hope of ray was generated for age-related macular degeneration (AMD), which is an eye disorder where people lose their eyesight owing to age. Using the Deep neural network (DNN) and the AI program GoogLeNet, an effort was made to expedite quality assurance assessments of stem cells used to treat AMD (Schaub et al., 2020). The biosafety and bio-efficacy of stem cell therapies are other significant issues that need to be taken care of to treat diseases. Ashraf et al. used ML algorithms to review this aspect by comparing data of cancer cells and stem cells under various conditions to characterize morphological and phenotypic differences. The generated data is utilized in DL algorithms mainly convolutional neural network (CNN), an artificial neural network (ANN)-based algorithm, to ensure biosafety and bioefficacy of stem cell-based therapy (Ashraf et al., 2021). This chapter aims to peek through the glass using AI-based algorithms to assess the potential application of stem cell research in medical therapy.

2. Types of stem cells and their therapeutic applications

Stem cells are undifferentiated cells, initially not designed for a specific purpose, and can be converted into specific cell types based on the body's requirements. Initially, stem cells are nondividing and nonspecific cells until they receive messages from the body to repair or grow new tissues. These cells can be divided into embryonic stem cells (ESC's) and

Computational Biology for Stem Cell Research. https://doi.org/10.1016/B978-0-443-13222-3.00007-1
Copyright © 2024 Elsevier Inc. All rights reserved, including those for text and data mining, AI training, and similar technologies.

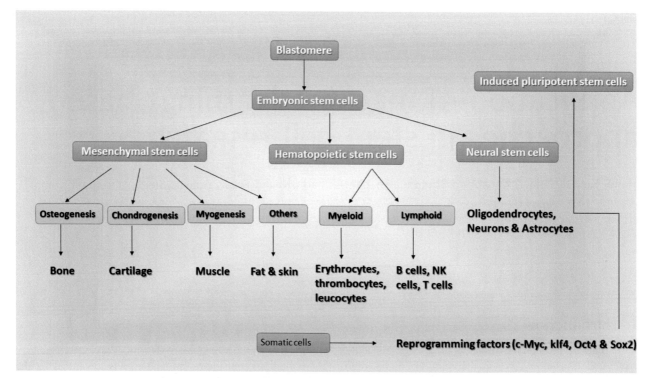

FIGURE 6.1 Stem cell differentiation.

adult stem cells (ASCs), based on their origin (Fig. 6.1). The stem cells that come from the inner cell layer (mass) of the blastocyst are considered ESCs and can differentiate into any cell type, while the cells derived from a variety of tissues, viz., bone marrow, liver, heart, etc. are adult or tissue-specific or somatic stem cells. Although ASCs have less potential for differentiation than ESCs, they are nonetheless employed in regenerative medicine because of the ethical issues concerning the use of ESCs. The therapeutic potential of different types of stem cells may use to handle previously untreated diseases such as genetic disorders, autoimmune diseases, or degenerative disorders (Fig. 6.2).

2.1 Embryonic stem cell therapy

ESCs have the great potential to predict toxicity in humans and are an important in vitro model for drug testing (van Dartel et al., 2010). According to earlier investigations, ESCs may be used as a screening platform to find low-molecular-weight drugs that influence endogenous stem cell populations and heal injured tissues. Additionally, an array of human embryonic stem cell (hESC) cell lines, for example, H1, H7, H9, and mouse embryonic stem cell (mESC) cell lines, D3, are available for screening of drugs (Lou & Liang, 2011). For instance, ventricular cardiomyocytes that form the muscular walls of the heart are needed in conditions such as myocardial infarction or chronic heart failure. The embryoid bodies that are created with ESCs can produce these cells (Fijnvandraat et al., 2003). Additionally, utilizing neuronal tissue derived from mESCs, motor neurons that engraft and restore function have been created in a rat model of spinal cord injury and hemiplegia (Chiba et al., 2003). In several cases, mESC-derived dopaminergic neurons have been utilized to repair motor dysfunctions in a rodent model of Parkinson's disease (Kim et al., 2002). Various cell types that have different applications in the field of medicine have been derived from hESCs, namely neural tissue, insulin-secreting cells, cardiomyocytes, hematopoietic cells, endothelial cells, osteoblasts, and hepatocytes (Lerou & Daley, 2005). In another study, the potential of ESCs is used to heal liver damage that occurs due to toxins or medications or some hereditary conditions, and/or infections. This could be accomplished through the transplantation of ESCs-derived hepatocytes, in which highly specific cytochrome P450 (CYP450), hepatocyte nuclear factor-alpha (HNF-α), and other molecules repopulate damaged liver tissues, and produce a regenerated liver, which may be used as a model for drug testing (Tolosa et al., 2015). A notable improvement in glucose levels was observed when ESCs were used to create a therapy option for diabetic patients, particularly in type I diabetes mellitus (T1DM) and type II diabetes mellitus (T2DM). After transplantation, the pancreatic progenitor cells (PPCs) produced from ESCs differentiated into beta cells, released insulin, and showed positive expression of pancreatic markers including PDX1, GCK, and GLUT2 (Bruin et al., 2015; Salguero-Aranda et al., 2016).

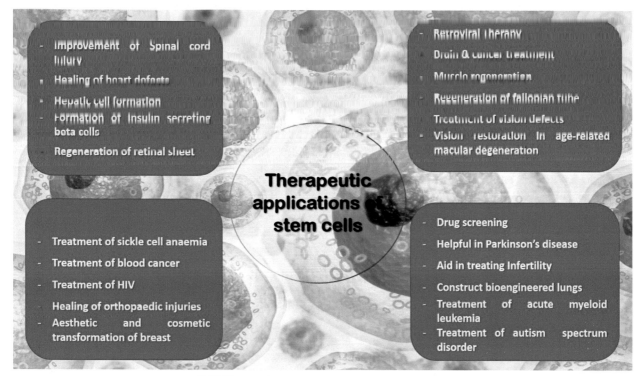

FIGURE 6.2 Therapeutic applications of stem cell therapy.

2.2 Induced pluripotent stem cell therapy

Scientists Yamanaka and Takahashi effectively converted multipotent ASCs into a pluripotent state, leading to the discovery of induced pluripotent stem cells (iPSCs) (Takahashi & Yamanaka, 2006). In several investigations, iPSCs specific to a particular disease were derived from patients suffering from neurodegenerative diseases, such as Parkinson's disease, Alzheimer's disease, amyotrophic lateral sclerosis, and spinal muscular atrophy. Since the neurons created from these iPSCs replicated similar pathological characteristics, they can be used to test the effectiveness of drugs and can be used as a prototype model for cell replacement. cell replacement (Jung et al., 2012). An excellent illustration is Parkinson's disease, where gene therapy aims to supply the deficient chemicals, dopamine, and other components, to the affected cells. Scientists have successfully introduced the helpful genes needed for the basal ganglia of the brain's dopaminergic pathways to produce, store, and absorb dopamine. The ability of transplanted differentiated cells to survive, engraft, and differentiate into adult rodent dopaminergic neurons was demonstrated in studies employing human iPSCs obtained from Parkinson patients, with positive functional benefits (Tailor et al., 2012). In a study, Hayashi et al. showed that iPSCs may be used to generate functional sperm and egg cells in a rat model, which could be a feasible solution to address problems associated with infertility issues (Hayashi et al., 2011). Additionally, using iPSC technology in conjunction with genome editing tools and genome-wide association analyses (GWAS) can aid in understanding the genetic defects linked to a certain disease, such as insulin resistance (Elsayed et al., 2021).

2.3 Induced tissue-specific stem cell therapy

Induced tissue-specific stem (iTS) cells are created through tissue-specific selection followed by partial reprogramming of somatic cells using transient overexpression of reprogramming factors via plasmids. Mouse pancreatic cells that are capable of self-renewal and that express the transcription factor, Pdx1, unique to pancreatic tissues, have been used to generate iTS cells (Saitoh et al., 2016). When iTS cells are implanted in naked mice, there are no reports of teratoma formation. The iTS cells are found to be superior for therapeutic applications in comparison with their counterparts; ESCs or iPSCs, due to their lower association with tumorigenicity. Since the iTS cells are partially reprogrammed, their distinct methylation patterns are somewhat more conserved compared with ESCs or iPSCs. As a result, because they preserve the donor tissue's epigenetic memory, iTS cells can be considered as a possible candidate for cell replacement treatment (Mukherjee et al., 2021).

2.4 Fetal stem cell and adipose tissue—derived stem cell therapy

Fetal stem cells are pluripotent cells with the capacity to divide, multiply, and develop into any form of cell. Fetal stem cells can be obtained from aborted fetal tissues and extraembryonic structures such as amniotic fluid, umbilical cord, Wharton's jelly, and placenta (Marcus & Woodbury, 2008). In utero transplantation (IUT) of allogenic stem cells is still a glimmer of hope for treating postnatal disorders. In such a situation, fetal stem cells have a definite competitive advantage over ASCs (O'Donoghue & Fisk, 2004). Autologous hematopoietic stem cells (HSCs) obtained from fetus are first isolated, transduced in vitro, and then transplanted back to the fetus in ex vivo gene therapy to treat numerous disease conditions (Campagnoli et al., 2002). Numerous studies suggest that human fetal mesenchymal stem cells (hfMSCs) could be used to treat musculoskeletal conditions such as osteoporosis, osteoarthritis, and osteogenesis (Song et al., 2022).

Adipose tissue—derived stem cells (ADSCs) are multipotent mesenchymal cells that can be differentiated into adipocytes, chondrocytes, myocytes, osteoblasts, and neurocytes among other cell lineages. ADSCs appear to be the most beneficial, among ASCs, in terms of their applicability in cell-based therapies and tissue engineering. They have been used extensively in clinical settings on patients, especially for reconstructive, esthetic, and cosmetic objectives such as breast augmentation and facial contouring (Bacakova et al., 2018). Because of their immunomodulatory and immunosuppressive properties, ADSCs have been successfully used for the treatment of inflammatory and autoimmune illnesses, including rheumatoid arthritis, Crohn's disease, and multiple sclerosis, through intravenous infusion of these stem cells (Ceccarelli et al., 2020). According to a study, a combined therapy using ADSCs, platelet-rich plasma, and endorectal flaps has also been used to treat patients with refractory perineal Crohn's disease (Wainstein et al., 2018). The simultaneous transplantation of ADSCs and human cells secreting neurotrophic factors improved the therapy of multiple sclerosis in a rat model (Razavi et al., 2018). In addition, bioengineered lungs were created on scaffolds made by decellularizing nontransplantable lungs utilizing ADSCs in a dynamic bioreactor (Farré et al., 2018).

2.5 Hematopoietic stem cell therapy

HSCs are clinically useful as they are multipotent cells that can differentiate into all other types of blood cells, including myeloid- and lymphoid-lineage cells. Clinical trials conducted based on hematopoietic stem cell transplantation (HSCT) to treat a range of illnesses, including leukemia, cardiac abnormalities, autoimmune diseases, and hepatic and neurologic issues. HSCs extracted from bone marrow or peripheral blood have been used for infusion in the specific trials (Müller et al., 2016). HSCT is just one of many ground-breaking HSC-based treatment options including cell-based and gene therapy. Leukemia, among many other illnesses, calls for HSC-based therapy in particular. Acute myeloid leukemia (AML) is one of the most aggressive types of cancer, associated with a high mortality rate. The initial line of treatment for AML is chemotherapy, which lowers the risk of relapsing, but a considerable amount of functional immune cells are damaged in this process. As a result, HSCT is used for all types of AML because it can replace functional immune cells that are lost during chemotherapy (Lee & Hong, 2020).

Allogenic HSCT procedures are proving beneficial for curing single-gene illnesses such as sickle cell anemia and β-thalassemia, which were previously thought to be incurable. Moreover, a large number of metabolic disorders, including Hurler syndrome and adrenoleukodystrophy, may also be benefitted by using HSCT techniques in a similar manner (Talib & Shepard, 2020). Allogenic HSCT is also used to treat several disorders, namely leukemia, including AML, acute lymphocytic leukemia (ALL), chronic myeloid leukemia, and chronic lymphocytic leukemia. The allogenic transplantation technique also shows a promising result for myeloproliferative disorders such as non-Hodgkin's lymphoma (NHL). Autologous HSCT is the most frequent option available to treat multiple myeloma cases where patients' own HSCs are transplanted (Kanate et al., 2012). Additionally, patients treated with this technology following chemotherapy sessions significantly outlive patients treated only with chemotherapy.

2.6 Mesenchymal stem cell therapy

Mesenchymal stem cells (MSCs) are ASCs with multilineage differentiation and self-renewal capability. MSCs are often referred to as mesenchymal stromal cells (Hmadcha et al., 2020). Although the term "MSCs" was once solely used to refer to stromal cells in the bone marrow, it has now been expanded to include cells from almost all postnatal connective tissues, such as synovium, adipose tissue, dental pulp, etc. (Bianco et al., 2008; Chamberlain et al., 2007). MSCs-based therapies have been used in several preclinical investigations over the past few decades to treat a wide range of illnesses, including neurological disorders, diabetes, heart ischemia, and diseases related to cartilage and bone (Eridani, 2014). MSCs have the innate capacity to move toward the injured tissues and encourage tissue regeneration through the release of paracrine substances with pleiotropic effects (Akimoto et al., 2013). Human MSCs (hMSCs) derived from the umbilical cord showed tremendous benefits by enhancing ventricular function in a pig model of myocardial ischemia (Liu et al., 2010). Advanced therapies in cases such as brain cancer glioblastoma have demonstrated the employment of engineered MSCs to be useful in chasing down evasive tumor cells in the brain and delivering specific payloads, such as TNF-alpha-related apoptosis induced ligand (TRAIL) (Cheetu et al., 2018). Due to its antiapoptotic, antifibrosis, chemoattractive, and chondrogenesis actions, the MSC secretome also has a significant impact on osteoarthritis and cartilage protection (Mancuso et al., 2019).

3. Overview of machine learning—based model building

There is a need of intelligent algorithms, implementation of which can improve the quality of healthcare provided to the general public. ML and DL algorithms fall under the umbrella term of intelligent algorithms. These algorithms will assist machines to learn on their own, come to decisions, and recognize patterns (Amisha et al., 2019; Mukherjee et al., 2021). Both ML and DL techniques have a wide range of applications in stem cell research.

3.1 Machine learning algorithms

ML algorithms are designed to train machines to perform various actions such as predictions, recommendations, and estimations based on available labeled or unlabeled data sets. ML algorithm—based learning enables computers to mimic human behavior by training them with the help of past experiences. The ML algorithm can be broadly divided into supervised, unsupervised, reinforcement, and semisupervised learning based on different learning experiences. In supervised learning, machines are trained on well-labeled data. The term "labeled data" refers to input data that has already been assigned the appropriate output. To forecast and categorize the test samples, the underlying mathematical model will first learn its parameters from the labeled samples, while unsupervised learning does not require labeled data to train models. Environment feedback is a component of reinforcement learning. It is, therefore, not entirely unsupervised. Since no label samples are available for training in this method, it cannot be considered supervised learning either. Instead of relying on manually adjusted parameters or previously fixed processing steps, ML learns the processing rules from model examples. ML algorithms mainly clustering and regression algorithms have been widely used in dealing with medical imaging, disease diagnosis, and treatment of diseases including stem cell—based treatments. While the clustering algorithms include grid-based clustering, hierarchical clustering, partitioning clustering, and coclustering, the regression algorithm includes linear regression, decision tree, ANNs, and hierarchical and ensemble methods (Mukherjee et al., 2021).

The regression algorithm is another type of ML algorithm that involves the prediction of the association of one or more independent variables with a dependent (target) and independent (predictor) variable. ANN is a more sophisticated ML algorithm with a wide range of modifications that generally aims to emulate the operation of biological neural networks. This algorithm's input, hidden, and output nodes are all different, and each layer has a higher level of processing abstraction than before. The nodes (sometimes referred to as neurons) within each succeeding layer of the multilayer perceptron, a traditional ANN version, are all coupled to one another between levels. Each neuron uses the inputs from the previous layer, which are weighted independently, to produce an output after carrying out a few straightforward mathematical calculations (Sotoudeh et al., 2019).

The clustering algorithm uses an unsupervised learning approach to cluster unlabeled datasets based on similarities and dissimilarities. Since it is part of unsupervised learning, no supervision is provided to the algorithm, and prediction is done solely based on characteristics of given training data. For instance, in genetic interaction, clustering algorithms aim to solve an optimization problem by placing an unidentified target gene in the relationship map based on preset criteria and a predetermined cost function.

3.2 Deep learning algorithms

DL algorithms are a variation of ANNs that use many ("deep") layers to increase complexity. A neural network's architecture can vary in complexity, giving rise to a large variety of ANNs that may be applied to DL. Since the layer of data travels solely in the forward direction, the term "feedforward" is used to describe the typical neural network. CNN is another commonly used DL algorithm applied to biological datasets. The initial layers of CNN compare each component of an image to small subimages. Each node's output to the succeeding layer is decided by how closely an area of the image resembles the characteristic that each node holds. The classification of the entire image is carried out by a typical fully connected neural network after the convolution layers (Sotoudeh et al., 2019).

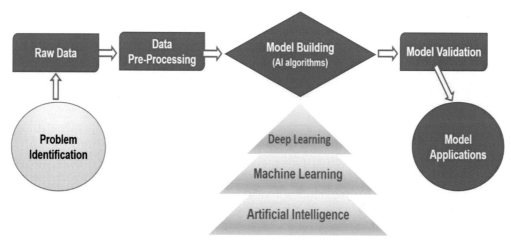

FIGURE 6.3 General schema for building a prediction model.

3.3 Model building and model evaluation metrics

Building a ML model is a multistep process. The first step of model building is to identify problems (such as biological problems) that need to be solved by intelligent techniques. The subsequent steps required data collection and data processing. The model is built by choosing an appropriate ML algorithm. Before deploying the model on untested data, its performance must be examined and validated. The validation step's major goal is to reduce the amount of behavior change that occurs once the model is applied to the practical dataset. The complete data are divided into three datasets: training, validation, and testing. The complete process of model building is given in Fig. 6.3.

Model evaluation involves comparing a model's performance to actual data to determine how well it performs. Before including them in the production environment, the model must be validated by taking into account the elements and components. The best model validation testing procedures involve creating training and test sets from the data. Further, checking the models' validity for various combinations of the training and test sets using the same data source. Moreover, creating the classification matrices also provides an inside look at the model's mathematical viability (Ramakrishna et al., 2020).

4. Applications of machine learning in stem cell–based therapies

ML is a data-driven approach that enables machines to learn from real-world data and develop intelligent algorithms capable of solving problems of predictive analysis and classification without explicit programming. ML algorithms are extensively applied in different areas of stem cell–based research. The clinical use of human iPSCs provides a potential solution to problems related to immune rejection in place of hESCs and does not raise the same ethical concerns as ESCs when employed in regenerative medicine. The applicability of iPSCs enables advancements in biological research. In one of their research, Nishino et al. created a learning model based on linear classification to discriminate between somatic cells, iPSCs, ESCs, and embryonal carcinoma cells (ECCs) based on their DNA methylation signatures (Nishino et al., 2021). The classification model after training was able to distinguish different cell types with an accuracy of 94.23%. Additionally, the learned models' component analysis revealed the distinct epigenetic signature of iPSCs, which differs from that of ESCs. This classification algorithm-based study highlights the iPSC-specific methylation variations present on chromosomes 7, 8, 12, and 22.

Cardiomyocyte is the cell present in the heart and responsible for the contraction and relaxation of heart muscles. It is difficult to extract human cardiomyocytes as their renewal capability is lost in the adult stage. Typically, the electrophysiological systems that are used to assess the characteristics of cardiomyocytes are expensive, challenging to operate, and could result in unfavorable cellular responses (Laflamme et al., 2007). The use of iPSC-derived cardiomyocytes in the development of cardiovascular medications and safety evaluation has a lot of potential in cell therapy. It takes a lot of physical labor to identify abnormal Ca^{2+} transients to understand and evaluate the functionality of cardiomyocytes. So, an analytical pipeline is developed by combining a classification-based ML algorithm and analytical algorithm that can automatically detect Ca^{2+} transient signals to differentiate hiPSC-derived cardiomyocytes (Hwang et al., 2020). The Ca^{2+} transient data of 254 cells are divided into training and test datasets. Training data consist of 200 cell signals having 1893 peaks and test data having 54 cell signals (454 peaks). The three processes comprise manual evaluation, enhanced analytical technique, and SVM approach to assess the normality of studied signals. The data generated from mentioned peak measurement methods in combination with other parameters associated with cell variability are used to train cell level SVM classifier to detect cell abnormalities. The performance of the cell based SVM model is evaluated using the leave one-out cross-validation method. The evaluation metrics show 89.9% accuracy, 94.7% sensitivity, and 83.3% specificity on training data, and 87% accuracy, 88.9% sensitivity, and 83.3% for test data. Juhola et al. used ML classifiers to study iPSC derived cardiomyocytes using their calcium transient signals (Juhola et al., 2021). With the help of this information, signals were divided into three groups: baseline, adrenaline, and dantrolene according to their differential effect on calcium levels. First, the peak recognition algorithm was used to identify patterns of transient signals. After that, 12 peak variables were calculated for each signal peak. The dantrolene group was further divided into responder, semiresponder, and nonresponder categories based on the percentage of anomalies that dantrolene could diminish while still reflecting the architecture of the calcium peaks. The dantrolene group was further divided into a responder, semiresponder, and nonresponder categories based on the percentage of anomalies that dantrolene could diminish while also reflecting the morphology of the calcium peaks. The random forest and least square SVM outperforms other studied ML algorithms in identifying patterns of Ca^{2+} transient signals.

Several studies have applied clustering algorithms in stem cell research to identify different clusters based on their specific characteristics. For instance, Chen et al. used microarray-based gene expression data of ESCs and applied multiple clustering ML techniques, such as K-means, hierarchical clustering, self-organizing maps (SOMs), and partition around medoids (PAMs) clustering, to discover genes based on their distinct signature patterns (Chen et al., 2002). This comparative study showed that there is partial consistency in the identification of distinct clusters by applying different clustering algorithms. This study provides a roadmap for using efficient clustering algorithms for extracting meaningful biological information from microarray-based gene expression data of ESCs. A study conducted by Maddah and Loewke has defined an automated process that includes segmenting beating zones, calculating a reliable beating signal, modeling and quantifying the signal, and hierarchical clustering of cardiomyocytes derived from stem cells (Maddah & Loewke, 2014). In contrast to other imaging-based techniques, this hierarchical clustering-based technique provides therapeutic applications by recording heartbeat patterns and arrhythmias throughout normal and sick cells with different densities.

In a similar kind of study, Zhu et al. proposed an advanced clustering algorithm based on the single-cell structure entropy (SSE) minimization principle to identify cell populations based on cell type differentiation (Zhu et al., 2019). The algorithm is applied to eight single-cell RNA sequencing (scRNA-seq) test datasets including stem cell data. The performance of the SSE algorithm is compared with other techniques: nonnegative matrix factorization (NMF), single-cell interpretation via multi-kernel learning (SIMLR), and structural entropy (SE) minimization principle. Compared with studied methods, the SSE algorithm can identify biologically diverse cells into various clusters with high accuracy.

5. Applications of deep learning in stem cell–based therapies

With an expanded number of hidden layers, the innovative DL idea has become more sophisticated and, as a result, even more advanced. ML algorithms rely on manual extraction of underlying features in the dataset, which is a laborious and time-consuming approach. The manual feature extraction is replaced in DL approaches, where features are automatically extracted by using hidden neural networks. ANN approach mimics behavior of human neurons in terms of processing information. It consists of multiple layers of fully connected neural networks having an input layer, neural hidden layers, and an output layer with an activation function. The development of ANN enables the identification of cell types without molecular labeling. The use of ANNs in the diagnosis of acute graft-vs-host disease (aGVHD) following unrelated donor-HSCT for thalassemia disease has been the subject of one study conducted by Caocci et al. group. The prognostic model is constructed using two separate algorithms: logistic regression (LR) and ANN (Caocci et al., 2010). The initial data consist of 24 independent variables taken from 78 patients. When comparing the prognosis performance of the two techniques, the sensitivity of ANN was higher (83.3%) compared with LR (21.7%) in predicting aGVHD in patients. Additionally, ANN's mean specificity for detecting the avoidance of aGVHD in patients who had not experienced symptoms was 90.1%, whereas LR's mean specificity was 80.5%.

The development of supervised/unsupervised ML models is cumbersome, time-consuming, error-prone, and needs manual curation for iPSCs related functional studies. Thus, there is an urgent need for automation for examining iPSC colonies. Nowadays, CNNs are one technique that currently offers an automated platform for analyzing iPSC colonies (Ramakrishna et al., 2020). CNNs, a subset of multilayered neural networks, are used in DL to rectify data properties and can significantly aid in the identification of images and the diagnosis of diseases. CNNs can differentiate iPSCs colonies with high accuracy based on their morphologic and textural changes. As a result, CNNs may one day be used to build DL tasks that address a variety of issues in stem cell research.

In one study, human urinal cells were used to create iPSCs, and CNN was used for semisupervised segmentation and colony recognition (Fan et al., 2017). To pinpoint the location and borders of colonies, CNN recognizes visual information from sparsely labeled data. The hidden Markov model (HMM) was used in this study to forecast the colony formation time window and growth phase while cells undergo reprogramming. HMM is a popular learning approach that can process the raw data of time-dependent multiple images without preprocessing. The existence of mouse-derived iPSC and urine cells iPSC was verified using a computer vision technique that included a created binary image, a brightfield image, colony-picking king options, and manual cross-referencing. The outcomes were next contrasted with those of hESCs. Notably, the computer vision technique discovered worked well for detecting iPSCs in mice as well as humans.

Kavitha et al. in their study used a vector-based CNN (V-CNN) algorithm to extract features from the iPSC colony to identify their distinct characteristics (Kavitha et al., 2017). Later on, the robustness of the V-CNN model is compared with SVM classifiers to identify feature-based differences in healthy and unhealthy iPSC colonies. The evaluation metrics show a high-level accuracy of 90% for V-CNN models compared with SVM models (75%–77%). The proposed V−CNN model shows superiority over SVM models and, thus, provides a reliable and robust automated framework for classifying iPSCs colonies.

Chu et al. identified the early possibility of hiPSCs formation and predicted their morphology using multiple deep learning models (Chu et al., 2023). Images taken from cell culture by using time-lapse brightfield microscopy were used to look for signs that they might become hiPSCs in the initial stages of reprogramming. The CNN model was trained on images taken at early stage of hiPSC formation. Later on, the hiPSC probability images obtained from the CNN model were used as an input for recurrent neural network (RNN) models to generate predicted hiPSC images to identify morphological features of future hiPSC colonies. The proposed study has the potential to enhance the efficiency of culturing hiPSC cell lines.

In one of the studies conducted by Aida et al., a conditional generative adversarial network (CGAN) was employed to create AI models for locating regions of cancer stem cells (CSCs) to find unrecognized morphological characteristics (Aida et al., 2020). The phase-contrast images obtained from tagging CSCs with Nanog-green fluorescent protein were used to train AI models. The choice of images for training boosted several criteria for evaluating segmentation quality, leading to a novel method for identifying malignant cells. However, the performance of the model used in the study was less effective at identifying Nanog-expressing cells compared with green fluorescent protein (GFP) fluorescence analysis, but it might be improved and used as a CSC diagnostic technique soon. CNN simultaneously recognized the nucleus, bleeding, mitosis, and cell shape. The best image assessment outcomes were obtained when mouse embryonic fibroblast feeder cells were present. Phase-contrast images were used to demonstrate how the AI model represented CSCs using GFP fluorescence.

6. Evolution of nature-inspired computing and its applications in stem cell−based therapies

More recently, the conversion of high-throughput data into understanding stem cell research has relied heavily on computational methods. Recent years have seen the emergence of computational stem cell research that integrates high-throughput molecular stem cell data. For the first time, when Alan Turing studied the role of morphogen concentration in defining the patterns of virtual embryo formation by writing a computer code, computational methods and their applications have a significant impact on developmental biology (AM TURING, 1952). Before the development of OMICs, the study of theoretical mechanisms of morphogenesis is the only application of computational methods in stem cell biology. The in silico methods were used to model the dynamics of ASCs and to study their positional information during embryogenesis (Lander, 2007).

Stem cell research has changed dramatically since techniques for creating iPSCs were discovered (Takahashi & Yamanaka, 2006). As a result, there are more prospects for developmental research, tissue regeneration technologies, cell transplantation, and human illness research. Rapid advancements in single-cell sequencing techniques have made it possible to identify and characterize cellular behaviors across tissues under differential conditions (van Dijk et al., 2018). The analysis of single-cell data overcomes technical constraints, such as gene dropouts and poor capture rates, by taking into account a large number of individual samples and enables the identification of cellular subpopulations at an unheard-of resolution (del Sol & Jung, 2021). It is now possible to create advanced computational models that catch the aggregate behavior of genes at cellular as well as molecular levels thanks to the mass production of these multiomics single-cell data. This creates the perfect framework for addressing critical issues in the stem cell field. The nature-inspired intelligent

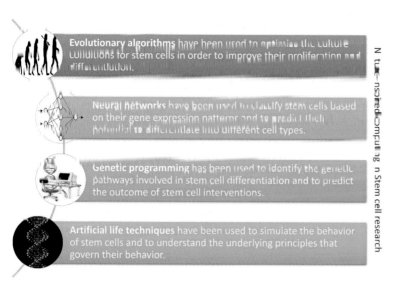

FIGURE 6.4 Nature-inspired computing in stem cell research.

computing (NIC) techniques imitate behavior of biological systems. NIC-based methods offer fresh approaches to comprehend, describe, and analyze natural complexity and include problem-solving strategies based on concepts of biological processes. Nowadays, NIC algorithms are becoming useful in providing solutions to complex optimization problems of stem cell—based therapies (Fig. 6.4).

In one study, Egri et al. used biologically inspired algorithms to control the production of stem cells (Egri et al., 2020). Bioinspired reinforcement learning algorithm learns from the interactions based on positive and negative feedback systems and improves the controller of the current platform. The study showed that bioinspired reinforcement-based learning may be a more viable alternative way, compared with traditional stem cell processing methods.

Computational techniques that depend on the reconstruction of cell-based gene regulatory networks (GRNs) are beneficial for simulating cellular conversion and identifying the best combinations of conversion parameters (Barbuti et al., 2020). These network-based strategies are crucial, but they rely on the precise identification of individual cell-based GRNs, which is not always possible, especially for recently discovered cell subtypes. Therefore, the best tool for studying gene interactions in single cells is scRNA-seq. The pdiversity of expression of genes across cells and their coexpression in specific cells can be studied by this method in a single experiment (del Sol & Jung, 2021). As a result, network-based approaches for cellular conversion can be implemented. GRN patterns connected to various states of pluripotency have been discovered particularly utilizing machine learning approaches. This information based on GRNs is critical for developing novel methods to control the ability of stem cells to differentiate into particular cell types. Single-cell-based GRN inference techniques have made it possible to optimize several cell conversion protocols for differentiation, trans-differentiation, and reprogramming. The evaluation of GRN-based inferences concerning their reprogramming paths led to the discovery of a combination of small drugs that improved the effectiveness of the conversion of embryonic fibroblasts to pluripotent stem cells.

7. Conclusion and future aspects

Stem cell—based regenerative therapies have a great deal of potential to improve patient care and speed up healing. However, there are also negatives brought on by inefficient production, drawn-out and complicated processes, and mistakes made by humans as a result of excessive human effort. It is crucial to comprehend the genetic factors that affect how an organ develops, including its size, shape, and orientation (Polak, 2010). A potent solution for a greater comprehension of such systems may be provided by AI-driven models and productive algorithms. These models might automate and reduce human error factors in the creation of stem cell—based regenerative therapies. Given their exceptional skills in image categorization, detection, and segmentation, CNN algorithms have been implemented to extract vital information from medical images. CNNs have completely changed the way of morphological examination of stem cells. For instance, CNNs can accurately recognize iPSCs utilizing microscopic pictures instead of the more traditional methods of manual molecular labeling. This discovery has significantly improved the use of iPSCs in stem cell—based research. As a result,

ML- and DL-driven methodologies have become essential components of stem cell research. Due to the data-driven nature of these methodologies, results will be flawed if there is insufficient or error-prone data. Despite the fact that algorithms have improved over the past few years, accurate outcomes still require manual intervention by experts. Soon, further improvements in algorithms could successfully advance the development of stem cell—based regenerative treatments and assist with decisions for medical professionals.

References

Ahuja, A. S. (2019). The impact of artificial intelligence in medicine on the future role of the physician. *PeerJ, 7*, e7702. https://doi.org/10.7717/peerj.7702

Aida, S., Okugawa, J., Fujisaka, S., Kasai, T., Kameda, H., & Sugiyama, T. (2020). Deep learning of cancer stem cell morphology using conditional generative adversarial networks. *Biomolecules, 10*(6). https://doi.org/10.3390/biom10060931

Akimoto, K., Kimura, K., Nagano, M., Takano, S., To'a Salazar, G., Yamashita, T., & Ohneda, O. (2013). Umbilical cord blood-derived mesenchymal stem cells inhibit, but adipose tissue-derived mesenchymal stem cells promote, glioblastoma multiforme proliferation. *Stem Cells and Development, 22*(9), 1370—1386. https://doi.org/10.1089/scd.2012.0486

AM TURING, F. (1952). The chemical basis of morphogenesis. *Sciences-cecm.usp.br.*

Amisha, Malik, P., Pathania, M., & Rathaur, V. K. (2019). Overview of artificial intelligence in medicine. *Journal of Family Medicine and Primary Care, 8*(7), 2328—2331. https://doi.org/10.4103/jfmpc.jfmpc_440_19

Ashraf, M., Khalilitousi, M., & Laksman, Z. (2021). Applying machine learning to stem cell culture and differentiation. *Current Protocols, 1*(9), e261. https://doi.org/10.1002/cpz1.261

Bacakova, L., Zarubova, J., Travnickova, M., Musilkova, J., Pajorova, J., Slepicka, P., Kasalkova, N. S., Svorcik, V., Kolska, Z., Motarjemi, H., & Molitor, M. (2018). Stem cells: Their source, potency and use in regenerative therapies with focus on adipose-derived stem cells - a review. *Biotechnology Advances, 36*(4), 1111—1126. https://doi.org/10.1016/j.biotechadv.2018.03.011

Barbuti, R., Gori, R., Milazzo, P., & Nasti, L. (2020). A survey of gene regulatory networks modelling methods: From differential equations, to Boolean and qualitative bioinspired models. *Journal of Membrane Computing, 2*(3), 207—226. https://doi.org/10.1007/s41965-020-00046-y

Bianco, P., Robey, P. G., & Simmons, P. J. (2008). Mesenchymal stem cells: Revisiting history, concepts, and assays. *Cell Stem Cell, 2*(4), 313—319. https://doi.org/10.1016/j.stem.2008.03.002

Bruin, J. E., Saber, N., Braun, N., Fox, J. K., Mojibian, M., Asadi, A., Drohan, C., O'Dwyer, S., Rosman-Balzer, D. S., Swiss, V. A., Rezania, A., & Kieffer, T. J. (2015). Treating diet-induced diabetes and obesity with human embryonic stem cell-derived pancreatic progenitor cells and antidiabetic drugs. *Stem Cell Reports, 4*(4), 605—620. https://doi.org/10.1016/j.stemcr.2015.02.011

Campagnoli, C., Bellantuono, I., Kumar, S., Fairbairn, L. J., Roberts, I., & Fisk, N. M. (2002). High transduction efficiency of circulating first trimester fetal mesenchymal stem cells: Potential targets for in utero ex vivo gene therapy. *BJOG: An International Journal of Obstetrics and Gynaecology, 109*(8), 952—954. https://doi.org/10.1111/j.1471-0528.2002.t01-1-02011.x

Caocci, G., Baccoli, R., Vacca, A., Mastronuzzi, A., Bertaina, A., Piras, E., Littera, R., Locatelli, F., Carcassi, C., & La Nasa, G. (2010). Comparison between an artificial neural network and logistic regression in predicting acute graft-vs-host disease after unrelated donor hematopoietic stem cell transplantation in thalassemia patients. *Experimental Hematology, 38*(5), 426—433. https://doi.org/10.1016/j.exphem.2010.02.012

Ceccarelli, S., Pontecorvi, P., Anastasiadou, E., Napoli, C., & Marchese, C. (2020). Immunomodulatory effect of adipose-derived stem cells: The cutting edge of clinical application. *Frontiers in Cell and Developmental Biology, 8*, 236. https://doi.org/10.3389/fcell.2020.00236

Chamberlain, G., Fox, J., Ashton, B., & Middleton, J. (2007). Concise review: Mesenchymal stem cells: Their phenotype, differentiation capacity, immunological features, and potential for homing. *Stem Cells, 25*(11), 2739—2749. https://doi.org/10.1634/stemcells.2007-0197

Chen, G., Jaradat, S. A., Banerjee, N., Tanaka, T. S., Ko, M. S. H., & Zhang, M. Q. (2002). Evaluation and comparison of clustering algorithms in analyzing ES cell gene expression data. *Statistica Sinica, 12*(1), 241—262.

Chiba, S., Iwasaki, Y., Sekino, H., & Suzuki, N. (2003). Transplantation of motoneuron-enriched neural cells derived from mouse embryonic stem cells improves motor function of hemiplegic mice. *Cell Transplantation, 12*(5), 457—468. https://doi.org/10.3727/000000003108747019

Chu, S.-L., Sudo, K., Yokota, H., Abe, K., Nakamura, Y., & Tsai, M.-D. (2023). Human induced pluripotent stem cell formation and morphology prediction during reprogramming with time-lapse bright-field microscopy images using deep learning methods. *Computer Methods and Programs in Biomedicine, 229*, 107264. https://doi.org/10.1016/j.cmpb.2022.107264

van Dartel, D. A., Pennings, J. L., de la Fonteyne, L. J., van Herwijnen, M. H., van Delft, J. H., van Schooten, F. J., & Piersma, A. H. (2010). Monitoring developmental toxicity in the embryonic stem cell test using differential gene expression of differentiation-related genes. *Toxicological Sciences, 116*(1), 130—139. https://doi.org/10.1093/toxsci/kfq127

del Sol, A., & Jung, S. (2021). The importance of computational modeling in stem cell research. *Trends in Biotechnology, 39*(2), 126—136. https://doi.org/10.1016/j.tibtech.2020.07.006

van Dijk, D., Sharma, R., Nainys, J., Yim, K., Kathail, P., Carr, A. J., Burdziak, C., Moon, K. R., Chaffer, C. L., Pattabiraman, D., Bierie, B., Mazutis, L., Wolf, G., Krishnaswamy, S., & Pe'er, D. (2018). Recovering gene interactions from single-cell data using data diffusion. *Cell, 174*(3), 716—729. https://doi.org/10.1016/j.cell.2018.05.061. e727.

Egri, P., Csáji, B. C., Kis, K. B., Monostori, L., Váncza, J., Ochs, J., Jung, S., König, N., Schmitt, R., Brecher, C., Pieske, S., & Wein, S. (2020). Bio-inspired control of automated stem cell production. *Procedia CIRP, 88*, 600—605. https://doi.org/10.1016/j.procir.2020.05.105

Elsayed, A. K., Vimalraj, S., Nandakumar, M., & Abdelalim, E. M. (2021). Insulin resistance in diabetes: The promise of using induced pluripotent stem cell technology. *World Journal of Stem Cells, 13*(3), 221–235. https://doi.org/10.4252/wjsc.v13.i3.221

Eridani, S. (2014). Types of human stem cells and their therapeutic applications. *Stem Cell Discovery, 4*, 15–20. https://doi.org/10.4236/scd.2014.42003

Fan, K., Zhang, S., Zhang, Y., Lu, J., Holcombe, M., & Zhang, X. (2017). A machine learning assisted, label-free, non-invasive approach for somatic reprogramming in induced pluripotent stem cell colony detection and prediction. *Scientific Reports, 7*(1), 13496, https://doi.org/10.1038/s41598-017-13680-x

Paré, R., Otero, J., Almendros, I., & Navajas, D. (2018). Bioengineered lungs: A challenge and an opportunity. *Archivos de Bronconeumología, 54*(1), 31–39. https://doi.org/10.1016/j.arbres.2017.09.002

Fijnvandraat, A. C., Lekanne Deprez, R. H., & Moorman, A. F. M. (2003). Development of heart muscle-cell diversity: A help or a hindrance for phenotyping embryonic stem cell-derived cardiomyocytes. *Cardiovascular Research, 58*(2), 303–312. https://doi.org/10.1016/s0008-6363(03)00248-3

Hayashi, K., Ohta, H., Kurimoto, K., Aramaki, S., & Saitou, M. (2011). Reconstitution of the mouse germ cell specification pathway in culture by pluripotent stem cells. *Cell, 146*(4), 519–532. https://doi.org/10.1016/j.cell.2011.06.052

Hmadcha, A., Martin-Montalvo, A., Gauthier, B. R., Soria, B., & Capilla-Gonzalez, V. (2020). Therapeutic potential of mesenchymal stem cells for cancer therapy. *Frontiers in Bioengineering and Biotechnology, 8*, 43. https://doi.org/10.3389/fbioe.2020.00043

Hole, K. J., & Ahmad, S. (2019). Biologically driven artificial intelligence. *Computer, 52*(8), 72–75. https://doi.org/10.1109/MC.2019.2917455

Hwang, H., Liu, R., Maxwell, J. T., Yang, J., & Xu, C. (2020). Machine learning identifies abnormal Ca2+ transients in human induced pluripotent stem cell-derived cardiomyocytes. *Scientific Reports, 10*(1), 16977. https://doi.org/10.1038/s41598-020-73801-x

Juhola, M., Penttinen, K., Joutsijoki, H., & Aalto-Setälä, K. (2021). Analysis of drug effects on iPSC cardiomyocytes with machine learning. *Annals of Biomedical Engineering, 49*(1), 129–138. https://doi.org/10.1007/s10439-020-02521-0

Jung, Y. W., Hysolli, E., Kim, K. Y., Tanaka, Y., & Park, I. H. (2012). Human induced pluripotent stem cells and neurodegenerative disease: Prospects for novel therapies. *Current Opinion in Neurology, 25*(2), 125–130. https://doi.org/10.1097/WCO.0b013e3283518226

Kanate, A. S., Kharfan-Dabaja, M. A., & Hamadani, M. (2012). Controversies and recent advances in hematopoietic cell transplantation for follicular non-hodgkin lymphoma. *Bone Marrow Research, 2012*, 897215. https://doi.org/10.1155/2012/897215

Kavitha, M. S., Kurita, T., Park, S. Y., Chien, S. I., Bae, J. S., & Ahn, B. C. (2017). Deep vector-based convolutional neural network approach for automatic recognition of colonies of induced pluripotent stem cells. *PLoS One, 12*(12), e0189974. https://doi.org/10.1371/journal.pone.0189974

Kim, J.-H., Auerbach, J. M., Rodríguez-Gómez, J. A., Velasco, I., Gavin, D., Lumelsky, N., Lee, S.-H., Nguyen, J., Sánchez-Pernaute, R., Bankiewicz, K., & McKay, R. (2002). Dopamine neurons derived from embryonic stem cells function in an animal model of Parkinson's disease. *Nature, 418*(6893), 50–56. https://doi.org/10.1038/nature00900

Laflamme, M. A., Chen, K. Y., Naumova, A. V., Muskheli, V., Fugate, J. A., Dupras, S. K., Reinecke, H., Xu, C., Hassanipour, M., Police, S., O'Sullivan, C., Collins, L., Chen, Y., Minami, E., Gill, E. A., Ueno, S., Yuan, C., Gold, J., & Murry, C. E. (2007). Cardiomyocytes derived from human embryonic stem cells in pro-survival factors enhance function of infarcted rat hearts. *Nature Biotechnology, 25*(9), 1015–1024. https://doi.org/10.1038/nbt1327

Lander, A. D. (2007). Morpheus unbound: Reimagining the morphogen gradient. *Cell, 128*(2), 245–256. https://doi.org/10.1016/j.cell.2007.01.004

Lee, J. Y., & Hong, S. H. (2020). Hematopoietic stem cells and their roles in tissue regeneration. *International Journal of Stem Cells, 13*(1), 1–12. https://doi.org/10.15283/ijsc19127

Lerou, P. H., & Daley, G. Q. (2005). Therapeutic potential of embryonic stem cells. *Blood Reviews, 19*(6), 321–331. https://doi.org/10.1016/j.blre.2005.01.005

Liu, C. B., Huang, H., Sun, P., Ma, S. Z., Liu, A. H., Xue, J., Fu, J. H., Liang, Y. Q., Liu, B., Wu, D. Y., Lü, S. H., & Zhang, X. Z. (2016). Human umbilical cord-derived mesenchymal stromal cells improve left ventricular function, perfusion, and remodeling in a porcine model of chronic myocardial ischemia. *Stem Cells Translational Medicine, 5*(8), 1004–1013. https://doi.org/10.5966/sctm.2015-0298

Lou, Y.-j., & Liang, X.-g. (2011). Embryonic stem cell application in drug discovery. *Acta Pharmacologica Sinica, 32*(2), 152–159. https://doi.org/10.1038/aps.2010.194

Maddah, M., & Loewke, K. (2014). Automated, non-invasive characterization of stem cell-derived cardiomyocytes from phase-contrast microscopy. In *Medical Image Computing and Computer-Assisted Intervention, 17* pp. 57–64). https://doi.org/10.1007/978-3-319-10404-1_8. 25333101.

Mancuso, P., Raman, S., Glynn, A., Barry, F., & Murphy, J. M. (2019). Mesenchymal stem cell therapy for osteoarthritis: The critical role of the cell secretome. *Frontiers in Bioengineering and Biotechnology, 7*, 9. https://doi.org/10.3389/fbioe.2019.00009

Marcus, A. J., & Woodbury, D. (2008). Fetal stem cells from extra-embryonic tissues: Do not discard. *Journal of Cellular and Molecular Medicine, 12*(3), 730–742. https://doi.org/10.1111/j.1582-4934.2008.00221.x

Mukherjee, S., Yadav, G., & Kumar, R. (2021). Recent trends in stem cell-based therapies and applications of artificial intelligence in regenerative medicine. *World Journal of Stem Cells, 13*(6), 521–541. https://doi.org/10.4252/wjsc.v13.i6.521

Müller, A. M., Huppertz, S., & Henschler, R. (2016). Hematopoietic stem cells in regenerative medicine: Astray or on the path? *Transfusion Medicine and Hemotherapy, 43*(4), 247–254. https://doi.org/10.1159/000447748

Nishino, K., Takasawa, K., Okamura, K., Arai, Y., Sekiya, A., Akutsu, H., & Umezawa, A. (2021). Identification of an epigenetic signature in human induced pluripotent stem cells using a linear machine learning model. *Human Cell, 34*(1), 99–110. https://doi.org/10.1007/s13577-020-00446-3

O'Donoghue, K., & Fisk, N. M. (2004). Fetal stem cells. *Best Practice & Research Clinical Obstetrics & Gynaecology, 18*(6), 853–875. https://doi.org/10.1016/j.bpobgyn.2004.06.010

Polak, D. J. (2010). Regenerative medicine. Opportunities and challenges: A brief overview, *Journal of The Royal Society Interface, 7*(Suppl. 1_6), 3777–3781. https://doi.org/10.1098/rsif.2010.0362.focus

Ramakrishna, R. R., Abd Hamid, Z., Wan Zaki, W. M. D., Huddin, A. B., & Mathialagan, R. (2020). Stem cell imaging through convolutional neural networks: Current issues and future directions in artificial intelligence technology. *PeerJ, 8*, e10346. https://doi.org/10.7717/peerj.10346

Razavi, S., Ghasemi, N., Mardani, M., & Salehi, H. (2018). Co-transplantation of human neurotrophic factor secreting cells and adipose-derived stem cells in rat model of multiple sclerosis. *Cell Journal, 20*(1), 46–52. https://doi.org/10.22074/cellj.2018.4777

Saitoh, I., Sato, M., Soda, M., Inada, E., Iwase, Y., Murakami, T., Ohshima, H., Hayasaki, H., & Noguchi, H. (2016). Tissue-specific stem cells obtained by reprogramming of non-obese diabetic (NOD) mouse-derived pancreatic cells confer insulin production in response to glucose. *PLoS One, 11*(9), e0163580. https://doi.org/10.1371/journal.pone.0163580

Salguero-Aranda, C., Tapia-Limonchi, R., Cahuana, G. M., Hitos, A. B., Diaz, I., Hmadcha, A., Fraga, M., Martín, F., Soria, B., Tejedo, J. R., & Bedoya, F. J. (2016). Differentiation of mouse embryonic stem cells toward functional pancreatic β-cell surrogates through epigenetic regulation of Pdx1 by nitric oxide. *Cell Transplantation, 25*(10), 1879–1892. https://doi.org/10.3727/096368916x691178

Schaub, N. J., Hotaling, N. A., Manescu, P., Padi, S., Wan, Q., Sharma, R., George, A., Chalfoun, J., Simon, M., Ouladi, M., Simon, C. G., Jr., Bajcsy, P., & Bharti, K. (2020). Deep learning predicts function of live retinal pigment epithelium from quantitative microscopy. *Journal of Clinical Investigation, 130*(2), 1010–1023. https://doi.org/10.1172/jci131187

Sheets, K. T., Bagó, J. R., & Hingtgen, S. D. (2018). Delivery of cytotoxic mesenchymal stem cells with biodegradable scaffolds for treatment of postoperative brain cancer. *Methods in Molecular Biology, 1831*, 49–58. https://doi.org/10.1007/978-1-4939-8661-3_5

Song, I., Rim, J., Lee, J., Jang, I., Jung, B., Kim, K., & Lee, S. (2022). Therapeutic potential of human fetal mesenchymal stem cells in musculoskeletal disorders: A narrative review. *International Journal of Molecular Sciences, 23*(3). https://doi.org/10.3390/ijms23031439

Sotoudeh, H., Shafaat, O., Bernstock, J. D., Brooks, M. D., Elsayed, G. A., Chen, J. A., Szerip, P., Chagoya, G., Gessler, F., Sotoudeh, E., Shafaat, A., & Friedman, G. K. (2019). Artificial intelligence in the management of glioma: Era of personalized medicine. *Frontiers in Oncology, 9*, 768. https://doi.org/10.3389/fonc.2019.00768

Srinivasan, M., Thangaraj, S. R., Ramasubramanian, K., Thangaraj, P. P., & Ramasubramanian, K. V. (2021). Exploring the current trends of artificial intelligence in stem cell therapy: A systematic review. *Cureus, 13*(12), e20083. https://doi.org/10.7759/cureus.20083

Tailor, J., Andreska, T., & Kittappa, R. (2012). From stem cells to dopamine neurons: Developmental biology meets neurodegeneration. *CNS & Neurological Disorders: Drug Targets, 11*(7), 893–896. https://doi.org/10.2174/1871527311201070893

Takahashi, K., & Yamanaka, S. (2006). Induction of pluripotent stem cells from mouse embryonic and adult fibroblast cultures by defined factors. *Cell, 126*(4), 663–676. https://doi.org/10.1016/j.cell.2006.07.024

Talib, S., & Shepard, K. A. (2020). Unleashing the cure: Overcoming persistent obstacles in the translation and expanded use of hematopoietic stem cell-based therapies. *Stem Cells Translational Medicine, 9*(4), 420–426. https://doi.org/10.1002/sctm.19-0375

Tolosa, L., Caron, J., Hannoun, Z., Antoni, M., López, S., Burks, D., Castell, J. V., Weber, A., Gomez-Lechon, M.-J., & Dubart-Kupperschmitt, A. (2015). Transplantation of hESC-derived hepatocytes protects mice from liver injury. *Stem Cell Research & Therapy, 6*(1), 246. https://doi.org/10.1186/s13287-015-0227-6

Wainstein, C., Quera, R., Fluxá, D., Kronberg, U., Conejero, A., López-Köstner, F., Jofre, C., & Zarate, A. J. (2018). Stem cell therapy in refractory perineal Crohn's disease: Long-term follow-up. *Colorectal Disease.* https://doi.org/10.1111/codi.14002

Zhu, X., Li, H. D., Xu, Y., Guo, L., Wu, F. X., Duan, G., & Wang, J. (2019). A hybrid clustering algorithm for identifying cell types from single-cell RNA-seq data. *Genes, 10*(2). https://doi.org/10.3390/genes10020098

Chapter 7

Stem cell therapy in the era of machine learning

Asif Adil[1], Mohammed Asger[1], Musharaf Gul[2], Akib Mohi Ud Din Khanday[3] and Rayees Ahmad Magray[4]

[1]Department of Computer Sciences, Baba Ghulam Shah Badshah University, Rajouri, Jammu and Kashmir, India; [2]Departmetn of Zoology, Hemvati Nandan Bahuguna Gharwal University (A Central University), Srinagar, Uttarakhand, India; [3]Department of Computer Science and Software Engineering-CIT, United Arab Emirates University, Al Ain, United Arab Emirates; [4]Department of Environmental Studies, Government Degree College, Kupwara, Jammu and Kashmir, India

1. Introduction

Stem cells are the undifferentiated cells present in the human body (Zakrzewski et al., 2019). In contrast to other cell types (somatic cells), stem cells have the potential to both self-renew as undifferentiated cells and, under the correct circumstances, develop into nearly any other cell type in the body (Kumar et al., 2010). They are distinguished by the following three qualities: self-renewing (the power to replicate indefinitely); potent (the ability to divide into one or more cell types); and clonogenic (the potential to create identical copies or replicas) (Kumar et al., 2010). The stem cells are able to replicate, either naturally or in a lab, to produce what are termed daughter cells. These daughter cells may either continue to divide and form more stem cells or differentiate into specialized cells (e.g., blood cells, brain cells, heart muscle cells, and bone cells) with a more defined role in the body (Shende & Devlekar, 2021).

Numerous studies in the past several decades have shown that this fascinating adaptability has important therapeutic significance (Duncan & Valenzuela, 2017; Genc et al., 2019; Salybekov et al., 2021; Sugaya & Vaidya, 2018; Tatulian, 2022; Zakrzewski et al., 2019). Consequently, stem cell treatment is increasingly employed as a therapeutic technique for several disorders, including cancer, Alzheimer's disease (AD) (Yamanaka, 2020). The 2007 Nobel Prize in Physiology or Medicine was shared by Mario R. Capecchi, Sir Martin J. Evans, and Oliver Smithies for their studies that provided a framework for harnessing mouse embryonic stem cells to insert targeted mutations into the mouse genome (Khandpur et al., 2021). The field of stem cell research has seen a considerable change toward a more data-driven and methodical approach with the emergence of machine learning (ML).

ML is an area of artificial intelligence (AI) that focuses on the development of algorithms that can "learn" from data and make inferences or decisions without being explicitly programmed to do so (Merkin et al., 2022). Many industries, including medicine, have started using it to boost their diagnostic, therapeutic, and pharmaceutical capabilities. With the biological experiments being conducted on a larger scale worldwide, the data thus accumulated provides an opportunity to use computational tools like ML and big data to dig deeper and find answers to previously unswered biological questions (Adil et al., 2021). ML has been used for a variety of medical tasks, including diagnosis, diagnosis confirmation, and prognosis. Medical imaging scans, such as computerized tomography (CT) and magnetic resonance imaging (MRI's), have been analyzed using ML algorithms to better diagnose illnesses such as cancer (Litjens et al., 2017). Furthermore, ML has been used in drug discovery and development, which has resulted in identifying the novel therapeutic targets, the prediction of medication effectiveness, and toxicity (Duvenaud et al., 2015).

The use of ML algorithms in stem cell research has been seen as having the most potential in the field of cell reprogramming. The specific genetic factors necessary for efficient cell reprogramming have been found using ML techniques (Kim et al., 2015). This has enabled researchers to reprogram cells more quickly and effectively, which has the potential to significantly enhance regenerative medicine. ML is also used for the identification and characterization of stem cells. Off late, ML has allowed for the automated analysis of vast volumes of cell imaging and gene expression data, allowing for the precise identification and classification of stem cells (Polo et al., 2012).

Computational Biology for Stem Cell Research. https://doi.org/10.1016/B978-0-443-13222-3.00004-6
Copyright © 2024 Elsevier Inc. All rights reserved, including those for text and data mining, AI training, and similar technologies.

In addition to these specialized uses, ML is also being used to evaluate enormous volumes of data from stem cell research (Loh & Guo, 2018). This has enabled academics to find previously difficult or impossible-to-detect patterns and trends in the data. This has resulted in fresh insights into stem cell biology and has the potential to significantly enhance our knowledge of how stem cells work.

The current study takes a look at stem cell research's present position in the age of ML, paying special attention to the applications of ML to the field and the possible consequences of this emerging technology.

2. Current state of stem cell therapies

Stem cells are a specific class of undifferentiated cells found in the body that can give rise to a large variety of other cell types. Their prospects in regenerative medicine have been the focus of substantial research owing to their importance in tissue growth and repair. Broadly, stem cells are categorized into embryonic stem cells (ESCs) and adult stem cells (ASCs) (Prentice, 2019; Zhang et al., 2020). ESCs are formed from the inner cell mass of the blastocyst, an early stage of embryonic development, and are capable of generating all cell types in the body. ASCs, on the other hand, are distributed in diverse tissues across the body and have a relatively limited capacity for differentiation (Grochowski et al., 2018; Nguyen et al., 2016).

Significant progress has been made in stem cell research in recent years. The development of "induced pluripotent stem cells (iPSCs)" (Coronnello & Francipane, 2022), for instance, has transformed the study of stem cells by enabling scientists to make ESC-like cells from adult cells without the ethical difficulties connected with the use of ESCs. iPSCs have been utilized to simulate diseases, create new therapeutics, and investigate the molecular basis for development and differentiation. Stem cell therapy is a rapidly developing field with a lot of potential for the treatment of various diseases and conditions. The following are some of the current trends in stem cell therapy:

Regenerative medicine: Regenerative medicine is an emerging field that employs a variety of methods to heal or replace diseased or damaged cells, tissues, or organs (Atala, 2011). Spinal cord injuries, cardiovascular disease, and diabetes might all be significantly impacted by advancements in this area (Atala, 2011; Goradel et al., 2018). Stem cell therapy, which utilizes stem cells to regenerate damaged tissues, is an illustration of the potential of regenerative medicine (Han et al., 2020) Stem cell therapy has already demonstrated promise in the treatment of a number of diseases, including cardiovascular disease and age-related macular degeneration (Han et al., 2020).

Clinical trials: The number of clinical trials using stem cells is growing, with a focus on safety and efficacy (Hoang et al., 2022; Trounson & McDonald, 2015). Clinical trials are essential for the advancement of stem cell therapies, as they assess the viability and efficacy of treatment. Numerous clinical trials are actively investigating the use of stem cells to treat diverse ailments, including cardiovascular disease, neurological disorders, and cancer (Goradel et al., 2018; Pérez López & Otero Hernández, 2012).

Autologous stem cells: The stem cells that are extracted from the patient's own body (autologous) are gaining popularity since they pose less of a threat of rejection and immunological reactions (Al Hamed et al., 2019; Figueiredo et al., 2021; Moreau et al., 2011). With their reduced risk of triggering an immune response, autologous stem cells show great promise as a regenerative medicine approach. When used for tissue regeneration, autologous stem cells reduce the requirement for immunosuppressant medicines and improve the likelihood of a positive outcome (Genc et al., 2019). This ability of not getting rejected by the body makes it an advantageous stem cell therapy in the modern era.

Gene editing: Gene editing is a rapidly developing field that has the potential to revolutionize the treatment of genetic conditions. The use of gene editing techniques, such as CRISPR/Cas9 (Doudna & Charpentier, 2014; Gasperini et al., 2019), in stem cells is particularly promising, as stem cells have the ability to differentiate into any type of cell in the body and have the potential to regenerate damaged tissues and organs. In stem cell therapy, the gene editing is being used in the following ways:

(A) **Correction of genetic mutations:** Gene editing techniques are being used to correct genetic mutations that cause diseases, such as sickle cell anemia, thalassemia, and cystic fibrosis (Rouanet et al., 2013; Sebastiano et al., 2011; Theodoris et al., 2021). By correcting these mutations in stem cells, researchers aim to develop treatments that will cure these diseases.

(B) **Generation of iPSCs:** Adult cells that have been reprogrammed to a condition similar to embryonic development are referred to as induced pluripotent stem cells (iPSCs). iPSCs are generated via the use of gene editing methods, and then they may be differentiated into any type of cell found in the body. This eliminates ethical problems as well as the possibility of cell rejection (Okita et al., 2011; Su et al., 2013). The iPSCs have the potential to be used for regenerative medicine, as well as for the development of personalized medicine, where treatments are tailored to the specific needs and genetic makeup of each patient (Lien et al., 2023).

(C) Gene therapy: Gene therapy is a type of treatment that involves introducing a functional gene into the body to correct a genetic mutation. Gene therapy using stem cells has the potential to treat a range of genetic conditions, including inherited diseases, cancer, and immune disorders (De Masi et al., 2020). Gene therapy using stem cells has the advantage of being able to treat the entire body, as the stem cells can differentiate into any type of cell in the body (Rah, 2011).

(D) Enhanced therapeutic outcomes: Gene editing in stem cells can be used to enhance therapeutic outcomes, by increasing the effectiveness of treatments such as drug therapies or cell-based therapies (De Luca et al., 2019; Sugaya & Vaidya, 2018). By incorporating gene editing techniques into stem cell therapies, researchers are able to improve the efficacy of treatments and provide better outcomes for patients.

iPSCs: The iPSCs have emerged as a promising tool in the field of stem cell therapy. iPSCs are generated by reprogramming adult cells, such as skin or blood cells, into an ESC-like state using a combination of genetic factors (Takahashi & Yamanaka, 2006). This technology has the potential to revolutionize the study of stem cells by allowing scientists to generate patient-specific stem cells for use in personalized medicine.

The capacity of iPSCs to create cells that are genetically identical to the patient eliminates the danger of rejection that is often associated with the use of ESCs derived from other people. This is one of the most significant advantages offered by iPSCs. Because of this, iPSCs are a promising candidate for the development of cell replacement treatments, such as those used in the treatment of neurodegenerative conditions including Parkinson's disease and Alzheimer's disease (Zhang et al., 2020; Vuidel et. al., 2022).

In recent years, the use of iPSCs in the development of new therapies has become a rapidly growing area of research. For example, researchers have used iPSCs to generate functional dopamine neurons for the treatment of Parkinson's disease (Bang et al., 2018; Liu & Cheung, 2020) and to study the molecular basis of degenerative diseases such as Huntington's disease (Choi et al., 2018; Maucksch et al., 2013).

However, before the full potential of iPSCs can be realized in the clinic, there are still a number of difficulties that need to be solved. For instance, the production of iPSCs might be a process that takes a lot of time and requires a lot of labor, and the efficiency of reprogramming is still rather low. In addition, the production of iPSCs is associated with a broad variety of ethical and safety issues. These issues include the potential for the development of tumors as well as the chance of unintentional genetic modifications occurring in the cells that have been reprogrammed. In spite of these impediments, there is optimism for the future of iPSC research, and it appears that significant progress will be made in this area over the course of the next several years. Scientists all around the world now have the opportunity to examine iPSCs on a molecular level as they have never been able to do before because of the widespread availability of genomic and epigenomic data. Because of this, they are able to create novel techniques for directing the differentiation of iPSCs and gain new insights into the intricate processes that regulate the behavior of iPSCs. In addition, developments in the engineering of iPSCs are allowing researchers to produce iPSCs with specific features and to employ iPSCs as delivery vehicles for therapeutic genes.

3. Machine learning applications in different biological and medicine fields

Developing ML algorithms have become a rapidly growing field in biology and medicine, with a growing number of applications being developed and adopted. These are being applied to analyze large and complex biological and medical datasets, providing new insights into the functioning of biological systems and the underlying mechanisms of disease (Chen et al., 2022).

A number of ML algorithms are being utilized to analyze large datasets of patient data, such as electronic health records, to identify risk factors for specific diseases and to predict patient outcomes. The outcomes of cardiovascular disease and cancer patients have recently been predicted using ML algorithms (Hussein et al., 2019; Krittanawong et al., 2020; Lu et al., 2021). Additionally, ML is being used to analyze genomic data to identify new targets for drug development and to improve our understanding of the underlying mechanisms of disease. ML algorithms have been used to identify novel targets for cancer therapy and to understand the molecular basis of neurological diseases (Rafique et al., 2021).

In stem cell therapy, it is becoming an increasingly important tool offering new prospects for advancing our understanding of stem cell biology and for developing new therapeutic strategies. The capability of ML algorithms to analyze large and complex datasets is particularly valuable in stem cell research, where vast amounts of data are generated from experiments and from patient studies. It can also be used to identify novel stem cell populations and to classify stem cells based on their properties, such as differentiation potential or function (Qu-Petersen et al., 2002; Semina et al., 2022). Additionally, these algorithms can be used to predict stem cell behavior and to identify new targets for stem cell based therapies (Shende & Devlekar, 2021).

The potential benefits of ML in stem cell therapy are vast, and new advancements are made frequently. For instance, algorithms based on ML are used to establish novel methods for reprogramming adult cells into stem cells, which may be utilized to generate patient-specific cells for regenerative medicine and to model disease in the lab (Coronnello & Francipane, 2022). Additionally, ML algorithms are being used to identify new targets for stem cell–based therapies, such as the identification of novel small molecules that can control stem cell behavior (Salybekov et al., 2021).

4. Current trends in machine learning–based stem cell therapy

Off late, ML applicability in stem cell research has found quite a good space in the scientific communities worldwide. We performed a publication trend analysis for the MeSH terms on PubMed (Fig. 7.1). For Fig. 7.1B, the following search query was used:

(((Stem Cell Therapy) AND (Machine Learning)) OR ((iPSC AND Machine Learning)) OR ((cell Therapy AND Deep learning))) [MeSH Major Topic].

A total of 218 papers were found relevant for analysis, and further processing led to the inclusion of 17 papers for the review process. As can be seen in Fig. 7.1B, there is a steep rise in the research focusing on implementing the ML for stem cell research.

The process of selecting appropriate stem cells for a given treatment is one of the key applications of ML in the field of stem cell therapy. This is one of the primary areas where ML is being applied. The traditional techniques of choosing stem cells frequently took a significant amount of time, involved subjectivity, and were inaccurate. Researchers are able to assess a vast dataset including stem cell properties and construct prediction models with the help of ML algorithms. These models can determine which stem cells are the most effective for a certain treatment. Heo et al. (2019) used ML algorithms to predict the efficacy of stem cell–based therapies for stroke. Another application of ML in stem cell therapy is in the field of regenerative medicine. ML algorithms are being utilized to design 3D scaffolds for stem cell growth and regeneration. The algorithms can predict the optimal conditions for stem cell proliferation and differentiation, resulting in improved tissue regeneration. Franks et al. (2020) revealed that ML algorithms can predict stem cell behavior in different microenvironments, leading to improved stem cell–based therapies for various medical conditions.

ML is also being utilized in the development of new stem cell–based drugs. Traditional methods of drug development are often time-consuming, expensive, and lack efficiency. With ML algorithms, researchers can analyze large amounts of data to identify potential drug targets and predict their efficacy (Kusumoto et al., 2022).

Furthermore, ML algorithms are also being utilized to optimize the delivery of stem cells to the affected area. Traditional methods of stem cell delivery were often limited by the lack of efficient delivery methods. With ML algorithms, researchers can analyze different delivery methods and predict the optimal delivery method for a particular treatment (Shende & Devlekar, 2021). Using ML techniques, we can improve stem cell–based therapy for cardiac disorders by maximizing their delivery to the heart (Krittanawong, 2017).

In addition to this, Table 7.1 presents a breakdown of a few leading studies in stem cell research and ML. The table gives a holistic view of the research contribution, target diseases, ML type used, i.e., classic ML algorithms (such as random forest classifier, support vector machine, logistic regression, etc.), fuzzy clustering, deep learning and neural

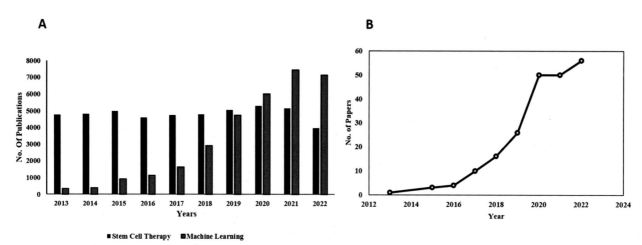

FIGURE 7.1 (A) Year wise distribution of research articles for the search keywords "stem cell therapy" and "machine learning" on PubMed. (B) The number of papers having used/addressing both machine learning and stem cell therapy.

TABLE 7.1 Contribution of different authors in the field of stem cell and stem cell therapy using ML, AI, and/or deep learning.

Target disease(s)	Data used	ML type used	Contribution	Limitation	References
Various	Multimodal	ML	Predicted stem cell donor availability	Clinical intervention, good number of false positives and false negatives, data integration	Li et al. (2020)
Organ regeneration	Clinical trial and animal studies	Neural nets	Critical information prediction such as stem cell dosage for cartilage repair, percentage of defected area, tissue repair, etc.	Clinical intervention required, less accuracy, data integration	Fredik Liu et al. (2021)
Scleroderma	Peripheral blood cells of SCOT patients	ML	Measured the stem cell transplant measurement. Hematopoietic stem cell transplant responses well in comparison with the cyclophosphamide treatment receivers	Smaller sample size	Franks et al. 2020)
	iPSC-RPEs image data	Deep learning	Designed a multislice tensor deep learning model to identify the delineation efficacy of iPSC-RPEs	Work focuses only on single cell type and arguably may not show good accuracy in a multicellular type environment	Lien et al. (2023)
Heart valve disease	RNA-seq assays	ML	Developed ML models to correct the dysregulated gene networks in diseases using iPSCs. Discovered molecules with restorative effects on gene networks	Uses KNN as a classifier, which does not work well as the dimensions of the data increase and the data load is increased	Theodori et al. (2021)
Ovarian cancer	RNA-seq assay	Deep learning	Developed a deep learning—based tool for simulating the development of artificially iPSCs and artificial natural killer cells	Uses simulated data for validation	Esmail and Danter (2021)
Lung cancer	RNA-seq data	Unsupervised fuzzy clustering	Identified the relationship between gene expression and clinical symptoms; identified the biomarkers	Sparsity of data can lead to ML errors	Bhuvaneswal et al. (2021)
Cancer	Image data	Deep learning	Developed a tool for visualizing and analyzing the cancer metastasis in mouse bodies	Deep learning has promise but still at infancy. Inapplicability of study to humans	Pan et al. 2019)
Gastric cancer	RNA-seq data	ML	ML-based intratumoral heterogeneity (cellular subtypes; study of gastric cancer	*	Xiang et al. 2021)

Note: *in the table indicates that no limitation was either mentioned in the study's discussion area, or after SO analysis, no significant limitation was found.

networks, etc. The limitation in the table were recorded either as per the limitations mentioned by the original authors of the study or by examining the strengths and opportunities of the paper. The later we call "SO analysis" of the paper.

The trend of implementing and developing ML-based methods for stem cells related studies has proven fruitful, as is evident through Table 7.1. The limitations indicate the potential areas of research, which can be explored for either improvement or developing the new methodology.

5. Conclusion and future prospects

The field of stem cell therapy is evolving rapidly, with new trends and developments emerging regularly. Further studies are required to fully comprehend the potential of stem cell therapies and to deliver safe and effective treatments to patients. The applications of ML in stem cell therapy have proved to be a game changer. It has led to the development of more efficient, accurate, and cost-effective stem cell—based therapies. Additionally, the inclusion of DL in stem cell research has also opened up new avenues of simulating the process of stem cell for therapeutic usage. The trend of utilizing ML in stem cell therapy is expected to continue and will lead to the development of even more advanced stem cell—based therapies in the future. Integration of big data and AI and ML can play a significant role in advancing stem cell research and improving treatment outcomes.

Author contribution

AA and MA conceived the idea. AA, MG, and AMK performed literature survey. AA and RM performed statistical analysis. All authors cowrote the manuscript and discussed the results.

References

Adil, A., Kumar, V., Jan, A. T., & Asger, M. (2021). Single-cell transcriptomics: current methods and challenges in data acquisition and analysis. *Frontiers in Neuroscience, 15*, 591122. https://doi.org/10.3389/fnins.2021.591122

Al Hamed, R., Bazarbachi, A. H., Malard, F., Harousseau, J. L., & Mohty, M. (2019). Current status of autologous stem cell transplantation for multiple myeloma. *Blood Cancer Journal, 9*(4), 1—10. https://doi.org/10.1038/s41408-019-0205-9

Atala, A. (2011). Tissue engineering of human bladder. *British Medical Bulletin, 97*(1), 81—104. https://doi.org/10.1093/BMB/LDR003

Bang, J. S., Choi, N. Y., Lee, M., Ko, K., Lee, H. J., Park, Y. S., Jeong, D., Chung, H. M., & Ko, K. (2018). Optimization of episomal reprogramming for generation of human induced pluripotent stem cells from fibroblasts. *Animal Cells and Systems, 22*(2), 132. https://doi.org/10.1080/19768354.2018.1451367

Bhuvaneswari, M. S., Priyadharsini, S., Balaganesh, N., Theenathayalan, R., & Hailu, T. A. (2022). Investigating the lung adenocarcinoma stem cell biomarker expressions using machine learning approaches. *BioMed Research International*. https://doi.org/10.1155/2022/3518190

Krittanawong, C., Zhang, H. J., Wang, Z., Aydar, M., & Kitai, T. (2017). Artificial intelligence in precision cardiovascular medicine. *Journal of the American College of Cardiology, 69*(21), 2657—2664.

Chen, B., Zhou, X., Yang, L., Zhou, H., Meng, M., Wu, H., Liu, Z., Zhang, L., & Li, C. (2022). Glioma stem cell signature predicts the prognosis and the response to tumor treating fields treatment. *CNS Neuroscience and Therapeutics, 28*(12), 2148—2162. https://doi.org/10.1111/CNS.13956

Choi, K. A., Choi, Y., & Hong, S. (2018). Stem cell transplantation for Huntington's diseases. *Methods, 133*, 104—112. https://doi.org/10.1016/j.ymeth.2017.08.017

Coronnello, C., & Francipane, M. G. (2022). Moving towards induced pluripotent stem cell-based therapies with artificial intelligence and machine learning. *Stem Cell Reviews and Reports, 18*(2), 559—569. https://doi.org/10.1007/S12015-021-10302-Y

De Luca, M., Aiuti, A., Cossu, G., Parmar, M., Pellegrini, G., & Robey, P. G. (2019). Advances in stem cell research and therapeutic development. *Nature Cell Biology, 21*(7), 801—811. https://doi.org/10.1038/S41556-019-0344-Z

De Masi, C., Spitalieri, P., Murdocca, M., Novelli, G., & Sangiuolo, F. (2020). Application of CRISPR/Cas9 to human-induced pluripotent stem cells: From gene editing to drug discovery. *Human Genomics, 14*(1), 1—12. https://doi.org/10.1186/S40246-020-00276-2/TABLES/2

Doudna, J. A., & Charpentier, E. (2014). The new frontier of genome engineering with CRISPR-Cas9. *Science, 346*(6213). https://doi.org/10.1126/SCIENCE.1258096

Duncan, T., & Valenzuela, M. (2017). Alzheimer's disease, dementia, and stem cell therapy. *Stem Cell Research & Therapy, 8*(1), 1—9. https://doi.org/10.1186/S13287-017-0567-5

Duvenaud, D., Maclaurin, D., Aguilera-Iparraguirre, J., Gómez-Bombarelli, R., Hirzel, T., Aspuru-Guzik, A., & Adams, R. P. (2015). Convolutional networks on graphs for learning molecular fingerprints. *Advances in Neural Information Processing Systems, 28*.

Esmail, S., & Danter, W. R. (2022). Stem-cell based, machine learning approach for optimizing natural killer cell-based personalized immunotherapy for high-grade ovarian cancer. *FEBS Journal, 289*(4), 985—998. https://doi.org/10.1111/FEBS.16214

Figueiredo, F. C., Glanville, J. M., Arber, M., Carr, E., Rydevik, G., Hogg, J., Okonkwo, A., Figueiredo, G., Lako, M., Whiter, F., & Wilson, K. (2021). A systematic review of cellular therapies for the treatment of limbal stem cell deficiency affecting one or both eyes. *Ocular Surface, 20*, 48—61. https://doi.org/10.1016/J.JTOS.2020.12.008

Franks, J. M., Martyanov, V., Wang, Y., Wood, T. A., Pinckney, A., Crofford, L. J., Keyes-Elstein, L., Furst, D. E., Goldmuntz, E., Mayes, M. D., McSweeney, P., Nash, R. A., Sullivan, K. M., & Whitfield, M. L. (2020). Machine learning predicts stem cell transplant response in severe scleroderma. *Annals of the Rheumatic Diseases, 79*(12), 1608–1615. https://doi.org/10.1136/ANNRHEUMDIS-2020-217033

Fredrik Liu, Y. Y., Lu, Y., Oh, S., & Conduit, G. J. (2020). Machine learning to predict mesenchymal stem cell efficacy for cartilage repair. *PLoS Computational Biology, 16*(10). https://doi.org/10.1371/JOURNAL.PCBI.1008275

Gasperini, M., Hill, A. J., McFaline-Figueroa, J. L., Martin, B., Kim, S., Zhang, M. D., Jackson, D., Leith, A., Schreiber, J., Noble, W. S., Trapnell, C., Ahituv, N., & Shendure, J. (2019). A genome-wide framework for mapping gene regulation via cellular genetic screens. *Cell, 176*(1-2), 377–390.e19. https://doi.org/10.1016/j.cell.2018.11.029

Genç, B., Bozan, H. R., Genç, S., & Genc, K. (2019). Stem cell therapy for multiple sclerosis. *Advances in Experimental Medicine and Biology, 1084*, 145–174. https://doi.org/10.1007/5584_2018_247

Goradel, N. H., Hour, F. G., Negahdari, B., Malekshahi, Z. V., Hashemzehi, M., Masoudifar, A., & Mirzaei, H. (2018). Stem cell therapy: A new therapeutic option for cardiovascular diseases. *Journal of Cellular Biochemistry, 119*(1), 95–104. https://doi.org/10.1002/JCB.26169

Grochowski, C., Radzikowska, E., & Maciejewski, R. (2018). Neural stem cell therapy-Brief review. *Clinical Neurology and Neurosurgery, 173*, 8–14. https://doi.org/10.1016/J.CLINEURO.2018.07.013

Han, F., Wang, J., Ding, L., Hu, Y., Li, W., Yuan, Z., Guo, Q., Zhu, C., Yu, L., Wang, H., Zhao, Z., Jia, L., Li, J., Yu, Y., Zhang, W., Chu, G., Chen, S., & Li, B. (2020). Tissue engineering and regenerative medicine: Achievements, future, and sustainability in asia. *Frontiers in Bioengineering and Biotechnology, 8*. https://doi.org/10.3389/FBIOE.2020.00083

Heo, J. N., Yoon, J. G., Park, H., Kim, Y. D., Nam, H. S., & Heo, J. H. (2019). Machine learning-based model for prediction of outcomes in acute stroke. *Stroke, 50*(5), 1263–1265. https://doi.org/10.1161/STROKEAHA.118.024293

Hoang, D. M., Pham, P. T., Bach, T. Q., Ngo, A. T. L., Nguyen, Q. T., Phan, T. T. K., Nguyen, G. H., Le, P. T. T., Hoang, V. T., Forsyth, N. R., Heke, M., & Nguyen, L. T. (2022). Stem cell-based therapy for human diseases. *Signal Transduction and Targeted Therapy, 7*(1). https://doi.org/10.1038/S41392-022-01134-4

Hussein, S., Kandel, P., Bolan, C. W., Wallace, M. B., & Bagci, U. (2019). Lung and pancreatic tumor characterization in the deep learning era: Novel supervised and unsupervised learning approaches. *IEEE Transactions on Medical Imaging, 38*(8), 1777–1787. https://doi.org/10.1109/TMI.2019.2894349

Khandpur, S., Gupta, S., & Gunaabalaji, D. R. (2021). Stem cell therapy in dermatology. *Indian Journal of Dermatology, Venereology and Leprology, 87*(6), 753–767. https://doi.org/10.25259/IJDVL_19_20

Kim, J. B., Ziller, M. J., Wu, G. C., & Huang, S. (2015). Genomic and epigenomic landscapes of adult somatic cells. *Cell, 161*(7), 1677–1689.

Krittanawong, C., Virk, H. U. H., Bangalore, S., Wang, Z., Johnson, K. W., Pinotti, R., Zhang, H. J., Kaplin, S., Narasimhan, B., Kitai, T., Baber, U., Halperin, J. L., & Tang, W. H. W. (2020). Machine learning prediction in cardiovascular diseases: A meta-analysis. *Scientific Reports, 10*(1), 1–11. https://doi.org/10.1038/s41598-020-72685-1

Kumar, R., Sharma, A., Pattnaik, A. K., & Varadwaj, P. K. (2010). Stem cells: An overview with respect to cardiovascular and renal disease. *Journal of Natural Science, Biology and Medicine, 1*(1), 43. https://doi.org/10.4103/0976-9668.71674

Kusumoto, D., Yuasa, S., & Fukuda, K. (2022). Induced pluripotent stem cell-based drug screening by use of artificial intelligence. *Pharmaceuticals, 15*(5), 562. https://doi.org/10.3390/PH15050562

Lien, C.-Y., Chen, T.-T., Tsai, E.-T., Hsiao, Y.-J., Lee, N., Gao, C. E., Yang, Y.-P., Chen, S.-J., Yarmishyn, A. A., Hwang, D.-K., Chou, S.-J., Chu, W.-C., Chiou, S.-H., & Chien, Y. (2023). Recognizing the differentiation degree of human induced pluripotent stem cell-derived retinal pigment epithelium cells using machine learning and deep learning-based approaches. *Cells, 12*(2), 211. https://doi.org/10.3390/CELLS12020211

Li, Y., Masiliune, A., Winstone, D., Gasieniec, L., Wong, P., Lin, H., Pawson, R., Parkes, G., & Hadley, A. (2020). Predicting the availability of hematopoietic stem cell donors using machine learning. *Biology of Blood and Marrow Transplantation: Journal of the American Society for Blood and Marrow Transplantation, 26*(8), 1406–1413. https://doi.org/10.1016/J.BBMT.2020.03.026

Litjens, G., Kooi, T., Bejnordi, B. E., Setio, A. A. A., Ciompi, F., Ghafoorian, M., van der Laak, J. A. W. M., van Ginneken, B., & Sánchez, C. I. (2017). A survey on deep learning in medical image analysis. *Medical Image Analysis, 42*, 60–88. https://doi.org/10.1016/J.MEDIA.2017.07.005

Liu, Z., & Cheung, H. H. (2020). Stem cell-based therapies for Parkinson disease. *International Journal of Molecular Sciences, 21*(21), 1–17. https://doi.org/10.3390/IJMS21218060

Loh, K. M., & Guo, X. (2018). Machine learning in single-cell genomics. *Nature Reviews Genetics, 19*(11), 659–670.

Lu, J., Wang, L., Bennamoun, M., Ward, I., An, S., Sohel, F., Chow, B. J. W., Dwivedi, G., & Sanfilippo, F. M. (2021). Machine learning risk prediction model for acute coronary syndrome and death from use of non-steroidal anti-inflammatory drugs in administrative data. *Scientific Reports, 11*(1). https://doi.org/10.1038/S41598-021-97643-3

Maucksch, C., Vazey, E. M., Gordon, R. J., & Connor, B. (2013). Stem cell-based therapy for Huntington's disease. *Journal of Cellular Biochemistry, 114*(4), 754–763. https://doi.org/10.1002/JCB.24432

Merkin, A., Krishnamurthi, R., & Medvedev, O. N. (2022). Machine learning, artificial intelligence and the prediction of dementia. *Current Opinion in Psychiatry, 35*(2), 123–129. https://doi.org/10.1097/YCO.0000000000000768

Moreau, P., Avet-Loiseau, H., Harousseau, J. L., & Attal, M. (2011). Current trends in autologous stem-cell transplantation for myeloma in the era of novel therapies. *Journal of Clinical Oncology, 29*(14), 1898–1906. https://doi.org/10.1200/jco.2010.32.5878

Nguyen, P. K., Rhee, J. W., & Wu, J. C. (2016). Adult stem cell therapy and heart failure, 2000 to 2016: A systematic review. *JAMA Cardiology, 1*(7), 831–841. https://doi.org/10.1001/JAMACARDIO.2016.2225

Okita, K., Matsumura, Y., Sato, Y., Okada, A., Morizane, A., Okamoto, S., Hong, H., Nakagawa, M., Tanabe, K., Tezuka, K. I., Shibata, T., Kunisada, T., Takahashi, M., Takahashi, J., Saji, H., & Yamanaka, S. (2011). A more efficient method to generate integration-free human iPS cells. *Nature Methods, 8*(5), 409−412. https://doi.org/10.1038/NMETH.1591

Pan, C., Schoppe, O., Parra-Damas, A., Cai, R., Todorov, M. I., Gondi, G., von Neubeck, B., Böğürcü-Seidel, N., Seidel, S., Sleiman, K., Veltkamp, C., Förstera, B., Mai, H., Rong, Z., Trompak, O., Ghasemigharagoz, A., Reimer, M. A., Cuesta, A. M., Coronel, J., … Ertürk, A. (2019). Deep learning reveals cancer metastasis and therapeutic antibody targeting in the entire body. *Cell, 179*(7), 1661−1676.e19. https://doi.org/10.1016/J.CELL.2019.11.013

Pérez López, S., & Otero Hernández, J. (2012). Advances in stem cell therapy. *Advances in Experimental Medicine and Biology, 741*, 290−313. https://doi.org/10.1007/978-1-4614-2098-9_19

Polo, J. M., Liu, S., & Wernig, M. (2012). Cell reprogramming: iPSCs in the new millennium. *Nature Reviews Genetics, 13*(11), 671−682.

Prentice, D. A. (2019). Adult stem cells: Successful standard for regenerative medicine. *Circulation Research, 124*(6), 837−839. https://doi.org/10.1161/CIRCRESAHA.118.313664

Qu-Petersen, Z., Deasy, B., Jankowski, R., Ikezawa, M., Cummins, J., Pruchnic, R., Mytinger, J., Cao, B., Gates, C., Wernig, A., & Huard, J. (2002). Identification of a novel population of muscle stem cells in mice: Potential for muscle regeneration. *The Journal of Cell Biology, 157*(5), 851. https://doi.org/10.1083/JCB.200108150

Rafi, M. A. (2011). Gene and stem cell therapy: Alone or in combination? *BioImpacts: BI, 1*(4), 213. https://doi.org/10.5681/BI.2011.030

Rafique, R., Islam, S. M. R., & Kazi, J. U. (2021). Machine learning in the prediction of cancer therapy. *Computational and Structural Biotechnology Journal, 19*, 4003−4017. https://doi.org/10.1016/J.CSBJ.2021.07.003

Rouanet, S., Warrick, E., Gache, Y., Scarzello, S., Avril, M. F., Bernerd, F., & Magnaldo, T. (2013). Genetic correction of stem cells in the treatment of inherited diseases and focus on xeroderma pigmentosum. *International Journal of Molecular Sciences, 14*(10), 20019. https://doi.org/10.3390/IJMS141020019

Salybekov, A. A., Wolfien, M., Kobayashi, S., Steinhoff, G., & Asahara, T. (2021). Personalized cell therapy for patients with peripheral arterial diseases in the context of genetic alterations: Artificial intelligence-based responder and non-responder prediction. *Cells, 10*(12). https://doi.org/10.3390/CELLS10123266

Sebastiano, V., Maeder, M. L., Angstman, J. F., Haddad, B., Khayter, C., Yeo, D. T., Goodwin, M. J., Hawkins, J. S., Ramirez, C. L., Batista, L. F. Z., Artandi, S. E., Wernig, M., & Joung, J. K. (2011). In situ genetic correction of the sickle cell anemia mutation in human induced pluripotent stem cells using engineered zinc finger nucleases. *Stem Cells (Dayton, Ohio), 29*(11), 1717−1726. https://doi.org/10.1002/STEM.718

Semina, S. E., Alejo, L. H., Chopra, S., Kansara, N. S., Kastrati, I., Sartorius, C. A., & Frasor, J. (2022). Identification of a novel ER-NFκB-driven stem-like cell population associated with relapse of ER+ breast tumors. *Breast Cancer Research, 24*(1), 1−19. https://doi.org/10.1186/S13058-022-01585-1/FIGURES/7

Shende, P., & Devlekar, N. P. (2021). A review on the role of artificial intelligence in stem cell therapy: An initiative for modern medicines. *Current Pharmaceutical Biotechnology, 22*(9), 1156−1163. https://doi.org/10.2174/1389201021666201007122524

Su, R. J., Baylink, D. J., Neises, A., Kiroyan, J. B., Meng, X., Payne, K. J., Tschudy-Seney, B., Duan, Y., Appleby, N., Kearns-Jonker, M., Gridley, D. S., Wang, J., Lau, K. H. W., & Zhang, X. B. (2013). Efficient generation of integration-free iPS cells from human adult peripheral blood using BCL-XL together with Yamanaka factors. *PLoS One, 8*(5). https://doi.org/10.1371/journal.pone.0064496

Sugaya, K., & Vaidya, M. (2018). Stem cell therapies for neurodegenerative diseases. *Advances in Experimental Medicine and Biology, 1056*, 61−84. https://doi.org/10.1007/978-3-319-74470-4_5

Takahashi, K., & Yamanaka, S. (2006). Induction of pluripotent stem cells from mouse embryonic and adult fibroblast cultures by defined factors. *Cell, 126*(4), 663−676. https://doi.org/10.1016/J.CELL.2006.07.024

Tatulian, S. A. (2022). Challenges and hopes for Alzheimer's disease. *Drug Discovery Today, 27*(4), 1027−1043. https://doi.org/10.1016/J.DRUDIS.2022.01.016

Theodoris, C. V., Zhou, P., Liu, L., Zhang, Y., Nishino, T., Huang, Y., Kostina, A., Ranade, S. S., Gifford, C. A., Uspenskiy, V., Malashicheva, A., Ding, S., & Srivastava, D. (2021). Network-based screen in iPSC-derived cells reveals therapeutic candidate for heart valve disease. *Science (New York, N.Y.), 371*(6530). https://doi.org/10.1126/SCIENCE.ABD0724

Trounson, A., & McDonald, C. (2015). Stem cell therapies in clinical trials: Progress and challenges. *Cell Stem Cell, 17*(1), 11−22. https://doi.org/10.1016/J.STEM.2015.06.007

Vuidel, A., Cousin, L., Weykopf, B., Haupt, S., Hanifehlou, Z., Wiest-Daesslé, N., Segschneider, M., Lee, J., Kwon, Y.-J., Peitz, M., Ogier, A., Brino, L., Brüstle, O., Sommer, P., & Wilbertz, J. H. (2022). High-content phenotyping of Parkinson's disease patient stem cell-derived midbrain dopaminergic neurons using machine learning classification. *Stem Cell Reports, 17*(10), 2349−2364. https://doi.org/10.1016/J.STEMCR.2022.09.001

Xiang, R., Song, W., Ren, J., Wu, J., Fu, J., & Fu, T. (2021). Identification of stem cell-related subtypes and risk scoring for gastric cancer based on stem genomic profiling. *Stem Cell Research & Therapy, 12*(1). https://doi.org/10.1186/S13287-021-02633-X

Yamanaka, S. (2020). Pluripotent stem cell-based cell therapy-promise and challenges. *Cell Stem Cell, 27*(4), 523−531. https://doi.org/10.1016/J.STEM.2020.09.014

Zakrzewski, W., Dobrzyński, M., Szymonowicz, M., & Rybak, Z. (2019). Stem cells: Past, present, and future. *Stem Cell Research & Therapy, 10*(1), 1−22. https://doi.org/10.1186/S13287-019-1165-5/FIGURES/8

Zhang, F. Q., Jiang, J. L., Zhang, J. T., Niu, H., Fu, X. Q., & Zeng, L. L. (2020). Current status and future prospects of stem cell therapy in Alzheimer's disease. *Neural Regeneration Research, 15*(2), 242−250. https://doi.org/10.4103/1673-5374.265544

Chapter 8

Computational and stem cell biology: Challenges and future perspectives

Rajiv Kumar[1], Agnieszka Maria Jastrzębska[2], Magali Cucchiarin[3], Neelam Chhillar[4] and Mitrabasu Chhillar[5]

[1]University of Delhi, New Delhi, Delhi, India; [2]Warsaw University of Technology, Faculty of Materials Science and Engineering, Warsaw, Poland; [3]Center of Experimental Orthopaedics, Saarland University Medical Center and Saarland University, Homburg, Germany; [4]Department of Biochemistry, School of Medicine, DY Patil University, Navi Mumbai, Maharashtra, India; [5]Naval Dockyard, Mumbai, Maharashtra, India

1. Introduction

Current knowledge of stem cells quickly intensifies computational biology and the present capability of researchers to engage them for disease modeling, drug development, and regeneration. Many examples illustrate fundamental ideas that have generally governed computational biology and are still relevant today, although nowadays computation is playing a more prominent role in stem cell biology (Bian & Cahan, 2016). The value of computing models for developing hypotheses and the necessity of iterating between modeling and laboratory testing give a better understanding developed by computational biologists who work collaboratively across disciplines, even with laboratory scientists. Significant difficulties in this field were highlighted in the hope that they would encourage more people to learn about this fascinating field. In this context, computational biology enters the picture (Wang, Neu, et al., 2019). The following concepts were underlined, including disease modeling, stem cell bioengineering, regulatory networks, existing challenges in computational and stem cell biology, upcoming challenges of single-cell data analysis, existing challenges in computational and stem cell biology, upcoming challenges of single-cell data analysis, lineage tracking, genomic and proteomic cell type, and trajectory inference to illustrate how computation might be used in stem cell biology. The most important issues, which lie ahead while considering substantial computational biology advancements over the past two decades, have been focused on the underlined topics. Furthermore, progress in computational biology is not purely computational; it is linked to experiments because of the intimate linkage between computational biology and experimental research (Ul-Haq & Madura, 2015). Recent data analysis reveals and confirms significant improvements in biological computing, but there is still a massive amount of effort required.

Multiscale biological modeling, analysis of vast gene expression and proteomics information, and other techniques allowed for the construction of detailed and prospective models for cellular pathways. Furthermore, new strategies were proposed for insinuating connections and can be marked as the first steps to cell modeling (Edgar et al., 2020). Models of neurons' electrical activity have been developed on another level, aiding in identifying genes associated with various diseases. It has been possible to view molecules in motion and disentangle the networks that control how cells function. For example, finding the right targets for targeted cancer therapy is challenging. Additionally, creating computer systems for surveillance systems, mapping the biological networks and routes responsible for cancer initiation, growth, and dissemination, predicting performance and epigenetic malfunction in diseases from the structure of molecules, determining the underlying mechanisms of oncogenic mutations and the cellular network that are rewired in cancer, and performing accurately, effectively, and entirely in dynamic models have been underlined for further investigations. Moreover, challenging obstacles were also identified. This ability is lost as cells differentiate. For instance, a liver cell can only divide to generate new liver cells, and it cannot transform into a lung or brain cell (Wang, Sun, et al., 2019). Stem cells are precious in medicine because of their pluripotency. They could replace cells lost to spinal cord injuries, repopulate Parkinson's disease-related neurons, and regenerate pancreatic insulin-producing cells used to treat type I diabetes (Zhou & Melton, 2018). Despite making great strides over the past decade, multiscale biological modeling still needs much work. Fusion approaches spanning sizes, sources, and disciplines are crucial features required for achieving this goal. By incorporating

Copyright © 2024 Elsevier Inc. All rights reserved, including those for text and data mining, AI training, and similar technologies.

data from various sources, including cross-linking, fluorescence resonance energy transfer (FRET), crystallography, electron microscopy, nuclear magnetic resonance (NMR), and small-angle X-ray scattering, hybrid methods combine information from various sources (Seffernick & Lindert, 2020). The creation of validation processes for models is also crucial. Advance models built on the integration of experimental data might be anticipated. These must follow a precise protocol to maintain quality control if these models are to be stored in a public archive, which is currently a common goal. Finally, it is becoming increasingly important to study the dynamics of substantial integrated models to comprehend the regulation and function of large complexes in cells (Tieu et al., 2012). Until now, it has been challenging to understand how specific molecules behave and how they associate. Further, a tremendously complicated layer is added to the system's complexity by the mechanics of such extensive relationships. This problem is aggravated by the potential for major sections of the molecules to be distorted as well as the numerous temporal posttranslational alterations that take place in a variety of combinations to indicate different functions.

Dynamics conformational ensembles perform as a mediating factor in genetics. The secrets of life can be solved with the aid of solid concepts such as the free energy landscape, which have been transferred from physics and chemistry. On a different level, improved methods for analyzing tumor mutations and knowledge of the duplicated pathways that can take over within cancer may contribute to current and actionable findings. These approaches forecast the development and resistance of cancer (Wechman et al., 2020). Such anticipatory thinking has the potential to be very effective, providing the oncologist with a cancer prognosis and mechanical knowledge, which in turn allows for more judicious treatment choices (Rajiv, 2021). Last but not least, the community must address the issue of how to link global phenotypes such as cancer's genetic makeup to genetics.

To overcome persistent difficulties in computational and stem cell biology, we outline a computational strategy that integrates gene expression analysis, comprehension skills from proteomic pathway informatics, and cell signaling models to designate critical transitional phases of developing cells at high resolution (Kumar et al., 2021b). For instance, our network models link sparse gene signatures with related but dissimilar biological processes to elucidate the molecular mechanisms that regulate cell fate transitions. These findings use an integrative system approach to pinpoint new regulatory nodes and processes that are advantageous for cell engineering, which can be used to address the problems by resolving existing challenges in computational and stem cell biology, single-cell data analysis, disease modeling, computational biology, stem cell bioengineering, genomics, and proteomics (Ottoboni et al., 2020). We believe that the stem cell community should emphasize the significance of sponsoring and assisting in the creation of computational tools.

In addition, the society ought to highlight the duty to improve and use tools in a specific manner that can promote impartial and fair interpretation (Kumar, 2021a). Researchers should work to spot and stop the improper use of computational and mathematical tools, much as it is our responsibility as a stem cell community to recognize the inappropriate use of stem cells (Fig. 8.1). When creating techniques, much attention was paid to interacting with multimodal information and avoiding a static image of cell type (Bosch, 2009).

In conclusion, the following points can be underlined from the aforementioned discussion: (1) the stem cell society as a whole is accountable for upholding more robust standards for data collection and deposition, which are urgently needed; (2) the ongoing cell atlas initiatives should place stem cells more prominently; and where it is practical, these studies should be connected with existing stem cell resources where comprehensive characterization of the lines has already been done; and (3) a precise and systematic data gathering process will become even more critical than the amount of data gathered and developed since the potential for development of generative tools will expand our understanding of stem cell biology.

2. Current challenges and contests: Crossing hurdles in stem cell and computational biology

The study of biological extent data is conducted through the interdisciplinary field of bioinformatics, which combines computer science techniques with mathematical and statistical modeling. Currently, new techniques explore the biological features to search for instant applications in biological fields such as investigation and innovation used in exploring disease profiles. These methodologies comprehensively analyze a vast amount of biological and clinical data. Network inference, molecular oncology, genomics, and medical genomics are just a few biological issues researchers are investigating (Hamey et al., 2017). Understanding the mechanistic underpinnings of human diseases will be accomplished by combining experimental and computational data. In the fields of molecular biology and customized medicine, the key aspects are (1) bioinformatics with high throughput, utilizing computer learning and probabilistic techniques, (2) high-throughput sequencing data processing for both bulk and single-cell data, (3) high-throughput bioinformatics initiatives made use

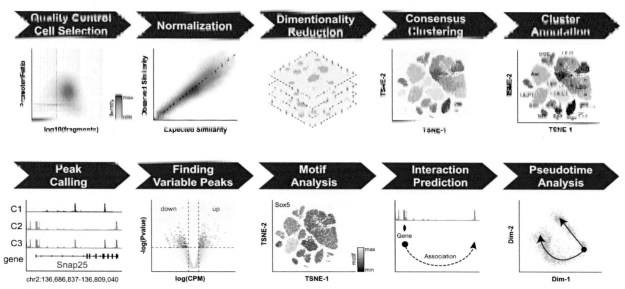

FIGURE 8.1   Overview of the SnapATAC analysis workflow in a schematic. *Reprinted (adapted) with permission from ref. Fang et al. (2021) Copyright (2021) Bing Ren et al. Nature communications, Nature.*

of the ability to track gene activity to paint a clearer picture of the underlying biology, (4) machine learning statistics such as nonparametric procedures, Gaussian processes, and generative models (Frisch et al., 2016), (5) statistical machine learning programs aiming to apply cutting-edge deep learning techniques to get beyond the challenges of interpreting large amounts of data, (6) computational biology applied in molecular biology, individualized healthcare, cancer research, and cancer immunology, (7) biobank and health data analysis, (8) clinical data (a special kind of data received via high-throughput molecular profiling and computational techniques by exploring clinical data), and (9) applications for the control of various biological events also used in the analysis of transcription of genes (Rajiv, 2021e). Thus, in initiatives involving the control of gene transcription, we wish to investigate the as-yet-undiscovered code contained in the DNA sequence that is in charge of cell-specific gene activation. Overall, modern biology in general and stem cell biology in particular now depend on the computational or systemic biological processing of experimental data. This is obvious when looking at the statistical and bioinformatic analysis of molecular high-throughput data (Rajiv, 2021h). Without computational techniques, the complexity and enormous amount of data are just unmanageable. But it is also becoming clearer that theoretical methodologies, such as mathematical modeling and computer simulation, can significantly contribute to the creation of new insights in other areas of stem cell biology and the profile of disease models by dissolving continued contests.

Bioengineering chiefly enjoys the challenges that lie in the areas of stem cells and computational biology. When viewed under a microscope, they appear to be works of art: a red dapple with green crescents, angular green dabs, and deep blue and purple patches. These, however, are cells that researchers have grown and studied and that have been highlighted with fluorescent dyes and antibodies. In the labs of computer science and biological engineering, researchers conduct research on synthetic biology. Now, scientists "train" stem cells to advance more multifaceted assemblies, such as pancreatic, and liver tissues, by manipulating the genetic code. Photos of the tissues they generated in the lab at a microscopic scale have evidenced. For example, each type of cell that forms a tissue contributes uniquely to the tissue's functionality (Lowe et al., 2007). When necessary, beta cells in pancreatic tissue secrete insulin to control the body's blood sugar levels. Scientists are working to develop human-induced pluripotent stem cells (hiPSCs) into functional beta cells. For instance, the cells can be capable of producing insulin but unable to secrete it, or they might be capable of doing so but not in the proper quantities. In the lab, functioning beta cells might be created, especially when utilized alongside other types of support cells. They might eventually be able to help diabetics whose own beta cells are not functioning properly, which is a difficult issue (Rather et al., 2021). The testified practice is a plate of stem cells with added chemicals that drive the cells to undergo alterations and differentiate to produce specific types of cells. But in most cases, this leads to an undifferentiated cell population that is ill-equipped to perform its intended job. Current research employs a novel approach, such as genetic circuits to "train" stem cells to develop into elusive beta cells. The idea of a genetic circuit may seem strange, and various questions come into existence that reflect the current challenges, including, the question "can you program a cell like a computer?" (Wang et al., 2018). However, scientists have found that despite their differences, cells and computers are incredibly similar in that they both require an input, some form of pattern recognition, and an output. When anyone types a specific command into a computer, the device will carry it out.

Creating and inserting DNA that directs a stem cell to change its color once it develops into a particular cell type can be done (McBride et al., 2003). Thus, rather than typing words into a cell, genetic information fragments can be used to program it. With the help of earlier research, biologists now have access to various capabilities, including enzymes that can cut DNA segments at specific locations and sew them back together (Rajiv, 2021d). The researchers must create a new program and tools based on computational biology for the cells and review previous work to make an excellent educated guess regarding the combination and temporal sequence of genetic components, specifically genes and promoters, which will cause the cell to do what the researchers want. Using a virus, they deliver the DNA to the cells and incorporate it into the genome after creating the genetic program (Rajiv et al., 2020). This approach has a few significant benefits. In other words, they can train several stem cell clusters in various ways, then combine them, as researchers put it: "With genetic circuits, they can acquire control at the single-cell level." According to their programming, each cell would evolve into one of the various cell types required for an entire tissue. When building DNA, taking care of the cells, or doing numerous tests to examine cellular activity and tissue structure, researchers always balance the project to develop pancreatic beta cells with independent research on cell signaling in complex liver tissue construction. These issues are incredibly fascinating, and biology may quickly become incredibly difficult. Engineering complexity is a topic that interests researchers a lot (Kaushik et al., 2016). Let's analyze biological diversity, understand it, and see what we might develop with it to finally solve problems relating to medicine, rather than trying to reduce the problem. This new approach to computational stem cell biology effectively discovers and addresses these or similar challenges in a close collaboration involving experimentalists and theorists.

Current challenges can be resolved by combining and enhancing the use of theoretical concepts and by increasing the visibility of these approaches both inside and outside any research center. Encouraging a methodological debate in the disciplines of mathematical modeling, bioinformatics, and biometry extended to stem cell–related topics can be possible, but enforcement communication is missing among researchers conducting experimental research (Rea et al., 2019). Human pluripotent stem cells (hPSCs) can completely alter how genes are discovered in developing mammals using high-throughput screening technologies. For example, hPSCs can be used to create progenitor communities for cell-based therapies to explore the causes of human blood cell disorders. With a better consideration of the signaling pathways in hematopoiesis and their impact on gene expression, we can further improve our ability to design novel, clinically relevant methods and address current knowledge gaps in adult hematopoietic stem cells (HSCs) (Rajiv and Gulia, 2021) (Fig. 8.2). Computational and stem cell biology investigate molecular processes that control hematopoietic formation in healthy and pathological settings (Rajiv, 2021). This chapter explores the current understanding of the various cellular and molecular mechanisms investigated by computational and stem cell biology.

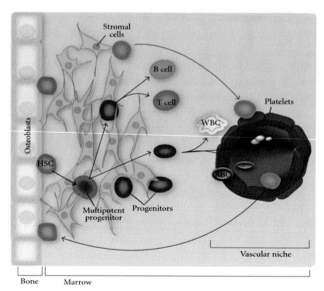

FIGURE 8.2 The hematopoietic stem cell (HSC) niche's quantitative characteristics. A structural representation of the niche is shown in the left panel, and a mathematical model for controlling the fate of hematopoietic stem cells is shown in the right panel in schematic form. The stem cell population dynamics may be influenced by signaling pathways and population counts, which are included in the model. *Reprinted (adapted) with permission from ref. Tieu et al. (2012) Copyright (2012) Mary E. Sehl et al. Stem Cells International, Hindwai.*

Investigating results through stem cell biology in the generation of blood from hPSCs and recent developments in computational techniques and stem cell biology may enhance the current understanding of human hematopoietic advancement. Recently, genetic disorders and the ineffectiveness of gene editing, computational, and stem cell biology methods have explored appropriate models for researching human genetic diseases. However, newly innovated techniques, including computational and stem cell biology tools, make hPSCs ideal for investigating the effect of harmful mutations and obtaining cells that have been gene corrected for use in autologous cell therapy (Rajiv, 2021c). Current stem cell applications for hiPSCs in genetic illnesses were discussed, including cardiovascular diseases and neurodegenerative disorders (Gingold et al., 2016). The following causes are mentioned later and include numerous significant issues and current challenges that can be resolved using computational and stem cell biology tools and bioinformatic techniques, including how can the enormous amount of diverse experimental data such as measurements of the proteome, metabolome, and transcriptome as well as imaging data be examined systematically? And how can the retrieved information be appropriately incorporated into the analysis? What connections exist between various regulatory elements (transcription factors, cell−cell interactions, and signaling molecules), and what causal relationships do they share? How many events at the cellular and tissue levels, along with stem cells' differentiation capacity, can be accounted for by intracellular and molecular mechanisms? Can one (potentially) categorize and forecast the functional features of stem cells based on their molecular or phenomenological traits? Are the existing ideas, conceptions, and assumptions about different stem cell systems' features quantitatively in line with the results of experiments? These challenges can be resolved by applying computational stem cell biology tools and bioinformatic techniques and will contribute to upcoming discoveries (Zhang, 2011).

Congenital disorders and in vitro differentiation bias are both affected by the recognition of source code and noncoding genetic mutations that alter the tendency of variation into distinct families. Finding ways to use single-cell omics, proteomics, phosphoproteomics, and epigenetics to better understand stem cells and their differentiation pathways is another difficulty. Few reported technologies are still working toward single-cell resolution.

Intercellular signaling networks, a biological network distinct from gene regulatory networks (GRNs), anticipate the network of tissue-specific cell−cell interactions necessary for tissue homeostasis and development using statistical frameworks and knowledge of ligand−receptor complexes from the past. Identifying the fundamental clinical relevance necessary for sustaining tissue homeostasis and fostering tissue regeneration will be greatly aided in the future by statistical models of cell signaling pathways. Moreover, by looking at the epithelial cell−cell images of the organs and the cell−cell interactome of diseased or wounded tissues, dysregulated interactions can be revealed that can direct the generation of methods for regaining tissue homeostasis (Petersen et al., 2017). With the use of imaging-based techniques for spatial transcriptome restructuring and these simulations of tissue-specific cell signaling pathways, it will be possible to characterize the entire interactome in a substantially resolved manner and, as a result, establish a more accurate prediction. There was another topic we glossed over. It is a description of cell-based modeling, a method being used by in computational methods to investigate how cells cooperate and evolve (Duffy et al., 2019). In the lengthy history of cell-based modeling of stem cells and advances, predictions of embryonic toxicity and various interpretations of signaling pathways that spatially shape the limb have both been explored. Thus, by combining the information gathered from single-cell sequencing, we expect to create more precise and potent cell-based modeling systems that will be applied to develop multicellular behavior and function dynamically.

3. Disease modeling and computational biology: Current challenges

Disease models based on human stem cells have considerable potential to increase the current knowledge of how human illness is initiated and prolonged. These models can represent diseases in vitro, using people with genetic disorders as a source and modifying them via genome editing and various differentiation techniques. However, numerous issues have prevented stem cell−based methodologies for in vitro disease modeling from reaching their full potential. Ex vivo cell- and tissue-based models play a crucial role in the workflow of pathophysiology investigations, drug discovery, the creation of individualized therapeutic approaches, assay development, and disease modeling (Rajiv, 2021i). Because ex vivo stem cell models have superior predictive potential, they have been used in scientific and pharmacological research for these goals. In simple words, it is now possible to develop and perform research on cellular function disease models even more swiftly and successfully. The development of separation and in vitro approaches for cultivating autologous human embryonic stem cells (ESCs) and the standardization of procedures for developing patient-derived iPSCs are among such approaches. Traditional shorter-term monolayer cultures frequently fail to accurately represent these factors since many genetically predisposed diseases take time to emerge in certain tissue microenvironments (Acheson et al., 2010). These difficulties must be overcome, mainly when using animal models that only partially mimic the complex diseases prevalent in humans. Herein, we highlight how cutting-edge genome editing technologies and computational biology in human stem cell and organoid cultures, particularly intestinal, retina, and kidney organoids, offer genetically defined models that mimic disease symptoms. In addition, the shortcomings of two-dimensional cultivation systems were eliminated, and tissue shape and function imitation were enhanced. The ability of stem cells and scientific computing to produce more complex systems, such as scaffold cell models, tissue and organs, or organs-on-a-microfluidics, has been innovated. Last but not least, the development of genome editing and gene therapy innovations significantly impacted the creation of more accurate stem cell disease models and the development of successful therapeutic approaches (Pulido-Reyes et al., 2017). This section discusses significant developments in stem cell–based models, emphasizing current trends, assistance, and limitations (Rajiv, 2021g). It was stressed that experimental biology must be integrated with computational methods that can describe complex biological entities and produce results or predictions in the setting of personalized medicine, such as neurons and cardiomyocytes (Fig. 8.3).

A new era of disease modeling has been made possible by recent advancements in cellular reprogramming (Takahashi, 2014). This topical collection will present the most recent findings in drug development, disease modeling, and stem cell–based therapy. In addition to improving our knowledge on stem cell differentiation, the causes of disease, the efficacy and toxicity of medications, and the potential for tissue regeneration using stem cells, this section will be a crucial source while developing innovative technologies for reprogramming cells (de Magalhães & Ocampo, 2022). We discuss high-caliber articles on the following subjects: disease modeling, stem cell biology, drug development, and stem cell–based therapy, as well as read reviews, research papers, editorials, case studies, letters to the editor, and technical reports, which cover (1) stem cell–based therapy and disease modeling, (2) stem cell differentiation routes and mechanisms, (3) maturation of stem cell–derived somatic cells, (4) the developmental processes' mechanisms, (5) mechanisms of stem cell–mediated tissue regeneration, (6) three-dimensional (3D) organoids models and 3D-printed tissues, (7) stem cell–based drug screening for toxicity and efficacy, and (8) precision medicine using stem cells. The study of molecular genetics has made use of computing for more than 50 years. More recently, the conversion of high-throughput sequencing data into an understanding of both embryonic and stem molecular genetics has relied heavily on computational methodologies. In recent years, a brand-new area of computational biology and medicine has emerged that combines molecular data with architecture models of stem cell properties. In addition, single-cell transcriptomics is anticipated to give us a better grasp on the development of and our capacity to engineer cell fate, which is covered in this section's description of a new field paying particular attention (Ali et al., 2022). For example, type 2 diabetes' consequence is diabetic cardiomyopathy, which is believed to be influenced by genetics and lifestyle factors (Rajput et al., 2021).

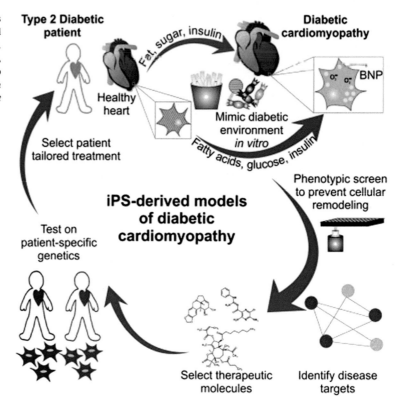

FIGURE 8.3 A human pluripotent stem cell line was employed to generate an organoid kidney. Differentiated nephrons are observable using immunofluorescence. NPHS1, yellow glomeruli, LTL, pink proximal tubules, and DTLs, pink distal tubules/collecting ducts make up these nephrons (CDH1, green). *Reprinted (adapted) with permission from ref. Drawnel et al. (2014) Copyright (2014) Roberto Iacone et al. Cell Reports, Elsevier Inc.*

Human iPSC-derived cardiomyocytes can be utilized to develop in vitro models of the illness that take environmental and genetic factors into account. First, morphological and functional disorders were seen in a diabetic cardiomyopathy surrogate caused by the analytical laboratories of diabetes. The influence of genetics has been evaluated by obtaining cardiomyocytes from two diabetic individuals with varying disease development. The patient specific cells recreate the cardiomyopathic phenotype primarily, with a degree of severity based on their initial clinical status. Thus, to uncover drugs that shield the cardiomyocyte morphology in vitro from the stress of diabetes, these models are subjected to several screening stages (Drawnel et al., 2014). In this study, a patient iPSC model of a complicated metabolic disorder has been demonstrated to signify the potency of this technology for developing and evaluating treatment plans for a condition of growing clinical importance. Consequently, computer models are employed to analyze statistical data by offering a mechanistic understanding of biological mechanisms and by producing fresh predictions that might direct experimental investigation (Xu et al., 2019). Hence, the use of computational models enables the resolution of several biological issues and problems in the fields of stem cell studies and regenerative medicine.

These models have been demonstrated to help direct experimental research, especially in developing cellular conversion techniques. Nevertheless, given the advancements in single-cell technologies and their recent debut, which can now provide a significant amount of data of various kinds, the time is right for creating more complex models. In this sense, we can now create more accurate models of biological processes at multiple scales of biological complexity. When predicting cell destiny determinants in vivo, for instance, a model that considers stemming cell−niche interactions is preferable (Das & Baker, 2016). To advance regenerative medicine tactics, including in vivo reprogramming, stem cell transplantation, and rejuvenation, these models can be beneficial in overcoming the limits of present experimental protocols (Kohn et al., 2021). At several regulatory levels, valid omics data may now be produced in recognition of the widespread adoption of high-throughput sequencing technology. Integrative computer models help unravel the intricate interactions between these unified parameter layers by methodically evaluating these vast amounts of biomedical data.

Network modeling of intricate gene−gene interfaces has been effectively applied to the study of disease-related dysregulation and the prediction of potential therapeutic targets to reverse the diseased phenotype in the context of human disorders (Kumar & Gulia, 2021). These computer network models have also become a promising tool for analyzing the mechanisms of developmental processes such as cellular differentiation, transdifferentiation, and reprogramming (Gulia, 2021). Here, current developments and acknowledged limitations in computer modeling of cellular systems have been featured. A particular focus is on highlighting the contribution of computer modeling to our comprehension of stem cell biology and the intricate multifactorial characteristics of human disorders and their management. Additionally, scientific modeling of human ailments makes it possible to identify illnesses' cellular and molecular causes and create new treatments for them. Animal models for various diseases have their restrictions, but specific PSCs can self-renew and specialize into practically any cell type. The ability to model human diseases utilizing cultured PSCs has fundamentally changed how we study complex, early-onset, and late-onset diseases and monogenic, epigenetic, and complex disorders (Gingold et al., 2016). Many methods are employed to develop these disease models, using either patient-specific iPSCs or other stem cells, opening up new opportunities for developing models and their application in drug discovery.

Modern modeling attempts can be supported by developments in biosystems theory, making them more therapeutically helpful. Researchers face surprising findings and problems while studying complicated diseases that they had not anticipated when focusing on fundamental biological processes. With this aim, we think that it makes more sense and is more manageable in the immediate term to research groups of individuals with similar disease patterns than to try to fully comprehend individualized dynamics. An illness model can be employed to understand the underlying mechanisms of the disease's cellular, biochemical, and metabolic phenotypes and generate new treatments to lessen the sickness. In general, the most advantageous or complex traits that influence the choice of illnesses are identified and highlighted for future modeling. PSC-derived cells are currently employed in evaluating numerous conditions and in forthcoming pharmacological treatments (Choi et al., 2022). Furthermore, disease models based on human stem cells have considerable potential to increase our knowledge of human illness. These models can represent diseases in vitro by using people with genetic disorders as a source of information and manipulating them using genome editing and various differentiation techniques. However, several issues have prevented stem cell−based in vitro disease modeling from reaching its full potential. Many genetically programmed diseases take time to develop and occur in certain tissue microenvironments. These factors are frequently insufficiently modeled using more familiar, shorter-term monolayer cultures. These difficulties must be overcome, mainly when using animal models that only partially mimic the complex diseases prevalent in humans.

Herein, a few queries have been replied to, including how cutting-edge genome editing technologies in human stem cell and organoid cultures, particularly the example of intestinal organoids, offer genetically defined models that mimic disease symptoms (Blutt & Estes, 2022). Because stem cell research is one of biology's fields with some of the quickest growth and the most promising future applications, to consolidate and explain some of the most recent developments, this section summarizes the most current results in adult stem cell, iPSC, ESC, and malignant stem cell research. Studies on CRISPR/Cas9 genome engineering, the dynamic characterization of iPSC-derived 3D brain organoids, the development of iPSC-derived cardiac tissue and astrocytes, and nonviral minicircle vector-based gene change of stem cells have all been discussed (Acheson et al., 2010). Effective differentiation methods for chondrocytes, adipocytes, and neurons were also described. As well as being described, many stem cell uses are used in disease modeling. Additionally, this volume emphasizes stem cells' most recent advances and uses in disease therapy and fundamental science. The identification of disease causes and medication responses has become more and more reliant on omics-based methods. Because diseases and medication reactions are often coexpressed and planned in important omics data exchanges, it is not always possible to draw meaningful conclusions from the usual methods for collecting omics data from isolated single layers. Furthermore, it costs money to develop new treatments for diseases, and drugs can have side effects that harm people.

The absence of easily accessible techniques for tissue differentiating continued to be one of the main barriers inhibiting the broader implementation of this approach in cancer research, despite PSC technology's versatility as a model for many cancer etiologies. As more epithelial types become empirically available, new developments in directed distinguishing methodologies and expanding 3D organoids biological methods open significant potential in disease research, facilitating disease modeling in neuroscience, cardiology, and epithelial regeneration, where PSC models are now frequently used (Rajiv, 2021k). Novel PSC-derived organoids techniques are expected to fill in the gaps to manufacture several cancer cell types that are now unavailable, such as breast, prostate, and ovary. To sum up, PSCs will eventually be used in fascinating ways in the realm of cancer biology. We anticipate the widespread adoption of PSC approaches into the toolkits of basic and clinical researchers who want to effectively recreate disease, define changes resulting from cancer-associated mutations, or enhance their current investigations with a human model. Dynamic modeling seeks to create mathematical models that can forecast the course of the disease and the effects of drugs. Model parameters were calculated using data from the clinic. Dynamic models may include new features that were derived from the possible participants identified by static modeling (Yin et al., 2019). It is possible to effectively intervene in the progression of the disease by using optimal medicine dosages created by control algorithms, which also suggests that the predictive model used in static modeling will once again be accurate. It is a potent tool for analyzing diseases and creating treatments since it combines static and dynamic modeling and performs better than other learning techniques at making static modeling predictive, but the model's generality is reduced because the underlying assumptions cannot apply in practice. Using learning algorithms with precise predictions of future testing data is always preferable (Shimoni et al., 2016). The intricate design objectives (such as control with restrictions) cannot be met by simple control algorithms regarding drug administration, and complex regulator algorithms may not be time-efficient despite aiming at more control intents. It is desirable to have an optimization method that computes the required drug dosage effectively while managing many objectives.

Disease modeling by expanding drugs and goals offers alternatives for (1) finding new targets for the same drug, (2) looking into various medications for the same ailment, and (3) reducing off-target side effects for a safe therapy. For safety reasons, side effects should be avoided, but this does not mean that all pharmaceuticals that cause them should be withdrawn. Because chemotherapy treatments destroy healthy cells, the pharmacological agents that selectively target cancer cells were selected. It is possible to achieve a treatment that is both more effective therapeutically and less impaired using control theory in conjunction with drug delivery. The scientific "culture" of the various theoretical biology subdisciplines was found to differ significantly during the workshop talks. This distinction was particularly pronounced when comparing various model-building strategies used in computational systems biology and computational neuroscience. Formal approaches to model design and data-driven model development are highlighted in the former (Chiao, 2020). On the other hand, formal model selection approaches are essentially nonexistent in computational neuroscience, with the notable exception of computational neuroimaging. Contemporary research has looked at the antecedents of these distinctions between the two subdisciplines.

In conclusion, developing hierarchical modeling and selecting suitable models are top goals for advancing complex illness modeling. There is a need for a platform (network and/or international initiatives) where control theory, physics, and applied mathematics can encourage method development across many domains of computational biology. Drug combinations, repurposing, and precision medicine are three areas where computational systems based on biology-based pharmacological predictions have shown promises. Utilizing known omics data, static networks of chemical bonding can forecast estimate pairs by transferring messages through its nodes and edges (Sabzpoushan, 2020).

The emerging contributors to the network of molecular interactions make it feasible to fully understand illness pathogenesis and therapy response. As a result, therapeutic molecules with greater therapeutic effectiveness can be used without producing unintended side effects. Additionally, possible patient-specific controls can be found to explain the variation in individualized treatment. Dynamic modeling is essential since the predicted models could fail because of in vivo changing behavior.

4. Differentiating stem cells, desired cell types, and organoids: Challenges and future aspects

Clinical research has a lot of potential for recent developments in stem cell engineering. It is difficult to use stem cells for translational drug discovery screening due to the difficulty in controlling the environment of the cells and the scarcity of high-throughput techniques. By actively building and assessing 3D interactions between platforms using microfluidic techniques, such as organ-on-a-chip and the process of reconstructing platforms, researchers have lately made significant strides in stem cell engineering (Rai et al., 2020). The future and practical advantages of stem cell–based organs-on-a-chip or organoid-on-a-chip systems for drug screening and disease modeling were also highlighted, along with high-throughput analytical techniques for microfluidic 3D cell culture (Aghamiri et al., 2020). To give tailored treatments, "human modeling" systems can benefit from the core mechanistic foundation provided by human organoids (Aghamiri et al., 2020). Although there have been significant advances in the study of organogenesis, in vitro, modeling still does not completely capture all of the biological functions of the human body.

To replicate the human body's morphogenetic and molecular routes and procedures, many biological models, such as 2D/3D cell culture and associated animal models, have been developed (Ballav et al., 2022). However, these models have run into significant problems that have limited their practical use. Among these, stem cell research holds considerable promise for elucidating the workings of illness and providing potential treatments by simulating human tissues. With the help of such technologies, future organoid differentiation processes may be optimized, resulting in better spatial organization (Boehnke et al., 2016). These technologies may enable high-throughput investigations of organoid architecture. Organoids are human PSC−based 3D cell culture platforms that replicate the properties of growing tissue by containing cell types that are localized in tissue. Given the challenge of gaining access to the growing human cerebral cortex and the intricacy of neurological illnesses, neural organoids are an exceptionally creative scientific development. Neural organoids have become an essential technique for mimicking features of the developing human brain that are poorly represented in animal models (Aleynik et al., 2014). Organoids also hold promise for research into the peculiar cellular, molecular, and dynamic generation that support neurological diseases. Organoids may also be a source of cell-based therapies for disorders or lesions of the brain and a testing ground for medicines in human cells. Despite the promising qualities of organoids, their widespread application is restricted by several obstacles that still need to be overcome. For example, the nonexistence of high-fidelity cell types, constrained maturation, atypical brain structure, and a lack of realization can minimize their dependability for specific applications.

Organoid platforms have created great opportunities for innovative therapeutic therapies by improving disease modeling and in vitro organogenesis. In addition to developing novel biological models and methods for studying and controlling tissue renewal, organoid simulations can reproduce pathological conditions and potentially generate more precise drug screening techniques and patient-specific treatments. Current solutions to enhance tissue cell suspension cultures are valuable because they can mimic complex niche interactions in nature (Rajiv, 2022). Organoids offer a more sophisticated in vitro tool, allowing for the performance of physiologically relevant investigations that cannot be carried out in animals or people. Organoids can be used more effectively by improving control over environmental stimuli and providing a previously unheard-of possibility to observe and control cellular function recognition to the vast array of bioengineering techniques currently available. A new level of control that enables fine-tuning organoids using genetic circuits and genome editing is the ability to affect how cells internally digest external information (Driehuis & Clevers, 2017). The transition from easily standard-sable in vitro cell monolayer model systems to multiphenotype designs coupled with a variety of bioengineering tools will cause a standardization nightmare, even if it is undoubtedly vital to continue to grow this collection of bioengineering methods. It is also critical to consider how to produce and manipulate organoids in a way that effectively couples appropriate procedures with unmet demands. As has previously been the case with iPSC-derived products, this will definitely make it more difficult for groups to encourage comparability across systems. It is necessary to take into account the challenges of simultaneously maintaining organoids in growth at sizes ranging from microscopic to macroscopic (Yin et al., 2016). Also, the intrinsic capability of the system is individually change each parameter. Numerous significant biological and medicinal investigations have already shown the value of organoid systems in generating new knowledge and advancing the development of remedies for diseases that had previously been considered intractable.

Organoids are effective platforms for investigating particular illnesses and developing specialized therapies. This new technology has enhanced the probability that drugs would transition into preclinical therapeutics and recreate the complexity of organs by providing a variety of approaches for human ailment modeling, drug development, diagnosis, tissue engineering, and regenerative medicine. In this study, we first give a brief description of the evolving organoid technology, then we list the practical uses it has been put to, and finally we discuss the challenges and limitations that 3D organoids encounter (Qian et al., 2019). We conclude by arguing that human organoids offer a fundamental mechanistic underpinning for "human modeling" systems that make personalized drug recommendations. This section focuses on the numerous types of human organoid models, in vitro organoid creation techniques, and organoids. The potential of organoids as a kind of therapy is discussed in the article, along with biological methods to increase 3D organoid complexity. In contrast to incomplete review articles that have previously been published in the literature and that only looked at a small portion of organoid technology, the current work examines this technology in depth and critically from various angles. We briefly cover the uses of organoid technology in this chapter, as well as the difficulties and restrictions that 3D organoids face, before outlining the development of the technology's history and summarizing its faithful applications. Evolutionary and stem cell biology is being revolutionized by the ability of organoids to produce intricate 3D structures resembling organs.

We are currently laying the groundwork for the future of cell treatment using organoids, including applications for personalized medicine, drug screening, and translational applications of organoids. However, despite a few preclinical studies supporting their efficacy, a clinical transformation of organoids into tissue engineering medicines has been put on hold because there are no reliable, repeatable, and scalable production methods compatible with current pharmacological standards, and there are not. Joaquim Vives and colleagues offer a verified bioprocess design in this issue of stem cell research and cell therapy for the mass synthesis of human pancreatic organoids from cadaveric tissue in compliance with existing good manufacturing practices(Vives & Batlle-Morera, 2020). Additionally, given that this type of medicine differs from previous medicinal goods due to its complex composition and the biological nature of the active constituent, the authors offer a set of standards for beginning materials and essential quality attributes of finished products relevant to other advances (Chueng et al., 2016). Although it is still challenging to produce functioning cells that secrete insulin on a large scale, approaches like the one Katerina Bittenglova and colleagues showed help advance the therapeutic application of organoids for treating type 1 diabetes and pave the way for future clinical uses of organoids in degenerative illnesses (Bittenglova et al., 2021).

Human PSCs can now differentiate into "organoids," or self-organizing tissues, according to recent advances in stem cell research. These organoids' cells and the way they are arranged within them exhibit striking similarities to tissues found in living organisms, which may one day pave the road to regenerative medicine. Examples include kidney organoids made from human iPSCs, which are astonishingly highly organized and contain every type of cell found in the human kidney. Although immediate applications such as nephrotoxicity screening, the modeling of kidney formation and renal disorders, or "kidney-on-a-chip" concepts are already beginning to take shape, more work will be required to address problems such as the lack of cell maturation, restricted cell diversity, vascularization, and functionality.

This section will cover the most recent kidney organoid generation techniques, microfluidic bioengineering methods, and modeling of renal disorder phenotypes in kidney organoids produced from patient-derived PSCs. Although it is currently difficult to study development, toxicity, genetics, and viral disease using primary tissue, liver cell types produced from iPSCs may be able to. An additional advantage of patient individuality might be significant in any of these fields. Many iPSC diversification strategies concentrate on 3D or organotypic transformation due to their advantage of more accurately mimicking in vivo systems, such as developing tissue-like structures and inflammasomes/cross-talk among various kinds of cells (Rajiv, 2021f). Ultimately, these simulations might be employed in therapeutic settings in conjunction with, or perhaps instead of, animal simulations. It will also be necessary to collaboratively develop imaging technologies to make it possible to visualize organotypic segments and submodels.

Numerous liver simulators that have been discussed in the literature are referred to as "organoids" in this sentence. These frameworks range from straightforward spheres or cysts composed of a single cell type, typically hepatocytes, to those made up of several cell types mixed during divergences, such as hepatic stellate cells, epithelial cells, and intercellular cells, which regularly contribute to enhanced hepatic genetic makeup. Because they are much more straightforward than ex vivo tissue cultures, they lack versatility and broad applicability. These enable the evaluation of particular functionalities or readouts, such as enhanced phenotypes, protein synthesis, or medication metabolism. They are frequently created in the liver field rather than formed as a single organ, in contrast to other organoid models such as the brain, kidneys, lung, and intestine. Accessibility to organotypic liver surrogates will significantly improve disease, toxicity, and drug discovery models by merging domains such as a microfluidic microchip with organoids. This would pave the way toward novel treatments. These surrogates contain various cell types, interact, and have an architecture similar to that found in the living body. In the current world, metabolic illnesses such as obesity, diabetes mellitus, and cardiovascular disease pose a serious threat to health. Yet, efforts to understand underlying mechanisms and create logical therapies are hampered by the absence of suitable human model systems (Rajiv, 2021j). A significant development is the creation of cellular models that mimic the histological, molecular, and physiological characteristics of human organs. This is made possible by breakthroughs in stem cell and organoid technologies. Human stem cells and organoids offer unheard-of systems for researching the causes of underlying metabolic disorders, especially when combined with significant advancements in gene editing techniques. Here, the development and application of organoids and metabolic cell types produced from stem cells in studying metabolic diseases, including obesity and liver disorders, were covered.

We specifically go over the drawbacks of using animal models and the benefits of using stem cells and organoids, particularly their use in treating metabolic illnesses. Additionally, medication action mechanisms were elaborated, gum producing the effectiveness and toxicity of current therapy, searching for novel treatments, and seeking tailored therapeutics. We emphasize the potential to transform the method to discover therapies for metabolic illnesses by integrating stem cell–derived organoids with gene editing and genetic analysis.

Thanks to current organoid culture technology, adult and PSCs can self-organize into multiple (3D) constructions that precisely imitate the look and function of in vivo organs. Particularly, liver organoids made from PSCs of humans (PSC-LOs) can be applied in preclinical settings such as disease modeling, drug research, and regenerative medicine. By carefully regulating the cellular microenvironment, the organoid technologies are combined with cutting-edge bioengineering toolkits such as micromachining, and 3 D biomaterials to enhance the functional strength of the character of human PSC-LOs and increase the effectiveness of hepatic differentiation (Nikolova & Chavali, 2019). Hepatic transitional epithelium, endothelial, and mesenchymal cells must be united to sustain the hepatocellular capabilities of in vitro liver organoids for an extended period. As a result, the biological function and scalability of human PSC-LOs have been enhanced.

Bioengineering approaches have been applied to identify different zonal hepatocyte populations in liver organoids for capturing diverse illnesses. Customized liver organoids made from distinct PSCs will therefore be useful for tissue regeneration and future in vitro clinical studies. This chapter will focus on the most current developments in bioengineering methods for liver biomaterial fermentation procedures (Rouwkema et al., 2011). These systems promote in vitro drug development, provide cell therapy materials for transplantation, and offer a timely and essential study to mimic disease pathogenesis. Without a doubt, 3D organoid systems can offer fresh opportunities for developing organ processes while addressing important problems in the clinical fields, especially in light of some important limitations in bioreactors and animal models. Because of the 3D structure of organoids and improved cellular assembly with genuine tissues, in vitro, human models have a significant potential to be more accurate. The roles of cell genomics and proteomics can be explored through single reactions and elevated characterizations of chromosome configurations such as chromatin domains and transcriptional regulatory elements (Zhang, Liu, et al., 2020).

Through modeling many degenerative disorders, including genetic, infectious, neuropsychiatric, and cancerous diseases, organoids are used to screen medications and assess their toxicity (Kumar et al., 2021a). This knowledge would change how we think about how organs work and are affected by disease, changing how we create drugs and think about tailored and regenerative treatments. Powerful in vitro systems known as brain organoids have been produced recently owing to advancements in 3D cultures that take advantage of PSCs' complex "assemblies" which can disclose cell capacity for self-organization. Many facets of in vivo human brain growth and pathologies are reproduced by these 3D tissues (Lancaster et al., 2017). These in vitro systems allow the creation of increasingly complex "assemblies," which can disclose cell diversity, microcircuits, and cell–cell interactions inside their 3D organization when combined with enhanced differentiation techniques. A review and discussion of the ways that human brain organoids that have helped stimulate developmental diseases and demystify the intricacies of brain development are presented here (Reiss, 2009).

The topic of brain organoids is still emerging, and there has been discussion of difficult concerns that are related to this. The brain hepatocytes are self-assembling, 3D aggregates that use PSCs with cell types, and cytoarchitecture like those in the developing human brain. They have consequently evolved into reducing model systems that can be used to study the development and disorders of the neural network (Chen et al., 2020). Brain organoids closely mirror several crucial aspects of the early stages of human brain development at the biochemical, cellular, structural, and system levels. But not all features of brain development are completely mirrored, including the formation of several pyramidal neuronal divisions, gyrification, and the establishment of complex neural circuits. Here, we discuss the applications of brain biomaterial techniques in disease modeling and highlight recent advancements in this field. We also emphasize elements that can now be recapitulated and those that cannot, and we compare modern organoid systems to the developing human brain. Finally, we address future directions for developing modern brain organoid technologies to broaden their applicability (Lin et al., 2018).

The development of 3D organoids, however, still faces numerous challenges. Organotypic simulations of patient-specific tumors have transformed our understanding of disease heterogeneity and its implications for individualized care. These advancements may be in part attributable to the ability of organoid models to maintain the genetic, biochemical, morphological, and pharmacokinetic traits of the original tumor in vitro while offering unmatched genomic and environmental manipulation (Yellapantula et al., 2019). Despite recent developments in organoid techniques, the methods now used for cancer organoid growth are inherently unpredictable and irreplaceable due to several nonstandard components, including cancer tissue origins and subsequent analysis, media formulations, and a 3D animal-derived matrix. Given the potential of cancer organoids to faithfully reproduce the intra- and intertumoral biological heterogeneity related to patient-specific malignancies, it is essential to eliminate the unfavorable technical variability associated with cancer organoid culture to establish repeatable platforms that hasten the translation of findings into patient treatment. Here, the prospects for standardizing the newest cancer organoid systems and the recent multidisciplinary breakthroughs, obstacles, and successes were discussed (Skaga et al., 2019). Major challenges include controlling personality to produce mature organoids, lowering the high variability of organoid inoculating, increasing physiologically required documents and dimensions, widening the organoid lifespan to produce accurate output cells in vitro, and fictionalizing the physiological intricacies of native organs by implementing the main biological subsystems such as the nervous system and vasculature (Fig. 8.4). The field of bioengineering must adopt a particularly successful multidisciplinary strategy to overcome these restrictions. For instance, artificial biomaterials have recently offered well-defined platforms to overcome such constraints

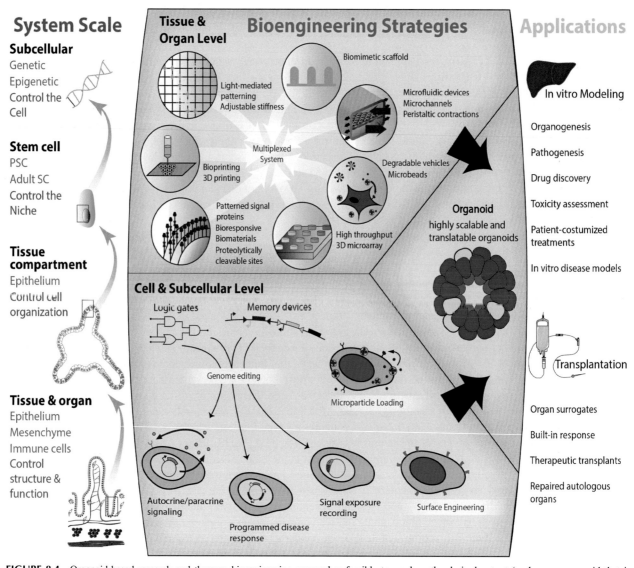

FIGURE 8.4 Organoid-based research and therapy: bioengineering approaches feasible to produce the desired output (such as an organoid that is particular to a tissue or an in vitro disease model) with the proper combination and sequence of the input signals that the niche transmits to the cell. Several bioengineering techniques can be used to alter these signals and track appropriate responses. According to the same logic, after the biology of organoids has been clarified, it may be possible to use fresh information to develop artificial niches. Combining several bioengineering methods that imitate particular niche components will allow the niche to be manufactured. *Reprinted (adapted) with permission from ref. Oswald and Baranov (2018) Copyright (2015) Oren Levyet al. Cell Stem Cell, Elsevier Inc.*

using logically designed 3D matrices that mimic biological tissues' ECM. Numerous studies are progressively focusing on the therapeutic advantages of organoids in regenerative medicine, gene therapy, personalized medicine, and drug development. However, it is far more challenging to translate 3D organoids into medical applications (Takahashi, 2019). Later on, additional authorized organoids will be available for clinical trials.

The creation of potent in vitro systems known as brain organoids has been made possible by advances in 3D culturing that take advantage of PSCs capacity for self-organization. These 3D tissues accurately represent various in vivo elements of human brain growth and pathologies (Marton & Pașca, 2020). These in vitro systems enable the creation of increasingly complex "assemblies," which can reveal cell diversity, simulate their effects, and establish cell—cell contacts within their 3D organization when combined with enhanced differentiation techniques.

5. Future prospective and outlook

The current branch of computational stem cell biology, which combines system-level stem cell models with high-throughput molecular data, has only recently come into existence. Single-cell multimodal assay development provides a powerful tool for investigating several facets of cellular heterogeneity, opening fresh insights into growth, tissue homeostasis, and disease. More recently, the conversion of high-throughput data into understanding, including developmental and stem cell biology, has relied heavily on computational methodologies (Bosch, 2009). New concepts and definitions contextualize current challenges and suggest new concepts for a better perspective. A thorough study has demonstrated how biology and computer engineering can help one another by resolving issues, leading to an enhanced comprehension of biological processes' current and future views while also advancing algorithm design (Yang et al., 2021). Unfortunately, it is not uncommon for convergence between biology and engineering to be difficult, especially given the unique cultural differences between these two communities. Moreover, with the use of organoid platforms, in vitro organogenesis and clinical diagnostics have advanced, opening promising possibilities for creating novel treatments. The dorsal brain, choroid plexus, hippocampus, ventral telencephalon, and crypt-villus structures resembling those in the intestinal epithelial lining are only a few examples of important features of many tissues and organ subregions that have been effectively duplicated in 3D organoid models (Salle, 2021). Organoid models can be used to simulate disease states, create more accurate drug screening platforms, and develop medical therapies, as well as create new biological systems and tools for researching and manipulating tissue regeneration. Spheroid cultures and tissue explants are beneficial because they can replicate the intricate niche interactions found in nature. Organoids offer a more sophisticated in vitro tool, allowing for the performance of physiologically relevant investigations that cannot be carried out in animals or people (Zhang, Jin, et al., 2020). Organoids can be used more effectively by improving control over environmental stimuli and providing a previously unheard-of possibility to observe and control cellular function thanks to the vast array of bioengineering techniques currently available. Genome editing and genetic circuitry can be used to fine-tune organoids by altering how cells inwardly process external inputs. With that said, there are still difficulties. A standardization nightmare will result from the switch from easy characteristics of the system in vitro cell monolayer preclinical studies to multiphenotype models integrated with multiple bioengineering technologies (Almodovar et al., 2015). Determining the best approaches to link relevant procedures with unmet medical requirements in the creation and management of organoids is therefore vital. We must continue building this library of bioengineering techniques. Similar to how it has been with iPSC-derived products, this will undoubtedly make it challenging for groups to compare outcomes across systems (Balbi & Vassalli, 2020). The challenges of instantly and easily trying to control organoids from microscopic to macroscopic levels while maintaining them in culture, as well as the experiments of using bioengineering techniques without attempting to tamper with the system's ability to adjust each parameter automatically, must be taken into account. Nevertheless, numerous fundamental biological and therapeutic investigations have already shown the value of organoid systems in advancing new information and bringing us closer to treatments for illnesses that previously seemed incurable.

To highlight some of the current unmet needs and demonstrate how pervasive computation is in this subject, we only briefly touch on a few topics at the intersection of computational biology and stem cell biology in this chapter. Throughout our examination of the application of computational biology in stem cell biology, we identified a set of key ideals or guiding principles. Although not comprehensive and in some respects applicable to computational biology more generally, this list does encompass the most popular and, in our opinion, most significant concepts that can address current and future issues. Regenerative medicine's long-term goal is still a way off. However, it is anticipated that soon, disease modeling and drug screening will be able to benefit from kidney cell cultures or perhaps the more complex renal organoids (Kitano, 2017). Right now, kidney organoids can include up to 100 nephrons and grow to a diameter of around 8 mm after about 3 weeks in culture. The issue cannot be resolved by simply increasing the number of nephrons because the kidney also needs a particular histological architecture, an integrated collecting duct architecture, and a functional exit route to the bladder to properly reabsorb fluids, amino acids, electrolytes, and other nutrients. This is still a work in progress. As a result, creating an entirely functionally adequate replacement organ still presents a significant problem (Ali et al., 2022). One potential method is introducing decellularized scaffolds made from human kidneys into which iPSC-derived kidney cells have been infused. Human kidneys have been de- and recellularized utilizing a variety of techniques. It is possible that producing certain renal cell types to employ in cell therapy to transplant back into the damaged kidney could be beneficial (Zorova et al., 2018).

Early research indicates that this is the case, with signals of lessened damage appearing when human-derived renal progenitors were transplanted into animal models that were susceptible to harm. The ability of these cells to function over the long term and whether they will do so in the presence of chronic renal damage is still uncertain (Luiz et al., 2014). The use of organ-on-a-chip technology based on microfluidics and kidney cell types derived from iPSCs may be more practical. Prospects still include such choices. All of these possibilities, however, are now possible thanks to the discovery of the technology to transfer a PSC state to a kidney endpoint. Additionally, it provides the first step toward a more profound recognition of the molecular underpinnings of a human being's normal kidney development, an area with a wealth of untapped knowledge (Rajiv, 2021b). The fate of stem cells is controlled by a complex interplay of biophysical and biochemical components found in the natural stem cell niche. By adopting design cues from native stem cell microenvironments, engineering strategies to maintain stemness and direct transformation ex vivo have been established.

Cell−cell interactions, microscale structure, matrix mechanics, and biochemistry are only a few of the crucial regulators of stem cell fate that hydrogel materials can control (Lovett et al., 2020). This level of control offers the possibility of designing hydrogel niches to produce biomimetic tissue structures, boosts the reproducibility and throughput of stem cell growth, and makes it easier to create stem cells for therapeutic use. Regulating plasticity is a practical way to prevent epithelial-to-mesenchymal transition (EMT), or the recently identified gross cell fate transitions of one cancer type into another, which are becoming recognized as an escape route. In pathologic situations, such as those indicated by fibrosis or noncancerous metaplasia, a framework for imagining how cell fate plasticity might be encouraged or prevented has been offered. This underlines how crucial it is to identify and comprehend every cell type implicated in a specific pathology, whether they are represented by typical cell types or unique pathogenic inflammatory responses that do not typically exist (Kumar, 2021b).

Numerous data points should be available in this direction owing to emerging technologies, particularly single-cell transcriptomics. The risk of promoting pathologic cell plasticity, such as cancer, is present with all therapies, including those that promote positive plasticity. Any cell modulation therapy must also be done correctly to only affect the intended tissues and not interfere with the ratios of unintended tissues or other vital cell types. An overview of recent challenges that occurred during the discoveries of this emerging discipline (computational and stem cell biology) has also been presented, as have essence considerations elaborated on the implications that single-cell transcriptomics is predicted to have better knowledge of expansion with better capacity to design cell fate for improved prospects. Different subtopics, including existing challenges in crossing hurdles in computational and stem cell biology and upcoming challenges in single-cell data analysis, disease modeling, computational biology, stem cell bioengineering, genomics, and proteomics, can be used to address those problems, which stem cell and computational biology are facing now and are discussing here with their future aspects.

An overview of recent challenges during the discoveries of this emerging discipline (computational and stem cell biology) has also been presented. Essential considerations have elaborated on the implications that single-cell transcriptomics is predicted to have in terms of a better expansion of knowledge and a better capacity to design cell fate for improved prospects. Different subtopics, including existing challenges in crossing hurdles in computational and stem cell biology and upcoming challenges in single-cell data analysis, disease modeling, computational biology, stem cell bioengineering, genomics, and proteomics, can be used to address the problems that stem cell and computational biology are currently facing.

6. Conclusion

By combining omics data from the transcriptional, transcriptional, proteomic, and metabolic layers, systems biology predicts likely reaction mechanisms and addresses well-discussed problems. The constructed models improve the therapeutic effectiveness of drugs when paired with known drug reactions by utilizing already-existing treatments and combining pharmacological substances without off-target effects. Drug administration laws are created to balance toxicity and efficiency based on the developed computational models. To model disease causes and treatment responses, an overview of therapeutic systems and an examination of interconnections connecting genes, protein, and medication molecules were presented (Palleria et al., 2013). Combining computational and predictive models with the drug administration outlined in control rules can enhance therapeutic performance (Ye et al., 2021). PSC disease models have begun finding use in understanding disease biology and creating innovative treatments in addition to successfully reproducing disease symptoms (Kumar & Gulla, 2021). Applications of PSC's to melanoma will be beneficial for (1) modeling disease population groups, (2) elucidating pathophysiologic mechanisms, (3) predicting survival in patients, (4) identifying potential biomarkers and targeted therapies, (5) finding haploinsufficient genes by properly functional mapping of illness-associated gen chromosome loss, and (6) applying random drug testing to identify possible drugs to rescue particular disease phenotypes. PSC modeling eliminates numerous drawbacks of existing model systems, such as the difficulty of obtaining patient samples, the heterogeneity of tumors, and interspecies variations (Adasme et al., 2022). In conclusion, hPSC-derived organoids offer potent tools for drug development and recapitulation of the pathogenic states in cancer. For example, clarifying the early stages of tumor growth is particularly difficult since cancer cell lines and tumor-derived mouse models already have pronounced levels of genetic abnormalities.

Biomolecules are dynamic, constantly switching between forms with different energies, rather than being static. These concepts provide insight into the mechanisms underlying the self-organization of biomolecules into ensembles of 3D conformation that are relevant for specific functional purposes (Karimi, 2018). Another fundamental goal is to create effective, high-affinity medications. Any research advancement or challenge, whether it is met or only aspirationally pursued, is significant. This list does, however, demonstrate the breadth of the difficulties that computational biology as a field must overcome. It is difficult to say whether and how much computational biology researchers have contributed to each development and challenge because there is no valuable method for measuring a journal's impact on a discipline. However, it can be difficult to tell the difference between fresh theories that could result in significant advancements and speculative claims.

To improve biological community, computational biology researchers seek publications that cover all aspects of the field (Tisoncik et al., 2012). This suggested technique describes fresh notions about having substantial new insights into biological processes and method papers providing new procedures for handling important problems that have been demonstrated to produce or can deliver new biological insights. When and how to be optimized, computational approaches should provide precise instructions.

Multiple parameters affect the analysis results in most computational biology programs. Reproducibility should be ensured while applying and using computational approaches. Data must be characterized and kept to be later reanalyzed to facilitate data integration and enable computational phenotypic inference (Bornot et al., 2014). This approach is generally followed because it is widely acknowledged that FAIR (findable (F), accessible (A), interoperable (I), and reusable (R)) data management has advantages, and because most publications now demand that genome-scale data be deposited in open repositories. The repeatability of computational analytics has been hampered, too, by the varying degrees to which publications demand that code be made publicly available.

We value close, win−win interactions between computational and evolutionary biologists, where ideas and information are shared in both directions. Due to the incorporation of computational, statistical, and mathematical factors into the experimental design and the biological input provided by the experimentalists, which ensures the computational models' applicability and insight, these deep links will lead to more fruitful investigations (Robertson et al., 2018). Such in-depth interconnections are extremely desirable because they make it possible for people with unusual origins to enter the field, such as those in physics, mathematics, or even other scientific computing, and also because they enable them to accumulate and work on implementing that knowledge specific to stem cell biology. The problem of using computational methods to advance our understanding of biological systems is something that researchers are working together to tackle (Lalani et al., 1997). Identification of a stem cell or transformed cell type was a complicated task. The implicit presumption that cells in the body are static structures emerges more from our perception of cellular systems than from any inherent reality about those processes. The fingerprints identified through theoretical analysis of omics data must be harmonized with the situation, which allows for cell kinds based on form and studied as histopathology.

Computing-based labeling of a group of cells can be done in several ways, including (1) annotation based on prior knowledge, such as the presence of a validated bloodline marker, (2) resemblance to collected and compiled benchmark datasets, (3) computer metadata, and, (4) interpretation with prior knowledge. Utilizing markers is the most popular method of determining a cell type (Newman et al., 2019). Marker-based cluster labeling encourages integrity with subsequent investigations because these identifiers can also be used for cell separation and sorting. It was a remarkable success on the first conference of computational stem cell biology to bring together researchers from different cultural backgrounds and with diverse computerized stem cell research implementations. The considerations brought to light openings and obstacles for computational regenerative medicine that demand concerted effort from the entire stem cell community. Realizing a comprehensive grasp of the molecular and biological mechanisms controlling cell differentiation is a severe obstacle to stem cell engineering (Rajiv, 2021a).

Author contributions

Dr. Rajiv Kumar drafted the work and figures. Prof. Agnieszka M. Jastrzebska, and Prof. Magali Cucchiarini contributed by formatting all the sections. Dr. Neelam Chhillar, Mitrabasu Chhillar, and Dora I. Medina revised it critically for important intellectual content. In the end, all the authors approved the final version.

Availability of data and materials

Where applicable, the reference section contains appropriate citations.

References

Acheson, A., Sunshine, J. L., Rutishauser, U., Ailor, E., Takahashi, N., Tsukamoto, Y., Masuda, K., Rahman, B. A., Jarvis, D. L., Lee, Y. C., Betenbaugh, M. J., Angata, K., Chan, D., Thibault, J., Fukuda, M. N., Huckaby, V., Ranscht, B., Terskikh, A., Marth, J. D., … Roth, J. (2010). Molecular basis for polysialylation: A novel polybasic polysialyltransferase domain (PSTD) of 32 amino acids unique to the alpha 2,8-polysialyltransferases is essential for polysialylation. *Journal of Biological Chemistry, 276*(1), 9443–9448. http://www.ncbi.nlm.nih.gov/pubmed/9134419.

Adasme, M. F., Bolz, S. N., Al-Fatlawi, A., & Schroeder, M. (2022). Decomposing compounds enables reconstruction of interaction fingerprints for structure-based drug screening. *Journal of Cheminformatics, 14*(1). https://doi.org/10.1186/s13321-022-00592-w

Aghamiri, S., Rabiee, N., Ahmadi, S., Rabiee, M., Bagherzadeh, M., & Karimi, M. (2020). Microfluidics: Organ-on-a-chip. In *Biomedical applications of microfluidic devices*. https://doi.org/10.1016/B978-0-12-818791-3.00001-2

Aleynik, A., Gernavage, K. M., Mourad, Y. S., Sherman, L. S., Liu, K., Gubenko, Y. A., & Rameshwar, P. (2014). Stem cell delivery of therapies for brain disorders. *Clinical and Translational Medicine, 3*(1). https://doi.org/10.1186/2001-1326-3-24

Ali, M., Ribeiro, M. M., & del Sol, A. (2022). Computational methods to identify cell-fate determinants, identity transcription factors, and niche-induced signaling pathways for stem cell research. *Methods in Molecular Biology, 2471*. https://doi.org/10.1007/978-1-0716-2193-6_4

Almodovar, J., Castilla Casadiego, D. A., & Ramos Avilez, H. V. (2015). Polysaccharide-based biomaterials for cell-material interface. In , *Vol. 5. Cell and material interface: Advances in tissue engineering, biosensor, implant, and imaging technologies* (pp. 230–244).

Balbi, C., & Vassalli, G. (2020). Exosomes: Beyond stem cells for cardiac protection and repair. *Stem Cells, 38*(11). https://doi.org/10.1002/stem.3261

Ballav, S., Jaywant Deshmukh, A., Siddiqui, S., Aich, J., & Basu, S. (2022). *Two-dimensional and three-dimensional cell culture and their applications.* https://doi.org/10.5772/intechopen.100382

Bian, Q., & Cahan, P. (2016). Computational tools for stem cell biology. *Trends in Biotechnology, 34*(12). https://doi.org/10.1016/j.tibtech.2016.05.010

Bittenglova, K., Habart, D., Saudek, F., & Koblas, T. (2021). The potential of pancreatic organoids for diabetes research and therapy. *Islets, 13*(5–6). https://doi.org/10.1080/19382014.2021.1941555

Blutt, S. E., & Estes, M. K. (2022). Organoid models for infectious disease. *Annual Review of Medicine, 73*. https://doi.org/10.1146/annurev-med-042320-023055

Boehnke, K., Iversen, P. W., Schumacher, D., Lallena, M. J., Haro, R., Amat, J., Haybaeck, J., Lieba, S., Lange, M., Schäfer, R., Regenbrecht, C. R. A., Reinhard, C., & Velasco, J. A. (2016). Assay establishment and validation of a high-throughput screening platform for three-dimensional patient-derived colon cancer organoid cultures. *Journal of Biomolecular Screening, 21*(9). https://doi.org/10.1177/1087057116650965

Bornot, A., Blackett, C., Engkvist, O., Murray, C., & Bendtsen, C. (2014). The role of historical bioactivity data in the deconvolution of phenotypic screens. *Journal of Biomolecular Screening, 19*(5), 696–706. https://doi.org/10.1177/1087057113518966

Bosch, T. C. G. (2009). Hydra and the evolution of stem cells. *BioEssays, 31*(4). https://doi.org/10.1002/bies.200800183

Chen, A., Guo, Z., Fang, L., & Bian, S. (2020). Application of fused organoid models to study human brain development and neural disorders. *Frontiers in Cellular Neuroscience, 14*. https://doi.org/10.3389/fncel.2020.00133

Chiao, J. Y. (2020). Computational theory of mind. In *Philosophy of computational cultural neuroscience*. https://doi.org/10.4324/9780429327674-10

Choi, S., Choi, J., Cheon, S., Song, J., Kim, S. Y., Kim, J., Nam, D. H., Manzar, G., Kim, S. M., Kang, H. S., Kim, K. K., Jeong, S. H., Lee, J. H., Park, E. K., Lee, M., Lee, H. A., Kim, K. S., Park, H. J., Oh, W. K., … Kim, E. M. (2022). Pulmonary fibrosis model using micro-CT analyzable human PSC–derived alveolar organoids containing alveolar macrophage-like cells. *Cell Biology and Toxicology, 38*(4). https://doi.org/10.1007/s10565-022-09698-1

Chueng, S.-T. D., Yang, L., Zhang, Y., & Lee, K.-B. (2016). Multidimensional nanomaterials for the control of stem cell fate. *Nano Convergence*. https://doi.org/10.1186/s40580-016-0083-9

Das, S., & Baker, A. B. (2016). Biomaterials and nanotherapeutics for enhancing skin wound healing. *Frontiers in Bioengineering and Biotechnology*. https://doi.org/10.3389/fbioe.2016.00082

Drawnel, F. M., Boccardo, S., Prummer, M., Delobel, F., Graff, A., Weber, M., Gérard, R., Badi, L., Kam-Thong, T., Bu, L., Jiang, X., Hoflack, J. C., Kiialainen, A., Jeworutzki, E., Aoyama, N., Carlson, C., Burcin, M., Gromo, G., Boehringer, M., … Iacone, R. (2014). Disease modeling and phenotypic drug screening for diabetic cardiomyopathy using human induced pluripotent stem cells. *Cell Reports, 9*(3). https://doi.org/10.1016/j.celrep.2014.09.055

Driehuis, E., & Clevers, H. (2017). CRISPR/Cas 9 genome editing and its applications in organoids. *American Journal of Physiology - Gastrointestinal and Liver Physiology, 312*(3). https://doi.org/10.1152/ajpgi.00410.2016

Duffy, F., Maheshwari, N., Buchete, N. V., & Shields, D. (2019). Computational opportunities and challenges in finding cyclic peptide modulators of protein–protein interactions. *Methods in Molecular Biology, 2001*, 73–95. https://doi.org/10.1007/978-1-4939-9504-2_5. Humana Press Inc.

Edgar, L., Pu, T., Porter, B., Aziz, J. M., La Pointe, C., Asthana, A., & Orlando, G. (2020). Regenerative medicine, organ bioengineering and transplantation. *British Journal of Surgery, 107*(7). https://doi.org/10.1002/bjs.11686

Fang, R., Preissl, S., Li, Y., Hou, X., Lucero, J., Wang, X., Motamedi, A., Shiau, A. K., Zhou, X., Xie, F., Mukamel, E. A., Zhang, K., Zhang, Y., Behrens, M. M., Ecker, J. R., & Ren, B. (2021). Comprehensive analysis of single cell ATAC-seq data with SnapATAC. *Nature Communications, 12*(1). https://doi.org/10.1038/s41467-021-21583-9

Frisch, M. J., Trucks, G. W., Schlegel, H. B., Scuseria, G. E., Robb, M. A., Cheeseman, J. R., Scalmani, G., Barone, V., Mennucci, B., Petersson, G. A., Nakatsuji, H., Caricato, M., Li, X., Hratchian, H. P., Izmaylov, A. F., Bloino, J., Zheng, G., Sonnenberg, J. L., Hada, M., & ..., ... Fox, D. J. (2016). *Gaussian 09, revision E. 01.* Wallingford CT: Gaussian Inc. (Gaussian Inc).

Gingold, J., Zhou, R., Lemischka, I. R., & Lee, D. F. (2016). Modeling cancer with pluripotent stem cells. *Trends in Cancer, 2*(9). https://doi.org/10.1016/j.trecan.2016.07.007

Gulia, R. K., & K. (2021). Cell machine's transduction machinery, and cell signalling defects. Small tools and nano-bio interface for influential regenerative remedies. *Journal of Cell Signaling, 6*(5), 1—14.

Hamey, F. K., Nestorowa, S., Kinston, S. J., Kent, D. G., Wilson, N. K., & Gottgens, B. (2017). Reconstructing blood stem cell regulatory network models from single-cell molecular profiles. *Proceedings of the National Academy of Sciences of the United States of America, 114*(23). https://doi.org/10.1073/pnas.1610609114

Karimi, T. (2018). Molecular mechanism of self-organization in biological systems. In *Molecular mechanisms of autonomy in biological systems.* https://doi.org/10.1007/978-3-319-91824-2_3

Kaushik, S. N., Kim, B., Cruz Walma, A. M., Choi, S. C., Wu, H., Mao, J. J., Jun, H. W., & Cheon, K. (2016). Biomimetic microenvironments for regenerative endodontics. *Biomaterials Research.* https://doi.org/10.1186/s40824-016-0061-7

Kitano, H. (2017). Biological complexity and the need for computational approaches. In *History, philosophy and theory of the life sciences* (Vol. 20). https://doi.org/10.1007/978-3-319-47000-9_16

Kohn, S., Leichsenring, K., Kuravi, R., Ehret, A. E., & Böl, M. (2021). Direct measurement of the direction-dependent mechanical behaviour of skeletal muscle extracellular matrix. *Acta Biomaterialia, 122.* https://doi.org/10.1016/j.actbio.2020.12.050

Kumar, R. (2021a). Computational explorations and interpretation of bonding in metallopharmaceutical for a thorough understanding at the quantum level. *Sustainable Chemical Engineering.* https://doi.org/10.37256/sce.212021828

Kumar, R. (2021b). Emerging role of neutrophils in wound healing and tissue repair: The routes of healing. *Biomedical Journal of Scientific & Technical Research, 36*(4), 28687—28688.

Kumar, R., Chhikara, B. S., Gulia, K., & Chhillar, M. (2021a). Review of nanotheranostics for molecular mechanisms underlying psychiatric disorders and commensurate nanotherapeutics for neuropsychiatry: The mind knockout. *Nanotheranostics, 5*(3), 288—308. https://doi.org/10.7150/ntno.49619

Kumar, R., Chhikara, B. S., Gulia, K., & Chhillar, M. (2021b). Cleaning the molecular machinery of cellsviaproteostasis, proteolysis and endocytosis selectively, effectively, and precisely: Intracellular self-defense and cellular perturbations. *Molecular Omics, 17*(1), 11—28. https://doi.org/10.1039/d0mo00085j. Royal Society of Chemistry.

Kumar, R., & Gulia, K. (2021). The convergence of nanotechnology-stem cell, nanotopography-mechanobiology, and biotic-abiotic interfaces: Nanoscale tools for tackling the top killer, arteriosclerosis, strokes, and heart attacks. *Nano Select, 2*(4), 655—687. https://doi.org/10.1002/nano.202000192

Lalani, I., Bhol, K., & Ahmed, A. R. (1997). Interleukin-10: Biology, role in inflammation and autoimmunity. *Annals of Allergy, Asthma and Immunology, 79*(6), 469—484. https://doi.org/10.1016/S1081-1206(10)63052-9. American College of Allergy, Asthma and Immunology.

Lancaster, M. A., Corsini, N. S., Wolfinger, S., Gustafson, E. H., Phillips, A. W., Burkard, T. R., Otani, T., Livesey, F. J., & Knoblich, J. A. (2017). Guided self-organization and cortical plate formation in human brain organoids. *Nature Biotechnology, 35*(7). https://doi.org/10.1038/nbt.3906

Lin, B., Srikanth, P., Castle, A. C., Nigwekar, S., Malhotra, R., Galloway, J. L., Sykes, D. B., & Rajagopal, J. (2018). Modulating cell fate as a therapeutic strategy. *Cell Stem Cell, 23*(3). https://doi.org/10.1016/j.stem.2018.05.009

Lovett, M. L., Nieland, T. J. F., Dingle, Y. T. L., & Kaplan, D. L. (2020). Innovations in 3D tissue models of human brain physiology and diseases. *Advanced Functional Materials, 30*(44). https://doi.org/10.1002/adfm.201909146

Lowe, A. W., Olsen, M., Hao, Y., Lee, S. P., Lee, K. T., Chen, X., van de Rijn, M., & Brown, P. O. (2007). Gene expression patterns in pancreatic tumors, cells and tissues. *PLoS One, 2*(3). https://doi.org/10.1371/journal.pone.0000323

de Magalhães, J. P., & Ocampo, A. (2022). Cellular reprogramming and the rise of rejuvenation biotech. *Trends in Biotechnology, 40*(6). https://doi.org/10.1016/j.tibtech.2022.01.011

Luiz, G., Noelio, D., Anielle, S., João, M. M., Ana, G. B. M., Carlos, U. V., Patricia, F., Yara Maia, P. S., Ana, P. F. J. A., & Isabela, G. (2014). Frontiers of biology in human diseases: Strategies for biomolecule's discovery, nanobiotechnologies and biophotonics. *BMC Proceedings, 8*(Suppl 4), O9. https://bmcproc.biomedcentral.com/articles/10.1186/1753-6561-8-S4-O9.

Marton, R. M., & Paşca, S. P. (2020). Organoid and assembloid technologies for investigating cellular crosstalk in human brain development and disease. *Trends in Cell Biology, 30*(2). https://doi.org/10.1016/j.tcb.2019.11.004

McBride, W. H., Iwamoto, K. S., Syljuasen, R., Pervan, M., & Pajonk, F. (2003). The role of the ubiquitin/proteasome system in cellular responses to radiation. *Oncogene, 22*(37). https://doi.org/10.1038/sj.onc.1206676. REV. ISS. 3.

Newman, A. M., Steen, C. B., Liu, C. L., Gentles, A. J., Chaudhuri, A. A., Scherer, F., Khodadoust, M. S., Esfahani, M. S., Luca, B. A., Steiner, D., Diehn, M., & Alizadeh, A. A. (2019). Determining cell type abundance and expression from bulk tissues with digital cytometry. *Nature Biotechnology, 37*(7). https://doi.org/10.1038/s41587-019-0114-2

Nikolova, M. P., & Chavali, M. S. (2019). Recent advances in biomaterials for 3D scaffolds: A review. *Bioactive Materials, 4.* https://doi.org/10.1016/j.bioactmat.2019.10.005

Oswald, J., & Baranov, P. (2018). Regenerative medicine in the retina: From stem cells to cell replacement therapy. *Therapeutic Advances in Ophthalmology, 10.* https://doi.org/10.1177/2515841418774433

Ottoboni, L., von Wunster, B., & Martino, G. (2020). Therapeutic plasticity of neural stem cells. *Frontiers in Neurology, 11.* https://doi.org/10.3389/fneur.2020.00148

Palleria, C., Di Paolo, A., Giofrè, C., Caglioti, C., Leuzzi, G., Siniscalchi, A., De Sarro, G., & Gallelli, L. (2013). Pharmacokinetic drug-drug interaction and their implication in clinical management. *Journal of Research in medical Sciences, 18*(7).

Petersen, F., Yue, X., Riemekasten, G., & Yu, X. (2017). Dysregulated homeostasis of target tissues or autoantigens - a novel principle in autoimmunity. *Autoimmunity Reviews, 16*(6). https://doi.org/10.1016/j.autrev.2017.04.006

Pulido-Reyes, G., Leganes, F., Fernández-Piñas, F., & Rosal, R. (2017). Bio-nano interface and environment: A critical review. *Environmental Toxicology and Chemistry, 36*(12). https://doi.org/10.1002/etc.3924

Qian, X., Song, H., & Ming, G. L. (2019). Brain organoids: Advances, applications and challenges. *Development (Cambridge), 146*(8). https://doi.org/10.1242/dev.166074

Rai, N., Singh, A. K., Singh, S. K., Gaurishankar, B., Kamble, S. C., Mishra, P., Kotiya, D., Barik, S., Atri, N., & Gautam, V. (2020). Recent technological advancements in stem cell research for targeted therapeutics. *Drug Delivery and Translational Research, 10*(4). https://doi.org/10.1007/s13346-020-00766-9

Rajiv, K. (2021). Molecular entities of antimicrobial drugsand resistance mechanism underlying: Bioaccessibility, bioavailability, and bioaccumulation. *International Journal of Bioorganic and Medicinal Chemistry, 1*(1), 17—19.

Rajiv, K. (2021a). Architecting and tailoring of cell repair molecular machinery: Molecule-by-Molecule and atom-by-atom. *Austin Journal of Pharmacology and Therapeutics, 9*(1), 1128—1129.

Rajiv, K. (2021b). Biomedical applications of nanoscale tools and nano-bio interface: A blueprint of physical, chemical, and biochemical cues of cell mechanotransduction machinery. *Biomedical Research and Clinical Reviews, 4*(2), 1—4, 64.

Rajiv, K. (2021c). Cell shrinkage, cytoskeletal pathologies, and neurodegeneration: Myelin sheath formation and remodeling. *Archives of Medical and Clinical Research, 1*(1), 1—5.

Rajiv, K. (2021d). DNA looping initiating types of machinery of transcription, recombination, and replication: An experimental, and theoretical insight. *Global Journal of Medical Research: B Pharma, Drug Discovery, Toxicology & Medicine, 21*(2), 1—4. https://medicalresearchjournal.org/index.php/GJMR/article/view/2449.

Rajiv, K. (2021e). Elucidation of the origin of autoimmune diseases via computational multiscale mechanobiology and extracellular matrix remodeling: Theories and phenomenon of immunodominance. *Current Medical and Drug Research, 5*(1). Art. ID 215 (2021).

Rajiv, K. (2021f). Host-environment interface, host defense, and mast cell: Autoimmunity, allergy, inflammation, and immune response. *JSM Clinical Pharmaceutics, 5*(1), 1—4. https://www.jscimedcentral.com/ClinicalPharmaceutics/clinicalpharmaceutics-5-1018.pdf.

Rajiv, K. (2021g). Macrophage subtypes, phenotypes, inflammatory molecules, cytokines, and atherosclerotic lesions - atherosclerosis, metabolic diseases, and pathogenesis, the therapeutic challenges. *Journal of Clinical & Experimental Pathology, 11*(11S), 1—3.

Rajiv, K. (2021h). Molecular profiling and progression of malignant melanoma: Nanomedicine and immunotherapeutic remedies for diagnosis, treatment, and therapy. *Clinical Oncology, 6*(18846), 1—3.

Rajiv, K. (2021i). Physiology, coagulation cascade: Inherited disorders, and the molecular phenomenon of alterations in hemostasis. *Journal of Clinical Haematology, 2*(2), 62—64.

Rajiv, K. (2021j). Repair of the molecular machinery of the cell at the nanoscale. *Journal of Biomedical Research & Studies, 1*(1), 1—4.

Rajiv, K. (2021k). Traumatic brain injury: Mechanistic insight on pathophysiological mechanisms underlying, neurotransmitters, and potential therapeutic targets. *Medical and Clinical Reviews, 7*(8), 1—3. https://medical-clinical-reviews.imedpub.com/traumatic-brain-injury-mechanisticinsight-on-pathophysiological-mechanismsunderlying-neurotransmitters-andpotential-therapeutic-ta.pdf.

Rajiv, K. (2022). A mechanistic insight on pathophysiological mechanisms of inflammatory diseases and potential therapeutic targets. *Journal of Scientific Research and Biomedical Informatics, 3*(1), 1—4.

Rajiv, K., & Gulia, K. (2021). Cytokine storm and signaling pathways: Pathogenesis of SARS-CoV-2 Infection, managing and treatment strategies. *Biomedical Journal of Scientific & Technical Research, 35*(3), 27754—27758.

Rajiv, K., Gulia, K., Chaudhary, M. P., & Shah, M. A. (2020). SARS-CoV-2, influenza virus and nanoscale particles trapping, tracking and tackling using nanoaperture optical tweezers: A recent advances review. *Journal of Materials NanoScience, 7*(2), 79—92. https://pubs.thesciencein.org/journal/index.php/jmns/article/view/220.

Rajput, A., Sharma, R., & Bharti, R. (2021). Pharmacological activities and toxicities of alkaloids on human health. *Materials Today: Proceedings, 48.* https://doi.org/10.1016/j.matpr.2021.09.189

Rather, M. A., Khan, A., Alshahrani, S., Rashid, H., Qadri, M., Rashid, S., Alsaffar, R. M., Kamal, M. A., & Rehman, M. U. (2021). Inflammation and alzheimer's disease: Mechanisms and therapeutic implications by natural products. *Mediators of Inflammation, 2021* https://doi.org/10.1155/2021/9982954

Rea, R., De Angelis, M. G., & Baschetti, M. G. (2019). Models for facilitated transport membranes: A review. *Membranes, 9*(2). https://doi.org/10.3390/membranes9020026

Reiss, A. L. (2009). Childhood developmental disorders: An academic and clinical convergence point for psychiatry, neurology, psychology and pediatrics. *The Journal of Child Psychology and Psychiatry and Allied Disciplines, 50*(1–2). https://doi.org/10.1111/j.1469-7610.2008.02016.x

Robertson, S. N., Campsie, P., Childs, P. G., Madsen, F., Donnelly, H., Henriquez, F. L., Mackay, W. G., Salmerón-Sánchez, M., Tsimbouri, M. P., Williams, C., Dalby, M. J., & Reid, S. (2018). Control of cell behaviour through nanovibrational stimulation: Nanokicking. *Philosophical Transactions of the Royal Society A: Mathematical, Physical and Engineering Sciences.* https://doi.org/10.1098/rsta.2017.0290

Rouwkema, J., Gibbs, S., Lutolf, M. P., Martin, I., Vunjak-Novakovic, G., & Malda, J. (2011). In vitro platforms for tissue engineering: Implications for basic research and clinical translation. *Journal of Tissue Engineering and Regenerative Medicine, 5*(8). https://doi.org/10.1002/term.414

Sabzpoushan, S. H. (2020). A system biology-based approach for designing combination therapy in cancer precision medicine. *BioMed Research International.* https://doi.org/10.1155/2020/5072697

Salle, V. (2021). Coronavirus-induced autoimmunity. *Clinical Immunology, 226.* https://doi.org/10.1016/j.clim.2021.108694. Academic Press Inc.

Seffernick, J. T., & Lindert, S. (2020). Hybrid methods for combined experimental and computational determination of protein structure. *Journal of Chemical Physics, 153*(24). https://doi.org/10.1063/5.0026025

Shimoni, S., Bar, I., Meledin, V., Derazne, E., Gandelman, G., & George, J. (2016). Circulating endothelial progenitor cells and clinical outcome in patients with aortic stenosis. *PLoS One, 11*(2). https://doi.org/10.1371/journal.pone.0148766

Skaga, E., Kulesskiy, E., Fayzullin, A., Sandberg, C. J., Potdar, S., Kyttälä, A., Langmoen, I. A., Laakso, A., Gaál-Paavola, E., Perola, M., Wennerberg, K., & Vik-Mo, E. O. (2019). Intertumoral heterogeneity in patient-specific drug sensitivities in treatment-naïve glioblastoma. *BMC Cancer, 19*(1). https://doi.org/10.1186/s12885-019-5861-4

Takahashi, K. (2014). Cellular reprogramming. *Cold Spring Harbor Perspectives in Biology, 6*(2). https://doi.org/10.1101/cshperspect.a018606

Takahashi, T. (2019). Organoids for drug discovery and personalized medicine. *Annual Review of Pharmacology and Toxicology, 59.* https://doi.org/10.1146/annurev-pharmtox-010818-021108

Tieu, K. S., Tieu, R. S., Martinez-Agosto, J. A., & Sehl, M. E. (2012). Stem cell niche dynamics: From homeostasis to carcinogenesis. *Stem Cells International.* https://doi.org/10.1155/2012/367567

Tisoncik, J. R., Korth, M. J., Simmons, C. P., Farrar, J., Martin, T. R., & Katze, M. G. (2012). Into the eye of the cytokine storm. *Microbiology and Molecular Biology Reviews, 76*(1), 16–32. https://doi.org/10.1128/mmbr.05015-11

Ul-Haq, Z., & Madura, J. D. (2015). Frontiers in computational chemistry. *Frontiers in Computational Chemistry, 1.* https://doi.org/10.2174/97816810816701170301. Elsevier Inc.

Vives, J., & Batlle-Morera, L. (2020). The challenge of developing human 3D organoids into medicines. *Stem Cell Research and Therapy, 11*(1). https://doi.org/10.1186/s13287-020-1586-1

Wang, X., Neu, C. P., & Pierce, D. M. (2019). Advances toward multiscale computational models of cartilage mechanics and mechanobiology. *Current Opinion in Biomedical Engineering, 11*, 51–57. https://doi.org/10.1016/j.cobme.2019.09.013. Elsevier B.V.

Wang, J., Sun, M., Liu, W., Li, Y., & Li, M. (2019). Stem cell-based therapies for liver diseases: An overview and update. *Tissue Engineering and Regenerative Medicine, 16*(2). https://doi.org/10.1007/s13770-019-00178-y

Wang, G., Xiao, S., Chen, X., & Li, X. (2018). Application of genetic algorithm in automatic train operation. *Wireless Personal Communications, 102*(2). https://doi.org/10.1007/s11277-017-5228-6

Wechman, S. L., Emdad, L., Sarkar, D., Das, S. K., & Fisher, P. B. (2020). Vascular mimicry: Triggers, molecular interactions and in vivo models. *Advances in Cancer Research, 148.* https://doi.org/10.1016/bs.acr.2020.06.001

Xu, X., An, H., Zhang, D., Tao, H., Dou, Y., Li, X., Huang, J., & Zhang, J. (2019). A self-illuminating nanoparticle for inflammation imaging and cancer therapy. *International Journal of Agricultural and Statistical Sciences, 14*(2). https://doi.org/10.1126/sciadv.aat2953

Yang, B. A., Westerhof, T. M., Sabin, K., Merajver, S. D., & Aguilar, C. A. (2021). Engineered tools to study intercellular communication. *Advanced Science, 8*(3). https://doi.org/10.1002/advs.202002825. John Wiley and Sons Inc.

Ye, T., Li, F., Ma, G., & Wei, W. (2021). Enhancing therapeutic performance of personalized cancer vaccine via delivery vectors. *Advanced Drug Delivery Reviews, 177.* https://doi.org/10.1016/j.addr.2021.113927

Yellapantula, V., Hultcrantz, M., Rustad, E. H., Wasserman, E., Londono, D., Cimera, R., Ciardiello, A., Landau, H., Akhlaghi, T., Mailankody, S., Patel, M., Medina-Martinez, J. S., Arango Ossa, J. E., Levine, M. F., Bolli, N., Maura, F., Dogan, A., Papaemmanuil, E., Zhang, Y., & Landgren, O. (2019). Comprehensive detection of recurring genomic abnormalities: A targeted sequencing approach for multiple myeloma. *Blood Cancer Journal, 9*(12). https://doi.org/10.1038/s41408-019-0264-y

Yin, X., Mead, B. E., Safaee, H., Langer, R., Karp, J. M., & Levy, O. (2016). Engineering stem cell organoids. *Cell Stem Cell, 18*(1). https://doi.org/10.1016/j.stem.2015.12.005

Yin, A., Moes, D. J. A. R., van Hasselt, J. G. C., Swen, J. J., & Guchelaar, H. J. (2019). A review of mathematical models for tumor dynamics and treatment resistance evolution of solid tumors. *CPT: Pharmacometrics and Systems Pharmacology, 8*(10). https://doi.org/10.1002/psp4.12450

Zhang, L. (2011). Development and application of cheminformatics approaches to facilitate drug discovery and environmental toxicity assessment [The University of North Carolina at Chapel Hill]. In *ProQuest Dissertations and theses*. https://search.proquest.com/docview/923797093? accountid=8359.

Zhang, C., Jin, M., Zhao, J., Chen, J., & Jin, W. (2020). Organoid models of glioblastoma: Advances, applications and challenges. *American Journal of Cancer Research, 10*(8).

Zhang, M., Liu, Y., & Chen, Y. G. (2020). Generation of 3D human gastrointestinal organoids: Principle and applications. *Cell Regeneration, 9*(1). https://doi.org/10.1186/s13619-020-00040-w

Zhou, Q., & Melton, D. A. (2018). Pancreas regeneration. *Nature, 557*(7705). https://doi.org/10.1038/s41586-018-0088-0

Zorova, L. D., Popkov, V. A., Plotnikov, E. J., Silachev, D. N., Pevzner, I. B., Jankauskas, S. S., Zorov, S. D., Babenko, V. A., & Zorov, D. B. (2018). Functional significance of the mitochondrial membrane potential. *Biochemistry (Moscow) Supplement Series A: Membrane and Cell Biology*. https://doi.org/10.1134/S1990747818010129

Application of genomic and proteomic approaches in stem cell research

Chapter 9

Single-cell transcriptome profiling in unraveling distinct molecular signatures from cancer stem cells

Dibyabhaba Pradhan[1,3] and Usha Agrawal[2]

[1]Indian Biological Data Centre, Regional Centre for Biotechnology, Faridabad, Haryana, India; [2]ICMR-National Institute of Pathology, Safdarjung Hospital Campus, New Delhi, Delhi, India; [3]Department of Bioinformatics, Centralized Core Research Facility, All India Institute of Medical Sciences, New Delhi, Delhi, India

1. Introduction

Despite impressive remissions from cancer achieved through targeted therapy, cancer patients remain at a risk of relapse often due to intratumoral heterogeneity (McGranahan & Swanton, 2015), which has a differential treatment response to distinct tumor subpopulation—the cancer stem cells (CSCs). The rare CSCs are observed to exist in tumors; a subpopulation of them are resistant to therapy and even persist during remission (Magee et al., 2012; Tehranchi et al., 2010; Woll et al., 2014). However, CSCs often could not be characterized in remission cases by large being outnumbered by their normal tissue counterparts (Mustjoki et al., 2013). Therefore, it is imperative that elimination of all CSCs should be sufficient to cure cancer. Nevertheless, the challenge remains in identifying the distinct signature of the CSCs in various cancers. Single-cell gene-expression profiling holds promises in capturing tumor cell heterogeneity, in particular, the distinct gene expression signature in rare CSCs populations (McGranahan & Swanton, 2015; Tehranchi et al., 2010).

2. Cancer stem cells

The characteristics feature of stem cells includes capacity for self-renewal and the ability to differentiate into diverse specialized cell types. Embryonic stem cells (ESCs) can differentiate all tissue during embryonic development, while adult stem cells help in replenishing and repairing adult tissues. CSCs have both stem cell—like and tumorigenic properties, i.e., have the capability of self-renewal and differentiation as well as develop tumors upon transplantation in animal host (Yu et al., 2012). CSCs can initiate tumor, promote drug resistance, and relapse cancer and metastasis (Wen & Tang, 2016). The enhanced capacity of CSCs to repair DNA damage, low levels of reactive oxygen species (ROS), and slow proliferation deems them to be resistant to chemotherapy and radiation therapy (Reya et al., 2001). Therefore, distinguishing CSCs in heterogenous tumor tissue is of paramount importance in cancer therapeutics. CSCs can often be distinguished from other cell types based on cell surface markers (CD44, CD24, and CD133). Further, the gene expression pattern of CSCs can be compared with gene expression pattern of other tumor cell types to delineate CSCs unique gene expression signature in various cancer.

3. Single-cell transcriptomics

Whole transcriptome analysis has been one of the forefront techniques to compare genome-wide gene expression signatures in diverse experimental conditions. Microarray and bulk RNA-seq are primarily used to compare average cellular expressions. These methods have the potentials to identify disease biomarkers and potential drug targets, if the research design does not have the goal of looking for cellular heterogeneity. Bulk RNA-seq in fact has been a good choice for

Copyright © 2024 Elsevier Inc. All rights are reserved, including those for text and data mining, AI training, and similar technologies.

comparative transcriptomics for several disease conditions and drug response studies. However, in tumor microenvironment, the heterogenous cell types such as CSCs have been identified as one of the reasons for cancer relapse (McGranahan & Swanton, 2015). Furthermore, different cancer cell types can have aberrant gene expression pattern, which gets unrecognized in bulk RNA-seq.

Single-cell transcriptomics (scRNA-seq) enables the identification of each cell types, classifying them into subpopulations, and then defines genome-wide transcription for each of the detected "cell types," thus defining cellular heterogeneity (Buettner et al., 2015). It allows the transcriptomes of individual cells to be compared such that transcriptional similarities and differences within cell populations could be determined in a specific tumor sample. In addition, scRNA-seq helps in delineating characteristics of gene expression such as gene splicing patterns, expression of single alleles, and identification of coregulated gene modules.

The growing interest of scientific communities to study cancer heterogeneity at single-cell precision and different cell types has led to development of single-cell transcriptomics profiling raw data acquisition databases as well as powerful bioinformatics analysis tools that could perform preprocessing, quality control, analysis, and downstream analysis. Standard practices in scRNA-seq analysis and their implementations by various research groups have been illustrated herein to demonstrate successful application of scRNA-seq in detecting distinct CSC signature.

3.1 Databases for storing scRNA-seq raw data

High-throughput sequencing raw data including scRNA-seq are deposited in FASTQ format in European Nucleotide Archive (ENA) (ENA Browser, n.d.), NCBI Sequence Read Archive (SRA) (Home - SRA - NCBI, n.d.) and DDBJ Sequence Read Archive (DRA) (Sequence Read Archive, 2023) as per the International Nucleotide Sequence Database Collaboration (INSDC) standard. NCBI Gene Expression Omnibus (GEO) (Barrett et al., 2013) and EBI ArrayExpress (BioStudies, n.d.) are two specific databases that store gene expression count matrix for all kinds of whole transcriptome studies (microarray and RNA-seq). In addition, there are scRNA-seq specific databases developed by Bioinformatics research groups. EBI Single Cell Expression Atlas (Expression Atlas Update: Gene and Protein Expression in Multiple Species | Nucleic Acids Research | Oxford Academic, n.d.), PanglaoDB (Franzén et al., 2019), CancerSEA (Yuan et al., 2019), and Cancer Single-Cell Expression Map (CancerSCEM) (Zeng et al., 2022) are few renowned databases that archive single cell sequencing raw data. CancerSEA and CancerSCEM are the two cancer-specific scRNA databases that store raw data as well as provide functionality on exploring cell types in various functional states. These databases provide a means to visually inspect heterogeneity in gene expression in different cancer cell types.

3.1.1 CancerSEA

CancerSEA is developed by systematically acquiring scRNA data from cancer-related scRNA-seq datasets in human from SRA, GEO, and ArrayExpress. The database collated 41,900 cancer single cells from 25 cancer types. Overall 14 crucial functional states of cancer cells are identified, and their corresponding gene signature was built from literature and known databases (Yuan et al., 2019). A set of 166 gene signature was built to define stemness properties of cancer cell (Pinto et al., 2018). The genes involved in stemness in specific cancer and their correlation (positive/negative) can be explored comprehensively in CancerSEA along with corresponding protein coding gene (PCGs)/lncRNA expression profiles. An example of relevance of CCL5 across 14 functional states shows that the gene has positive correlation for stemness functionality in various cancers (Fig. 9.1).

3.1.2 Cancer single-cell expression map

CancerSCEM consists of 208 cancer samples across 28 studies and 20 human cancer types. The database contains 33 cell types and 638,341 high-quality cells after filtering low-quality cells with abnormal gene expression levels or high mitochondria RNA percentages (Zeng et al., 2022). The database provides gene-level and sample-level analysis functionalities. It uses cell-specific makers to distinguish one cell type from other. Fourteen genes, viz., CD34, ITGA5, PROM1, CD105, VCAM1, CD164, THY1, KIT, ACE, CMAH, ABCG2, CD41, ALDH1A1, and BMI1 are used as markers to detect stem cell—like cells in tumor samples (HSCs, hematopoietic stem cells) (Zeng et al., 2022). A basic analysis report for single-cell sample AML-029-01-1E is demonstrated in Fig. 9.2. The results show presence of HSCs in the tumor samples (Fig. 9.2B). It would be interesting to observe change in HSCs population after patient receives therapy. Further, it would be also imperative to note change in HSCs gene expression pattern after patient receives therapy for a period of time. Single-cell analysis on a set of sample of same cancer type pre- and posttherapy using a standard analysis protocol could help in identifying distinct CSC signature in various cancers.

Apart from CancerSEA and CancerSCEM, three databases known to integrate scRNA-seq data for cancer are given as Table 9.1.

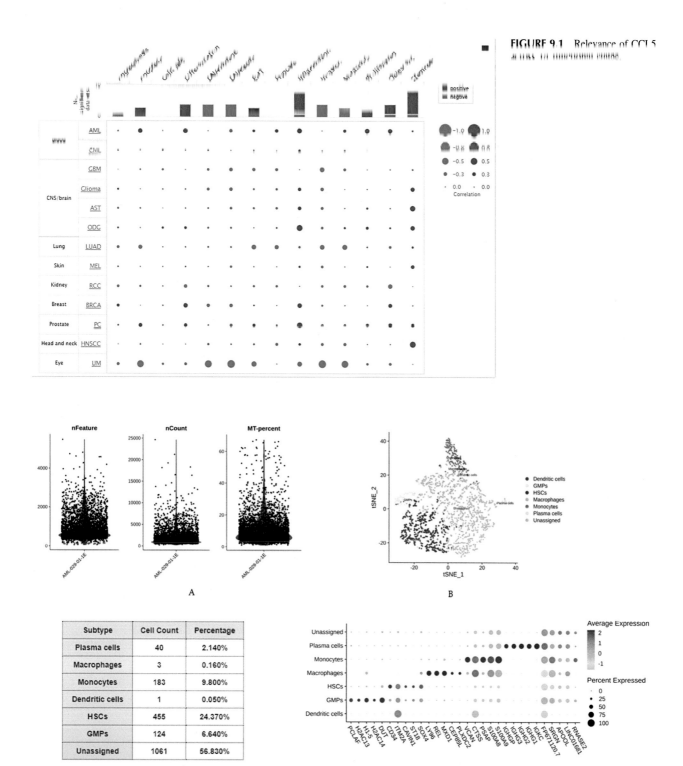

FIGURE 9.2 The single-cell analysis report of sample AML-029-01-1E (A) plot showing features, counts, and mitochondria (MT) percent, (B) unsupervised clustering of 1867 cells (resolution = 1, PCs = 1:20, MT < 10%). (C) cell type, cell count, and percentage, (D) thirty top DEGs in different cell types.

4. Single-cell data analysis workflow

A standard protocol need to be followed to detect stem cell signature in various cancer types. The reads generated from 10X Genomics platform can be well processed and analyzed using Cell Ranger v5.0. The reads generated from Smart-Seq2 or Drop-Seq can be processed with zUMIs v2.9.7 (Parekh, 2017/2023) and alignment using STAR (Dobin, 2014/2023) or HISAT2 (Kim et al., 2019). Once unique molecular identifier (UMI) counts are obtained, cell quality control, principal

TABLE 9.1 Cancer-specific scRNA-seq databases.

Sl. no	Database name	Description	Reference
1	Integrated iMMUnoprofiling of large adaptive CANcer patient cohorts (IMMUcan) scDB (https://immucanscdb.vital-it.ch)	It is created by screening literature databases (PubMed, bioRvix)—from 144 datasets—78 filtered datasets and 56 cancer indications are available to connect cell types and gene expression patterns to specific clinical patterns.	Camps et al. (2023)
2	Tumor Immune Single-hub 2 (TISCH2) (http://tisch.comp-genomics.org/)	It is a scRNA-seq database with 190 datasets and 6,297,320 cells that enable exploring tumor microenvironment (TME) in 27 cancer types.	Han et al. (2023)
3	Deeply Integrated Single-Cell omics data (DISCO) (https://www.immunesinglecell.org/)	DISCO offers a platform to explore published single-cell data and compare with researchers own data. It has subatlas for specific diseases including cancer.	Li et al. (2022)

component analysis (PCA) dimension reduction, t-distributed stochastic neighbor embedding (t-SNE), and uniform manifold approximation and projection (UMAP) clustering and cell type—based clustering—based input marker genes can be achieved by Seurat v3.2.3. The marker gene list can be a list populated from literature search or generated from in-house research or from specific databases/tools, viz., StemMapper, scCancer v2.2.0, CopyKAT v1.0.4, and SingleR v1.4.1 (Pinto et al., 2018). Differential gene expression analysis for CSCs can be performed using FindMarkers' function in Seurat. The downstream analysis such as pathway and gene ontology enrichment, oncogene and tumor suppressor genes detection, receptor genes, ligands, receptor-ligand interactions, and survival analysis can be obtained/performed using specialized databases or tools or bioconductor packages as well as TCGA bulk RNA-seq data (Fig. 9.3).

The single-cell analysis involves steps such as raw data alignment, quality control and normalization, feature selection, dimensionality reduction, and visualization. The alignment of raw reads with reference sequence helps in quantification of UMIs. The quality control of quantified UMIs performed based on counts per barcode, genes per barcode, and fraction of

Single cell transcriptome data analysis

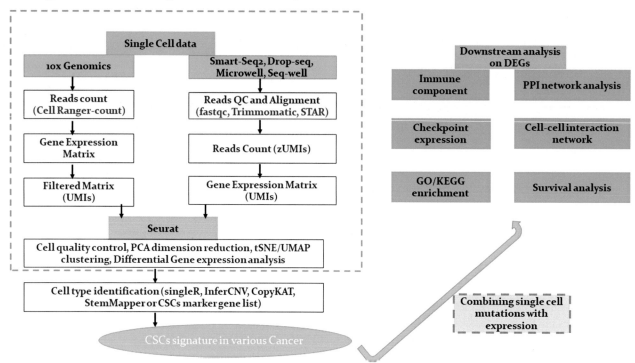

FIGURE 9.3 Overview of analyses for scRNA-seq data.

counts from mitochondrial genes per barcode. Further, the counts are normalized to correct relative gene expression abundances between cells by scaling the count data. The most common normalization approach is count depth scaling, referred to as "counts per million" (CPM). There are several normalization methods, namely, SAMstrt, BASiCS, GRM, scran, SCnorm, and Linnorm, which have been used to reduce noise in scRNA-seq data (Lytal et al., 2020). Upon quality control and normalization, the genes containing useful information, i.e., genes most variable in an identified or clustered cell population are picked. Dimension reduction techniques, viz., PCA, t-SNE, and UMAP, are implemented to cluster counts data into different cell types. The scRNA-seq results are visualized through 2D or 3D clusters, heatmaps, and violin plots.

5. Single-cell studies on detecting distinct molecular signature in cancer

Single-cell transcriptomics with high-sensitivity mutation detection method was applied on 2000 CSCs from patients with chronic myeloid leukemia (CML). It showed a subgroup of CML-CSCs (BCR-ABL + CSCs) with a distinct molecular signature that selectively persisted during prolonged tyrosine kinase inhibitor (TKI) therapy led to frequent relapse upon treatment discontinuation (Giustacchini et al., 2017). Single-cell sequencing of bladder CSCs, and subsequent CRISPR/Cas9-based analysis on ARID1A, GPRC5A, and MLL2 genes revealed mutations on these three genes drives self-renewal property of human bladder CSCs (Yang et al., 2017). In both cases, it was challenging to identify mutations from single-cell data due to low coverage. Therefore, specific marker or genes were chosen evaluating stem cell properties (Table 9.2).

TABLE 9.2 List of publications in single cell for delineating the tissue heterogeneity in cancer.

Sl. no	Title	Description	Reference
1	Single-cell RNA sequencing highlights intratumor heterogeneity and intercellular network featured in adamantinomatous craniopharyngioma	Integrative analysis of 44,038 single-cell transcriptome profiles showed four major eoplastic cell states with distinctive expression signatures in adamantinomatous craniopharyngioma	Jiang et al. (2023)
2	Comprehensive molecular phenotyping of ARID1A-deficient gastric cancer reveals pervasive epigenomic reprogramming and therapeutic opportunities	Tumor microenvironment change due to ARID1A inactivation was observed across GC molecular subtypes through single cell RNA-seq	Xu et al. (2023)
3	Molecular phenotypic linkage between N6-methyladenosine methylation and tumor immune microenvironment in hepatocellular carcinoma	The study validated credible linkage between m^6A methylation pattern and tumor immune microenvironment in hepatocellular carcinoma	Zhang et al. (2023)
4	Transcriptomic intratumor heterogeneity of breast cancer patient-derived organoids may reflect the unique biological features of the tumor of origin	Transcriptomic intratumor heterogeneity in breast cancer patient-derived organoids has been confirmed through this study.	Saeki et al. (2023)
5	Risk modeling of single-cell transcriptomes reveals the heterogeneity of immune infiltration in hepatocellular carcinoma	The study linked conventional prognostic gene signature to the immune microenvironment and cellular heterogeneity at the single-cell level	Wang et al. (2023)
6	Single-cell profiling of γδ hepatosplenic T cell lymphoma unravels tumor cell heterogeneity associated with disease progression	The study revealed heterogenous microenvironment underlying pathogenesis of hepatosplenic T cell lymphoma	Song et al. (2023)
7	Single-cell characterization of myeloma and its precursor conditions reveals transcriptional signatures of early tumorigenesis	The study uncovered early-occurring transcriptional changes in myeloma that remained unnoticed through bulk RNA-seq studies	Boiarsky et al. (2022)
8	Single-cell profiling reveals molecular basis of malignant phenotypes and tumor microenvironments in small bowel adenocarcinomas	The study provided information on distinct tumor microenvironment cells in small bowel adenocarcinomas	Yang et al. (2022)
9	Mesenchymal and adrenergic cell lineage states in neuroblastoma possess distinct immunogenic phenotypes	Distinct immunogenic phenotypes of neuroblastoma have been delineated through the study along with the immunogenic potential of the mesenchymal lineage	Sengupta et al. (2022)
10	The heterogeneous immune landscape between lung adenocarcinoma and squamous carcinoma revealed by single-cell RNA sequencing	The study depicted distinct immune landscape in transcriptome profiles of lung adenocarcinoma and squamous carcinoma	Wang et al. (2022)

6. Future perspectives

Several cancer-specific single-cell raw data repositories along with robust analysis protocols have given novel insights on role of CSCs with respect to cancer relapse. In fact, CSCs markers available in public domain are pretty useful in detecting CSCs population. Further, large-scale study on CSCs in pre- and posttherapeutic interventions in various cancer types would be helpful in understanding clonal heterogeneity in cancer remissions and relapse. In addition, as only a subpopulation of CSCs are often responsible for cancer relapse, single-cell genomics-based mutation detection on same samples would definitely improve our understanding on distinct signature governing tumor reoccurrence.

Authors contribution

UA planned that chapter, DP prepared draft manuscript. UA gave further insight to the manuscript. Both authors read and finalized the manuscript.

References

Barrett, T., Wilhite, S. E., Ledoux, P., Evangelista, C., Kim, I. F., Tomashevsky, M., Marshall, K. A., Phillippy, K. H., Sherman, P. M., Holko, M., Yefanov, A., Lee, H., Zhang, N., Robertson, C. L., Serova, N., Davis, S., & Soboleva, A. (2013). NCBI GEO: Archive for functional genomics data sets—update. *Nucleic Acids Research, 41*(D1), D991–D995. https://doi.org/10.1093/nar/gks1193

BioStudies. (n.d.). BioStudies < The European Bioinformatics Institute < EMBL-EBI. Retrieved February 13, 2023, from https://www.ebi.ac.uk/biostudies/arrayexpress.

Boiarsky, R., Haradhvala, N. J., Alberge, J.-B., Sklavenitis-Pistofidis, R., Mouhieddine, T. H., Zavidij, O., Shih, M.-C., Firer, D., Miller, M., El-Khoury, H., Anand, S. K., Aguet, F., Sontag, D., Ghobrial, I. M., & Getz, G. (2022). Single cell characterization of myeloma and its precursor conditions reveals transcriptional signatures of early tumorigenesis. *Nature Communications, 13*(1), 7040. https://doi.org/10.1038/s41467-022-33944-z

Buettner, F., Natarajan, K. N., Casale, F. P., Proserpio, V., Scialdone, A., Theis, F. J., Teichmann, S. A., Marioni, J. C., & Stegle, O. (2015). Computational analysis of cell-to-cell heterogeneity in single-cell RNA-sequencing data reveals hidden subpopulations of cells. *Nature Biotechnology, 33*(2), 155–160. https://doi.org/10.1038/nbt.3102

Camps, J., Noël, F., Liechti, R., Massenet-Regad, L., Rigade, S., Götz, L., Hoffmann, C., Amblard, E., Saichi, M., Ibrahim, M. M., Pollard, J., Medvedovic, J., Roider, H. G., & Soumelis, V. (2023). Meta-analysis of human cancer single-cell RNA-seq datasets using the IMMUcan database. *Cancer Research, 83*(3), 363–373. https://doi.org/10.1158/0008-5472.CAN-22-0074

Dobin, A. (2023). *STAR 2.7.10b [C]* (Original work published 2014) https://github.com/alexdobin/STAR.

ENA Browser. (n.d.). Retrieved February 13, 2023, from https://www.ebi.ac.uk/ena/browser/home.

Expression Atlas update: Gene and protein expression in multiple species | Nucleic Acids Research | Oxford Academic. (n.d.). Retrieved February 13, 2023, from https://academic.oup.com/nar/article/50/D1/D129/6438036?login=false.

Franzén, O., Gan, L.-M., & Björkegren, J. L. M. (2019). PanglaoDB: A web server for exploration of mouse and human single-cell RNA sequencing data. *Database: The Journal of Biological Databases and Curation*, baz046. https://doi.org/10.1093/database/baz046, 2019.

Giustacchini, A., Thongjuea, S., Barkas, N., Woll, P. S., Povinelli, B. J., Booth, C. A. G., Sopp, P., Norfo, R., Rodriguez-Meira, A., Ashley, N., Jamieson, L., Vyas, P., Anderson, K., Segerstolpe, Å., Qian, H., Olsson-Strömberg, U., Mustjoki, S., Sandberg, R., Jacobsen, S. E. W., & Mead, A. J. (2017). Single-cell transcriptomics uncovers distinct molecular signatures of stem cells in chronic myeloid leukemia. *Nature Medicine, 23*(6), 692–702. https://doi.org/10.1038/nm.4336

Han, Y., Wang, Y., Dong, X., Sun, D., Liu, Z., Yue, J., Wang, H., Li, T., & Wang, C. (2023). TISCH2: Expanded datasets and new tools for single-cell transcriptome analyses of the tumor microenvironment. *Nucleic Acids Research, 51*(D1), D1425–D1431. https://doi.org/10.1093/nar/gkac959

Home—SRA - NCBI. (n.d.). Retrieved February 13, 2023, from https://www.ncbi.nlm.nih.gov/sra.

Jiang, Y., Yang, J., Liang, R., Zan, X., Fan, R., Shan, B., Liu, H., Li, L., Wang, Y., Wu, M., Qi, X., Chen, H., Ren, Q., Liu, Z., Wang, Y., Zhang, J., Zhou, P., Li, Q., Tian, M., … Xu, J. (2023). Single-cell RNA sequencing highlights intratumor heterogeneity and intercellular network featured in adamantinomatous craniopharyngioma. *Science Advances, 9*(15), eadc8933. https://doi.org/10.1126/sciadv.adc8933

Kim, D., Paggi, J. M., Park, C., Bennett, C., & Salzberg, S. L. (2019). Graph-based genome alignment and genotyping with HISAT2 and HISAT-genotype. *Nature Biotechnology, 37*(8), 907–915. https://doi.org/10.1038/s41587-019-0201-4

Li, M., Zhang, X., Ang, K. S., Ling, J., Sethi, R., Lee, N. Y. S., Ginhoux, F., & Chen, J. (2022). Disco: A database of deeply integrated human single-cell omics data. *Nucleic Acids Research, 50*(D1), D596–D602. https://doi.org/10.1093/nar/gkab1020

Lytal, N., Ran, D., & An, L. (2020). Normalization methods on single-cell RNA-seq data: An empirical survey. *Frontiers in Genetics, 11*. https://www.frontiersin.org/articles/10.3389/fgene.2020.00041.

Magee, J. A., Piskounova, E., & Morrison, S. J. (2012). Cancer stem cells: Impact, heterogeneity, and uncertainty. *Cancer Cell, 21*(3), 283–296. https://doi.org/10.1016/j.ccr.2012.03.003

McGranahan, N., & Swanton, C. (2015). Biological and therapeutic impact of intratumor heterogeneity in cancer evolution. *Cancer Cell, 27*(1), 15–26. https://doi.org/10.1016/j.ccell.2014.12.001

Mustjoki, S., Richter, J., Barbany, G., Ehrencrona, H., Fioretos, T., Gedde-Dahl, T., Gjertsen, B. T., Hovland, R., Hernesniemi, S., Josefsen, D., Koskenvesa, P., Dybedal, I., Markevärn, B., Olofsson, T., Olsson-Strömberg, U., Rapakko, K., Thunberg, S., Stenke, L., Simonsson, B., … Nordic

CML Study Group (NCML SG). (2013). Impact of malignant stem cell burden on therapy outcome in newly diagnosed chronic myeloid leukemia patients. *Leukemia, 27*(7), 1320–1326. https://doi.org/10.1038/leu.2013.19

Pacelli, O. (2023). Welcome to LGMA (Original work published 2019) https://github.com/edpm.club/aVMh.

PINH, L. P., Mayhugh, R. S. R, Mauno, R, Hivers, H. V., Mahoney, S., Annabir, R. P., Nayana, L., Thalif, T., & Thishak, M. T. (2018). StemMapper: A curated gene expression database for stem cell lineage analysis. *Nucleic Acids Research, 46*(D1), D788–D793. https://doi.org/10.1093/nar/gkx921

Reya, T., Morrison, S. J., Clarke, M. F., & Weissman, I. L. (2001). Stem cells, cancer, and cancer stem cells. *Nature, 414*(6859), 105–111. https://doi.org/10.1038/35102167

Saeki, S., Kumegawa, K., Takahashi, T., Tang, L., Osako, T., Tada, M., Otsuji, K., Miyata, K., Yamakawa, K., Suzuki, J., Sukimoto, Y., Ozaki, Y., Takano, T., Sano, T., Noda, T., Ohno, S., Yao, R., Ueno, T., & Maruyama, R. (2023). Transcriptomic intratumor heterogeneity of breast cancer patient-derived organoids may reflect the unique biological features of the tumor of origin. *Breast Cancer Research: BCR, 25*(1), 21. https://doi.org/10.1186/s13058-023-01617-4

Sengupta, S., Das, S., Crespo, A. C., Cornel, A. M., Patel, A. G., Mahadevan, N. R., Campisi, M., Ali, A. K., Sharma, B., Rowe, J. H., Huang, H., Debruyne, D. N., Cerda, E. D., Krajewska, M., Dries, R., Chen, M., Zhang, S., Soriano, L., Cohen, M. A., … George, R. E. (2022). Mesenchymal and adrenergic cell lineage states in neuroblastoma possess distinct immunogenic phenotypes. *Nature Canada, 3*(10), 1228–1246. https://doi.org/10.1038/s43018-022-00427-5

Sequence read archive.(February 14, 2023). https://www.ddbj.nig.ac.jp/dra/index-e.html.

Song, W., Zhang, H., Yang, F., Nakahira, K., Wang, C., Shi, K., & Zhang, R. (2023). Single cell profiling of γδ hepatosplenic T-cell lymphoma unravels tumor cell heterogeneity associated with disease progression. *Cellular Oncology, 46*(1), 211–226. https://doi.org/10.1007/s13402-022-00745-x

Tehranchi, R., Woll, P. S., Anderson, K., Buza-Vidas, N., Mizukami, T., Mead, A. J., Astrand-Grundström, I., Strömbeck, B., Horvat, A., Ferry, H., Dhanda, R. S., Hast, R., Rydén, T., Vyas, P., Göhring, G., Schlegelberger, B., Johansson, B., Hellström-Lindberg, E., List, A., … Jacobsen, S. E. W. (2010). Persistent malignant stem cells in del(5q) myelodysplasia in remission. *New England Journal of Medicine, 363*(11), 1025–1037. https://doi.org/10.1056/NEJMoa0912228

Wang, L., Chen, Y., Chen, R., Mao, F., Sun, Z., & Liu, X. (2023). Risk modeling of single-cell transcriptomes reveals the heterogeneity of immune infiltration in hepatocellular carcinoma. *Journal of Biological Chemistry, 299*(3), 102948. https://doi.org/10.1016/j.jbc.2023.102948

Wang, C., Yu, Q., Song, T., Wang, Z., Song, L., Yang, Y., Shao, J., Li, J., Ni, Y., Chao, N., Zhang, L., & Li, W. (2022). The heterogeneous immune landscape between lung adenocarcinoma and squamous carcinoma revealed by single-cell RNA sequencing. *Signal Transduction and Targeted Therapy, 7*(1), 289. https://doi.org/10.1038/s41392-022-01130-8

Wen, L., & Tang, F. (2016). Single-cell sequencing in stem cell biology. *Genome Biology, 17*, 71. https://doi.org/10.1186/s13059-016-0941-0

Woll, P. S., Kjällquist, U., Chowdhury, O., Doolittle, H., Wedge, D. C., Thongjuea, S., Erlandsson, R., Ngara, M., Anderson, K., Deng, Q., Mead, A. J., Stenson, L., Giustacchini, A., Duarte, S., Giannoulatou, E., Taylor, S., Karimi, M., Scharenberg, C., Mortera-Blanco, T., … Jacobsen, S. E. W. (2014). Myelodysplastic syndromes are propagated by rare and distinct human cancer stem cells in vivo. *Cancer Cell, 25*(6), 794–808. https://doi.org/10.1016/j.ccr.2014.03.036

Xu, C., Huang, K. K., Law, J. H., Chua, J. S., Sheng, T., Flores, N. M., Pizzi, M. P., Okabe, A., Tan, A. L. K., Zhu, F., Kumar, V., Lu, X., Benitez, A. M., Lian, B. S. X., Ma, H., Ho, S. W. T., Ramnarayanan, K., Anene-Nzelu, C. G., Razavi-Mohseni, M., … Singapore Gastric Cancer Consortium. (2023). Comprehensive molecular phenotyping of ARID1A-deficient gastric cancer reveals pervasive epigenomic reprogramming and therapeutic opportunities. *Gut, gutjnl-2022-328332.* https://doi.org/10.1136/gutjnl-2022-328332

Yang, Z., Li, C., Fan, Z., Liu, H., Zhang, X., Cai, Z., Xu, L., Luo, J., Huang, Y., He, L., Liu, C., & Wu, S. (2017). Single-cell sequencing reveals variants in ARID1A, GPRC5A and MLL2 driving self-renewal of human bladder cancer stem cells. *European Urology, 71*(1), 8–12. https://doi.org/10.1016/j.eururo.2016.06.025

Yang, J., Zhou, X., Dong, J., Wang, W., Lu, Y., Gao, Y., Zhang, Y., Mao, Y., Gao, J., Wang, W., Li, Q., Gao, S., Wen, L., Fu, W., & Tang, F. (2022). Single-cell profiling reveals molecular basis of malignant phenotypes and tumor microenvironments in small bowel adenocarcinomas. *Cell Discovery, 8*(1), 92. https://doi.org/10.1038/s41421-022-00434-x

Yuan, H., Yan, M., Zhang, G., Liu, W., Deng, C., Liao, G., Xu, L., Luo, T., Yan, H., Long, Z., Shi, A., Zhao, T., Xiao, Y., & Li, X. (2019). CancerSEA: A cancer single-cell state atlas. *Nucleic Acids Research, 47*(Database issue), D900–D908. https://doi.org/10.1093/nar/gky939

Yu, Z., Pestell, T. G., Lisanti, M. P., & Pestell, R. G. (2012). Cancer stem cells. *The International Journal of Biochemistry & Cell Biology, 44*(12), 2144–2151. https://doi.org/10.1016/j.biocel.2012.08.022

Zeng, J., Zhang, Y., Shang, Y., Mai, J., Shi, S., Lu, M., Bu, C., Zhang, Z., Zhang, Z., Li, Y., Du, Z., & Xiao, J. (2022). CancerSCEM: A database of single-cell expression map across various human cancers. *Nucleic Acids Research, 50*(D1), D1147–D1155. https://doi.org/10.1093/nar/gkab905

Zhang, F., Bi, J., Liao, J., Zhong, W., Yu, M., Lu, X., Che, J., Chen, Z., Xu, H., Hu, S., Liu, X., & Guo, S. (2023). Molecular phenotypic linkage between N6-methyladenosine methylation and tumor immune microenvironment in hepatocellular carcinoma. *Journal of Cancer Research and Clinical Oncology.* https://doi.org/10.1007/s00432-023-04589-2

Chapter 10

The single-cell big data analytics: A game changer in bioscience

Sonali Rawat, Yashvi Sharma and Sujata Mohanty

Stem Cell Facility, DBT-Centre of Excellence for Stem Cell Research, All India Institute of Medical Sciences, New Delhi, Delhi, India

1. Introduction

For many years, the methodology in cell biology has remained largely unchanged. Scientists isolate cells and try to maintain them in vitro using their limited understanding of their natural conditions. They then study the cell's behavior under various conditions, changes over time, or responses to manipulation. This is particularly true in stem cell biology, where researchers have worked tirelessly to improve culture conditions and differentiation protocols based on a fragmented understanding of development processes. Despite this, stem cell therapies are already being used in clinics, but the range of cell types used is still limited. Recently, technology has advanced to safely and efficiently produce cell therapies, mainly focused on T cells, hematopoietic stem cells (HSCs), and pluripotent stem cells (PSCs). However, the lack of diversity in cell sources is preventing cell therapy from reaching its full potential and highlights the need for new methods to isolate or generate source cells.

In the past decade, numerous techniques for genome-wide profiling such as RNA sequencing and ChIP-seq have been created and utilized to study comprehensive alterations in gene expression, transcription factor attachment to chromatin, and epigenetic modifications. However, these techniques often require more sample material than is obtainable from a single cell, limiting their usefulness in single-cell studies. Despite the significant progress achieved through these methods, profiling of single cells would be highly advantageous to the field. Since the discovery of mRNA in single cells in 2009, a plethora of techniques for genome-wide profiling of individual cells and their applications have been reported (Tang et al., 2009). The wide array of techniques available for measuring RNA in single cells, such as Smart-seq, CEL-seq, and Quartz-seq, can be attributed to the simplicity of capturing, amplifying, and sequencing mRNA to capture a cellular snapshot of its state. Besides evaluating single-cell gene expression, methods have also been devised to assess DNA variation, chromatin organization, chromatin accessibility, DNA–protein interactions, and DNA methylation in individual cells (Buenrostro et al., 2015; Cusanovich et al., 2015; Hashimshony et al., 2012; Ramsköld et al., 2012; Rotem et al., 2015; Sasagawa et al., 2013; Smallwood et al., 2014a). Stem cell biology, like many other fields, is being influenced by the advent of "big data" Firstly, the application of next-generation sequencing techniques on individual cells is leading to the identification of previously unknown and uncommon cell types, which have the potential for utilization in cell-based therapies. Secondly, data-driven tools can be used to predict changes in cell fate. For example, with the present tools, it is possible to anticipate the transcription factors required to convert human cells from one type to another. The integration of these advancements will create a novel data-oriented approach to the development of cell-based therapies.

As single-cell sequencing techniques have advanced quickly, the computational workflows for analyzing the data have also been improved to manage the increased pace of single-cell RNA sequencing (scRNA-seq) experiments (Andrews et al., 2021). The existing analysis pipeline comprises two primary components: preprocessing, which entails refining the data matrix by eliminating extraneous information via quality control, data correction, normalization, feature selection, dimensionality reduction, and downstream analysis at the cell and gene levels, employed to unearth biological understanding and describe the fundamental biological mechanism (Tugizimana et al., 2016). Computational biologists have devised several methods for each of these stages, each performing effectively in different tasks and circumstances, making it difficult to establish a universal workflow for the analysis of single-cell experiments.

Computational Biology for Stem Cell Research. https://doi.org/10.1016/B978-0-443-13222-3.00002-2
Copyright © 2024 Elsevier Inc. All rights are reserved, including those for text and data mining, AI training, and similar technologies.

115

This chapter of the book will cover the computational workflow for high-throughput data techniques and draw attention to the tools accessible for each stage. Additionally, it will elaborate on the significant technological improvements that have facilitated the analysis of vast amounts of data as an invaluable tool in biomedical research.

2. Techniques for producing single-cell big data in revolutionizing regenerative medicine healthcare

Over the years, multiple techniques for analyzing single cells have emerged (Fig. 10.1). These techniques generate high-throughput data from individual cells that can offer novel insights into stem cell biology and guide the formulation of new regenerative medicine approaches.

2.1 Single-cell RNA sequencing

Since its inception in 2009, scRNA-seq has garnered significant attention (Tang et al., 2009). Its appeal lies in the high-resolution, genome-wide evaluation of RNA molecules in individual cells. Previously, scRNA-seq was mainly utilized by specialized research teams, but it is now becoming more accessible to general researchers and clinicians, unlocking the potential for notable breakthroughs using this potent technique. After the discovery by Tang et al., other techniques were developed, including tag sequencing methods such as STRT, which was described by Islam et al. in 2011 (Islam et al., 2011; Tang et al., 2009). To apply the STRT method, individual cells are put into separate wells in a PCR plate containing lysis buffer. Reverse transcription reagents are then added to generate cDNA. Subsequently, each well is given a distinctive template-switching oligo (TSO) containing a particular sequence on the 3′ end and a universal primer sequence on the 5′. This prompts a process where the TSO and cDNA molecules merge, causing the integration of the sequence present at the TSO's end. The integration of these "barcode" sequences allows for the unbiased RNA sequencing of several cells concurrently via multiplexing (Gu et al., 2021).

The STRT-seq method, however, had a limitation in that only the 5′ end fragment was captured and sequenced after full-length cDNA amplification via template switching. To address this issue, Ramsköld et al. and Picelli et al. introduced the full-length SMART-seq and SMART-seq2 protocols (Picelli et al., 2013; Ramsköld et al., 2012). SMART-seq is a technique that provides better coverage of transcripts compared with traditional tag-based approaches, allowing for detailed analysis of alternative transcript isoforms and the identification of single-nucleotide polymorphisms. Brennecke et al. have also developed an improved version of the SMART-seq protocol called SMART-seq2 (Brennecke et al., 2013; Picelli et al., 2013). The study involved the use of the Fluidigm C1 system to rapidly and efficiently sequence 96 isolated cells in individual compartments without manual intervention. This was made possible by an integrated fluidic circuit method that allowed for passive sequencing. The SPLiT-seq technique, developed by Rosenberg et al., addressed the need for a large number of oligos in scRNA-seq by enabling low-cost transcriptional profiling of hundreds of thousands of fixed cells or nuclei in a single experiment (Carangelo et al., 2022).

There are multiple components involved in the scRNA-seq including single-cell isolation, sequencing procedure, quality assurance along with quality control, data normalization, data assimilation, genetic network construction. All these steps involve various kits and software. Single-cell isolation may involve techniques such as fluorescence-activated cell sorting (FACS) and LCM. Sequencing can be performed using techniques such as SPLiT-seq, STRT-seq, and Drop-seq.

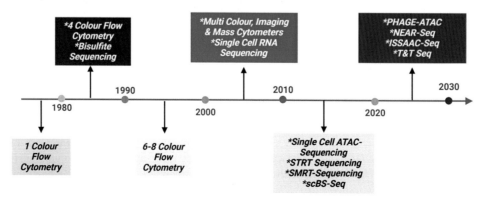

FIGURE 10.1 A representative illustration showing a timeline graph depicting the increasing significance of single cells in biomedical research.

Data normalization maybe done using DESeq2, SCnorm, SAMstrt, etc. Data assimilation may involve various steps for example, data imputation that can be done using technologies such as MAGIC, SAVER, and DrImpute. Several software packages may be involved in gene network creation such as METAMATH, PIDC, etc. to discover the correlations between genes (Chen et al., 2019). Despite facing common challenges such as non–single cell resolution, low sensitivity, premium, and hand-operative processes, scRNA-seq is gaining increasing importance and is expected to become a basic area of research in both exploration and therapies, even though it is still in its early stages.

2.2 Advancements in single-cell ATAC-seq for profiling chromatin accessibility

This method enables the examination of the epigenetic control of gene expression in individual cells. It has the potential to investigate the accessibility of chromatin in stem cells and discover regulatory components that regulate the behavior of stem cells. It was introduced in 2013, and ever since that, it is a widely used tool in chromatin accessibility analysis. This technique involved a bioengineered Tn5 transposase, which creates a staggered nick of nine base pairs in the chromatin and assigns the resulting sequences with high-throughput adaptor sequences. This results in a duplicate sequence of nine base pairs, which is followed by paired-end sequencing. It is an easy and efficient method, which has high specificity and requires a limited number of cells ($\sim$50,000). The primary objective of this method is to detect and describe single cells within a heterogenous cell population. This technique required four simple steps including quality control and alignment before the analysis, followed by core analysis, also called peak calling. This is proceeded by advanced analysis to identify motifs and nucleosomes. Finally, there is an assimilation of the data so generated with data from multiple omics for restructuring regulatory networks.

The first step before the analysis is the prealignment quality check, which checks the quality of the sequences, duplications, nucleotide content, etc. This gives the information to make alignment adjustments to identify duplications or biases. Further with this information, duplicated or low-quality data are removed.

The next step, which is peak calling, involves analysis, which is shape and number of reads based and also identifies transposase bias. This step makes the comparison of specific peaks concerning the background obtained to make specific recommendations. This leads to the analysis based on generated peaks in terms of shape and spread. Now with these data, one can work on the visualization of pathways and enriched genes via an amalgamation of the generated data from ATAC with the data generated from RNA-sequencing (Rosenberg et al., 2018).

Out of the various steps involved in ATAC-seq, a common platform for preanalysis is FastQC, which can help in envisaging the nucleotide content and scoring, duplications, and contamination in the data for assisting the removal of bases of reduced quality. Further, Bowtie2 is a software for alignment of the clarified data. The next major step is the peak calling, which can be done using the MACS2 which is a regular platform for peak calling for channeling of ENCODE ATAC-seq. Several other peak calling platforms could be HOMER, F-seq, and HMMRATAC. Further novel technologies such as sci-CAR, scNMT-seq, and Pi-ATAC have also been introduced for the assessment of transcriptomics, proteomics, and epigenomics from the same cell (Yan et al., 2020).

This method has been extensively utilized for the detection of genetic alterations in cancer cells as well, for example, Nobuchi et al. utilized ATAC for analysis of the resistance to radiotherapy in oral squamous carcinoma cells (Yan et al., 2020). Taavitsainen et al. identified drug resistance relative to chromatic landscapes in the case of prostate cancer using ATAC (Nobuchi et al., 2022). Likewise, this technique can be employed for the detection of heterogeneity in the induced pluripotent stem cells (iPSCs) of the patients to develop personalized therapies, and a lot of research is yet ongoing to mitigate the issues involved in this. Milani et al. devised a protocol for the freezing of human iPSCs−derived neuronal cells for making them suitable for ATAC analysis (Taavitsainen et al., 2021). This approach can also be used to investigate the patterns of chromatin reorganization during the differentiation of stem cells into various cell fates, as well as assess the suitability of particular types of stem cells for diverse clinical uses (Milani et al., 2016).

2.3 Single-cell bisulfite sequencing

This method is regarded as the most reliable approach for analyzing DNA methylation patterns in single cells. It can be employed to examine the epigenetic control of gene expression in stem cells and detect novel regulatory elements. One significant aspect of this method is that it offers quantitative data at the level of individual cells. The main approach to detecting methylation patterns involves the specific deamination of cytosine residues that are unmethylated to uracil by bisulfite salts, as cytosine residues that are methylated are considered to be protected. Several optimizations have been made in this technique, for example, Clark et al., stated that they have optimized the single-cell bisulfite sequencing (scBS-seq) to make it attuned with robotics to parallelly process up to 96 sup toes. scBS-seq requires the creation of a library, which starts with the step of cell lysis. This is followed by the bisulfite salts treatment followed by oligo tagging and amplification to create libraries. This library is then monitored in terms of quality checks and processed for data mining (Finkbeiner et al., 2022). The data input in sc-BS is usually accepted in the FASTA format. Then tools such as FASTAX-Toolkit, PRINTSEQ, etc. are used for processing of data for removal of errors in the raw data. This leads to data alignment using tools including MethylCoder, RMAP, GSNAP, etc., and this finally provides insights on the differentially methylated regions via statistical analyses (Shafi et al., 2018).

This technique can be employed for stem cell analysis as well wherein it can be used to identify the methylation patterns in the early and late passages of stem cells, which might give information about the upregulation and down-regulation of genes linked to the health and potency state of stem cells. It can also be utilized for identifying the variability in methylation patterns of differentiated adult stem cells as compared with undifferentiated stem cells to give insights into the quintessential genes for fate selection. Smallwood et al. utilized this technique for assessing the epigenetic heterogeneity of embryonic stem cells grown in serum media and 2i media and accurately measured about 48% of CpG methylation sites (Buenrostro et al., 2015; Rotem et al., 2015; Smallwood et al., 2014a).

2.4 Fluorescence-activated cell sorting and flow cytometry in stem cell research and analysis

This technique allows for the assessment of the properties of single cells, including the expression of specific markers, through the use of fluorescently labeled antibodies. It also allows for the charge-based separation of specific cell populations based on their properties, such as the expression of specific markers. FCM is utilized for the analysis and identification of the fluorescently tagged cells, while FACS requires a cell sorter that not only identifies the cells but also sorts them based on the charge for further analysis. The key highlight of flow cytometry (FC) is that it quantitatively provides data considering the properties of the single cells. The size distribution of cells can be inferred using this method, which can assist in separating various cell types in a mixed population. For instance, analyzing the size distribution of cells in whole blood can be helpful in distinguishing red blood cells, platelets, lymphocytes, monocytes, and granulocytes, as is done in FC. It can also help in the comparative evaluation of the granularity and internal complexity of the cells, along with measuring the fluorescent intensity when the fluorescently tagged cells are acquired via this instrument. Measurements are made separately on each cell and not as an average of the whole population; thereby, it is an important tool for the analysis of single cells. It majorly comprises three parts—fluidics, optics, and electronics. Fluidics is used for the acquisition of the cells into the cytometer and to concentrate them into a single cell stream so that the focusing of lasers is aligned to each cell. This process is based on the principle of hydrodynamic focusing using sheath fluid. Following this, the fluorescence is recorded via optics and converted into digital signals via electronics for data generation. This technique is widely used in stem cell analysis. For example, the ISHAGE assay is a popular method for quantifying CD34+ hematopoietic stem cells in whole blood. It is necessary for determining the number of stem cells in peripheral blood for hematopoietic stem cell transplants and umbilical cord blood banking (Clark et al., 2017; Smallwood et al., 2014b). Another noteworthy application of FC in stem cells is surface marker profiling, which is needed for the identity characterization of the mesenchymal stem cells being more than 95% positive for markers such as CD105, CD29, CD73, CD29, and HLA-ABC and being less than 2% positive for markers such as HLA-DR, CD34, and CD45. Similarly, it has also been employed for iPSCs identification, and it was found that CD44 was a negative marker in the fibroblast reprogramming of iPSCs (Luo et al., 2018). Therefore the expression of this marker may be checked via FC for iPSCs identification. Liu et al. utilized FACS for the sorting of muscle stem cells (Sutherland et al., 1996). This technique can also be furthered for the clinical release of stem cells and good manufacturing practices (de Morais et al., 2022). Not just stem cells, but also their derivates such as extracellular vesicles (EVs), which are nanoscale vesicles, are also being characterized using flow cytometers (Quintanilla et al., 2014). Flow cytometers are now also employing nanoscale measurement technologies wherein in terms of EV-based FC, diagnostics are utilizing this technique to identify the circulating EVs from different disease scenarios to locate molecular makers of different diseases and cancers (Liu et al., 2015).

There are several databases that offer readily available data for the single-cell analysis such as alona, SCRAT, ASAP, Granatum, iS-CellR PanglaoDB, ICARUS, and WASP (Franzén & Björkegren, 2020; Franzén et al., 2019; Garmire et al., 2021; Jiang et al., 2022; Patel, 2018). These techniques may help in a wide variety of applications, which can lead to the unraveling of information that remains hidden in bulk data processing (Fig. 10.2).

3. Unlocking the potential of stem cells through single-cell data analysis

Ever since scRNA-seq was first introduced, there has been a surge in interest in this novel technology. Its widespread appeal can be attributed, in part, to its capacity to offer a detailed characterization of RNA molecules at the individual cell level, with superior resolution and extensive genome-wide coverage. In the past, scRNA-seq was restricted to specialized research groups, but now it has become more widely available to researchers and clinicians. This has enabled important

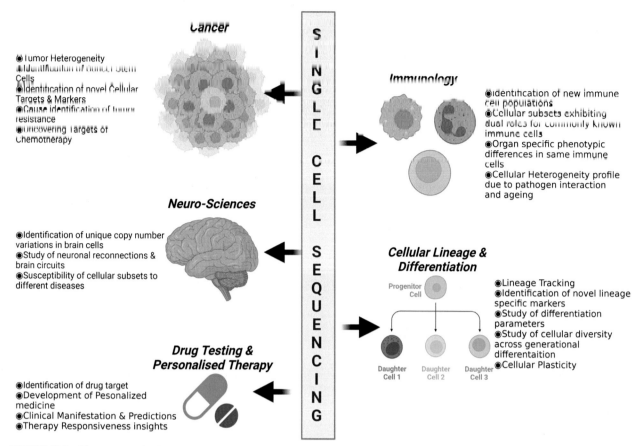

FIGURE 10.2 The representative image demonstrates the progress of single-cell sequencing, highlighting the availability of various databases that offer readily accessible data for single-cell analysis.

discoveries to be made using this powerful technique (Brocco et al., 2019; Camilleri et al., 2016; Haque et al., 2017; Janockova et al., 2021).

scRNA-seq is highly effective for analyzing the transcriptomes of individual cells and is commonly used for assessing the heterogeneity of cell populations. Its primary purpose is to compare transcriptional similarities and differences within cell populations, which is especially valuable for detecting rare cell populations that may have been overlooked by conventional experimental methods. While scRNA-seq was previously limited to specialized research groups, it has become more widely available to researchers and clinicians, leading to significant discoveries using this powerful approach (Fig. 10.3).

Analyzing single-cell big data involves several steps, which can be summarized as follows:

1. **Data preprocessing:** This step involves the processing and cleaning of raw data to remove any technical noise and batch effects. It includes quality control, normalization, and gene filtering.
2. **Dimensionality reduction**: Reducing the dimensionality of single-cell data is crucial for effective visualization and analysis, given the large number of dimensions involved. There are different techniques available to achieve this, including t-distributed stochastic neighbor embedding, principal component analysis, and uniform manifold approximation and projection.
3. **Cell clustering**: After reducing the dimensionality of single-cell data, it becomes feasible to group cells having similar gene expression patterns into clusters. Unsupervised machine learning techniques, such as k-means or hierarchical clustering, can be used to accomplish the grouping process.
4. **Cell type identification:** To determine the cell types present, the gene expression profiles of the cells within each cluster are compared to established gene expression patterns for various cell types.
5. **Differential gene expression analysis**: It is a useful tool for identifying genes that exhibit differential expression between distinct cell types or conditions.

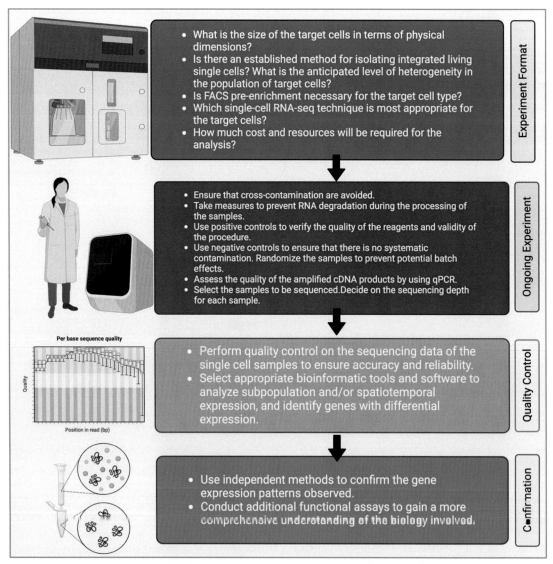

FIGURE 10.3 This diagram illustrates the key stages involved in a typical single-cell sequencing project. The project typically involves four main phases: experimental planning, experimental implementation, bioinformatic analysis, and confirmation of findings. The example given here is a single-cell RNA sequencing project. It is crucial to keep in mind that if any issues arise during the project, researchers need to backtrack to the previous stage to determine the root cause and make appropriate adjustments. In real-world projects, this process may need to be repeated several times.

6. **Trajectory analysis:** Trajectory analysis can be used to model the developmental trajectories of different cell types or to investigate cellular differentiation pathways.
7. **Functional analysis**: Functional analysis can be used to investigate the biological pathways and processes that are enriched in different cell types or conditions.
8. **Integration with other data types**: To gain a more in-depth comprehension of cellular function, single-cell data can be combined with other data types, including epigenomics, proteomics, and spatial transcriptomics data.
9. **Validation:** Finally, the results of the analysis should be validated using independent datasets or experimental validation.

In general, analyzing big data from single cells involves the integration of statistical, computational, and biological knowledge, as well as a comprehension of the underlying biological processes being investigated.

Tian et al. have proposed a new computational technique called scMelody, which is crucial for identifying cell types and heterogeneities in single-cell data through clustering. It utilizes an improved consensus-based clustering model and is capable of clustering single-cell DNA methylation data from various methods, including scBS-seq, scME-seq, scWGBS, scNOMe-seq, scTrio-seq, and snmC-seq. The authors demonstrated that scMelody outperforms other currently available methods in terms of clustering performance and scalability, which was demonstrated by analyzing both real and simulated datasets (Tian et al., 2022). Two instances where scMelody was implemented resulted in the discovery of new cell clusters from human hematopoietic cells and mouse neuron datasets.

Li and colleagues analyzed four scRNA-seq datasets from individuals with upper gastrointestinal cancer (UGIC) to identify a common subcluster of cancer stem cells (CSCs) (Li & Li, 2018). The researchers verified the identity of these cells by comparing them with scRNA-seq datasets from six different types of cancer, which included glioma, melanoma, breast cancer, stellate cell cancer, ovarian cancer, and osteosarcoma. The authors found a distinct population of CSCs shared across four scRNA-seq datasets from patients with UGIC. The UGIC-specific CSCs displayed upregulation of 33 genes and downregulation of 141 genes. These genes were distinct from those found in other cancers examined in the study and were found to be involved in inflammatory and WNT pathways. Even though scRNA-seq technology has accredit gene expression profiling at a single-cell level, analyzing large cohorts of samples remains a challenging task (Stuart & Satija, 2019). In data mining, an important matter that needs to be addressed is the merging of preexisting bulk transcriptomic datasets. Novel discoveries in disease research have been made by combining the strengths of single-cell and bulk approaches. The scientists used single-cell transcriptome data from patients to illustrate the effectiveness of single-cell technology in identifying essential cell types and significant genes, thereby broadening the potential clinical uses of this technology. They predict that by utilizing advanced computational analysis methods on single-cell datasets, more accurate and valuable biomarkers can be identified, resulting in remarkable advancements in the diagnosis and treatment of complex diseases.

4. The promising future of single-cell data in advancing stem cell biology research

Stem cells are distinguished by two properties that are frequently correlated: prolonged self-renewal capability and multilineage prospects. Embryonic development proceeds naturally from the totipotent zygote (capable of generating all types of cell) to pluripotent cells attributing to the three germ layers, which then diversified into tissues preserved by more limited stem and progenitor cells (Vickaryous & Hall, 2006; Shang et al., 2021). Adult stem cells competent of regrowing organs or parts of organs (e.g., hematopoietic, skeletal, gut) were investigated, with the possibility of immediate therapeutic trials (Tang et al., 2010). It quickly became clear that estimating how a cell state will change ("cell fate") under certain conditions is difficult, minimizing the relevance of several prevailing stem cell modification efforts. Amid this advancement, our efforts are still primarily vague and inadequate due to a lack of insight into the intricate working mechanisms. To verify efficiency and safety in the medical setting, we must depend on broad screening strategies, with subsequent improvement in efficiency, precision, and reliability. Complex single-cell assessments are now transforming our understanding of cell types and offer enormous potential for unraveling the molecular circuits that govern cell transformation and cell fate technology.

The future of single-cell data in stem cell science is very promising and holds the potential for many new advances in the field. Here are a few ways in which single-cell data are expected to impact stem cell biology in the future:

4.1 Improved understanding of stem cell behavior

The embryonic development process is still largely unknown to scientists, particularly in regard to when the blastomere undergoes changes in gene regulation and cell fate. Before the advent of scRNA-seq, conducting experiments on a limited number of cells in stem cell biology was difficult. Researchers had proposed that the differentiation of blastomeres began at the 8- or 16-cell stage, but there was no molecular evidence to support this hypothesis. The emergence of scRNA-seq has provided researchers with unique opportunities to examine the gene expression changes that occur during the differentiation of blastomeres. In vitro models of embryonic stem cells are regarded as the most effective means of investigating the self-renewal and differentiation potential of PSC. Through scRNA-seq, it has been possible to trace the derivation of mouse and human embryonic stem cells from the inner cell mass of the blastocyst under appropriate pluripotency-supporting conditions (Oostdyk et al., 2019; Tang et al., 2010). A research study employed scRNA-seq to examine the gene expression of more than 80,000 individual cells from 31 human brain organoids. This approach allowed the researchers to identify the expression of different cell markers in distinct cell clusters. By comparing their findings with various gene expression datasets from mice and humans, the researchers were able to identify characteristic genes of various cell types, including dopaminergic neurons, glutamatergic and gamma-aminobutyric neurons, and retinal cells such as photoreceptor müller glial, amacrine, bipolar, and retinal pigment epithelium cells (Quadrato et al., 2017). Treutlein et al. conducted a study that showcased how scRNA-seq can be utilized to identify novel stem cell types. They used Smart-seq to sequence 198 alveolar epithelial cells from mice to comprehend the cell type and hierarchy of the developing lung. Even though alveolar type 1 (AT1) and alveolar type 2 (AT2) cells are epithelial cell types, the characteristics of alveolar progenitor cells were previously unclear (Brazovskaja et al., 2019; Treutlein et al., 2014). Brunskill et al. employed Smart-seq to analyze 235 cells from mouse kidneys and establish gene expression patterns at various developmental stages of the kidney. Using single-cell microarray and Smart-seq, they detected the coexpression of Foxd1 and Six2 in single cells of metanephric mesenchyme, which are markers of stromal-committed cells and nephron-committed cells, respectively. This finding was subsequently validated through protein expression analysis (Brunskill et al., 2014).

4.2 Development of new therapies

The state of a cell can be considered as a vast collection of features, including more than 20,000 genes that encode proteins and RNAs, as well as various other molecules such as metabolites and lipids. Extracellular situations, which include cell neighbors, extracellular matrix interrelations, endocrine signaling, and many other factors, also influence cell state. Because we are still finding new facets of cell biology, it is difficult to list all of the elements. To anticipate and modify cell fates, we must first clearly describe cell state, which necessitates a broad coverage of various parameters, which is now feasible with the advancement of highly parallel single-cell metrics (multiomics). By combining living cells and materials, it is possible to manipulate cells in a more precise manner, with a focus on single-cell applications. This integration allows for the creation of surface-functionalized single cells that can perform new, nonnatural functions with great significance for various applications. These technological goals are crucial due to the growing global population, which is causing issues such as increased energy and environmental pollution, biomass consumption, and health problems. Currently, scRNA-seq data are the most widely used and established high-throughput technology for single-cell analysis (Denyer et al., 2019). However, to derive valuable insights for biological and medical applications, it is important to have advanced algorithms and software in the field of bioinformatics. These tools are necessary to accurately convert the raw scRNA-seq data into useful information and knowledge. Due to the stochastic nature of current scRNA-seq methods and the restricted amount of RNA in individual cells, it is not practical to capture and sequence all the messenger RNAs (mRNAs) present in a single cell (Hwang et al., 2018). Thus, to ensure the quality of scRNA-seq data is crucial in the analysis, and typical quality control measures include assessing the number of expressed genes and the percentage of mitochondrial genes. The primary goal of quality control is to identify and remove cells that are empty, doublets, or have poor-quality data. However, it is worth noting that the choice of quality control metrics may differ depending on the biological sample being studied, and there are no universally accepted thresholds for these metrics (Hong et al., 2022).

Cancer poses the greatest threat to human life due to its ability to replicate limitlessly, invade tissues, metastasize, and evade the immune system. The variety of cancer cells is a significant challenge in achieving complete cancer remission, and it is a fundamental aspect of nearly all human cancers. This diversity arises from chance genetic and epigenetic alterations in cancer cells from the same patient. The use of advanced technology has allowed scientists to develop a more comprehensive understanding of the molecular processes involved in the progression of cancer. This is particularly true when studying the range of different tumor cells and immune cells involved. As a result, scRNA-seq has become an important tool for exploring the characteristics and variety of both cancer and immune cells (Marusyk et al., 2012) (Fig. 10.4).

Xu and colleagues employed scRNA-seq to scrutinize the transcriptome profile of 96,796 individual cells obtained from 15 pairs of primary tumors and axillary lymph nodes. This examination resulted in the identification of nine subgroups of cancer cells (Xu et al., 2020). Analysis of the transcriptome data suggests that NECTIN2-TIGIT interactions between metastatic cancer cells and other cells in the tumor microenvironment play a role in promoting lymph node metastasis in breast cancer patients. Besides revealing the mechanisms behind therapy resistance, disease recurrence, gene mutations, migration, and invasion, the use of scRNA-seq technology has also facilitated research on gene regulatory networks in various forms of cancer (Serrano-Ron et al., 2021). It is essential to utilize scRNA-seq technology to conduct detailed analyses of tumors at the individual cell level due to the intricate and varied nature of the tumor microenvironment.

4.3 Better understanding of disease states

In cancerous tissues, it is typical to discover subsets of cells with different phenotypes and functional capabilities. The CSC theory posits that CSCs are a highly tumorigenic subset of cells that are responsible for many crucial biological properties of tumors, such as drug resistance, disease recurrence, self-renewal, tumor initiation, and metastasis. Despite significant research, there is still a contentious discussion about whether the present comprehension of CSCs is applicable to all types of tumors. By detecting very small nucleic acid sequences, scRNA-seq can help identify these cells and potentially lead to new discoveries and insights (Wen & Tang, 2016). Yang et al. utilized scRNA-seq technology to explore the genetic

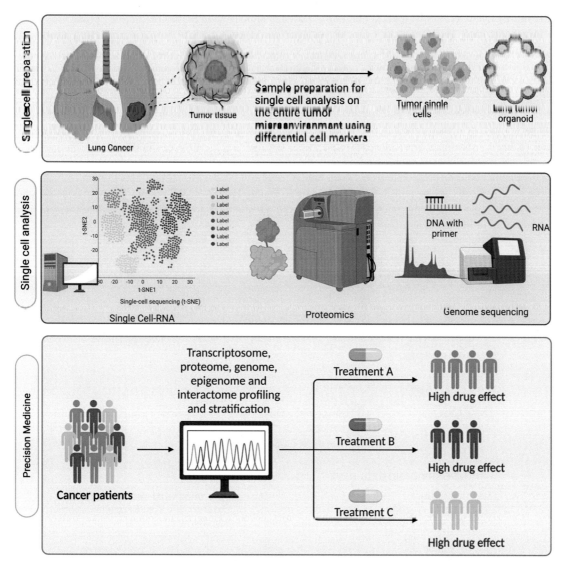

FIGURE 10.4   The representative illustration is a depiction of how personalized cancer treatment can be accomplished by utilizing single-cell technologies and databases. In other words, the illustration demonstrates how single-cell methods and information repositories can be leveraged to tailor cancer therapies to the unique characteristics of individual patients.

composition and source of bladder CSCs, which have not been fully understood. They conducted their analysis by studying 59 cells from three bladder cancer samples, including bladder CSCs, nonstem cells, bladder epithelial stem cells, and bladder epithelial nonstem cells. The results showed that bladder CSCs were clonally homogenous and likely derived from bladder epithelial stem cells or bladder epithelial nonstem cells, as revealed by phylogenetic analysis. Furthermore, the researchers discovered 21 essential genes responsible for modifications in bladder CSCs, among which were six genes that were previously unknown (PAWR, RGS9BP, MKL1, GPRC5A, PITX2, and ETS1) (Yang et al., 2017). The simultaneous presence of mutations in GPRC5A, ARID1A, and MLL2 genes was observed to amplify the self-renewal and tumor initiation capacities of bladder cancer nonstem cells, making them akin to stem cells. These results indicate the potential for developing new targeted therapies for bladder cancer. Gao and coworkers conducted an investigation in which they analyzed 1000 individual cells from 12 patients with triple-negative breast cancer (TNBC). They utilized a model of single-cell copy number alteration to gain insight into tumor development, diagnosis, and treatment. The team discovered that the most frequent tumor-derived indication in all cells was the basal-like 1 proliferation signal, which was associated with luminal progenitor-like cells that were actively dividing (Gao et al., 2016).

Karaayvaz et al. discovered that there are distinct cell subpopulations that are biologically identified by activation of glycosphingolipid metabolism and correlated with innate immunity processes, as evidenced by Convergence of Gene

Expression Profiles. They found that glycosphingolipid pathway signals were highly predictive of diagnosis and treatment outcomes in patients with TNBC. Furthermore, the researchers identified a number of potential therapeutic targets for TNBC, including S1PR1, which was found to be highly expressed in tumors (Karaayvaz et al., 2018).

Additionally, while single-cell analysis is a state-of-the-art technology, it still has several drawbacks. Generating individual cell samples, carrying out genome-wide amplification, sequencing the samples, and analyzing the resulting data all pose difficulties, including low coverage, biases, and errors that may differ depending on the sequencing platform and methodology employed. Hence, it is essential to enhance the procedures for preparing the samples, patterns of whole-genome amplification, sequence analysis, and data analysis algorithms to make progress in the field. There are also fundamental biological questions yet to be answered in the field of CSCs, such as identifying the driving genes behind cancer cell onset and development, understanding how these driving factors change during the occurrence of cancer cells, identifying the distinctions and relationships between varied range of CSCs within a single malignant cell, and determining which cell or group of cells are responsible for the tumor's origins. However, with technological advancements and the decreasing costs of single sequencing, these challenges can be adequately addressed, and the scRNA-seq innovation will ultimately contribute to human health (Fig. 10.5).

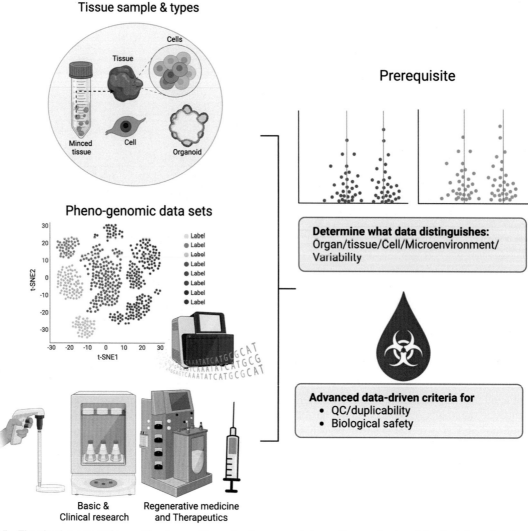

FIGURE 10.5 The visual representations outline the specific needs of cancer and stem cell research, shaped by the information they yield. Put differently, the distinct characteristics of stem cell studies are defined by the insights and data obtained from cancer and stem cell research.

4.4 Improved tissue engineering

Using biomaterial systems, it has become possible to create complex tissue behaviors in vitro that were not achievable using conventional two-dimensional culture systems. This opens up new opportunities for investigating both normal and disease-related developmental processes. We suggest that to determine the appropriate design parameters for biomaterial systems, it is necessary to identify the molecular mechanisms that underlie the observed emergent behaviors. This can be achieved by using advanced technologies in systems biology, including single-cell omics, genetic engineering, and high-content imaging (Cahan et al., 2021). The field of systems tissue engineering brings together the fields of tissue engineering and systems biology to explore the underlying mechanisms behind complex tissue behaviors. This approach takes into account the dynamic regulation and dysregulation of tissue development, the presence of single-cell heterogeneity and rare cell types, and the spatial distribution and structure of individual cells and cell types within tissues. By considering these factors, systems tissue engineering can aid in identifying the key factors contributing to intricate tissue behaviors (Hockemeyer & Jaenisch, 2016). The fusion of biological and materials data science in systems tissue engineering can expedite the identification of biomaterial design parameters, resulting in more effective basic scientific discoveries and applications. Cutting-edge molecular techniques are being employed to study native in vivo tissues, enabling the measurement of thousands or even millions of factors in one experiment. scRNA-seq is a technique that has revealed significant molecular phenotype heterogeneity among cells that are considered to be of the same conventional type, even within the same tissue compartment (Park et al., 2020). The use of advanced molecular tools has resulted in the generation of large datasets, which can reveal interconnected responses of spatially coordinated biological systems, rare cell behaviors within complex systems, and the dynamic nature of response networks. Engineered 3D hydrogels and microsystems can be used to facilitate the analysis of dynamics and to identify the main factors that determine tissue function. When cells are moved from 2D tissue culture plastic to 3D culture, particularly in systems that can be remodeled, their environment undergoes dynamic changes in terms of geometry, structure, and composition (Kang et al., 2021; Napolitano et al., 2007). scRNA-seq has uncovered molecular diversity within cells of the same type in a tissue compartment, but this is not enough to understand the complex cellular functions that determine tissue development and disease. To create predictive models and go beyond descriptive research, there is a need to connect systems-level molecular phenotypes with the underlying dynamic microenvironment that controls cell and tissue functions.

A systematic experimental approach is crucial to improve biomaterial microsystems and understand emergent tissue behaviors. This should include dynamic, spatial coordination, and rare cell factors that contribute to tissue functions. The use of available tools for systems-level analysis can lead to innovations in biomaterial platforms and integration with computational algorithms to identify design parameters, accelerating the pace of translation and improving understanding of biomaterial function. The future of single-cell data in stem cell biology is promising, and improvements in single-cell data analysis techniques are anticipated to enhance our knowledge of stem cell behavior and inform the development of new regenerative medicine approaches.

5. Conclusion

In summary, the future of stem cell biology will require the use of big data–driven techniques. Stem cell therapy has made significant progress over the past 60 years, but it has struggled to achieve widespread adoption of various cell types. The advent of single-cell biology has made this problem even more challenging. However, with the help of new data-driven tools, rare cell types with therapeutic potential are being discovered. Unfortunately, our current approaches may not be sufficient to create or maintain these cells, so we must turn to data-driven tools for a solution. By doing so, we can expand the range of indications in which existing cell therapies show promise.

Acknowledgments
The authors would like to acknowledge the Biorender software for figures.

References

Andrews, T. S., Kiselev, V. Y., McCarthy, D., & Hemberg, M. (2021). Tutorial: Guidelines for the computational analysis of single-cell RNA sequencing data. *Nature Protocols, 16*(1), 1–9.

Brazovskaja, A., Treutlein, B., & Camp, J. G. (2019). High-throughput single-cell transcriptomics on organoids. *Current Opinion in Biotechnology, 55*, 167–171.

Brennecke, P., Anders, S., Kim, J. K., Kołodziejczyk, A. A., Zhang, X., Proserpio, V., ... Heisler, M. G. (2013). Accounting for technical noise in single-cell RNA-seq experiments. *Nature Methods, 10*(11), 1093–1095.

Brocco, D., Lanuti, P., Simeone, P., Bologna, G., Pieragostino, D., Cufaro, M. C., Graziano, V., Peri, M., Di Marino, P., De Tursi, M., & Grassadonia, A. (November 18, 2019). Circulating cancer stem cell-derived extracellular vesicles as a novel biomarker for clinical outcome evaluation. *Journal of oncology, 2019.*

Brunskill, E. W., Park, J. S., Chung, E., Chen, F., Magella, B., & Potter, S. S. (2014). Single cell dissection of early kidney development: Multilineage priming. *Development, 141*(15), 3093−3101.

Buenrostro, J. D., Wu, B., Chang, H. Y., & Greenleaf, W. J. (2015). ATAC-seq: A method for assaying chromatin accessibility genome-wide. *Current Protocols in Molecular Biology, 109*(1), 21−29.

Cahan, P., Li, H., Morris, S. A., Da Rocha, E. L., Daley, G. Q., & Collins, J. J. (2014). CellNet: Network biology applied to stem cell engineering. *Cell, 158*(4), 903−915.

Camilleri, E. T., Gustafson, M. P., Dudakovic, A., Riester, S. M., Garces, C. G., Paradise, C. R., Takai, H., Karperien, M., Cool, S., Sampen, H. J., & Larson, A. N. (December 2016). Identification and validation of multiple cell surface markers of clinical-grade adipose-derived mesenchymal stromal cells as novel release criteria for good manufacturing practice-compliant production. *Stem Cell Research & Therapy, 7*(1), 1−6.

Carangelo, G., Magi, A., & Semeraro, R. (2022). From multitude to singularity: An up-to-date overview of scRNA-seq data generation and analysis. *Frontiers in Genetics, 13*, 994069.

Chen, G., Ning, B., & Shi, T. (2019). Single-cell RNA-seq technologies and related computational data analysis. *Frontiers in Genetics, 10*, 317.

Clark, S. J., Smallwood, S. A., Lee, H. J., Krueger, F., Reik, W., & Kelsey, G. (March 2017). Genome-wide base-resolution mapping of DNA methylation in single cells using single-cell bisulfite sequencing (scBS-seq). *Nature Protocols, 12*(3), 534−547.

Cusanovich, D. A., Daza, R., Adey, A., Pliner, H. A., Christiansen, L., Gunderson, K. L., ... Shendure, J. (2015). Multiplex single-cell profiling of chromatin accessibility by combinatorial cellular indexing. *Science, 348*(6237), 910−914.

Denyer, T., Ma, X., Klesen, S., Scacchi, E., Nieselt, K., & Timmermans, M. C. (2019). Spatiotemporal developmental trajectories in the Arabidopsis root revealed using high-throughput single-cell RNA sequencing. *Developmental Cell, 48*(6), 840−852.

Finkbeiner, C., Ortuño-Lizarán, I., Sridhar, A., Hooper, M., Petter, S., & Reh, T. A. (January 25, 2022). Single-cell ATAC-seq of fetal human retina and stem-cell-derived retinal organoids shows changing chromatin landscapes during cell fate acquisition. *Cell Reports, 38*(4), 110294.

Franzén, O., & Björkegren, J. L. (2020). alona: a web server for single-cell RNA-seq analysis. *Bioinformatics, 36*(12), 3910−3912.

Franzén, O., Gan, L. M., & Björkegren, J. L. (2019). PanglaoDB: A web server for exploration of mouse and human single-cell RNA sequencing data. *Database, 2019.*

Gao, R., Davis, A., McDonald, T. O., Sei, E., Shi, X., Wang, Y., ... Navin, N. E. (2016). Punctuated copy number evolution and clonal stasis in triple-negative breast cancer. *Nature Genetics, 48*(10), 1119−1130.

Garmire, D. G., Zhu, X., Mantravadi, A., Huang, Q., Yunits, B., Liu, Y., ... Garmire, L. X. (2021). GranatumX: A community-engaging, modularized, and flexible webtool for single-cell data analysis. *Genomics, Proteomics & Bioinformatics, 19*(3), 452−460.

Gu, H., Raman, A. T., Wang, X., Gaiti, F., Chaligne, R., Mohammad, A. W., ... Gnirke, A. (2021). Smart-RRBS for single-cell methylome and transcriptome analysis. *Nature Protocols, 16*(8), 4004−4030.

Haque, A., Engel, J., Teichmann, S. A., & Lönnberg, T. (2017). A practical guide to single-cell RNA-sequencing for biomedical research and clinical applications. *Genome Medicine, 9*(1), 1−12.

Hashimshony, T., Wagner, F., Sher, N., & Yanai, I. (2012). CEL-seq: Single-cell RNA-seq by multiplexed linear amplification. *Cell Reports, 2*(3), 666−673.

Hockemeyer, D., & Jaenisch, R. (2016). Induced pluripotent stem cells meet genome editing. *Cell Stem Cell, 18*(5), 573−586.

Hong, R., Koga, Y., Bandyadka, S., Leshchyk, A., Wang, Y., Akavoor, V., ... Campbell, J. D. (2022). Comprehensive generation, visualization, and reporting of quality control metrics for single-cell RNA sequencing data. *Nature Communications, 13*(1), 1688.

Hwang, B., Lee, J. H., & Bang, D. (2018). Single-cell RNA sequencing technologies and bioinformatics pipelines. *Experimental & Molecular Medicine, 50*(8), 1−14.

Islam, S., Kjällquist, U., Moliner, A., Zajac, P., Fan, J. B., Lönnerberg, P., & Linnarsson, S. (2011). Characterization of the single-cell transcriptional landscape by highly multiplex RNA-seq. *Genome Research, 21*(7), 1160−1167.

Janockova, J., Matejova, J., Moravek, M., Homolova, L., Slovinska, L., Nagyova, A., Rak, D., Sedlak, M., Harvanova, D., Spakova, T., & Rosocha, J. (January 2021). Small extracellular vesicles derived from human chorionic MSCs as modern perspective towards cell-free therapy. *International Journal of Molecular Sciences, 22*(24), 13581.

Jiang, A., Lehnert, K., You, L., & Snell, R. G. (2022). ICARUS, an interactive web server for single cell RNA-seq analysis. *Nucleic Acids Research, 50*(W1), W427−W433.

Kang, S. M., Kim, D., Lee, J. H., Takayama, S., & Park, J. Y. (2021). Engineered microsystems for spheroid and organoid studies. *Advanced Healthcare Materials, 10*(2), 2001284.

Karaayvaz, M., Cristea, S., Gillespie, S. M., Patel, A. P., Mylvaganam, R., Luo, C. C., ... Ellisen, L. W. (2018). Unravelling subclonal heterogeneity and aggressive disease states in TNBC through single-cell RNA-seq. *Nature Communications, 9*(1), 3588.

Li, W. V., & Li, J. J. (2018). An accurate and robust imputation method scImpute for single-cell RNA-seq data. *Nature Communications, 9*(1), 997.

Liu, L., Cheung, T. H., Charville, G. W., & Rando, T. A. (October 2015). Isolation of skeletal muscle stem cells by fluorescence-activated cell sorting. *Nature Protocols, 10*(10), 1612−1624.

Luo, Y., He, J., Xu, X., Sun, M. A., Wu, X., Lu, X., & Xie, H. (March 21, 2018). Integrative single-cell omics analyses reveal epigenetic heterogeneity in mouse embryonic stem cells. *PLoS Computational Biology, 14*(3), e1006034.

Marusyk, A., Almendro, V., & Polyak, K. (2012). Intra-tumour heterogeneity: A looking glass for cancer? *Nature Reviews Cancer, 12*(5), 323−334.

Milani, P., Escalante-Chong, R., Shelley, B. C., Patel-Murray, N. L., Xin, X., Adam, M., Mandefro, B., Sareen, D., Svendsen, C. N., & Fraenkel, E. (May 5, 2016). Cell freezing protocol suitable for ATAC-Seq on motor neurons derived from human induced pluripotent stem cells. *Scientific Reports, 6*(1).

de Morais, P. D., Cristina, C., Dias Alves Pinto, J., Wagner de Souza, K., Izu, M., Fernando da Silva Bouzas, L., & Henrique Paraguassú-Braga, F. (March 21, 2022). Validation of the single platform ISHAGE protocol for enumeration of CD34+ hematopoietic stem cells in mobilized cord blood in a Brazilian center. *Hematology, Transfusion and Cell Therapy, 44*, 45–55.

Napolitano, A. P., Chai, P., Dean, D. M., & Morgan, J. R. (2007). Dynamics of the self-assembly of complex cellular aggregates on micromolded nonadhesive hydrogels. *Tissue Engineering, 13*(8), 2087–2094.

Nobuchi, T., Ouito, T., Kawamatsu, A., Kurwaki, K., Nozaki, R., Kase, Y., Iroda, M., Ouito, M., Uno, T., & Uzawa, K. (March 15, 2022). Assay for transposase accessible chromatin with high throughput sequencing reveals radioresistance related genes in oral squamous cell carcinoma cells. *Biochemical and Biophysical Research Communications, 597*, 115–121.

Oostdyk, L. T., Shank, L., Jividen, K., Dworak, N., Sherman, N. E., & Paschal, B. M. (2019). Towards improving proximity labeling by the biotin ligase BirA. *Methods, 157*, 66–79.

Park, D., Lee, J., Chung, J. J., Jung, Y., & Kim, S. H. (2020). Integrating organs-on-chips: Multiplexing, scaling, vascularization, and innervation. *Trends in Biotechnology, 38*(1), 99–112.

Patel, M. V. (2018). iS-CellR: a user-friendly tool for analyzing and visualizing single-cell RNA sequencing data. *Bioinformatics, 34*(24), 4305–4306.

Picelli, S., Björklund, Å. K., Faridani, O. R., Sagasser, S., Winberg, G., & Sandberg, R. (2013). Smart-seq2 for sensitive full-length transcriptome profiling in single cells. *Nature Methods, 10*(11), 1096–1098.

Quadrato, G., Nguyen, T., Macosko, E. Z., Sherwood, J. L., Min Yang, S., Berger, D. R., ... Arlotta, P. (2017). Cell diversity and network dynamics in photosensitive human brain organoids. *Nature, 545*(7652), 48–53.

Quintanilla, R. H., Jr., Asprer, J. S., Vaz, C., Tanavde, V., & Lakshmipathy, U. (January 9, 2014). CD44 is a negative cell surface marker for pluripotent stem cell identification during human fibroblast reprogramming. *PLoS One, 9*(1), e85419.

Ramsköld, D., Luo, S., Wang, Y. C., Li, R., Deng, Q., Faridani, O. R., ... Sandberg, R. (2012). Full-length mRNA-Seq from single-cell levels of RNA and individual circulating tumor cells. *Nature Biotechnology, 30*(8), 777–782.

Rosenberg, A. B., Roco, C. M., Muscat, R. A., Kuchina, A., Sample, P., Yao, Z., ... Seelig, G. (2018). Single-cell profiling of the developing mouse brain and spinal cord with split-pool barcoding. *Science, 360*(6385), 176–182.

Rotem, A., Ram, O., Shoresh, N., Sperling, R. A., Goren, A., Weitz, D. A., & Bernstein, B. E. (2015). Single-cell ChIP-seq reveals cell subpopulations defined by chromatin state. *Nature Biotechnology, 33*(11), 1165–1172.

Sasagawa, Y., Nikaido, I., Hayashi, T., Danno, H., Uno, K. D., Imai, T., & Ueda, H. R. (2013). Quartz-seq: A highly reproducible and sensitive single-cell RNA sequencing method, reveals non-genetic gene-expression heterogeneity. *Genome Biology, 14*(4), 1–17.

Serrano-Ron, L., Cabrera, J., Perez-Garcia, P., & Moreno-Risueno, M. A. (2021). Unraveling root development through single-cell omics and reconstruction of gene regulatory networks. *Frontiers in Plant Science, 12*, 661361.

Shafi, A., Mitrea, C., Nguyen, T., & Draghici, S. (2018). A survey of the approaches for identifying differential methylation using bisulfite sequencing data. *Briefings in Bioinformatics, 19*(5), 737–753.

Shang, F., Yu, Y., Liu, S., Ming, L., Zhang, Y., Zhou, Z., ... Jin, Y. (2021). Advancing application of mesenchymal stem cell-based bone tissue regeneration. *Bioactive Materials, 6*(3), 666–683.

Smallwood, S. A., Lee, H. J., Angermueller, C., Krueger, F., Saadeh, H., Peat, J., Andrews, S. R., Stegle, O., Reik, W., & Kelsey, G. (2014b). Single-cell genome-wide bisulfite sequencing for assessing epigenetic heterogeneity. *Nature Methods, 11*(8), 817–820.

Smallwood, S. A., Lee, H. J., Angermueller, C., Krueger, F., Saadeh, H., Peat, J., ... Kelsey, G. (2014a). Single-cell genome-wide bisulfite sequencing for assessing epigenetic heterogeneity. *Nature Methods, 11*(8), 817–820.

Stuart, T., & Satija, R. (2019). Integrative single-cell analysis. *Nature Reviews Genetics, 20*(5), 257–272.

Sutherland, D. R., Anderson, L., Keeney, M., Nayar, R., & Chin-Yee, I. A. (June 1996). The ISHAGE guidelines for CD34+ cell determination by flow cytometry. *Journal of hematotherapy, 5*(3), 213–226.

Taavitsainen, S., Engedal, N., Cao, S., Handle, F., Erickson, A., Prekovic, S., Wetterskog, D., Tolonen, T., Vuorinen, E. M., Kiviaho, A., & Nätkin, R. (September 6, 2021). Single-cell ATAC and RNA sequencing reveal pre-existing and persistent cells associated with prostate cancer relapse. *Nature Communications, 12*(1), 5307.

Tang, F., Barbacioru, C., Bao, S., Lee, C., Nordman, E., Wang, X., ... Surani, M. A. (2010). Tracing the derivation of embryonic stem cells from the inner cell mass by single-cell RNA-Seq analysis. *Cell Stem Cell, 6*(5), 468–478.

Tang, F., Barbacioru, C., Wang, Y., Nordman, E., Lee, C., Xu, N., ... Surani, M. A. (2009). mRNA-Seq whole-transcriptome analysis of a single cell. *Nature Methods, 6*(5), 377–382.

Tian, Q., Zou, J., Tang, J., Liang, L., Cao, X., & Fan, S. (2022). scMelody: An enhanced consensus-based clustering model for single-cell methylation data by reconstructing cell-to-cell similarity. *Frontiers in Bioengineering and Biotechnology, 10*.

Treutlein, B., Brownfield, D. G., Wu, A. R., Neff, N. F., Mantalas, G. L., Espinoza, F. H., ... Quake, S. R. (2014). Reconstructing lineage hierarchies of the distal lung epithelium using single-cell RNA-seq. *Nature, 509*(7500), 371–375.

Tugizimana, F., Steenkamp, P. A., Piater, L. A., & Dubery, I. A. (2016). A conversation on data mining strategies in LC-MS untargeted metabolomics: Pre-processing and pre-treatment steps. *Metabolites, 6*(4), 40.

Vickaryous, M. K., & Hall, B. K. (2006). Human cell type diversity, evolution, development, and classification with special reference to cells derived from the neural crest. *Biological Reviews, 81*(3), 425–455.

Wen, L., & Tang, F. (2016). Single-cell sequencing in stem cell biology. *Genome Biology, 17*(1), 1–12.

Xu, G., Liu, Y., Li, H., Liu, L., Zhang, S., & Zhang, Z. (2020). Dissecting the human immune system with single cell RNA sequencing technology. *Journal of Leukocyte Biology, 107*(4), 613–623.

Yang, Z., Li, C., Fan, Z., Liu, H., Zhang, X., Cai, Z., … Wu, S. (2017). Single-cell sequencing reveals variants in ARID1A, GPRC5A and MLL2 driving self-renewal of human bladder cancer stem cells. *European Urology, 71*(1), 8–12.

Yan, F., Powell, D. R., Curtis, D. J., & Wong, N. C. (2020). From reads to insight: A hitchhiker's guide to ATAC-seq data analysis. *Genome Biology, 21*, 1–16.

Chapter 11

Unravelling the genomics and proteomics aspects of the stemness phenotype in stem cells

Sorra Sandhya[1], Kaushik Kumar Bharadwaj[1], Joyeeta Talukdar[2] and Debabrat Baishya[1]

[1]Department of Bioengineering and Technology, Gauhati University, Guwahati, Assam, India; [2]Department of Biochemistry, All India Institute of Medical Sciences (AIIMS), New Delhi, Delhi, India

1. Introduction

Stem cells (SCs) are pluripotent cells with a cellular phenotype, i.e., the ability to self-renew and differentiate into multiple adult cell types in response to extracellular signals (Orkin, 2011). Through extracellular signaling networks that balance protooncogenes (which encourage self-renewal), gate-keeping tumour suppressors (which limit self-renewal), and caring tumor suppressors (which protect genomic integrity), SCs maintain self-renewal (He et al., 2009). In response to homeostasis and tissue repair, SCs differentiate through signaling networks that regulate gene expression (Das et al., 2020). However, significant heterogeneity has been observed in the gene expression level of SCs, which has made it harder to comprehend the fundamentals of stem cell (SC) biology (Hough et al., 2014). Hence, it is very challenging to identify which gene or protein signaling networks account specifically for SCs' self-renewal or differentiation at the molecular level.

Advanced SC research allows the in vitro production of high-quality clinical-grade SCs for cell-mediated therapy or translational research. However, their quantitative similarity or differences to their primary cell types remain largely unresolved. Basically, adult stem cells (ASCs) and embryonic stem cells (ESCs) are two main categories of SCs. Both ASCs and ESCs have different sets of assets and disadvantages in translational research and regenerative therapy. ASCs-mediated therapy such as bone marrow (BM)—originated hematopoietic stem cells (HSCs) transplants has been used to treat leukemia since the first successful BM transplant was done in 1957 in a patient with acute leukemia (Thomas et al., 1957). ASCs-mediated therapy is somehow successful because of its less ethical concern, the feasibility of autologous transplantation, which results in no immunological rejection, the capability for differentiation into a certain specific lineage of cells, and a lower chance of tumour formation. Thus, ASCs are acknowledged as the gold standard in cell-mediated therapy (Aly, 2020; Prentice, 2019). So far, more than 3000 clinical trials using ASCs have been registered in the WHO (World Health Organization) International Clinical Trials Registry, which include several blood-related disorders, malignancies, immune system issues, spinal cord injuries, brain trauma, stroke, Parkinson's disease, juvenile diabetes, blindness, and so forth (Aly, 2020).

On the other hand, the application of ESCs-mediated therapy in human trials to cure disease may require more extensive study along with special safety precautions. Due to the hidden dangers of genetic aberration, tumour development, immunological rejection of transplanted tissue, difficulty in regenerating homogenous cell populations after differentiation, and ethical debate surrounding the death of embryos for research, ESCs-mediated therapy is controversial (Ahn et al., 2010; Draper et al., 2004; Maitra et al., 2005; Rebuzzini et al., 2015; Vogel, 2005). Despite the controversy, scientists are optimistic that human-derived ESCs (hESCs) hold promise for cell-mediated therapies in the future, as well as an in vitro ESCs-based platform to screen various drugs (Ilic & Ogilvie, 2017). Significant advancements have been made in SCs research over a span of nearly two decades since the first hESCs were obtained (Ilic & Ogilvie, 2017; Thomson et al., 1998). So far, hESCs have been used as ESCs-mediated therapy for various diseases and injuries, such as Parkinson's disease, spinal cord injury, intrauterine adhesion, age-related mascular degeneration, autoimmune, cardiovascular, and neurological diseases, and type 1 diabetes (Cubillo et al., 2018; Ilic & Ogilvie, 2017; Yamanaka, 2020). Current

Computational Biology for Stem Cell Research. https://doi.org/10.1016/B978-0-443-13222-3.00028-9
Copyright © 2024 Elsevier Inc. All rights reserved, including those for text and data mining, AI training, and similar technologies.

excitement over SC findings is the use of ESCs such as human-induced pluripotent stem cells (hiPSCs) in translational medicine after the initial generation of iPSCs in vitro by inducing OCT3/4, SOX2, KIF4, and MYC transcriptional factors in mouse foetal and adult fibroblasts (Takahashi et al., 2007; Takahashi & Yamanaka, 2006; Yu et al., 2007). Since then, other organizations have worked to develop iPSCs as an iPSCs-mediated therapy, and some of them have already advanced to the clinical trials stage for a number of illnesses such as Parkinson's disease, muscular degeneration, patient transfusion, graft versus host disease, etc. (Yamanaka, 2020). Although iPSCs have the potential for unlimited proliferation, ongoing practical concerns such as their inherent tumorigenicity, immunogenicity, and heterogeneity limit their clinical application (Ortuno-Costela et al., 2019; Yamanaka, 2020).

Therefore, it is important to first understand the basic fundamental mechanisms involved in the cellular phenotype of SCs and then utilize their functional aspects. To overcome these issues, scientists initiated the Multi-Omics Platform (MOP) to investigate an in-depth phenotypic study of SCs by underlying extracellular signaling networks based on gene/DNA (Genomics), RNA (Transcriptomics), protein (Proteomics), and functional level (Metabolomics) (Fig. 11.1A). This MOP reveals high-throughput genome, transcriptome, and proteome data profiling of SCs regulation, which may identify possible therapeutic usage of SCs. Thus, in this chapter, we summarize the techniques used to characterize the genomic and proteomic analysis of SCs. And then we intended a systematic genomics and proteomics analysis to understand the stemness phenotype in SCs, especially in ESCs to accelerate clinical applications of ESCs as ESCs-mediated therapy in the future. Our findings contribute to the mechanistic insight into regulatory signaling networks that underlie the in vitro maintenance of ESCs pluripotency and self-renewal to prevent differentiation for further therapeutic application.

2. Genomic and proteomic aspects to understand stem cell phenotype

The "whole genome" of an organism's DNA can be studied by the genomic approach, which reveals a complete set of genetic information. The human "whole genome" is estimated to be between 20,000 and 25,000 genes, according to the International Human Genome Sequencing Consortium (Roth, 2019). The proteomics approach, on the other hand, entails the identification and measurement of every protein present in a cell, tissue, or organism (Aslam et al., 2017; Yu et al., 2010). The most recent statistics from the neXtProt database, which were published on August 18, 2022, revealed that 20,359 proteins exist in the human body. A total of 90.41% of the data correspond to PE1 (Experimental evidence at the protein level), 5.57% to PE2 (Experimental evidence at the transcript level), 0.69% to PE3 (Protein inferred by homology), 0.06% to PE4 (Protein predicted), and 2.99% to PE5 (Protein uncertain). However, there are now 1557 proteins without a function annotated and 1343 missing proteins. Thus, enhancing our understanding of the global gene and protein expression profiles of SCs will have significant implications for the therapeutic application of SCs-mediated therapy (Hipp et al., 2010). Therefore, in this section, we have presented a comprehensive table summarising the different genomics and proteomics types based on their analytical findings and applications (Fig. 11.2). Then, in order to characterize SCs at the genome and proteome levels, we discussed in more detail some popular genomic and proteomic methods/tools. Based on types of genomic and proteomic analysis, one can potentially analyze the structural, functional, comparative, differential, and mutational status of genes as well as proteins involved in SC phenotype (Fig. 11.2).

2.1 Genomic platform to characterize SCs phenotype

The main purpose of characterizing the gene regulation profile of SCs is to monitor the genomic integrity of the cells and track the expression of proteins associated with pluripotency. To date, various genomic tools have been developed to study SCs phenotype. One of the most common platforms for evaluating the gene expression profile of SCs is Affymetrix Gene Chip technology or high-density synthetic oligonucleotide microarray (Hipp et al., 2010; Lipshutz et al., 1999; Lockhart et al., 1996). Approximately one million 25 base nucleotide sequences make up this tiny platform, which enables quantitative study of 47,000 transcripts and variations, including 38,500 levels of human gene expression (HG-U133 Plus 2.0, www.affymetrix.com) (Hipp et al., 2010). Other techniques, such as quantitative hybridization of a high-density cDNA microarray (Hipp et al., 2010; Yang & Speed, 2002), serial analysis of gene expression (SAGE) (Rao & Stice, 2004; Velculescu et al., 1995), subtractive hybridization (SH) (Duguid et al., 1988; Rao & Stice, 2004; Wu et al., 1994) and representational difference analysis (RDA) (Lisitsyn et al., 1993; Rao & Stice, 2004) techniques, have been utilized for greater understanding of SCs regulation as well as characterizing the SCs' gene expression profile. Oligonucleotide and cDNA microarray technologies have emerged as leading methods for analysing global gene expression patterns between different populations. Despite their distinct strengths, both technologies offer high-throughput data, making them valuable tools in this field. These techniques were also effectively used to clarify the cellular and molecular mechanisms behind the

A.

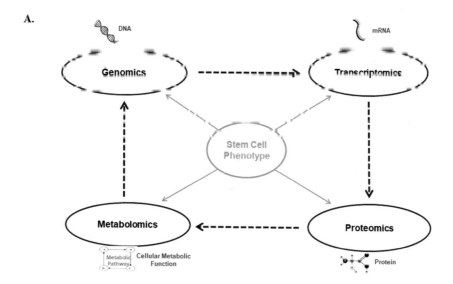

B.   *Evolution of Genomic Platform*

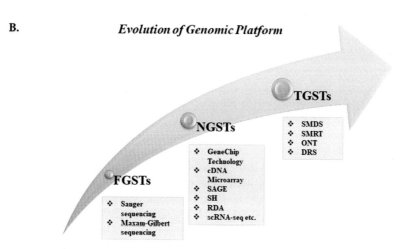

C.   *Analytical Proteomic platforms*

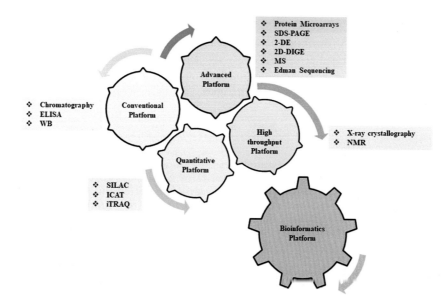

132

initial differentiation process and maintenance of pluripotency in human ESCs (Rao & Stice, 2004), as well as to analyze gene expression profiling of various SCs such as Mesenchymal SCs (MSCs), hematopoietic SCs (HSCs) etc (Table 11.1).

However, the variation and cellular heterogeneity in SC regulatory gene expression levels led to the development of next-generation sequencing technologies (NGSTs), also known as single-cell sequencing or second-generation sequencing analysis, that demonstrate single-cell genome analysis using ultra-high speed sequencing (Bode et al., 2021; Chen et al., 2020; Hough et al., 2014; Hwang et al., 2018; Wen & Tang, 2016). Hence, the development of NGSTs ends the era of first-generation sequencing technologies (FGSTs) dominated by Sanger sequencing methods (Sanger & Coulson, 1975; Wen & Tang, 2022) and Maxam-Gilbert sequencing (Kchouk et al., 2017) (Fig. 11.1B). Next-generation sequencing (NGS) is an enormously parallel or extensive sequencing technology employed for DNA and RNA analysis at the molecular level that has revolutionized the fields of genomics and clinical research (Behjati & Tarpey, 2013; Liu et al., 2017; Sandhya Verma, 2021). The NGS technologies can evaluate hundreds of thousands to billions of 25−800 nucleotide-long sequences within days in a low-cost manner compared to Sanger sequencing (Wen & Tang, 2022). The NGS-based tools also include single-cell genome (to analyze DNA sequencing at the single-cell level), epigenome (to determine cell identity and function with chromatin organization and characterize regulatory elements), and transcriptome (screening of Single-cell RNA-sequencing [scRNA-seq] technologies). These techniques serve as powerful tools to investigate cellular heterogeneity within a "homogeneous" SC population and are useful to identify distinct phenotypic cell types (Bode et al., 2021; Holliday, 2006; Hwang et al., 2018; Navin et al., 2011). Using these technologies, encouraging discoveries have been made in the SC fields, which include an in-depth study of pluripotent stem cells (PSCs) such as ESCs, tissue-derived SCs, and cancer stem cells (CSCs), a subtype of cancer cells (Chen et al., 2020; Hwang et al., 2018). For example, the derivation of human and mouse ESCs from the blastocyst's inner cell mass has been traced by scRNA-seq analysis (Chen et al., 2020; Mantsoki et al., 2016; Tang et al., 2010; Yan et al., 2013). The scRNA-seq analysis also reveals three cell-lineages such as pluripotent epiblast cells, extra-embryonic ectoderm cells, and primitive endoderm cells of the human blastocyst (Blakeley et al., 2015). Recent studies reported that a distinct dynamic gene expression behavior of sex chromosomes during early human embryogenesis (Moreira de Mello et al., 2017; Zhou et al., 2019) and different patterns of gene expression in human preimplantation embryos' glucose metabolism have been observed by scRNA-seq analysis (Zhao et al., 2019). Thus, NGSTs play an important role in evaluating fundamental insights into ESCs phenotype.

Furthermore, despite recent advancements in the NGS platform, third-generation sequencing technologies (TGSTs) have advanced SCs research (Fig. 11.1B). For example, by using third-generation sequencing (TGS)−based long-read nanopore sequencing technology, researchers have shed light on the biological process of microhomology-mediated break-induced replication in human PSCs with copy number variations (CNVs) of chromosome 20q11.21 (Halliwell et al., 2021). The TGS platform includes single-molecule DNA sequencing (SMDS), nanopore direct RNA sequencing (DRS), Oxford nanopore technology (ONT), and single-molecule real-time (SMRT) sequencing technologies (Mehdi Kchouk et al., 2017; Ozsolak, 2012; Wen & Tang, 2022). However, the utilization of these TGSTs in SC research is not well documented yet.

Additionally, web-based tools such as Stembase and StemChecker online databases are also very worthwhile for reanalyzing the data generated by scientists for further exploration (Table 11.2). The Stembase database consists of microarrays on SCs and their derivatives, which is useful to search for novel SC markers and genes that have unique functions in SCs (Porter et al., 2007). The StemChecker is a collection of stemness signatures based on transcription factor gene sets that are useful to discover and explore stemness signatures in gene sets (Pinto et al., 2015). Thus, our findings describe the basic technologies of the genomic platforms that contributed to SC phenotype analysis and how these platforms evolved from FGSTs to TGSTs in the past two decades.

FIGURE 11.1   Schematic diagram representing the Multi-Omics Platform (MOP) to understand stem cell phenotypes along with analytical techniques used as Genomic and Proteomic platforms. (A) The image represents interlinked MOP analysis, which includes genomics (gene/DNA level), transcriptomics (RNA level), proteomics (protein level), and metabolomics (Functional Level) analysis, will provide high-throughput genome, transcriptome, and proteome data profiling of SCs regulation that may reveal snapshots of various aspects of the molecular and cellular phenotype of the SCs. (B) the image represents the evaluation of genomic platforms from First-generation sequencing technologies (FGSTs) to Next-generation sequencing technologies (NGSTs) and then NGSTs to Third-generation sequencing technologies (TGSTs) based on analytical techniques utilized to analyse genomic integrity of SCs in the last 2 decades. (C) The image shows the total 5 analytical proteomic platforms, such as the conventional, advanced, quantitative, high-throughput and bioinformatics platforms divided based on techniques utilized to track protein expression of SCs. Importantly, a bioinformatics platform has been established to handle and store an enormous quantity of proteomic data collected from the other four platforms. This Bioinformatic platform is also useful for predicting the 3D structure of a protein, rapid analysis of protein−protein interactions, domain and motif analysis, mass spectroscopy data analysis, and investigating evolutionary relationships (Perez-Riverol et al., 2015; Vihinen, 2001).

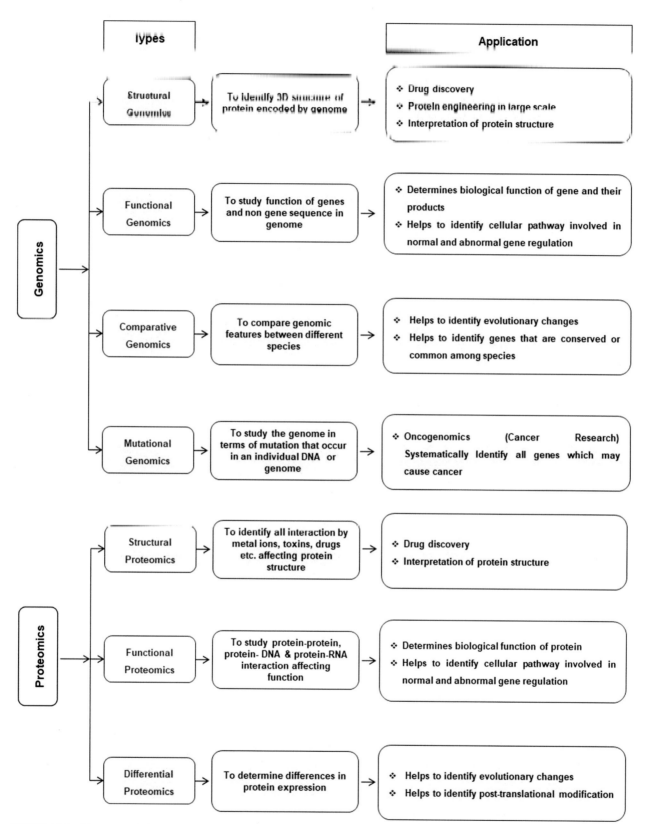

FIGURE 11.2   Diagrammatic flowchart showing the different types of Genomics and Proteomics and their applications. A genomic study will offer an analysis of the gene's structural, functional, comparative, and mutational aspects. Proteomics studies, on the other hand, interpret protein structural, functional, and differential proteomic analyses. Both the genomics and proteomics analysis of genes and proteins are very useful for drug discovery, the interpretation of protein structure, the identification of evolutionary changes, etc.

TABLE 11.1 An overview of the genetic techniques/tools used to identify distinct gene expression in particular stem cell populations.

Sl.no.	Techniques/tools	Utilization of the technique	Stem cell population
1.	Affymetrix gene chip technology	Utilized to assess the quantitative, parallel measurements from gene expression profiling and SC DNA sequencing (Hipp et al., 2010; Lipshutz et al., 1999; Lockhart et al., 1996)	hESCs (Gao et al., 2013; Rao & Stice, 2004; Suarez-Farinas et al., 2005), mESCs (Glover et al., 2006), MSCs (Kubo et al., 2009), DP-MSCs (Kim et al., 2011), BM-MSCs (Kim et al., 2011), WJ-MSCs (Gao et al., 2013)
2.	Quantitative hybridization of a high-density cDNA microarray	Used to identify genes that are differentially expressed between two separate cell populations and provide high-throughput data to compare global gene expression patterns between different populations (Hipp et al., 2010; Yang & Speed, 2002)	hESCs (Kim et al., 2006), hHSCS (Kim et al., 2006), hMSCs (Brendel et al., 2005; Kim et al., 2006), DP-MSCs (Kang et al., 2016; Lee et al., 2022), PL-MSCs (Lee et al., 2023), UC-MSCs (Kang et al., 2016; Lee et al., 2023), stem cell lines (Piscaglia et al., 2007)
3.	Serial analysis of gene expression	Used to obtain access to a deeper understanding of SCs regulation and the quantitative and simultaneous analysis of many transcripts (Rao & Stice, 2004; Velculescu et al., 1995)	hESCs (Hirst et al., 2007; Richards et al., 2006), BM-MSCs (Silva et al., 2003)
4.	Subtractive hybridization	Utilized for isolating single-stranded cDNA from nanogram quantities of double-stranded cDNA (Duguid et al., 1988; Rao & Stice, 2004; Wu et al., 1994)	ESCs (Ryan et al., 2001; Tanaka et al., 2002), M-HSCs (Huttmann et al., 2006; Terskikh et al., 2003)
5.	Representational difference analysis	Used to distinguish between probes that check for polymorphisms between two people and probes linked to viral genomes present in human DNA as single copies (Lisitsyn et al., 1993; Rao & Stice, 2004)	HSCs (Degar et al., 2001)
6.	Single-cell RNA-sequencing technologies	To investigate cellular heterogeneity within a "homogeneous" stem cell population and are useful to identify distinct phenotypic cell types (Bode et al., 2021; Holliday, 2006; Hwang et al., 2018; Navin et al., 2011)	hESCs and mESCs (Biase et al., 2014; Chen et al., 2020; Hwang et al., 2018; Mantsoki et al., 2016; Tang et al., 2010; Yan et al., 2013), PSCs (Shi et al., 2015; Wen & Tang, 2016)
7.	Oxford nanopore technology	Used to determine the nucleotide in a DNA sequence so that structural genomic variations and repetition content can be resolved more effectively (Mehdi Kchouk et al., 2017; Wen & Tang, 2022)	PSCs (Halliwell et al., 2021)
8.	Single-molecule DNA sequencing	Used to analyze unmanipulated, natural DNA molecules in a massively parallel manner (Ozsolak, 2012)	—
9.	Nanopore direct RNA sequencing	Used to analyze order of nucleotides in continuous native RNA strands (Jain et al., 2022; Ozsolak, 2012)	ESCs (Rao & Stice, 2004), HSCs (Zeng et al., 2019)
10.	Single-molecule real-time sequencing	Used to analyze long-read lengths (exceeding several kilobases) of repetitive regions of complex genomes (Mehdi Kchouk et al., 2017; Ozsolak, 2012; Wen & Tang, 2022)	—

bone marrow (BM); dental pulp (DP); embryonic stem cells (ESCs); hematopoietic stem cells (HSCs); human (h); mesenchymal stem cells (MSCs); mouse (m); murine (M); periodontal ligament (PL); pluripotent stem cells (PSCs); umbilical cord (UC); Wharton's jelly (WJ).

TABLE 11.2 Summary of web based genomic techniques/tools used to characterize stemness phenotype in stem cells.

Sl. no.	Techniques/tools	Utilization of the technique
1.	Stembase database	Helpful for finding novel stem cell markers and genes with specific roles in stem cell regulation (Porter et al., 2007)
2.	StemChecker database	Very useful to discover and explore stemness signatures in gene sets (Pinto et al., 2015)
3.	Microarray and weighted gene correlation analysiss bioinformatics analysis.	Helpful for locating gene clusters with high correlation and for locating potential biomarkers or treatment targets (Langfelder & Horvath, 2008)

2.2 Proteomic platform to characterize stem cell phenotype

Genomic integrity, such as variation and cellular heterogeneity of the SCs phenotype, can be determined by interrogating gene expression levels in a single cell through the genomic platform (Mantsoki et al., 2016). However, coupling protein with gene expression data and tracking the expression of proteins associated with the pluripotency of SCs phenotype is also necessary for understanding how SCs behavior is regulated (van Hoof et al., 2012). Hence, the proteomic analysis of SCs should be considered to comprehend the basic functional aspects of genes involved in SC phenotype at the protein level. Importantly, the proteomics platform considers the most significant and applicable dataset to characterize a biological system (Aslam et al., 2017). Proteomic analysis is much more complicated than genomic analysis since proteins are effector molecules that initiate biological function and their expression levels depend on corresponding mRNA levels (Aslam et al., 2017; Lander et al., 2001).

Proteomic analysis of SCs can be evolved by using five analytical proteomic platforms (Fig. 11.1C and Table 11.3). First, the conventional or traditional platform uses chromatography-based techniques such as ion exchange chromatography, affinity chromatography, gas chromatography, enzyme-linked immunosorbent assay (ELISA), and western blot (WB) (Hage et al., 2012; Jungbauer & Hahn, 2009; Munoz et al., 2014). This platform is designed for the extraction, purification, and identification of proteins (Aslam et al., 2017). For instance, employing lectin affinity chromatography of glycopeptides, Gerardo and his colleague discovered 119 lectin-bound glycopeptides as possible protein glycobiomarkers of ESCs. The combined datasets from ESCs and embryoid bodies (EBs) have also shown 293 distinct N-linked glycopeptide sequences (from 180 glycoproteins) (Alvarez-Manilla et al., 2010). Similarly, Saifun et al. identified 417 proteins from human adipose tissue–derived mesenchymal SCs (ADSCs) and 417 proteins from mouse ADSCs by utilizing liquid chromatography (LC) with tandem mass spectrometry (Nahar et al., 2018). Liyuan et al. (Zhang et al., 2012) utilized precolumn derivatization high-performance LC (HPLC) with fluorescence detection to determine amino acids such as leucine, isoleucine, proline, valine, methionine, tryptophan, phenylalanine, and lysine in mouse ESCs. The second platform is an advanced one that uses mass spectrometry (MS), two-dimensional gel electrophoresis (2-DE), two-dimensional differential gel electrophoresis (2D-DIGE), protein microarrays or clips based on analytical, functional, and reverse-phase protein analysis, gel-based approaches, and Edman sequencing methods. Here, gel-based and MS techniques were specially allotted for the characterization of proteins, and Edman Sequencing was allotted for amino acid sequence analysis of proteins (Aslam et al., 2017). Third, the quantitative platform consists of stable isotope labeling with amino acids in cell culture (SILAC) (Ong & Mann, 2007), isotope-coded affinity tag (ICAT) labeling (Shiio & Aebersold, 2006), and isobaric tag for relative and absolute quantitation (iTRAQ) techniques (Wiese et al., 2007) that allow sequence identification and accurate quantification of proteins in complex mixtures. Fourth, a high-throughput proteomic platform such as X-ray crystallography (Smyth & Martin, 2000) and magnetic resonance spectroscopy (MRS) or nuclear magnetic resonance (NMR) spectroscopy (Semmler, 2018) together with matrix-assisted laser desorption/ionization–time of flight (MALDI-TOF) MS will provide the three-dimensional (3D) structure of a protein, which may also provide insight into its biological function (Aslam et al., 2017). Therefore, with the support of the aforementioned four platforms, a huge amount of high-throughput proteomic data is collected, and finally, a "bioinformatic proteomic platform" has been established to handle and store an enormous quantity of data. This bioinformatic platform is specially designed to predict the 3D structure of a protein, analyze MS data, analyze protein–protein interactions quickly, and analyze protein domains and motifs. This platform is also useful for investigating evolutionary relationships (Perez-Riverol et al., 2015; Vihinen, 2001). Thus, proteome analysis may provide structural and functional information about proteins that are involved in the SCs phenotype.

TABLE 11.3 Summary of Proteomics Platforms that can be used for analyzing proteins associated with Stem cells regulation.

Sl. no.	Platform	Techniques/tools	Utilization of the technique
1.	The conventional platform	• Exchange chromatography • Affinity chromatography • Gas chromatography • ELISA • WB	Used for the extraction, purification and identification of SCs protein (Aslam et al., 2017; Hage et al., 2012; Jungbauer & Hahn, 2009; Munoz et al., 2014)
2.	The advanced platform	• Protein microarrays • SDS-PAGE • 2-DE • 2D-DIGE • MS • Edman sequencing techniques	Used for the characterization of protein and Edman sequencing was allotted for amino-acid sequence analysis of protein (Aslam et al., 2017)
3.	The quantitative platform	• SILAC • ICAT • iTRAQ	Used to determine protein sequences and precisely measure their quantities in complicated mixes (Ong & Mann, 2007; Shiio & Aebersold, 2006; Wiese et al., 2007)
4.	The high-throughput proteomic platform	• X-ray crystallography • MRS or NMR spectroscopy • MALDI-TOF mass spectrometry	Used to reveal 3D structure of a protein (Semmler, 2018; Smyth & Martin, 2000)
5.	The bioinformatic platform	• String database • CORUM • Uniprot • KEGG • Reactome	Used for quick protein–protein interaction analysis, protein domain and motif analysis, and MS data analysis. Used to determine the 3D structure of a protein. The investigation of evolutionary links can also be aided by this platform (Perez-Riverol et al., 2015; Vihinen, 2001)

3. Mechanistic insights into the stemness phenotype of stem cells through genomic and proteomic platform

The stemness phenotype of SCs deals with the molecular process underlying the fundamental properties of SCs such as self-renewal and differentiation (Saei Arezoumand et al., 2017). SCs maintain the stemness phenotype by highly organized and strictly controlled signaling network systems that control the regulation of various stemness-associated genes such as NANOG, OCT4, SOX2, REX1, NOTCH1, NESTIN, etc. (Gonzalez-Garza et al., 2018) and thus retain a balance between proliferation, quiescence, and regeneration (Mushtaq et al., 2020). This phenomenon of SCs laid the foundation for cell based therapies for several diseases, such as Parkinson's condition, spinal cord harm, age related macular degeneration, osteoarthritis, etc., which cannot be cured by conventional medicines (Ilic & Ogilvie, 2017; Loo & Wong, 2021; Mahla, 2016; Yamanaka, 2020). And thus, SC-based therapy became the frontier of regenerative medicine.

Despite significant progress in SC research, SC-mediated therapy raises sharp ethical and political controversies. These controversies mainly concern patient consent, destruction of embryos, creation of embryos, legal and ethical conflict of using SCs lines, as well as the risk and benefits of experimental intervention in clinical trials, etc. (Lo & Parham, 2009). Therefore, it is important to apply SC-mediated therapy in an ethically suitable way, particularly for ESC-mediated therapy, particularly for ESC-mediated therapy, which raises a number of serious ethical questions, including those related to the likelihood of a genetic anomaly, tumor development, immune rejection, difficulty obtaining uniform cell populations followed by differentiation, and destruction of embryos (Ahn et al., 2010; Draper et al., 2004; Maitra et al., 2005; Rebuzzini et al., 2015; Vogel, 2005).

Additionally, it has been noticed that pluripotent ESCs and CSCs have similar traits such as rapid proliferation, a need for metabolism, and the ability to keep themselves in an undifferentiated condition for an extended period of time (Hadjimichael et al., 2015). Based on studies of biomarkers, gene signatures, and epigenetic regulator signaling networks, Hadjimichael et al. (2015) documented common stemness regulators in ESCs and CSCs. Importantly, it has also been documented that in vitro cultured pluripotent ESCs when injected into immunodeficient mice cause benign tumors and teratomas, suggesting that ESCs have inherent tumorigenic potential (Hadjimichael et al., 2015; Solter, 2006). Normally, the pluripotency of ESCs is maintained by the master stemness regulator genes OCT4, SOX2, and NANOG (Boyer et al., 2005). Similarly, CSCs obtained from epithelial tumors exhibit ESC-like gene signatures plus MYC gene upregulation (Ben-Porath et al., 2008; Wong et al., 2008). Additionally, we also demonstrated that the MYC-mediated HIF2α stemness phenotype via NANOG and SOX2 enhances the stemness phenotype in CSCs that promotes tumor formation in the mouse model (Das et al., 2019). Thus, these studies suggest that the stemness phenotype is also associated with cancer progression.

Therefore, to utilize SCs as successful cell-mediated therapies, it is important to understand the basic molecular and cellular signaling networks that maintain stemness property in SCs as well as diseases associations with stemness mechanisms. However, obtaining a fair understanding of how cell signaling networks maintain stemness as well as pluripotency in SCs has been a crucial step due to a number of internal factors such as culture conditions, tools used to explore cell signaling, identification of specific pathways, and variations in experimental design (Dalton, 2013). Signaling networks usually have diverse effects depending on their activation (Dalton, 2013). Thus, in this section, we reviewed the molecular signature of stemness in SCs, especially in the case of ESCs, through genomic and proteomic approaches and described how these approaches take SC research forward.

3.1 Molecular signatures of stemness in embryonic stem cells through genomics

The core cell mass of the blastocyst is where ESCs are found. They exhibit a stemness phenotype, which includes the capacity for self-renewal and differentiation into all three germ layers (ectoderm, mesoderm, and endoderm) (Chen et al., 2022). Through a variety of pluripotent regulator molecules, including transcriptional regulator factors, soluble extracellular signaling molecules, epigenetic factors, and micro-RNA molecules that prevent spontaneous differentiation, ESCs were able to maintain self-renewal and pluripotency in vitro (Cerulo et al., 2014; Fritsch & Singer, 2008; Kimura et al., 2004; Swain et al., 2020). The core stemness regulator factors SOX2, OCT4, and NANOG play a significant role in the propagation of undifferentiated ESCs in culture and thus maintain ESCs at their stemness state (Boyer et al., 2005; Hassiotou et al., 2013; Huang et al., 2015; Singh et al., 2007; Swain et al., 2020). Importantly, these stemness regulator factors of ESCs are also associated with carcinogenesis (Das et al., 2019; Swain et al., 2020). Through a cooperative interaction, these core stemness regulator factors along with extracellular signaling pathways such as bone morphogenetic protein (BMP), NODAL/ACTIVIN, leukemia inhibitor factor (LIF), WNT, SMAD1, SMAD2/3, ERBB2/ERBB3, signal transducer and activator of transcription 3 (STAT3), and transcription factor 3 (TCF3) signaling, form the core

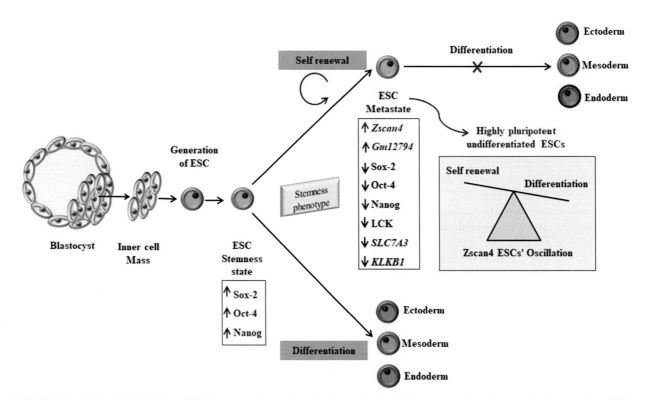

FIGURE 11.3 Mechanistic insight into ESC stemness and metastate. The inner cell mass of blastocytes is the main source for the generation of ESCs. These ESCs exhibit a stemness phenotype, i.e., the ability to self-renew and differentiate into ectoderm, mesoderm, and endoderm germ layers (Chen et al., 2022). The core transcription factors SOX2, OCT4, and NANOG maintain ESCs in their stemness state and play an important role in pluripotency, self-renewal, and propagation of undifferentiated ESCs in culture (Chen et al., 2022). This highly pluripotent undifferentiated condition of ESCs is maintained by transient activation of Zscan4, which oscillates between self-renewal and differentiation balance and maintains their developmental potency in long-term cell culture, marked as "ESC metastate or Zscan4 ESCs' oscillation" (Amano et al., 2013; Cerulo et al., 2014). More specifically, Gm12794, a leukaemia inhibitor factor (LIF) dependent not-canonical pluripotency signature that specially marks Zscan4 ESCs' Oscillation, is essential for ESCs self-renewal maintenance (Cerulo et al., 2014). However, these highly pluripotent Zscan4 ESCs don't express transiently high levels of pluripotency-associated genes such as NANOG, SOX2, OCT4, LCK, SLC7A3 and KLKB1, which prevent differentiation in ESC propagation (Kim et al., 2014).

transcriptional regulatory network that regulates self-renewal and pluripotency in ESCs (Boyer et al., 2005; Chen et al., 2022; Cole et al., 2008; Hough et al., 2014; Peerani et al., 2007; Singh et al., 2012; Vallier et al., 2009; Wang et al., 2007; Xu et al., 2008). In long-term cell culture, this undifferentiated condition of ESCs is also marked by a "metastate" that oscillates between self-renewal and differentiation equilibrium (Fig. 11.3) (Amano et al., 2013; Cerulo et al., 2014). ESC "metastate" is maintained by transient activation of Zscan4 (zinc finger and SCAN domain containing protein 4), which controls telomere length and genomic integrity in ESCs (Amano et al., 2013; Zalzman et al., 2010). Moreover, Gm12794, an LIF-dependent noncanonical pluripotency signature that specifically marks ESC "metastate" or Zscan4 ESCs' oscillation, is essential for ESCs self-renewal maintenance in vitro (Cerulo et al., 2014). In this regard, Luigi et al. suggested that this novel Zscan4ESC "metastate" may be enhanced by supplementing LIF-dependent high-pluripotency culture conditions attained through the addition of two more inhibitors (2i), which collectively block extracellular signal–regulated kinase (ERK) and glycogen synthase kinase-3 (Gsk-3) signaling (Cerulo et al., 2014; Marks et al., 2012). However, these highly pluripotent Zscan4 ESCs do not express transiently high levels of pluripotency in the NANOG gene (Singh et al., 2007), indicating that ESC cultures consist of multiple but not overlapping levels of pluripotency (Cerulo et al., 2014). More specifically, using microarray and weighted gene correlation analysis (WGCNA), Jeffrey and his team discovered a new set of self-renewal factors in the same year. These factors include LCK, KLKB1, SLC7A3, RHOJ, ZEB2, and ADAM12 (Kim et al., 2014). They demonstrated a potential functional impact of lymphocyte-specific protein tyrosine kinase (LCK) and suggested that downregulation of NANOG, SOX2, OCT4, LCK, SLC7A3, and KLKB1 prevents differentiation in ESCs propagation (Kim et al., 2014) (Fig. 11.3). These findings suggest that the self-renewal of undifferentiated ESCs is controlled through the up- and downregulation of a specific set of genes that balance the oscillation between self-renewal and differentiation (Fig. 11.3).

Regardless of the extensive use of ESCs in research, the lack of a clear understanding of the stemness associated pluripotency state of ESCs during in vitro culture limits their use for therapeutic purposes. It is very difficult to observe cell fate decisions and insight into signaling networks that maintain the pluripotency of SCs due to the variability of gene expression in ESCs and population-wise, lineage priming, or coexpression of lineage specific and pluripotency genes in culture (Enver et al., 2009; Hough et al., 2014). Interestingly, the temporary pluripotent state of ESCs, which is ready to undergo specification towards extraembryonic, and subsequently embryonic cell lineages, has inherent heterogeneity (Hough et al., 2014). However, lineage differentiation of SCs can be repressed by epigenetic factors, when self-renewal has driven strongly, and consequently, self-renewal and pluripotency may each be controlled separately (Hough et al., 2014). Additionally, a significant portion of hESC lines have major detrimental structural variants, finer-scale structural variants, and single-nucleotide variants (SNVs), which compromise the safety of hESC-derived therapy in clinical individuals (Merkle et al., 2022). Within 12 hours of fertilisation, mouse embryonic stem cells (ESCs) sustain global demethylation of both the maternal and paternal genomes, resulting in a highly unclosed chromatin state that serves as a definition of heterogeneity. This is followed by an increase in closeness following the late zygote stage (Guo et al., 2017). Thus, heterogeneity in gene expression in ESCs initiates an unstable regulatory mechanism for self-renewal during in vitro culture.

To overcome ESCs heterogeneity issues during in vitro culture, recent attempts have been made to generate pluripotent ESCs with a self-renewal phenotype in vitro (Fig. 11.4) (Chen et al., 2022). The use of trophoblast cells, such as mitomycin C−treated inactive mouse embryo fibroblasts (MEFs) in the feeder layer during in vitro culture, prevents the differentiation of mouse ESCs (Conner, 2001). These trophoblastic cells secrete the cytokine LIF, and then LIF

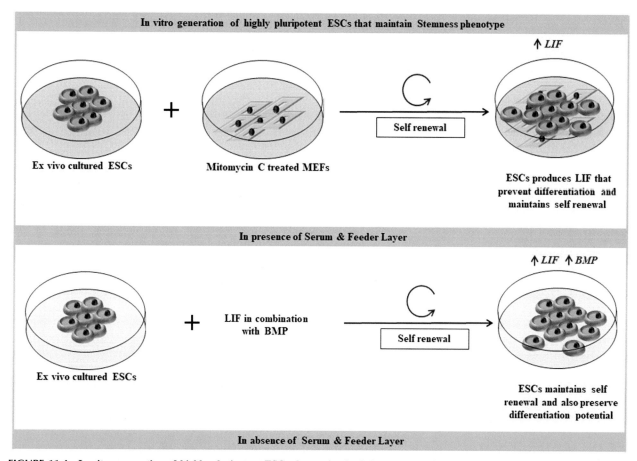

FIGURE 11.4 In vitro generation of highly pluripotent ESCs that maintained the stemness phenotype. The use of trophoblast cells such as mitomycin C-treated inactive mouse embryo fibroblasts (MEFs) feeder layer during in vitro culture prevents the differentiation of mouse ESCs (Conner, 2001). In the presence of serum, these trophoblastic cells secrete leukaemia inhibitor factor (LIF), which directs mouse ESCs self-renewal by altering the function of trophoblasts and preventing differentiation (Chen et al., 2022; Stewart et al., 1992; Williams et al., 1988). However, in the absence of serum and feeder layer, LIF supplementation in combination with BMP maintains the self-renewal of ESCs and also preserves multilineage differentiation, chimaera colonization, and germline transmission properties (Ying, Nichols et al., 2003; Ying, Stavridis et al., 2003).

directs mESCs self-renewal by altering the function of the trophoblast and preventing differentiation in the presence of serum (Chen et al., 2022; Stewart et al., 1992; Williams et al., 1988). LIF maintains the pluripotency of ESCs by activating the JAK (janus kinase)/STAT3 signaling network through binding to its receptor gp130/LIFR (Matsuda et al., 1999; Niwa et al., 1998). However, LIF supplementation in combination with BMP maintains ESCs self-renewal in the absence of serum and the feeder layer and also preserves their multilineage differentiation, chimaera colonization, and germ line transmission potentials (Ying, Nichols, et al., 2003; Ying, Stavridis, et al., 2003). Additionally, in the presence of LIF, BMP activates the SMAD pathway, which in turn activates the inhibitors of differentiation (Id) genes. Id proteins then repress the lineage-specific regulator factors, maintaining ESC self-renewal via the LIF/STAT3 signaling network (Ying, Nichols, et al., 2003). Interestingly, physical interaction between SMAD1 and NANOG blocks BMP-induced mesodermal differentiation of ESCs (Suzuki et al., 2006). Thereby, these findings strongly indicate the stemness phenotype of ESCs can be maintained by proper in vitro culture conditions that prevent the differentiation process for long-term culture.

3.2 Molecular signatures of stemness in embryonic stem cells through proteomics

The controlled stemness networks in ex vivo cultured ESCs greatly encourage research and application of ESCs in the field of regenerative medicine (Young, 2011). Therefore, an intracellular signaling network that accounts for the stemness phenotype based on the proteome-wide scale will be worthwhile for SCs-mediated therapy. Consequently, large-scale proteome-wide analysis of ESCs using advanced MS-based analytical techniques has been observed. Previously, Dormeyer and his colleagues conducted an extensive plasma membrane (PM) proteomics study to identify 237 and 219 PM proteins in human ESCs (HUES-7 cell lines) and human embryonal carcinoma cells (hECCs) (NT2/D1 cell lines) (Dormeyer, van Hoof, Braam, et al., 2008). It has been reported that it is very difficult to study the intracellular signaling network of PM proteins, especially those embedded in the lipid bilayer of cells, due to the hydrophobic nature of transmembrane domains; thus, identifying the molecular signature of PM proteins in cells may provide beneficial insights into the regulation of their biological activities (Dormeyer, van Hoof, Mummery, et al., 2008; Van Hoof et al., 2008). They revealed NPC1, FZD2, FZD6, FZD7, LRP6, and SEMA4D proteins expressed differently in hESCs and hECCs associated with the hedgehog and WNT pathway, which play a vital role in self-renewal as well as in cancer growth. The advantage of these studies was that it provided a useful guide for identifying PM proteins from ESCs by analyzing the proteome profile of human ESCs (Dormeyer, van Hoof, Braam, et al., 2008; Dormeyer, van Hoof, Mummery, et al., 2008). They also revealed a significant number of receptors, transporters, signal transducers, and cell−cell adhesion proteins, as well as well-known stemness-mediated cell surface markers such as ALP, CD9, and CTNNB. Further, Mi-Young et al. identified REX1 (reduced expression protein 1), a canonical pluripotency signature in undifferentiated human ESCs, by performing a comparative proteomic analysis (Son et al., 2015). Downregulation of REX1 has been observed in differentiated cells (Wang et al., 2006). They were also able to identify 140 differentially expressed proteins (DEPs) in human ESCs (control), compare transcriptome data with REX1 knockdown (KD) hESCs, and suggest a highly interconnected network between REX1 KD-mediated DEPs predominantly associated with ribosome-mediated translation and regulation of mitochondrial activity. Thus, understanding the pluripotency of ESCs at the molecular level should illuminate the fundamental landscape of cellular reprogramming. In this context, Jianlong and his coworkers explored the protein intrinsic signaling network in which NANOG operates ESCs pluripotency (Wang et al., 2006). NANOG, in coordination with OCT4 (Babaie et al., 2007; Nichols et al., 1998) and SOX2 (Avilion et al., 2003), maintains ESCs in the pluripotent or "ground state" (Ying et al., 2008). Hence, the "ground state" of ESCs maintained by tightly controlled intrinsic signaling networks, including interplay between key transcriptional regulators of stemness such as NANOG, OCT4, KIF-4, SOX2, and c-MYC (Babaie et al., 2007; Mossahebi-Mohammadi et al., 2020; Parisi & Russo, 2011). Importantly, ESCs' "ground state" can also be preserved via extrinsic signaling networks, such as the interaction between LIF and BMI (Fig. 11.4) (Chen et al., 2022; Matsuda et al., 1999; Stewart et al., 1992; Williams et al., 1988; Ying, Nichols, et al., 2003; Ying, Stavridis, et al., 2003). On the other side, fibroblast growth factor (FGF) acts as a key extrinsic signaling regulating SC pluripotency (Mossahebi-Mohammadi et al., 2020). Activation of the ERK signaling network stimulated by FGF triggers the transition from pluripotent self-renewal to differentiation (Mossahebi-Mohammadi et al., 2020). Thus, inhibiting the network of mitogen-activated protein kinase (MAPK)/ERK signaling network that controls FGF signaling may thereby limit ESCs differentiation and increase ESC stability and stemness (Kunath et al., 2007). Recently, Matteo and his team revealed human embryonic development by exploring FGF signaling by performing scRNA-seq of human embryos (Mole et al., 2021). They discovered extraembryonic hypoblast cells that operate as the anterior signaling centre to pattern the epiblast and express antagonist effects on BMP, NODAL, and WNT signaling networks (Mole et al., 2021). Additionally, earlier research demonstrated that SMAD3 cooccupies the genome with OCT4 in ESCs to promote the pluripotency state and that transforming growth factor beta (TGF-β) signalling, mediated by SMAD2 and SMAD3 (SMADmad2/3), plays a role in this (Mullen et al., 2011; Mullen & Wrana, 2017). Thus, to support the stemness phenotype of ESCs, these extracellular signalling networks are made up of a variety of proteins, including BMP, LIF, NODAL/ACTIVIN, JAK/STAT, FGF, TGF, and TCF3. These proteins also cooccupy the genome with cell type—specific core stemness regulator factors OCT4, SOX2, and NANOG (Wang et al., 2006). Therefore, these findings revealed mechanistic insights into the stemness phenotype of ESCs through the core transcriptional regulatory network.

4. Conclusion

Stem cells maintain the stemness (self-renewal and differentiation) phenotype by dynamic cross-talk between cellular regulatory signaling networks, which includes genetics, epigenetics, and transcriptional (posttranscriptional and post-translational) regulation of stemness regulator genes. OCT4, SOX2, and NANOG, the stemness regulator genes, work in conjunction with other extracellular signaling networks such as BMP, LIF, NODAL/ACTIVIN, JAK/STAT, FGF, TGF, and TCF3 signalling to maintain the stemness phenotype in SCs and thereby maintain the balance between cell proliferation, quiescence, and regeneration. However, a lack of understanding of the basic biology of the stemness phenotype in SCs continues to cast doubt on the successful clinical application of these cells. Thus, we aim to provide a concise review of genomic and proteomic aspects for understanding the basics of the SC stemness phenotype. Taken together, our findings contribute to the mechanistic insight into regulatory signaling networks that underlie the in vitro maintenance of ESCs pluripotency and self-renewal to prevent the differentiation process. These observations and the in vitro maintenance of the stemness phenotype in ESCs greatly encourage the prospect of utilizing ex vivo cultured ESCs as a platform to screen the toxicity effects of various drugs as a future direction. However, the use of ex vivo cultured ESCs as ESCs-mediated therapy in clinical subjects remains limited. Long-term culture of ESCs may lead them to reprogram themselves into enhanced stemness phenotype, which may cause tumor (teratomas) or cancer development. Thus, the process of producing ESC lines is inefficient. Moreover, it is very unclear whether ex vivo cultured ESCs would be safe and feasible for autologous or analogous transplantation. Therefore, more in vivo studies and clinical trials are required. Nevertheless, ESC-mediated therapy has major ethical concerns since it is associated with a risk factor for female donors being consented to, and to acquire the inner cell mass, the embryo should be destroyed.

Acknowledgments
Sorra Sandhya would like to thank Indian Council of Medical Research (ICMR)-Research Associate (RA)-III fellowship (45/11/2020 DDI/BMS) for financial support.

Author contributions
S.S.: Conception and design, data analysis, figures and tables, interpretation, manuscript writing, and final approval of manuscript. K.K.B.: Assembly of data and manuscript writing, tables design, and preparation of manuscript. J.T.: Assembly of data and manuscript writing. D.B.: Conception and design, data analysis and interpretation, final approval of manuscript.

References

Ahn, S. M., Simpson, R., & Lee, B. (2010). Genomics and proteomics in stem cell research: The road ahead. *Anatomy & Cell Biology, 43*(1), 1—14. https://doi.org/10.5115/acb.2010.43.1.1

Alvarez-Manilla, G., Warren, N. L., Atwood, J., 3rd, Orlando, R., Dalton, S., & Pierce, M. (2010). Glycoproteomic analysis of embryonic stem cells: Identification of potential glycobiomarkers using lectin affinity chromatography of glycopeptides. *Journal of Proteome Research, 9*(5), 2062—2075. https://doi.org/10.1021/pr8007489

Aly, R. M. (2020). Current state of stem cell-based therapies: An overview. *Stem Cell Investigation, 7*, 8. https://doi.org/10.21037/sci-2020-001

Amano, T., Hirata, T., Falco, G., Monti, M., Sharova, L. V., Amano, M., ... Ko, M. S. (2013). Zscan4 restores the developmental potency of embryonic stem cells. *Nature Communications, 4*, 1966. https://doi.org/10.1038/ncomms2966

Aslam, B., Basit, M., Nisar, M. A., Khurshid, M., & Rasool, M. H. (2017). Proteomics: Technologies and their applications. *Journal of Chromatographic Science, 55*(2), 182—196. https://doi.org/10.1093/chromsci/bmw167

Avilion, A. A., Nicolis, S. K., Pevny, L. H., Perez, L., Vivian, N., & Lovell-Badge, R. (2003). Multipotent cell lineages in early mouse development depend on SOX2 function. *Genes & Development, 17*(1), 126—140. https://doi.org/10.1101/gad.224503

Babaie, Y., Herwig, R., Greber, B., Brink, T. C., Wruck, W., Groth, D., ... Adjaye, J. (2007). Analysis of Oct4-dependent transcriptional networks regulating self-renewal and pluripotency in human embryonic stem cells. *Stem Cells, 25*(2), 500—510. https://doi.org/10.1634/stemcells.2006-0426

Behjati, S., & Tarpey, P. S. (2013). What is next generation sequencing? *Archives of Disease in Childhood: Education and Practice Edition, 98*(6), 236−238. https://doi.org/10.1136/archdischild-2013-304340

Ben-Porath, I., Thomson, M. W., Carey, V. J., Ge, R., Bell, G. W., Regev, A., & Weinberg, R. A. (2008). An embryonic stem cell-like gene expression signature in poorly differentiated aggressive human tumors. *Nature Genetics, 40*(5), 499−507. https://doi.org/10.1038/ng.127

Biase, F. H., Cao, X., & Zhong, S. (2014). Cell fate inclination within 2-cell and 4-cell mouse embryos revealed by single-cell RNA sequencing. *Genome Research, 24*(11), 1787−1796. https://doi.org/10.1101/gr.177725.114

Blakeley, P., Fogarty, N. M., del Valle, I., Wamaitha, S. E., Hu, T. X., Elder, K., ... Niakan, K. K. (2015). Defining the three cell lineages of the human blastocyst by single-cell RNA-seq. *Development, 142*(18), 3151−3165. https://doi.org/10.1242/dev.123547

Bode, D., Cull, A. H., Rubio-Lara, J. A., & Kent, D. G. (2021). Exploiting single-cell tools in gene and cell therapy. *Frontiers in Immunology, 12*, 702636. https://doi.org/10.3389/fimmu.2021.702636

Boyer, L. A., Lee, T. I., Cole, M. F., Johnstone, S. E., Levine, S. S., Zucker, J. P., ... Young, R. A. (2005). Core transcriptional regulatory circuitry in human embryonic stem cells. *Cell, 122*(6), 947−956. https://doi.org/10.1016/j.cell.2005.08.020

Brendel, C., Kuklick, L., Hartmann, O., Kim, T. D., Boudriot, U., Schwell, D., & Neubauer, A. (2005). Distinct gene expression profile of human mesenchymal stem cells in comparison to skin fibroblasts employing cDNA microarray analysis of 9600 genes. *Gene Expression, 12*(4−6), 245−257. https://doi.org/10.3727/000000005783992043

Cerulo, L., Tagliaferri, D., Marotta, P., Zoppoli, P., Russo, F., Mazio, C., ... Falco, G. (2014). Identification of a novel gene signature of ES cells self-renewal fluctuation through system-wide analysis. *PLoS One, 9*(1), e83235. https://doi.org/10.1371/journal.pone.0083235

Chen, T., Li, J., Jia, Y., Wang, J., Sang, R., Zhang, Y., & Rong, R. (2020). Single-cell sequencing in the field of stem cells. *Current Genomics, 21*(8), 576−584. https://doi.org/10.2174/1389202921999200624154445

Chen, G., Yin, S., Zeng, H., Li, H., & Wan, X. (2022). Regulation of embryonic stem cell self-renewal. *Life (Basel), 12*(8). https://doi.org/10.3390/life12081151

Cole, M. F., Johnstone, S. E., Newman, J. J., Kagey, M. H., & Young, R. A. (2008). Tcf3 is an integral component of the core regulatory circuitry of embryonic stem cells. *Genes & Development, 22*(6), 746−755. https://doi.org/10.1101/gad.1642408

Conner, D. A. (2001). Mouse embryo fibroblast (MEF) feeder cell preparation. *Current Protocols in Molecular Biology, 23.* https://doi.org/10.1002/0471142727.mb2302s51 (Chapter 23), Unit 23 22.

Cubillo, E. J., Ngo, S. M., Juarez, A., Gagan, J., Lopez, G. D., & Stout, D. A. (2018). Embryonic stem cell therapy applications for autoimmune, cardiovascular, and neurological diseases: A review. *AIMS Cell and Tissue Engineering, 1*(3), 191−223. https://doi.org/10.3934/celltissue.2017.3.191

Dalton, S. (2013). Signaling networks in human pluripotent stem cells. *Current Opinion in Cell Biology, 25*(2), 241−246. https://doi.org/10.1016/j.ceb.2012.09.005

Das, D., Fletcher, R. B., & Ngai, J. (2020). Cellular mechanisms of epithelial stem cell self-renewal and differentiation during homeostasis and repair. *Wiley Interdisciplinary Reviews: Developmental Biology, 9*(1), e361. https://doi.org/10.1002/wdev.361

Das, B., Pal, B., Bhuyan, R., Li, H., Sarma, A., Gayan, S., ... Felsher, D. W. (2019). MYC regulates the HIF2alpha stemness pathway via Nanog and Sox2 to maintain self-renewal in cancer stem cells versus non-stem cancer cells. *Cancer Research, 79*(16), 4015−4025. https://doi.org/10.1158/0008-5472.CAN-18-2847

Degar, B. A., Baskaran, N., Hulspas, R., Quesenberry, P. J., Weissman, S. M., & Forget, B. G. (2001). The homeodomain gene Pitx2 is expressed in primitive hematopoietic stem/progenitor cells but not in their differentiated progeny. *Experimental Hematology, 29*(7), 894−902. https://doi.org/10.1016/s0301-472x(01)00661-0

Dormeyer, W., van Hoof, D., Braam, S. R., Heck, A. J., Mummery, C. L., & Krijgsveld, J. (2008). Plasma membrane proteomics of human embryonic stem cells and human embryonal carcinoma cells. *Journal of Proteome Research, 7*(7), 2936−2951. https://doi.org/10.1021/pr800056j

Dormeyer, W., van Hoof, D., Mummery, C. L., Krijgsveld, J., & Heck, A. J. (2008). A practical guide for the identification of membrane and plasma membrane proteins in human embryonic stem cells and human embryonal carcinoma cells. *Proteomics, 8*(19), 4036−4053. https://doi.org/10.1002/pmic.200800143

Draper, J. S., Smith, K., Gokhale, P., Moore, H. D., Maltby, E., Johnson, J., ... Andrews, P. W. (2004). Recurrent gain of chromosomes 17q and 12 in cultured human embryonic stem cells. *Nature Biotechnology, 22*(1), 53−54. https://doi.org/10.1038/nbt922

Duguid, J. R., Rohwer, R. G., & Seed, B. (1988). Isolation of cDNAs of scrapie-modulated RNAs by subtractive hybridization of a cDNA library. *Proceedings of the National Academy of Sciences of the United States of America, 85*(15), 5738−5742. https://doi.org/10.1073/pnas.85.15.5738

Enver, T., Pera, M., Peterson, C., & Andrews, P. W. (2009). Stem cell states, fates, and the rules of attraction. *Cell Stem Cell, 4*(5), 387−397. https://doi.org/10.1016/j.stem.2009.04.011

Fritsch, M. K., & Singer, D. B. (2008). Embryonic stem cell biology. *Advances in Pediatrics, 55*, 43−77. https://doi.org/10.1016/j.yapd.2008.07.006

Gao, L. R., Zhang, N. K., Ding, Q. A., Chen, H. Y., Hu, X., Jiang, S., ... Zhu, Z. M. (2013). Common expression of stemness molecular markers and early cardiac transcription factors in human Wharton's jelly-derived mesenchymal stem cells and embryonic stem cells. *Cell Transplantation, 22*(10), 1883−1900. https://doi.org/10.3727/096368912X662444

Glover, C. H., Marin, M., Eaves, C. J., Helgason, C. D., Piret, J. M., & Bryan, J. (2006). Meta-analysis of differentiating mouse embryonic stem cell gene expression kinetics reveals early change of a small gene set. *PLoS Computational Biology, 2*(11), e158. https://doi.org/10.1371/journal.pcbi.0020158

Gonzalez-Garza, M. T., Cruz-Vega, D. E., Cardenas-Lopez, A., de la Rosa, R. M., & Moreno-Cuevas, J. E. (2018). Comparing stemness gene expression between stem cell subpopulations from peripheral blood and adipose tissue. *American Journal of Stem Cells, 7*(2), 38−47.

Guo, F., Li, L., Li, J., Wu, X., Hu, B., Zhu, P., ... Tang, F. (2017). Single-cell multi-omics sequencing of mouse early embryos and embryonic stem cells. *Cell Research, 27*(8), 967−988. https://doi.org/10.1038/cr.2017.82

Hadjimichael, C., Chanoumidou, K., Papadopoulou, N., Arampatzi, P., Papamatheakis, J., & Kretsovali, A. (2015). Common stemness regulators of embryonic and cancer stem cells. *World Journal of Stem Cells, 7*(9), 1150—1184. https://doi.org/10.4252/wjsc.v7.i9.1150

Hage, D. S., Anguizola, J. A., Bi, C., Li, R., Matsuda, R., Papastavros, E., ... Zheng, X. (2012). Pharmaceutical and biomedical applications of affinity chromatography: Recent trends and developments. *Journal of Pharmacy Biomedicine Analytical, 69*, 93—105. https://doi.org/10.1016/j.jpba.2012.01.004

Halliwell, J. A., Baker, D., Judge, K., Quail, M. A., Oliver, K., Betteridge, E., ... Barbaric, I. (2021). Nanopore sequencing indicates that tandem amplification of chromosome 20q11.21 in human pluripotent stem cells is driven by break induced replication. *Stem Cells and Development, 30*(11), 378—386. https://doi.org/10.1089/scd.2021.0013

Haagiotou, E., Hepworth, A. R., Beltran, A. S., Mathews, M. M., Stuebe, A. M., Hartmann, P. E., ... Blancafort, P. (2013). Expression of the pluripotency transcription factor OCT4 in the normal and aberrant mammary gland. *Frontiers in Oncology, 3*, 79. https://doi.org/10.3389/fonc.2013.00079

He, S., Nakada, D., & Morrison, S. J. (2009). Mechanisms of stem cell self-renewal. *Annual Review of Cell and Developmental Biology, 25*, 377—406. https://doi.org/10.1146/annurev.cellbio.042308.113248

Hipp, J. A., Hipp, J. D., Atala, A., & Soker, S. (2010). Functional genomics: New insights into the 'function' of low levels of gene expression in stem cells. *Current Genomics, 11*(5), 354—358. https://doi.org/10.2174/138920210791616680

Hirst, M., Delaney, A., Rogers, S. A., Schnerch, A., Persaud, D. R., O'Connor, M. D., ... Marra, M. A. (2007). LongSAGE profiling of nine human embryonic stem cell lines. *Genome Biology, 8*(6), R113. https://doi.org/10.1186/gb-2007-8-6-r113

Holliday, R. (2006). Epigenetics: A historical overview. *Epigenetics, 1*(2), 76—80. https://doi.org/10.4161/epi.1.2.2762

Hough, S. R., Thornton, M., Mason, E., Mar, J. C., Wells, C. A., & Pera, M. F. (2014). Single-cell gene expression profiles define self-renewing, pluripotent, and lineage primed states of human pluripotent stem cells. *Stem Cell Reports, 2*(6), 881—895. https://doi.org/10.1016/j.stemcr.2014.04.014

Huang, G., Ye, S., Zhou, X., Liu, D., & Ying, Q. L. (2015). Molecular basis of embryonic stem cell self-renewal: From signaling pathways to pluripotency network. *Cellular and Molecular Life Sciences, 72*(9), 1741—1757. https://doi.org/10.1007/s00018-015-1833-2

Huttmann, A., Duhrsen, U., Heydarian, K., Klein-Hitpass, L., Boes, T., Boyd, A. W., & Li, C. L. (2006). Gene expression profiles in murine hematopoietic stem cells revisited: Analysis of cDNA libraries reveals high levels of translational and metabolic activities. *Stem Cells, 24*(7), 1719—1727. https://doi.org/10.1634/stemcells.2005-0486

Hwang, B., Lee, J. H., & Bang, D. (2018). Single-cell RNA sequencing technologies and bioinformatics pipelines. *Experimental & Molecular Medicine, 50*(8), 1—14. https://doi.org/10.1038/s12276-018-0071-8

Ilic, D., & Ogilvie, C. (2017). Concise review: Human embryonic stem cells-what have we done? What are we doing? Where are we going? *Stem Cells, 35*(1), 17—25. https://doi.org/10.1002/stem.2450

Jain, M., Abu-Shumays, R., Olsen, H. E., & Akeson, M. (2022). Advances in nanopore direct RNA sequencing. *Nature Methods, 19*(10), 1160—1164. https://doi.org/10.1038/s41592-022-01633-w

Jungbauer, A., & Hahn, R. (2009). Ion-exchange chromatography. *Methods in Enzymology, 463*, 349—371. https://doi.org/10.1016/S0076-6879(09)63022-6

Kang, C. M., Kim, H., Song, J. S., Choi, B. J., Kim, S. O., Jung, H. S., ... Choi, H. J. (2016). Genetic comparison of stemness of human umbilical cord and dental pulp. *Stem Cells International*, 3453890. https://doi.org/10.1155/2016/3453890

Kchouk, M., Gibrat, J.-F., & Elloumi, M. (2017). Generations of sequencing technologies: From first to next generation. *Biology and Medicine, 9*(3), 1—8. https://doi.org/10.4172/0974-8369.1000395

Kim, J. J., Khalid, O., Namazi, A., Tu, T. G., Elie, O., Lee, C., & Kim, Y. (2014). Discovery of consensus gene signature and intermodular connectivity defining self-renewal of human embryonic stem cells. *Stem Cells, 32*(6), 1468—1479. https://doi.org/10.1002/stem.1675

Kim, S. H., Kim, Y. S., Lee, S. Y., Kim, K. H., Lee, Y. M., Kim, W. K., & Lee, Y. K. (2011). Gene expression profile in mesenchymal stem cells derived from dental tissues and bone marrow. *Journal of Periodontal & Implant Science, 41*(4), 192—200. https://doi.org/10.5051/jpis.2011.41.4.192

Kim, C. G., Lee, J. J., Jung, D. Y., Jeon, J., Heo, H. S., Kang, H. C., ... Kim, H. S. (2006). Profiling of differentially expressed genes in human stem cells by cDNA microarray. *Molecules and Cells, 21*(3), 343—355.

Kimura, H., Tada, M., Nakatsuji, N., & Tada, T. (2004). Histone code modifications on pluripotential nuclei of reprogrammed somatic cells. *Molecular and Cellular Biology, 24*(13), 5710—5720. https://doi.org/10.1128/MCB.24.13.5710-5720.2004

Kubo, H., Shimizu, M., Taya, Y., Kawamoto, T., Michida, M., Kaneko, E., ... Kato, Y. (2009). Identification of mesenchymal stem cell (MSC)-transcription factors by microarray and knockdown analyses, and signature molecule-marked MSC in bone marrow by immunohistochemistry. *Genes to Cells, 14*(3), 407—424. https://doi.org/10.1111/j.1365-2443.2009.01281.x

Kunath, T., Saba-El-Leil, M. K., Almousailleakh, M., Wray, J., Meloche, S., & Smith, A. (2007). FGF stimulation of the Erk1/2 signalling cascade triggers transition of pluripotent embryonic stem cells from self-renewal to lineage commitment. *Development, 134*(16), 2895—2902. https://doi.org/10.1242/dev.02880

Lander, E. S., Linton, L. M., Birren, B., Nusbaum, C., Zody, M. C., Baldwin, J., ... International Human Genome Sequencing, C. (2001). Initial sequencing and analysis of the human genome. *Nature, 409*(6822), 860—921. https://doi.org/10.1038/35057062

Langfelder, P., & Horvath, S. (2008). WGCNA: An R package for weighted correlation network analysis. *BMC Bioinformatics, 9*, 559. https://doi.org/10.1186/1471-2105-9-559

Lee, H. J., Jeon, M., Kim, Y. H., Kim, S. O., & Lee, K. E. (2023). Comparative gene expression analysis of stemness between periodontal ligament and umbilical cord tissues in humans. *Journal of Dental Science, 18*(1), 211—219. https://doi.org/10.1016/j.jds.2022.06.005

Lee, K. E., Kang, C. M., Jeon, M., Kim, S. O., Lee, J. H., & Choi, H. J. (2022). General gene expression patterns and stemness of the gingiva and dental pulp. *Journal of Dental Science, 17*(1), 284−292. https://doi.org/10.1016/j.jds.2021.02.012

Lipshutz, R. J., Fodor, S. P., Gingeras, T. R., & Lockhart, D. J. (1999). High density synthetic oligonucleotide arrays. *Nature Genetics, 21*(Suppl. l), 20−24. https://doi.org/10.1038/4447

Lisitsyn, N., Lisitsyn, N., & Wigler, M. (1993). Cloning the differences between two complex genomes. *Science, 259*(5097), 946−951. https://doi.org/10.1126/science.8438152

Liu, Y. J., Zhang, F., Liu, H. D., & Sun, X. (2017). The application of next-generation sequencing techniques in studying transcriptional regulation in embryonic stem cells. *Yi Chuan, 39*(8), 717−725. https://doi.org/10.16288/j.yczz.16-390

Lo, B., & Parham, L. (2009). Ethical issues in stem cell research. *Endocrine Reviews, 30*(3), 204−213. https://doi.org/10.1210/er.2008-0031

Lockhart, D. J., Dong, H., Byrne, M. C., Follettie, M. T., Gallo, M. V., Chee, M. S., ... Brown, E. L. (1996). Expression monitoring by hybridization to high-density oligonucleotide arrays. *Nature Biotechnology, 14*(13), 1675−1680. https://doi.org/10.1038/nbt1296-1675

Loo, S. J. Q., & Wong, N. K. (2021). Advantages and challenges of stem cell therapy for osteoarthritis (Review). *Biomedical Reports, 15*(2), 67. https://doi.org/10.3892/br.2021.1443

Mahla, R. S. (2016). Stem cells applications in regenerative medicine and disease therapeutics. *International Journal of Cell Biology*, 6940283. https://doi.org/10.1155/2016/6940283

Maitra, A., Arking, D. E., Shivapurkar, N., Ikeda, M., Stastny, V., Kassauei, K., ... Chakravarti, A. (2005). Genomic alterations in cultured human embryonic stem cells. *Nature Genetics, 37*(10), 1099−1103. https://doi.org/10.1038/ng1631

Mantsoki, A., Devailly, G., & Joshi, A. (2016). Gene expression variability in mammalian embryonic stem cells using single cell RNA-seq data. *Computational Biology and Chemistry, 63*, 52−61. https://doi.org/10.1016/j.compbiolchem.2016.02.004

Marks, H., Kalkan, T., Menafra, R., Denissov, S., Jones, K., Hofemeister, H., ... Stunnenberg, H. G. (2012). The transcriptional and epigenomic foundations of ground state pluripotency. *Cell, 149*(3), 590−604. https://doi.org/10.1016/j.cell.2012.03.026

Matsuda, T., Nakamura, T., Nakao, K., Arai, T., Katsuki, M., Heike, T., & Yokota, T. (1999). STAT3 activation is sufficient to maintain an undifferentiated state of mouse embryonic stem cells. *The EMBO Journal, 18*(15), 4261−4269. https://doi.org/10.1093/emboj/18.15.4261

Merkle, F. T., Ghosh, S., Genovese, G., Handsaker, R. E., Kashin, S., Meyer, D., ... Eggan, K. (2022). Whole-genome analysis of human embryonic stem cells enables rational line selection based on genetic variation. *Cell Stem Cell, 29*(3), 472−486. https://doi.org/10.1016/j.stem.2022.01.011

Mole, M. A., Coorens, T. H. H., Shahbazi, M. N., Weberling, A., Weatherbee, B. A. T., Gantner, C. W., ... Zernicka-Goetz, M. (2021). A single cell characterisation of human embryogenesis identifies pluripotency transitions and putative anterior hypoblast centre. *Nature Communications, 12*(1), 3679. https://doi.org/10.1038/s41467-021-23758-w

Moreira de Mello, J. C., Fernandes, G. R., Vibranovski, M. D., & Pereira, L. V. (2017). Early X chromosome inactivation during human preimplantation development revealed by single-cell RNA-sequencing. *Scientific Reports, 7*(1), 10794. https://doi.org/10.1038/s41598-017-11044-z

Mossahebi-Mohammadi, M., Quan, M., Zhang, J. S., & Li, X. (2020). FGF signaling pathway: A key regulator of stem cell pluripotency. *Frontiers in Cell and Developmental Biology, 8*, 79. https://doi.org/10.3389/fcell.2020.00079

Mullen, A. C., Orlando, D. A., Newman, J. J., Loven, J., Kumar, R. M., Bilodeau, S., ... Young, R. A. (2011). Master transcription factors determine cell-type-specific responses to TGF-beta signaling. *Cell, 147*(3), 565−576. https://doi.org/10.1016/j.cell.2011.08.050

Mullen, A. C., & Wrana, J. L. (2017). TGF-Beta family signaling in embryonic and somatic stem-cell renewal and differentiation. *Cold Spring Harbor Perspectives in Biology, 9*(7). https://doi.org/10.1101/cshperspect.a022186

Munoz, N., Kim, J., Liu, Y., Logan, T. M., & Ma, T. (2014). Gas chromatography-mass spectrometry analysis of human mesenchymal stem cell metabolism during proliferation and osteogenic differentiation under different oxygen tensions. *Journal of Biotechnology, 169*, 95−102. https://doi.org/10.1016/j.jbiotec.2013.11.010

Mushtaq, M., Kovalevska, L., Darekar, S., Abramsson, A., Zetterberg, H., Kashuba, V., ... Kashuba, E. (2020). Cell stemness is maintained upon concurrent expression of RB and the mitochondrial ribosomal protein S18-2. *Proceedings of the National Academy of Sciences of the United States of America, 117*(27), 15673−15683. https://doi.org/10.1073/pnas.1922535117

Nahar, S., Nakashima, Y., Miyagi-Shiohira, C., Kinjo, T., Kobayashi, N., Saitoh, I., ... Fujita, J. (2018). A comparison of proteins expressed between human and mouse adipose-derived mesenchymal stem cells by a proteome analysis through liquid chromatography with tandem mass spectrometry. *International Journal of Molecular Sciences, 19*(11). https://doi.org/10.3390/ijms19113497

Navin, N., Kendall, J., Troge, J., Andrews, P., Rodgers, L., McIndoo, J., ... Wigler, M. (2011). Tumour evolution inferred by single-cell sequencing. *Nature, 472*(7341), 90−94. https://doi.org/10.1038/nature09807

Nichols, J., Zevnik, B., Anastassiadis, K., Niwa, H., Klewe-Nebenius, D., Chambers, I., ... Smith, A. (1998). Formation of pluripotent stem cells in the mammalian embryo depends on the POU transcription factor Oct4. *Cell, 95*(3), 379−391. https://doi.org/10.1016/s0092-8674(00)81769-9

Niwa, H., Burdon, T., Chambers, I., & Smith, A. (1998). Self-renewal of pluripotent embryonic stem cells is mediated via activation of STAT3. *Genes & Development, 12*(13), 2048−2060. https://doi.org/10.1101/gad.12.13.2048

Ong, S. E., & Mann, M. (2007). Stable isotope labeling by amino acids in cell culture for quantitative proteomics. *Methods in Molecular Biology, 359*, 37−52. https://doi.org/10.1007/978-1-59745-255-7_3

Orkin, S. H. (2011). Genome medicine: Stem cells, genomics and translational research. *Genome Medicine, 3*(6), 34. https://doi.org/10.1186/gm250

Ortuno-Costela, M. D. C., Cerrada, V., Garcia-Lopez, M., & Gallardo, M. E. (2019). The challenge of bringing iPSCs to the patient. *International Journal of Molecular Sciences, 20*(24). https://doi.org/10.3390/ijms20246305

Ozsolak, F. (2012). Third-generation sequencing techniques and applications to drug discovery. *Expert Opinion on Drug Discovery, 7*(3), 231−243. https://doi.org/10.1517/17460441.2012.660145

Parisi, S., & Russo, T. (2011). Regulatory role of Klf5 in early mouse development and in embryonic stem cells. *Vitamins and Hormones, 87*, 381−397. https://doi.org/10.1016/B978-0-12-386015-6.00037-8

Pevrani, E., Rao, B. M., Bauwens, C., Yin, T., Wood, G. A., Nagy, A., … Zandstra, P. W. (2007). Niche-mediated control of human embryonic stem cell self-renewal and differentiation. *The EMBO Journal, 26*(22), 4744−4755. https://doi.org/10.1038/sj.emboj.7601896

Perez-Riverol, Y., Alpi, E., Wang, R., Hermjakob, H., & Vizcaino, J. A. (2015). Making proteomics data accessible and reusable: Current state of proteomics databases and repositories. *Proteomics, 15*(5−6), 930−949. https://doi.org/10.1002/pmic.201400302

Pinto, J. P., Kalathur, R. K., Oliveira, D. V., Barata, T., Machado, R. S., Machado, S., … Futschik, M. E. (2015). StemChecker: A web-based tool to discover and explore stemness signatures in gene sets. *Nucleic Acids Research, 43*(W1), W72−W77. https://doi.org/10.1093/nar/gkv529

Piscaglia, A. C., Shupe, T., Gasbarrini, A., & Petersen, B. E. (2007). Microarray RNA/DNA in different stem cell lines. *Current Pharmaceutical Biotechnology, 8*(3), 167−175. https://doi.org/10.2174/138920107780906478

Porter, C. J., Palidwor, G. A., Sandie, R., Krzyzanowski, P. M., Muro, E. M., Perez-Iratxeta, C., & Andrade-Navarro, M. A. (2007). StemBase: A resource for the analysis of stem cell gene expression data. *Methods in Molecular Biology, 407*, 137−148. https://doi.org/10.1007/978-1-59745-536-7_11

Prentice, D. A. (2019). Adult stem cells. *Circulation Research, 124*(6), 837−839. https://doi.org/10.1161/CIRCRESAHA.118.313664

Rao, R. R., & Stice, S. L. (2004). Gene expression profiling of embryonic stem cells leads to greater understanding of pluripotency and early developmental events. *Biology of Reproduction, 71*(6), 1772−1778. https://doi.org/10.1095/biolreprod.104.030395

Rebuzzini, P., Zuccotti, M., Redi, C. A., & Garagna, S. (2015). Chromosomal abnormalities in embryonic and somatic stem cells. *Cytogenetic and Genome Research, 147*(1), 1−9. https://doi.org/10.1159/000441645

Richards, M., Tan, S. P., Chan, W. K., & Bongso, A. (2006). Reverse serial analysis of gene expression (SAGE) characterization of orphan SAGE tags from human embryonic stem cells identifies the presence of novel transcripts and antisense transcription of key pluripotency genes. *Stem Cells, 24*(5), 1162−1173. https://doi.org/10.1634/stemcells.2005-0304

Roth, S. C. (2019). What is genomic medicine? *Journal of the Medical Library Association, 107*(3), 442−448. https://doi.org/10.5195/jmla.2019.604

Ryan, K. M., Hayes, I., Hampson, L., Heyworth, C. M., Clark, A., Wootton, M., … Graham, G. J. (2001). Differentiating embryonal stem cells are a rich source of haemopoietic gene products and suggest erythroid preconditioning of primitive haemopoietic stem cells. *Journal of Biological Chemistry, 276*(12), 9189−9198. https://doi.org/10.1074/jbc.M008354200

Saei Arezoumand, K., Alizadeh, E., Pilehvar-Soltanahmadi, Y., Esmaeillou, M., & Zarghami, N. (2017). An overview on different strategies for the stemness maintenance of MSCs. *Artificial Cells, Nanomedicine and Biotechnology, 45*(7), 1255−1271. https://doi.org/10.1080/21691401.2016.1246452

Sandhya Verma, R. K. G. (2021). Next-generation sequencing: An expedition from workstation to clinical applications. In K. R. A. N. Dey (Ed.), *Translational bioinformatics in healthcare and medicine* (pp. 29−47). Sciencedirect.

Sanger, F., & Coulson, A. R. (1975). A rapid method for determining sequences in DNA by primed synthesis with DNA polymerase. *Journal of Molecular Biology, 94*(3), 441−448. https://doi.org/10.1016/0022-2836(75)90213-2

Semmler, W. (2018). Nuclear magnetic resonance: Pioneers of discovery. *Radiologe, Der, 58*(6), 590−594. https://doi.org/10.1007/s00117-018-0385-5

Shi, J., Chen, Q., Li, X., Zheng, X., Zhang, Y., Qiao, J., … Duan, E. (2015). Dynamic transcriptional symmetry-breaking in pre-implantation mammalian embryo development revealed by single cell RNA-seq. *Development, 142*(20), 3468−3477. https://doi.org/10.1242/dev.123950

Shiio, Y., & Aebersold, R. (2006). Quantitative proteome analysis using isotope-coded affinity tags and mass spectrometry. *Nature Protocols, 1*(1), 139−145. https://doi.org/10.1038/nprot.2006.22

Silva, W. A., Jr., Covas, D. T., Panepucci, R. A., Proto-Siqueira, R., Siufi, J. L., Zanette, D. L., … Zago, M. A. (2003). The profile of gene expression of human marrow mesenchymal stem cells. *Stem Cells, 21*(6), 661−669. https://doi.org/10.1634/stemcells.21-6-661

Singh, A. M., Hamazaki, T., Hankowski, K. E., & Terada, N. (2007). A heterogeneous expression pattern for Nanog in embryonic stem cells. *Stem Cells, 25*(10), 2534−2542. https://doi.org/10.1634/stemcells.2007-0126

Singh, A. M., Reynolds, D., Cliff, T., Ohtsuka, S., Mattheyses, A. L., Sun, Y., … Dalton, S. (2012). Signaling network crosstalk in human pluripotent cells: A Smad2/3-regulated switch that controls the balance between self-renewal and differentiation. *Cell Stem Cell, 10*(3), 312−326. https://doi.org/10.1016/j.stem.2012.01.014

Smyth, M. S., & Martin, J. H. (2000). x ray crystallography. *Molecular Pathology, 53*(1), 8−14. https://doi.org/10.1136/mp.53.1.8

Solter, D. (2006). From teratocarcinomas to embryonic stem cells and beyond: A history of embryonic stem cell research. *Nature Reviews Genetics, 7*(4), 319−327. https://doi.org/10.1038/nrg1827

Son, M. Y., Kwak, J. E., Kim, Y. D., & Cho, Y. S. (2015). Proteomic and network analysis of proteins regulated by REX1 in human embryonic stem cells. *Proteomics, 15*(13), 2220−2229. https://doi.org/10.1002/pmic.201400510

Stewart, C. L., Kaspar, P., Brunet, L. J., Bhatt, H., Gadi, I., Kontgen, F., & Abbondanzo, S. J. (1992). Blastocyst implantation depends on maternal expression of leukaemia inhibitory factor. *Nature, 359*(6390), 76−79. https://doi.org/10.1038/359076a0

Suarez-Farinas, M., Noggle, S., Heke, M., Hemmati-Brivanlou, A., & Magnasco, M. O. (2005). Comparing independent microarray studies: The case of human embryonic stem cells. *BMC Genomics, 6*, 99. https://doi.org/10.1186/1471-2164-6-99

Suzuki, A., Raya, A., Kawakami, Y., Morita, M., Matsui, T., Nakashima, K., … Izpisua Belmonte, J. C. (2006). Nanog binds to Smad1 and blocks bone morphogenetic protein-induced differentiation of embryonic stem cells. *Proceedings of the National Academy of Sciences of the United States of America, 103*(27), 10294−10299. https://doi.org/10.1073/pnas.0506945103

Swain, N., Thakur, M., Pathak, J., & Swain, B. (2020). SOX2, OCT4 and NANOG: The core embryonic stem cell pluripotency regulators in oral carcinogenesis. *Journal of Oral and Maxillofacial Pathology, 24*(2), 368−373. https://doi.org/10.4103/jomfp.JOMFP_22_20

Takahashi, K., Tanabe, K., Ohnuki, M., Narita, M., Ichisaka, T., Tomoda, K., & Yamanaka, S. (2007). Induction of pluripotent stem cells from adult human fibroblasts by defined factors. *Cell, 131*(5), 861−872. https://doi.org/10.1016/j.cell.2007.11.019

Takahashi, K., & Yamanaka, S. (2006). Induction of pluripotent stem cells from mouse embryonic and adult fibroblast cultures by defined factors. *Cell, 126*(4), 663−676. https://doi.org/10.1016/j.cell.2006.07.024

Tanaka, T. S., Kunath, T., Kimber, W. L., Jaradat, S. A., Stagg, C. A., Usuda, M., … Ko, M. S. (2002). Gene expression profiling of embryo-derived stem cells reveals candidate genes associated with pluripotency and lineage specificity. *Genome Research, 12*(12), 1921−1928. https://doi.org/10.1101/gr.670002

Tang, F., Barbacioru, C., Bao, S., Lee, C., Nordman, E., Wang, X., … Surani, M. A. (2010). Tracing the derivation of embryonic stem cells from the inner cell mass by single-cell RNA-Seq analysis. *Cell Stem Cell, 6*(5), 468−478. https://doi.org/10.1016/j.stem.2010.03.015

Terskikh, A. V., Miyamoto, T., Chang, C., Diatchenko, L., & Weissman, I. L. (2003). Gene expression analysis of purified hematopoietic stem cells and committed progenitors. *Blood, 102*(1), 94−101. https://doi.org/10.1182/blood-2002-08-2509

Thomas, E. D., Lochte, H. L., Jr., Lu, W. C., & Ferrebee, J. W. (1957). Intravenous infusion of bone marrow in patients receiving radiation and chemotherapy. *New England Journal of Medicine, 257*(11), 491−496. https://doi.org/10.1056/NEJM195709122571102

Thomson, J. A., Itskovitz-Eldor, J., Shapiro, S. S., Waknitz, M. A., Swiergiel, J. J., Marshall, V. S., & Jones, J. M. (1998). Embryonic stem cell lines derived from human blastocysts. *Science, 282*(5391), 1145−1147. https://doi.org/10.1126/science.282.5391.1145

Vallier, L., Mendjan, S., Brown, S., Chng, Z., Teo, A., Smithers, L. E., … Pedersen, R. A. (2009). Activin/Nodal signalling maintains pluripotency by controlling Nanog expression. *Development, 136*(8), 1339−1349. https://doi.org/10.1242/dev.033951

Van Hoof, D., Heck, A. J., Krijgsveld, J., & Mummery, C. L. (2008). Proteomics and human embryonic stem cells. *Stem Cell Research, 1*(3), 169−182. https://doi.org/10.1016/j.scr.2008.05.003

van Hoof, D., Krijgsveld, J., & Mummery, C. (2012). Proteomic analysis of stem cell differentiation and early development. *Cold Spring Harbor Perspectives in Biology, 4*(3). https://doi.org/10.1101/cshperspect.a008177

Velculescu, V. E., Zhang, L., Vogelstein, B., & Kinzler, K. W. (1995). Serial analysis of gene expression. *Science, 270*(5235), 484−487. https://doi.org/10.1126/science.270.5235.484

Vihinen, M. (2001). Bioinformatics in proteomics. *Biomolecular Engineering, 18*(5), 241−248. https://doi.org/10.1016/s1389-0344(01)00099-5

Vogel, G. (2005). Cell biology. Ready or not? Human ES cells head toward the clinic. *Science, 308*(5728), 1534−1538. https://doi.org/10.1126/science.308.5728.1534

Wang, J., Rao, S., Chu, J., Shen, X., Levasseur, D. N., Theunissen, T. W., & Orkin, S. H. (2006). A protein interaction network for pluripotency of embryonic stem cells. *Nature, 444*(7117), 364−368. https://doi.org/10.1038/nature05284

Wang, L., Schulz, T. C., Sherrer, E. S., Dauphin, D. S., Shin, S., Nelson, A. M., … Robins, A. J. (2007). Self-renewal of human embryonic stem cells requires insulin-like growth factor-1 receptor and ERBB2 receptor signaling. *Blood, 110*(12), 4111−4119. https://doi.org/10.1182/blood-2007-03-082586

Wen, L., & Tang, F. (2016). Single-cell sequencing in stem cell biology. *Genome Biology, 17*, 71. https://doi.org/10.1186/s13059-016-0941-0

Wen, L., & Tang, F. (2022). Recent advances in single-cell sequencing technologies. *Precision Clinical Medicine, 5*(1), pbac002. https://doi.org/10.1093/pcmedi/pbac002

Wiese, S., Reidegeld, K. A., Meyer, H. E., & Warscheid, B. (2007). Protein labeling by iTRAQ: A new tool for quantitative mass spectrometry in proteome research. *Proteomics, 7*(3), 340−350. https://doi.org/10.1002/pmic.200600422

Williams, R. L., Hilton, D. J., Pease, S., Willson, T. A., Stewart, C. L., Gearing, D. P., … Gough, N. M. (1988). Myeloid leukaemia inhibitory factor maintains the developmental potential of embryonic stem cells. *Nature, 336*(6200), 684−687. https://doi.org/10.1038/336684a0

Wong, D. J., Liu, H., Ridky, T. W., Cassarino, D., Segal, E., & Chang, H. Y. (2008). Module map of stem cell genes guides creation of epithelial cancer stem cells. *Cell Stem Cell, 2*(4), 333−344. https://doi.org/10.1016/j.stem.2008.02.009

Wu, G., Su, S., & Bird, R. C. (1994). Optimization of subtractive hybridization in construction of subtractive cDNA libraries. *Genetic Analysis Techniques and Applications, 11*(2), 29−33. https://doi.org/10.1016/1050-3862(94)90057-4

Xu, R. H., Sampsell-Barron, T. L., Gu, F., Root, S., Peck, R. M., Pan, G., … Thomson, J. A. (2008). NANOG is a direct target of TGFbeta/activin-mediated SMAD signaling in human ESCs. *Cell Stem Cell, 3*(2), 196−206. https://doi.org/10.1016/j.stem.2008.07.001

Yamanaka, S. (2020). Pluripotent stem cell-based cell therapy-promise and challenges. *Cell Stem Cell, 27*(4), 523−531. https://doi.org/10.1016/j.stem.2020.09.014

Yang, Y. H., & Speed, T. (2002). Design issues for cDNA microarray experiments. *Nature Reviews Genetics, 3*(8), 579−588. https://doi.org/10.1038/nrg863

Yan, L., Yang, M., Guo, H., Yang, L., Wu, J., Li, R., … Tang, F. (2013). Single-cell RNA-Seq profiling of human preimplantation embryos and embryonic stem cells. *Nature Structural & Molecular Biology, 20*(9), 1131−1139. https://doi.org/10.1038/nsmb.2660

Ying, Q. L., Nichols, J., Chambers, I., & Smith, A. (2003). BMP induction of Id proteins suppresses differentiation and sustains embryonic stem cell self-renewal in collaboration with STAT3. *Cell, 115*(3), 281−292. https://doi.org/10.1016/s0092-8674(03)00847-x

Ying, Q. L., Stavridis, M., Griffiths, D., Li, M., & Smith, A. (2003). Conversion of embryonic stem cells into neuroectodermal precursors in adherent monoculture. *Nature Biotechnology, 21*(2), 183−186. https://doi.org/10.1038/nbt780

Ying, Q. L., Wray, J., Nichols, J., Batlle-Morera, L., Doble, B., Woodgett, J., … Smith, A. (2008). The ground state of embryonic stem cell self-renewal. *Nature, 453*(7194), 519−523. https://doi.org/10.1038/nature06968

Young, R. A. (2011). Control of the embryonic stem cell state. *Cell, 144*(6), 940−954. https://doi.org/10.1016/j.cell.2011.01.032

Yu, J., Vodyanik, M. A., Smuga-Otto, K., Antosiewicz-Bourget, J., Frane, J. L., Tian, S., ... Thomson, J. A. (2007). Induced pluripotent stem cell lines derived from human somatic cells. *Science, 318*(5858), 1917—1920. https://doi.org/10.1126/science.1151526

Yu, L.-R., Stewart, N. A., & Veenstra, T. D. (2010). Proteomics: The deciphering of the functional genome. In H. W. Geoffrey Ginsburg (Ed.), *Proteomics. The deciphering of the functional genome* (pp. 89—96). Elsevier.

Zalzman, M., Falco, G., Sharova, L. V., Nishiyama, A., Thomas, M., Lee, S. L., ... Ko, M. S. (2010). Zscan4 regulates telomere elongation and genomic stability in ES cells. *Nature, 464*(7290), 858—863. https://doi.org/10.1038/nature08882

Zeng, Y. C., He, J., Bai, Z., Li, Z., Gong, Y., Liu, C., ... Liu, B. (2019). Tracing the first hematopoietic stem cell generation in human embryo by single-cell RNA sequencing. *Cell Research, 29*(11), 881—894. https://doi.org/10.1038/s41422-019-0228-6

Zhang, L., Li, Y., Zhou, H., Li, L., Wang, Y., & Zhang, Y. (2012). Determination of eight amino acids in mice embryonic stem cells by pre-column derivatization HPLC with fluorescence detection. *Journal of Pharmacy Biomedicine Analytical, 66*, 356—358. https://doi.org/10.1016/j.jpba.2012.03.014

Zhao, D. C., Li, Y. M., Ma, J. L., Yi, N., Yao, Z. Y., Li, Y. P., ... Wu, L. Q. (2019). Single-cell RNA sequencing reveals distinct gene expression patterns in glucose metabolism of human preimplantation embryos. *Reproduction, Fertility and Development, 31*(2), 237—247. https://doi.org/10.1071/RD18178

Zhou, Q., Wang, T., Leng, L., Zheng, W., Huang, J., Fang, F., ... Kristiansen, K. (2019). Single-cell RNA-seq reveals distinct dynamic behavior of sex chromosomes during early human embryogenesis. *Molecular Reproduction and Development, 86*(7), 871—882. https://doi.org/10.1002/mrd.23162

Chapter 12

Cutting-edge proteogenomics approaches to analyze stem cells at the therapeutic level

Saifullah Afridi[1,2], Tabassum Zahra[1], Umar Nishan[1] and Daniel C. Hoessli[3]

[1]Department of Chemistry, Kohat University of Science and Technology, Kohat, Khyber Pakhtunkhwa, Pakistan; [2]Department of Allied Health Sciences, Faculty of Life Sciences, Sarhad University of Science & Information Technology (SUIT), Mardan Campus, Mardan, Khyber Pakhtunkhwa, Pakistan; [3]Dr. Panjwani Center for Molecular Medicine and Drug Research, International Center for Chemical and Biological Sciences, University of Karachi, Karachi, Sindh, Pakistan

1. Stem cell characterization

Stem cells are a crucial element in developing, maintaining, and repairing tissues in the human body. However, stem cells have also been implicated in the development of cancers. Cancer stem cells (CSCs) are a subpopulation of cells with a self-renewal capacity that allows tumor initiation, growth, and recurrence. CSCs are also responsible for the resistance of tumors to chemotherapy and radiotherapy by expressing different proteogenomic markers (Kreso & Dick, 2014). Therefore, stem cell culture characterization has two main purposes: tracking the expression of the protein and monitoring the genomic integrity associated with their pluripotency (Fig. 12.1). To ascertain that cells did not undergo any chromosomal changes due to chromosomal loss, gain, or changes in epigenetic profiles, genomic analysis monitoring of cultured stem cells is crucial. As Joseph et al. (2020), Avinash et al. (2017), Marrazzo et al. (2021) did when exploring methods for stem cells, isolation have been explored for therapeutic usage in wound healing and microbial growth inhibition during infection. Similarly, proteomic testing confirmed that cultured cells express all the elements required to maintain the pluripotency of stem cells. Examining the important genetic and protein markers CD44, CD90, STRO-1, and CD146 to establish the condition of the differentiated cells provides an accurate cell type identification (Sebastião et al., 2021). The primary techniques for stem cell analysis are listed in this section, including karyotyping, PCR, RT-qPCR, epigenetic profiling, single-nucleotide polymorphism (SNP) analysis, western blotting, enzyme-linked immunosorbent assay (ELISA), DELFIA, flow cytometry (fluorescence-activated cell sorting [FACS]: Afridi et al., 2015) immunohistochemistry (IHC), biomarker analysis, mass spectrometry (MS), and artificial intelligence (AI) (Arjmand et al., 2022; Avinash et al., 2017; Noell et al., 2018; Tsujimoto et al., 2022). For instance, Noell (2018) and Arjmand et al. (2022) highlighted the main signaling protein markers to target CSCs metabolic processes and their application in respiratory medicine. Broadly, they are divided into three groups.

1.1 Genetic approaches

For stem cell analysis, different genetic techniques are used. These include karyotyping, SNP analysis, fluorescence in situ hybridization (FISH), comparative genomic hybridization (CGH), next-generation sequencing (NGS), and epigenetic profiling. Embryonic stem cells (ESCs) and induced pluripotent stem cells (iPSCs) in particular are more susceptible to genetic instability and should be monitored periodically for chromosomal changes. Different types and classes of stem cells are usually characterized by diverse marker combinations. Alongside with viable stem cell lines, iPSCs screening from adult somatic cells reprogramming or tracking stem cells' development in regenerative medicine and gene therapy for inherited disorders, different marker combinations are utilized (Mahla, 2016; Staal et al., 2019).

Computational Biology for Stem Cell Research. https://doi.org/10.1016/B978-0-443-13222-3.00031-9
Copyright © 2024 Elsevier Inc. All rights reserved, including those for text and data mining, AI training, and similar technologies.

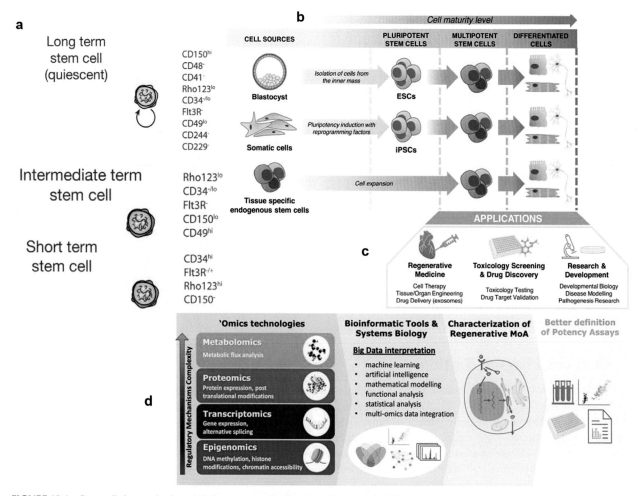

FIGURE 12.1   Stem cell characterizations. (A) shows types and cell markers for stem cells. (B) represents the source and nature of stem cells. Pluripotent cells (ESCs and iPSCs) differentiate into multipotent stem cells. (C) highlights stem cells' diverse therapeutic applications. (D) represents the technologies used to dissect stem cells at therapeutic levels. *Adopted and modified from Sebastião et al. (2021).*

1.1.1 Karyotyping

Karyotyping is a technique used to define cellular chromosome morphology and numbers. Differences in morphology include chromosome size, banding patterns, and position of centromeres. Cultured stem cells can accrue changes including chromosomal abnormalities and genomic alterations that may lead to deviations in cellular functions and gene expression, increasing the risk of tumorigenic stem cell formation. Therefore, checking the stem cell culture for any chromosomal abnormalities, especially those destined for therapeutic application, is essential (Joseph et al., 2020). A stem cell line should typically be karyotyped every 10—15 passages to check for chromosomal deletions, insertions, duplications, centromere loss, or translocations. Chromosomes in a metaphase cell are stained in traditional karyotyping methods to create different banding patterns (Wang, Li, et al., 2022). Giemsa staining, often called as G-band karyotyping, is the most widely used approach. Other techniques include C-banding (constitutive heterochromatin staining), T-banding (telomeric staining), R-banding (reverse Giemsa staining), and Q-banding (quinacrine staining). Changes in banding patterns are also widely used (Lund et al., 2012; Mahla, 2016).

1.1.2 Fluorescence in situ hybridization

FISH is used to hybridize fluorescently labeled DNA overlapping fragments to metaphase or interphase chromosomes. The desired cell chromosomes are chemically/thermally denatured in formalin fixed on the surface and followed by fluorescence microscopy (Dekel-Naftali et al., 2012). Spectral karyotyping (SKY) and multiplex fluorescence in situ hybridization (M-FISH) serve for the detection of chromosomal changes. Briefly, both techniques use fluorescently labeled DNA in hybridization-based combinations. Each chromosome has a distinct spectral signature as a result of labeling with a variety of fluorophores specifically tailored for that chromosome. After the detection of chromosomes' distinctive signatures, a pseudo-colored visualization aid is added and is suits for a presentation visualization. The chromosome regions are then collected and processed separately by the M-FISH and SKY procedures. These two approaches can identify translocations between and within the chromosomes with greater precision than traditional karyotyping. They are unable to identify duplications, deletions, and inversions of about 5 Mb. The primary drawback of FISH for the study of stem cells is the requirement of full-length chromosomal probes. Most commonly, FISH is used for other techniques to seek confirmation using chromosome-specific region probes (Liao et al., 2021; Lu et al., 2021). Fast FISH is usually used to detect aneuploidies, insertions, and deletions ≥20 kb that are found in ESCs and iPSCs.

1.1.3 Comparative genomic hybridization

CGH can detect both copy number variations (CNVs) and chromosomal ploidy. In this technique, different probes are used for DNA labeling from both the reference and the stem cell population, and both DNAs are first denatured and then hybridized to metaphase chromosomes, followed by fluorescence measurements to note any differences between the reference and test samples in their fluorescent signals. Without requiring the exact sequence, this technique finds and identifies CNVs (Dekel-Naftali et al., 2012; Neavin et al., 2021) The most prominent disadvantage of this technique is its lack of sensitivity when applied to whole metaphase chromosomes with a ~5 Mb limit of detection. The use of genomic clones based on microarrays increased the efficiency and detection sensitivity. Usually, CGH does not detect inversions, minor gains, and losses, or balanced translocations. Using arrays for stem cell populations complicates the interpretation of the results in the presence of repetitive regions. To analyze variability among individuals in some regions of chromosomes, the best reference is a DNA sample extracted from early stem cell passage or other stem cells (Dekel-Naftali et al., 2012; Elliott et al., 2010).

1.1.4 Single-nucleotide polymorphism analysis

Single base-pair mutations, often referred to as SNPs, are among the most common genetic variants in DNA. SNPs may accumulate in stem cells over time, which may result in physiological and phenotypic alterations that affect cell development or survival. These genetic modifications could lead to aberrantly increased tumorigenicity or decreased pluripotency. With oligonucleotide arrays and SNP-based high-density genotyping, it is possible to pinpoint the ancestry of a stem cell line and keep track of its genetic integrity. SNPs can be identified using digital PCR, DNA sequencing, and microarrays (Steventon-Jones et al., 2022). Loss of heterozygosity (LOH) in chromosome pairs can be detected by SNP arrays. As a chromosome pair comes from two different parents, many SNPs differences occur. Normally with passages, cultured ESCs are prone to lose heterozygosity that can be linked with an upsurge in tumorigenic potential. Presently, LOH chromosome specific stem cells are being used as cancer models. Similarly, unbalanced translocations, aneuploidy, duplications, and deletions can be detected with SNP arrays. Low degrees of mosaicism may make interpretations difficult, and inversions are not recognized. Similar to comparative genomic hybridization, reference material selection is crucial to assess variations between people. SNP detection and comparative genomic hybridization can be easily integrated into an array (Neavin et al., 2021; Steventon-Jones et al., 2022).

1.1.5 Epigenetic profiling

Due to the epigenetic status of stem cells, their pluripotency is relevant over a broad spectrum. By making DNA accessible to the transcriptional machinery, histone modification and DNA methylation control gene expression (Wen et al., 2020; Yokobayashi et al., 2021). By processing DNA under carefully regulated conditions with bisulfite, which causes uracils to develop from cytosines while leaving methylated cytosines unaltered, DNA methylation patterns can be analyzed. Using a PCR, chip array, or DNA sequencing, it is possible to examine the global methylation patterns conversion in the DNA. The DNA binds to histones by cross-linking; histone modification patterns can be analyzed via chromatin immunoprecipitation (ChIP). Using the appropriate protein or histone-specific antibodies, the cross-linked chromatin can be purified for analysis of the cross-linked regions by real-time qPCR, microarrays, or DNA sequencing (Trapp et al., 2021). Initially, new stem cell type can be categorized by probing the epigenetic status of some key genes, such as Dlk1/MEG3, Notch 1, Xist, H19, Oct4, and PWS/AS (Azizi et al., 2022; Fortress et al., 2023; Masih et al., 2023). Meanwhile, quantitative chromatin structure assessment in cultured cells using specific epigenetics tools can be done. First in the absence or presence of nuclease, chromatin is digested, purified, quantitated and then the genomic DNA assessed via real-time PCR by linking epigenetically silenced or expressed genes (Lopez-Lozano et al., 2022). Epigenetic profiling of stem cells has shown that epigenetic memory in iPSCs can be retained to some extent. As cells during development differentiate, patterns of different gene expression also change; however, during dedifferentiation, it varies in different stem cells situations (He et al., 2020; Wen et al., 2020). Therefore, newly produced iPSC must be screened for epigenetic memory retention analysis. For instance, Fortress et al. (2023) used these techniques for stem cell characterization in retinal disease with stem cell therapy.

1.1.6 Digital, RT, and qRT-PCR

To evaluate the gene regulation profile of any cell type, RT-digital PCR and reverse transcription quantitative PCR (RT-qPCR) offer sensitive, quick, and quantitative approaches. If pluripotency is maintained, it depends on how much Oct4, SOX2, NANOG transcriptional factors, and other regulators are present (Azizi et al., 2022; Ma et al., 2021). RT-qPCR multiplexed panel assays offer an effective method to identify and measure transcription factors, kinases, and other molecules that are important for both preserving pluripotency and cell differentiation. Having the ability to perform multiplex transcriptome analysis increases the sensitivity of assessing the health of cells (Refat et al., 2021). Even with lesser amounts of nucleic acid, digital PCR can increase measurement accuracy. Refat et al. (2021) used these key techniques for stem cell application in diabetic patients to correct pulmonary dysfunction. Droplet digital PCR enables absolute transcript measurement and rare sequence finding. In addition to assisting with transcriptome research, digital PCR is remarkably accurate for finding SNPs and CNVs.

1.1.7 Next-generation sequencing

The term "NGS" refers to high-throughput DNA sequencing methods not based on the Sanger procedures (Fig. 12.2) (Shendure & Ji, 2008). NGS sequences millions or billions of DNA fragments simultaneously, thus increasing throughput significantly and reducing the requirement for fragment-cloning techniques, which are frequently employed in the Sanger sequencing method. NGS technology can be used for a widespread variety of sequencing ranging from individual cell genes to entire genomes (Merkle et al., 2022). Similarly, NGS-based microarrays, RNA-seq, single transcripts, and full transcriptome analysis are also frequently used and produce enormous amounts of data that can be applied to different stem cell analysis (Kim et al., 2021; Kumar et al., 2021; Liao et al., 2021). For instance, Kim et al. (2021) used next-generation sequencing to track stem cells treatment in acute myeloid leukemias.

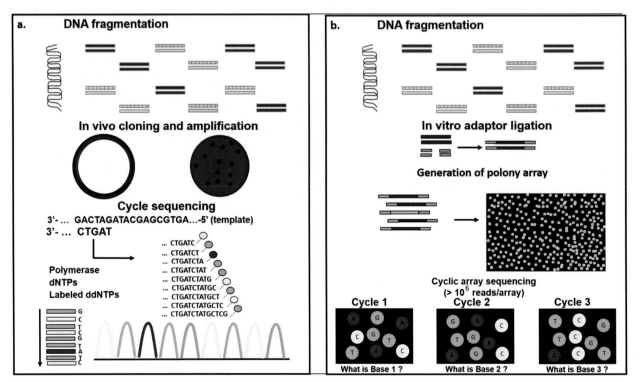

FIGURE 12.2 Difference between Sanger and next-generation sequencing. (A) Sanger-based sequencing steps. (B) NGS-based sequencing steps. *Adopted from Jay Shendure and Hanlee Ji.* Nature-Biotechnology *27, 1135−1145 (2008).*

1.2 Proteomic approaches

For the proteomic analysis of stem cells, a series of different protein-based techniques are used (Pinada et al., 2021; Venkatesh et al., 2021). These include enzyme-linked immunoassay (ELISA), dissociation-enhanced lanthanide fluorescence immunoassay (DELFIA), FACS, immunocytology, IHC, and MS (Guerin et al., 2021; Nowzari et al., 2021; Červenka et al., 2021).

1.2.1 Flow cytometry

For the examination of stem cells, flow cytometry is widely used. Cell sorting is used to physically divide various stem cell populations, while fluorescently labeled monoclonal antibodies serve to differentiate particular cell populations. For instance, iPSCs can be distinguished from other cells that have been or have not been reprogrammed or not have been reprogrammed as a somatic cell population (Belotti et al., 2021; Simard et al., 2023). Cell sorting is frequently used for stem cell culture expansion and maintenance. Sorting can help to keep the population homogenous by removing undesirable cell types. Cellular phenotypes can be identified using a variety of well-characterized cell surface markers, which can also be used to divide mixed populations (DeVilbiss et al., 2021; Schoof et al., 2021). As markers, stem cells also express a number of cell surface proteins for pluripotency and lineage characterization. For instance, the keratin sulfate antigens TRA-1−81 and TRA-1−60 as well as the glycolipid antigens SSEA-3 and SSEA-4 are often used for sorting and detecting ESC populations. Similarly, on the basis of transcription factors expression such as Oct-3/4 and NANOG, fixed stem cells can also be sorted (Morita et al., 2022). When stem cells differentiate the expression of additional markers and loss of these markers can be used to monitor the population's lineages and differentiation (Klyuchnikov et al., 2021; Meyfour et al., 2021). For instance, Belotti et al. (2021) used this technique to analyze stem cell therapeutic potential in multiple myeloma patients' treatment.

1.2.1.1 Immunocyto- and histochemistry

To confirm whether proteins are expressed in stem cells or not, IHC is frequently performed. Prior to fluorescence imaging, antibodies targeting particular proteins are used for the cells' (often fixed) treatment. With the help of this procedure, subcellular localization and relative expression levels of the target proteins, and the number of cells expressing specific proteins can be easily determined (Amirian et al., 2022; Nowzari et al., 2021). IHC is another technique used to detect stem cells in tissues, including when they engraft after transplantation (Azizi et al., 2022; Niknejad et al., 2021). It is possible to transduce stem cells expressing a protein that can be monitored in tissue- and organism-specific experimental contexts. IHC has the capacity to distinguish between transplanted and any resident stem cells by imaging the target protein that is not expressed in a particular tissue or organism (Dembitskaya et al., 2022; Song et al., 2020; Yang et al., 2020). Using antibody sandwich assays with anti GFP (green fluorescent protein) antibody offer great signal amplification. The GFP tag can be detected either indirectly by using an anti-GFP antibody or directly by green fluorescence analysis. Amirian, Azizi et al. (2022), Song et al. (2020) used these key techniques to analyze different cells types, their correlations with disease treatment, and progression including nonobstructive azoospermia and cancer.

1.2.2 ELISA and DELFIA assays

In vitro biodetection immunoassay is an important analytical tool, which relies on the highest specific molecular recognition between an antigen and antibody. Immunoassays have several uses in disease diagnosis, food safety inspection, and environmental monitoring due to their great advantages of high sensitivity and excellent specificity. Currently, a number of immunoassay-based technologies have been developed by simply fusing the traditional immunoassay format with a number of signal transducers, including colorimetry, chemiluminescence, fluorescence, electrochemistry, light scattering, chirality, and magnetic signals. Due to its greater sensitivity, fluorescence immunoassay is a desirable replacement for conventional colorimetric and chemiluminescence immunoassays. One of the most sensitive fluorescence bioanalysis techniques is the DELFIA (dissociation-enhanced lanthanide (Ln^{3+}) fluoroimmunoassay). In the DELFIA assay, the particular immunological recognition of targets with a Eu^{3+-} based luminous signal transducer is applied by using antibodies that have been tagged with europium (Eu^{3+}) chelates as molecular probes (Allicotti, Borras, Pinilla et al., 2003; Singh, 2021). The nonluminescent Eu^{3+} chelate tagged on the antibody is induced to change into a highly luminescent Eu^{3+} micelle in a typical DELFIA, which results in a greatly enhanced luminosity via the "antenna effect" from the enhancer solution. Additionally, the DELFIA approach can significantly boost sensitivity by employing the time-resolved luminescence/fluorescence (TRL/TRF) methodology, which entirely eliminates background signal noise coming from scattered light and short-lived autofluorescence.

Bead-based multiplex immunoassays can measure the phosphorylation statuses of many proteins involved in signal transduction as well as monitor proteins, cytokines, and chemokines simultaneously in any stem cell culture type (Allicotti et al., 2003).

1.2.3 Western blotting

Western blotting is an important technique used to assess the efficacy of transfection as well as to analyze the mechanism of stem cell differentiation. Regulation of gene expression either introduced or silenced by antisense RNA or RNAi molecule. Quantitation, and degree of protein expression by western blotting is frequently combined with measurements of transcript levels by RT-qPCR. Zhu et al. (2021) analyzed the MSC treatment outcome in COVID-19 patients using this technique. The impact of transfection on the expression of downstream proteins can easily be examined with accuracy using western blots (Du et al., 2022; Zhu et al., 2021).

1.2.4 Mass spectrometry—based proteomic biomarker analysis

In stem cells comparative protein levels, posttranslational modifications and protein interactions can easily be analyzed using MS (Prantl et al., 2020). With the help of MS analysis, protein and peptide fingerprinting can easily reveal the biological pathways regulation tracking and analysis (Liang et al., 2021, pp. 159—179; Vaňhara et al., 2020). Using this method, stem cell differentiation, lineage-specific characteristics, and pluripotency can be assessed. The capability of MS techniques permits the examination of stem cells biological reaction, pathways, and signaling networks monitoring in real-time analysis. It has been recently used for PBMC-based metabolite detection in multiple myeloma stem cell transplant patients (Derman et al., 2021; Vaňhara et al., 2020; Zhu et al., 2021).

1.2.5 CyTOF or mass cytometry analysis

CyTOF also named mass cytometry is a powerful technique used to analyze different cell populations systematically. Conceptually, metal-conjugated antibodies are used for cell markers tagging detection (Zhang, Warden, Li, Ding, 2020). CyTOF is working on the mass spectrometer time-of-flight (TOF) principle with exceptionally precise measurement of the mass-to-charge ratio of an ion. The first metal is ionized within known electric field strength followed by time and distance measurement of ion to reach the detector (Fig. 12.3). Then the detected ions are collected to examine the expression of the associated antibodies, permitting the cell type determination by the tagged antibodies to be recognized. It therefore helps to define any cell population with high sensitivity at the supramolecular level. Despite being usually considered a variation-based flow cytometry, CyTOF varies from FACS by many features including sensitivity, dimensionality, cell throughput and background noise (Derman et al., 2021; DeVilbiss et al., 2021). CyTOF has benefited greatly from clever transition and fluorescence readout to heavy metal-tagged antibody probes-based detection. The dimensionality increase represents a real revolution for analysis. Irrespective of normal FACS of three to seven color fluorescence signals overlapping issues, CyTOF has the capability up to 45 parameters analysis simultaneously with a reliable quantification (Mohamad & Capitano, 2022). Theoretically using isotopic discrimination, CyTOF can distinguish between ≥100 different elemental masses, indicating its capability even at further higher multiplex applications in near future. It is therefore possible to screen stem cells and system-wide CyTOF profiling with minimal cross-talk and compensation constriction (Wang, Xu, et al., 2022; Zhang et al., 2020).

1.3 Artificial intelligence

As a new emerging field, AI has revolutionized various fields, including stem cell research, enabling researchers to accelerate the discovery of new treatments for complex diseases (Mukherjee et al., 2021; Ye et al., 2020). The potential applications of AI in stem cell research are numerous and include identifying new targets for drug development, improving cell culture and differentiation methods, and predicting the efficacy of various drug candidates (Coronnello et al., 2022, pp. 1—11). For instance, one of the major challenges in stem cell research is identifying the optimal conditions for culturing and differentiating stem cells. This process can be time-consuming and requires careful monitoring of various parameters, such as pH, temperature, and nutrient levels. However, AI algorithms can analyze large datasets to identify the optimal conditions for stem cell culture and differentiation, reducing the time and cost required for this process (Dembitskaya et al., 2022; Ouyang et al., 2022). As recently showed, AI algorithms could optimize the differentiation of human iPSCs into cardiac cells (Coronnello et al., 2022, pp. 1—11). The researchers used AI to identify the optimal growth factors combination and signaling molecules required for efficient differentiation. This study highlights the potential of AI in optimizing cell differentiation protocols, which could have significant implications for regenerative medicine.

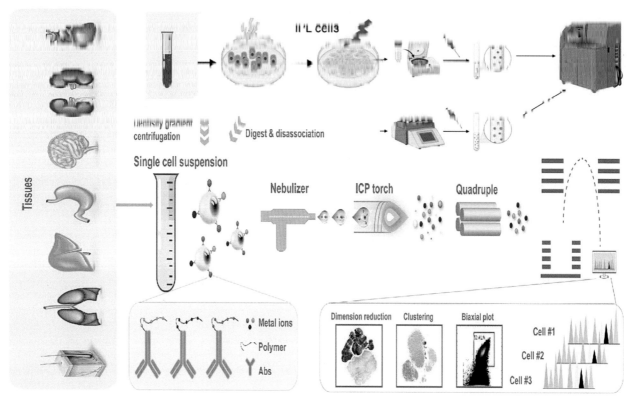

FIGURE 12.3   CyTOF workflow principle. Desired cell populations at the single-cell suspension level are tagged with metal-conjugated antibodies followed by metal ionization, detection, and analysis. *Adopted and Modified from Zhang et al. (2020).*

Another area where AI is making significant contributions to stem cell research in drug discovery. By using AI algorithms to analyze large datasets of molecular structures, researchers can identify new drug candidates and predict their efficacy in vitro and in vivo. For instance, AI algorithms could be used to predict the efficacy of various drugs on human iPSC-derived cardiomyocytes and T cell populations (Afridi et al., 2021; Shad et al., 2018; Topno et al., 2021). They trained a deep learning algorithm using data from over 3×10^4 individual cells and used it to predict the effects of various drugs on the cells. Their finding suggests that AI could be used to accelerate drug discovery and reduce the time and cost required for clinical trials. Meanwhile, AI is also being used to identify new targets for drug development. By analyzing large datasets of gene expression profiles, researchers can identify new signaling pathways and unique genes that are involved in stem cell differentiation and disease development (Kusumoto et al., 2022). AI algorithms could be used to identify novel genes that regulate the differentiation of neural stem cells. They used a machine learning algorithm to identify genes that were highly expressed during the early stages of differentiation and then validated these findings using genetic manipulation techniques (Coronnello et al., 2022, pp. 1−11; Topno et al., 2021). This study highlights the potential of AI in identifying new targets for stem cell−based drug development, which could have significant implications for the treatment of several diseases.

2. Conclusion

In this chapter, we highlighted the most important key techniques used to investigate stem cells at transcriptional and translational levels. Using these techniques can analyze the target stem cell population with full genes and protein profiling linked with pathways, histone modification, receptor guidance, coexpressions, cotransport, biodegradation, biosensing, differential analysis, cell fate, and their final destination. Meanwhile, the potential applications of AI in stem cell research and recent advances in this field demonstrate its significant impact. AI algorithms are being used to optimize stem cell culture and differentiation protocols, accelerate drug discovery, and identify new targets for drug development. These advances have the potential to revolutionize stem cell regenerative medicine and could lead to the discovery of new treatments for complex diseases.

Author contributions
SA and TZ wrote the manuscript, and UN and DCH critically evaluated, modified, and finalized the manuscript.

References

Afridi, S., Adnan, M., Hameed, M. W., Khalil, A. W., Iqbal, Z., Hoessli, D. C., … Zhang, X. J. B.c. (2021). Small organic molecules accelerate the expansion of regulatory T cells. *Bioorganic Chemistry, 111*, 104908.

Afridi, S., Shaheen, F., Roetzschke, O., Shah, Z. A., Abbas, S. C., Siraj, R., … communications, b. r. (2015). A cyclic peptide accelerates the loading of peptide antigens in major histocompatibility complex class II molecules. *Biochemical and Biophysical Research Communications, 456*(3), 774—779.

Allicotti, G., Borras, E., Pinilla, C. J.o. I., & Immunochemistry. (2003). A time-resolved fluorescence immunoassay (DELFIA) increases the sensitivity of antigen-driven cytokine detection. *Journal of Immunoassay and Immunochemistry, 24*(4), 345—358.

Amirian, M., Azizi, H., Hashemi Karoii, D., & Skutella, T. J. S. R. (2022). VASA protein and gene expression analysis of human non-obstructive azoospermia and normal by immunohistochemistry. *Immunocytochemistry, and Bioinformatics Analysis, 12*(1), 17259.

Arjmand, B., Hamidpour, S. K., Alavi-Moghadam, S., Yavari, H., Shahbazbadr, A., Tavirani, M. R., … Larijani, B. J. F.i. p. (2022). Molecular docking as a therapeutic approach for targeting cancer stem cell metabolic processes. *Frontiers in Pharmacology, 13*.

Avinash, K., Malaippan, S., & Dooraiswamy, J. N. J. I.j. o. s. c. (2017). Methods of isolation and characterization of stem cells from different regions of oral cavity using markers: a systematic review. *International Journal of Stem Cells, 10*(1), 12—20.

Azizi, H., Karoii, D. H., & Skutella, T. (2022). *Protein and gene expression analysis in the differentiation of spermatogonia stem cells into functional mature neurons by immunohistochemistry, immunocytochemistry, and bioinformatics analysis.*

Belotti, A., Ribolla, R., Cancelli, V., Villanacci, A., Angelini, V., Chiarini, M., … Ferrari, S. J. C. M. (2021). Predictive role of diffusion-weighted whole-body MRI (DW-MRI) imaging response according to MY-RADS criteria after autologous stem cell transplantation in patients with multiple myeloma and combined evaluation with MRD assessment by flow cytometry. *Cancer Medicine, 10*(17), 5859—5865.

Červenka, J., Tylečková, J., Kupcová Skalníková, H., Vodičková Kepková, K., Poliakh, I., Valeková, I., … Pánková, T. J. F.i. c. n. (2021). Proteomic characterization of human neural stem cells and their secretome during in vitro differentiation. *Frontiers in Cellular Neuroscience, 14*, 612560.

Coronnello, C., Francipane, M. G. J. S. C. R., & Reports. (2022). *Moving towards induced pluripotent stem cell-based therapies with artificial intelligence and machine learning.*

Dekel-Naftali, M., Aviram-Goldring, A., Litmanovitch, T., Shamash, J., Reznik-Wolf, H., Laevsky, I., … Hourvitz, A. J. E. j. o. h. g (2012). Screening of human pluripotent stem cells using CGH and FISH reveals low-grade mosaic aneuploidy and a recurrent amplification of chromosome 1q. *European Journal of Human Genetics, 20*(12), 1248—1255.

Dembitskaya, Y., Boyce, A., Idziak, A., Pourkhalili, A., Le Bourdeelles, G., Girard, J., … de Amezaga, A. O. (2022). *Shadow imaging for panoptical visualization of brain tissue in vivo.*

Derman, B. A., Stefka, A. T., Jiang, K., McIver, A., Kubicki, T., Jasielec, J. K., & Jakubowiak, A. J. J. B. C. J. (2021). Measurable residual disease assessed by mass spectrometry in peripheral blood in multiple myeloma in a phase II trial of carfilzomib, lenalidomide, dexamethasone and autologous stem cell transplantation. *Blood Cancer Journal, 11*(2), 19.

DeVilbiss, A. W., Zhao, Z., Martin-Sandoval, M. S., Ubellacker, J. M., Tasdogan, A., Agathocleous, M., … Morrison, S. J. J. E. (2021). Metabolomic profiling of rare cell populations isolated by flow cytometry from tissues. *eLife, 10*, e61980.

Du, Y., Liu, Y., Zhou, Y., Zhang, P. J. S. C. R., & Therapy. (2022). Knockdown of CDC20 promotes adipogenesis of bone marrow-derived stem cells by modulating β-catenin. *Stem Cell Research & Therapy, 13*(1), 1—13.

Elliott, A. M., Hohenstein Elliott, K. A., & Kammesheidt, A. J. M. B. (2010). High resolution array-CGH characterization of human stem cells using a stem cell focused microarray. *Molecular Biotechnology, 46*, 234—242.

Fortress, A. M., Miyagishima, K. J., Reed, A. A., Temple, S., Clegg, D. O., Tucker, B. A., … Ludwig, T. E. (2023). Stem cell sources and characterization in the development of cell-based products for treating retinal disease: An NEI Town Hall report. *BioMed Central.*

Guerin, C. L., Guyonnet, L., Goudot, G., Revets, D., Konstantinou, M., Chipont, A., … Reports. (2021). Multidimensional proteomic approach of endothelial progenitors demonstrate expression of KDR restricted to CD19 cells. *Stem Cell Reviews and Reports, 17*, 639—651.

He, Q., Wang, L., Zhao, R., Yan, F., Sha, S.Cui, C., … (2020). Mesenchymal stem cell-derived exosomes exert ameliorative effects in type 2 diabetes by improving hepatic glucose and lipid metabolism via enhancing autophagy. *Stem Cell Research & Therapy, 11*(1), 1—14.

Joseph, A., Baiju, I., Bhat, I. A., Pandey, S., Bharti, M., Verma, M., … Saikumar, G. J. J.o. C. P. (2020). Mesenchymal stem cell-conditioned media: A novel alternative of stem cell therapy for quality wound healing. *Journal of Cellular Physiology, 235*(7—8), 5555—5569.

Kim, H.-J., Kim, Y., Kang, D., Kim, H. S., Lee, J.-M., Kim, M., & Cho, B.-S. J. B. C. J. (2021). Prognostic value of measurable residual disease monitoring by next-generation sequencing before and after allogeneic hematopoietic cell transplantation in acute myeloid leukemia. *Blood Cancer Journal, 11*(6), 109.

Klyuchnikov, E., Christopeit, M., Badbaran, A., Bacher, U., Fritzsche-Friedland, U., von Pein, U. M., … Kröger, N. J. E.j. o. h. (2021). Role of pre-transplant MRD level detected by flow cytometry in recipients of allogeneic stem cell transplantation with AML. *European Journal of Haematology, 106*(5), 606—615.

Kreso, A., & Dick, J. E. J. C.s. c. (2014). Evolution of the cancer stem cell model. *Cell Stem Cell, 14*(3), 275—291.

Kumar, A., Bantilan, K., Jacob, A., Park, A., Schoninger, S., & Sauter, C. (2021). Noninvasive monitoring of mantle cell lymphoma by immunoglobulin gene next-generation sequencing in a phase 2 study of sequential chemoradioimmunotherapy followed by autologous stem-cell rescue. *Clin Lymphoma Myeloma Leuk, 21*(4), 230—237.e212.

Kusumoto, D., Yuasa, S., & Fukuda, K. J. P. (2022). Induced pluripotent stem cell-based drug screening by use of artificial intelligence. *Pharmaceuticals (Basel), 15*(5), 562.

Liang, Y., Truong, T., Zhu, Y., & Kelly, R. T. J. L. S. C. M. (2021). In-depth mass spectrometry-based single-cell and immunoblot proteomics.

Liao, J., Suen, H. C., Luk, A. C. S., Yang, L., Lee, A. W. T., Qi, H., & Lee, T. L. J. P. G. (2021). Transcriptomic and epigenomic profiling of young and aged spermatogonial stem cells reveals molecular targets regulating differentiation. *PLOS Genetics, 17*(7). e1009369.

Lopez Lozano, A. P., Arevalo-Niño, K., Gutierrez-Puente, Y., Montiel-Hernandez, J. L., Urrutia-Baca, V. H., Del Angel-Mosqueda, C., ... Medicine, F. (2022). SSEA-4 positive dental pulp stem cells from deciduous teeth and their induction to neural precursor cells. *Head & Face Medicine, 18*(1), 9.

Lu, Y., Liu, M., Yang, J., Weissman, S. M., Pan, X., Katz, S. G., & Wang, S. J. C. D. (2021). Spatial transcriptome profiling by MERFISH reveals fetal liver hematopoietic stem cell niche architecture. *Nature, 7*(1), 47.

Lund, R. J., Nikula, T., Rahkonen, N., Närvä, E., Baker, D., Harrison, N., ... Lahesmaa, R. J. S. c. r. (2012). High-throughput karyotyping of human pluripotent stem cells. *Stem Cell Research & Therapy, 9*(3), 192−195.

Ma, H., Bell, K. N., Loker, R. N. J. M. T.-M., & Development, C. (2021). qPCR and qRT-PCR analysis: Regulatory points to consider when conducting biodistribution and vector shedding studies. *Molecular Therapy — Methods & Clinical Development, 20*, 152−168.

Mahla, R. S. J. I.j. o. c. b. (2016). Stem cells applications in regenerative medicine and disease therapeutics. *International Journal of Cell Biology*, 2016.

Marrazzo, P., Pizzuti, V., Zia, S., Sargenti, A., Gazzola, D., Roda, B., ... Alviano, F. J. A. (2021). Microfluidic tools for enhanced characterization of therapeutic stem cells and prediction of their potential antimicrobial secretome. *Antibiotics (Basel), 10*(7), 750.

Masih, K. E., Gardner, R. A., Chou, H.-C., Abdelmaksoud, A., Song, Y. K., Mariani, L., ... Adebola, S. O. J. B. A. (2023). A stem cell epigenome is associated with primary nonresponse to CD19 CAR T-cells in pediatric acute lymphoblastic leukemia. *Bloodadvances*, 2022008977.

Merkle, F. T., Ghosh, S., Genovese, G., Handsaker, R. E., Kashin, S., Meyer, D., ... Pato, M. J. C.s. c. (2022). Whole-genome analysis of human embryonic stem cells enables rational line selection based on genetic variation. *Cell Stem Cell, 29*(3), 472−486. e477.

Meyfour, A., Pahlavan, S., Mirzaei, M., Krijgsveld, J., Baharvand, H., Salekdeh, G. H. J. C., & Sciences, M. L. (2021). The quest of cell surface markers for stem cell therapy. *Cellular and Molecular Life Sciences, 78*, 469−495.

Mohamad, S. F., & Capitano, M. L. (2022). Utilizing CyTOF to examine hematopoietic stem and progenitor phenotype. In *Hematopoietic stem cells: Methods and protocols* (pp. 113−126). Springer.

Morita, Y., Kishino, Y., Fukuda, K., & Tohyama, S. J. C. P. (2022). Scalable manufacturing of clinical-grade differentiated cardiomyocytes derived from human-induced pluripotent stem cells for regenerative therapy. *Cell Proliferation, 55*(8), e13248.

Mukherjee, S., Yadav, G., & Kumar, R. J. W. Jo. S. C. (2021). Recent trends in stem cell-based therapies and applications of artificial intelligence in regenerative medicine. *World Journal of Stem Cells, 13*(6), 521.

Neavin, D., Nguyen, Q., Daniszewski, M. S., Liang, H. H., Chiu, H. S., Wee, Y. K., ... Lidgerwood, G. E. J. G.b. (2021). Single cell eQTL analysis identifies cell type-specific genetic control of gene expression in fibroblasts and reprogrammed induced pluripotent stem cells. *Genome Biology, 22*(1), 1−19.

Niada, S., Giannasi, C., Magagnotti, C., Andolfo, A., & Brini, A. T. J. Jo. P. (2021). Proteomic analysis of extracellular vesicles and conditioned medium from human adipose-derived stem/stromal cells and dermal fibroblasts. *Journal of Proteomics, 232*, 104069.

Niknejad, P., Azizi, H., & Sojoudi, K. J. C. R. (2021). POU5F1 protein and gene expression analysis in neonate and adult mouse testicular germ cells by immunohistochemistry and immunocytochemistry. *Cellular Reprogramming, 23*(6), 349−358.

Noell, G., Faner, R., & Agustí, A. J. E. R. R. (2018). From systems biology to P4 medicine: Applications in respiratory medicine. *European Respiratory Review, 27*(147).

Nowzari, F., Wang, H., Khoradmehr, A., Baghban, M., Baghban, N., Arandian, A., ... Najarasl, M. J. F.i. V. S. (2021). Three-dimensional imaging in stem cell-based researches. *Frontiers in Veterinary Science, 8*, 657525.

Ouyang, Q., Yang, W., Wu, Y., Xu, Z., Hu, Y.Hu, N., ... (2022). Multi-labeled neural network model for automatically processing cardiomyocyte mechanical beating signals in drug assessment. *Biosensors and Bioelectronics, 209*, 114261.

Prantl, L., Eigenberger, A., Klein, S., Limm, K., Oefner, P. J., Schratzenstaller, T., ... Surgery, R. (2020). Shear force processing of lipoaspirates for stem cell enrichment does not affect secretome of human cells detected by mass spectrometry in vitro. *Plastic and Reconstructive Surgery, 146*(6), 749e−758e.

Refat, M. S., Hamza, R. Z., Adam, A. M. A., Saad, H. A., Gobouri, A. A., Al-Harbi, F. S., ... El-Megharbel, S. M. J. P. O. (2021). Quercetin/zinc complex and stem cells: A new drug therapy to ameliorate glycometabolic control and pulmonary dysfunction in diabetes mellitus: Structural characterization and genetic studies. *PLoS One, 16*(3). e0246265.

Schoof, E. M., Furtwängler, B., Üresin, N., Rapin, N., Savickas, S., Gentil, C., ... Porse, B. T. J. N.c. (2021). Quantitative single-cell proteomics as a tool to characterize cellular hierarchies. *Nature Communications, 12*(1), 3341.

Sebastião, M. J., Serra, M., Gomes-Alves, P., & Alves, P. M. J. C. O.i. B. (2021). Stem cells characterization: OMICS reinforcing analytics. *Current Opinion in Biotechnology, 71*, 175−181.

Shad, K. F., Salman, S., Afridi, S., Tariq, M., & Asghar, S. (2018). Introductory chapter: Ion channels. In *Ion channels in health and sickness*. IntechOpen.

Shendure, J., & Ji, H. J. N.b. (2008). Next-generation DNA sequencing. *Nature Biotechnology, 26*(10), 1135−1145.

Simard, C., Fournier, D., & Trepanier, P. J. I. Jo. L. H. (2023). Validation of a rapid potency assay for cord blood stem cells using phospho flow cytometry: The IL-3-pSTAT5 assay. *International Journal of Laboratory Hematology, 45*(1), 46−52.

Singh, A. J. N.m. (2021). Towards resolving proteomes in single cells. *Nature Methods, 18*(8), 856.

Song, G., Shi, Y., Zhang, M., Goswami, S., Afridi, S., Meng, L., ... Zhang, J. J. C. D. (2020). Global immune characterization of HBV/HCV-related hepatocellular carcinoma identifies macrophage and T-cell subsets associated with disease progression. *Cell Discovery, 6*(1), 90.

Staal, F. J., Aiuti, A., & Cavazzana, M. J. F.i. p. (2019). Autologous stem-cell-based gene therapy for inherited disorders: State of the art and perspectives. *Frontiers in Pediatrics, 7*, 443.

Steventon-Jones, V., Stavish, D., Halliwell, J. A., Baker, D., & Barbaric, I. J. C. P. (2022). Single nucleotide polymorphism (SNP) arrays and their sensitivity for detection of genetic changes in human pluripotent stem cell cultures. *Current Protocols, 2*(11), e606.

Topno, R., Singh, I., Kumar, M., & Agarwal, P. J. B.c. (2021). Integrated bioinformatic analysis identifies UBE2Q1 as a potential prognostic marker for high grade serous ovarian cancer. *BMC Cancer, 21*, 1–13.

Trapp, A., Kerepesi, C., & Gladyshev, V. N. J. N. A. (2021). Profiling epigenetic age in single cells. *Nature Aging, 1*(12), 1189–1201.

Tsujimoto, H., Katagiri, N., Ijiri, Y., Sasaki, B., Kobayashi, Y., Mima, A., ... Osafune, K. J. P.o. (2022). In vitro methods to ensure absence of residual undifferentiated human induced pluripotent stem cells intermingled in induced nephron progenitor cells. *PLoS One, 17*(11). e0275600.

Vaňhara, P., Moráň, L., Pečinka, L., Porokh, V., Pivetta, T., Masuri, S., ... Havel, J. (2020). Intact cell mass spectrometry for embryonic stem cell biotyping. In *Mass spectrometry in life sciences and clinical laboratory*. IntechOpen.

Venkatesh, S., Baljinnyam, E., Tong, M., Kashihara, T., Yan, L., Liu, T., ... Physiology, C. (2021). Proteomic analysis of mitochondrial biogenesis in cardiomyocytes differentiated from human induced pluripotent stem cells. *American Journal of Physiology-Regulatory, Integrative and Comparative Physiology, 320*(4), R547–R562.

Wang, L.-B., Li, Z.-K., Wang, L.-Y., Xu, K., Ji, T.-T., Mao, Y.-H., ... Zhao, Q. J. S. (2022). A sustainable mouse karyotype created by programmed chromosome fusion. *Science, 377*(6609), 967–975.

Wang, Y., Xu, B., & Xue, L. J. E. (2022). Applications of CyTOF in brain immune component studies. *Engineering, 16*, 187–197.

Wen, Z., Mai, Z., Zhu, X., Wu, T., Chen, Y., & Geng, D. (2020). Mesenchymal stem cell-derived exosomes ameliorate cardiomyocyte apoptosis in hypoxic conditions through microRNA144 by targeting the PTEN/AKT pathway. *Stem Cell Research & Therapy, 11*(1), 1–17.

Yang, H., Ye, S., Goswami, S., Li, T., Wu, J., Cao, C., ... Chen, Y. J. I.j. o. c. (2020). Highly immunosuppressive HLADRhi regulatory T cells are associated with unfavorable outcomes in cervical squamous cell carcinoma. *International Journal of Cancer, 146*(7), 1993–2006.

Ye, K., Takemoto, Y., Ito, A., Onda, M., Morimoto, N., Mandai, M., ... Osakada, F. J. S. R. (2020). Reproducible production and image-based quality evaluation of retinal pigment epithelium sheets from human induced pluripotent stem cells. *Scientific Reports, 10*(1), 14387.

Yokobayashi, S., Yabuta, Y., Nakagawa, M., Okita, K., Hu, B., Murase, Y., ... Yamamoto, T. J. C.r. (2021). Inherent genomic properties underlie the epigenomic heterogeneity of human induced pluripotent stem cells. *Cell Reports, 37*(5), 109909.

Zhang, T., Warden, A. R., Li, Y., Ding, X. J. C., & Medicine, T. (2020). Progress and applications of mass cytometry in sketching immune landscapes. *Clinical and Translational Medicine, 10*(6), e206.

Zhu, R., Yan, T., Feng, Y., Liu, Y., Cao, H., Peng, G., ... Hou, W. J. C.r. (2021). Mesenchymal stem cell treatment improves outcome of COVID-19 patients via multiple immunomodulatory mechanisms. *Cell Research, 31*(12), 1244–1262.

Chapter 13

Cheminformatics, metabolomics, and stem cell tissue engineering: A transformative insight

Rajiv Kumar[1], Magali Cucchiarin[2], Agnieszka Maria Jastrzębska[3], Gerardo Caruso[4], Johannes Pernaa[5] and Zarrin Minuchehr[6]

[1]University of Delhi, New Delhi, Delhi, India; [2]Center of Experimental Orthopaedics, Saarland University Medical Center and Saarland University, Homburg, Germany; [3]Warsaw University of Technology, Faculty of Materials Science and Engineering, Warsaw, Poland; [4]Department of Biomedical and Dental Sciences and Morphofunctional Imaging, Unit of Neurosurgery, University of Messina, Messina, Italy; [5]The Unit of Chemistry Teacher Education, Department of Chemistry, Faculty of Science, University of Helsinki, Helsinki, Finland; [6]National Institute of Genetic Engineering and Biotechnology (NIGEB), Tehran, Iran

1. Introduction

Mechanotransduction, biomicroelectromechanical systems, cellular mechanics and cell signaling, cell—matrix interactions, biosystems and biomaterial engineering, microfluidics, gene chips, computational biology, and tissue structure—function visualization are only a few topics covered under the themes of stem cell tissue engineering, metabolomics, and cheminformatics (Rajiv, 2021a). In stem cell tissue engineering and regeneration, stem cells play key roles and operate as supporting cells to regulate healing, growth, and metabolism via the regulation of tissue homeostasis (Kumar et al., 2021). The quantitative and dependable regulation of distinct stem cells fate and behavior, as well as their population, in vitro and in vivo, is a crucial objective of regenerative medicine and bioengineering (Edgar et al., 2020). Stem cells, which are usually termed undistinguishable cells of a multicellular being and have a scaled ability for persistent self-renewal with the capacity to develop into specific cell types, are the central features of this attempt. Stem cells are essential for multicellular organisms to grow from a single cell during embryogenesis and tissue homeostasis (Richter et al., 2001). For research, cell-based therapeutics, and drug testing, stem cells hold the potential to deliver an endless resource of human tissue. It will be necessary to exert this control over the creation of increasingly complex tissue-like structures in addition to precisely controlling stem cell self-renewal and differentiation to realize this promise (Nava et al., 2012). Additionally, stem cells serve as a source of target cells with the ability to differentiate into many distinct types.

Many chemical informatics operations are made easier by cell-based representations, which are derivations of coordinate-based spaces. These images typically experience the "curse of dimensionality." This chapter suggests networks as an appealing paradigm for describing chemical space because they avoid numerous drawbacks of coordinate and cell-based representations, such as the curse of dimensionality (Hao et al., 2020). Computational modeling is an effective method for describing biological assemblies and producing forecasts (Angerer et al., 2017). Cellular measurements and application methods (microarray profiling, sequencing, ChIP-Seq, ChIP-chip, ChIP-Seq of histone modification, CLIP-Seq, RIP-Seq, Bisulfite-Seq, proteomics, Y2H, and DNase hypersensitivity mapping), cellular entity (DNA: SNPs, microRNA, protein—DNA interaction, protein abundance MS-based methods, protein—RNA interaction, epigenetic, DNA methylation, histone modification), and posttranslational modification (Ribo-Seq/ribosome profiling, gene regulatory regions) were utilized for cellular measurements. These tools and techniques have been innovated in various methods that can be incorporated into the nomenclature for big data in biology (Bian & Cahan, 2016). Models can be developed to understand biological association.

Computational Biology for Stem Cell Research. https://doi.org/10.1016/B978-0-443-13222-3.00001-0
Copyright © 2024 Elsevier Inc. All rights reserved, including those for text and data mining, AI training, and similar technologies.

Spectroscopy, spectrometry, cheminformatics, and metabolomics were established during an early instrumental revolution, which can be thought of as an evolved type of chemical analysis. This was monitored by the expansion of compound libraries and database systems, as well as high-throughput and high-performance liquid chromatography (LC)/ gas chromatography (GC)−mass spectrometry (MS) (Ramraje et al., 2020). Metabolomics platforms and technologies used to communicate operations and accelerate data analysis. While providing ontologies and annotations, computational biology deals a circumstantial methodology to efficient depiction of metabolite outlines from a dataset (Tang et al., 2019). Particularly in the fields of medicinal chemistry and chemical biology, the idea of chemical space is becoming more and more significant in many areas of chemical research (Dobson, 2004). Computational stem cell biology will help understand genetics and epigenetic deformities, leading to disease (Zhang, 2011). In this chapter, we further explore key technical steps in acquiring, processing, and analyzing metabolomics data. Software solutions for acquisition and preprocessing data across metabolomics platforms have been developed, for instance, software packages (MS-DIAL, XCMS, MZmine2, Mnova, speaq 2.0, Metaboetabo Analyst, rDolphin, BATMAN, rNMR) with selected features (built-in DIA analysis, annotation and visualization, statistical analysis, batch mode, user-friendly, retention time correction, deconvolution, single suite for processing and visualization, statistical analysis, visualization, peak picking and grouping); multivariate statistical functions, modules for integrative data analysis, enhances reactive oxygen intermediates (ROIs), visualization of nuclear magnetic resonance (NMR) signals (GC/LC/MS, LC/MS, NMR, 1D NMR, and ^{1}H-NMR) have used at various platforms.

Data mining, arithmetical methods, and metabolomics data can be used to understand cell and tissue functions (Fernandes et al., 2019). Bioinformatics analysis of stem cells, computational stem cell biology and evolving natural science, computational tools for stem cell biology, gene expression analysis, computational evolutionary biology, biosensors, novel gene regulatory networks, drug discovery, cellular computing, and biorobotics are among the key areas in which cheminformatics, metabolomics, and stem cell tissue engineering have already proven their worth, and thus, there is a need for a transformative insight, as the author has portrayed her.

2. Computational tools and stem cell biology

Computational science is a discipline of study that uses computers and software engineering to understand processes and visualize assemblies. It entails the use of computational tools to represent and replicate biological frameworks as well as to clarify experimental findings, typically on a large scale. Computational stem cell biology combines molecular data with systems-level modeling (Acheson et al., 2010). Single-cell transcriptomics is predicted to influence expansion and control the fate of individual cells. Exploring the enormous potential of embryonic stem (ES) cells and making key discoveries have been made possible by the combination of computational biology and large-scale biology. A thorough and up-to-date introduction and reference for the computational biology of ES cells are provided in this chapter by the compilation of reviews and articles from many areas of stem cell research and computational biology (Kumar, 2021).

Single-cell computational biology and genomics are revolutionizing stem cell biology by understanding cell processes such as separation, improvement, repair, reinvention, and pluripotency. It is crucial to recognize and consider critical executive traits and transcriptional paths to unambiguously decode these processes at a subatomic level (McBride et al., 2003). Two developments that have lately emerged demand an explanation of the ultramodern in computational stem cell biological science. First, far-reaching molecular statistics and systems-level models of stem cell activity and function have started to be successfully combined via systems biology and network biology methodologies. Second, new-fangled technologies that enable single-cell genome-wide molecular outlining are now advanced. The ability to profile numerous high-quality objects in a single test provided by current high-throughput atomic profiling advances offers an effective means of addressing these questions (Karn, 2013). However, the data generated by such developments are translated via bioinformatics. More recently, computational approaches such as ScoreCard (theoretically conceivable to include new cell types in the lineage scorecard prediction), PluriTest (hypothetically imaginable to prolong to global DNA methylation and RNA-Seq), TeratoScore (tentatively imaginable to lengthen to RNA-Seq, qPCR), KeyGenes, CellNet (Bulk, and single cell RNA-Seq), Mogrify, Heinaniemi (RNA-Seq), D'Alessio, Lang (landscape could be assembled with additional genes, histone modification data, microRNAs), Crespo, and Davis have been used judgmentally during the interpretation of high-throughput statistics into understanding of developing stem cell biology (Bian & Cahan, 2016). Thus, a few subjects covered in this chapter are in-depth studies of epigenome, genome, transcriptome, proteome, and regulatory networks of ES cells, theories of computational biology and methods employed in these studies, and recently released databases and online resources for bioinformatics study. For cluster-based analysis, researchers apply various resources for high-performance computing outputs. Running multiprocessor applications and batch scripting explain the basic structure of computer programs and scripts with a focus on Python scripting, create analytical pipelines to complete

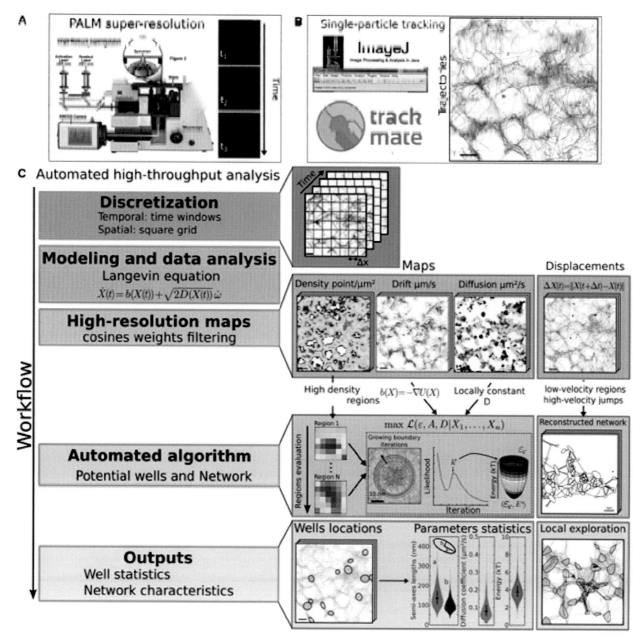

FIGURE 13.1 Pipeline for high-throughput SPT analysis. Automated workflow for high-throughput trajectory analysis. (A) A single-particle experiment's acquisition device and raw data. (B) Using traditional software, such as Trackmate, which is accessible as an image J plugin, raw data from (A) are converted into trajectories. (C) A diagram illustrating how the high-throughput analysis employed in this work was organized. Using a square grid and temporal binning, trajectories are first segmented both spatially and chronologically (time-windows analysis). Then, using the Langevin equation to analyze the trajectories, researchers may create high-resolution maps of the motion of the local trajectories. To find possible wells and recreate the network, automated methods remove high-density, low-velocity portions of the maps. Finally, the outputs are statistics for the reconstructed network and well locations. These traits make it possible to study how local nanoscale environmental exploration paths interact with their surroundings. *Reprinted (adapted) with permission from Ref. Parutto et al. (2021), Copyright (2022) David Holcman et al. Cell Reports Methods, Elsevier.*

challenging jobs, describe the creation, concerned processes used in basic operations for databases, script database procedures to find information, explore data, and gather research data (Fig. 13.1).

Authors possess a fundamental comprehension of creation, manipulation, and formats for research graphics and develop practical knowledge of computer vision and a fundamental grasp of artificial intelligence. Alternately, using quantitative estimates from a mathematical model, mathematical modeling can connect the measures between a theory and measurable facts (Edler et al., 2002). Since it is typically hard to directly observe particular cell fate outcomes in vivo and can only be done with multimodal data of cell clones, researching cell fate dynamics is made more difficult. Because the data and hypotheses cannot be openly likened, traditional arithmetical examinations cannot be used, just as experimental data alone cannot be used to confirm biological models (Gasteiger & Engel, 2018).

For researchers and clinicians working in the area of stem cell research, as well as for learners and healthcare professionals with an interest in regenerative medicine, bioinformatics, developmental biology, and computational biology, this chapter offers an outline of current findings on computational methodologies for stem cell tissue engineering (Larsen et al., 2006). Here, in addition, we explain and outline how to narrate the comments behind an assumption (cell destiny conclusions) individually as a stochastic model, how to scientifically assess such a rule-based model using methodical computation or stochastic simulations of the model's Master equation, and how to forecast results of clonal figures for particular hypotheses (Lei et al., 2014). To evaluate models, authors also provide two methods for directly comparing these predictions with the clonal data.

To successfully apply tissue engineering in therapeutic settings, it is crucial to master various complexity levels. Biological complexity may be studied more quantitatively, and recognitions to computer modeling tools have been reported (Kitano, 2017). To be more precise, computational tools can aid in (1) quantifying and optimizing tissue-engineered products, such as by modifying scaffold design to enhance microenvironmental signals or the selection criteria to increase homogeneity of chosen cell populations; (2) evaluating and optimizing tissue engineering processes, such as by modifying bioreactor design to enhance quality and quantity of finished products; and (3) evaluating impact in vivo. This section's contributors cover a range of future thoughts on connected research initiatives. This chapter focuses on metabolomics and stem cell tissue engineering, using computer modeling and cheminformatics as a framework for understanding fundamental processes, such as cell attachment and migration, nutrition transport and usage, matrix formation, local cell–cell interactions, and cell population dynamics (Muzzarelli et al., 1999). Major advancements can be made by merging current experimental and computational methodologies. Moreover, a comprehensive view of these essential factors will be essential to establishing new lines of inquiry in tissue engineering research. Last but not least, contemporary issues in tissue engineering and regenerative medicine have been discussed and might be resolved using the underlined techniques here for stem cell engineering specifically. Overall, cheminformatics, tissue engineering, and stem cell research were analyzed in metabolomics data acquisition and applied in the engineering of stem cell niche (Lu et al., 2011) to govern and manipulate the stem cell microenvironment, landscape of G protein–coupled receptor signaling, regulate ESC fate, regenerate pancreatic β-cell, and produce stem cells used as clinical and preclinical agents, in cardiac tissue regeneration and adipose tissue repair.

3. Computational approaches, genomics and proteomics

To more accurately anticipate cellular conversion factors, cell–cell interactions, and cell identity transcription factors that are crucial for tissue regeneration and homeostasis, single-cell-based mechanistic models must be developed (Lustig et al., 2021). As a modeling framework (probabilistic model, logical model, continuous model, and descriptive model), we employed model types (undirected graph, directed graph, Boolean network, dynamic Bayesian network, regression-based equation, and linear ordinary differential equation) in several applications (extrapolation of cell–cell communication networks from ligand–receptor pair appearance, inference of Boolean logic rules of gene regulatory networks (GRNs) established biological importance) (forecast of tissue cell–cell interactions, identification of cooperative TF regulation, and prophecy of cell conversion features and reprogramming efficiency) (del Sol & Jung, 2021). Underlined models will help experimentalists in creating stem cell transplantation and gene therapy regimens, which will hasten the development of novel regenerative medicine treatments. Tissue engineering combines scaffolds, cells, and chemicals to create functional tissues. With the use of tissue engineering, damaged cells or entire organs can be repaired, preserved, or improved (Stouder & Gallagher, 2013). Although the Food and Drug Administration (FDA) has approved a number of artificial tissues, as well as synthetic cartilage and skin, their use in treating human patients is presently circumscribed.

A subset of tissue regeneration that also includes tissue engineering is the research of self-healing, in which the body uses its own mechanisms and, sometimes, the help of external biological matter to rebuild cells, tissues, and organs. For example, clinical application of musculoskeletal tissue-engineered constructions necessitates improved biological and mechanical functionality, which can only be attained through multidisciplinary research (Chen et al., 2022). Tissue engineering and tissue regeneration have become interchangeable to focus on therapies for complex diseases. Since proteomics and genomics methods are more deep-rooted in the quest for disease biomarkers, the discipline of metabolomics is fast developing and seems to be catching up with them (Senevirathna & Asakawa, 2021). However, a more thorough review of how to manage missing data is necessary to define the bases, protocols, and standard operating procedures. To achieve this, it is required to use statistical analysis to compare the effects of imputation of incomplete data among analytical metabolomics platforms and by the kind of biological matrix (Van Steendam et al., 2013). The author presents a discussion on cell signaling via soluble components during direct cell–cell connection and cell–extracellular matrix interface (Gulia, 2021). Highlighted research transfers information gained from one tissue to another (Fig. 13.2A and B).

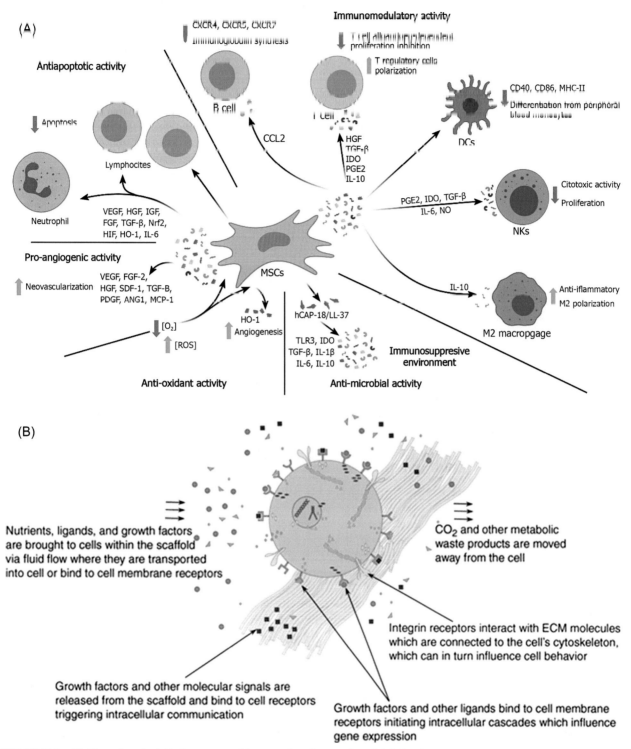

FIGURE 13.2   (A). Illustration of soluble factors secreted by mesenchymal stem cells and their functions. *ANG*1, Angiogenesis 1; *CCL*, CC-chemokine ligand; *CXCR*, C-X-C chemokine receptor type; *DCs*, dendritic cells; *FGF*, fibroblast growth factor; *hCAP*, human cathelicidin anti-microbial peptide; *HGF*, Hepatocyte growth factor; *HIF*, Hypoxiahypoxia-inducible factor; *HO*-1, heme oxygenase; *IDO*, indoleamine 2,3-dioxygenase; *IGF*, insulin-like growth factor; *IL*, interleukin; *MCP*-1, monocyte chemotactic protein-1; *MSCs*, mesenchymal stem cells; *NKs*, natural killer cells; *NO*, nitric oxide; *Nrf*2, nuclear factor erythroid-related factor 2; *PDGF*, platelet-derived growth factor; *PGE*2, prostaglandin E2; *ROS*, reactive oxygen species; *SDF*, stromal cell-derived factor; *TGF-β*, transforming growth factor-−β; *VEGF*, vascular endothelial growth factor. (B) Diagram showing how cells and scaffolds are related. In a tissue engineering scaffold, a cell is surrounded by a complex microenvironment. Transport of nutrients carries ligands, growth factors, and other signals that may attach to cell receptors. In addition, the deteriorating scaffold can discharge chemical messengers that attach to membrane receptors and affect intracellular communication and cellular functions including gene transcription. Through integrin receptors, cells also adhere to the scaffold. Integrin receptors are intimately linked to the cytoskeleton of the cell and transmit additional information to the cell, changing how the cell functions. *(A) Reprinted (adapted) with permission from Ref. González-González et al. (2020), Copyright (2020), Baishideng Publishing Group inc. Open Access under the terms of the Creative Commons CC-BY License; (B) Reprinted (adapted) with permission from Ref. Evans et al. (2006), Copyright (2006), Elsevier B.V.*

Thus, a global picture of an understudied system was aided by the insertion and incorporation of additional circumstantial biological equivalents such as genomics and proteomics (Lal et al., 2018). Currently, to provide good coverage and optimize chemical identification, matching experimental spectra data necessitates more searches of numerous separate database resources.

To achieve the maximum level of accuracy, records should ideally be manually annotated and corrected. The resources in question should encompass both spectrum and compound chemical features, as well as biological activities compiled from various sources. Moreover, to accurately identify new compounds, structural elucidation is difficult, and time-consuming operations can be made easier by applying computationally assisted tools and algorithms (Xu et al., 2021). Improvements in metabolite spectral archives, bioinformatics tools, and algorithms are expected. These types of investigations will influence the development of new drugs and tailored medical interventions, as well as advance our understanding of biology and disease etiology (Rajiv, 2021f).

New technologies offer opportunities for training future scientists, such as SWOT analysis and 3D printing. In the study, engineering was viewed as a theoretical link that connects expertize (cheminformatics databases and software), chemical science (molecular visualizations), and arithmetic (graph theory) into a cohesive, informatively worthwhile source (Pernaa, 2022). Cheminformatics offers excellent opportunities for theoretical learning, according to the analysis's major finding. It is a field of chemistry study that focuses on developing practical artifacts, such as databases, software, and visualizations. This is ideal for a theoretical approach since it enables an engineering-based methodology that assures apprentices of playing a dynamic and imaginative part. Because of cheminformatics' multidisciplinary character, the key difficulty is a high content and requisite knowledge (Wishart, 2016). Training and the design of a collaborative learning environment can easily help in tackling this type of problem. The development of enlightening cheminformatics is quiet in its infancy, but it appears to be a very capable area for assisting the theoretical education of chemistry (Pernaa, 2022). The shift from cancer science to genomics has had significant effects on cancer biology, leading to new treatments. The development of both competitive and "understandable" single-cell platforms has resulted in a flood of single-cell OMIC statistics. This discipline is now coping with it (Kumar et al., 2021). Further, researchers assert that the study of computational stem cell science will result in a paradigm-shifting comprehension of biological studies and improve our power to accurately and effectively change cell fate by merging single-cell data and computational modeling methodologies (Rajiv, 2021b). But there are several unresolved issues.

4. Cheminformatics, tissue engineering, and stem cell research

Although cheminformatics was initially developed as a tool to aid in developing and discovering drug processes, it now has a significant impact on various fields in stem cell biology, biochemistry, and tissue engineering. The purpose of this section is to introduce readers to the topic of cheminformatics and its role in tissue engineering and demonstrate its features, showing how it not only has much in common with the science of bioinformatics but can also greatly improve much of what is now being done in that field (Schaduangrat et al., 2020). Cheminformatics is responsible for organizing, processing, gathering, and transformation of lab data to develop pharmaceuticals. Cheminformatics technologies assist medicinal chemists in understanding the intricate structures of chemical molecules.

The goal of cheminformatics is to identify novel chemical entities and create new molecules, which can be used to treat diseases. The underlined computational modeling subfield is illustrated in this section with examples (Kumar et al., 2013). Blood arteries, the trachea, cartilage, and bone are only a few of the basic tissue engineering applications. In this part that provides examples of the first two domains, the main emphasis is on (the optimization of) mechanical signals, fluid flow encountered, mass movement by the cells in scaffolds and bioreactors, as well as the optimization of the cell population itself.

Modeling can help understand and control complex relationships between cells and microenvironment. To apply these models to practice, various scientific computing methods are required, as well as fresh equipment that could be used to empirically verify the findings of the computations (Chang et al., 2016). Even though cheminformatics is a rapidly growing discipline, it is still relatively new and has little public recognition. Additionally, there is not much historical study on the application of cheminformatics in teaching. Most academics, although it is incorporated, are concentrating on chemistry software over and done with data handling and visualization.

A team of researchers reported fundamental chemistry readings that incorporate diverse cheminformatics skills (Jirát et al., 2013). To help researchers with visual impairments, a team of researchers created an adaptive cheminformatics system that reads chemical compound names aloud (Kamijo et al., 2016). Moreover, to promote learning of biochemistry, Lohning et al. examined current perceptions of the use of 3D printing and cheminformatics software. The usage of cheminformatics software, according to their study, improved learning outcomes in terms of molecular interactions. By examining how cheminformatics fits within the STEM education paradigm, this part advances prior knowledge in same area as reported earlier. An illustration of a cheminformatics-based training package was developed for this aim. The results of the recently mentioned studies **indicate** that databases, software, and websites are heavily emphasized in cheminformatics research. Because of Lohning et al. educational cheminformatics study, 3D printing technique was used to exemplify the engineering component. Other researchers were also developed a realistic context for STEM learning in chemistry using 3D printing (Vernengo et al., 2020). Molecular visuals illustrate chemistry, 3D printing, cheminformatics, and graph theory (Fig. 13.3) The choice of graph theory was made because of how frequently it is used in the field of cheminformatics, for instance, almost all cheminformatics software supports graph theory-based computer-based 2D and 3D visualizations, according to Rassokhin (2020).

Authors understand that papers focusing on the fundamentals of chemical characterization and cheminformatics would be helpful to beginners in the subject, which can range from undergraduates to seasoned chemists originating from empirical subfields of chemistry. This is because chemical information is becoming increasingly important in the practices of chemistry, biology, and related sciences (Wishart, 2016). Visualization is essential for conducting nondisruptive studies with great spatial resolution (Acheson et al., 2010). Two-photon microscopy based on laser pulses is emerging as a major tool for cell evaluation and modification due to its great depth resolution, depth of penetration through every specimen, and low cell damage. A potent tool for stem cell research is a nondisturbing single-cell observation and manipulation approach.

Tissue engineering is a claim of engineering and biology principles and methods to fundamental comprehension of the structure/function interactions in healthy and sick tissues and the creation of biological alternatives to sustain, improve, or restore utility (Kumar & Gulia, 2021). Extended-connectivity fingerprints that describe molecules quantitatively and popular deep learning architectures have right shape and are very effective in cheminformatics, RNNs (recurrent neural networks), and RNN function well with SMILES (simplified molecular-input line-entry system). These tools have been applied to represent sequences that could be used to locate molecules with desired attributes (such as a specific solubility or toxicity). Graph convolutional networks (GCNs) have been developed to resolve this issue by taking molecule graphs and their features as input, which can frequently be used when working with molecules. Publicly available molecule datasets (ChemSpider: ChEMBL: RDKit) have also been developed and recommended.

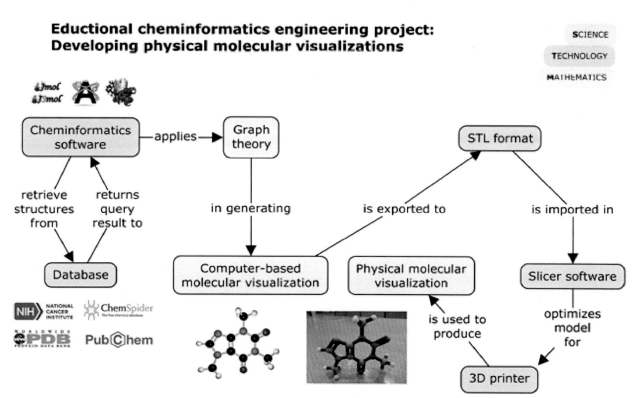

FIGURE 13.3 An illustration of molecular visualization engineering project's relation to the STEM education framework. *Reprinted (adapted) with permission from Ref. Pernaa (2022) Copyright (2022), American Chemical Society and Division of Chemical Education, Inc.*

In this part, we highlight current advancements in cell and tissue engineering in relation to the revolution in imaging techniques. The understanding of pluripotent stem cells (PSCs), ES cells and multipotent stem cells, and hematopoietic stem cells (HSCs) has significantly advanced during the past decade (Walter et al., 2019). Nonetheless, it is difficult to research the evolving potential of underlined stem cells since these types of cells live in complicated cellular contexts, making it difficult to fully understand features of their delivery, survival, migration, proliferation, differentiation, and engraftment using only a few snapshots of their environment or location or genetic markers. So, it is vitally desirable to have trustworthy imaging techniques to keep an eye on or track the stem cells' progress (Erickson et al., 2020). Imaging tools enable long-term monitoring of stem cells. Researchers can study cell performance in the setting of an alive organism thanks to confocal and multiphoton microscopy, numerous noninvasive imaging technologies, and time-lapse imaging technology (Pancholi & Dave, 2022). The understanding of stem cell biology has now advanced because of newfound information. The use of most recent imaging techniques for studying HSCs and PSCs, as well as the difficulties that still lie ahead, has been covered in this chapter. Surgical reconstruction, machinery, organ transplantation, and metabolic supplements are all options for treating organ or tissue loss in healthcare.

5. Metabolomics data acquisition and preprocessing

An untargeted metabolomics study's data preparation phase is one of the most crucial parts of the procedure. To ensure that high-quality datasets are created, each step of the process for data preparation, from the data's "raw" format to a table of metabolic properties, must be carried out using simple, repeatable processes. Metabolomics has become a popular discipline for various fields of inquiry, including tailored drugs, because metabolic reactions are more similar to the phenotype (Liew et al., 2021). The newest in omics sciences is metabolic, which tries to measure and describe metabolites, which are typically byproducts of biological reactions and are tiny chemical molecules (1500 Da) on cells, tissue, or biofluids (De San-Martin et al., 2020). Data processing is essential for metabolomics to analyze, recognize, and quantify endogenous and exogenous metabolites. Because they are so closely related to functional endpoints and the phenotypes of the organisms, the metabolites are essential parts of biological systems and are extremely revealing about their functional condition (Lei et al., 2014). This section's goal is to outline data processing tools that are accessible, enabling even a novice researcher to provide trustworthy results without having to go through a protracted parameterization procedure. One of the primary metabolomics analytical platforms is NMR spectroscopy, along with MS (Rajiv, 2021e). Because of the technological advancements in the area of NMR spectroscopy, it is now possible to identify and quantify numerous metabolites in a single sample of biofluids in an untargeted and nondestructive manner (van Dongen et al., 2002). The typical procedures are needed to transform complicated datasets collected from MS and untargeted NMR spectroscopy-based metabolomics research into tables of features suitable for statistical analysis. Using a variety of analytical methods such as NMR spectroscopy and MS, recent developments in metabolic outlining procedures enable overall profiling of metabolites in species, tissues, or cells (Petukhov, 2009). Because of the technological and structural complexity of the raw data collected by this equipment, statistically relevant information cannot be easily extracted. Preprocessing is the process of converting data from equipment such as GC/LC-MS and NMR spectra into a practice that may be used for subsequent analysis and biological interpretation. Basic baseline correction, scaling, quantification, normalization, detection, and peak alignment are the main areas of focus for common data preparation methods used during analysis. The advancement of the use of metabolomics in the area of biomarker discovery has been made possible by the combination of biofluid NMR spectra with pattern recognition techniques (Ash et al., 2019). According to the number of studies that have been published, metabolomics is becoming increasingly significant in diagnostics, such as when defining pathological status or discovering biomarkers (Rajiv, 2021g). With a focus on metabolomics of cerebrospinal fluid and identification of multiple sclerosis biomarkers, author highlights advancements in data collection and multivariate analysis of NMR-based metabolomics data in this study (Walter et al., 2019). Starting with preprocessing the raw data, an informatics pipeline is used to make logic of untargeted GC-MS or LC-MS data. To contextualize untargeted metabolomics data, results from data preprocessing are statistically analyzed and then mapped to metabolic pathways. LC-MS and GC-MS data preparation has focused on the development of the ADAP suite of computational techniques. It comprises two independent computational algorithms that, from corresponding raw LC-MS and GC-MS data, derive compound-relevant information (Fig. 13.4). MS and separation techniques are used for metabolomic and proteomic research, but data processing is a challenge. A summary of the widely used outfits in the data preprocessing pipeline is given. Thus, recent years have seen the emergence of numerous software applications for data preprocessing (Bonfitto et al., 2021).

Here, we also include MZmine 2, a new version of the well-known open-source data processing toolset. Extracted ion chromatogram creation, peak identification, spectral deconvolution, and alignment are computational processes. The two

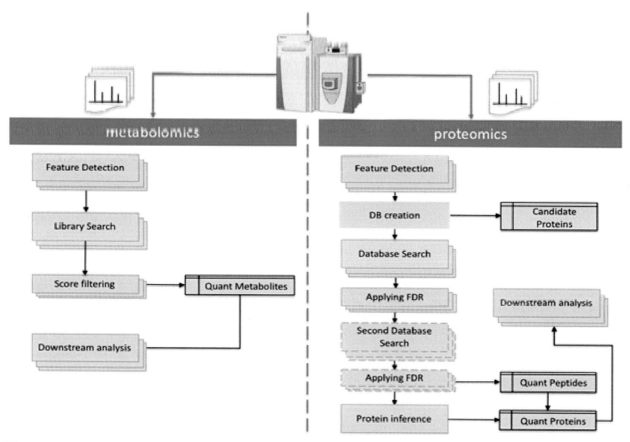

FIGURE 13.4 Processes in bioinformatics for proteomics and MS. The proteomics lane (right) shows the label-free database search analysis workflow, which entails the phases of MS1 spectra feature identification, protein database creation, protein inference, database search, and statistical analysis at the end. A typical spectrum search workflow is the metabolomics process. *Reprinted (adapted) with permission from Ref. Perez-Riverol and Moreno (2020) Copyright (2019), John Wiley & Sons, Inc.*

procedures were integrated into cross-platform and graphical MZmine 2 frameworks (modular outline for processing, displaying, and analyzing MS-based molecular profile data), and cytosolic adhesion and degranulation-promoting adapter protein (ADAP)−specific graphical user interfaces were created for ADAP to be used easily. In summary, we show how to use ADAP to preprocess GC-MS and LC-MS data and highlight the significance of algorithmic ideas underlying important phases in the two workflows.

Moreover, scalable data examination in proteomics and metabolomics have been performed using workflows, and biocontainers including metabolomics, mzQuality, Fluxomics stationary ^{13}C-MS iso2flux, ecometabolomics, LC-MS XCMS 3.0, proteogenomics, metaproteomics, peptide and protein quantification workflow, proteogenomics database (pgdb) engines, IPAW proteogenomics and other main tools (CAMERA, XCMS, W4M multivariate, OpenMS, MetFrag), and univariate statistical analysis tools (metaQuantome, ms-vetfc, Ecomet, XCMS, CAMERA, Iso2flux, Escher metabolic pathway viewer, SearchGUI, msconvert, CustomProDB, PeptideShaker, MSGF+, MetaProteomeAnalyzer, Unipept, OpenMS, and pypgatk) (Perez-Riverol & Moreno, 2020).

6. Manipulation of stem cells and tissue regeneration

Adipocytes, vascular smooth muscle cells, adipose-derived stem cells (ASCs), and vascular endothelial cells are all found in fat tissue. ASCs can develop into many lineages, just as human bone marrow-derived stem cells (BMSCs) can (neuronal cells, fibroblast, adipose cells, endothelial cells, chondrocytes, osteoblasts, cardiomyocytes, and myocytes). Additionally, it has been demonstrated that in long-term cultures, they exhibit immunoprophylaxis and genetic stability. ASCs can be extracted in huge numbers with little invasiveness, unlike BMSCs, though. Cells may be useful for tissue engineering and reconstructing prescriptions. The paradigm shift brought about by stem cell−based treatments for the healing and renewal of numerous organs, and tissues may pave the way for creating new therapeutic modalities for treating a range of diseases.

Although ES cells and prompted PSCs have a lot of potential applications, there are a number of limitations on their utilization due to genetic engineering, cell laws, and ethical issues (Lele et al., 2019). Contrarily, adult stem cells are derived from autologous tissue. These cells are more readily accessible and do not raise ethical or immunological issues (Gilfillan & Beaven, 2011). Human adipose tissue is a novel source of MSCs for mesenchymal tissue regeneration due to its potential for adipogenic, myogenic, neurogenic, osteogenic, and chondrogenic potential. ASCs have been employed as a more efficient therapeutic option based on the findings of our earlier studies because of their molecular characteristics, differentiation potential, aptitude for wound healing, and potential application in cell-based therapies and tissue engineering (Rajiv, 2021h). Adipose tissue regeneration and angiogenesis have been demonstrated to be aided by the therapeutic use of adipose stem/progenitor cells (Acheson et al., 2010), identifying adipocyte-releasing factors for adipose tissue regeneration.

Injection of precursors induces dramatic neogenesis in perichondrium and periosteum. ASC clinical applications will be successful if the milieu is carefully designed. Then, the cell distribution strategy is used to prevent unexpected behavior and generate maximum potential, and the target diseases are chosen. Adipose tissue is a renewable type of tissue known as ASCs, which secrete vital growth factors to reduce inflammation, regulate the immune system, speed up wound healing, and target wounded areas (Rozwadowska & Kurpisz, 2018). ASCs are a subset of a diverse stromal cell population isolated from adipose tissue, and scientists are looking into ways to make adipose tissue a suitable cell source in large quantities.

7. Stem cell microenvironment and cardiac tissue regeneration: engineered approaches

Tissue engineering has been pursued to create biomimetic cardiac tissue since the 1990s with the goal of regenerating the heart (Bursac, 2009). However, cardiomyocytes created from PSCs have little in common with those of the adult heart. Better knowledge of human stem cell biology and improvements in process engineering have made it possible to obtain an endless supply of cells, particularly cardiomyocytes, for generating functional cardiac muscle. Cardiac tissue engineering techniques using biomaterial scaffolds, including cells and growth factors, are proving to be quite effective at repairing and regenerating the heart (Kumar & Gulia, 2021). Stem cells can be used to create cardiac microengineered tissue structures, using methods such as injections, spheroids, hydrogels, scaffolds, and bioprinting. To maintain stemness and/or differentiate into cardiac-specific lineages, the stem cell microenvironment is critical (Bazil et al., 2010). Autologous stem cell—based tissue engineering technologies can restore injured cardiac tissue. Throughout cardiac tissue engineering platforms, special attention is paid to the functions of the extracellular matrix in controlling stem cells' physiological responses.

Medical micro- and nanoscale engineering techniques can generate physiologically appropriate stem cell microenvironments to better understand stem cell activities and recent stem cell therapy (Fig. 13.5A and B). The realization of these developments will affect several therapeutic applications. Thus, the restoration of heart tissue is possibly the most immediate need (Rajiv, 2021d).

Despite significant progress in controlling biochemical regulatory factors to produce physiologically realistic in vivo stem cell niches, more synergy of cutting-edge techniques is required that hold promise for elucidating the significance of numerous physical cues, including stem cell differentiation into cardiac cells, the development of bioengineered cardiac tissue grafts, and the electromechanical coupling of these cells (Edgar et al., 2020). Consequently, numerous physiologically significant micro- and nanoengineering initiatives were developed to resolve these problems. Cardiovascular tissue has been created from stem cells during differentiation, when it was controlled by structural cues and transcription factors. Most of this chapter, however, focuses on highlighting cutting-edge and unorthodox microscale engineering methods that have been used for stem cell—based cardiac tissue creation. These methods include microfluidic, mechanical, electrical, topographic, biomaterial, and optical stimulation (Ghafar-Zadeh et al., 2011). Biomaterials and stem cells can help regenerate damaged myocardium. Additionally, the development of appropriate scaffolding biomaterials and the regulation of the stem cells' microenvironment have been made possible by nanoenabled techniques (nanomaterials and nanofeatured surfaces) (Piard et al., 2015). Stem cell—based regenerative therapies require repeatable outcomes to ensure well-being and efficiency in patients with heart failure.

The most difficult task is determining and choosing the best type of stem cell for cardiac renewing remedy. Human PSCs become a popular cell source for producing cardiomyocytes (CMs), with potential uses in drug discovery, disease modeling, and cutting-edge cell treatments (Naryzhnaya et al., 2019). The growth of CMs originating from stem cells was better understood using lessons from embryology. But it is strongly advised to create a population with CMs with a

(A)

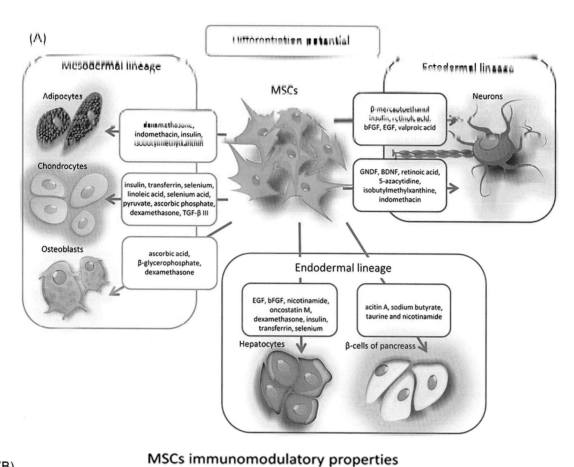

(B)

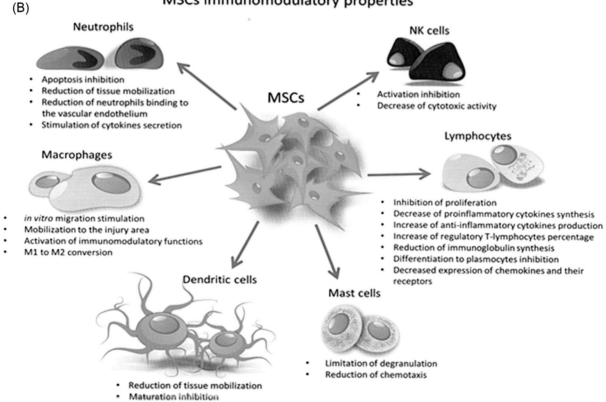

FIGURE 13.5   (A) The ability of mesenchymal stem cells to differentiate. (B) The mesenchymal stem cells' potential for immunomodulation is shown schematically *(A) Reprinted (adapted) with permission from Ref. Andrzejewska et al. (2019), John Wiley and Sons, inc. According to the terms of the Creative Commons CC-BY Licence, it is an open access source.*

consistent cardiac subtype, adult development, and functional characteristics. Additionally, obstacles related to cancer, immunological rejection, arrhythmogenesis, and graft cell death requirement are to be solved in clinical exercise. Novel techniques are developed to improve cardiac regenerative medicine.

8. Future prospective and present outlook

Computational biology, cheminformatics, metabolomics, and stem cell tissue engineering are among the potent approaches that can illustrate biological arrangements and create forecasts across many geographical and temporal scales to provide revolutionary insight in the field of single-cell big data. The development of simulations that connect many heights of biological components, including intracellular connections, the activities of cell populations, and cellular activity, is made possible by the continually growing amount of data. Prototypes of stem cell—niche interactions will assist in developing fresh methodologies for stem cell transformation. Increased interaction between experimentalists and computer scientists will facilitate stem cell investigation and clinical regenerative remedies. Direct reprogramming is a potential strategy for developing cell and tissue engineering therapies.

Metabolomics technologies identify and quantify thousands of metabolites in biological samples, reducing the cost of measurement for each sample. Cellular reprogramming allows direct conversion of mature cell types. This makes the direct reprogramming method a promising strategy for creating various cellular and tissue engineering treatments. Direct reprogramming uses multidisciplinary approaches to detect changes in stem cell identity. According to the author's perspective, new experimental analyses of engineered tissues will be driven by predictive metabolomics, and new computational science analysis frameworks will be motivated by the needs of tissue engineering.

9. Conclusion

Nowadays, omics technologies are expanding quickly and offer complete molecular data to thoroughly analyze biological systems. Omics data collection and single-cell technologies enable computational biology for stem cell research. Computational biology, metabolomics, and stem cell tissue engineering could advance regenerative medicine. Deeper understanding of biological mechanisms is needed to improve cellular reprogramming. Multiscale modeling can help develop stem cell therapies (Rajiv, 2021c). Single cell—based computational biology, metabolomics, and stem cell tissue engineering can improve in vitro cell creation. Direct reprogramming of cells involves combining experimental and computational techniques, including stem cell tissue engineering, computational biology, metabolomics, and cheminformatics. Metabolomics quantifies metabolites and metabolic pathways related to cellular energy status, stem cell proliferation, suitability, and fate decisions. Further, these engineered methods and methodologies have been used in the innovation of clinical and preclinical agents for the regulating stem cell microenvironment, cardiac tissue regeneration, manipulating stem cell activities, and adipose tissue repair. The collective approaches of metabolomics, cheminformatics, computational biology, and stem cell tissue engineering provide a framework for stem cell—targeted cardiac tissue engineering. The most important element of this chapter is one of that "researchers will expose transdisciplinary fields of computational biology and stem cell engineering to have a thorough grasp of the confluence of tissue engineering, computational science, molecular and cell biology." Therefore, the scientists will practice and improve this new discipline. Sooner or later, they will cross traditional disciplinary barriers and be able to easily communicate in the cultures of tissue engineering, pharmacological and cellular biology, and computational methods. Experienced researchers in metabolomics, stem cell tissue engineering, and cheminformatics are needed.

Acknowledgments
For inspiration, the author (Rajiv Kumar) heartily acknowledges his younger brother Bitto.

Author contributions
Dr. Rajiv Kumar drafted the work. Prof. Magali Cucchiarini and Prof. Gerardo Caruso contributed by formatting all the sections. Dr. Johannes Pernaa and Dr. Zarrin Minuchehr revised it critically for important intellectual content. In the end, all the authors approved the final version.

References

Acheson, A., Sunshine, J. L., Rutishauser, U., Ailor, E., Takahashi, N., Tsukamoto, Y., Masuda, K., Rahman, B. A., Jarvis, D. L., Lee, Y. C., Betenbaugh, M. J., Angata, K., Chan, D., Thibault, J., Fukuda, M. N., Huckaby, V., Ranscht, B., Terskikh, A., Marth, J. D., et al. (2010). Molecular basis for polysialylation: A novel polybasic polysialyltransferase domain (PSTD) of 32 amino acids unique to the alpha 2,8-polysialyltransferases is essential for polysialylation. *Journal of Biological Chemistry, 276*(1), 9443–9448. http://www.ncbi.nlm.nih.gov/pubmed/9134419.

Andrzejewska, A., Lukomska, B., & Janowski, M. (2019). Concise review: Mesenchymal stem cells: From roots to boost. *Stem Cells, 37*(7). https://doi.org/10.1002/stem.3016

Angerer, P., Simon, L., Tritschler, S., Wolf, F. A., Fischer, D., & Theis, F. J. (2017). Single cells make big data: New challenges and opportunities in transcriptomics. *Current Opinion in Systems Biology, 4*, 85–91. https://doi.org/10.1016/j.coisb.2017.07.004. Elsevier Ltd.

Ash, J. R., Kuenemann, M. A., Rotroff, D., Motsinger-Reif, A., & Fourches, D. (2019). Cheminformatics approach to exploring and modeling trait-associated metabolite profiles. *Journal of Cheminformatics, 11*(1). https://doi.org/10.1186/s13321-019-0366-3

Bazil, J. N., Buzzard, G. T., & Rundell, A. E. (2010). Modeling mitochondrial bioenergetics with integrated volume dynamics. *PLoS Computational Biology.* https://doi.org/10.1371/journal.pcbi.1000632

Bian, Q., & Cahan, P. (2016). Computational tools for stem cell biology. *Trends in Biotechnology, 34*(12). https://doi.org/10.1016/j.tibtech.2016.05.010

Bonfitto, S., Casiraghi, E., & Mesiti, M. (2021). Table understanding approaches for extracting knowledge from heterogeneous tables. *Wiley Interdisciplinary Reviews: Data Mining and Knowledge Discovery, 11*(4). https://doi.org/10.1002/widm.1407

Bursac, N. (2009). Cardiac tissue engineering using stem cells. *IEEE Engineering in Medicine and Biology, 28*(2). https://doi.org/10.1109/MEMB.2009.931792

Chang, K. J., Redmond, S. A., & Chan, J. R. (2016). Remodeling myelination: Implications for mechanisms of neural plasticity. *Nature Neuroscience, 19*(2), 190–197. https://doi.org/10.1038/nn.4200. Nature Publishing Group.

Chen, Y., Wang, Y., Luo, S. C., Zheng, X., Kankala, R. K., Wang, S. B., & Chen, A. Z. (2022). Advances in engineered three-dimensional (3D) body articulation unit models. *Drug Design, Development and Therapy*, 16. https://doi.org/10.2147/DDDT.S344036

De San-Martin, B. S., Ferreira, V. G., Bitencourt, M. R., Pereira, P. C. G., Carrilho, E., de Assunção, N. A., & de Carvalho, L. R. S. (2020). Metabolomics as a potential tool for the diagnosis of growth hormone deficiency (Ghd): A review. *Archives of Endocrinology and Metabolism, 64*(6). https://doi.org/10.20945/2359-3997000000300

Dobson, C. M. (2004). Chemical space and biology. *Nature, 432*(7019). https://doi.org/10.1038/nature03192

van Dongen, M., Weigelt, J., Uppenberg, J., Schultz, J., & Wikström, M. (2002). Structure-based screening and design in drug discovery. *Drug Discovery Today, 7*(8). https://doi.org/10.1016/S1359-6446(02)02233-X

Edgar, L., Pu, T., Porter, B., Aziz, J. M., La Pointe, C., Asthana, A., & Orlando, G. (2020). Regenerative medicine, organ bioengineering and transplantation. *British Journal of Surgery, 107*(7). https://doi.org/10.1002/bjs.11686

Edler, L., Poirier, K., Dourson, M., Kleiner, J., Mileson, B., Nordmann, H., Renwick, A., Slob, W., Walton, K., & Würtzen, G. (2002). Mathematical modelling and quantitative methods. *Food and Chemical Toxicology, 40*(Issues 2–3). https://doi.org/10.1016/S0278-6915(01)00116-8

Erickson, M. A., Wilson, M. L., & Banks, W. A. (2020). In vitro modeling of blood-brain barrier and interface functions in neuroimmune communication. *Fluids and Barriers of the CNS, 17*(1). https://doi.org/10.1186/s12987-020-00187-3

Evans, N. D., Gentleman, E., & Polak, J. M. (2006). Scaffolds for stem cells. *Materials Today, 9*(12). https://doi.org/10.1016/S1369-7021(06)71740-0

Fernandes, M., Sanches, B., & Husi, H. (2019). Cheminformatics and computational approaches in metabolomics. In *Computational biology* (pp. 143–159). Codon Publications. https://doi.org/10.15586/computationalbiology.2019.ch9

Gasteiger, J., & Engel, T. (2018). Applied chemoinformatics: Achievements and future opportunities. In *Applied chemoinformatics*. https://doi.org/10.1002/9783527806539.ch6h

Ghafar-Zadeh, E., Waldeisen, J. R., & Lee, L. P. (2011). Engineered approaches to the stem cell microenvironment for cardiac tissue regeneration. *Lab on a Chip, 11*(18). https://doi.org/10.1039/c1lc20284g

Gilfillan, A. M., & Beaven, M. A. (2011). Regulation of mast cell responses in health and disease. *Critical Reviews in Immunology, 31*(6), 475–530. https://doi.org/10.1615/critrevimmunol.v31.i6.30

González-González, A., García-Sánchez, D., Dotta, M., Rodríguez-Rey, J. C., & Pérez-Campo, F. M. (2020). Mesenchymal stem cells secretome: The cornerstone of cell-free regenerative medicine. *World Journal of Stem Cells, 12*(12). https://doi.org/10.4252/wjsc.v12.i12.1529

Gulia, R. K. (2021). Cell mechanotransduction machinery, and cell signaling defects: Small tools and nano-bio interface for influential regenerative remedies. *Journal of Cell Signaling, 6*(5), 1–14.

Hao, Y., Cheng, S., Tanaka, Y., Hosokawa, Y., Yalikun, Y., & Li, M. (2020). Mechanical properties of single cells: Measurement methods and applications. *Biotechnology Advances, 45*. https://doi.org/10.1016/j.biotechadv.2020.107648

Jirát, J., Čech, P., Znamenáček, J., Šimek, M., Škuta, C., Vaněk, T., Dibuszová, E., Nič, M., & Svozil, D. (2013). Developing and implementing a combined chemistry and informatics curriculum for undergraduate and graduate students in the Czech Republic. *Journal of Chemical Education, 90*(3). https://doi.org/10.1021/ed3001446

Kamijo, H., Morii, S., Yamaguchi, W., Toyooka, N., Tada-Umezaki, M., & Hirobayashi, S. (2016). Creating an adaptive technology using a cheminformatics system to read aloud chemical compound names for people with visual disabilities. *Journal of Chemical Education, 93*(3). https://doi.org/10.1021/acs.jchemed.5b00217

Karn, T. (2013). High-throughput gene expression and mutation profiling: Current methods and future perspectives. *Breast Care, 8*(6). https://doi.org/10.1159/000357461

Kitano, H. (2017). Biological complexity and the need for computational approaches. In *History, philosophy and theory of the life sciences*. https://doi.org/10.1007/978-3-319-47000-9_16

Kumar, R. (2021). Computational explorations and interpretation of bonding in metallopharmaceutical for a thorough understanding at the quantum level. *Sustainable Chemical Engineering.* https://doi.org/10.37256/sce.2120211828

Kumar, R., & Gulia, K. (2021). The convergence of nanotechnology-stem cell, nanotopography-mechanobiology, and biotic-abiotic interfaces: Nanoscale tools for tackling the top killer, arteriosclerosis, strokes, and heart attacks. *Nano Select, 2*(4), 655−687. https://doi.org/10.1002/nano.202000192

Kumar, O. L., Rachana, S., & Rani, B. M. (2013). Modern drug design with advancement in QSAR : A review. *International Journal of Research Studies in Biosciences, 2*(1), 1−12. http://www.ijrbs.in.

Lal, C. V., Bhandari, V., & Ambalavanan, N. (2018). Genomics, microbiomics, proteomics, and metabolomics in bronchopulmonary dysplasia. *Seminars in Perinatology, 42*(7). https://doi.org/10.1053/j.semperi.2018.09.004

Larsen, M., Artym, V. V., Green, J. A., & Yamada, K. M. (2006). The matrix reorganized: Extracellular matrix remodeling and integrin signaling. *Current Opinion in Cell Biology, 18*(5). https://doi.org/10.1016/j.ceb.2006.08.009

Lei, J., Levin, S. A., & Nie, Q. (2014). Mathematical model of adult stem cell regeneration with cross-talk between genetic and epigenetic regulation. *Proceedings of the National Academy of Sciences of the United States of America.* https://doi.org/10.1073/pnas.1324267111

Lele, T. P., Brock, A., & Peyton, S. R. (2019). Emerging concepts and tools in cell mechanomemory. *Annals of Biomedical Engineering.* https://doi.org/10.1007/s10439-019-02412-z

Liew, F. F., Dutta, S., Sengupta, P., & Chhikara, B. S. (2021). Chemerin and male reproduction: 'a tangled rope' connecting metabolism and inflammation. *Chemical Biology Letters, 8*(4), 224−237. https://pubs.thesciencein.org/journal/index.php/cbl/article/view/270.

Lustig, A., Margi, R., Orlov, A., Orlova, D., Azaria, L., & Gefen, A. (2021). The mechanobiology theory of the development of medical device-related pressure ulcers revealed through a cell-scale computational modeling framework. *Biomechanics and Modeling in Mechanobiology, 20*(3), 851−860. https://doi.org/10.1007/s10237-021-01432-w. Springer Science and Business Media Deutschland GmbH.

Lu, P., Takai, K., Weaver, V. M., & Werb, Z. (2011). Extracellular Matrix degradation and remodeling in development and disease. *Cold Spring Harbor Perspectives in Biology, 3*(12). https://doi.org/10.1101/cshperspect.a005058

McBride, W. H., Iwamoto, K. S., Syljuasen, R., Pervan, M., & Pajonk, F. (2003). The role of the ubiquitin/proteasome system in cellular responses to radiation. *Oncogene, 22*(37). https://doi.org/10.1038/sj.onc.1206676

Muzzarelli, R. A., Mattioli-Belmonte, M., Pugnaloni, A., & Biagini, G. (1999). Biochemistry, histology and clinical uses of chitins and chitosans in wound healing. *EXS, 87,* 251−264. https://doi.org/10.1007/978-3-0348-8757-1_18

Naryzhnaya, N. V., Maslov, L. N., & Oeltgen, P. R. (2019). Pharmacology of mitochondrial permeability transition pore inhibitors. *Drug Development Research, 80*(8). https://doi.org/10.1002/ddr.21593

Nava, M. M., Raimondi, M. T., & Pietrabissa, R. (2012). Controlling self-renewal and differentiation of stem cells via mechanical cues. *Journal of Biomedicine and Biotechnology, 2012.* https://doi.org/10.1155/2012/797410

Pancholi, U. V., & Dave, V. (2022). Review of computational approaches to model transcranial direct current stimulations tDCS and its effectiveness. *Journal of Integrated Science and Technology, 10*(1), 1−10. http://pubs.iscience.in/journal/index.php/jist/article/view/1388.

Parutto, P., Heck, J., Lu, M., Kaminski, C., Avezov, E., Heine, M., & Holcman, D. (2021). High-throughput super-resolution single particle trajectory analysis reconstructs organelle dynamics and membrane Re-organization. *SSRN Electronic Journal.* https://doi.org/10.2139/ssrn.3985166

Perez-Riverol, Y., & Moreno, P. (2020). Scalable data analysis in proteomics and metabolomics using BioContainers and workflows engines. *Proteomics, 20*(9). https://doi.org/10.1002/pmic.201900147

Pernaa, J. (2022). Possibilities and challenges of using educational cheminformatics for STEM education: A SWOT analysis of a molecular visualization engineering project. *Journal of Chemical Education, 99*(3). https://doi.org/10.1021/acs.jchemed.1c00683

Petukhov, P. A. (2009). Cheminformatics approaches to virtual screening. *Journal of the American Chemical Society, 131*(9), 3407−3408. https://doi.org/10.1021/ja9007326

Piard, C. M., Chen, Y., & Fisher, J. P. (2015). Cell-laden 3D printed scaffolds for bone tissue engineering. *Clinical Reviews in Bone and Mineral Metabolism.* https://doi.org/10.1007/s12018-015-9198-5

Rajiv, K. (2021a). Architecting and tailoring of cell repair molecular machinery: Molecule-by-Molecule and atom-by-atom. *Austin Journal of Pharmacology and Therapeutics, 9*(1), 1128−1129.

Rajiv, K. (2021b). Cell shrinkage, cytoskeletal pathologies, and neurodegeneration: Myelin sheath formation and remodeling. *Archives of Medical and Clinical Research, 1*(1), 1−5.

Rajiv, K. (2021c). Elucidation of the origin of autoimmune diseases via computational multiscale mechanobiology and extracellular matrix remodeling: Theories and phenomenon of immunodominance. *Current Medical and Drug Research, 5*(1). Art. ID 215 (2021).

Rajiv, K. (2021d). Macrophage subtypes, phenotypes, inflammatory molecules, cytokines, and atherosclerotic lesions - atherosclerosis, metabolic diseases, and pathogenesis, the therapeutic challenges. *Journal of Clinical and Experimental Pathology, 11*(11S), 1−3.

Rajiv, K. (2021e). Metabolic and immune system interface: Immunometabolism, micro biota, and diseases. *Annals of Clinical and Medical Microbiology, 5*(1), 1−3.

Rajiv, K. (2021f). Molecular profiling and progression of malignant melanoma: Nanomedicine and immunotherapeutic remedies for diagnosis, treatment, and therapy. *Clinics Oncology, 6*(18846), 1−3.

Rajiv, K. (2021g). Traumatic brain injury: Mechanistic insight on pathophysiological mechanisms underlying, neurotransmitters, and potential therapeutic targets. *Medical and Clinical Reviews, 7*(8), 1−3. https://medical-clinical-reviews.imedpub.com/traumatic-brain-injury-mechanisticinsight-on-pathophysiological-mechanismsunderlying-neurotransmitters-andpotential-therapeutic-ta.pdf.

Rajiv, K. (2021h). Wound pathophysiology: Insights of Ca2+ signaling and cellular senescence mechanisms in healing, and regeneration. *Journal of Blood Disorders and Transfusion, 12*(4), 463.

Ramraje, G. R., Patil, S. D., Patil, P. H., & Pawar, A. R. (2020). A brief review on: Separation techniques chromatography. *Asian Journal of Pharmaceutical Analysis, 10*(4). https://doi.org/10.5958/2231-5675.2020.00041.1

Rassokhin, D. (2020). The C++ programming language in cheminformatics and computational chemistry. *Journal of Cheminformatics, 12*(1). https://doi.org/10.1186/s13321-020-0415-y

Richter, K. S., Harris, D. C., Daneshmand, S. T., & Shapiro, B. S. (2001). Quantitative grading of a human blastocyst: Optimal inner cell mass size and shape. *Fertility and Sterility, 76*(6). https://doi.org/10.1016/S0015-0282(01)02870-9

Rozwadowska, N., & Kurpisz, M. (2018). Myocardial infarction. In *A roadmap to nonhematopoietic stem cell-based therapeutics: From the bench to the clinic* (pp. 223–249). Elsevier. https://doi.org/10.1016/B978-0-12-811920-4.00009-4

Schaduangrat, N., Lampa, S., Simeon, S., Gleeson, M. P., Spjuth, O., & Nantasenamat, C. (2020). Towards reproducible computational drug discovery. *Journal of Cheminformatics, 12*(1). https://doi.org/10.1186/s13321-020-0408-x. BioMed Central Ltd.

Senevirathna, J. D. M., & Asakawa, S. (2021). Multi-omics approaches and radiation on lipid metabolism in toothed whales. *Life, 11*(4). https://doi.org/10.3390/life11040364

del Sol, A., & Jung, S. (2021). The importance of computational modeling in stem cell research. *Trends in Biotechnology, 39*(2). https://doi.org/10.1016/j.tibtech.2020.07.006

Stouder, M. D., & Gallagher, S. (2013). Crafting operational counterintelligence strategy: A guide for managers. *International Journal of Intelligence & Counter Intelligence, 26*(3). https://doi.org/10.1080/08850607.2013.780560

Tang, B., Pan, Z., Yin, K., & Khateeb, A. (2019). Recent advances of deep learning in bioinformatics and computational biology. *Frontiers in Genetics, 10*. https://doi.org/10.3389/fgene.2019.00214. Frontiers Media S.A.

Van Steendam, K., De Ceuleneer, M., Dhaenens, M., Van Hoofstat, D., & Deforce, D. (2013). Mass spectrometry-based proteomics as a tool to identify biological matrices in forensic science. *International Journal of Legal Medicine, 127*(2). https://doi.org/10.1007/s00414-012-0747-x

Vernengo, A. J., Grad, S., Eglin, D., Alini, M., & Li, Z. (2020). Bioprinting tissue analogues with decellularized extracellular matrix bioink for regeneration and tissue models of cartilage and intervertebral discs. *Advanced Functional Materials, 30*(44). https://doi.org/10.1002/adfm.201909044

Walter, J. E., Ayala, I. A., & Milojevic, D. (2019). Autoimmunity as a continuum in primary immunodeficiency. *Current Opinion in Pediatrics, 31*(6), 851–862. https://doi.org/10.1097/MOP.0000000000000833. Lippincott Williams and Wilkins.

Wishart, D. S. (2016). Introduction to cheminformatics. *Current protocols in bioinformatics, 2016*. https://doi.org/10.1002/0471250953.bi1401s53

Xu, J., Cao, X. M., & Hu, P. (2021). Perspective on computational reaction prediction using machine learning methods in heterogeneous catalysis. *Physical Chemistry Chemical Physics, 23*(19). https://doi.org/10.1039/d1cp01349a

Zhang, L. (2011). Development and application of cheminformatics approaches to facilitate drug discovery and environmental toxicity assessment [The University of North Carolina at Chapel Hill]. In *ProQuest dissertations and theses*. https://search.proquest.com/docview/923797093?accountid=8359.

Chapter 14

Advances in regenerative medicines based on mesenchymal stem cell secretome

Bhawna Sharma[1], Himanshu Sehrawat[1] and Vandana Gupta[2]

[1]Department of Biosciences, Jamia Millia Islamia, New Delhi, India; [2]Department of Microbiology, Ram Lal Anand College, University of Delhi, New Delhi, India

1. Introduction

Stem cells are the undifferentiated predecessor cells that are able to self-renew and differentiate into several types of cells. Broadly, they are of three types, namely (1) embryonic stem cells (ESCs) that are sourced from the early-stage embryos; (2) induced pluripotent stem cells (iPSCs), which are very similar to ESCs and are sourced from adult somatic cells (specifically skin and blood cells), have potential of rapid proliferation in vitro; (3) adult stem cells (ASCs), which are rare cell populations found throughout the postnatal life in the body and can differentiate into a restricted types of cell lineages. ASCs include hematopoietic stem cells (HSCs), neural stem cells (NSCs), and mesenchymal stem cells (MSCs) (Vizoso et al., 2017).

The MSCs in culture are defined through three criteria laid down by "Mesenchymal and Tissue Stem Cell Committee of the International Society for Cellular Therapy" based on their activities in the culture:

(a) Adhere to the plastic surface in standard cultures.
(b) ≥95% of MSC population should express cell differentiation markers (CD markers) 73, 90, and 105 while lack expression of HLA-DR (≤2% positive), CD markers 14 or 11b, 79a or 19, 34, and 45 on their surface (Fig. 14.1).
(c) Under the usual in vitro conditions, cells should be differentiated into chondroblasts, adipocytes, and osteoblasts (Dominici et al., 2006; Horwitz et al., 2005).

A cell's secretome is the collection of all molecules/factors secreted extracellularly by it. These factors can be lipids, soluble proteins (e.g., growth factors, chemokines, and cytokines), nucleic acids, and extracellular vesicles (EVs) among other things. Based on their origin, density, size, and surface markers EVs are subdivided into apoptotic bodies, MVs (microvesicles), ectosomes, oncosomes, exosomes, and exosome-like vesicles (Beer et al., 2017; Vizoso et al., 2017). EVs is a generic term, which is used for secreted vesicles. Exosome is a subtype of EV expressing tumor susceptibility gene 101, CD9, CD81, and CD63. Multivesicular bodies give rise to exosomes through inward budding, and these are smaller in size (30−100 nm) than other subtypes and involved in the transfer of functional cargos such as peptides, protein s, miRNAs, mRNAs, cytokines, etc. Individual cells have their specific secretome, which can change in response to variations in the pathological and physiological conditions. The utility of secretome in regenerative medicines could potentially avoid infection transmission, immune compatibility, and tumorigenicity. The secretome plays important roles in various cellular processes such as tissue repair, angiogenesis, inflammatory modulation, autoimmune response modulation, neuroprotection, and apoptotic and antiapoptotic effects (Xia et al., 2019).

2. MSC-Secretome: conditioned media from mesenchymal stem cells and exosomes

MSC-Secretome is the collective term, used for factors released from the cell. This includes cytokines, chemokines, growth factors, paracrine factors, proteasomes, antioxidants, antiinflammatory factors, microRNAs, exosomes, angiogenic factors,

Computational Biology for Stem Cell Research. https://doi.org/10.1016/B978-0-443-13222-3.00008-3
Copyright © 2024 Elsevier Inc. All rights reserved, including those for text and data mining, AI training, and similar technologies.

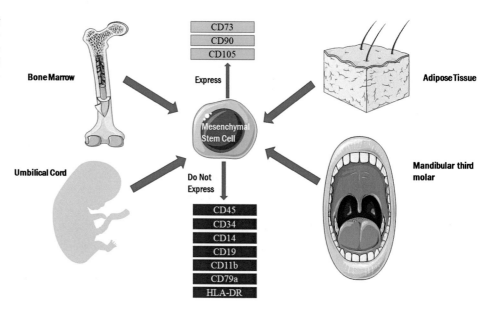

FIGURE 14.1 The origin of mesenchymal stem cells and their surface markers. *This figure was created using the Servier Medical Art Commons Attribution 3.0 Unported Licence, http://smart.servier.com, accessed April 16, 2023.*

etc. The paracrine actions of these factors are responsible for the therapeutic properties of the stem cells (Eleuteri & Fierabracci, 2019; Montero-Vilchez et al., 2021; Xia et al., 2019).

The conditioned medium (CM) is a culture medium in which stem cells are grown, and it contains the secretome of that particular type of stem cell. The sources of MSCs for the MSC-CM are different; bone marrow-MSCs, adipose tissue-MSCs, umbilical cord-MSCs, placenta, amnion, dental pulp, etc. The paracrine factors in the CM are vascular endothelial growth factor (VEGF), insulin-like growth factor-1 (IGF-1), IGF-2, stromal cell−derived factor-1 (SDF-1), and hepatocyte growth factor (HGF). The composition of the CM is dependent on the source cell, but in general, it is abundant in cytokines. These cytokines including IGF-binding protein-7, metalloproteinase-1, EGF, and other growth factors are linked to the migration of endothelial cell and angiogenesis (Montero-Vilchez et al., 2021).

MSC-derived exosomes (MSC-Exos) and MSC-derived extracellular vesicles (MSC-EVs) are paracrine mediators of cytokines, peptides, proteins, lipids, mRNAs, miRNAs, etc., from the MSCs to the recipient cells. These contribute to the communication between the cells and thus are very important in therapy. These are more advantageous in therapies compared with implanting MSCs as they act as a powerful tool in tissue homeostasis along with other benefits such as no proliferation unlike cells, and ease of production, preservation, and transfer (Eleuteri & Fierabracci, 2019; Nikfarjam et al., 2020). RNA populations are more abundant in EVs than other molecules. These RNA molecules are involved in the important effector functions of MSC-EVs and exosomes. miRNAs (miR-22, miR-27a, miR-125A-3p, miR-126-3p, miR-196a, miR-206) are involved in immunoregulatory functions, regeneration, cell cycle regulation, and apoptosis (Eleuteri & Fierabracci, 2019). Preservation of exosomes is necessary to use them as therapeutic agents and vehicles for drug delivery, in which lyophilization is more preferred than other ways of the storage. Lyophilized exosomes can be stored at room temperature without affecting DNA, RNA, and protein content. The function of proteins and their pharmacokinetics and physicochemical properties remain the same (Charoenviriyakul et al., 2018).

Cell-free therapies using MSC-Secretome have numerous advantages in the regenerative medicine over stem cell−based applications (Vizoso et al., 2017):

(a) Using MSC-Secretome can be a viable solution to address the safety concerns related to transplanted proliferative cells. This approach avoids issues such as generation of tumors, formation of emboli, immune compatibility, and transmission of infections.
(b) It can be tested on safety parameters, clinical dosage, and efficacy similar to their conventional pharmaceutical counterparts.
(c) It can be stored for a long period in a lyophilized form without losing effectiveness and does not require usage of potentially toxic cryoprotective agents.
(d) It is more cost-effective and practical solution for various clinical applications.
(e) It provides a convenient source of the production of bioactive factors in mass.

(f) The problem of the high cost and time for the upkeep of cultured of stem cells can be overcome through MSC-secretome therapies, and these are immediately available for treatment purposes.

(g) There is a flexibility of modifications in the biological products obtained by secretome for desired cell-specific effects.

MSC-Secretome is a new booster to regenerative medicine and has shown potential in many clinical applications such as the prevention of diseases such as tumors, cardiac dysfunction, type 1 diabetes, hair loss, joint osteoarthritis, and neurodegenerative diseases (Xia et al., 2019).

3. Effector biological functions of MSC-Secretome and their applications

3.1 Immunomodulation

MSCs affect major stages of the immune response, which include T cell effector stage; recognition and presentation of an antigen; and the T cells activation, proliferation, and differentiation. These show a suppressive effect on both adaptive and innate immunities in animal models according to some preclinical studies. Activation and function of the cells including cytotoxic and helper T cells of immune system is modulated by MSCs, and proliferation of CD4+ and CD8+ T cells is inhibited. MSC-CM (MSC-conditioned medium) shows antiinflammatory effects because of cytokines such as IL (interleukin)13, IL10, IL17E, IL27, IL12p70, IL1RA, NT-3, CNTF, IL18BP, TGFβ; and some proinflammatory cytokines such as IL9, IL8, IL6, and IL1b. MSCs balance the pro- and antiinflammatory cytokines release by inhibiting IFN-γ and TNFα (proinflammatory cytokines) while stimulating IL10 (antiinflammatory cytokines) (Vizoso et al., 2017).

Besides MSC-CM, MSC-Exos are also involved in immunomodulation as they contain several effector chemokines such as PGE2, IL-1Ra, IL10, HGF, IDO-1, and TGFβ. Some examples of the immunomodulation response performed by either the MSC-CM, MSC-Exos, MSC-Secretome, or all of them are as follows (Harrell, Fellabaum, et al., 2019):

1. The attenuation of IL-2 induces the proliferation of CD8+ and CD4+ T cells by MSC-Exos and MSC-CM. This happens due to Jak-1/Stat-5 dependent on the arrest of the cell cycle in G1 phase.
2. Thrombospondin 1 (TSP1) in the MSC-Secretome regulates the TGF-β/Smad signaling pathway activation. MSC-Exos increase the cytotoxic potential and proliferation of NK cells by suppressing this signaling pathway.
3. MSC-CM and MSC-Exos both induce the programmed cell death of activated PB-MNCs (peripheral blood mononuclear cells).
4. MSC-CM and MSC-Exos induce a regulatory phenotype in DCs, and as a result, these regulatory DCs induce the Tregs (regulatory T cells expressing $CD4^+CD25^+FoxP3^+$ markers) expansion through the production of KYN (kynurenine). This immunomodulation eventually enables the immunosuppressive environment generation in the inflamed tissues.
5. In the peripheral blood of the aplastic anemia patients, MSC-Exos alleviate the generation of IL-10-producing and Fox-P3-expressing Tregs and attenuate the IL-17-producing Th17 cells.
6. MSC-Exos that bear the IL-1-Ra diminish the skin inflammation and accelerate wound healing.
7. The cytotoxic and inflammatory potential of NKT cells and T lymphocytes is suppressed by the MSC-CM's administration in inducible nitric oxide synthase and indoleamine 2,3-dioxygenase 1 (IDO1)/kynurenine-dependent manner.

MSC-Exos with improved immunosuppressive activities can be sourced by MSCs that are primed with the TNF-α and IFN-γ. MSC-Exos show better therapeutic effects in the treatment of inflammatory and autoimmune disease (Harrell, Fellabaum, et al., 2019; Shen et al., 2021), such as the attenuation of ulcerative colitis is done by the MSC-EVs through modulation of function and phenotype of colonic macrophages (Harrell, Jovicic, et al., 2019). Chronic liver inflammation is also attenuated by MSC-EVs through the downregulation of the inflammatory cytokines and production of the TGF-β1 in Kupffer cells. Additionally, the alternatively activated M2 macrophage expansion is promoted by MSC-EVs involved in the repair and regeneration of tissues (Harrell, Jovicic, et al., 2019).

MSC-Secretome-based treatment of the inflammation and injury of many organs is effective. MSC-EVs reduce renal failures by inhibiting oxidative stress, autophagy, apoptosis, and necrosis in epithelial cells of the renal tubule. Moreover, MSC-EVs also suppress the detrimental immune response. The cardiomyocytes are efficiently protected from ischemic injury with the help of MSC-EVs. Delivery of miR-22 mediated by MSC-Exos exhibited improvement in cardiac function in ischemic cardiomyocytes. The repair and regeneration of eye's injured neurons is promoted by MSC-Exos. These deliver miR21, miR-146, and miR-17-92 into the injured retinal ganglion cells (RGCs). The neural cell death is prevented in the cerebral I/R injury by MSC-Exos through the inhibition of caspases 9 and 3. The MSC-sourced factor, pigment epithelium-derived Factor (PEDF), suppresses the caspase-3-driven apoptosis and as a result reduces the I/R-induced cerebral injury (Harrell, Jovicic, et al., 2019).

MSC-EVs are reported to be effective for many autoimmune diseases such as uveitis, type 1 diabetes, inflammatory bowel disease, multiple sclerosis, Sjogren's syndrome, and rheumatoid arthritis. The MSC-sourced Secretome, EVs, and Exos can be the potential new cell-free drugs in the autoimmune disorders treatment (Shen et al., 2021).

3.2 Antiapoptotic function

MSC-Secretome shows antiapoptotic effects as it prevents cell death by increasing expressions of antiapoptotic proteins and inhibitor proteins of apoptosis. It decreases the cleaved caspase-3 expression and inhibits the proapoptotic factor "Bax" and upregulates expression of antiapoptotic factor Bcl-2, in parenchymal cells. This prevents cell loss during ongoing inflammation (Harrell, Fellabaum, et al., 2019; Vizoso et al., 2017). Bcl-2 and Bcl-xL are apoptosis suppressors, and Bak, Bax, and Bad are apoptosis promoters. MSC-Secretome increases the ratio of Bcl-2 to Bax and inhibits the stimulation of caspase-3 in cardiomyocytes and thus prevents apoptosis (He et al., 2009).

MSC-Exos show hepatoprotective effects in acute liver failure by increasing the Bcl-xL (antiapoptotic gene) expression in injured hepatocytes (Harrell, Fellabaum, et al., 2019). MSC-Exos sourced from human menstrual blood attenuate the acute liver injury and suppress the caspase-3-mediated apoptosis of hepatocytes. It suppresses the NLRP3-dependent caspase-1 activation and inhibits the caspase-1-driven pyroptosis for hepatoprotective effects. Exosomal miR-233 induces the suppression and degradation of NLRP3 caspase-1-induced pyroptosis and NLRP3 mRNA, respectively in hepatocytes (Harrell, Jovicic, et al., 2019).

The MSC-Secretome antiapoptotic effects help in promoting myocardial regeneration. MSC-Exos prevent apoptosis of cardiomyocytes by suppression of caspase-3 activities, and by downregulation and upregulation of Bax and Bcl-2, respectively. The MSC-Exos-mediated delivery of antiapoptotic miRs, such as miR-210, miR-146a, miR-15, miR-126, miR-21, miR-22, etc., promote survival of cardiac stem cells. The miR-210 and miR-125b-5p both prevent p53- and Bak-1-driven apoptosis in cardiomyocytes and increase their survival (Harrell et al., 2020; Harrell, Fellabaum, et al., 2019). The miR-19a is responsible for antiapoptotic activities, induced by MSC-Exos in cardiomyocytes. It downregulates the proapoptotic phosphatase and tensin homolog (PTEN) activation and upsurges activation and phosphorylation of Akt, which as a result upregulates Bcl-2 protein. The MSC-EVs-derived miR21 is mainly responsible for cardioprotection (Harrell et al., 2020; Harrell, Jovicic, et al., 2019). The antiapoptotic miR-21 is present in MSC-EVs as well as MSC-CM and shows antiapoptotic properties (Sandonà et al., 2021).

The MSC-EVs show antiapoptotic activities in the case of AIH (autoimmune hepatitis). MSC-EVs released miR-21-5p that protects lung epithelial cells by inhibiting both extrinsic and intrinsic apoptotic pathways (Harrell, Jovicic, et al., 2019). Kyon and colleagues evaluated the antapoptotic activity of Wharton's Jelly—sourced MSCs in the C2C12, which is a cell line derived from skeletal myoblast of mouse. High levels of chemokine C motif ligand-1 are secreted by Wharton's Jelly—MSCs when cocultured with C2C12, which is responsible for antiapoptotic effects (Sandonà et al., 2021).

3.3 Neuroprotective and neurotrophic effects

Numerous studies have been performed to know neuroprotective and neurotrophic activities of the MSC-secretome, and many beneficial effects were observed. These effects include increase in the myelin sheath thickness, inflammatory site environment modulation, acceleration in regeneration and organization of fibers, and vascularization enhancement at the regenerating site. Many neurotrophic factors are present in the MSC-CM. There are mutually beneficial effects of growth factors released by NSCs and MSCs (Vizoso et al., 2017). NT-4/5, erythropoietin, ciliary neurotrophic factor, basic fibroblast growth factor, nerve growth factor, neurotrophin-3, vascular endothelial growth factor, brain-derived neurotrophic factor, and insulin-like growth factor-1 are common growth factors in the MSC secretome. Three groups of neurotrophic factors, based on the targeting receptors, are neurokines, neurotrophins, and growth factor-β (TGF-β) family. The cerebral dopamine neurotrophic factors exhibit neurotrophic and neuroprotective activities in PNS and CNS injuries (Caseiro et al., 2016).

MSC-EVs- and MSC-Exos-based therapies have been proved as effective treatment options in spinal cord injuries (SCIs) and ischemic brain damage. MSC-Exos increase the number of endothelial cells and neuroblasts in the brain's ischemic regions and promote axonal growth. MSC-Exo transfers miR-133b in neurons to target RhoA (Ras homolog gene family member A), which promotes the neurite outgrowth. Additionally, they also activate Stat-3 and Erk1/2 signaling pathways for neuron regeneration and promote recovery from SCI (Harrell, Fellabaum, et al., 2019). MSC-Exos-sourced PEDF delivery reduces the immune response—induced injury (Harrell, Jovicic, et al., 2019).

In amyotrophic lateral sclerosis (ALS), adipose tissue—derived MSCs (AT-MSCs) increase the level of glial-derived neurotrophic factors (GDNFs), which neuroprotect motor neurons by increasing their survival, functionality, and number. Exosomes derived from AT-MSC in the 0.2 μg/mL concentration protect the cell from H_2O_2-induced damage, and this has been proved in NSC-34 motor neuron cell line in which different incubation timing (2, 4, 6, 8, and 18 h) were tested with exosomes and H_2O_2. AT-MSC derived exosomes display neurogeneric and neuroprotective activities in ALS because of the presence of a protein called ribonuclease RNAse4, which is mutated in this disease. These exosomes also transfer SOD1 in mutated cells to destroy free superoxide radicals and improve response to oxidative stress (Bonafede et al., 2016; Sandonà et al., 2021).

The bone marrow-MSC-Exos (BM-MSC-Exos)-based therapy can be proved effective in the most common inflammatory disease of CNS, called multiple sclerosis (MS). These exosomes promote M2 phenotype polarization by inhibiting microglia to develop into an M1 phenotype and secrete antiinflammatory cytokines, e.g., TGF-β, TNF-α, and IL-10. MSC-Exos also alleviate conditions of experimental autoimmune encephalomyelitis (EAE) through the transfer of mRNA, miRNA, and proteins. However, MSC-EVs show alleviation in EAE conditions through the transfer of tolerance molecules, such as galecin-1 (GAL-1), TGF-β, and PD-L1 (Shen et al., 2021).

The "conditioned medium–human amniotic mesenchymal stem cell" inhibits monocyte-derived dendritic cells differentiation through the induction of antiproliferative effects on T cells, reduction in Th1 and Th17 populations and as a result protects from traumatic brain injuries. The release of taurine, spermidine, lysine, and alpha-aminoadipic-acid shows neuroprotective effects (Giampà et al., 2019). MSC-Secretome-based therapeutics also hold promise in the treatment of spinal cord injuries (Herbert et al., 2022).

3.4 Wound healing and tissue repair

Wound healing is a complex and constantly evolving process, which is characterized by a high degree of organization and strict regulation. It involves a variety of mechanisms at the cellular, humoral, and molecular levels. Wound healing is divided into four main phases (Fig. 14.2):

A. The homeostasis phase is the first stage of wound healing. It begins immediately after injury lasts for some hours. On damage to the skin, the activated platelets start aggregating to stop the bleeding. In the end, the platelet clumps convert into a blood clot by thrombin, generating a fibrin mesh.

B. During the inflammatory phase, neutrophils and monocytes (later macrophages) focus on phagocytosis of bacteria and degradation of necrotic tissue and debris, and the wound bed provides foundation of the new tissue. Neutrophils release

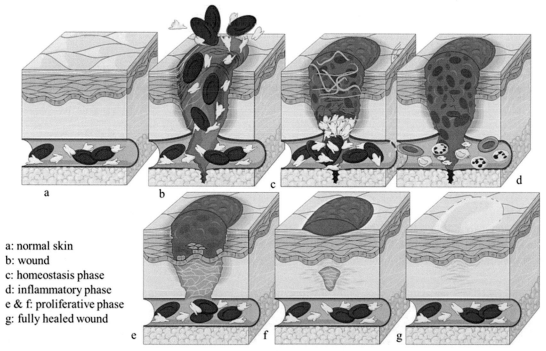

a: normal skin
b: wound
c: homeostasis phase
d: inflammatory phase
e & f: proliferative phase
g: fully healed wound

FIGURE 14.2 The stages of wound healing. *This figure was created using the Servier Medical Art Commons Attribution 3.0 Unported Licence, http:// smart.servier.com, accessed April 16, 2023.*

multiple soluble mediators such as highly potent antimicrobial substances and proteinases and mediators such as TNF-α, IL-1β, and IL-6, which work in amplifying the inflammatory response. Macrophages act as antigen-presenting cells (APCs) and phagocytose pathogens during wound repair while secreting several potent growth factors such as TGF-α, VEGF, platelet-derived growth factor (PDGF), etc. to upregulate proliferation of the cells.

C. The proliferation phase starts approximately 3−10 days after the injury. In this phase, the granulation tissue covers the wound surface and creates new blood vessels to supply oxygen to the fresh granulation tissue. Fibroblasts produce collagen, components of the extracellular matrix (ECM) such as fibronectin, hyaluronic acid, and glycosaminoglycans. The ECM acts as a scaffold for cell adhesion and permits the expansion, movement, and differentiation of the cells.

D. The remodeling a.k.a. the maturation phase can start anytime between 21 days to 1 year postinjury. In the final remodeling phase, the angiogenesis decline, with decrease in the flow of blood to the wound, the inflammatory phase slowly resolves, the metabolic activity in wound gradually reduces until it stops, and in the end, scar tissue is formed. The new tissue steadily recovers strength and flexibility, the type III collagen is remodeled to stronger collagen type I, and finally the wound closes completely (Reinke & Sorg, 2012).

Researchers have developed novel treatment methods for nonhealing wounds, one of which is stem cell−based therapy. Self-renewal property of stem cells can be used for tissue repair as they can secrete numerous bioactive molecules that accelerate wound healing. However, using living cells poses several challenges such as biosafety, their immune compatibility, possible tumorigenicity, risk of infection, difficulty in storage, and expensive cost, thus limiting their use (Ibrahim et al., 2022).

Research, however, suggests that the secreted molecules by stem cells can generate higher therapeutic impacts than using actual cells. Stem cell secretomes gained considerable interest when they improved the healing process by accelerating the rate of angiogenesis (Janockova et al., 2021; Zhang et al., 2015), reducing inflammation, and stimulating proliferation of fibroblast and keratinocytes in the management of skin wound, as demonstrated in the in vitro as well as in vivo experiments (Ibrahim et al., 2022).

Bermudez et al. reported the implications of MSCs in wound healing and tissue repair (Bermudez et al., 2015) Bermudez induced dry eye in rats via lacrimal gland removal and caused disruption of epithelial layer in cornea with NaOH, hence inducing corneal ulcers along with inflammation in the eye. Corneal ulcers are commonly caused due to using contact lenses, trauma, chemical injuries, infections, and keratoconjunctivitis sicca (commonly referred as dry eye syndrome). After causing injury, dry eyes were treated with "conditioned media from human uterine cervical stem cells" (CM-hUESCs). Following treatment, a remarkable improvement was seen in epithelial regeneration, and mRNA expression levels were reduced for TNF-α and inflammatory protein 1 alpha (MIP-1α) and in corneal macrophage when treated with CM-hUESCs or sodium hyaluronate eye drops. Also, an increase in the levels of fibroblast growth factor 6 and 7, inhibitors of metalloproteinases 1 and 2, and hepatocyte growth factor was also seen. These proteins probably mediate the healing effects (Bermudez et al., 2015).

The activities of MSCs, like the secretion of mediators of antiinflammation and angiogenesis clubbed together with their differentiation capabilities, make them quite important in wound healing and tissue repair. However, due to the inconsistencies in delivery protocols among different MSC populations, it is hard for researchers to standardize the delivery route, system, and time. Using exosomes can provide an alternative to these issues by their cell-free nature. Still, their inadequate production poses a challenge to their widespread clinical use (Bian et al., 2022).

3.5 Regulation of angiogenesis

Restoring the blood vessels is an essential process in wound healing. MSCs play an important role in angiogenesis, considering that abnormalities and insufficiency in vessel growth can lead to several diseases, as seen in atherosclerotic ailments and disorders related to wound healing. Several researches have proved the importance of MSC-Secretome in angiogenesis (Janockova et al., 2021). For instance, MSC populations derived from bone marrow, adipose tissue, amniotic fluid, and Wharton jelly umbilical vein were found to promote multiplication and migration of endothelial cells, not only stimulating tube formation but also preventing their apoptosis as well in vitro (Burlacu et al., 2013; Vizoso et al., 2017).

In vivo angiogenesis and cutaneous wound healing was observed on injection of human umbilical cord−MSCs (huc-MSC) sourced exosomes in two doses (80 and 160 µg/mL) to rat model with second-degree burns. The Wnt signaling pathway is essential for angiogenesis and needed by the endothelial cell of the vascular system for their proliferation, migration, remodeling, and maturation. Therefore, activation of Wnt/βcatenin in the endothelial cells can be a possible mechanism for repairing tissues by exosomes in improving angiogenesis (Janockova et al., 2021; Zhang et al., 2015).

The uptake of human adipose-derived mesenchymal stem cells (adMSC-Exos) sourced exosomes by endothelial cells significantly stimulated angiogenesis in vitro as well as in vivo. The presence of miR-125a within adMSC-Exos was responsible for this effect by targeting the "transmembrane region of delta-like 4 (DLL4), which is known to inhibit angiogenesis, and repressing its expression. Furthermore, miR-125a was found to enhance the formation of endothelial tip cells, thereby modulating endothelial cell angiogenesis. adMSC-Exo with its proangiogenic properties, thus, emerges as a promising candidate for therapeutic tissue repair (Liang et al., 2016). This suggests that MSC-Exos have similar healing properties as their cellular counterparts in angiogenesis. All these reports present MSC-Exos as an attractive choice for treating skin wounds and represent an avenue for a cell free therapy (El Toothy et al., 2017).

3.6 Antitumor and antimicrobial effects

EVs derived from MSCs demonstrated comparable or superior effectiveness as compared with MSCs in reducing inflammation and promoting healing in various preclinical models of lung injury. Systemically administered MSC-EVs from porcine was demonstrated to be safe and inhibitory to influenza virus replication in a mixed swine "influenza-induced lung injury model," suggesting that EVs may possibly provide viable therapeutics for virus-induced lung injury (Masterson et al., 2021).

In a study by Park et al. (2019), MVs released from human bm-MSCs were tested for their therapeutic efficacy. It exhibited considerable improvement in alveolar fluid clearance (AFC) and reduction in bacterial load in the damaged alveoli in an ex vivo human model of bacterial pneumonia. Their bactericidal effect was more conspicuous when MSCs were treated with a TLR3 agonist poly (I:C) before the isolation of MVs. The success of MVs in clearing the infection is possibly due to its angiopoietin-1 content; a protein that is known for its antipermeability, antiinflammatory, and endothelial protective attributes. Despite some limitations, the study established that MVs can be used as a practical alternative to live cells in treating acute lung injury. MVs offer several benefits including but not limited to their easy isolation and storage, thus avoiding the need for using dimethyl sulfoxide (DMSO) for its cryopreservation and a bone marrow transplantation facility. Using cells that are alive presents a challenge due to their tumorigenic potential that may be avoided, live cells may also cause hemodynamic instability with administration. Pretreating cells can modify the MVs and improve their therapeutic effects. However, due to their low production, they cannot be used in clinical use right now (Park et al., 2019).

4. Clinical studies with secretome from mesenchymal stem cells

Several clinical trials have been conducted to address various attributes of MSCs-based therapeutics such as feasibility, safety, and efficacy. Some of these trials are presented in Table 14.1.

A number of clinical trials on use of MSC-Secretome are conducted in critically ill COVID-19 patients after approval by FDA, indicating the possibility of this approach in treating diseases such as COVID-19 where immune dysregulation contributes significantly to the pathogenesis of the disease (Chouw et al., 2022; Rossello-Gelabert et al., 2022; Tan et al., 2022).

The future of MSC-EVs in regenerative medicine is promising as is evident by the preclinical and clinical studies for repairing and regenerating damaged tissues such as the heart, lungs, liver, pancreas, bones, skin, cornea, blood diseases, cancer therapeutics, etc. However, the clinical application of MSC-EVs requires careful evaluation of issues such as scalability, isolation, quantification, tissue targeting, biodistribution, characterization, safety, mode of administration, delivery vehicles, and dosage efficacy. Further investigation is necessary to enhance their efficacy, scalability, and potency for clinical applications (Gemayel et al., 2023; Williams et al., 2023).

5. Applications of computational tools in stem cell–based therapies

As the number of cells and cell lines to be explored increases, so does the demand for advancement in a fully automated, computerized approach to accelerate analysis of the said data. Research in computational methods will be indispensable in advancing medical applications in the future. These methods can be used to differentiate altered cells from normal ones, study pathophysiology of diseases, and screen drugs. Every year dozens of datasets for experimental regenerative biology are published, but international standards, high-quality, and the tools to analyze the said datasets are still lacking creating a hurdle to effectively use them to produce meaningful patterns (Juhola et al., 2015).

Teixeira et al. demonstrated how using a stirred suspension bioreactor controlled by a computer can improve the neuroregulatory properties of the MSC-Secretome in humans. A dynamically cultured secretome can help us in

TABLE 14.1 Clinical trials undertaken to evaluate MSC-Secretome-based therapeutics.

Condition	Cell source/secretome	Result/s Expected result (if study is in a trial)	References
Chronic kidney disease	UC	Administration od cord-blood-MSC-EVs in grade III–IV CKD patients was found to be safe and ameliorated the inflammation and improved the kidney function	(Nassar et al., 2016)
Liver fibrosis	BM	Fibrosis is reduced in vivo by MSC-Exos through the Wnt/β-catenin pathway. The downstream gene expression in liver fibrosis tissue and hepatic stellate cells is inhibited, as exosomes reduce the expression of β-catenin, PPARγ, Wnt10b, and Wnt3a.	(Janockova et al., 2021)
Wound healing and angiogenesis	UC	In vivo, the tissue repair mechanism is activated by exosomes through the delivering Wnt4 to activate Wnt/β-catenin signaling, improving angiogenesis in the repair of skin burn injury.	(Janockova et al., 2021)
		In vitro, exosomes elevate endothelial cell formation, migration, and tube formation.	
Alzheimer's disease	AP	The levels of Aβ40 and Aβ42 are significantly decreased by exosome transfer to N2a cells, whereas exosomes also secrete enzymatically active neprilysin.	(Janockova et al., 2021)
Acute myocardial infarction	AP	The apoptosis in myocardial cells is reduced by exosomes through their sub-jection to oxidative stress in vitro.	(Janockova et al., 2021)
Traumatic brain injury	BM	Exosomes improve functional recovery, reduce neuroinflammation, and pro-mote neurovascular remodeling in rats after traumatic brain injury.	(Janockova et al., 2021)
Multiple sclerosis	SHED-CM	Reduced demyelination, inflammation, and axon injury.	(El Moshy et al., 2020)
Diabetic polyneuropathy	Dental pulp MSC-CM	Showed angiogenic, neuroprotective, and antiinflammatory responses.	(El Moshy et al., 2020)
Coronavirus	MSC-Exos	Phase 1 trial NCT—NCT04276987.	(Ma et al., 2020)
		Inhalation of MSC-derived exosomes treats severe coronavirus pneumonia.	
Metastatic pancreatic adeno-carcinoma	MSC-Exos with KRAS G12D siRNA	Phase 1 trial NCT—NCT03608631 iExosomes will treat patients with meta-static cancer of pancreas, having KrasG12D mutation.	(Ma et al., 2020)
Diabetes mellitus type 1	MSC-Exos	Phase 2 and 3 trials NCT—NCT02138331.	(Ma et al., 2020)
		β-cell mass in T1DM is targeted by exosomes and microvesicles therapy.	
Macular holes	MSC-Exos	Early phase 1 trial NCT—NCT03437759.	(Ma et al., 2020)
		Healing of MHs by exosomes.	
Stroke	BM	Exosomes enhance neurogenesis, angiogenesis, and neurite remodeling. It also significantly increases the synaptophysin immunoreactive area of the ischemic boundary zone.	(Janockova et al., 2021)
Parkinson's disease	SHED-MVs and EXs	They stimulate the neurite outgrowth of neurons, followed by the inhibition of neuron apoptosis.	(El Moshy et al., 2020)
Critical-sized tongue defect in rats	Gingival MSC-EXs	They promote the recovery of tongue lingual papillae, and reinnervation and regeneration of taste buds.	(El Moshy et al., 2020)

AT, adipose tissue (AT); BM, bone marrow; NCT, National Clinical Trial; SHED-CM, stem cells from human exfoliated deciduous teeth-conditioned medium; T1DM, type I diabetes mellitus; UC, umbilical cord (UC); WJ, Wharton's jelly.

expanding many cells in one vessel while monitoring and controlling variables such as temperature, pH, and dissolved oxygen. It can also be used to easily scale up the process later to bigger systems. The secretome from cells induced more human neural progenitor cells to differentiate into neurons than their static conditions counterpart suggesting an improved approach for neural proliferation and differentiation allowing new therapeutic options in the future (Teixeira et al., 2016).

It is important to understand how apoptosis signaling pathways work to predict tumor growth under the influence of substances that can induce apoptosis such as MSC-Secretome. To understand this, Hendrata and Sudiono (2016) developed a multiscale model in which they integrated apoptotic signaling pathways with endothelial microenvironmental dynamics and cellular interaction. Their results were supported by Sandra et al. (2014) in their studies using HeLa cells. The simulation results indicated secretome's effectiveness in suppressing avascular tumor growth. Thus, their model provides a tool in the prediction of MSC-Secretome effects in tumor growth in cultured cells as well as in tumor spheroid experiment. The model, though comprehensive, has certain limitations. Due to unavailability of data for some of the parameters, it uses untested estimated values and can only capture tumor growth during the avascular stage. In future, the model needs to encompass three dimensions for improving exactness and include angiogenesis simulation to evaluate secretome efficacy during vascular growth and not just avascular stage (Hendrata & Sudiono, 2016).

Further, it is fundamental to understand the influence of genetic components on the development of an organ with respect to its size, shape, and orientation. And although we now understand the mechanism of most of the regenerative models along with their gene regulations, their step-by-step dynamics to develop into a particular organ shape are lacking. This is where AI and constructive algorithms step in. They can provide us a better understanding of these mechanisms and make the developing regenerative medicines automated while removing human contact and thus inadvertent errors (Mukherjee et al., 2021).

6. Conclusion

Secretions from the MSCs, called MSC-Secretome, in the culture media where stem cells are grown, possess therapeutic properties because of the presence of the factors such as EVs, peptides, chemokines, cytokines, miRNAs, etc. MSC-Secretome is replacing stem cell−based therapies because of the ease of use and fewer side effects. It has shown potential in many clinical applications such as cardiac dysfunction, diabetes, osteoarthritis, etc. The secretome with different cell sources such as bone marrow, adipose tissue, amnion, placenta, dental pulp, and umbilical cord has been observed with different effector biological functions. MSC-Secretome effects include immunomodulation, inflammatory responses, wound healing and repair, neuroprotection and neurotrophic effects, apoptotic and antiapoptotic effects, antitumor effects, antimicrobial effects, and angiogenesis. Clinical studies indicate that MSC-Secretome is a new booster in regenerative medicines. Quite a few clinical trials are being conducted to check the effectiveness of MSC-Secretome-based therapeutics. This chapter provides an insight into the biological effects of MSC-Secretome and the promise it holds in regenerative medicine through immunomodulation, antiapoptosis, neuroprotection and neurotrophism, wound healing and tissue repair, angiogenesis, and antitumor and antimicrobial activities.

MSC-Secretome is rich in EVs, peptides, chemokines, cytokines, miRNAs, etc. and has been demonstrated to possess therapeutic properties. MSC-Secretome offers benefits over stem cell−based regenerative therapies such as ease of use and fewer side effects. The biological effector functions of MSC-Secretome include immunomodulation, downregulation of the inflammatory responses, wound healing and repair, neuroprotection and neurotrophic effects, angiogenesis, apoptotic and antiapoptotic effects, and antitumor and antimicrobial activities. Use of MSC-Secretome has shown potential in many clinical applications such as cardiac dysfunction, diabetes, osteoarthritis, cancers, viral infections, immune disorders, etc.

The secretome derived from different cells has been observed to have different effector biological functions. Bone marrow−derived MSCs are most tested in clinical trials against many diseases, but iPSC-MSCs are more efficient because of the active differentiation and robust proliferation capacities, along with their capacity to generate from either patients or allogeneic sources.

MSC-derived products such as MSC-CM and MSC-EVs exhibit efficacy in clinical trials for use in various diseases including lung disorders, kidney diseases, type 1 diabetes, graft-versus-host disease (GvHD), etc. MSC-CM can reduce the lung injury severity, as shown in many in vitro and in vivo studies. The effect of MSCs-EV on improved clearance rate of alveolar fluid in human lung tissues rejected for transplantation has been demonstrated on an ex vivo lung perfusion model. MSC-EVs are demonstrated to be of benefit in patients of type 1 diabetes and steroid refractory GvHD in clinical trials.

Stem cell secretome provides an attractive alternative for improved and quick treatment of patients with its ease of use and fewer side effects than the cell-based therapies. However, the production is inefficient and slow and requires experienced personnel, posing a difficulty to the clinical use of MSC-Secretome currently. Various in vitro and in vivo studies followed by clinical studies indicate that MSC-Secretome will provide a new paradigm to regenerative medicines in coming years. However, the future research should be focused to understand site-specific homing of MSCs and differentiation, along with research on finding more biomarkers for the isolation of source-specific MSCs and their derived products to revolutionize the field of regenerative medicines.

Acknowledgments

Parts of the figures were drawn by using pictures from Servier Medical Art. Servier Medical Art by Servier is licensed under a Creative Commons Attribution 3.0 Unported License (https://creativecommons.org/licenses/by/3.0/).

References

Beer, L., Mildner, M., & Ankersmit, H. J. (2017). Cell secretome based drug substances in regenerative medicine: When regulatory affairs meet basic science (review of cell secretome based drug substances in regenerative medicine: When regulatory affairs meet basic science). *Annals of Translational Medicine, 5*(7), 170.

Bermudez, M. A., Sendon-Lago, J., Eiro, N., Treviño, M., Gonzalez, F., Yebra-Pimentel, E., Giraldez, M. J., Macia, M., Lamelas, M. L., Saa, J., Vizoso, F., & Perez-Fernandez, R. (2015). Corneal epithelial wound healing and bactericidal effect of conditioned medium from human uterine cervical stem cells. *Investigative Ophthalmology and Visual Science, 56*(2), 983–992.

Bian, D., Wu, Y., Song, G., Azizi, R., & Zamani, A. (2022). The application of mesenchymal stromal cells (MSCs) and their derivative exosome in skin wound healing: A comprehensive review. *Stem Cell Research and Therapy, 13*(1), 24.

Bonafede, R., Scambi, I., Peroni, D., Potrich, V., Boschi, F., Benati, D., Bonetti, B., & Mariotti, R. (2016). Exosome derived from murine adipose-derived stromal cells: Neuroprotective effect on in vitro model of amyotrophic lateral sclerosis. *Experimental Cell Research, 340*(1), 150–158.

Burlacu, A., Grigorescu, G., Rosca, A.-M., Preda, M. B., & Simionescu, M. (2013). Factors secreted by mesenchymal stem cells and endothelial progenitor cells have complementary effects on angiogenesis in vitro. *Stem Cells and Development, 22*(4), 643–653.

Caseiro, A. R., Pereira, T., Ivanova, G., Luís, A. L., & Maurício, A. C. (2016). Neuromuscular regeneration: Perspective on the application of mesenchymal stem cells and their secretion products. *Stem Cells International, 9756973,* 2016.

Charoenviriyakul, C., Takahashi, Y., Nishikawa, M., & Takakura, Y. (2018). Preservation of exosomes at room temperature using lyophilization. *International Journal of Pharmaceutics, 553*(1–2), 1–7.

Chouw, A., Milanda, T., Sartika, C. R., Kirana, M. N., Halim, D., & Faried, A. (2022). Potency of mesenchymal stem cell and its secretome in treating COVID-19. *Regenerative Engineering and Translational Medicine, 8*(1), 43–54. https://doi.org/10.1007/s40883-021-00202-5

Dominici, M., Le Blanc, K., Mueller, I., Slaper-Cortenbach, I., Marini, F., Krause, D., Deans, R., Keating, A., Prockop, D., & Horwitz, E. (2006). Minimal criteria for defining multipotent mesenchymal stromal cells. The International Society for Cellular Therapy position statement. *Cytotherapy, 8*(4), 315–317.

El Moshy, S., Radwan, I. A., Rady, D., Abbass, M. M. S., El-Rashidy, A. A., Sadek, K. M., Dörfer, C. E., & Fawzy El-Sayed, K. M. (2020). Dental stem cell-derived secretome/conditioned medium: The future for regenerative therapeutic applications. *Stem Cells International, 7593402,* 2020.

El-Tookhy, O. S., Shamaa, A. A., Shehab, G. G., Abdallah, A. N., & Azzam, O. M. (2017). Histological evaluation of experimentally induced critical size defect skin wounds using exosomal solution of mesenchymal stem cells derived microvesicles. *International Journal of Stem Cells, 10*(2), 144–153.

Eleuteri, S., & Fierabracci, A. (2019). Insights into the secretome of mesenchymal stem cells and its potential applications. *International Journal of Molecular Sciences, 20*(18). https://doi.org/10.3390/ijms20184597

Gemayel, J., Chaker, D., El Hachem, G., Mhanna, M., Salemeh, R., Hanna, C., Harb, F., Ibrahim, A., Chebly, A., & Khalil, C. (2023). Mesenchymal stem cells-derived secretome and extracellular vesicles: Perspective and challenges in cancer therapy and clinical applications. In *Clinical & translational oncology: Official publication of the federation of Spanish oncology societies and of the national cancer institute of Mexico.* Advance online publication. https://doi.org/10.1007/s12094-023-03115-7

Giampà, C., Alvino, A., Magatti, M., Silini, A. R., Cardinale, A., Paldino, E., Fusco, F. R., & Parolini, O. (2019). Conditioned medium from amniotic cells protects striatal degeneration and ameliorates motor deficits in the R6/2 mouse model of Huntington's disease. *Journal of Cellular and Molecular Medicine, 23*(2), 1581–1592.

Harrell, C. R., Fellabaum, C., Jovicic, N., Djonov, V., Arsenijevic, N., & Volarevic, V. (2019). Molecular mechanisms responsible for therapeutic potential of mesenchymal stem cell-derived secretome. *Cells, 8*(5). https://doi.org/10.3390/cells8050467

Harrell, C. R., Jovicic, N., Djonov, V., Arsenijevic, N., & Volarevic, V. (2019). Mesenchymal stem cell-derived exosomes and other extracellular vesicles as new remedies in the therapy of inflammatory diseases. *Cells, 8*(12). https://doi.org/10.3390/cells8121605

Harrell, C. R., Jovicic, N., Djonov, V., & Volarevic, V. (2020). Therapeutic use of mesenchymal stem cell-derived exosomes: From basic science to clinics. *Pharmaceutics, 12*(5). https://doi.org/10.3390/pharmaceutics12050474

He, A., Jiang, Y., Gui, C., Sun, Y., Li, J., & Wang, J.-A. (2009). The antiapoptotic effect of mesenchymal stem cell transplantation on ischemic myocardium is enhanced by anoxic preconditioning. *Canadian Journal of Cardiology, 25*(6), 353–358.

Hendrata, M., & Sudiono, J. (2016). A computational model for investigating tumor apoptosis induced by mesenchymal stem cell-derived secretome. *Computational and Mathematical Methods in Medicine, 4910603,* 2016.

Herbert, F. J., Bharathi, D., Suresh, S., David, E., & Kumar, S. (2022). Regenerative potential of stem cell-derived extracellular vesicles in spinal cord injury (SCI). *Current Stem Cell Research and Therapy, 17*(3), 280–293.

Horwitz, E. M., Le Blanc, K., Dominici, M., Mueller, I., Slaper-Cortenbach, I., Marini, F. C., Deans, R. J., Krause, D. S., Keating, A., & International Society for Cellular Therapy. (2005). Clarification of the nomenclature for MSC: The international society for cellular therapy position statement. *Cytotherapy, 7*(3), 393–395.

Ibrahim, R., Mndlovu, H., Kumar, P., Adeyemi, S. A., & Choonara, Y. E. (2022). Cell secretome strategies for controlled drug delivery and wound-healing applications. *Polymers, 14*(14), 2929.

Jakovljevic, J., Slavkova, L., Harvanova, D., Spakova, T., & Rosocha, J. (2021). New therapeutic approaches of mesenchymal stem cells-derived exosomes. *Journal of Biomedical Science, 28*(1), 39.

Juhola, M., Penttinen, K., Joutsijoki, H., Varpa, K., Saarikoski, J., Rasku, J., Siirtola, H., Honen, K., Tanskanen, J., Hyttinen, H., Hyttinen, J., & Aalto-Setälä, K. (2015). Signal analysis and classification methods for the calcium transient data of stem cell-derived cardiomyocytes. *Computers in Biology and Medicine, 61*, 1–7.

Liang, X., Zhang, L., Wang, S., Han, Q., & Zhao, R. C. (2016). Exosomes secreted by mesenchymal stem cells promote endothelial cell angiogenesis by transferring miR-125a. *Journal of Cell Science, 129*(11), 2182–2189.

Masterson, C. H., Ceccato, A., Artigas, A., dos Santos, C., Rocco, P. R., Rolandsson Enes, S., Weiss, D. J., McAuley, D., Matthay, M. A., English, K., Curley, G. F., & Laffey, J. G. (2021). Mesenchymal stem/stromal cell-based therapies for severe viral pneumonia: Therapeutic potential and challenges. *Intensive Care Medicine Experimental, 9*(1), 1–21.

Ma, Z.-J., Yang, J.-J., Lu, Y.-B., Liu, Z.-Y., & Wang, X.-X. (2020). Mesenchymal stem cell-derived exosomes: Toward cell-free therapeutic strategies in regenerative medicine. *World Journal of Stem Cells, 12*(8), 814–840.

Montero-Vilchez, T., Sierra-Sánchez, Á., Sanchez-Diaz, M., Quiñones-Vico, M. I., Sanabria-de-la-Torre, R., Martinez-Lopez, A., & Arias-Santiago, S. (2021). Mesenchymal stromal cell-conditioned medium for skin diseases: A systematic review. *Frontiers in Cell and Developmental Biology, 9*, 654210.

Mukherjee, S., Yadav, G., & Kumar, R. (2021). Recent trends in stem cell-based therapies and applications of artificial intelligence in regenerative medicine. *World Journal of Stem Cells, 13*(6), 521–541.

Nassar, W., El-Ansary, M., Sabry, D., Mostafa, M. A., Fayad, T., Kotb, E., Temraz, M., Saad, A.-N., Essa, W., & Adel, H. (2016). Umbilical cord mesenchymal stem cells derived extracellular vesicles can safely ameliorate the progression of chronic kidney diseases. *Biomaterials Research, 20*, 21.

Nikfarjam, S., Rezaie, J., Zolbanin, N. M., & Jafari, R. (2020). Mesenchymal stem cell derived-exosomes: A modern approach in translational medicine. *Journal of Translational Medicine, 18*(1), 449.

Park, J., Kim, S., Lim, H., Liu, A., Hu, S., Lee, J., Zhuo, H., Hao, Q., Matthay, M. A., & Lee, J.-W. (2019). Therapeutic effects of human mesenchymal stem cell microvesicles in an ex vivo perfused human lung injured with severe *E. coli* pneumonia. *Thorax, 74*(1), 43–50.

Reinke, J. M., & Sorg, H. (2012). Wound repair and regeneration. European surgical research. Europaische chirurgische forschung. *Recherches Chirurgicales Europeennes, 49*(1), 35–43.

Rossello-Gelabert, M., Gonzalez-Pujana, A., Igartua, M., Santos-Vizcaino, E., & Hernandez, R. M. (2022). Clinical progress in MSC-based therapies for the management of severe COVID-19. *Cytokine AND Growth Factor Reviews, 68*, 25–36. https://doi.org/10.1016/j.cytogfr.2022.07.002

Sandonà, M., Di Pietro, L., Esposito, F., Ventura, A., Silini, A. R., Parolini, O., & Saccone, V. (2021). Mesenchymal stromal cells and their secretome: New therapeutic perspectives for skeletal muscle regeneration. *Frontiers in Bioengineering and Biotechnology, 9*, 652970.

Sandra, F., Sudiono, J., Sidharta, E. A., Sunata, E. P., Sungkono, D. J., Dirgantara, Y., & Chouw, A. (2014). Conditioned media of human umbilical cord blood mesenchymal stem cell-derived secretome induced apoptosis and inhibited growth of HeLa cells. *The Indonesian Biomedical Journal, 6*(1), 57–62.

Shen, Z., Huang, W., Liu, J., Tian, J., Wang, S., & Rui, K. (2021). Effects of mesenchymal stem cell-derived exosomes on autoimmune diseases. *Frontiers in Immunology, 12*, 749192.

Tan, M. I., Alfarafisa, N. M., Septiani, P., Barlian, A., Firmansyah, M., Faizal, A., Melani, L., & Nugrahapraja, H. (2022). Potential cell-based and cell-free therapy for patients with COVID-19. *Cells, 11*(15), 2319. https://doi.org/10.3390/cells11152319

Teixeira, F. G., Panchalingam, K. M., Assunção-Silva, R., Serra, S. C., Mendes-Pinheiro, B., Patrício, P., Jung, S., Anjo, S. I., Manadas, B., Pinto, L., Sousa, N., Behie, L. A., & Salgado, A. J. (2016). Modulation of the mesenchymal stem cell secretome using computer-controlled bioreactors: Impact on neuronal cell proliferation, survival and differentiation. *Scientific Reports, 6*(1), 1–14.

Vizoso, F. J., Eiro, N., Cid, S., Schneider, J., & Perez-Fernandez, R. (2017). Mesenchymal stem cell secretome: Toward cell-free therapeutic strategies in regenerative medicine. *International Journal of Molecular Sciences, 18*(9), 1852.

Williams, T., Salmanian, G., Burns, M., Maldonado, V., Smith, E., Porter, R. M., Song, Y. H., & Samsonraj, R. M. (2023). Versatility of mesenchymal stem cell-derived extracellular vesicles in tissue repair and regenerative applications. *Biochimie, 207*, 33–48. https://doi.org/10.1016/j.biochi.2022.11.011

Xia, J., Minamino, S., Kuwabara, K., & Arai, S. (2019). Stem cell secretome as a new booster for regenerative medicine. *Bioscience Trends, 13*(4), 299–307.

Zhang, B., Wu, X., Zhang, X., Sun, Y., Yan, Y., Shi, H., Zhu, Y., Wu, L., Pan, Z., Zhu, W., Qian, H., & Xu, W. (2015). Human umbilical cord mesenchymal stem cell exosomes enhance angiogenesis through the Wnt4/β-catenin pathway. *Stem Cells Translational Medicine, 4*(5), 513–522.

Chapter 15

Paradigms of omics in bioinformatics for accelerating current trends and prospects of stem cell research

Santosh Kumar Behera[1,a], Seeta Dewali[2,a], Netra Pal Sharma[2], Satpal Singh Bisht[2], Amrita Kumari Panda[3], Sanghamitra Pati[4] and Sunil Kumar[5]

[1]Department of Biotechnology, NIPER, Ahmedabad, Gujrat, India; [2]Department of Zoology, Kumaun University, Nainital, Uttarakhand, India; [3]Department of Biotechnology, Sant Gahira Guru University, Ambikapur, Chhattisgarh, India; [4]ICMR-Regional Medical Research Centre, Bhubaneswar, Odisha, India; [5]ICAR-Indian Agricultural Statistical Research Institute, New Delhi, Delhi, India

1. Introduction

Stem cell research represents one of the fields of science developing fastest and has a couple of up-and-coming applications in the future. Embryonic stem cells (ESCs) are pluripotent stem cells formed from the inner cell mass of a blastocyst, an early-stage embryo. Because ESCs may develop into any type of cell in the body, they are extremely helpful for research and future medical applications (Liu et al., 2020). ESCs are cells found in the blastocyst of developing embryos and are regarded as the "gold" standard for pluripotency (Babu & Krishnamoorthy, 2012; Evans, 2011). Due of their capacity for diversification into various lineages and replenish themselves, in several clinical trials, stem cells were investigated for a variety of ailments, such as immunological, cardiovascular, and neurological problems. Adult stem cells, ESCs, induced pluripotent stem cells (iPSCs), and cancer stem cells (CSCs) are all extensively used in therapeutic applications and fundamental scientific study. Each of these stem cell types has distinct properties and applications. These ESCs, iPSCs, and CSCs have considerable application in "personalized" remedy, regenerative medicines, embryonic development, and cell differentiation research (Nestor & Noggle, 2013). Adult stem cells, particularly adipose tissue-derived stem cells (ADSCs), are present throughout the body and serve important roles in tissue homeostasis and repair. ADSCs are produced solely from adipose (fat) tissue. Adult stem cells, including ADSCs, replace cells that are lost or destroyed as a result of disease, injury, or regular physiological turnover. When tissues are damaged or cells are lost, ADSCs can be triggered and differentiated to make new cells and aid in tissue regeneration. They can develop into all three types of germ layer cells and self-renew forever (Eguizabal et al., 2019).

Age-, disease-, or injury-related damage to tissues and organs may be repaired or replaced by regenerative cell treatment. Stem cells have enormous potential as a source of cells for regenerative cell treatment, drawing more and more interest from fundamental scientists, doctors, and the general public (He et al., 2022). These stem cells are being created for many therapeutic uses that are expanding quickly. Adult stem cells can be employed to treat patients' cells, and immune rejection, ethics, and carcinogenesis are not contentious topics (Yamada et al., 2020). Therefore, they benefit from being accepted by all patients and are often used in research trials. ESCs and iPSCs are used safely and effectively to treat various disorders, including macular degeneration, spinal cord injury, and myocardial infarction (Martin, 2017).

Stem cells are helpful resources for diagnosing and treating illness and valuable resources for treatment. Notably, recent advancements in iPSCs have ushered in a new period of ailment modeling. iPSCs can be isolated, multiplied, and transformed into ailment-related cell types as like nerve cells can be cultivated in laboratories as two-dimensional monolayers. Neuronal culture techniques enable researchers to examine the behavior, function, and characteristics of

---

[a]Equal contribution.

Computational Biology for Stem Cell Research. https://doi.org/10.1016/B978-0-443-13222-3.00020-4
Copyright © 2024 Elsevier Inc. All rights reserved, including those for text and data mining, AI training, and similar technologies.

neurons outside of the body or incorporated into three-dimensional tissue cultures that are self-organized. With the goal to better comprehend illness causes, test treatment therapies, self-organized three-dimensional tissue cultures can be used (Schutgens et al., 2020).

The chronology of significant breakthroughs and developments in fundamental science and stem cell—based treatment for therapeutic use (Fig. 15.1). In the wake of the groundbreaking discovery of pluripotent human embryonic stem cells (hESCs) in November 1998, research became a hot topic in the medical community (Thomson et al., 1998). The first significant development in regenerative medicine occurred with the first description of "stem cells" in 1888. In 1902, hematopoietic progenitor cells were first identified. The first bone marrow transplant was performed in 1939 to cure aplastic anemia. Advancement of stem cell—based treatment has seen several breakthroughs and significant achievements thanks to the conversion of fundamental research into preclinical investigations and clinical trials (Hoang et al., 2022). Stem cells were later discovered, and the progress of stem cell—based rehabilitation for treating illnesses followed their isolation in 1991.

2. A brief overview of current therapeutic uses of stem cell treatment

Regenerative medicine technique is the most optimistic field in contemporary discipline of science. The field of regenerative medicine is quickly expanding, with the ultimate goal of restoring health by repairing or replacing damaged body parts, such as organs, tissues, or cells. Being grateful to the active dedication of the research communities in evaluating potential applications for a variety of conditions, such as neurodegenerative disorders and diabetes, the prospect of stem cell treatment can be a substitute to traditional drug-based therapy, which are a continuing and dynamic topic. While drug-based therapies have long been the mainstay of medical therapy, advances in scientific study and technology are opening

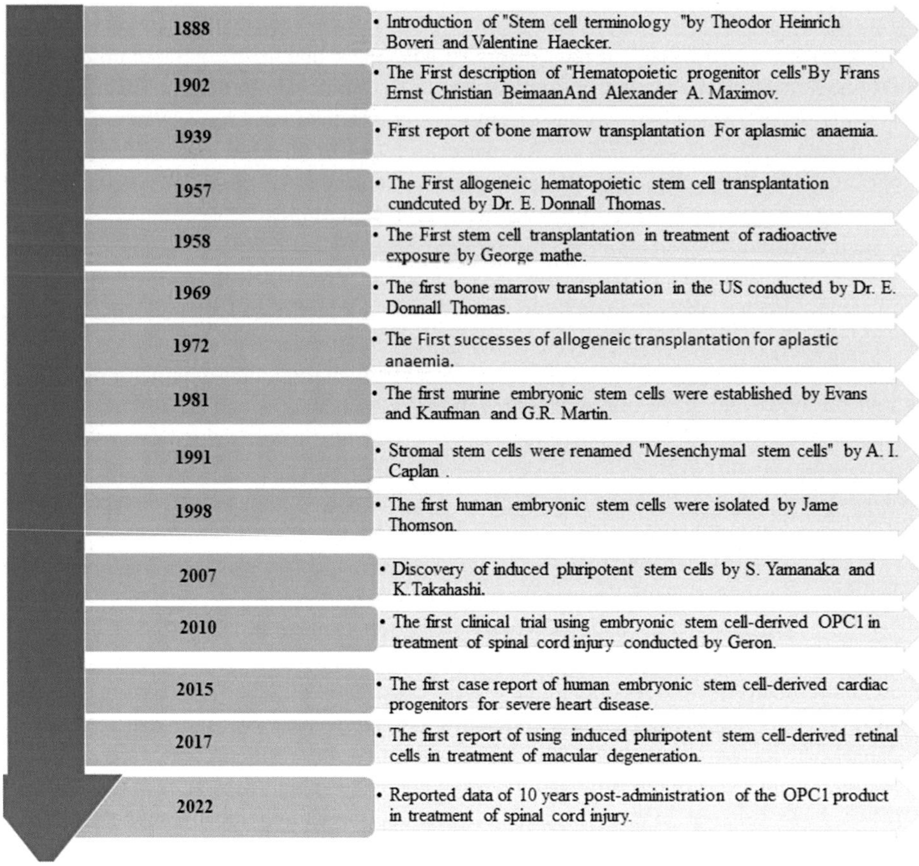

1888	Introduction of "Stem cell terminology "by Theodor Heinrich Boveri and Valentine Haecker.
1902	The First description of "Hematopoietic progenitor cells"By Frans Ernst Christian BeimaanAnd Alexander A. Maximov.
1939	First report of bone marrow transplantation For aplasmic anaemia.
1957	The First allogeneic hematopoietic stem cell transplantation cundcuted by Dr. E. Donnall Thomas.
1958	The First stem cell transplantation in treatment of radioactive exposure by George mathe.
1969	The first bone marrow transplantation in the US conducted by Dr. E. Donnall Thomas.
1972	The First successes of allogeneic transplantation for aplastic anaemia.
1981	The first murine embryonic stem cells were established by Evans and Kaufman and G.R. Martin.
1991	Stromal stem cells were renamed "Mesenchymal stem cells" by A. I. Caplan .
1998	The first human embryonic stem cells were isolated by Jame Thomson.
2007	Discovery of induced pluripotent stem cells by S. Yamanaka and K.Takahashi.
2010	The first clinical trial using embryonic stem cell-derived OPC1 in treatment of spinal cord injury conducted by Geron.
2015	The first case report of human embryonic stem cell-derived cardiac progenitors for severe heart disease.
2017	The first report of using induced pluripotent stem cell-derived retinal cells in treatment of macular degeneration.
2022	Reported data of 10 years post-administration of the OPC1 product in treatment of spinal cord injury.

FIGURE 15.1 Brief history of regenerative medicine or stem cell—based therapies.

the way for new and creative therapeutic choices (Chari et al., 2018). Recent studies documenting the successful use of regenerative medicine in patients have increased optimism that such reformative approaches might someday be used to treat various difficult conditions (Madl et al., 2018). Clinical trials involving stem cell-based therapeutics have advanced very exponentially in recent years. Some of these trials notably impacted several diseases (Pérez López & Otero Hernández, 2012), neurodegenerative disorders such as amyotrophic lateral sclerosis, and multiple sclerosis, ocular diseases, *spinal cord injury,* therapies for the treatment of diabetes (Aly, 2020). New possibilities for treating untreatable diseases and disorders have become available because of recent developments in stem cell research. Two types of stem cells that have currently become popular in multidisciplinary field are hPSCs and multipotent mesenchymal stem cells (MSCs). Human pluripotent stem cells (hPSCs) are self-renewing cell types that have the extraordinary capacity to differentiate into all three germ layers: ectoderm, mesoderm, and endoderm. These three germ layers give rise to the many tissues and organs that make up the human body. The International Society for Cell and Gene Therapy defines MSCs as multipotent progenitor cells with the self-renewal capability and transformed into mesenchymal lineages (Hoang et al., 2022). This chapter explains the outcomes of various recently concluded and ongoing clinical trials in human treatments (Table 15.1).

3. Characterization of stem cells using OMICS

Cell identification is particularly crucial when dealing with the ability of an organism or a cell to change its physical or biochemical characteristics in response to environmental stimuli of stem cells. The proliferation and obligation of MSCs, hematopoietic stem cells, and brain stem cells to explicit types of cells during culture may be affected by adaptation to environmental conditions (such as oxygen concentration) (Hawkins, 2013). Furthermore, cell culture setups expose cells to various stresses, nutritional gradients, and metabolite gradients, impacting biology of stem cell (Rodrigues et al., 2011). For instance, compared with usual culture conditions, iPSC-CMs (murine iPSC-derived cardiomyocytes) grew better after exposed to an amalgamation of low concentration of oxygen and irregular mixing in bioreactor (Correia et al., 2014).

Identity is frequently established by establishing characteristic cell shape and the biomarkers expression by using cytometry and immune cytochemistry (Tan et al., 2022). The routine identification of the origin of stem cell lines is frequently made using short tandem repeat (STR) profiling (Kerrigan et al., 2011). Genomic changes in PSC-derived cells raise the possibility of cancers when administered to patients. Techniques including whole genome/exome sequencing, molecular cytogenetic technique, i.e., FISH (fluorescence in situ hybridization), and karyotyping can be used to check genetic integrity of cells (Jo et al., 2020).

4. Transcriptomic and proteomic methods in stem cell characterization

Numerous essential genes and pathways tangled in stem cell self-renewal, proliferation, and differentiation were discovered using transcriptome analysis (Kosenko et al., 2022). Noncoding RNA is essential for differentiation and pluripotency. Some noncoding RNAs, such as Oct4 and Nanog, interact directly with the transcription factors (TFs) that regulate pluripotency. Other noncoding RNAs also operate via activating the epigenetic machinery (Gutierrez-Cruzet al., 2022).

It is challenging for the whole-genome mRNA to represent the protein dynamics of the stem cells because of posttranscriptional activity. The relative quantification of proteins and peptides, subcellular localization, protein—protein interactions, and posttranslational changes are all revealed by proteomic analysis (Atanasoaie, 2022). Proteomics research looked at changes in the cytoplasmic protein composition to find new markers and map protein—protein interactions of stem cell development phases. Proteomics has also pinpointed essential proteins in CSCs (Besson et al., 2011).

Cell signaling is significantly influenced by protein phosphorylation. Several proteins' phosphorylation states vary dynamically as stem cells differentiate. We must comprehend the phosphorylation landscape for more effective stem cell treatments to comprehend differentiation and proliferation (Detering et al., 2022). A few recently reported examples of omics technologies in stem cell characterization are summarized in Table 15.2.

5. Multiomics collaboration

The possibility of implementing these omics technologies using computational methods to simulate an integrated system seen in vivo has increased due to DNA and RNA sequencing, protein and metabolite measurement, and other advances (Nguyen et al., 2022). Rarely have numerous omics research on stem cells been combined into one study. One method is to build links between various omics datasets (metabolites, genes, and proteins) using an a priori enrichment analysis utilizing concepts from gene ontology (Mathe et al., 2018). However, "gene ontology" is connected to incomplete a priori

TABLE 15.1 In several clinical investigations, adipose tissue (AT)–derived mesenchymal stem cells (MSCs), bone marrow (BM)–derived MSCs, and umbilical cord (UC)–derived MSCs have been examined for the therapy of various human medical conditions.

S. no.	Disease	Significant sources of human mesenchymal stem cells (MSC)	No. of MSC-treated patients	Efficacy	References
1	Acute ischemic stroke	AT-MSC	4 patients	In comparison with placebo, there was not a significant difference in the mRS and NIHSS assessments.	de Celis-Ruiz et al. (2022)
2	Autism spectrum disorders	Bone marrow (BM)–derived MSCs	254 patients	94.48% of children showed improvement on the ISAA (Indian Scale for Assessment of Autism) and 95.27% of kids exhibited an improved score on the CARS (Childhood Autism Rating Scale) after transplantation. These improvements were seen in 86 (86/86) patients who underwent FDG-PET CT scans to measure improved brain activity.	Sharma et al. (2020)
3	Autism spectrum disorders	Umbilical cord (UC)–derived MSCs	12 patients	Six out of twelve partakers reported change on two ASD-specific criteria.	Sun et al. (2020)
4	Cerebral palsy	UC-derived MSCs	19 patients	ADL, CFA, and GMFM-88 scores significantly upgraded in comparison with the control group and prior to transplanting.	Gu et al. (2020)
4	Ischemic stroke	BM-derived MSCs	16 patients	The Basel index, mRS (Modified Rankin Scale) and NIHSS (National Institutes of Health Stroke Scale) did not change even after 2 years. the primary motor cortex's activity and motor performance could be enhanced with MSC-based therapy.	Jaillard et al. (2020)
5	Ischemic stroke	BM-derived MSCs	31 patients	The MSC group showed enhancement in motor functions. According to neuroimaging research, the corticospinal tract and posterior limb of the internal capsule fractional anisotropy did not decrease in the MSC group but dramatically decreased in the control group 90 days after infusion. Ipsilon and interhemispheric connections improved dramatically in the MSC group.	Lee et al. (2022)
6	Spinal cord injury	UC-derived MSCs	41 patients	41 improvements in the ASIA score, pinprick score, light touch, IANR-SCIFRS total score, and sphincter score compared with prior transplantation.	Yang et al. (2021)
7	Spinal cord injury	BM-derived MSCs	11 patients	Despite the lack of statistical significance, neurological impairment in 10 patients improved after MSC infusion.	Vaquero et al. (2018)
8	Respiratory disease ARDS	UC	9 patients	Reduced inflammation and improved clinical results	Gorman et al. (2021)
9	Respiratory disease BPD	UCB	33 patients	There was no appreciable difference between the groups in the primary outcomes. The effects of MSC transplantation on individuals with severe BPD were remarkable.	Ahn et al. (2021)

10	COVID-19	UC	SpO2 level increases, dyspnea and inflammatory cytokines r the serum decrease, and the CT score improves.	5 patients	Hashemian et al (2021)
11	COVID-19	UC	Decreased inflammatory response, elevated CRP, and improved CT score.	31 patients	Guo et al. (2020)
12	COVID-19	UC	Lessening of the pulmonary nodule or lung lesion.	65 patients	Shi et al. (2021)
13	COVID-19	UC	Increase patient survival and decrease inflammatory cytokine production.	12 patients	Lanzoni et al. (2021)
14	COVID-19	UC	Reduction of CRP, inflammatory cytokines, and neutrophil extracellular traps, immune homeostasis, and upkeep of SARS-CoV-2-specific antibodies.	29 patients	Zhu et al. (2021)
15	Type 1 diabetes	Adipose tissue (AT)	Baseline C-peptide and HbA1C levels have significantly improved.	7 patients	Dantas et al. (2021)
16	Type 2 diabetes	BM	HbA1c values decreased slightly in the first 3 years after administration, but after 6 months, they climbed and reverted to normal.	25 patients	Nguyen et al. (2021)
17	Premature ovarian Failure	BM-MSC	The improvement of menopausal symptoms and the rise in basal estrogen levels.	2 patients	Igboe et al. (2020)
18	Skin burns	BM-MSC & UC-MSC	Regarding of burn extent (%) and hospitalization time, the BM-MSC and UC-MSC groups both had considerably better healing rates than the traditionally treated group.	10 patients	Abo-Elkheir et al (2017)
19	Severe ischemic heart failure	BM	Enhancements in myocardial mass, LVEF, and LVESV. After a year of follow-up, there has been a significant decrease in the quality-of-life score and the amount of scar tissue. Also a substantial reduction in angina hospitalizations after 4 years of continuation.	30 patients	Mathiesen et al. (2015)
20	Ischemic heart failure	BM	Clinical results, such as MACE, have improved. However, there has been no change in LVEF, left ventricular, scar size. 6MWT, or peak oxygen consumption, which did not vary among clusters.	25 patients	Bolli et al. (2021)
21	Chronic ischemic heart disease	UC	LVEF, MLHF, and NYHA improvements and reduction in infarct size.	2 patients	Hu et al (2020)

TABLE 15.2 Few examples of "omics" approaches for stem cell characterization.

Approach	Types of stem cells	Outcomes	References
Single-cell transcriptomics	Leukemic stem cells (LSCs)	Identification of LSC-specific gene expression, understanding the biology of LSCs, and creating tailored treatments depend heavily on the identification of LSC-specific gene expression. This study proposed that CD96 is a specific LSC marker in few patients.	Velten et al. (2021)
Single nucleus (sn) RNAseq	Human pituitary stem cells	Illustrate gene expression and chromatin accessibility programs in pediatric, adult, and aged pituitaries cell lineages.	Zhang et al. (2021)
Single-cell transcriptomics	Hair follicle stem cells (HFSCs)	Five important HFSC populations should be separated from one another using markers that are known to be connected to both the bulge and the outer root sheath.	Chovatiya et al. (2021)
Single-cell RNA-sequencing	Limbal stem cells	Based on their unique expression pattern, the novel LSC markers TSPAN7 and SOX17 were identified and verified.	Li et al. (2021)
Proteomics LC-MS/MS in SWATH-MS mode	Human neural stem cells	VEGF signaling, Wnt signaling, and HIF-1 signaling pathways were identified as key pathways.	Cervenka et al. (2021)
Proteomics	Feline adipose-derived mesenchymal stem cells	228 exosome-specific proteins and 239 secretome proteins have been discovered.	Villatoro et al. (2021)

knowledge. The alternative strategy is coanalyzing data from many omics platforms, producing an independent metabolic network. These networks are prone to mistakes, and Ingenuity Pathway Analysis, Omix, Vanted, MetScape, Prometra, and Paintomics are the tools to manage omics data (Suravajhala et al., 2016).

6. Potency, omics approaches, and knowledge of the mechanism of action

According to appropriate tests or clinical evidence, potency is the unique ability of a substance to produce a specific therapeutic effect. Products made from stem cells are very complicated, and their effectiveness depends on several elements, such as the tissue's origin, the culture's circumstances, the age of cell donors, and multimorbidity (Zakrzewski et al., 2019). Regulators have elastic controlling strategy for stem cell–based treatment because they are aware of the inherent complexity of these products. The clinical effectiveness of the product should be reflected in the potency tests used for examining CQAs for product release, which is not an easy task given that the clinical efficacy of stem cell treatments is highly reliant on product attributes and particular clinical applications (George, 2011). The use of remestemcel-L allogenic generated MSCs meant for treating pediatrics with steroid-refractory acne is one recent instance demonstrating the difficulty of creating good potency tests for stem cell products (Kurtzberg et al., 2014).

Omics techniques enable multiplex target studies as well as untargeted approaches, resulting in more systematic and impartial insights that frequently lead to the development of MOAs and biomarkers. At the same time, most of the analytics suggested here are targeted-based approaches that involve studying a small number of hypothesized or known molecules (Xiao et al., 2022). Untargeted omics techniques enable systematic manipulation of cellular constituents such as transcriptomics, proteomics, and metabolomics as shown in Fig. 15.2. All these methods give advance knowledge for homeostasis of stem cell processes, communications with tissue environmental elements in endogenic places, paracrine signaling–mediated rejuvenation, communication between cell–cell, and molecular mechanisms of development (Lane et al., 2014).

Several variables, including lineage and maturity stage, significantly impact the highly dynamic gene expression profile during stem cell cultivation. RNA sequencing (RNA-seq), which aids the understanding of representative tags for each mRNA in a taster sample, has primarily supplanted hybridization-based microarrays (gene chips) in modern times (Wagner & Klein, 2020).

Along with mRNA-based transcriptomic investigations, microRNA (miRNA) research is becoming a significant component of the characterization of stem cells. MicroRNA shows a perilous regulating function in the creation and

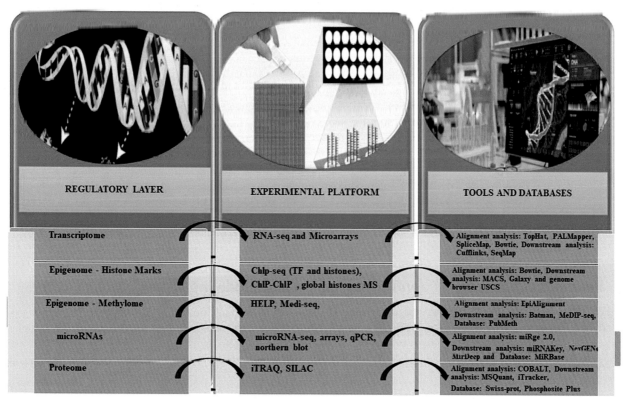

FIGURE 15.2 Dynamic system-level investigations are represented schematically, together with the experimental setup, many regulatory layers, databases and tools for the research.

differentiation of iPSCs as well as the differentiation, proliferation, and migration of MSCs (Ong et al., 2015). Different stem cell populations have released a variety of miRNAs linked to tissue regeneration. Epigenomic profiling is crucial for the study of stem cells and their potential applications in various areas such as DNA methylation, and histone modifications influence the gene expression regulation (Ramzan et al., 2021). Studies based on proteomic data provide more reliable insights into cell phenotype; meanwhile not every transcript is translated into a working protein. By using liquid chromatography, protein extraction, enzymatic digestion, peptide separation, and detection can be done, whereas mass spectrometry (MS) makes up a typical proteomic procedure toward identifying proteins (Angel et al., 2012).

Advanced quantitative proteomics techniques such as sequential window acquisition of all theoretical mass spectra (SWATH) allow the identification and quantification of huge numbers of proteins in a single experimentation (Sidoli et al., 2015). An illustration would be an understanding dynamic of protein during stem cell differentiation, maturity, or damage response. Proteomics can also access posttranslational modifications, as like glycosylation, which have been connected to essential elements of stem cell activity (Wang et al., 2014).

Besides levels of gene and protein expression, metabolism is known to have a significant role in regulating stem cell behavior. The amount of cell maturity significantly impacts the metabolism of stem cells. For instance, PSCs perform anaerobic glycolysis, whereas lineage-committed cells change their profile to become more oxidative, which is associated with more mature mitochondria (Rigaud et al., 2020).

MS and nuclear magnetic resonance (NMR) are the two main analytical methods for metabolic analysis, used to measure either extracellular or intracellular metabolites (Emwas et al., 2019). ^{13}C labeling can be used with NMR- and MS-based techniques for greater chemical specificity. Metabolites, instead of transcripts or proteins, exhibit various physicochemical characteristics (Chokkathukalam et al., 2014). LC is often used to separate these compounds depending on the metabolites' polarity, hydrophobicity, and other chemical characteristics (Buszewski & Noga, 2012).

High-throughput technologies that supply large volumes of data generate omics data. As a result, bioinformatics tools are necessary for all types of 'omics methods for development, analyses, and understand experimental information. The use of artificial aptitude to assess several iPSC-derived retinal pigment epithelial cell properties was a current achievement in this omics field (Blankenburg et al., 2009).

7. Importance of computational modeling in stem cell research

Recent high-throughput approaches produce a significant amount of stem cell data. A key objective of bioinformatics is to capture, handle, integrate, and analyze the massive volume of multidimensional data generated to deliver new and deeper insights into the underlying mechanisms. The production of large amounts of omics data makes it easier to construct computational models for stem cell research in addition to processing and analyzing massive datasets, even though computational models have been suggested in recent years. Computer models may direct experimental research by producing unique predictions and new insights into biological systems (Berger et al., 2013). Computational approaches have had a substantial impact on developmental biology program in computer to simulate attentions on morphogen that can influence pattern formation in embryo (Turing, 1952). To address pertinent issues in stem cell research, we may create models of systems biology at various degrees of complication in cell, tissue, and even organ. To better understand cellular differentiation and cellular conversion, for instance, cellular models such as gene regulatory network (GRN) models can be used (Yin et al., 2016). These models can also predict the critical TFs and cell communication pathway molecules regulating these progressions. Moreover, tissue-level models, such as those built on networks of cell—cell interactions, can help clarify fundamental ideas governing rejuvenation and create estimates of important cell communication measures (Weidemuller et al., 2021).

8. Research on stem cells: applications of several modeling types

Critical issues in stem cell study may be demonstrated using computational modeling. However, the research topic significantly impacts the modeling framework selection (Table 15.3). Cell conversion factors may be found using Boolean network models of GRNs. These models often feature regulatory interactions between a sizable number of TFs, even though they unable to necessitate the implication of kinetic parameters. Additionally, this modeling approach identifies the

TABLE 15.3 scRNA-seq-based mechanistic models.

Structured approach or methodology	Type of model	Application	Implication or significance
Descriptive model	**Simple graph or undirected graph**	Time-series data analysis and statistical inference for inferring gene coexpression networks	Prioritization of functional modules method of choosing which parts of a system should be created, implemented, or concentrated on based on their relative relevance or priority
Descriptive model	**Digraph or directed graph**	Using ligand—receptor pairs' expression, cell-to-cell signaling pathway networks can be inferred	Cell—cell interactions Prediction
Logical model	**Boolean logic networks Boolean network**	Coexpression modules and predicted TF-binding sites are used to infer a Boolean model of GRNs	Cell conversion factors Prediction
Logical model	**Boolean logic networks Boolean network**	A **Boolean logic networks** of GRNs is inferred using Granger causality analysis of time-series data	Regulatory interactions
			Governing cellular trajectories Prediction
Logical model	**Boolean logic networks Boolean network**	Using GRNs to infer Boolean logic rules	Cooperative transcription factor regulation Identification
Continuous model	Regression-based	Ridge regression and partial correlation are used to infer a quantitative model of GRNs	Prediction of gene expression dynamics during differentiation
Probabilistic model	Dynamic Bayesian network	Using prior knowledge, infer a probabilistic GRN model	Prediction of reprogramming efficiency

ideal combinations of TFs directing cellular conversion by allowing the deduction of cooperativity between TFs in regulation (Pratapa et al., 2020).

The active nature of gene expression throughout biological progressions including cellular differentiation is best predicted by continuous GRN models based on regular variance equations. It stands conceivable to deduce the controlling link between numerous genes from time-series data relating, for instance, cellular diversity using linear ordinary differential equations (Stroher et al., 2019). Furthermore, by perturbing specific TFs or signaling pathways, these comparisons may be used to expect the constant active behavior of gene expression quantitatively. One of the factors affecting the effectiveness of cellular reprogramming is randomness in expression of gene patterns and controlling communications in cell, which probabilistic methods as like probabilistic Boolean networks could possibly predict. As a result, these models may be used to rank the best TF combinations that, when perturbed, effectively cause cellular reprogramming. It is not necessary to determine the receptor−ligand binding affinity to infer directed graphs that depict cell−cell communication networks (Flottmann et al., 2012). The topology of cell−cell communication networks may be identified using this descriptive model to explain cells interaction to ensure function of tissue in homeostasis. This method also enables the detection of pertinent cell−cell communications that may impede tissue regeneration in illness (Kirouac et al., 2009).

9. Bioinformatics in stem cell research

An important task in bioinformatics is classifying high-level possible genes from high-throughput data. Analyzing matching groups of important features enhances thoughtful of the fundamental mechanisms of a biological method, giving important insights into the biology behind a specific phenotype, condition, or state. To cope with a considerable amount of data, the field of "stem cell bioinformatics" has developed as a distinct branch of the science of bioinformatics. In terms of stem cells, stem cell bioinformatics includes any in silico data analysis that supports the planning of studies and the interpretation of findings, for example, viewing/interpreting microarray data in a GRN of pluripotency (Anup, 2017).

It is challenging to track down high-quality, relevant examples of individual stem cells or developmental stages. In the case of public stem cell transcriptome research, platforms such as Stemformatics utilize QC measures that result in a failure rate of 30% (Choi et al., 2019).

Before the creation of omics, computer programs in developing and stem cell biology were primarily used to investigate the proposed process of morphogenesis by modeling the development of positional evidence throughout embryo development (Lander, 2007) and modeling adult stem cell self-renewal dynamics (Till et al., 1964). Around 15 years ago, massive genome sequencing initiatives began. Since then, the processing of enormous datasets has become the primary use of computer techniques in stem cell biology (Chinwalla et al., 2002; Lander et al., 2001). Two developments in recent years call for a summary of current knowledge in stem cell biology. One, systems-level demonstrating of stem cell behavior and work has begun to be synthesized using systems and network biology methodologies. The second is the development of methods that provide comprehensive molecular profiling of a single cell's genome.

10. Stem cell research mechanistic modeling frameworks

The characterization of the link is described in relationships of the natural method that generated the experimental data in mechanistic models, which describe a biological system based on a known or hypothesized interaction between genes (Bordbar et al., 2014). Thus, the direct integration of physical, biotic, and biochemical aspects of biological structures is a distinguishing feature of mechanistic models. Various scientific outlines, with logical and continuous models, have been developed in recent decades to depict a biological scheme at multiple steps of detail (Del Sol & Jung, 2021). The selected modeling framework has an impact on the defined biological reality's correctness, the capacity to simulate system behavior, and the capacity for prediction. A collection of rules that control the dynamics of the system serves as the foundation for conceptual models, which provide an outline for illustrating the association between genes. Each entity in the model has the ability to attain one of two states, denoted by the words "active" or "inactive," depending on the demands of its regulators. In stem cell research, using logical models to provide answers has been useful. GRN model of embryonic blood development was inferred from scRNA-seq data to predict the TFs controlling the differentiation of mesodermal progenitors into endothelium and primitive blood cells (Moignard et al., 2015). A GRN framework for blood-forming differentiating themselves, which integrates gene control, internal signaling pathways, and external ligand−receptor interactions to maintain the physical state of cell populations, emerged from the literature in a similar manner (Browaeys et al., 2020).

Researchers have been able to use continuous models to more precisely model biological systems thanks to the growing availability of single-cell data. This framework, as opposed to reasonable models, may offer a measurable description of biological schemes and does not call for categorizing items as "active" or "inactive." As a result, the system's temporal dynamics are depicted on a continuous time period, enabling a direct comparison between experimental results and the model state, to predict essential TFs involved in transdifferentiation, hematopoietic lineage specification, and stem cell differentiation. However, continuous models have other applications besides predicting TF. For instance, mechanical characteristics influencing cellular development were found by using a stem cell—niche interaction model. To help cellular differentiation processes, predictions regarding the culture's length and stiffness were created using this model.

11. Types of modeling in stem cell research

Graphs are made up of vertices and edges that reflect their connections, such as interactions between cells or coexpression of genes. The graph can be either directed or undirected depending on the kind of relationship it represents. It is important to note that graphs only reflect the facts; they do not impose modeling assumptions (Xulvi-Brunet & Li, 2010). Boolean networks are models that are built on graphs where individual vertex has a Boolean value assigned to it that represents its current states. These condition states are determined by the Boolean functions that are formalized as the states of other network vertices. Boolean networks' primary premise is that vertices' activity can be divided into two categories: active and inactive (Brim et al., 2023). This assumption is supported in the context of GRNs since the transcription rate acts like a Hill function, resembling a Boolean classification (Bhaskaran & Nair, 2015; Bocci et al., 2023). In this case, Boolean functions represent how various TFs work together to regulate a gene. Time-series or perturbation data are often needed for the identification of Boolean functions. However, the expanding scRNA-seq data set representing various cell subpopulations can be used. Continuous models, as opposed to Boolean networks, link vertices to steady states, like the constant gene expression levels.

Regression is frequently used to evaluate the relationship between vertices in this context. Connecting a purpose to a linear amalgamation of its results, such as gene expression to transcription rate, is a frequent modeling technique called a linear ordinary differential equation. The assumption of linearity forbids the consideration of nonadditive effects, such as competition TF regulation, although frequently acceptable in the framework of GRNs. Time-series data that illustrate the biological variation across time are often needed to infer continuous models. Stochastic processes can be simulated by temporal dynamic Bayesian networks (DBNs) (Bergen et al., 2020; Hecker et al., 2009). The provisional needs between the vertices of a DBN are represented by its probabilistic interactions. DBNs' central tenet is the reasonable assumption that, in the framework of gene regulation, an event may only influence subsequent events in the future and not in the past. Inferring DBNs requires time-series measurements, just like inferring continuous models.

12. Computational modeling applications in regenerative medicine

A potential method in the realm of regenerative medicine is stem cell rejuvenation. The goal of stem cell rejuvenation is to extend the life and function of the body's already-existing stem cells to promote tissue regeneration and repair. Disruption of endogenous stem cells' signaling pathways brought on by the adverse effects of a sick or old niche is one of the causes of stem cell malfunction (Bergen et al., 2020). When examining the intricate connections between stem cells and their niche environment, computational models are useful tools in stem cell research. These models can aid in the discovery of niche-specific stimuli and signaling pathways that affect stem cell behavior and result in dysregulation of stem cell function in aging and disease (Kalamakis et al., 2019).

Using phosphor-proteomics data and, if available, scRNA-seq data, for instance, representations that integrate signaling and transcriptional control systems might help identify the signaling molecules required for the steady upkeep of the disease (Del Sol & Jung, 2021). The following effects of cell-specific signaling signals on the cell transcriptional program that creates various stem cell phenotypes can be captured by this integrated model. In cases of aging or degenerative disorders, specific manipulation of the anticipated signaling biomolecules can also be employed to prevent the niche effect of stem cell upgrading.

13. Omics-based methods for managing the manufacturing of biomaterials

Recent advancements in biotechnology, notably in omics methods, have made it possible to functionally read out genome of cell, cell epigenome, cell transcriptome, cell proteome, and metabolome at high throughput. These omics-based methods can also be used to understand how biomaterials react in vivo, a field that now extensively uses estimated computational techniques rather than semiquantitative imaging-based techniques (Kersey et al., 2023).

The creation of biomaterials constituents to promote tissue renaissance using the body's innate healing capacities has been the focus of recent advancements in regenerative medicine (Mackay et al., 2021). To regulate coordinated reactions of the immune system, attract native progenitor, and encourage tissue renaissance, these biomaterials use biophysical and pharmacological cues. The creation of diverse biomaterials such as metallic, polymeric, ceramic, and composite biomaterials regulates cellular and molecular functions and promotes tissue renaissance (Zahid et al., 2022). In response to physiological cues, these designed biomaterials can release biologically active ions. All of these material-driven events have the power to draw in endogenous cells and influence the immune system. By incorporating biophysical and pharmacological signals inside these synthetic biomaterials, extracellular matrix (ECM) deposition may be regulated and regeneration of tissues can be promoted (Tang & Liu, 2023). To create the next generation of bioresponsive materials as material capabilities advance, assessment of short and enduring cellular and biomolecular interactions is essential (Behere & Ingavle, 2022). This thorough knowledge of cellular and molecular mechanisms will make it possible to quickly translate intended biomaterials for a variety of biological applications, such as medication delivery, immunological engineering, and regenerative medicine.

Traditional characterization approaches need to be translated to the clinic after being further investigated regarding how newly generated biomaterials interact with cells. Various experiments are carried out to measure metabolic activity, cell viability, proliferation, differentiation, and other factors for in vitro characterizations. To forecast the biological response, metabolites are expressed and characterized (Kersey et al., 2023). To measure the level of RNA and protein, conventional techniques such as Western blotting and quantitative polymerase chain reaction (qPCR) are used (Yue et al., 2023).

These low-throughput screening methods, however, only give a partial and biased picture of cellular behavior since they rely on "trial-and-error" methods (Sari et al., 2022). The cellular response, which includes molecular processes and regulatory networks known to cause complicated reactions, including immunogenicity and tumorigenicity, is crucially not assessed by current methods. Conventional tests also fall short in their ability to identify minor changes in how cells function and estimate systemic biological effects following clinical use. These problems result in material failures during the clinical trial stage and make it difficult for researchers to select from hundreds of potential biomaterials (Kersey et al., 2023).

14. Challenges in computational analysis in stem cell research

The execution of "omics" is not easy, frequently demanding expensive instrumentation and highly skilled personnel. Computational stem cell researchers must improve their hard skills (Computer programming, statistics, etc.) and soft skills (learning wet lab techniques, developing of transdisciplinary environment among biologists, physicists, system biologists, and many more) to create suitable tool and interpret complex samples in stem cell biology. To understand the results of computational models, machine learning skills are essential. Similarly, many tools have a specific algorithm pattern, which needs a better understanding of computational attributes. Storage of large amounts of big data generated from stem cell research and the cost of data handling, extraction, and visualization methodologies are another significant challenges for computational stem cell researchers.

15. Conclusion

There is an intriguing parallel between the current status of stem cell research and the evolution of cancer biology to cancer genetics. In fact, the emergence of "computational stem cell biology" denotes a comparable transformational stage in which computational methods and big data analysis are increasingly incorporated into stem cell research. Our understanding of stem cells and their intricate biochemistry could be fundamentally altered by this combination. In fact, there is an abundance of data in the domain of single-cell OMICs (omics) due to the development of "do-it-yourself" and commercial single-cell systems. Insights into cellular heterogeneity and molecular dynamics have never been more accessible because to the development of high-throughput, single-cell data, which has revolutionized the fields of genomics, transcriptomics, and proteomics. The multiomics era integrates data from various platforms, improving our understanding of stem cell biology, modeling human diseases, and developing novel therapeutic opportunities for regenerative remedies. The field of computational stem cell biology has enormous potential for providing developmental biology with paradigm-shifting deep understandings and for improving our capacity to precisely and efficiently control mechanism of cell. Researchers can better grasp the intricate mechanisms underpinning stem cell formation and differentiation by combining single-cell data with predictive modeling tools. We expect computational stem cell biology to develop a quantifiable background for defining cell type identification, and the biology's mechanisms of lineage commitment and maturation, which control gene expression and cellular differentiation, are intricate. There are some important mechanisms and elements that contribute to lineage commitment and maturation, even though the molecular reasoning behind these processes is still being researched and understood. This approach will provide the foundation for understanding how genetic and epigenetic anomalies interfere with normal development and lead to disease states. Before translating into effective clinical stem cell–based therapies, the interpretations generated from untargeted omics tools must be confirmed with preclinical trials. Consistent and reproducible 'omics-based analytical tools will support upcoming regulatory frameworks accelerating translation to therapy time gap.

Acknowledgements
SKB, SD, and AKP wrote the paper. NPS and SKB designed the figures with support from SD and AKP. SP and SPS supervised the project as well as edited the manuscript. All authors read and approved the final manuscript.

References

Abo-Elkheir, W., Hamza, F., Elmofty, A. M., Emam, A., Abdl-Moktader, M., Elsherefy, S., & Gabr, H. (2017). Role of cord blood and bone marrow mesenchymal stem cells in recent deep burn: A case-control prospective study. *American Journal of Stem Cells, 6*(3), 23–35.

Ahn, S. Y., Chang, Y. S., Lee, M. H., Sung, S. I., Lee, B. S., Kim, K. S., ... Park, W. S. (2021). Stem cells for bronchopulmonary dysplasia in preterm infants: A randomized controlled phase II trial. *Stem Cells Translational Medicine, 10*(8), 1129–1137.

Aly, R. M. (2020). Current state of stem cell-based therapies: An overview. *Stem Cell Investigation, 15*, 7–8. https://doi.org/10.21037/sci-2020-001

Angel, T. E., Aryal, U. K., Hengel, S. M., Baker, E. S., Kelly, R. T., Robinson, E. W., & Smith, R. D. (2012). Mass spectrometry-based proteomics: Existing capabilities and future directions. *Chemical Society Reviews, 41*(10), 3912–3928. https://doi.org/10.1039/c2cs15331a

Anup Som. (2017). In V. Verma, M. P. Singh, & M. Kumar (Eds.), *Bioinformatics strategies for stem cell research in: Stem cells from culture dish to clinic.* Nova Science Publishers, Inc, ISBN 978-1-53612-732-4. © 2017.

Atanasoai, I. (2022). *Technological advancements to study RBP and RNA biology.*

Babu, P. B. R., & Krishnamoorthy, P. (2012). Applications of bioinformatics tools in stem cell research: An update. *Journal of Pharmacy Research, 5*(9), 4863–4866.

Behere, I., & Ingavle, G. (2022). In vitro and in vivo advancement of multifunctional electrospun nanofiber scaffolds in wound healing applications: Innovative nanofiber designs, stem cell approaches, and future perspectives. *Journal of Biomedical Materials Research Part A, 110*(2), 443–461.

Bergen, V., Lange, M., Peidli, S., Wolf, F. A., & Theis, F. J. (2020). Generalizing RNA velocity to transient cell states through dynamical modeling. *Nature Biotechnology, 38*(12), 1408–1414.

Berger, B., Peng, J., & Singh, M. (2013). Computational solutions for omics data. *Nature Reviews Genetics, 14*(5), 333–346. https://doi.org/10.1038/nrg3433

Besson, D., Pavageau, A. H., Valo, I., Bourreau, A., Bélanger, A., Eymerit-Morin, C., Moulière, A., Chassevent, A., Boisdron-Celle, M., Morel, A., Solassol, J., Campone, M., Gamelin, E., Barré, B., Coqueret, O., & Guette, C. (2011). A quantitative proteomic approach of the different stages of colorectal cancer establishes OLFM4 as a new nonmetastatic tumor marker. *Molecular and Cellular Proteomics: MCP, 10*(12), M111. https://doi.org/10.1074/mcp.M111.009712, 009712.

Bhaskaran, S., & Nair, A. S. (2015). Hill equation in modeling transcriptional regulation. *Systems and Synthetic Biology*, 77–92.

Blankenburg, M., Haberland, L., Elvers, H. D., Tannert, C., & Jandrig, B. (2009). High-throughput omics technologies: Potential tools for the investigation of influences of EMF on biological systems. *Current Genomics, 10*(2), 86–92. https://doi.org/10.2174/138920209787847050

Bocci, F., Jia, D., Nie, Q., Jolly, M. K., & Onuchic, J. (2023). *Theoretical and computational tools to model multistable gene regulatory networks.* arXiv preprint arXiv:2302.07401.

Bolli, R., Mitrani, R. D., Hare, J. M., Pepine, C. J., Perin, E. C., Willerson, J. T., ... Cardiovascular Cell Therapy Research Network (CCTRN). (2021). A phase II study of autologous mesenchymal stromal cells and c-kit positive cardiac cells, alone or in combination, in patients with ischaemic heart failure: The CCTRN CONCERT-HF trial. *European Journal of Heart Failure, 23*(4), 661–674.

Bordbar, A., Monk, J. M., King, Z. A., & Palsson, B. O. (2014). Constraint-based models predict metabolic and associated cellular functions. *Nature Reviews Genetics, 15*(2), 107–120.

Brim, L., Pastva, S., Šafránek, D., & Šmijáková, E. (2023). Temporary and permanent control of partially specified Boolean networks. *Biosystems, 223*, 104795.

Browaeys, R., Saelens, W., & Saeys, Y. (2020). NicheNet: Modeling intercellular communication by linking ligands to target genes. *Nature Methods, 17*, 159–162. https://doi.org/10.1038/s41592-019-0667-5

Buszewski, B., & Noga, S. (2012). Hydrophilic interaction liquid chromatography (HILIC)- a powerful separation technique. *Analytical and Bioanalytical Chemistry, 402*(1), 231–247. https://doi.org/10.1007/s00216-011-5308-5

Cervenka, J., Tyleckova, J., Kupcova Skalníkova, H., Vodickova Kepkova, K., Poliakh, I., Valekova, I., ... Vodicka, P. (2021). Proteomic characterization of human neural stem cells and their secretome during in vitro differentiation. *Frontiers in Cellular Neuroscience, 14*, 612560.

Chari, S., Nguyen, A., & Saxe, J. (2018). Stem cells in the clinic. *Cell Stem Cell, 22*, 781–782.

Chinwalla, A. T., Cook, L. L., Delehaunty, K. D., Fewell, G. A., Fulton, L. A., Fulton, R. S., ... Zody, M. C. (2002). Initial sequencing and comparative analysis of the mouse genome. *Nature, 420*, 520–562.

Choi, J., Pacheco, C. M., Moshergen, R., Korn, O., Chen, T., Nagpal, I., Englent, S., Angel, P., Wu, & Wells, C., ... (2019). Bioinformatics, visualize and download clustered stem cell data. Nucleic Acids Research, (NDI), D841–D840.

Chokkathukalam, A., Kim, D. H., Barrett, M. P., Breitling, R., & Creek, D. J. (2014). Stable isotope-labeling studies in metabolomics: New insights into structure and dynamics of metabolic networks. *Bioanalysis, 6*(4), 511–524. https://doi.org/10.4155/bio.13.348

Chovatiya, G., Chuwalewala, S., Walter, L. D., Cosgrove, B. D., & Tumbar, T. (2021). High resolution single-cell transcriptomics reveals heterogeneity of self-renewing hair follicle stem cells. *Experimental Dermatology, 30*(4), 457–471

Correia, C., Serra, M., Espinha, N., Sousa, M., Brito, C., Burkert, K., ... Alves, P. M. (2014). Combining hypoxia and bioreactor hydrodynamics boosts induced pluripotent stem cell differentiation towards cardiomyocytes. *Stem Cell Review and Reports, 10*, 786–801

Dantas, J. R., Araùjo, D. B., Silva, K. R., Souto, D. L., Pereira, M. D. F. C., Luiz, R. R., ... Rodacki, M. (2021). Adipose tissue-derived stromal/stem cells+ cholecalciferol: A pilot study in recent-onset type 1 diabetes patients. *Archives of Endocrinology and Metabolism, 65*, 342–351.

de Celis-Ruiz, E., Fuentes, B., Alonso de Leciñana, M., Gutiérrez-Fernández, M., Borobia, A. M., Gutiérrez-Zúñiga, R., ... Díez-Tejedor, E. (2022). Final results of allogeneic adipose tissue-derived mesenchymal stem cells in acute ischemic stroke (AMASCIS): A phase II, randomized, double-blind, placebo-controlled, single-center, pilot clinical trial. *Cell Transplantation, 31*, 09636897221083863.

Del Sol, A., & Jung, S. (2021). The importance of computational modeling in stem cell research. *Trends in Biotechnology, 39*(2), 126–136.

Detering, N. T., Schüning, T., Hensel, N., & Claus, P. (2022). The phospho-landscape of the survival of motoneuron protein (SMN) protein: Relevance for spinal muscular atrophy (SMA). *Cellular and Molecular Life Sciences, 79*(9), 1–20.

Eguizabal, C., Aran, B., Chuva de Sousa Lopes, S. M., Geens, M., Heindryckx, B., Panula, S., ... Veiga, A. (2019). Two decades of embryonic stem cells: A historical overview. *Human Reproduction Open, 2019*(1), hoy024.

Emwas, A. H., Roy, R., McKay, R. T., Tenori, L., Saccenti, E., Gowda, G., Raftery, D., Alahmari, F., Jaremko, L., Jaremko, M., & Wishart, D. S. (2019). NMR spectroscopy for metabolomics research. *Metabolites, 9*(7), 123. https://doi.org/10.3390/metabo9070123

Evans, M. (2011). Discovering pluripotency: 30 years of mouse embryonic stem cells. *Nature Reviews in Molecular Cell Biology, 12*(10), 680–686.

Flöttmann, M., Scharp, T., & Klipp, E. (2012). A stochastic model of epigenetic dynamics in somatic cell reprogramming. *Frontiers in Physiology, 3*, 216. https://doi.org/10.3389/fphys.2012.00216

George, B. (2011). Regulations and guidelines governing stem cell-based products: Clinical considerations. *Perspectives in Clinical Research, 2*(3), 94–99. https://doi.org/10.4103/2229-3485.83228

Gorman, E., Shankar-Hari, M., Hopkins, P., Tunnicliffe, W. S., Perkins, G. D., Silversides, J., ... O'Kane, C. M. (2021). Repair of acute respiratory distress syndrome by stromal cell administration (REALIST) trial: A phase 1 trial. *EClinicalMedicine, 41*, 101167.

Gu, J., Huang, L., Zhang, C., Wang, Y., Zhang, R., Tu, Z., ... Liu, L. (2020). Therapeutic evidence of umbilical cord-derived mesenchymal stem cell transplantation for cerebral palsy: A randomized, controlled trial. *Stem Cell Research and Therapy, 11*, 1–12.

Guo, Z., Chen, Y., Luo, X., He, X., Zhang, Y., & Wang, J. (2020). Administration of umbilical cord mesenchymal stem cells in patients with severe COVID-19 pneumonia. *Critical Care, 24*, 1–3.

Gutierrez-Cruz, J. A., Maldonado, V., & Melendez-Zajgla, J. (2022). Regulation of the cancer stem phenotype by long non-coding RNAs. *Cells, 11*(15), 2352.

Hashemian, S. M. R., Aliannejad, R., Zarrabi, M., Soleimani, M., Vosough, M., Hosseini, S. E., ,,, Baharvand, H. (2021). Mesenchymal stem cells derived from perinatal tissues for treatment of critically ill COVID-19-induced ARDS patients: A case series. *Stem Cell Research and Therapy, 12*(1), 1–12.

Hawkins, K. E., Sharp, T. V., & Mckay, T. R. (2013). The role of hypoxia in stem cell potency and differentiation. *Regenerative Medicine, 8*, 771–782.

He, X., Wang, Q., Zhao, Y., Zhang, H., Wang, B., Pan, J., ... Wang, D. (2020). Effect of intramyocardial grafting collagen scaffold with mesenchymal stromal cells in patients with chronic ischemic heart disease: A randomized clinical trial. *JAMA Network Open, 3*(9), e2016236.

He, X., Zhu, Y., Ma, B., Xu, X., Huang, R., Cheng, L., & Zhu, R. (2022). Bioactive 2D nanomaterials for neural repair and regeneration. *Advanced Drug Delivery Reviews*, 114379.

Hecker, M., Lambeck, S., Toepfer, S., Van Someren, E., & Guthke, R. (2009). Gene regulatory network inference: Data integration in dynamic models—a review. *Biosystems, 96*(1), 86–103.

Hoang, D. M., Pham, P. T., Bach, T. Q., Ngo, A., Nguyen, Q. T., Phan, T., Nguyen, G. H., Le, P., Hoang, V. T., Forsyth, N. R., Heke, M., & Nguyen, L. T. (2022). Stem cell-based therapy for human diseases. *Signal Transduction and Targeted Therapy, 7*(1), 272. https://doi.org/10.1038/s41392-022-01134-4

Igboeli, P., El Andaloussi, A., Sheikh, U., Takala, H., ElSharoud, A., McHugh, A., ... Al-Hendy, A. (2020). Intraovarian injection of autologous human mesenchymal stem cells increases estrogen production and reduces menopausal symptoms in women with premature ovarian failure: Two case reports and a review of the literature. *Journal of Medical Case Reports, 14*(1), 1–11.

Jaillard, A., Hommel, M., Moisan, A., Zeffiro, T. A., Favre-Wiki, I. M., Barbieux-Guillot, M., ... ISIS-HERMES Study Group. (2020). Autologous mesenchymal stem cells improve motor recovery in subacute ischemic stroke: A randomized clinical trial. *Translational Stroke Research, 11*, 910–923.

Jo, H. Y., Han, H. W., Jung, I., Ju, J. H., Park, S. J., Moon, S., Geum, D., Kim, H., Park, H. J., Kim, S., Stacey, G. N., Koo, S. K., Park, M. H., & Kim, J. H. (2020). Development of genetic quality tests for good manufacturing practice-compliant induced pluripotent stem cells and their derivatives. *Scientific Reports, 10*(1), 3939. https://doi.org/10.1038/s41598-020-60466-9

Kalamakis, G., Brüne, D., Ravichandran, S., Bolz, J., Fan, W., Ziebell, F., ... Martin-Villalba, A. (2019). Quiescence modulates stem cell maintenance and regenerative capacity in the aging brain. *Cell, 176*(6), 1107–1119.

Kerrigan, L., & Nims, R. W. (2011). Authentication of human cell-based products: The role of a new consensus standard. *Regenerative Medicine, 6*, 255–260.

Kersey, A. L., Nguyen, T. U., Nayak, B., Singh, I., & Gaharwar, A. K. (2023). Omics-based approaches to guide the design of biomaterials. *Materials Today, 64*, 98–120.

Kirouac, D. C., Madlambayan, G. J., Yu, M., Sykes, E. A., Ito, C., & Zandstra, P. W. (2009). Cell-cell interaction networks regulate blood stem and progenitor cell fate. *Molecular Systems Biology, 5*, 293. https://doi.org/10.1038/msb.2009.49

Kosenko, A., Salame, T. M., Friedlander, G., & Barash, I. (2022). Macrophage-secreted CSF1 transmits a calorie restriction-induced self-renewal signal to mammary epithelial stem cells. *Cells, 11*(18), 2923.

Kurtzberg, J., Prockop, S., Teira, P., Bittencourt, H., Lewis, V., Chan, K. W., Horn, B., Yu, L., Talano, J. A., Nemecek, E., Mills, C. R., & Chaudhury, S. (2014). Allogeneic human mesenchymal stem cell therapy (remestemcel-L, Prochymal) as a rescue agent for severe refractory acute graft-versus-host disease in pediatric patients. *Biology of Blood and Marrow Transplantation, 20*(2), 229–235. https://doi.org/10.1016/j.bbmt.2013.11.001

Lander, E. S., Linton, L. M., Birren, B., Nusbaum, C., Zody, M. C., Baldwin, J., … Morgan, M. J. (2001). Initial sequencing and analysis of the human genome. *Nature, 409*, 860–921.

Lander, A. D. (2007). Morpheus unbound: Reimagining the morphogen gradient. *Cell, 128*, 245–256.

Lane, S. W., Williams, D. A., & Watt, F. M. (2014). Modulating the stem cell niche for tissue regeneration. *Nature Biotechnology, 32*(8), 795–803. https://doi.org/10.1038/nbt.2978

Lanzoni, G., Linetsky, E., Correa, D., Messinger Cayetano, S., Alvarez, R. A., Kouroupis, D., … Ricordi, C. (2021). Umbilical cord mesenchymal stem cells for COVID-19 acute respiratory distress syndrome: A double-blind, phase 1/2a, randomized controlled trial. *Stem Cells Translational Medicine, 10*(5), 660–673.

Lee, J., Chang, W. H., Chung, J. W., Kim, S. J., Kim, S. K., Lee, J. S., … Bang, O. Y. (2022). Efficacy of intravenous mesenchymal stem cells for motor recovery after ischemic stroke: A neuroimaging study. *Stroke, 53*(1), 20–28.

Li, D. Q., Kim, S., Li, J. M., Gao, Q., Choi, J., Bian, F., … Chen, R. (2021). Single-cell transcriptomics identifies limbal stem cell population and cell types mapping its differentiation trajectory in limbal basal epithelium of human cornea. *Ocular Surface, 20*, 20–32.

Liu, G., David, B. T., Trawczynski, M., & Fessler, R. G. (2020). Advances in pluripotent stem cells: History, mechanisms, technologies, and applications. *Stem Cell Reviews and Reports, 16*(1), 3–32.

Mackay, B. S., Marshall, K., Grant-Jacob, J. A., Kanczler, J., Eason, R. W., Oreffo, R. O., & Mills, B. (2021). The future of bone regeneration: Integrating AI into tissue engineering. *Biomedical Physics and Engineering Express, 7*(5), 052002.

Madl, C. M., Heilshorn, S. C., & Blau, H. M. (2018). Bioengineering strategies to accelerate stem cell therapeutics. *Nature, 557*, 335–342.

Martin, U. (2017). Therapeutic application of pluripotent stem cells: Challenges and risks. *Frontiers of Medicine, 4*, 229. https://doi.org/10.3389/fmed.2017.00229

Mathé, E., Hays, J. L., Stover, D. G., & Chen, J. L. (2018). The omics revolution continues: The maturation of high-throughput biological data sources. *Yearbook of Medical Informatics, 27*(1), 211–222. https://doi.org/10.1055/s-0038-1667085

Mathiasen, A. B., Qayyum, A. A., Jørgensen, E., Helqvist, S., Fischer-Nielsen, A., Kofoed, K. F., … Kastrup, J. (2015). Bone marrow-derived mesenchymal stromal cell treatment in patients with severe ischaemic heart failure: A randomized placebo-controlled trial (MSC-HF trial). *European Heart Journal, 36*(27), 1744–1753.

Moignard, V., Woodhouse, S., Haghverdi, L., Lilly, A. J., Tanaka, Y., Wilkinson, A. C., … Göttgens, B. (2015). Decoding the regulatory network of early blood development from single-cell gene expression measurements. *Nature Biotechnology, 33*(3), 269–276. https://doi.org/10.1038/nbt.3154

Nestor, M. W., & Noggle, S. A. (2013). Standardization of human stem cell pluripotency using bioinformatics. *Stem Cell Research and Therapy, 4*, 37.

Nguyen, L. T., Hoang, D. M., Nguyen, K. T., Bui, D. M., Nguyen, H. T., Le, H. T., … Bui, A. V. (2021). Type 2 diabetes mellitus duration and obesity alter the efficacy of autologously transplanted bone marrow-derived mesenchymal stem/stromal cells. *Stem Cells Translational Medicine, 10*(9), 1266–1278.

Nguyen, N., Jennen, D., & Kleinjans, J. (2022). Omics technologies to understand drug toxicity mechanisms. *Drug Discovery Today*, 103348.

Ong, S. G., Lee, W. H., Kodo, K., & Wu, J. C. (2015). MicroRNA-mediated regulation of differentiation and trans-differentiation in stem cells. *Advanced Drug Delivery Reviews, 88*, 3–15. https://doi.org/10.1016/j.addr.2015.04.004

Pérez López, S., & Otero Hernández, J. (2012). Advances in stem cell therapy. *Advances in Experimental Medicine and Biology, 741*, 290–313.

Pratapa, A., Jalihal, A. P., Law, J. N., Bharadwaj, A., & Murali, T. M. (2020). Benchmarking algorithms for gene regulatory network inference from single-cell transcriptomic data. *Nature Methods, 17*(2), 147–154. https://doi.org/10.1038/s41592-019-0690-6

Ramzan, F., Vickers, M. H., & Mithen, R. F. (2021). Epigenetics, microRNA and metabolic syndrome: A comprehensive review. *International Journal of Molecular Sciences, 22*(9), 5047. https://doi.org/10.3390/ijms22095047

Rigaud, V., Hoy, R., Mohsin, S., & Khan, M. (2020). Stem cell metabolism: Powering cell-based therapeutics. *Cells, 9*(11), 2490. https://doi.org/10.3390/cells9112490

Rodrigues, C. A., Fernandes, T. G., Diogo, M. M., da Silva, C. L., & Cabral, J. M. (2011). Stem cell cultivation in bioreactors. *Biotechnology Advances, 29*(6), 815–829.

Sari, B., Isik, M., Eylem, C. C., Bektas, C., Okesola, B. O., Karakaya, E., … Derkus, B. (2022). Omics technologies for high-throughput-screening of cell-biomaterial interactions. *Molecular Omics, 18*(7), 591–615.

Schutgens, F., & Clevers, H. (2020). Human organoids: Tools for understanding biology and treating diseases. *Annual Review of Pathology: Mechanisms of Disease, 15*, 211–234.

Sharma, A. K., Gokulchandran, N., Kulkarni, P. P., Sane, H. M., Sharma, R., Jose, A., & Badhe, P. B. (2020). Cell transplantation as a novel therapeutic strategy for autism spectrum disorders: A clinical study. *American Journal of Stem Cells, 9*(5), 89.

Shi, L., Huang, H., Lu, X., Yan, X., Jiang, X., Xu, R., ... Wang, F. S. (2021). Effect of human umbilical cord derived mesenchymal stem cells on lung damage in severe COVID-19 patients: A randomized, double-blind, placebo-controlled phase 2 trial. *Signal Transduction and Targeted Therapy, 6*(1), 58.

Sidoli, S., Lin, S., Xiong, L., Bhanu, N. V., Karch, K. R., Johansen, E., Hunter, C., Mollah, S., & Garcia, B. A. (2015). Sequential Window acquisition of all theoretical mass Spectra (SWATH) analysis for characterization and quantification of histone post-translational modifications. *Molecular and Cellular Proteomics, 14*(9), 2420–2428. https://doi.org/10.1074/mcp.O114.046102

Strober, B. J., Elorbany, R., Rhodes, K., Krishnan, N., Tayeb, K., Battle, A., & Gilad, Y. (2019). Dynamic genetic regulation of gene expression during cellular differentiation. *Science, 364*(6447), 1287–1290. https://doi.org/10.1126/science.aaw0040

Sun, J. M., Dawson, G., Franz, L., Howard, J., McLaughlin, C., Kistler, B., ... Kurtzberg, J. (2020). Infusion of human umbilical cord tissue mesenchymal stromal cells in children with autism spectrum disorder. *Stem Cells Translational Medicine, 9*(10), 1137–1146.

Suravajhala, P., Kogelman, L. J., & Kadarmideen, H. N. (2016). Multi-omic data integration and analysis using systems genomics approaches: Methods and applications in animal production, health and welfare. Genetics, selection, evolution. *GSE, 48*(1), 38. https://doi.org/10.1186/s12711-016-0217-x

Tan, J. P., Liu, X., & Polo, J. M. (2022). Establishment of human induced trophoblast stem cells via reprogramming of fibroblasts. *Nature Protocols*, 1–21.

Tang, R. Z., & Liu, X. Q. (2023). Biophysical cues of in vitro biomaterials-based artificial extracellular matrix guide cancer cell plasticity. *Materials Today Bio*, 100607.

Thomson, J. A., Itskovitz-Eldor, J., Shapiro, S. S., Waknitz, M. A., Swiergiel, J. J., Marshall, V. S., & Jones, J. M. (1998). Embryonic stem cell lines derived from human blastocysts. *Science, 282*, 1145–1147.

Till, J. E., McCulloch, E. A., & Siminovitch, L. (1964). A stochastic model of stem cell proliferation, based on the growth of spleen colony-forming cells. *Proceedings of the National Academy of Sciences, 51*(1), 29–36.

Turing, A. M. (1952). The chemical basis of morphogenesis. *Philosophical Transactions of the Royal Society of London Series B Biological Sciences, 237*, 37–72.

Vaquero, J., Zurita, M., Rico, M. A., Aguayo, C., Bonilla, C., Marin, E., ... Rodríguez, B. (2018). Intrathecal administration of autologous mesenchymal stromal cells for spinal cord injury: Safety and efficacy of the 100/3 guideline. *Cytotherapy, 20*(6), 806–819.

Velten, L., Story, B. A., Hernández-Malmierca, P., Raffel, S., Leonce, D. R., Milbank, J., ... Steinmetz, L. M. (2021). Identification of leukemic and pre-leukemic stem cells by clonal tracking from single-cell transcriptomics. *Nature Communications, 12*(1), 1–13.

Villatoro, A. J., Martín-Astorga, M. D. C., Alcoholado, C., Sánchez-Martín, M. D. M., & Becerra, J. (2021). Proteomic analysis of the secretome and exosomes of feline adipose-derived mesenchymal stem cells. *Animals, 11*(2), 295.

Wagner, D. E., & Klein, A. M. (2020). Lineage tracing meets single-cell omics: Opportunities and challenges. *Nature Reviews Genetics, 21*(7), 410–427. https://doi.org/10.1038/s41576-020-0223-2

Wang, Y. C., Peterson, S. E., & Loring, J. F. (2014). Protein post-translational modifications and regulation of pluripotency in human stem cells. *Cell Research, 24*(2), 143–160. https://doi.org/10.1038/cr.2013.151

Weidemuller, P., Kholmatov, M., Petsalaki, E., & Zaugg, J. B. (2021). Transcription factors: Bridge between cell signaling and gene regulation. *Proteomics, 21*(23–24), e2000034. https://doi.org/10.1002/pmic.202000034

Xiao, Y., Bi, M., Guo, H., & Li, M. (2022). Multi-omics approaches for biomarker discovery in early ovarian cancer diagnosis. *EBioMedicine, 79*, 104001. https://doi.org/10.1016/j.ebiom.2022.104001

Xulvi-Brunet, R., & Li, H. (2010). Co-Expression networks: Graph properties and topological comparisons. *Bioinformatics, 26*(2), 205–214.

Yamada, Y., Nakamura-Yamada, S., Konoki, R., & Baba, S. (2020). Promising advances in clinical trials of dental tissue-derived cell-based regenerative medicine. *Stem Cell Research and Therapy, 11*, 175.

Yang, Y., Pang, M., Du, C., Liu, Z. Y., Chen, Z. H., Wang, N. X., ... Rong, L. M. (2021). Repeated subarachnoid administrations of allogeneic human umbilical cord mesenchymal stem cells for spinal cord injury: A phase 1/2 pilot study. *Cytotherapy, 23*(1), 57–64.

Yin, X., Mead, B. E., Safaee, H., Langer, R., Karp, J. M., & Levy, O. (2016). Engineering stem cell organoids. *Cell Stem Cell, 18*(1), 25–38. https://doi.org/10.1016/j.stem.2015.12.005

Yue, Y., Ge, Z., Guo, Z., Wang, Y., Yang, G., Sun, S., & Li, X. (2023). Screening of lncRNA profiles during intramuscular adipogenic differentiation in longissimus dorsi and semitendinosus muscles in pigs. *Animal Biotechnology*, 1–11.

Zahid, A. A., Chakraborty, A., Shamiya, Y., Ravi, S. P., & Paul, A. (2022). Leveraging the advancements in functional biomaterials and scaffold fabrication technologies for chronic wound healing applications. *Materials Horizons, 9*(7), 1850–1865.

Zakrzewski, W., Dobrzyński, M., Szymonowicz, M., & Rybak, Z. (2019). Stem cells: Past, present, and future. *Stem Cell Research & Therapy, 10*(1), 68. https://doi.org/10.1186/s13287-019-1165-5

Zhang, Z., Zamojski, M., Smith, G. R., Willis, T. L., Yianni, V., Mendelev, N., ... Ruf-Zamojski, F. (2021). *Single nucleus pituitary transcriptomic and epigenetic landscape reveals human stem cell heterogeneity with diverse regulatory mechanisms*. bioRxiv.

Zhu, R., Yan, T., Feng, Y., Liu, Y., Cao, H., Peng, G., ... Zhao, R. C. (2021). Mesenchymal stem cell treatment improves outcome of COVID-19 patients via multiple immunomodulatory mechanisms. *Cell Research, 31*(12), 1244–1262.

Chapter 16

Transcriptomic profiling—based identification of biomarkers of stem cells

Swati Sharma, Daizy Kalpdev and Ankit Choudhary

Department of Pharmacology, All India Institute of Medical Sciences (AIIMS), New Delhi, Delhi, India

1. Introduction

The transcriptome is a whole set of transcripts that can be found within the cell, and understanding it will not only help in the interpretation of the genome's functional elements but also reveal the cell's and tissue's molecular constituents. This is the best tool to precisely measure transcript level and their isoforms in any type of cells including stem cells. The demand and rise in the high-throughput technologies has led to the revolution in computational field and thereby increase in the computational power subsequently, which made the analysis or characterization of data easy and efficient.

1.1 Global picture of cell function

The inheritance of genetic information from genes to the function of cells and the organisms exerting the biological effect comprises of a two-step process, viz., transcription and translation. The central dogma reveals the transcription of DNA into RNA, which is subsequently translated into protein. The whole of all RNA molecules set in a cell or in an organism thus makes up the transcriptome. It is interesting to note that the transcribed RNAs may include the entire range of coding and noncoding short-lasting subunits, serving diverse functions such as structural (rRNAs—ribosomal assembly), transporters (tRNA) along with regulatory (microRNAs—miRNAs, or long noncoding RNAs—lncRNAs) among others (Small et al., 2010). It is evident that the noncoding RNAs also play an important role in life-threatening diseases such as cancer (Kaur et al., 2020), cardiovascular diseases (Pedrotty et al., 2012), and recently neurological disorders (Lake et al., 2021).

The advent of postgenomic era has led to the emergence of various sophisticated omics techniques including transcriptomics, proteomics, and metabolomics. A transcriptome refers to the sum of all transcribed RNA, obtained either from a specific tissue or cell at any functional or developmental stage. This technique specializes in the understanding of a transcriptome, besides its identification and quantification, specific to a tissue or cell, and is the most widely used omics technology among all others (Byron et al., 2016). It is not only restricted to mRNAs, but it also provides important information about other noncoding RNAs. It also lays a detailed picture of all transcripts and is useful for predicting novel biomarkers and their underlying mechanisms to offer predictive diagnosis and targeted prevention (Casamassimi et al., 2017).

1.2 Profiling of a transcriptome

The profiling of a transcriptome exhibits a snapshot of the expression profile of actively expressed genes and transcripts under various conditions. It reveals the change in behavior of a cell in response to pathogens, diseases, or any environmental stress via gene expression profiling, which is extremely useful in quantifying and determining the phenotypic changes in the mutant by losing or gaining-of-function (Cappola & Margulies, 2011). Thus, due to having various applications in the domains including diagnosis of diseases (Casamassimi et al., 2017), identification of novel biomarkers (Penn-Nicholson et al., 2020), risk assessment associated with any new drugs (Bourdon-Lacombe et al., 2015) or environmental chemicals (Taubes et al., 2021), and also suggesting a holistic strategy of predictive, preventive, and personalized medicine (Wang et al., 2021), this technique besides other omics technologies has gained wide popularity in the

Computational Biology for Stem Cell Research. https://doi.org/10.1016/B978-0-443-13222-3.00034-4
Copyright © 2024 Elsevier Inc. All rights reserved, including those for text and data mining, AI training, and similar technologies.

203

diverse areas of biomedical researches. Further, with this technique, it is also possible to generate randomized RNA transcripts in vitro, to identify the sequences having enzymatic or therapeutic applications. In addition to the identification of thousands of signature genes associated with a disease, the sequencing of RNA can also be applied to discover novel gene structures, single-nucleotide polymorphisms (SNPs), and allele-specific expression (Battle et al., 2014). The transcriptome-based analysis also enables to establish a comparison between specific samples or conditions, most commonly, healthy and disease states, where the differences may have aroused due to changes in the molecular functions or external environmental factors. In the context of stem cells, the study of a transcriptome may be useful to understand the processes and underlying mechanisms of fundamental properties such as cell proliferation or differentiation, or embryonic development. For example, transcriptomics analysis in the case of cancer can enable classification beyond its anatomical location and histopathology and therefore, could be a great source for identifying gene targets for treatment and also, can establish benchmarks to predict tumor prognosis and therapy-based response (Blumenberg, 2019).

Single-cell transcriptome analysis is a propitious tool, which is used in stem cell biology to deal with cellular heterogeneity. The cell-to-cell variations are the characteristics of stem cell populations, and while using transcriptomic approach, these variations are masked because of bulky cells. The heterogeneity in umbilical cord—derived mesenchymal stem cells has been studied using single-cell transcriptome analysis by Huang et al. (2019) and finds positive results.

2. Timeline of main milestones in technologies for gene expression

The early approaches to study whole transcriptomes used various traditional techniques employed for gene expression profiling including RT-PCR, Northern blotting, and nuclease protection assay, but had bottlenecks in terms of analysis of a limited number of genes parallelly. However, at later stages, more advanced techniques such as differential display, subtractive hybridization, representational difference analysis (RDA), expressed sequence tags, cDNA fragment fingerprinting, serial analysis of gene expression (SAGE), and RNA sequencing (RNA-Seq) started flourishing that also revealed the identity of many unknown differentially expressed genes alongside known ones. The microarray encompassing probes against each annotated gene present in the genome and biochips with a high degree of parallelization made complete sequencing of the organisms possible and has become commercially available for its utilization in various researches (Stahl et al., 2011).

Though the dominant contemporary technique such as microarray is a powerful yet systematic technique that allows the sequencing of abundances of defined sets of transcriptomes of different types of biological samples using the hybridization method quite rapidly (Nelson, 2001), mostly, its design and functional applications were produced to focus on the known genes; however, the detection of multiple transcripts formed by alternative splicing using them was difficult (Heller, 2002). On the other hand, RNA-Seq gained popularity over microarrays by covering a broader range of expression values and providing a digital measure of count data, and scaling linearity at extreme values (Wang et al., 2009). Further, using RNA-Seq, it is also possible to extract information related to RNA splice events, which remains undetected by microarrays (Mortazavi et al., 2008). Areas such as the separation of alternative isoforms amid various populations or in an incipient species of a genetic nonmodel organism are another aspect yet to be explored using this high-end technique. Another drawback of using microarrays is the introduction of biases due to cross-hybridization trends while measuring the gene expression levels. This problem also exists for RNA-Seq in case of ambiguous alignment of the reads.

RNA-Seq relies on next-generation sequencing (NGS), which is a high-throughput sequencing method, which runs on different platforms including 454 system (Roche), Illumina's Illumina, and SOLiD technology system. In contrast to first-generation sequencing methods, RNA-Seq is cost-effective and has a stronger ability to characterize shorter reads and sequencing of thousands to millions of DNA molecules parallelly in a very shorter period, thereby overcoming most of the drawbacks of microarrays to a greater extent and bringing more deeper insights for transcriptome research (Lowe et al., 2017). A comparative description of various sequencing techniques based on their principle along with their advantages and disadvantages is shown in Table 16.1. The further section will discuss in detail about setting up of the experimental conditions, working methodology, and steps to be followed while conducting the transcriptome or RNA-Seq analysis and profiling. A schematic workflow and steps involved in RNA-Seq analysis and transcriptome profiling are shown in Fig. 16.1.

3. Working methodology

3.1 Setting up the experiment for transcriptome analysis

Before setting up an experiment for transcriptome analysis, it is important to define specific goals and assess their feasibility in terms of the available budget and methodology to be employed to achieve them. The basic questions to be

TABLE 16.1 Comparison of various sequencing techniques.

Technology	DNA sequencing		Microarray	RNA sequencing				
	First-generation sequencing	Next-generation sequencing		Pacibo	Ion Torrent system	454 system (Roche)	Illumina system	SOLiD system
Method	Dideoxychain-termination method (Sanger sequencing)	High-throughput sequencing	Hybridization	Single molecule in real time	Ion semiconductor	Pyrosequencing	Sequencing by synthesis	Sequencing by ligation
Signal	Digital signal	Digital signal	Fluorescence signal	Digital signal				
Resolution	Single bp	Single bp	Several-100bp	High resolution 3 kb read length	High resolution 200 bp read length	High resolution 700 bp read length	High resolution 50–250 bp read length	High resolution 50 bp read length
Nucleic acid required	High content	Low content	High content	Low content				
Advantages	Low cost, high precision, short turn around	Study of whole genomes and exomes, broad adaptability	Relatively inexpensive, advanced informatics and statistical tools available	Fast method and longest read length	Comparatively less expensive and fast method	Fast method and long read size	Cost-effective, high sequence yield, and more accuracy	Low cost per base
Disadvantages	Small sequence size, single species sequencing	Expensive and turnaround time, more complex data	High background, low sensitivity and limited dynamic range (probe-dependent)	Expensive and low yield at high accuracy	Create homopolymer errors	Expensive runs and create homopolymer errors	Expensive equipment and requires high DNA concentrations	Slower method, issues with palindromic sequence

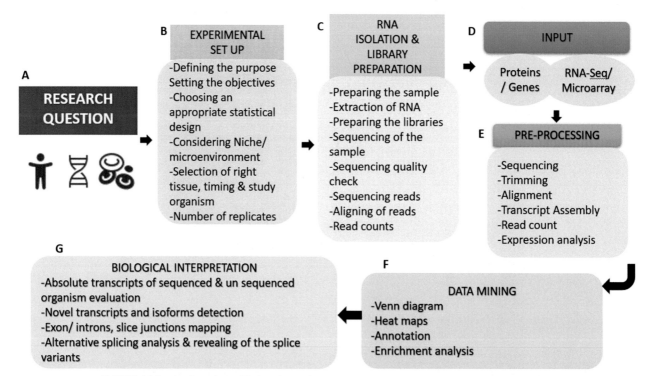

FIGURE 16.1 A schematic workflow and steps involved in RNA-Seq analysis and transcriptome profiling.

considered as early in the planning phase, while designing an experiment, include aspects of your interest in the transcriptome with a special focus on protein-coding mRNA or regulatory noncoding RNA, whether resources of sequencing available in the public platform, information about the sequence coverage, which is needed to study and best suiting sequencing technology, do my research include characterization of the transcriptome to compare expression between treatment and control group? How I am going to statistically determine differentially expressed genes by referring to the fold-change expression difference at a particular level of expression? What statistical tool shall I select to draw important inferences with some general assumptions (Auer & Doerge, 2010)?. It is noteworthy that the statistical model used for the analysis of RNA-Seq data has shifted its focus from analysis of a single sample to more complex statistical designs, which include linear regression models (Hardcastle & Kelly, 2010; Robinson & Oshlack, 2010).

Also, the expression of genes is highly sensitive to environmental conditions and the microniche in which the biological sample is originally placed. Even, the stages of the physiological development of a particular tissue, including embryonic development, reproductive stage, or the course of a day (circadian rhythms), can dramatically affect the transcript abundance and its isoform identity of a choice of tissue in observation (Nolte et al., 2009). While comparing between groups, the variance should be kept low to avoid systematic differences (e.g., at a variable time in a day, physiological state) within the sampling regime, and one should focus on the different areas of interest (e.g., divergence in population and its effect on gene expression) For studying the transcripts of unicellular organisms, pooling up of several individuals will be required due to limited yield of RNA material obtained from them for analysis, and therefore, the building of tissue-specific transcriptome may not work, whereas, in the case of large animals, the accessible tissues such as blood or skin are generally chosen, where choosing the right type of tissue, its developmental time stage, and environmental conditions needs to be premediated before designing RNA-Seq analysis experimental setup. The physiological pathways of the phenotype of interest to monitor the genes, which could serve as biomarkers, are considered an initiating point (Baran-Gale et al., 2018). For instance, Ducrest et al. (2008) contributed to the understanding of evolution traits by studying the hormonal control of pigmentation exhibited by brain tissue from the hypothalamus.

The other important steps to conduct a successful RNA-Seq experiment for a transcriptome analysis include sample preparation, extraction of RNA, and library preparation, which are discussed in the following.

3.2 Sample preparation

The limited stability of RNA and special precautions to be taken to avoid any form of contamination in the sample makes it vulnerable to fragmentation and loss due to RNase activity. Thus, working on fresh freezing-shock tissue stored in liquid nitrogen has been considered to be the most dependable method to prevent any tissue losses. Further, the buffers such as RNA Later and Trizol that are commercially available are used to store RNA at room temperature for some time.

3.3 Extraction of RNA and quality check

Different extraction protocols are required to extract RNA for sequencing, depending upon the focal species of RNA. For example, extraction of smaller RNA molecules of <200 bp is a difficult affair when mRNA extraction is performed through typical LiCl precipitation or any other commercially available kits. The extraction of miRNAs is done using different methods, and at the same time, the assessment of RNA quality is the first step performed for measuring gene expression effectively. Since, a transcriptome is constituted with a large fraction of depleted rRNA or poly-A enrichment rRNA, which needs to be removed before proceeding to the next step, i.e., library preparation. Eventually, polyadenylated mRNA molecules need to be developed by capturing them on magnetic beads coated with oligo dT- or other membranes to avoid chances of bias that could be introduced in sequencing at 3′ ends (Künstner et al., 2010).

Synthesis of cDNA: a high-quality yield of RNA isolated from the sample is generally copied onto a more stable template of double-stranded copy DNA (cDNA), which usually meets the required criteria of various sequencing methods in high-throughput RNA-Seq. This is an enzymatic reaction, which utilizes a hybridization technique where oligo-dT primer is hybridized onto the poly-A mRNA tail or by binding with random hexamer primers (Dong & Chen, 2013).

3.4 Library preparation

The next critical step in RNA-Seq is library preparation, which is platform specific and may be single-ended or paired-ended. The synthesized cDNA is shredded into smaller fragments serving as sequencing templates. A single-end strategy calls for partial sequencing of cDNA fragments (from one end), while in a paired-end strategy, short sequences are sequenced from both ends. The paired-end strategy is more suitable for obtaining an initial assembly of the transcriptome as well as for recognizing an isoform. However, it should be noted that neither too large (<300 bp) nor too short insert sizes are preferred for sequencing as the former is responsible for the loss of small size fraction of transcripts and the latter for contamination of the adapter, which needs a read removal and can complicate the analysis part (Ozsolak & Milos, 2011).

The amplification of cDNA, which is required in an appreciable amount as a starting material for most sequencing platforms, is achieved by PCR-based transcriptome amplification (Raz et al., 2011). The length of the template and the content of its sequence decide the efficiency of the PCR, which may show a biased result, having established a nonlinear correlation between the initial concentration of a gene before and after the PCR. Distortions in the final data of PCR should be checked and could be avoided by using a few amplification cycles in the starting, and if still present, it could be dealt with PCR duplication. In RNA-Seq, an estimate of actual gene expression is determined by the number of reads, and in highly expressed genes, there are fair chances of occurrence of duplications and its removal would downward bias in expression estimate. Therefore, appropriate mathematical prediction models are used to adjust projected duplicated reads to PCR artifacts. Mostly, PCR artifacts are indicated by the presence of identical read pairs having variation in their insert size between read pairs, which is a result due to their lesser probability of random duplication. This can be corrected by using standardized spike-in data controls that integrate most of the PCR-sensitive parameters (Jiang et al., 2011).

Further, to initiate the transcriptome characterization, the gene expression levels across genes need to be homogenized through library normalization, which is a costly and sensitive process (Künstner et al., 2010; Vijay et al., 2013). Though it provides a complete and comprehensive description of the transcriptome, it cannot be directly used for the quantification of gene expression.

4. Various platforms and strategies in RNA-Seq

There are numerous sequencing platforms commonly available for performing RNA-Seq, which include single-molecule real time sequencing by Pacific Biosciences, pyrosequencing-based 454 system by Roche, sequencing by synthesis from

Illumine, sequencing by ligation SOLiD system, while some others are in their developmental stages such as ion semi-conductor by Ion Torrent Sequencing, nanopore sequencing, combinatorial probe anchor synthesis, etc. (comparative description of various sequencing platforms is shown in Table 16.1) (Eid et al., 2009; Merriman & Rothberg, 2012; Raz et al., 2011). The most suitable platform among all is selected for RNA-Seq based on price/base pair, error rate, resolution, and read length. For creating de novo transcriptome assembly, longer and paired-end reads may be helpful. The end product of sequencing is revealed in the form of a number of accurately aligned reads in a gene, which determines the quantitative and inferential power of the gene expression.

It is worth mentioning that error profiles may exist and differ while using different sequencing platforms and should be considered in data interpretation. To overcome this, standard error probabilities/base is measured by the unit called phred quality cores, which was originally used in Sanger sequencing methods (Ewing & Green, 1998).

Further, an overall gene expression is performed with the characterization of the entire transcriptome and alternatively spliced isoforms allowing broader sequence coverage individually. The raw data generated by any sequencing platform in a single run require disk storage of a few gigabytes and can be analyzed using suitable data storage and bioinformatic tools and processes. Moreover, a de novo transcriptome assembly would ideally require a high-speed computing facility along with fast storage system (eigh-core machine with 32 GB of RAM), while at least 1 terabyte of storage or cloud computing facilities in the absence of sufficient storage space would be required for an experimental design of moderately sized RNA-Seq. Moreover, many downstream analyses could be carried out using different online tools such as Galaxy platform, R programming, bioconductor suite of R packages (www.bioconductor.org), which are easily available in the public domain (Goecks et al., 2010).

Furthermore, familiarity with various file formats of cross-platforms such as .fasta, .fastq, .sam, .bam, .vcf, .gtf, or .gff is advisable, as these may help perform basic tasks such as the elimination of adapter, quantification of duplicates, and maintenance of summary statistics of quality score. Preprocessing of quality control raw data is needed before feeding them into downstream analysis such as mapping or assembly.

4.1 Characterizing a transcriptome

The obtained reads of transcripts for a high-quality genome from RNA-Seq data are now identifiable by mapping them to a known genome or by de novo assembly, which is an expensive and laborious process. The mapping of spliced isoforms shows better inference as compared with mapping by de novo tools, whereas mapping assemblies carry all artifacts along with them.

4.2 Defining the variant

The critical molecular/biomarkers developed from the functional genome of transcribed DNA are an important aspect of RNA-Seq used in various transcriptomic types of research. Highly flexible data tools such as GATK pipeline (DePristo et al., 2011), along with others, are available for variant calling, which differs in the quality of data usage, i.e., base pair quality, population allele frequency, and substitution rates. However, it is extremely difficult to judge diploid genotypes due to allele-specific expression traits of transcriptome data.

Soon after the development of molecular markers through variant calling, quantification of gene transcripts is the next step to be followed, where the same consideration—distant mapping or de novo applies to evaluate the read counts. The read count of a sequenced transcriptome is defined as the number of reads mapping to a reference transcript, which is either assembled to a distant reference genome or de novo. Simulation strategy is again helpful and better represented of the two, for carrying out differential expression analysis (Vijay et al., 2013).

4.3 Mapping and alignment

To perform an accurate read quantification, the right strategy needs to be chosen and applied for sequence alignment of millions of short query sequenced fragments having some sequencing errors on a reference transcriptome. In case of a nonmodel species, where reads are aligned to distant references, it involves specific mapping and aligner tools (Fonseca et al., 2012). For example, Vijay et al. (2013) showed that the hybrid strategy of stampy can handle sequence distance divergence above 15%. Another challenge in this step is to deal with ambiguity associated with read mapping (Treangen & Salzberg, 2012). The greater tendency of having similarities among regions of the reference, which could be due to an increase in different parameters such as variation in the copy number, multigene families, domains repetition, etc., shows a lesser confidence value to place a read at a given location. This problem can be dealt with the following options: (1) discarding ambiguously mapped reads while retaining only the uniquely mapped reads; (2) retaining all matches lying within the range of general quality cutoff, thereby increasing the number of mapped reads as compared to the raw reads; (3) scoring using an effective algorithm for alignment to evaluate most favorable alignment and equal distribution of reads randomly across good loci; and (4) use of software tools and packages such as RSEM (Li & Dewey, 2011) and TopHat (Trapnell et al., 2009) to segregate the relative proportion of ambiguous reads based on their probabilistic inference.

Additionally, special packages such as ERANGE (Mortazavi et al., 2008) and TopHat spliced-read mapper (Trapnell et al., 2009) are available for mapping transcriptome data to a genomic reference that can handle inference of alternative splicing.

4.4 Assigning a gene name

Another crucial step is drawing meaningful results from the sequenced data in terms of gene function and its cross-interaction with RNA-Seq experimental setup, which are later compared among different studies. Subsequently, a gene name is assigned during mapping to annotated genomes references; however, an orthologous genes assignment from (distantly) related genomes is not fully determined in the absence of information about a sequenced gene, provided by contigs, particularly in the case of de novo assemblies. For closely related species, tools such as NUCmer and PROmer, which work on suffix tree—based methods, and for distantly related species, systems such as BLAST-based ortholog detection are available to complete the task (Vijay et al., 2013).

A relative statistical measurement of transcript abundance per transcript read counts, after mapping, is used to determine orders of magnitude across a large range of expression levels. It can be elaborated as transcript abundance of steady-state mRNA, and the number of reads should be directly proportional to each other, i.e., gene transcript of a gene having twice the amount of cellular concentration of transcript of another gene should have twice the number of reads.

4.5 Data normalization

There are several advantages of normalizing the raw data. It is helpful while comparing the transcripts of varying lengths, as it is assumed that more reads will be covered by longer transcripts as compared with a shorter transcript with an equal expression profile. But transcript length is not usually considered in statistical models when a comparison of the same transcript is made across different treatments (differential expression), which is particularly used for inferring differential expression.

Normalization is also done to compare the quality between two samples and in turn to compare gene expression profiles between both samples. This can be captured using RPKM or FPKM measures, which are used to evaluate reads/fragments per kilobase of transcript per million reads. For example, quantiles of mappable reads (read counts divided by a total number of mapping reads) can be used to solve a situation where genes from individual A (by isolating RNA from individual A), which is sequenced twice the depth coverage of individual B, show a higher expression level, despite having a same relative concentration in the cell (Bullard et al., 2010). Also, the comparability across different samples, transcripts, protocols, and platforms can be assured by using standardized spike-in RNA controls with known concentration and fixed length and GC content (Jiang et al., 2011). Furthermore, normalization methods such as trimmed mean normalization (software-edgeR) (Robinson & Oshlack, 2010) would be used to focus on the issues due to carry-over effects of expressed genes, exhibited from some genes to others. It is interesting to note that invariant internal control (housekeeping) genes have recently been in use for normalization and serve as standard for quantitative PCR analyses.

4.6 Differential gene expression

The nature of data acquired from the quantitative counts of transcripts is analyzed by suitable statistical models, which allow differential expression analyses and compare transcripts based on their expression levels and samples across the different treatment groups. Generally, count data with their statistical properties are described by Poisson process, but the errors inflated while preparing the library and mapping them cause variance in the read counts beyond expectation for an overdispersion in Poisson distribution. Software packages such as DESeq, edgeR, baySeq, and NOIseq are widely used to perform this task alongside the development of more novel and refined statistical methods to best estimate the overdispersion.

4.7 Alternative splicing

The biological reality of alternative splicing can be revealed by either reconstructing the sets of transcript isoforms de novo to create a transcriptome inventory or making use of distantly related available genomes (Vijay et al., 2013). The former approach is error-prone, and therefore, the latter is used to infer transcript isoforms of full length obtained from spliced reads and also to resolve the differences of unique as well as shared parts of isoforms. Till now, only a few attempts have been made to develop software to characterize de novo isoform (Grabherr et al., 2011). The other software tools that are used to conduct isoform-specific differential expression analyses include MMSEQ (Turro et al., 2011), Solasc (Richard et al., 2010), and Cufflinks (Trapnell et al., 2012). However, efforts are now being made to analyze the expression differences between exons directly, bypassing the reconstruction of isoforms (Anders et al., 2012).

4.8 Gene function and interaction

One of the main goals of conducting successful RNA-seq-based research is to identify a set of contrasting genes that differ across the treatments or populations. However, the specific aspects such as the overrepresentation of identified genes that are critical in a targeted metabolic pathway and the significant role played by them in that organism need to be addressed apparently. The gene ontology databases are prominently used for comparing the role of identified genes across species against a known reference (Ashburner et al., 2000; Harris et al., 2004). The genes of interest are further mapped directly to purported metabolic pathways (KeGG pathways) (Ogata et al., 1999) or different protein—protein interaction networks (Leskinen et al., 2012; Szklarczyk et al., 2011).

5. Conclusion and future perspectives

A plentiful research studies focusing on the identification of potential biomarkers in different diseased states (Kaur et al., 2020; Konieczna et al., 2014; Park et al., 2019; Pedrotty et al., 2012; Wang et al., 2021; Yang et al., 2020) for characterizing various transcriptomes using the high-throughput tool harnessing the power of next-generation sequencing are existing. The transcriptomic analysis of stem cell has been researched and found to be a propitious tool to unveil various research questions. Various studies (Billing et al., 2016; Choi et al., 2020; Pan et al., 2019; Wong et al., 2020; Zhang et al., 2022) focusing on stem cells transcriptomic analysis have been done and giving promising results to the researchers (Table 16.2). Creating a whole-genome assembly and functional genome annotation using RNA is now an achievable

TABLE 16.2 Recent developments and aspects of transcriptomic research in stem cell and various diseases.

S. No.	Description	Reference and PMID
	Transcriptomic-based research studies in stem cell biology and various diseases	
1.	Choi et al. (2020) performed a transcriptomics of tonsil-derived mesenchymal stem cells and found out the potential biomarkers of senescent cells.	Choi et al. (2020) PMID: 32807231
2.	Billing et al. (2016) performed characterization of human mesenchymal stem cells by transcriptomic and proteomic analysis to find out source-specific cellular markers.	Billing et al. (2016) PMID: 26857143
3.	Pan et al. (2019) identified biomarkers by network analysis of transcriptomic data stemness indices of bladder cancer to control cancer stem cell characteristics.	Pan et al. (2019) PMID: 31334127
4.	Wong et al. (2020) decoded the mesenchymal stem cells differentiation into mesangial cell by transcriptomic analysis.	Wong et al. (2020) PMID: 32635896
5.	Zhang et al. (2022) revealed the mesenchymal stem cells cellular heterogeneity by performing single-cell transcriptomic analysis.	Zhang et al. (2022) PMID: 35123072
6.	Konieczna et al. (2014) understood the lipid metabolism of rats in peripheral blood mononuclear cells by identifying early transcriptome-based biomarkers.	Konieczna et al. (2014) PMID: 24343050
7.	Park et al. (2019) identified glioblastoma multiomics and prognostic subtype signature using transcriptomic profiling.	Park et al. (2019) PMID: 31332251
8.	Kaur et al. (2020) used large-scale transcriptomic data to identify hepatocellular carcinoma platform independent diagnostic biomarker panel.	Kaur et al. (2020) PMID: 31998366
9.	Yang et al. (2020) did the high-throughput drug and biomarker discovery transcriptome profiling.	Yang et al. (2020) PMID: 32117438

objective (Ellegren et al., 2012). The technology has also humanized the concepts such as dose—response and compensation in a particular genetic model (Woll & Bryk, 2011). Further, with the conceiving and annotation of the human ENCODE project (The ENCODE Project Consortium, 2011), soon it will be possible to draw and infer epigenetic information that would be useful in characterizing the elements governing regulatory functions and uncover the areas such as allele-specific expression and posttranscriptional modifications. It will be feasible to edit RNA for carrying out research studies in the area of population based genetics (Skelly et al., 2011).

Therefore, this approach is extremely useful for personalized medicine and individualized cancer patient therapies, due to possibility of molecular characterization of tissue at various stages of its development.

Moreover, the utilization of system biology techniques has led to the discovery of genes' cross interactions (Chin delevitch et al., 2012). Additionally, the quantification and validation of mRNA specific to individual cells can now be achieved through the use of emerging tools (Larsson et al., 2010). Furthermore, in-situ hybridization experiments, along with the advancement of standardized methodological tools, will be employed at various stages of the RNA-seq workflow. The improvements in sequencing technology will be instrumental in providing further inspiration for conducting research and allow us to undertake studies to compare numerous organisms, tissues, and even environmental conditions based on their transcriptomes.

Various challenges have also been associated with the transcriptomic analysis whether it is the challenge of isolating sufficient amount of good-quality RNA or performing computational analysis. The computational analysis in transcriptomic has been challenging because of its high cost, requirement of high storage discs, and analysis of spatially resolved transcriptomic data.

Acknowledgments

The figures in the manuscript are developed using Microsoft Power Point.

Author's contributions

Dr. Swati Sharma contributed to the development and overall writing of the manuscript.

Ms Daizy Kalpdev and Mr Ankit contributed for the compilation of transcriptomic studies related to stem cells and performed technical formatting of the manuscript and corrections in manuscript wherever required.

References

Anders, S., Reyes, A., & Huber, W. (2012). Detecting differential usage of exons from RNA-seq data. *Genome Research, 22*(10), 2008−2017. https://doi.org/10.1101/gr.133744.111

Ashburner, M., Ball, C. G., Blake, J. B., Botstein, D., Butler, H., Cherry, J. M., Davis, A. P., Dolinski, K., Dwight, S. S., Eppig, J. T., Harris, M. P., Hill, D. J., Issel-Tarver, L., Kasarskis, A., Lewis, S. R., Matese, J. C., Richardson, J. D., Ringwald, M., Rubin, G. J., & Sherlock, G. (2000). Gene ontology: Tool for the unification of biology. *Nature Genetics, 25*(1), 25−29. https://doi.org/10.1038/75556

Auer, P. L., & Doerge, R. W. (2010). Statistical design and analysis of RNA sequencing data. *Genetics, 185*(2), 405−416. https://doi.org/10.1534/genetics.110.114983

Baran-Gale, J., Chandra, T., & Kirschner, K. (2018). Experimental design for single-cell RNA sequencing. *Briefings in Functional Genomics, 17*(4), 233−239. https://doi.org/10.1093/bfgp/elx035

Battle, A., Mostafavi, S., Zhu, X., Potash, J. B., Weissman, M. M., McCormick, C., Haudenschild, C. C., Beckman, K. B., Shi, J., Mei, R., Urban, A. E., Montgomery, S. B., Levinson, D. F., & Koller, D. (2014). Characterizing the genetic basis of transcriptome diversity through RNA-sequencing of 922 individuals. *Genome Research, 24*(1), 14−24. https://doi.org/10.1101/gr.155192.113

Billing, A. M., Ben Hamidane, H., Dib, S. S., Cotton, R. J., Bhagwat, A. M., Kumar, P., Hayat, S., Yousri, N. A., Goswami, N., Suhre, K., Rafii, A., & Graumann, J. (2016). Comprehensive transcriptomic and proteomic characterization of human mesenchymal stem cells reveals source specific cellular markers. *Scientific Reports, 6*(1), 21507. https://doi.org/10.1038/srep21507

Blumenberg, M. (2019). Introductory chapter: Transcriptome analysis. In M. Blumenberg (Ed.), *Transcriptome analysis [Internet]*. London: IntechOpen.

Bourdon-Lacombe, J., Moffat, I. D., Deveau, M., Husain, M., Auerbach, S. M., Krewski, D., Thomas, R. S., Bushel, P. R., Williams, A., & Yauk, C. L. (2015). Technical guide for applications of gene expression profiling in human health risk assessment of environmental chemicals. *Regulatory Toxicology and Pharmacology, 72*(2), 292−309. https://doi.org/10.1016/j.yrtph.2015.04.010

Bullard, J. B., Purdom, E., Hansen, K. D., & Dudoit, S. (2010). Evaluation of statistical methods for normalization and differential expression in mRNA-Seq experiments. *BMC Bioinformatics, 11*(1). https://doi.org/10.1186/1471-2105-11-94

Byron, S. A., Van Keuren-Jensen, K., Engelthaler, D. M., Carpten, J. D., & Craig, D. (2016). Translating RNA sequencing into clinical diagnostics: Opportunities and challenges. *Nature Reviews Genetics, 17*(5), 257−271. https://doi.org/10.1038/nrg.2016.10

Cappola, T. P., & Margulies, K. B. (2011). Functional genomics applied to cardiovascular medicine. *Circulation, 124*(1), 87−94. https://doi.org/10.1161/circulationaha.111.027300

Casamassimi, A., Federico, A., Rienzo, M., Esposito, S., & Ciccodicola, A. (2017). Transcriptome profiling in human diseases: New advances and perspectives. *International Journal of Molecular Sciences, 18*(8), 1652. https://doi.org/10.3390/ijms18081652

Chindelevitch, L., Ziemek, D., Enayetallah, A., Randhawa, R., Sidders, B., Brockel, C., & Huang, E. S. (2012). Causal reasoning on biological networks: Interpreting transcriptional changes. *Bioinformatics, 28*(8), 1114–1121. https://doi.org/10.1093/bioinformatics/bts090

Choi, D. H., Oh, S.-Y., Choi, J. K., Lee, K. E., Lee, J. Y., Park, Y. J., Jo, I., & Park, Y. S. (2020). A transcriptomic analysis of serial-cultured, tonsil-derived mesenchymal stem cells reveals decreased integrin α3 protein as a potential biomarker of senescent cells. *Stem Cell Research & Therapy, 11*(1), 359. https://doi.org/10.1186/s13287-020-01860-y

DePristo, M. A., Banks, E., Poplin, R., Garimella, K., Maguire, J., Hartl, C., Philippakis, A. A., Del Angel, G., Rivas, M. A., Hanna, M., McKenna, A., Fennell, T., Kernytsky, A., Sivachenko, A., Cibulskis, K., Gabriel, S., Altshuler, D., & Daly, M. J. (2011). A framework for variation discovery and genotyping using next-generation DNA sequencing data. *Nature Genetics, 43*(5), 491–498. https://doi.org/10.1038/ng.806

Dong, Z., & Chen, Y. (2013). Transcriptomics: Advances and approaches. *Science China Life Sciences, 56*(10), 960–967. https://doi.org/10.1007/s11427-013-4557-2

Ducrest, A., Keller, L., & Roulin, A. (2008). Pleiotropy in the melanocortin system, coloration and behavioural syndromes. *Trends in Ecology & Evolution, 23*(9), 502–510. https://doi.org/10.1016/j.tree.2008.06.001

Eid, J., Fehr, A., Gray, J., Luong, K., Lyle, J., Otto, G., Peluso, P. R., Kingan, S. B., Baybayan, P., Bettman, B., Bibillo, A., Bjornson, K., Chaudhuri, B., Christians, F., Cicero, R. L., Clark, S., Dalal, R. V., deWinter, A. D., Dixon, J., … Turner, S. T. (2009). Real-time DNA sequencing from single polymerase molecules. *Science, 323*(5910), 133–138. https://doi.org/10.1126/science.1162986

Ellegren, H., Smeds, L., Burri, R., Olason, P. I., Backström, N., Kawakami, T., Künstner, A., Mäkinen, H., Nadachowska-Brzyska, K., Qvarnström, A., Uebbing, S., & Wolf, J. B. (2012). The genomic landscape of species divergence in Ficedula flycatchers. *Nature, 491*(7426), 756–760. https://doi.org/10.1038/nature11584

ENCODE Project Consortium. (2012). An integrated encyclopedia of DNA elements in the human genome. *Nature, 489*(7414), 57–74. https://doi.org/10.1038/nature11247

Ewing, B., & Green, P. (1998). Base-calling of automated sequencer traces using phred. II. Error probabilities. *Genome Research, 8*(3), 186–194.

Fonseca, N. A., Rung, J., Brazma, A., & Marioni, J. C. (2012). Tools for mapping high-throughput sequencing data. *Bioinformatics, 28*(24), 3169–3177. https://doi.org/10.1093/bioinformatics/bts605

Goecks, J., Nekrutenko, A., & Taylor, J. (2010). Galaxy: A comprehensive approach for supporting accessible, reproducible, and transparent computational research in the life sciences. *GenomeBiology.com (London. Print), 11*(8), R86. https://doi.org/10.1186/gb-2010-11-8-r86

Grabherr, M., Haas, B. J., Yassour, M., Levin, J. Z., Thompson, D. A., Amit, I., Adiconis, X., Fan, L., Raychowdhury, R., Zeng, Q., Chen, Z., Mauceli, E., Hacohen, N., Gnirke, A., Rhind, N., Di Palma, F., Birren, B. W., Nusbaum, C., Lindblad-Toh, K., … Regev, A. (2011). Full-length transcriptome assembly from RNA-Seq data without a reference genome. *Nature Biotechnology, 29*(7), 644–652. https://doi.org/10.1038/nbt.1883

Hardcastle, T. J., & Kelly, K. A. (2010). baySeq: Empirical Bayesian methods for identifying differential expression in sequence count data. *BMC Bioinformatics, 11*(1). https://doi.org/10.1186/1471-2105-11-422

Harris, M. A., Clark, J. W., Ireland, A., Lomax, J., Ashburner, M., Foulger, R. E., Eilbeck, K., Lewis, S. R., Marshall, B. J., Mungall, C. J., Richter, J., Rubin, G. J., Blake, J. A., Bult, C. J., Dolan, M. E., Drabkin, H. A., Eppig, J. T., Hill, D. J., Ni, L. M., … White, R. (2004). The Gene Ontology (GO) database and informatics resource. *Nucleic Acids Research, 32*(90001), 258D–261. https://doi.org/10.1093/nar/gkh036

Heller, M. (2002). DNA microarray technology: Devices, systems, and applications. *Annual Review of Biomedical Engineering, 4*(1), 129–153. https://doi.org/10.1146/annurev.bioeng.4.020702.153438

Huang, Y., Li, Q., Zhang, K., Hu, M., Wang, Y., Du, L., Lin, L., Li, S., Sorokin, L., Melino, G., Shi, Y., & Wang, Y. (2019). Single cell transcriptomic analysis of human mesenchymal stem cells reveals limited heterogeneity. *Cell Death & Disease, 10*(5), 368. https://doi.org/10.1038/s41419-019-1583-4

Jiang, L., Schlesinger, F., Davis, C. A., Zhang, Y., Li, R., Salit, M. L., Gingeras, T. R., & Oliver, B. G. (2011). Synthetic spike-in standards for RNA-seq experiments. *Genome Research, 21*(9), 1543–1551. https://doi.org/10.1101/gr.121095.111

Kaur, H., Dhall, A., Kumar, R., & Raghava, G. P. S. (2020). Identification of platform-independent diagnostic biomarker panel for hepatocellular carcinoma using large-scale transcriptomics data. *Frontiers in Genetics, 10*. https://doi.org/10.3389/fgene.2019.01306

Konieczna, J., Sánchez, J., van Schothorst, E. M., Torrens, J. M., Bunschoten, A., Palou, M., Picó, C., Keijer, J., & Palou, A. (2014). Identification of early transcriptome-based biomarkers related to lipid metabolism in peripheral blood mononuclear cells of rats nutritionally programmed for improved metabolic health. *Genes and Nutrition, 9*(1), 366. https://doi.org/10.1007/s12263-013-0366-2

Künstner, A., Wolf, J. B. W., Backström, N., Whitney, O., Balakrishnan, C. N., Day, L. B., Edwards, S. V., Janes, D. E., Schlinger, B. A., Wilson, R. K., Jarvis, E. D., Warren, W. C., & Ellegren, H. (2010). Comparative genomics based on massive parallel transcriptome sequencing reveals patterns of substitution and selection across 10 bird species. *Molecular Ecology, 19*, 266–276. https://doi.org/10.1111/j.1365-294x.2009.04487.x

Lake, J., Storm, C. S., Makarious, M. B., & Bandres-Ciga, S. (2021). Genetic and transcriptomic biomarkers in neurodegenerative diseases: Current situation and the road ahead. *Cells, 10*(5), 1030. https://doi.org/10.3390/cells10051030

Larsson, C., Grundberg, I., Söderberg, O., & Nilsson, M. (2010). In situ detection and genotyping of individual mRNA molecules. *Nature Methods, 7*(5), 395–397. https://doi.org/10.1038/nmeth.1448

Leskinen, P. K., Laaksonen, T., Ruuskanen, S., Primmer, C. R., & Leder, E. H. (2012). The proteomics of feather development in pied flycatchers (Ficedulahypoleuca) with different plumage coloration. *Molecular Ecology, 21*(23), 5762–5777. https://doi.org/10.1111/mec.12073

Li, B., & Dewey, C. N. (2011). RSEM: Accurate transcript quantification from RNA-seq data with or without a reference genome. *BMC Bioinformatics, 12*(1), https://doi.org/10.1186/1471-2105-12-323

Lowe, R., Shirley, N., Bihlmaier, M. R., Pham, J. K., & Shinde, T. (2017). Transcriptomics technologies. *PLoS Computational Biology, 13*(5), e1005457. https://doi.org/10.1371/journal.pcbi.1005457

Merriman, B., & Rothberg, J. M. (2012). Progress in Ion Torrent semiconductor chip based sequencing. *Electrophoresis, 33*(23), 3397–3417. https://doi.org/10.1002/elps.201200424

Mortazavi, A., Williams, B. G., McCue, K., Schaeffer, L., & Wold, B. J. (2008). Mapping and quantifying mammalian transcriptomes by RNA-Seq. *Nature Methods, 5*(7), 621–628. https://doi.org/10.1038/nmeth.1226

Nimmon, M. J. (2001). Microarrays have arrived: Gene expression tool matures. *Journal of the National Cancer Institute, 93*(7), 492–494. https://doi.org/10.1093/jnci/93.7.492

Nolte, A. W., Renaut, S., & Bernatchez, L. (2009). Divergence in gene regulation at young life history stages of whitefish (*Coregonus* sp.) and the emergence of genomic isolation. *BMC Evolutionary Biology, 9*, 59. https://doi.org/10.1186/1471-2148-9-59

Ogata, H., Goto, S., Sato, K., Fujibuchi, W., Bono, H., & Kanehisa, M. (1999). KEGG: Kyoto encyclopedia of genes and genomes. *Nucleic Acids Research, 27*(1), 29–34. https://doi.org/10.1093/nar/27.1.29

Ozsolak, F., & Milos, P. M. (2011). RNA sequencing: Advances, challenges and opportunities. *Nature Reviews Genetics, 12*(2), 87–98. https://doi.org/10.1038/nrg2934

Pan, S., Zhan, Y., Chen, X., Wu, B., & Liu, B. (2019). Identification of biomarkers for controlling cancer stem cell characteristics in bladder cancer by network analysis of transcriptome data stemness indices. *Frontiers in Oncology, 9*, 613. https://doi.org/10.3389/fonc.2019.00613

Park, S., Shim, J. K., Yoon, S. J., Kim, H., Chang, J. H., & Kang, S. M. (2019). Transcriptome profiling-based identification of prognostic subtypes and multi-omics signatures of glioblastoma. *Scientific Reports, 9*(1). https://doi.org/10.1038/s41598-019-47066-y

Pedrotty, D. M., Morley, M., & Cappola, T. P. (2012). Transcriptomic biomarkers of cardiovascular disease. *Progress in Cardiovascular Diseases, 55*(1), 64–69. https://doi.org/10.1016/j.pcad.2012.06.003

Penn-Nicholson, A., Mbandi, S. K., Thompson, E. G., Mendelsohn, S. C., Suliman, S., Chegou, N. N., Malherbe, S. T., Darboe, F., Erasmus, M., Hanekom, W. A., Bilek, N., Fisher, M., Kaufmann, S. H. E., Winter, J., Murphy, M. M., Wood, R., Morrow, C., Van Rhijn, I., Moody, D. B., ... Team, C. I. (2020). RISK6, a 6-gene transcriptomic signature of TB disease risk, diagnosis and treatment response. *Scientific Reports, 10*(1). https://doi.org/10.1038/s41598-020-65043-8

Raz, T., Causey, M., Jones, D., Kieu, A., Letovsky, S., Lipson, D., Thayer, E., Thompson, J., & Milos, P. (2011). RNA sequencing and quantitation using the helicos genetic analysis system. *Methods in Molecular Biology (Clifton, N.J.)., 733*, 37–49. https://doi.org/10.1007/978-1-61779-089-8_3

Richard, H., Schulz, M. H., Sultan, M., Nürnberger, A., Schrinner, S., Balzereit, D., Dagand, E., Rasche, A., Lehrach, H., Vingron, M., Haas, S. A., & Yaspo, M. L. (2010). Prediction of alternative isoforms from exon expression levels in RNA-Seq experiments. *Nucleic Acids Research, 38*(10), e112. https://doi.org/10.1093/nar/gkq041

Robinson, M. D., & Oshlack, A. (2010). A scaling normalization method for differential expression analysis of RNA-seq data. *GenomeBiology.com (London. Print), 11*(3), R25. https://doi.org/10.1186/gb-2010-11-3-r25

Skelly, D. A., Johansson, M., Madeoy, J., Wakefield, J., & Akey, J. M. (2011). A powerful and flexible statistical framework for testing hypotheses of allele-specific gene expression from RNA-seq data. *Genome Research, 21*(10), 1728–1737. https://doi.org/10.1101/gr.119784.110

Small, E. I., Frost, R., & Olson, E. N. (2010). MicroRNAs add a new dimension to cardiovascular disease. *Circulation, 121*(8), 1022–1032. https://doi.org/10.1161/circulationaha.109.889048

Stahl, F., Hitzmann, B., Mutz, K., Landgrebe, D., Lübbecke, M., Kasper, C., Walter, J., & Scheper, T. (2011). Transcriptome analysis. In *Springer eBooks* (pp. 1–25). Springer Nature. https://doi.org/10.1007/10_2011_102

Szklarczyk, D., Franceschini, A., Kuhn, M., Simonovic, M., Roth, A., Minguez, P., Doerks, T., Stark, M., Muller, J., Bork, P., Jensen, L. J., & Von Mering, C. (2011). The STRING database in 2011: Functional interaction networks of proteins, globally integrated and scored. *Nucleic Acids Research, 39*(Database), D561–D568. https://doi.org/10.1093/nar/gkq973

Taubes, A., Nova, P., Zalocusky, K. A., Kosti, I., Bicak, M., Zilberter, M., Hao, Y., Yoon, S. H., Oskotsky, T., Pineda, S., Chen, B., Jones, E., Choudhary, K., Grone, B. P., Balestra, M. E., Chaudhry, F., Paranjpe, I., De Freitas, J. K., Koutsodendris, N., ... Huang, Y. (2021). Experimental and real-world evidence supporting the computational repurposing of bumetanide for APOE4-related Alzheimer's disease. *Nature Aging, 1*(10), 932–947. https://doi.org/10.1038/s43587-021-00122-7

Trapnell, C., Pachter, L., & Salzberg, S. L. (2009). TopHat: Discovering splice junctions with RNA-seq. *Bioinformatics, 25*(9), 1105–1111. https://doi.org/10.1093/bioinformatics/btp120

Trapnell, C., Roberts, A., Goff, L. A., Pertea, G., Kim, D., Kelley, D. E., Pimentel, H., Salzberg, S. L., Rinn, J. L., & Pachter, L. (2012). Differential gene and transcript expression analysis of RNA-seq experiments with TopHat and Cufflinks. *Nature Protocols, 7*(3), 562–578. https://doi.org/10.1038/nprot.2012.016

Treangen, T. J., & Salzberg, S. L. (2012). Repetitive DNA and next-generation sequencing: Computational challenges and solutions. *Nature Reviews Genetics, 13*(1), 36–46. https://doi.org/10.1038/nrg3117

Turro, E., Su, S. Y., Gonçalves, Â., Coin, L. J., Richardson, S., & Lewin, A. (2011). Haplotype and isoform specific expression estimation using multi-mapping RNA-seq reads. *Genome Biology, 12*(2), R13. https://doi.org/10.1186/gb-2011-12-2-r13

Vijay, N., Poelstra, J. W., Künstner, A., & Wolf, J. B. W. (2013). Challenges and strategies in transcriptome assembly and differential gene expression quantification. A comprehensive *in silico* assessment of RNA-seq experiments. *Molecular Ecology, 22*(3), 620–634. https://doi.org/10.1111/mec.12014

Wang, Z. L., Gerstein, M., & Li, X. (2009). RNA-seq: A revolutionary tool for transcriptomics. *Nature Reviews Genetics, 10*(1), 57−63. https://doi.org/10.1038/nrg2484

Wang, H., Tian, Q., Zhang, J., Liu, H., Zhang, J., Cao, W., Zhang, X., Li, X., Wu, L., Song, M., Kong, Y., Wang, W., & Wang, Y. (2021). Blood transcriptome profiling as potential biomarkers of suboptimal health status: Potential utility of novel biomarkers for predictive, preventive, and personalized medicine strategy. *The EPMA Journal, 12*(2), 103−115. https://doi.org/10.1007/s13167-021-00238-1

Wolf, J. B. W., & Bryk, J. (2011). General lack of global dosage compensation in ZZ/ZW systems? Broadening the perspective with RNA-seq. *BMC Genomics, 12*(1). https://doi.org/10.1186/1471-2164-12-91

Wong, C.-Y., Chang, Y.-M., Tsai, Y.-S., Ng, W. V., Cheong, S.-K., Chang, T.-Y., Chung, I. F., & Lim, Y. M. (2020). Decoding the differentiation of mesenchymal stem cells into mesangial cells at the transcriptomic level. *BMC Genomics, 21*(1), 467. https://doi.org/10.1186/s12864-020-06868-5

Yang, X., Kui, L., Tang, M., Li, D., Wei, K., Chen, W., Miao, J., & Dong, Y. (2020). High-throughput transcriptome profiling in drug and biomarker discovery. *Frontiers in Genetics, 11*. https://doi.org/10.3389/fgene.2020.00019

Zhang, C., Han, X., Liu, J., Chen, L., Lei, Y., Chen, K., Si, J., Wang, T. Y., Zhou, H., Zhao, X., Zhang, X., An, Y., Li, Y., & Wang, Q. F. (2022). Single-cell transcriptomic analysis reveals the cellular heterogeneity of mesenchymal stem cells. *Genomics, Proteomics & Bioinformatics, 20*(1), 70−86. https://doi.org/10.1016/j.gpb.2022.01.005

Chapter 17

Genomic and transcriptomic applications in neural stem cell therapeutics

Sushanth Adusumilli[1], Manvee Chauhan[2], Mahesh Mahadeo Mathe[1], Tapan Kumar Nayak[1] and Jayasha Shandilya[2]

[1]*Kusuma School of Biological Sciences, Indian Institute of Technology Delhi, New Delhi, Delhi, India;* [2]*Amity Institute of Molecular Medicine and Stem Cell Research, Amity University, Noida, Uttar Pradesh, India*

1. Introduction

The central nervous system (CNS) is a complex, yet highly organized ensemble of neurons, astrocytes, and oligodendrocytes. Anomaly in the distribution of brain cells and their network architecture are hallmarks of neurodevelopmental, neurodegenerative, and neuropsychiatric disorders including autism spectrum disorder (ASD), Alzheimer's disease, and schizophrenia (Braak & Braak, 1991; McCarthy, 2009; Stoner et al., 2014). Though most brain cells are postmitotic and terminally differentiated or quiescent, the nervous system shows considerable regenerative plasticity owing to niches of neural progenitor cells and a highly heterogenous population of neural stem cells (NSCs) (Kornblum, 2007).

Stem cells are functionally undifferentiated cells capable of self-renewal and differentiation into a variety of cell types. The embryonic stem cells are part of the inner cell mass (ICM) of developing preimplantation embryos, which are pluripotent and differentiate into cells of all the germ layers. The stem cell "potency" to differentiate into various cell types decreases with developing stages of the embryo (Ding et al., 2020). NSCs are multipotent stem cells that reside in the CNS (Kornblum, 2007). They are capable of self-renewal and give rise to neurons, astrocytes, and oligodendrocytes in the brain. Specific areas of the brain, mainly the subventricular zone (SVZ) and the subgranular zone (SGZ) in the dentate gyrus of hippocampus, are the active sites of neuronal cell regeneration from NSCs (Doetsch et al., 1999).

Most of the NSCs reside in the brain in a quiescent state. They can be reactivated to a proliferative state by external growth factors, nutrients, physical stimulation, injury, and/or drug administration (Lucassen et al., 2010; Puls et al., 2020). Aptly therefore, NSCs are increasingly viewed as potential therapeutic options in traumatic brain injury (Weston & Sun, 2018), spinal cord injury (Gregoire et al., 2015), Parkinson's disease (Taylor & Minger, 2005), Huntington's disease (Ryu et al., 2004), and neuropsychiatric disorders including autism (Mariani et al., 2015) and schizophrenia (Sacco et al., 2018). However, the pathogenesis of many of these disorders takes several years before their clinical symptoms manifest, thus making it difficult to establish causality with underlying genetic factors such as mutations, gene expression profiles, and epigenetic landscape. It has been proposed that one of the leading causes of neurodevelopmental disorders is alteration in gene expression profiles of NSCs, which in turn manifest as a loss of balance between NSC proliferation and differentiation (Gigek et al., 2015). For example, downregulation of metabotropic glutamate receptors (GRM7), implicated in attention deficit hyperexcitability disorder (ADHD) leads to increased proliferation of neural progenitor cells (NPCs) and reduced neural differentiation (Elia et al., 2011). Likewise, chromatin remodelers, transcription factors, oncogenes, and cytoskeletal proteins alter NSC homeostasis in the brain (Hovland et al., 2022; Sacco et al., 2018). Therefore, the major hurdle in the therapeutic application of NSCs in neuronal disorders is to have deeper understanding about the genomic, transcriptomic, and epigenetic state of the NSCs. Here, we present a comprehensive description of the methodology to study the genomic and transcriptomic state of the NSCs and discuss the hurdles in translating the current biological understanding of NSCs to therapeutic application in neurodegenerative and neuropsychiatric disorders.

Computational Biology for Stem Cell Research. https://doi.org/10.1016/B978-0-443-13222-3.00011-3
Copyright © 2024 Elsevier Inc. All rights are reserved, including those for text and data mining, AI training, and similar technologies.

2. Genomic approaches to study stem cells

The eukaryotic genome is organized as a complex nucleoprotein structure called chromatin, which is crucial for packaging of the genome inside the nucleus and regulation of fundamental DNA-dependent processes such as replication, repair, and transcription. The cellular and molecular heterogeneity apparent in the brain cells is a manifestation of the complex spatiotemporal regulation of gene expression. Gene expression landscape of brain cells is a function of variations in DNA sequences, covalent modification (methylation) of DNA, specific DNA-protein interaction, differential expression of regulatory RNA and proteins, turnover rates of different biomolecules, and overall, inherently stochastic gene expression rates. Cellular heterogeneity is a major challenge in the application of NSCs as a potential therapeutic option in neuronal disorders. Advent of high-throughput *-omics* approaches such as the next-generation sequencing (NGS) aimed at quantifying single cell or ensemble population's genome-wide expression status of mRNA and other noncoding RNAs have been of immense use in addressing the challenges of cellular heterogeneity.

2.1 Stem cells: promising tool for therapeutics

Stem cells have unique ability to regenerate themselves and differentiate into progenitors, which can further propagate to terminally differentiated cells. These cells are universally present in almost all tissues, starting from the embryonic development stages. They are characterized by the presence of protein markers such as cell surface cluster of differentiation antigens (CD9, CD24, CD90, and CD117), GPCRs (e.g., frizzled), transcriptions factors including octamer-binding protein 4 (Oct4), Sry-related high-mobility group (HMG) box-containing (Sox) family (Sox2), Kruppel-like factor (Klf) family, and enzymatic markers such as ATP-binding cassette (ABC) transporter G2, high telomerase activity, and alcohol dehydrogenase (ALDH) activity (Corti et al., 2006). In adult tissues, these cells reside in specialized regions known as "stem cell niche" (Corti et al., 2006). They are of many types, viz., embryonic stem cells, mesenchymal stem cells, hematopoietic stem cells, NSCs, etc. Different types of stem cells exhibit a wide spectrum of regeneration and differentiation abilities.

NSCs are multipotent in nature, which give rise to NPCs. The NPCs can further differentiate into neurons and glial cells via adult neurogenesis (Cervenka et al., 2020). NSCs are found in developing embryos as well as in specific regions in the adult brain. Embryonic NSCs have a higher degree of specialization and lower stemness than embryonic stem cells (ESCs). NSCs in adults are typically found in two specific niches: SVG and SGZ of dentate gyrus in the hippocampus (Doetsch et al., 1999).

2.1.1 Induction of neural stem cell differentiation

Dividing and differentiating NSCs have very distinct energy requirements. Mitochondrial cellular respiration fuels stem cell differentiation and self-renewal (Wanet et al., 2015). Mitochondrial dysfunction is associated with degenerative CNS disorders, which can be potentially reversed by exogenous stem cell transplantations. There are multiple ways through which NSC transplantation induces regeneration in animal models of CNS diseases by replenishing the damaged brain cells and delivering functional mitochondria to the target host cells through extracellular vesicles (Peruzzotti-Jametti et al., 2021).

Extensive studies have shown NSCs are heterogenous in nature in embryo as well as in adult mature brain (Llorens-Bobadilla et al., 2015). Therefore, categorization and functional analysis of adult NSCs will yield promising strategies in the replacement of damaged or diseased neural and glial cells. This may potentially open new avenues in the treatment of neurodegenerative diseases and spinal cord injuries. Reprogramming NSCs for therapeutic application, however, is still marred by significant hurdles. Better understanding of how NSCs are regulated in embryonic and in adult stem cell niche environments holds the key to our ability to reprogram them for wider therapeutic applications.

2.2 Applications of induced pluripotent stem cells in neurological disorders

The discovery that somatic cells could be converted into iPSCs has ushered in the limitless potential of this technology in stem cell therapeutics. The field has significantly advanced from the original experiments that employed the ectopic expression of four transcription factors (Oct-4, Klf-4, Sox-2, and c-Myc: Yamanaka factors) to generate stem cells from somatic cells (Takahashi & Yamanaka, 2006).

Extensive characterization studies have shown that iPSCs derived from diverse original somatic cell types share basic properties with human embryonic stem cells (hESCs) such as self-renewal, proliferation, and differentiation potential.

Further refinement and optimization of stem cell induction techniques and protocols such as induction of pluripotent stem cells in the absence of c-MYC (a protooncogene) has facilitated the generation of safer hiPSCs for research and medical applications (Nakagawa et al., 2008). Additionally, recent advances have shown successful application of chemical method utilizing small molecule cocktails for stem cell induction instead of the transgenes (Lin et al., 2009). Hence, it is now possible to generate patient-derived iPSCs models with better safety and efficacy to investigate human diseases (Fig. 17.1).

Currently, it is possible to successfully study several human neurological and neurodegenerative diseases using patient-derived iPSC models that were earlier limited by lack of suitable in vivo cellular models. Neurological diseases, such as autism, Alzheimer's disease, and Parkinson's disease, were early disease models developed using an iPSC approach.

2.2.1 Generation of neural progenitor cells from induced pluripotent stem cells

Application of protocols for enrichment of NPCs derived from iPSCs is promising substitute to tissue-derived primary cells used in medical research. Using specific protocols, it is possible to differentiate hiPSCs into various cells of the nervous system such as glia and different neuronal subtypes (motor, dopaminergic, and striatal). These iPSC-derived NPCs are being harnessed to study neuronal commitment to specialized neural cells bypassing the ethical issues linked with using hESCs.

Neurons derived from hiPSCs can be employed to study the underlying mechanisms of neurological disease pathogenesis and also identify biomarkers for early diagnosis and drug discovery efforts (Fig. 17.1). Patient-derived pluripotent stem cells have been successfully used to generate hiPSCs for disease modeling of neurodevelopmental disorders such as spinal muscular atrophy and Rett syndrome. Animal models utilizing hiPSC-derived cell transplantations are promising new strategies for studying neurological diseases in vivo and determine their therapeutic potential.

Significant advances have been made in generating hiPSCs using integration-free viruses or small molecules for the disease models of neurodevelopmental disorders such as schizophrenia and autism (Ronald & Hoekstra, 2011; Schizophrenia Working Group of the Psychiatric Genomics, 2014). This methodology has enabled the incorporation of complete genetic background of the donor in reprogrammed hiPSCs (Schlaeger et al., 2015). Current approaches in developing two-dimensional (2D) and three-dimensional (3D) brain organoids have significantly benefitted therapeutic targeting of neurodevelopmental disorders. Further ongoing research in directing these 3D cells to differentiate into specific brain regions such as cerebral cortex (Pasca et al., 2015) and midbrain (Jo et al., 2016) has brought a paradigm shift in the approaches to study the etiology and pathogenesis of neurobiological diseases such as schizophrenia and autism.

hiPSCs have been successfully used from autism patients to generate 3D brain organoids. High-throughput RNA sequencing (RNA-seq) approaches showed several genes associated with vascular development and lipid metabolism to be downregulated while genes associated with synaptic signaling, ligand-gated ion channel activity, and neural cell fate-associated transcription factors were overexpressed (Mariani et al., 2015) in such models.

Interesting mechanistic and genome wide transcriptomic studies in neurons derived from hiPSCs from schizophrenia patients have revealed significant differences from their normal counterpart. RNA-seq data from these neurons and NPCs showed differential gene expression signatures belonging to glutamate receptor signaling, synaptic transmission, insulin signaling, neuronal migration, and differentiation (Hoffman et al., 2017). This could account for reduced dendritic arborization and fewer neurites of schizophrenia hiPSC-derived neurons.

2.3 Epigenetic changes in neural stem cells

Epigenetic reprogramming of DNA and chromatin landscape in NSCs is a promising new strategy to regulate NSC differentiation and proliferation in combating neurodegenerative and neurodevelopment disorders (Lee et al., 2018; Won et al., 2016). Application of ascorbate (an agonist of histone demethylases) during iPSC reprogramming augments the expression of pluripotency genes, by removing several methylated histone H3 marks. This is especially useful in NSCs where addition of ascorbate was found to demethylate histone H3K9me3, H3K27me3 resulting in upregulation of dopaminergic neuron-specific genes. This increased the production of midbrain dopaminergic neurons (Miyamoto et al., 2017). Hence, NSCs and iPSCs are novel and rapidly evolving therapeutic tools of intervention in several neurological diseases and offer greater promise among the current therapies. Issues related to cellular heterogeneity can be addressed with genomic and transcriptomic approaches at the single cell level using specific markers to distinguish the stem cells and their subtypes. This will improve the efficacy and potency of stem cell—based therapies and generation of patient-derived disease models.

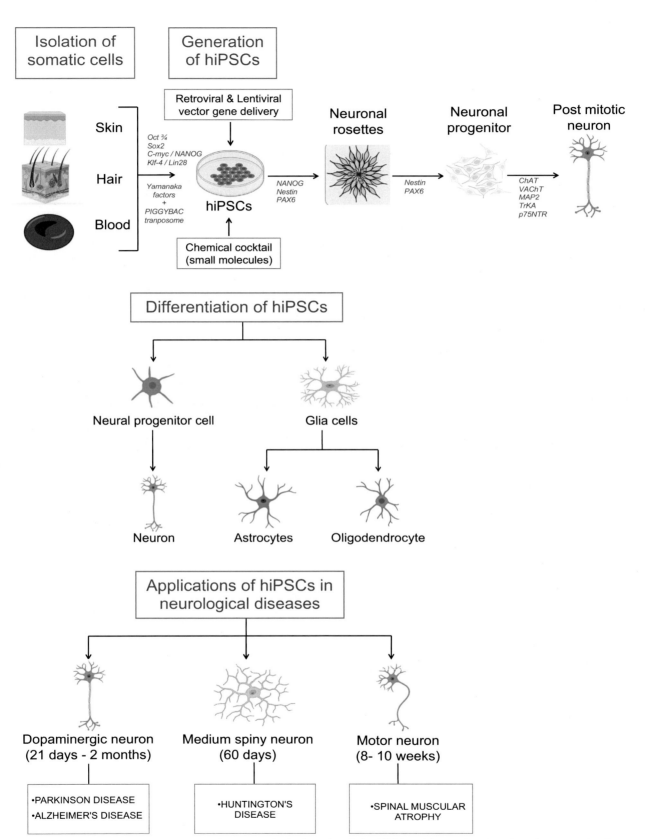

FIGURE 17.1 Schematic showing iPSC generation and application in patient-derived iPSC models. hiPSCs derived from diverse somatic cell types using various induction techniques ranging from retroviral based or integration-free virus-based gene delivery or alternatively using in combination with small molecules. Further differentiation of iPSCs into neural progenitor cells (NPCs). Using specialized protocols, the NSCs can be stimulated to generate distinct neuronal subtypes that can be used in therapeutic studies as well as in patient-derived iPSC disease modeling. *Note: Image was generated using BioRender.com*

3. Genomic methods in therapeutic application of neural stem cells

3.1 Genome wide association studies

Genome-wide association studies (GWAS) approach is applied to screen complete genomes of a large number of individuals in a population for finding associations between genetic variants with a disease or nondisease phenotype using statistical methods (Bale et al., 2009, Lee et al., 2018). To date, upwards of 3300 traits have been examined in more than 5700 GWAS (Watanabe et al., 2019), with number of participants ranging between several thousand to a million in some studies.

3.2 GWAS workflow

Briefly, GWAS involves data collection, data processing, and post-GWAS analyses to inform single-nucleotide polymorphisms (SNPs) on a genome (Uffelmann & Posthuma, 2021). Data collection involves selecting appropriate populations and performing genotyping using SNP arrays. Data processing involves quality control, imputation, association testing, accounting for false discovery, and statistical fine mapping. In the end, for the thousands of variants generated, potentially most important variants, called causal variants, are identified.

The next step is to derive functional associations with the causal variants. These involve finding out the affected gene(s) and determining the regulatory pathways and the biological processes impacted. Expression of quantitative trait is an amount of an mRNA or a protein. Chromosomal regions (loci) that explain variance in expression traits are known as eQTLs. Majority of these eQTLs are in noncoding regions of the genome including the cis- or trans-acting gene regulatory elements. eQTLs may be identified systematically by GWAS. However, most complex traits are polygenic. Therefore, algorithms have been developed to calculate polygenic risk score (PRS), which is calculated from GWAS of an individual genome using thousands of causal variant effects.

Conventionally, GWAS have used microarray panels (called SNP arrays) for genotyping. In these reference panels, unknown variants cannot be identified, rare variants are difficult to be identified due to low abundance, and population-

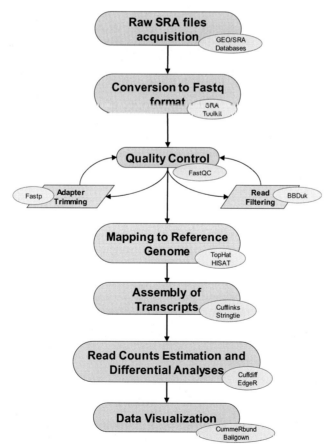

FIGURE 17.2A   Flowchart of whole RNA-seq pipeline. The gray elliptic nodes correspond to each critical step in the flowchart. The gray box nodes correspond to substeps in a critical step. The associated yellow nodes show the software involved in that step.

specific differences in variations lead to spurious associations (Korte & Farlow, 2013). A recent study did find that SNP arrays for well-represented populations are robust and cost-effective for determining common and rare variants, whereas sequencing may be a better alternative in case of poorly represented populations to improve quality of associations (Quick et al., 2020). Despite having these limitations, GWAS have been successful at generating testable predictions.

3.2.1 Applications of GWAS in neural stem cells

A recent study mapped publicly available insomnia GWAS data for European ancestry against enhancer maps of six brain regions. It was found that MADD (MAP kinase activating death domain), PPP2R3C (protein phosphatase 2 regulatory subunit), CASP9 (caspase 9), and PLEKHM2 (pleckstrin homology) genes were significantly prioritized along with notch signaling and glycerophospholipid metabolism pathways (Ding et al., 2018). The GWAS tool for complex traits used in the study prioritizes potential causal genes at given loci and finds tissues, cell types, and pathways enriched for prioritized genes. It was also found that insomnia-related genes were significantly enriched in NSCs. This study appears to generate new candidates in insomnia and opens new avenues for doing similar analyses against Alzheimer's, Parkinson's disease, schizophrenia, and other neurological disorders (Ding et al., 2018).

Within 3 years after miR137 was implicated in schizophrenia by psychiatric genomic consortium (Schizophrenia Working Group of the Psychiatric Genomics, 2014), empirical evidence was identified for its role in NSCs. Further, NOTCH4 has been experimentally identified as a schizophrenia risk gene in mouse NSCs (Zhang et al., 2021) having previously been implicated in insomnia (Ding et al., 2018). Another risk gene TMEM180 was identified by transcriptome-wide association studies (TWAS) through integration of GWAS and eQTL data, in which it was found that TMEM180 was downregulated in schizophrenia patients.

Further, an interesting study discovered that expression of an unannotated protein (TRANK1) was associated with SNP rs9834970, which results in severe mental illness. This effect was overcome by valproic acid (VPA) therapy in patient-derived iPSCs and its neural derivatives. VPA works on a cis-eQTL rs906482 (nearby SNP) that robustly impacts binding to transcription repressor CTCF, leading to TRANK1 downregulation (Jiang et al., 2019).

Bipolar (BP) disorder is a highly heritable, chronic psychiatric disorder involving recurrent manic and depressive episodes. Yet, only a small fraction of GWAS-identified risk loci are attributable to its heritability in a GWAS meta-analysis (Mullins et al., 2021; Kieseppa et al., 2004). To overcome the potential loss of true positives in GWAS due to the stringent significance threshold, a study designed to first perform whole-genome sequencing to identify the important rare variants and then GWAS was subsequently performed on a larger cohort using these variants, hence turning the process upside down (Truve et al., 2020).

Although GWAS have implicated disease-related pathways, more experimental characterization of predictions is necessary before these can act as starting points for drug development. Further, as whole-genome sequencing becomes more cost-effective, it would replace SNP arrays and whole-exome sequencing due to its ability of whole-genome-wide variation coverage (Uffelmann & Posthuma, 2021). It seems that while GWAS alone may be deficient, on integration with other high-throughput technologies, it becomes a powerful tool.

4. Transcriptomics

Transcriptomics is a powerful approach to evaluate cell-specific and tissue-wide gene expression patterns. It provides molecular insight into different biological networks and changes therein during different disorders. With the advent of NGS technologies, transcriptomics has taken large leaps in understanding overall gene regulation mechanisms as well as characterizing novel transcripts, fusion events, and even, de novo construction of new transcriptomes. The two main tools in transcriptomics are microarray and RNA-sequencing (RNA-seq). While microarrays are responsible for the genesis of transcriptomics approach, RNA-seq and its variants are the way forward due to its parallelization, high-throughput, sensitivity, and single base resolution. The only drawback of RNA-seq is that it is expensive and requires significant computational resources for data analysis. Nevertheless, with the significant reduction in costs of NGS technologies and establishment of high-performance computing facility, RNA-seq will be the future of transcriptomics. Here, we will elaborate the approach to RNA-seq data analysis in NSCs.

4.1 RNA-seq data analysis

The RNA-seq method can be split into two main sections: **cDNA-sequencing** and **data analysis**. cDNA-sequencing is the experimental section, whereas data analysis is the computational section.

4.1.1 cDNA-sequencing

It comprises the following steps:

a. Extraction and enrichment of RNA
b. cDNA library preparation
c. Fragmentation of cDNA/RNA library
d. NGS of the cDNA library

Every step is a complex procedure that can be discussed elaborately in separate sections. Since this is not our primary topic of focus, we will encourage our readers to explore the cited review article (Hrdlickova et al., 2017). The RNA quality and composition during extraction and library preparation protocols during sequencing has direct influence on the downstream data analysis. Two common modes of NGS are utilized for cDNA sequencing: single-end and paired-end. Single-end sequencing involves sequencing a cDNA in single direction, while paired-end sequencing involves sequencing in both directions.

4.1.2 Data analysis

RNA sequencing is followed by data analysis, which comprises of a series of defined steps that require computational biology and bioinformatics applications. Its primary objective is to attain the raw estimation of the counts of RNA transcripts corresponding to every region of the genome. Both coding (mRNA) and noncoding (miRNA, lncRNA etc.) RNA can be detected and estimated with high sensitivity and resolution. However, the detection of different types of RNA is dependent on several critical factors.

4.1.2.1 RNA detection and bias

rRNA constitutes 60%–80% of the total RNA of eukaryotic cells. Usually, extracted whole RNA is depleted of rRNA using specific probes (Archer et al., 2015). Insufficient depletion of rRNA causes sequencing bias toward rRNA sequences due to their overrepresentation and high levels of duplication among the total RNA. This leads to minimal diversity in sequenced transcriptome and failure in detection of low copy number RNAs, which comprises of most noncoding RNAs and many mRNAs. Similarly, enrichment of poly-A mRNA through oligo-dT priming introduces biases (Nam et al., 2002). Other factors include the total amount of starting RNA and the sequencing depth, i.e., total number of reads sequenced. The fragmentation and library preparation protocols can also introduce specific biases toward $3'/5'$ end of the sequences or cause accumulation of specific target sequences (Hrdlickova et al., 2017). The protocols are usually designed based on the RNA of interest that one wants to investigate. With suitable and optimized protocols, it is possible to accurately detect as low as 5–10 transcripts.

4.1.3 Data analysis pipeline

Fig. 17.2a shows a pipeline for RNA-seq data analysis following cDNA sequencing. The pipeline is more targeted toward RNA-seq of bulk RNA, specifically whole mRNA from a batch of cells (Shandilya et al., 2016). While the overall steps of the pipeline are valid for other variants of RNA-seq such as single-cell RNA-seq (scRNA-seq), targeted RNA-seq, small RNA-seq etc., it is imperative to understand that the detailed protocols and approaches in each step are different and can include additional steps. In fact, specialized software and algorithms have been developed for scRNA-seq that are unique and exclusive from the software for the analysis of bulk RNA (Table 17.1).

For RNA-seq data analysis, it is required to have basic proficiency with Linux architecture and bash scripting. Majority of the software employed in the pipeline run solely through command line prompts as they lack user-friendly interfaces. Small-scale RNA-seq data analysis can be performed on personal computers. But, for large-scale RNA-seq analyses, it is recommended to execute them on high-performance computing (HPC) systems to facilitate parallel computing. Handling of RNA-seq pipeline also requires basic programming in R/Python languages.

Each of the steps of the pipeline is described in the following. However, the details of the algorithms and command-line prompts of each software are out of the scope of this chapter.

4.1.3.1 Raw SRA file acquisition

The output of cDNA sequencing is in the *.sra* format. Every published sequencing data is hosted on the Sequence Read Archive (SRA) website (https://trace.ncbi.nlm.nih.gov/Traces/?view=run_browser&display=metadata) with an associated unique accession number (*SRRxx* ...). All the *.sra* files pertaining to a specific study are collected as a set and hosted on the

TABLE 17.1 List of software mentioned in RNA-seq pipeline section. For each software, the appropriate PubMed ID (PMID), weblink to its homepage, its role in the pipeline, and its compatibility with bulk/sc-RNAseq are mentioned.

Software	PMID	Link	Purpose	Compatibility
SRA Toolkit	–	https://github.com/ncbi/sra-tools/wiki/01.-Downloading-SRA-Toolkit	Fastq conversion	Bulk RNA-seq, scRNA-seq
FastQC	–	https://www.bioinformatics.babraham.ac.uk/projects/fastqc/	Quality control	Bulk RNA-seq, scRNA-seq
SinQC	27153613	https://morgridge.org/research/regenerative-biology/software-resources/sinqc/	Quality control	scRNA-seq
Fastp	30423086	https://github.com/OpenGene/fastp	Adapter trimming	Bulk RNA-seq
Trimmomatic	24695404	https://github.com/usadellab/Trimmomatic	Adapter trimming	Bulk RNA-seq
BBDuk	–	https://www.geneious.com/plugins/bbduk/	Read filtering	Bulk RNA-seq
SortMeRNA	23071270	https://bioinfo.lifl.fr/RNA/sortmerna/	Read filtering	Bulk RNA-seq
TopHat2	23618408	https://ccb.jhu.edu/software/tophat/index.shtml	Read alignment	Bulk RNA-seq, scRNA-seq
Bowtie2	22388286	https://bowtie-bio.sourceforge.net/bowtie2/index.shtml	Read alignment	Bulk RNA-seq, scRNA-seq
HISAT2	25751142	http://daehwankimlab.github.io/hisat2/	Read alignment	Bulk RNA-seq, scRNA-seq
STAR	23104886	https://github.com/alexdobin/STAR	Read alignment	Bulk RNA-seq, scRNA-seq
Cufflinks	20436464	http://cole-trapnell-lab.github.io/cufflinks/	Transcript assembly, quantification	Bulk RNA-seq, scRNA-seq
Stringtie	31842956	https://ccb.jhu.edu/software/stringtie/index.shtml	Transcript assembly, quantification	Bulk RNA-seq, scRNA-seq
edgeR	19910308	https://bioconductor.org/packages/release/bioc/html/edgeR.html	Quantification, visualization	Bulk RNA-seq
CummeRbund	–	http://compbio.mit.edu/cummeRbund/	Visualization	Bulk RNA-seq
Ballgown	25748911	https://www.bioconductor.org/packages/devel/bioc/vignettes/ballgown/inst/doc/ballgown.html	Visualization	Bulk RNA-seq
SAVER	29941873	https://github.com/mohuangx/SAVER	Transcript assembly, imputation, quantification	scRNA-seq
SAMstrt	23995393	https://github.com/shka/R-SAMstrt	Quantification	scRNA-seq
SCnorm	28418000	https://github.com/rhondabacher/SCnorm	Quantification	scRNA-seq
MAST	26653891	https://www.bioconductor.org/packages/release/bioc/html/MAST.html	Quantification	scRNA-seq

Gene Expression Omnibus (GEO) website (https://www.ncbi.nlm.nih.gov/geo/) with a specific series accession number (*GSExx...*). GEO additionally contains useful metadata about the library preparation protocols, sample conditions, sequencing platform, and analysis parameters provided the researchers upload that information. If one needs to reanalyze published datasets, the series accession number on GEO will be sufficient to retrieve every *.sra* file.

4.1.3.2 SRA to FASTQ format conversion

It is required to convert *.sra* files to *.fastq* for performing the analysis. *Fastq* files are a modified version of the universally known *Fasta* files that are used for denoting sequences. *Fastq* files contain additional information about the sequencing run of the sample as well as quality scores for each base of the sequence. The conversion is universally performed using the "SRA toolkit" (https://github.com/ncbi/sra-tools/wiki/01.-Downloading-SRA-Toolkit) software. Conversion of *.sra* files from single-end sequencing runs produces single *.fastq* files, whereas *.sra* files from paired-end sequencing runs are converted into a pair of *.fastq* files. FASTQ conversion is necessary since the subsequent steps require the per base quality score.

4.1.3.3 Quality control

"FastQC" (https://www.bioinformatics.babraham.ac.uk/projects/fastqc/) is a commonly used software to analyze each *.fastq* file as it evaluates a wide variety of parameters to check the sequencing quality. Parameters such as per sequence quality score, per base quality score, GC content, overrepresented sequences (>0.1% of total RNA), sequence duplication levels, and adapter contamination are assessed for quality control. While the overall quality scores cannot be corrected by any means other than resequencing, some aspects can be rectified for better accuracy in downstream analyses. The following steps can be taken to improve the data quality for RNA-seq analyses.

○ Adapter trimming

During NGS, specific adapters are attached to the ends of sequences to facilitate rapid sequencing. If the read length is too small, the adapter will get sequenced as part of the template cDNA strand. Since standard adapter sequences are used for most sequencing runs, they can be easily identified and trimmed from the ends of the reads. Fastp (https://github.com/OpenGene/fastp) and Trimmomatic (https://github.com/usadellab/Trimmomatic) are a couple of popular software used to detect and trim the adapter sequences.

○ Read filtering

Generally, many sequencing runs can have a small number of overrepresented sequences and acceptable levels of sequence duplication. This is expected as whole RNA does contain certain overexpressed genes as well as specific sequence repeats because of library preparation protocols and repeat RNA like rRNA. Mostly they can be ignored, but sometimes, it can be a source of contamination from viral sequences that can be verified by using the BLAST tool (https://blast.ncbi.nlm.nih.gov/Blast.cgi) for homologous sequences. If they are highly overexpressed, such sequences need to be filtered on a manually curated basis to reduce the noise during analysis. Tools such as BBDuk (https://www.geneious.com/plugins/bbduk/) and SortMeRNA (https://bioinfo.lifl.fr/RNA/sortmerna/) can be used to accurately filter reads by sequence.

It is better to avoid quality-based trimming and filtering of reads as it sacrifices sensitivity for accuracy (Del Fabbro et al., 2013). With the tools involved in subsequent steps considering the per base and sequence quality of reads for analysis, RNA-seq will benefit from higher sensitivity.

4.1.3.4 Read alignment or mapping

This is the most computationally expensive step in the entire analysis. The tens of millions of reads in a *.fastq* file are aligned or mapped to the reference genome or transcriptome of the source organism. There are many aligners developed for this purpose. TopHat (https://ccb.jhu.edu/software/tophat/index.shtml) is one of the oldest and widely used tools due to its established pipeline under the popular Tuxedo suite of software packages (Trapnell et al., 2012). HISAT (http://daehwankimlab.github.io/hisat2/) and STAR (https://github.com/alexdobin/STAR) are more recent and optimized aligners that are faster, efficient, and tailored for HPC systems (Dobin et al., 2013).

High alignment success rate (>85%) is essential for accurate gene/transcript count estimations and exhibits the quality of sequencing. Different parameters can be controlled such as number of mismatches, read gap lengths, intron lengths, and indels during mapping for accuracy (Kim et al., 2013). Further, a portion of the reads can be aligned to multiple places in the genome, due to palindromic sequences, highly similar gene families, and possible duplication events. This introduces significant variability in the overall analysis among genes with such traits (Robert & Watson, 2015). Normally, spliced reads and fusion events spanning exons are handled by separate aligners compared with single-exon reads. Spliced aligners are usually packaged together with nonspliced aligners. For example, TopHat (spliced) utilizes Bowtie (nonspliced) aligner (https://bowtie-bio.sourceforge.net/bowtie2/index.shtml) for mapping single-exon reads before handling spliced reads itself (Langmead & Salzberg, 2012). The final output of alignment step is exported as *.sam* (Sequence Alignment Map) or *.bam* (Binary Alignment Map) files.

4.1.3.5 Transcript assembly

The next step in the pipeline is the assembly of individual reads into full-length transcripts. Fragmentation of RNA/cDNA causes overlapping reads from each transcript. These overlapping reads will be merged sequentially to reconstruct full-length transcripts in a genome-guided manner. Cufflinks (http://cole-trapnell-lab.github.io/cufflinks/) is a suite of tools (part of the Tuxedo suite) that is utilized for this purpose (Trapnell et al., 2012). The assembled transcripts are outputted in a *.gtf* format.

4.1.3.6 Read count estimation and differential analysis

Finally, we arrive at the quantitative estimation of each gene/transcript. The raw metric for expression analysis is the direct count of the number of reads pertaining to a specific gene. The counting can be done at the mapped reads stage post-alignment or at the reconstructed transcripts stage postassembly, and accordingly, we require the *.sam/.bam* and *.gtf* files. While the count estimation of unique reads is only helpful for accurate assessment of expression of genes, count estimation of transcripts gives rise to additional information including alternate splicing and transcription start sites (TSS). However, since the source of a read from a common exon cannot be confidently assigned to a specific alternately spliced transcript, probabilistic approaches are employed for quantification of alternately spliced transcripts and TSS (Trapnell et al., 2012). Cufflinks utilizes the aforementioned approach.

As mentioned before, various types of biases are introduced into the analysis due to reasons such as library preparation and fragmentation, batch effects due to variability in cell sources and reagents across samples, technical variability in sequencing, etc. Another major source of bias is the transcripts themselves. A single 100 bp transcript will inherently fragment into lesser number of reads compared with a 1 kb transcript. Furthermore, a highly expressing transcript will be detected even with low sequencing depth compared with a transcript with low expression. Considering these sources of bias, different kinds of normalization methods for raw counts have been applied to minimize them.

Fragments per kilobase of exon per million mapped fragments (FPKM) is a well-known normalization method where the reads of a transcript are normalized by the sequencing depth as well as the reference transcript length. Another approach is to normalize by distribution. DESeq method measures the ratio of each read count to the geometric mean of all read counts in a sample (Love et al., 2014). The median of these ratios is chosen as a scaling factor for the entire read counts. Cuffdiff (a part of the Cufflinks suite) utilizes a modified version of the DESeq method. Trimmed mean of the M-values (TMM) involves choosing a sample as a reference and estimating fold changes and expression values of the remaining samples relative to the reference sample (Robinson & Oshlack, 2010). These are then trimmed from the read counts in the individual samples, and the mean of the trimmed reads is used as the scaling factor. EdgeR is a tool that relies on this method (Robinson et al., 2010). Selection of verified controls such as a set of housekeeping genes can also be employed as a reference for normalization. The choice of different normalization methods is highly dependent on the specific experimental protocols, sample characteristics, and finer technical details. Raw and normalized read counts are further analyzed to obtain differential gene expression data between different experimental conditions. Robust statistical testing of the raw or normalized counts is performed to correctly identify statistically significantly expressed genes.

4.1.3.7 Visualization of counts and analyses

CummeRbund (http://compbio.mit.edu/cummeRbund/) and Ballgown (https://www.bioconductor.org/packages/devel/bioc/vignettes/ballgown/inst/doc/ballgown.html) are tools that are designed to read the outputs of Cufflinks and Stringtie, respectively, and generate an array of plots for visualization of the analyzed data. Most of the plots, including bar graphs, heatmaps, scatterplots, regression analysis, and volcano plots, are generated using R programming language and environment. Hence, basic experience with R is essential for handling data. Tools such as EdgeR that are involved in differential analysis also facilitate the generation of required heatmaps, barplots, and scatterplots. However, tools such as Cufflinks and Stringtie are specialized in performing differential analysis and cannot plot data.

Fig. 17.2b illustrates an example RNA-seq analysis for hPSCs cells (Purkinje neurons). RNA-seq analysis provides highly valuable information on gene expression patterns, alternate splicing, gene fusion events, and even detection of novel transcripts across the entire genome. While rigorous statistical analysis can be applied to get the highest possible accuracy in the expression data, it cannot be conclusively ascertained whether a gene is showing significant changes in expression. Experimental validation by qPCR is necessary to confirm the findings of RNA-seq analysis. An accurate RNA-seq analysis with proper controls, however, may generate significant leads that can be investigated and verified experimentally with good success rate in different conditions.

4.2 Single-cell RNA-sequencing

Variations in biological properties among single cells manifest as cellular heterogeneity in cell populations. Cellular heterogeneity is highly relevant with regard to tumor microenvironment and cancer progression (Dagogo-Jack &

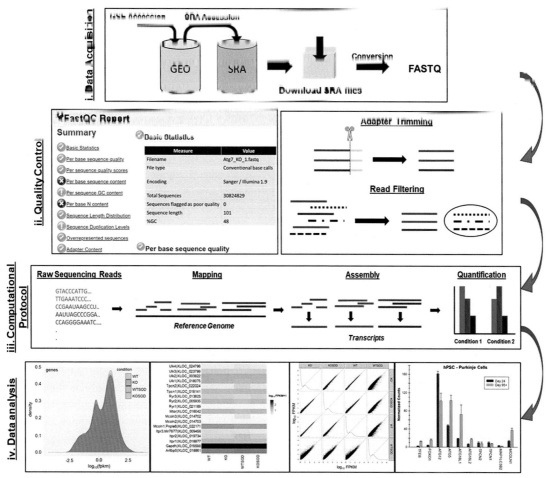

FIGURE 17.2B   RNA-seq general pipeline. The general process of RNA-seq analysis is depicted in the above figure. The entire process can be separated into four sections in a specific order: data acquisition, quality control, computational protocol, and data analysis. While the goal of each section in the general pipeline is the same for different kinds of RNA-seq, the individual steps in these sections and the algorithms involved are quite diverse for different kinds of RNA-seq. The steps detailed in this figure are most suitable for whole RNA-seq. Additionally, this pipeline requires the availability of a reference genome for the source organism. De novo analysis without a reference genome utilizes different pipelines and algorithms. (i) Data acquisition concerns the collection and conversion of raw sequencing data into FASTQ format. GSE accession numbers of datasets are gathered and logged into the GEO database to extract SRA accession numbers for raw sequencing data. Subsequently raw data is downloaded from SRA database using these SRA accession numbers and converted to FASTQ format. (ii) Quality control involves inspection of raw FASTQ data quality as well as preprocessing of data for accurate analyses. FastQC reports of raw data are generated by the widely used FastQC software that provides estimates of many quality control parameters. Preprocessing involves steps such as trimming of sequencing adapters from reads as well as identification and filtering of unwanted and contaminant reads from different sources. (iii) Computational protocol is the critical section that entails the reconstruction of short reads into full-length transcripts and accurate quantification of reads at the gene level. This comprises of three parts: mapping, assembly, and quantification. Mapping involves alignment of reads to the reference genome of the organism. It is essential for detection of every expressed gene as well as identification of splice junctions. Assembly involves merging of overlapping reads to form full-length transcripts. It provides an estimate of raw counts for each gene including total expression, alternate splicing events, and different transcription start sites (TSS). Quantification involves the normalization of the raw counts and elimination of biases introduced by various technical and biological factors. It is crucial for accurate estimation of gene expression patterns in different conditions. Each part is performed by specialized algorithms, and many algorithms have been developed that are employed by a whole host of software. (iv) Data analysis involves visualization and statistical analysis of expression data and identification of differentially expressed genes (DEGs) and other significant changes between conditions. The included plots from left to right correspond to density plots, heatmaps, scatterplots, and barplots. Density plots and scatterplots compare genome-wide expression patterns, whereas heatmaps and barplots portray expression patterns of specific gene sets.

Shaw, 2017; Lim et al., 2019). Investigation of the contributions of different cell populations in ensemble measurement has been crucial in other fields including stem cell biology, neurobiology, and immunology. It has been largely possible due to the emergence of technologies and sensitive assays that allow to capture properties of single cells. Single-cell RNA-seq (scRNA-seq) is a relatively recent method that quantifies gene expression in single cells. Since transcriptomic profile determines single-cell behavior in tissues, cellular heterogeneity is best addressed by scRNA-seq studies.

While the general pipeline for bulk RNA-seq and scRNA-seq is mostly similar, major differences arise in the specific protocols and bioinformatic tools (Chen et al., 2019). The pipeline for scRNA-seq will be briefly discussed with focus on highlighting the differences with that of bulk RNA-seq.

4.2.1 Single-cell isolation and library preparation

The experimental protocols of obtaining single cells, extracting single cell RNA and library preparation are quite complex compared with bulk RNA-seq. Many protocols have been developed that are very different in their approaches, technologies, and final transcript library. While discussion of different methods to harvest single cells and RNA isolation are beyond the scope of this chapter (reviewed in Hwang et al., 2018), it is important to consider specific aspects of library preparation. Based on the type of final transcript library obtained, they can be generally classified into three groups: full-length, 3'-end, and 5'-end transcript approach (Chen et al., 2019). It is generally thought that 3'-end or 5'-end approach has higher throughput and is less expensive to perform, whereas full-length approach leads to more sensitivity in detection of low abundance transcripts (Ziegenhain et al., 2017). While most methods involve enrichment of polyA mRNA, others involve isolation of noncoding RNAs such as miRNA, lncRNA, and circRNA. In any case, the average sample size of sequenced cells for scRNA-seq (every cell is a sample) is much higher compared with that of bulk RNA-seq.

4.2.2 Quality control

scRNA-seq results in noisy data compared with bulk RNA-seq due to extremely low RNA quantities ($\sim$fg to pg) and inefficient cell capture technologies. Broken or dead cells are often captured in large numbers, which need to be removed before processing the samples. RNA degradation also leads to lots of unmappable reads or underrepresentation of reads. Sequencing depth is another factor that introduces bias in the transcript library. These issues can be mitigated by careful cell capture, with stringent checks that omit broken, dead, or clumped cells. Some protocols involve introduction of external RNA control consortium (ERCC) controls, which enables the quantification of technical noise (Reid, 2005). ERCC controls involve addition of external reference RNA transcripts that can be used to calibrate the measurement of transcript counts. UMI (unique molecular identifier) tags are sometimes attached to the captured RNA for accurate identification and estimation of unique transcripts (Macosko et al., 2015). It is crucial to understand that ERCCs and UMIs are not compatible with all kinds of available scRNA-seq protocols.

Post-sequencing, FastQC software that is widely utilized for bulk RNA-seq can be employed for scRNA-seq as well. Nevertheless, other specialized methods such as SinQC have been developed (Jiang et al., 2016). Parameters such as mapping percentage postalignment, amount of reads mapping to ERCC controls, total number of genes detected, and proportion of mitochondrial RNA reads provide hints about the quality of cells and RNA sequenced.

4.2.3 Read alignment

This step is almost identical between bulk RNA seq and scRNA-seq. TopHat and HISAT can be used for scRNA-seq with high accuracy. The mapping percentage reveals the quality of RNA extracted and sequencing as mentioned before.

4.2.4 Transcript assembly and quantification

The workflow in this step is determined by the methods of cell isolation and library preparation. For protocols that follow the full-length transcript capture approach, the usual methods of bulk RNA-seq can be utilized for transcript assembly and quantification. Cufflinks and Stringtie software are quite appropriate for scRNA-seq analysis. If the protocol follows 3'/5'-end capture approach based on UMIs, specialized method, such as SAVER, is required for accurate reconstruction and quantification of transcripts (Huang et al., 2018).

4.2.5 Imputation

This is an additional step that is necessary for scRNA-seq. Normally, scRNA-seq data are riddled with plenty of missing reads due to insufficient amount of starting RNA during library preparation. The proportion of missing reads is far more significant in scRNA-seq compared with bulk RNA-seq. It causes high technical noise and cell-to-cell variability in detection of transcripts of many genes, leading to large impact on downstream analyses. Imputation is a strategy that involves substitution of missing reads with specific values. Many different methods have been developed for performing imputation in scRNA-seq. SAVER is one such method designed for performing imputation on scRNA-seq involving UMI-based protocols (Huang et al., 2018).

4.2.6 Normalization and differential analysis

Normalization approaches such as FPKM and TMM that are utilized for bulk RNA seq cannot be properly applied to scRNA-seq data due to abundance of missing reads. Different specialized approaches have been developed for normalization of scRNA-seq data. One way is to use ERCC controls as reference, against which all the other reads are normalized. SAMstrt software employs that approach (Katayama et al., 2013). Another approach is known as deconvolution where total expression values across pools of cells are used to perform normalization of data from individual cells. SCnorm is a normalization by distribution approach akin to DESeq2 or TMM, which is specifically designed for application in scRNA-seq data (Bacher et al., 2017)

Coming to the differential analysis of gene expression patterns, large sample sizes and unique normalization methods make the bulk RNA-seq tools such as Cuffdiff or EdgeR incompatible. Many tools such as MAST, Census, BCseq, etc. have been developed for performing differential analysis and they handle the missing reads, technical noise, and other variabilities with unique approaches (reviewed in Chen et al., 2019). Other downstream analysis including stratification of cell subpopulations, allelic variations, alternate splicing, and RNA editing can be performed with appropriate tools (Chen et al., 2019).

In summary, scRNA-seq is a powerful tool that enables us to study transcriptomic variations at a single-cell level across different conditions. However, there are many drawbacks at present. The protocols are highly complicated, inefficient, and expensive for large-scale studies. Due to the sheer diversity of cell capture and library preparation protocols, there is little consensus on the right approaches to be employed during data analysis. Optimization of protocols and customized analysis tailored to individual experiments present a cumbersome task for scientists in this field. Despite these drawbacks, scRNA-seq has a great potential to revolutionize transcriptomics and provide novel information that can benefit biological discovery and potentially novel therapeutics development. User-friendly bioinformatic tools, efficient protocols, and affordable sequencing technologies will further enable large-scale application of scRNA-seq in future.

5. Bottlenecks in -omics application in stem cell model for neurological disorders

There are several technical bottlenecks that limit the generation and successful application of hESC protocols, such as cellular heterogeneity that complicates the isolation and culturing of homogenous cell population needed for the generation of neuronal cell type. Further optimization and fine-tuning of protocols are warranted to improve the efficiency and efficacy of differentiating hiPSCs into different neural subtypes. Another challenge faced during hiPSCs differentiation studies is that they tend to differentiate more efficiently into cells that closely resemble the somatic cells from which they were initially derived (Kim et al., 2010). The inherent variation in the differentiation abilities of individual hiPSC clones adds further complexity in successful development of an in vitro model system. In this direction, identification and characterization of bona-fide markers associated with the various hiPSCs will remove the uncertainty in predicting its specific cell lineages (Bock et al., 2011).

There are additional issues with generating neuronal cell lineages as neuronal differentiation efficiency of hiPSCs is significantly lower than that of hESCs, while both transgene-derived or transgene-free hiPSCs have no differences in their abilities to generate nonneuronal subtypes. Additionally, spontaneous differentiation of hESC/hiPSCs results in the formation of heterogenous cell population. Hence, careful culturing techniques must be followed to prevent it. However, in recent times, significant advances have been achieved in neuronal differentiation from stem cells, and methods for stringent quality control and efficacy measurement have been developed. Use of high-throughput sequencing especially at the single-cell level has significantly furthered the field and has successfully addressed the challenges faced due to cellular heterogeneity. In a nutshell, the advances in transcriptomic methods will be a positive step ahead in the direction of successful and safe application of stem cell technology in treating neurological disorders.

Author contribution

Conceptualization: JS, TKN; Original manuscript preparation: JS, TKN, SA, MC, MM.

References

Archer, S. K., Shirokikh, N. E., & Preiss, T. (2015). Probe-directed degradation (PDD) for flexible removal of unwanted cDNA sequences from RNA-Seq libraries. *Current Protocols in Human, 85*, 11.15.11−11.15.36. https://doi.org/10.1002/0471142905.HG1115S85

Bacher, R., Chu, L. F., Leng, N., Gasch, A. P., Thomson, J. A.Stewart, R. M., … (2017). SCnorm: Robust normalization of single-cell RNA-seq data. *Nature Methods, 14*, 584−586. https://doi.org/10.1038/nmeth.4263

Bock, C., Kiskinis, E., Verstappen, G., Gu, H., Boulting, G., Smith, Z. D., ... Meissner, A. (2011). Reference maps of human ES and iPS cell variation enable high-throughput characterization of pluripotent cell lines. *Cell, 144*(3), 439–452. https://doi.org/10.1016/j.cell.2010.12.032

Braak, H., & Braak, E. (1991). Neuropathological stageing of Alzheimer-related changes. *Acta Neuropathologica, 82*(4), 239–259. https://doi.org/10.1007/BF00308809

Cervenka, J., Tyleckova, J., Kupcova Skalnikova, H., Vodickova Kepkova, K., Poliakh, I., Valekova, I., ... Vodicka, P. (2020). Proteomic characterization of human neural stem cells and their secretome during in vitro differentiation. *Frontiers in Cellular Neuroscience, 14*, 612560. https://doi.org/10.3389/fncel.2020.612560

Chen, G., Ning, B., & Shi, T. (2019). Single-cell RNA-seq technologies and related computational data analysis. *Frontiers in Genetics, 10*, 317. https://doi.org/10.3389/FGENE.2019.00317/BIBTEX

Corti, S., Locatelli, F., Papadimitriou, D., Donadoni, C., Salani, S., Del Bo, R., ... Comi, G. P. (2006). Identification of a primitive brain-derived neural stem cell population based on aldehyde dehydrogenase activity. *Stem Cells, 24*(4), 975–985. https://doi.org/10.1634/stemcells.2005-0217

Dagogo-Jack, I., & Shaw, A. T. (2017). Tumour heterogeneity and resistance to cancer therapies. *Nature Reviews Clinical Oncology, 15*, 81–94. https://doi.org/10.1038/nrclinonc.2017.166

Del Fabbro, C., Scalabrin, S., Morgante, M., & Giorgi, F. M. (2013). An extensive evaluation of read trimming effects on Illumina NGS data analysis. *PLoS One* (Vol. 8). https://doi.org/10.1371/JOURNAL.PONE.0085024

Ding, W. Y., Huang, J., & Wang, H. (2020). Waking up quiescent neural stem cells: Molecular mechanisms and implications in neurodevelopmental disorders. *PLoS Genetics, 16*(4), e1008653. https://doi.org/10.1371/journal.pgen.1008653

Ding, M., Li, P., Wen, Y., Zhao, Y., Cheng, B., Zhang, L., ... Zhang, F. (2018). Integrative analysis of genome-wide association study and brain region related enhancer maps identifies biological pathways for insomnia. *Progress in Neuro-Psychopharmacology and Biological Psychiatry, 86*, 180–185. https://doi.org/10.1016/j.pnpbp.2018.05.026

Dobin, A., Davis, C. A., Schlesinger, F., Drenkow, J., Zaleski, C.Jha, S., ... (2013). STAR: Ultrafast universal RNA-seq aligner. *Bioinformatics, 29*, 15. https://doi.org/10.1093/BIOINFORMATICS/BTS635

Doetsch, F., Caille, I., Lim, D. A., Garcia-Verdugo, J. M., & Alvarez-Buylla, A. (1999). Subventricular zone astrocytes are neural stem cells in the adult mammalian brain. *Cell, 97*(6), 703–716. https://doi.org/10.1016/s0092-8674(00)80783-7

Elia, J., Glessner, J. T., Wang, K., Takahashi, N., Shtir, C. J., Hadley, D., ... Hakonarson, H. (2011). Genome-wide copy number variation study associates metabotropic glutamate receptor gene networks with attention deficit hyperactivity disorder. *Nature Genetics, 44*(1), 78–84. https://doi.org/10.1038/ng.1013

Gigek, C. O., Chen, E. S., Ota, V. K., Maussion, G., Peng, H., Vaillancourt, K., ... Ernst, C. (2015). A molecular model for neurodevelopmental disorders. *Translational Psychiatry, 5*(5), e565. https://doi.org/10.1038/tp.2015.56

Gregoire, C. A., Goldenstein, B. L., Floriddia, E. M., Barnabe-Heider, F., & Fernandes, K. J. (2015). Endogenous neural stem cell responses to stroke and spinal cord injury. *Glia, 63*(8), 1469–1482. https://doi.org/10.1002/glia.22851

Hoffman, G. E., Hartley, B. J., Flaherty, E., Ladran, I., Gochman, P., Ruderfer, D. M., ... Brennand, K. J. (2017). Transcriptional signatures of schizophrenia in hiPSC-derived NPCs and neurons are concordant with post-mortem adult brains. *Nature Communications, 8*(1), 2225. https://doi.org/10.1038/s41467-017-02330-5

Hovland, A. S., Bhattacharya, D., Azambuja, A. P., Pramio, D., Copeland, J., Rothstein, M., & Simoes-Costa, M. (2022). Pluripotency factors are repurposed to shape the epigenomic landscape of neural crest cells. *Developmental Cell, 57*(19), 2257–2272 e2255. https://doi.org/10.1016/j.devcel.2022.09.006

Hrdlickova, R., Tolouc, M., & Tian, B. (2017). RNA-Seq methods for transcriptome analysis. *Wiley Interdisciplinary Reviews: RNA, 8*. https://doi.org/10.1002/wrna.1364

Huang, M., Wang, J., Torre, E., Dueck, H., Shaffer, S.Bonasio, R., ... (2018). SAVER: Gene expression recovery for single-cell RNA sequencing. *Nature Methods, 15*, 539–542. https://doi.org/10.1038/s41592-018-0033-z

Jiang, X., Detera-Wadleigh, S. D., Akula, N., Mallon, B. S., Hou, L., Xiao, T., ... McMahon, F. J. (2019). Sodium valproate rescues expression of TRANK1 in iPSC-derived neural cells that carry a genetic variant associated with serious mental illness. *Molecular Psychiatry, 24*(4), 613–624. https://doi.org/10.1038/s41380-018-0207-1

Jiang, P., Thomson, J. A., & Stewart, R. (2016). Quality control of single-cell RNA-seq by SinQC. *Bioinformatics, 32*, 2514–2516. https://doi.org/10.1093/BIOINFORMATICS/BTW176

Jo, J., Xiao, Y., Sun, A. X., Cukuroglu, E., Tran, H. D., Goke, J., ... Ng, H. H. (2016). Midbrain-like organoids from human pluripotent stem cells contain functional dopaminergic and neuromelanin-producing neurons. *Cell Stem Cell, 19*(2), 248–257. https://doi.org/10.1016/j.stem.2016.07.005

Katayama, S., Töhönen, V., Linnarsson, S., & Kere, J. (2013). SAMstrt: Statistical test for differential expression in single-cell transcriptome with spike-in normalization. *Bioinformatics, 29*, 2943–2945. https://doi.org/10.1093/BIOINFORMATICS/BTT511

Kieseppa, T., Partonen, T., Haukka, J., Kaprio, J., & Lonnqvist, J. (2004). High concordance of bipolar I disorder in a nationwide sample of twins. *American Journal of Psychiatry, 161*(10), 1814–1821. https://doi.org/10.1176/ajp.161.10.1814

Kim, K., Doi, A., Wen, B., Ng, K., Zhao, R., Cahan, P., ... Daley, G. Q. (2010). Epigenetic memory in induced pluripotent stem cells. *Nature, 467*(7313), 285–290. https://doi.org/10.1038/nature09342

Kim, D., Pertea, G., Trapnell, C., Pimentel, H., Kelley, R., & Salzberg, S. L. (2013). TopHat2: Accurate alignment of transcriptomes in the presence of insertions, deletions and gene fusions. *Genome Biology, 14*, 1–13. https://doi.org/10.1186/GB-2013-14-4-R36/FIGURES/6

Kornblum, H. I. (2007). Introduction to neural stem cells. *Stroke, 38*(Suppl. 2), 810–816. https://doi.org/10.1161/01.STR.0000255757.12198.0f. PMID: 17261715.

Korte, A., & Farlow, A. (2013). The advantages and limitations of trait analysis with GWAS: A review. *Plant Methods, 9*, 29. https://doi.org/10.1186/1746-4811-9-29

Langmead, B., & Salzberg, S. L. (2012). Fast gapped-read alignment with Bowtie 2. *Nature Methods, 9*, 357–359. https://doi.org/10.1038/nmeth.1923

Lee, J. J., Wedow, R., Okbay, A., Kong, E., Maghzian, O., Zacher, M., … Cesarini, D. (2018). Gene discovery and polygenic prediction from a genome-wide association study of educational attainment in 1.1 million individuals. *Nature Genetics, 50*(8), 1112–1121. https://doi.org/10.1038/s41588-018-0147-3

Lim, S. B., Yeo, T.Lee, W. D., … (2019). Addressing cellular heterogeneity in tumor and circulation for refined prognostication. *Proceedings of the National Academy of Sciences of the United States of America, 116*, 17957–17962. https://doi.org/10.1073/PNAS.1907904116/SUPPL_FILE/PNAS.1907904116.SAPP.PDF

Lin, T., Ambasudhan, R., Yuan, X., Li, W., Hilcove, S., Abujarour, R., … Ding, S. (2009). A chemical platform for improved induction of human iPSCs. *Nature Methods, 6*(11), 805–808. https://doi.org/10.1038/nmeth.1393

Llorens-Bobadilla, E., Zhao, S., Baser, A., Saiz-Castro, G., Zwadlo, K., & Martin-Villalba, A. (2015). Single-cell transcriptomics reveals a population of dormant neural stem cells that become activated upon brain injury. *Cell Stem Cell, 17*(3), 329–340. https://doi.org/10.1016/j.stem.2015.07.002

Love, M. I., Huber, W., & Anders, S. (2014). Moderated estimation of fold change and dispersion for RNA-seq data with DESeq2. *Genome Biology, 15*, 1–21. https://doi.org/10.1186/S13059-014-0550-8/FIGURES/9

Lucassen, P. J., Meerlo, P., Naylor, A. S., van Dam, A. M., Dayer, A. G., Fuchs, E., … Czeh, B. (2010). Regulation of adult neurogenesis by stress, sleep disruption, exercise and inflammation: Implications for depression and antidepressant action. *European Neuropsychopharmacology, 20*(1), 1–17. https://doi.org/10.1016/j.euroneuro.2009.08.003

Macosko, E. Z., Basu, A.Satija, R., … (2015). Highly parallel genome-wide expression profiling of individual cells using nanoliter droplets. *Cell, 161*, 1202–1214. https://doi.org/10.1016/J.CELL.2015.05.002

Mariani, J., Coppola, G., Zhang, P., Abyzov, A., Provini, L., Tomasini, L., … Vaccarino, F. M. (2015). FOXG1-dependent dysregulation of GABA/glutamate neuron differentiation in autism spectrum disorders. *Cell, 162*(2), 375–390. https://doi.org/10.1016/j.cell.2015.06.034

McCarthy, R. A. (2009). TYM and Alzheimer's disease. Age profile? *BMJ, 339*, b2831. https://doi.org/10.1136/bmj.b2831

Miyamoto, K., Tajima, Y., Yoshida, K., Oikawa, M., Azuma, R., Allen, G. E., … Yamada, M. (2017). Reprogramming towards totipotency is greatly facilitated by synergistic effects of small molecules. *Biology Open, 6*(4), 415–424. https://doi.org/10.1242/bio.023473

Mullins, N., Forstner, A. J., O'Connell, K. S., Coombes, B., Coleman, J. R. I., Qiao, Z., … Andreassen, O. A. (2021). Genome-wide association study of more than 40,000 bipolar disorder cases provides new insights into the underlying biology. *Nature Genetics, 53*(6), 817–829. https://doi.org/10.1038/s41588-021-00857-4

Nakagawa, M., Koyanagi, M., Tanabe, K., Takahashi, K., Ichisaka, T., Aoi, T., … Yamanaka, S. (2008). Generation of induced pluripotent stem cells without Myc from mouse and human fibroblasts. *Nature Biotechnology, 26*(1), 101–106. https://doi.org/10.1038/nbt1374

Nam, D. K., Lee, S.Zhou, G., … (2002). Oligo(dT) primer generates a high frequency of truncated cDNAs through internal poly(A) priming during reverse transcription. *Proceedings of the National Academy of Sciences of the United States of America, 99*, 6152. https://doi.org/10.1073/PNAS.092140899

Pasca, A. M., Sloan, S. A., Clarke, L. E., Tian, Y., Makinson, C. D., Huber, N., … Pasca, S. P. (2015). Functional cortical neurons and astrocytes from human pluripotent stem cells in 3D culture. *Nature Methods, 12*(7), 671–678. https://doi.org/10.1038/nmeth.3415

Peruzzotti-Jametti, L., Bernstock, J. D., Willis, C. M., Manferrari, G., Rogall, R., Fernandez-Vizarra, E., … Pluchino, S. (2021). Neural stem cells traffic functional mitochondria via extracellular vesicles. *PLoS Biology, 19*(4), e3001166. https://doi.org/10.1371/journal.pbio.3001166

Puls, B., Ding, Y., Zhang, F., Pan, M., Lei, Z., Pei, Z., … Chen, G. (2020). Regeneration of functional neurons after spinal cord injury via in situ NeuroD1-mediated astrocyte-to-neuron conversion. *Frontiers in Cell and Developmental Biology, 8*, 591883. https://doi.org/10.3389/fcell.2020.591883

Quick, C., Anugu, P., Musani, S., Weiss, S. T., Burchard, E. G., White, M. J., … Fuchsberger, C. (2020). Sequencing and imputation in GWAS: Cost-effective strategies to increase power and genomic coverage across diverse populations. *Genetic Epidemiology, 44*(6), 537–549. https://doi.org/10.1002/gepi.22326

Reid, L. H. (2005). Proposed methods for testing and selecting the ERCC external RNA controls. *BMC Genomics, 6*, 1–18. https://doi.org/10.1186/1471-2164-6-150/TABLES/10

Robert, C., & Watson, M. (2015). Errors in RNA-Seq quantification affect genes of relevance to human disease. *Genome Biology, 16*, 177. https://doi.org/10.1186/s13059-015-0734-x

Robinson, M. D., McCarthy, D. J., & Smyth, G. K. (2010). edgeR: A Bioconductor package for differential expression analysis of digital gene expression data. *Bioinformatics, 26*, 139–140. https://doi.org/10.1093/BIOINFORMATICS/BTP616

Robinson, M. D., & Oshlack, A. (2010). A scaling normalization method for differential expression analysis of RNA-seq data. *Genome Biology, 11*, 1–9. https://doi.org/10.1186/GB-2010-11-3-R25/FIGURES/3

Ronald, A., & Hoekstra, R. A. (2011). Autism spectrum disorders and autistic traits: A decade of new twin studies. *American Journal of Medical Genetics, Part B: Neuropsychiatric Genetics, 156B*(3), 255–274. https://doi.org/10.1002/ajmg.b.31159

Ryu, J. K., Kim, J., Cho, S. J., Hatori, K., Nagai, A., Choi, H. B., … Kim, S. U. (2004). Proactive transplantation of human neural stem cells prevents degeneration of striatal neurons in a rat model of Huntington disease. *Neurobiology of Disease, 16*(1), 68–77. https://doi.org/10.1016/j.nbd.2004.01.016

Sacco, R., Cacci, E., & Novarino, G. (2018). Neural stem cells in neuropsychiatric disorders. *Current Opinion in Neurobiology, 48,* 131–138. https://doi.org/10.1016/j.conb.2017.12.005

Sale, M. M., Mychaleckyj, J. C., & Chen, W. M. (2009). Planning and executing a genome wide association study (GWAS). *Methods in Molecular Biology, 590,* 403–418. https://doi.org/10.1007/978-1-60327-378-7_25

Schizophrenia Working Group of the Psychiatric Genomics, C. (2014). Biological insights from 108 schizophrenia-associated genetic loci. *Nature, 511*(7510), 421–427. https://doi.org/10.1038/nature13595

Schlaeger, T. M., Daheron, L., Brickler, T. R., Entwisle, S., Chan, K., Cianci, A., ... Daley, G. Q. (2015). A comparison of non-integrating reprogramming methods. *Nature Biotechnology, 33*(1), 58–63. https://doi.org/10.1038/nbt.3070

Shandilya, J., Gao, Y., Nayak, T. K., Roberts, S. G., & Medler, K. F. (2016). AP1 transcription factors are required to maintain the peripheral taste system. *Cell Death and Disease, 7*(10), e2433. https://doi.org/10.1038/cddis.2016.343

Stoner, R., Chow, M. L., Boyle, M. P., Sunkin, S. M., Mouton, P. R., Roy, S., ... Courchesne, E. (2014). Patches of disorganization in the neocortex of children with autism. *New England Journal of Medicine, 370*(13), 1209–1219. https://doi.org/10.1056/NEJMoa1307491

Takahashi, K., & Yamanaka, S. (2006). Induction of pluripotent stem cells from mouse embryonic and adult fibroblast cultures by defined factors. *Cell, 126*(4), 663–676. https://doi.org/10.1016/j.cell.2006.07.024

Taylor, H., & Minger, S. L. (2005). Regenerative medicine in Parkinson's disease: Generation of mesencephalic dopaminergic cells from embryonic stem cells. *Current Opinion in Biotechnology, 16*(5), 487–492. https://doi.org/10.1016/j.copbio.2005.08.005

Trapnell, C., Roberts, A.Goff, L., ... (2012). Differential gene and transcript expression analysis of RNA-seq experiments with TopHat and Cufflinks. *Nature Protocols, 7,* 562–578. https://doi.org/10.1038/nprot.2012.016

Truve, K., Parris, T. Z., Vizlin-Hodzic, D., Salmela, S., Berger, E., Agren, H., & Funa, K. (2020). Identification of candidate genetic variants and altered protein expression in neural stem and mature neural cells support altered microtubule function to be an essential component in bipolar disorder. *Translational Psychiatry, 10*(1), 390. https://doi.org/10.1038/s41398-020-01056-1

Uffelmann, E., & Posthuma, D. (2021). Emerging methods and resources for biological interrogation of neuropsychiatric polygenic signal. *Biological Psychiatry, 89*(1), 41–53. https://doi.org/10.1016/j.biopsych.2020.05.022

Wanet, A., Arnould, T., Najimi, M., & Renard, P. (2015). Connecting mitochondria, metabolism, and stem cell fate. *Stem Cells and Development, 24*(17), 1957–1971. https://doi.org/10.1089/scd.2015.0117

Watanabe, K., Stringer, S., Frei, O., Umicevic Mirkov, M., de Leeuw, C., Polderman, T. J. C., ... Posthuma, D. (2019). A global overview of pleiotropy and genetic architecture in complex traits. *Nature Genetics, 51*(9), 1339–1348. https://doi.org/10.1038/s41588-019-0481-0

Weston, N. M., & Sun, D. (2018). The potential of stem cells in treatment of traumatic brain injury. *Current Neurology and Neuroscience Reports, 18*(1), 1. https://doi.org/10.1007/s11910-018-0812-z

Won, H., de la Torre-Ubieta, L., Stein, J. L., Parikshak, N. N., Huang, J., Opland, C. K., ... Geschwind, D. H. (2016). Chromosome conformation elucidates regulatory relationships in developing human brain. *Nature, 538*(7626), 523–527. https://doi.org/10.1038/nature19847

Zhang, Y., Li, S., Li, X., Yang, Y., Li, W., Xiao, X., ... Luo, X. (2021). Convergent lines of evidence support NOTCH4 as a schizophrenia risk gene. *Journal of Medical Genetics, 58*(10), 666–678. https://doi.org/10.1136/jmedgenet-2020-106830

Ziegenhain, C., Vieth, B.Parekh, S., ... (2017). Comparative analysis of single-cell RNA sequencing methods. *Molecular Cell, 65,* 631–643. https://doi.org/10.1016/J.MOLCEL.2017.01.023

Stem cell network modeling and systems biology

Integration of multi-omic data to identify transcriptional targets during human hematopoietic stem cell erythroid differentiation

Meera Prasad[1,2,a], Avik Choudhuri[1,3,a], Song Yang[1,2,3], Emmet Flynn[1,2], Leonard I. Zon[1,2,3,4] and Yi Zhou[1,2,3]

[1]Stem Cell Program and Division of Hematology/Oncology, Boston Children's Hospital and Harvard Medical School, Boston, MA, United States; [2]Pediatric Hematology/Oncology Program of Boston Children's Hospital and Dana Farber Cancer Institute, Harvard Medical School, Boston, MA, United States; [3]Harvard Department of Stem Cell and Regenerative Biology, Harvard University, Cambridge, MA, United States; [4]Howard Hughes Medical Institute, Boston, MA, United States

1. Introduction

Gene expression describes how genetic information is decoded and manufactured into a functional gene product. Different cell types within an organism exhibit distinct biological functions, despite containing the same genetic material or DNA sequence. This, in part, can be attributed to the various ways gene expression in a cell is regulated. At a microscopic level, a DNA template is "transcribed" into RNA and then "translated" into a protein, the functional gene product. This process involves coordinating many dynamic events, which are subject to regulation at each step. These regulatory events happen on multiple levels, namely the transcription, the posttranscription mRNA modifications, the protein translation, and finally, the posttranslational protein modifications. Processes integral to multicellular organismal development occur through a series of gene regulatory events, cascading in a gene product that determines cell fate. Proper gene expression regulation is crucial to developing healthy living organisms (Lee & Young, 2013).

1.1 Coordinated action of transcription factors and enhancers regulates transcription

Transcriptional regulation of gene expression dictates tissue-specific gene expression within individual cell types in response to extracellular stimuli and is orchestrated by transcription factors (TFs). TFs are protein factors that can recognize and bind to unique DNA sequences, known as "motifs," to regulate gene transcription. TFs can act alone in activating gene expression, but they frequently work with specific regions that increase the transcriptional output, known as enhancers. Enhancers are cis-regulatory DNA sequences located upstream, downstream, or intronic relative to the gene body. The interaction between enhancers and TFs recruits additional regulators to bind and, together with geometric alteration of the chromatin architecture, forms a transcriptional "preinitiation complex"—an assembly of RNA polymerase II, cofactors, and other TFs—on the DNA motif close to the genes of interest (Spitz & Furlong, 2012) (Fig. 18.1A). This interaction creates a transcriptional "hotspot" nearby target gene(s) to activate or repress their transcription.

Gene transcription involves many molecular mechanisms that are tightly regulated. Malfunctioning at any step during the assembly of this large transcriptional machinery can certainly compromise this whole process, leading to pathological conditions of several diseases (Lee & Young, 2013; Spitz & Furlong, 2012). Understanding key molecular mechanisms

---

[a]Contributed equally.

Computational Biology for Stem Cell Research. https://doi.org/10.1016/B978-0-443-13222-3.00005-8
Copyright © 2024 Elsevier Inc. All rights reserved, including those for text and data mining, AI training, and similar technologies.

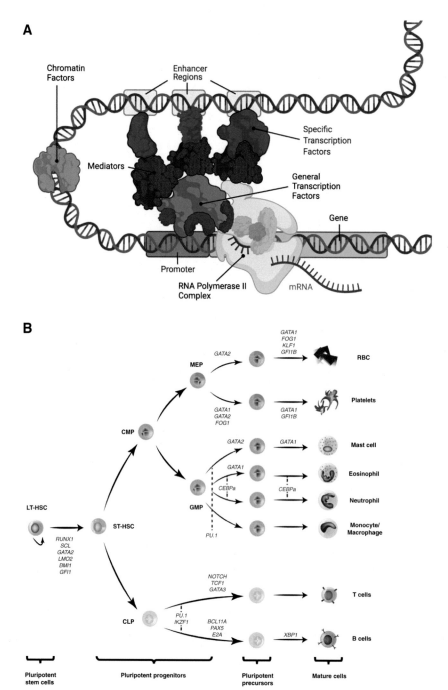

FIGURE 18.1   Overview of eukaryotic gene transcription regulation and representative TFs that determine the fate of hematopoietic stem and progenitor cells as they transition to distinct mature cells. (A) Gene transcription in a cell-specific manner is a tightly and well-orchestrated process. It involves structurally exposing and relaxing peculiar chromatin domains via histone and DNA modifications (not shown) and chromatin factor binding in a locus- and cell-specific manner. The domain-specific organizations facilitate interactions between different genome-coded DNA elements (motifs) such as enhancers and promoters. These interactions can be short and long distances between features coded on the same chromosome (cis) or even on different chromosomes (trans) (not shown). Ultimately, general TFs, mediators, cell- or lineage-specific TFs, and RNA polymerase II complex work coherently to transcribe encoded genes into mRNAs to produce proteins. (B) The specific steps during which individual factors are important, as determined by the loss-of-function studies, are as indicated. Abbreviated terms are explained as follows: *CLP*, common lymphoid progenitor; *CMPs*, common myeloid progenitors; *GMP*, granulocyte-macrophage progenitors; *LT-HSCs*, long-term hematopoietic stem cells; *MEP*, megakaryocyte-erythroid progenitor; *RBCs*, red blood cells; *ST-HSCs*, short-term hematopoietic stem cells.

causing human genetic disorders requires an in-depth examination of the interactions between TFs and enhancer DNA sequences and how they influence chromatin architecture to facilitate gene expression.

1.2 Hematopoiesis as a model system for studying multilevel genetic regulation

The blood production, also known as hematopoiesis, can be a useful paragon for understanding stem cell biology and its contribution to aging and disease, including cancer (Orkin, 1995). Fully mature blood cells have a limited lifespan such that throughout life, hematopoietic stem cells (HSC's) are responsible for replenishing the pool of multilineage progenitors and the individual hematopoietic lineage committed precursors. HSCs represent scarce cells that reside in adult mammals' bone marrow producing "progenitors" that eventually form single or several blood lineages. The progenitors differentiate into blood "precursors" committed to giving rise to mature blood cells, e.g., red blood cells (RBCs), megakaryocytes, myeloid cells (monocyte, macrophage, and neutrophil), and lymphocytes (Fig. 18.1B). HSCs can also self-renew to replenish the blood stem cell pool to ascertain lifelong maintenance of all blood cell lineages (Fig. 18.1B). Several hematopoietic TFs are key determinants of how HSCs will develop and choose their fate during embryogenesis and during programmed lineage-restricted differentiation in adults. Over several decades, functional insight into specific hematopoietic TFs has resulted from knockouts of genes encoding these TFs, as well as forced overexpression of these genes in mice and other model organisms (e.g., zebrafish, chicken, flies, and frogs). The crucial hematopoietic TFs are from all classes of DNA-binding proteins, rather than belong to members of a specific family. Functional requirements of many key TFs, as established through conventional gene targeting analysis, are summarized in Fig. 18.1B (Pevny et al., 1991; Simon et al., 1992; Tsai et al., 1994; Tsai & Orkin, 1997). Hematopoietic development is an extensively studied system, and in this chapter, the human hematopoietic development was selected as a model system, specifically, the maturation process of hematopoietic progenitors into RBCs, a process known as erythropoiesis, to gain detailed insight of the transcriptional regulation of gene expression.

HSCs and hematopoietic progenitor cells express unique cell surface proteins recognizable by specific antibodies directed to them. These membrane proteins are defined as the cell (or lineage) specific markers. They are used to separate stem and progenitor cells from more committed progenitors and other marrow cells through fluorescence-activated cell sorting (FACS) (Shields et al., 2015). One such marker is CD34 (AbuSamra et al., 2017; Kim et al., 2009), a glycoprotein, marking the hematopoietic cells at the stem-progenitor stage (HSPC) in humans and other mammalian species. The CD34$^+$ HSPCs (referred to as CD34$^+$ cells hereafter) play critical roles in maintaining and reconstituting the entire hematopoietic system in homeostasis and postinjury. They are often used clinically to restore the hematopoietic system through bone marrow transplantation for the treatment of numerous hematopoietic diseases and conditions, including recovery after chemotherapy or radiation. The functional importance of CD34$^+$ cells has drawn significant interest in identifying the genetic basis of CD34$^+$ cells, as exemplified by profiling their gene expression with increased scope. These efforts have identified numerous genes and pathways associated with the hematopoietic self-renewal and differentiation of the CD34$^+$ cells. Several studies have used CD34$^+$ cells to model human erythropoiesis (Kim et al., 2009; Lee et al., 2015; Ronzoni et al., 2008; Sankaran et al., 2008; Trompouki et al., 2011). Commonly, CD34$^+$ cells are freed from bone marrow by growth factors, isolated from either bone marrow or peripheral blood, and then differentiated in a cocktail of cytokines such as stem cell factor (SCF), IL-3, and erythropoietin (Epo). In vitro modeling of erythropoiesis with BFU-e assays in a semiliquid culturing system has enabled mechanistic studies of erythroid differentiation in both normal and pathological conditions. The field has gained extensive knowledge of genes critical for general RBC development but still does not fully understand how key erythroid TFs dictate required stage-specific gene expression in CD34$^+$ cells during erythroid differentiation. Further, the expression regulation of key TF genes associated with erythropoiesis of CD34$^+$ cells remains elusive.

1.3 High-throughput sequencing has paved the way for an integrative approach to understanding gene expression regulation

Functional genomics approaches have helped elucidate gene transcription regulation and the biological function of cells and tissues. The field began by focusing on one gene at a time during the Sanger sequencing era. Sanger sequencing (Maxam & Gilbert, 1977; Sanger & Coulson, 1975; Sanger et al., 1977) allowed sequencing of individually cloned DNA molecules at high accuracy and long length. Genes expression was determined by measuring single genes with known sequences by successive Northern assays (Krumlauf, 1994). Individual DNA regulatory elements and their interactive transcriptional and epigenetic regulatory factors were evaluated with gel mobility shift assays (Fried, 1989; Lane et al., 1992). Preliminary genome sequencing projects were made possible with the development of high-throughput Sanger sequencing platforms (International Human Genome Sequencing, 2004; Lander et al., 2001; Venter et al., 2001), which transformed functional genomic analyses from individual genes and loci to a genome-wide level. Comprehensive gene expression and genomic DNA element identification and quantification were realized using high-throughput microarray assays and direct sequence counting, such as SAGE and MPSS (Pariset et al., 2009).

Over the past few decades, several high-throughput sequencing technologies have become available to reveal specific cis- and trans-elements in transcription regulatory machinery in a genome-wide manner (Hollbacher et al., 2020) (Fig. 18.2A). Together, these methodologies are defined as next-generation sequencing (NGS) (Goodwin et al., 2016). Chromatin immunoprecipitation followed by sequencing (ChIP-seq) is powerful to explore protein−DNA interactions, such as TF binding, transcriptional complex assembly, and characterizing the epigenetic landscape by various histones and histone modifications. Additional burgeoned sequencing technologies include assay for transposase-accessible chromatin using sequencing (ATAC-seq) to measure genome-wide chromatin accessibility, and RNA sequencing (RNA-seq) to profile the transcriptome. All of these techniques and how they can be utilized to analyze transcriptional control of gene expression during human erythropoiesis are outlined in detail in the following sections.

1.4 Introduction to common next-generation sequencing methodologies

RNA-seq (Fig. 18.2A) is a technique used to view transcriptional changes, which have been influenced by an experimental perturbation (Stark et al., 2019). A large number of cells are used ($\sim$50,000), and RNA activity is assessed in the aggregate of this population of cells. In general, RNA is transcribed by RNA polymerase, fragmented by heat, synthesized to cDNA by reverse transcription, and then ligated with adapters to facilitate sequencing.

Chromatin accessibility is a key factor affecting cell function and is measured by ATAC-seq (Buenrostro et al., 2015) (Fig. 18.2A). Hyperactive Tn5 transposase cuts exposed areas of DNA and ligates adapters to be used later for amplifying those sections of DNA with PCR, and then sequenced with NGS. A large number (approximately 50,000) of cells are used, and how often a section of the genome gets cut by the hyperactive transposase serves as an analog for how open the chromatin is in that location in the population of cells measured. Further, how open the chromatin is in that section of the genome can be used to infer how much activity is going on in that area of the genome. This chromatin accessibility depends on many different factors, including degree of nucleosome occupancy and histone modifications.

ChIP-seq, or chromatin immunoprecipitation assay followed by next-generation sequencing (Fig. 18.2A), is a technique used to identify genome-wide binding sites of TFs and histones (Park, 2009). In ChIP-seq, DNA is cross-linked to proteins to bind those proteins to the genome. Chromatin is then fragmented with sonication, and antibodies against the protein of interest are used to isolate the parts of the genome bound by that protein. The protein and DNA complexes are then uncrosslinked, and the DNA is amplified and sequenced. ChIP-seq is useful for determining the activity of a specific TF or epigenetic elements in a population of cells.

In combination, these three techniques allow for a general assessment of the function of a population of cells from a genomic perspective. Note, bulk sequencing methods are a snapshot of what is occurring in a population of cells at a single time point.

2. Rationale

The sequential progression of blood progenitors to committed precursor cells and subsequent maturation into terminally differentiated cells with specialized functions is a highly controlled process (Andrews & Orkin, 1994; Kim & Bresnick, 2007). As previously discussed, transcriptional control is a central determinant of cellular differentiation, and TFs interact with DNA regulatory elements to establish cell type−specific gene expression programs. Crucial to establishing these transcriptional programs is an epigenetic landscape that supports lineage-specific gene expression. In the blood system, changes in chromatin architecture are one of the first signs of cellular differentiation. These changes are mediated by the occupancy of master regulatory TFs in lineage-specific regulatory regions that direct the expression of genes crucial to each stage of differentiation (Hnisz et al., 2017; Kornberg, 1999; Spitz & Furlong, 2012).

Uncovering epigenetic regulation of gene expression through TF binding and chromatin accessibility is a multistep and integrative process. Erythropoiesis is the commitment of hematopoietic progenitors into RBCs and is a well-established cell differentiation model (Dzierzak & Philipsen, 2013). Hence, the differentiation of human hematopoietic progenitor

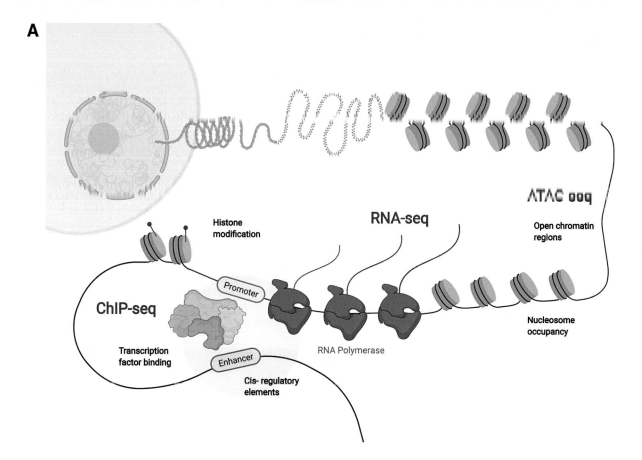

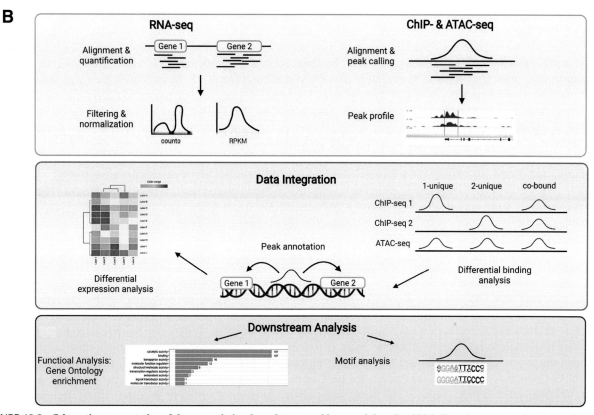

FIGURE 18.2 **Schematic representation of the transcriptional regulatory machinery and the related high-throughput sequencing approaches and bioinformatic analysis workflow to integrate data obtained from high-throughput sequencing approaches.** (A) ChIP-seq is utilized to study protein–DNA interactions, including the binding of TFs to enhancer and promoter DNA-regulatory regions, as well as histone modifications. ATAC-seq measures genome-wide chromatin open regions that are accessible to TFs or chromatin modifiers. RNA-seq captures the gene expression profile of the whole transcriptome. (B *top panel*) RNA-, ChIP-, and ATAC- seq reads are aligned to the reference genome and raw counts per gene or peak are calculated. Subsequent background correction and normalization steps determine the RPKM for a gene for RNA-seq, and the peak intensity for ChIP- and ATAC-seq. (B *middle panel*) Data integration steps include differential expression analysis for RNA-seq and differential binding analysis for ChIP-seq. Differential binding is correlated with ATAC-seq peaks to determine peak accessibility at TF-bound regions. Regions differentially accessible are annotated with their nearest gene and correlated with the differential gene expression profile from RNA-seq. (B *bottom panel*) Downstream analyses include GO-term pathway enrichment analysis for differentially expressed genes associated with differentially accessible regions, as well as motif analyses predicting additional TFs regulating the nearby genes.

cells (CD34[+]) into proerythroblasts (Ronzoni et al., 2008) as a paradigm was utilized to explore TF-mediated control of cell stage—specific gene expression. Erythropoiesis is modulated by the coordinated action of numerous lineage-specific master TFs, including GATA2 and GATA1 (Cantor & Orkin, 2002; Raghav & Gangenahalli, 2021). GATA2 acts primarily in the progenitor stages, while GATA1 dictates lineage commitment into proerythroblasts. GATA2 and GATA1 work in tandem to control differentiation stage—specific gene expression programs. GATA2 is interchanged with GATA1 both at the protein expression level and at the binding of enhancers important for erythroid commitment—a phenomenon defined as "GATA-switch"—to gradually establish erythroid-specific gene expression programs (Bresnick et al., 2010). Using the recently published dataset in the human CD34[+] erythroid differentiation system, which includes cell differentiation stage—matched ChIP-seq, RNA-seq, and ATAC-seq (Choudhuri et al., 2020), the genomic occupancy of GATA2 at a progenitor stage and GATA1 at an erythroid stage was correlated with the changes in gene expression and chromatin accessibility to explore GATA-switch-mediated erythroid commitment at the genomic level. In the following sections, a comprehensive genome-wide bioinformatic analysis for in-depth evaluation of interactions between TFs and DNA-regulatory regions that influence cell stage—specific gene expression is described in Fig. 18.2B. This approach includes collecting data for gene expression, chromatin binding of transcription factors, and chromatin accessibility (Fig. 18.2B, *top panel*), integrated data analysis of gene expression changes resulted from gene regulation by TF and chromatin accessibility (Fig. 18.2B, *middle panel*), and functional annotation of the detected changes (Fig. 18.2B, *bottom panel*). It allowed us to identify new erythroid cell—specific gene sets and their regulatory enhancers, as well as reveal hitherto unknown genetic mutations, which contribute to numerous human erythroid disorders including anemia, thalassemia, and red cell leukemia.

3. Experimental approach

3.1 Expansion and differentiation of CD34[+] cells to perform genome-wide assays by next-generation sequencing

Human CD34[+] cells, isolated from the healthy donors' peripheral blood mobilized with granulocyte colony-stimulating factor, were purchased from the Fred Hutchinson Cancer Research Center. The maintenance and differentiation of the CD34[+] cells follow the well-established protocols (Choudhuri et al., 2020; Sankaran et al., 2008; Trompouki et al., 2011). In summary, the cells were allowed to expand in StemSpan medium (Stem Cell Technologies Inc.) supplemented with StemSpan CC100 cytokine mix (Stem Cell Technologies Inc.) and 2% penicillin/streptomycin (P/S) for 6 days. After the expansion, the cells were treated for 2 h with human recombinant (hr) BMP4 (R&D) at a final concentration of 25 ng/mL to stimulate them and harvested for analyses corresponding to the progenitor stage (day 0, D0) time point. To study differentiation, the 6-day expanded cells were reseeded in the differentiation medium (StemSpan SFEM Medium with 2% P/S, 20 ng/ml hr-stem cell factor (SCF), 1 U/mL hr-erythropoietin (Epo), 5 ng/ml hr-interleukin-3 (IL-3), 2 mM dexamethasone, and 1 mM b-estradiol), at a density of $0.5-1.3 \times 10^6$ cells/mL and cultured for a total of 5 days prior to harvesting for the day 5 (D5) time point, which represents the erythroid stage after erythroid commitment. The harvested cells at D0 and D5 were subjected to NGS assays (ChIP-seq, RNA-seq and ATAC-seq) following library preparation and sequencing protocols as described in our recent publication (Choudhuri et al., 2020).

3.2 Preparing libraries to be sequenced by using illumina next-generation sequencing

NGS surpasses older sequencing modalities in that it allows massively parallel sequencing reactions to occur on a solid surface simultaneously (Alekseyev et al., 2018). All NGS technologies require the construction of sequencing libraries. The use of Illumina NGS sequencing platforms is the focus of the discussion (Fig. 18.3). A sequencing library is a collection of DNA fragments that are representative of the molecular signature of the cells. For RNA-seq libraries, mRNA molecules are enriched and fragmented before being converted to cDNA from all of the transcripts captured; for ATAC-seq, open chromatin regions are sampled; for ChIP-seq, genomic fragments attached to TFs of interest are analyzed. There are different protocols for library construction for each method. Here, three different approaches to enrich for protein-coding mRNA before cDNA synthesis and library construction are outlined (Bivens & Zhou, 2016; Hrdlickova et al., 2017), but the last method discussed was used to generate the data for this chapter (Fig. 18.3, *left panel*).

The first method to enrich mRNA is for samples with a small amount of biological material since it has the highest retention of very limited RNA. The total RNA in cells, which includes ribosomal RNA, transfer RNA, noncoding RNA,

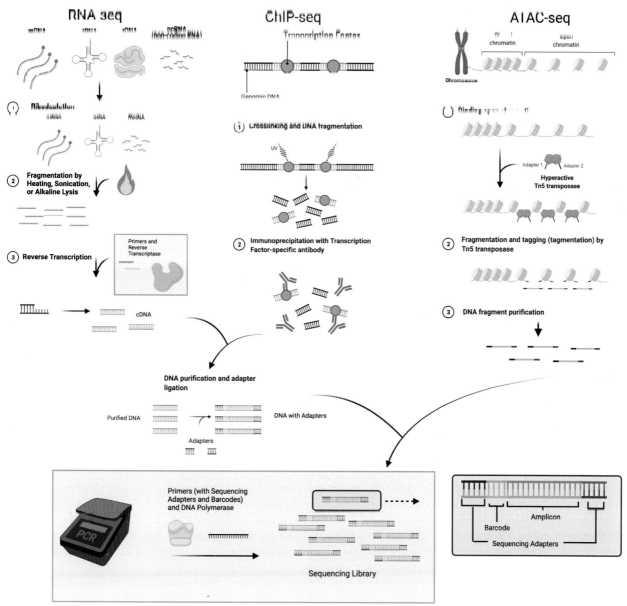

FIGURE 18.3 Schematic representation of library preparation for RNA-, ChIP-, and ATAC-seq is shown. Panels from *left* to *right* show the experimental flows of RNA-seq, ChIP-seq, and ATAC-seq library preparation. Only the ribo-depletion approach, the last of the three RNA-seq library preparation methods described in the body of the chapter is shown here. RNA, shown in *blue*, is isolated, ribo-depleted, and fragmented, resulting in RNA fragments shown in step 2. Reverse transcription creates strands of complementary DNA, or cDNA, shown in *yellow*. DNA purification and adapter ligation steps are shared by both RNA-seq and ChIP-seq. The *orange* fragments with *red/green* and *blue/purple* adapters are added separately on different ends of each double-strand DNA fragment. ChIP-seq begins with live cells with intact chromatin, shown as *yellow* DNA segments attached to *yellow* TFs (or protein of interest) and *blue* DNA segments without TFs. The DNA and binding proteins are cross-linked with formaldehyde and then fragmented by sonication. The DNA fragments attached to the TF are immunoprecipitated with antibodies against the TF, shown in *red*. The TF-associated DNA fragments are released from proteins by reverse cross-link and purified before the adapter ligation step. ATAC-seq also begins with live cells with open and closed chromatin, with histones shown as *blue* spheres and DNA shown as *blue* strands. Open chromatin regions are more accessible and more frequently bound by hyperactive Tn5 transposases, shown in *purple* with *blue* and *red* adapters, that enzymatically cut open regions and attach adapters to the ends of DNA fragments. The DNA fragments, shown with *green*, *blue*, and *red* segments, are purified and subject to a DNA amplification step common to those used in RNA-seq and ChIP-seq library processes. A PCR machine in *black* that combines DNA polymerase in *light blue* and sequencing adapters as PCR primers in *dark blue* is used to amplify the library DNA and add illumina sequencing primers to DNA fragments to complete the sequencing library construction. A single fragment is shown in the cut-out, consisting of the two *red/green* and *blue/purple* adapters, a barcode shown in *light blue*, and the DNA fragment, cDNA (RNA-seq), or genomic DNA (ChIP- and ATAC-seq) shown in *orange*.

and mRNA, is separated out from the rest of the cell material using standard methods. The primary RNA of interested for RNA-seq is mRNA. Most mRNA and very few other RNA contain a long stretch of adenosines (As) at their 3′-ends, which are referred to as poly-A$^+$ tails. Oligo-dT primers, which selectively bind to these poly-A$^+$ tails, are used with reverse transcriptase during reverse transcription (RT) to produce complementary DNA, or cDNA, from the poly-adenylated RNA. RNases are introduced to degrade the remaining RNA before cDNA purification. The isolated cDNA is then fragmented through enzymatic digestion or sonication, adapters with barcodes are ligated to the ends allowing for sample multiplexing during sequencing, and the cDNA is amplified by PCR using sequencing primers to create the sequencing library.

The second method for mRNA enrichment is used for samples with a large amount of biological material. This method increases the probability of detecting low-abundance mRNA sequences compared with other methods described. It requires more starting materials to ensure the recovery of a sufficient amount of mRNA during the isolation step. After the total RNA is isolated, oligo-dT primers attached to magnetic beads are hybridized to the polyA$^+$ RNA to retain them in the subsequent washing steps. The unbound RNA is then washed away with a wash buffer allowing polyA$^+$ RNAs to remain bound to the beads. The beads with bound polyA$^+$ RNAs are separated from unbound RNAs by centrifugation or magnets, and the hybrids on the beads are denatured in an elution buffer, allowing the bound polyA$^+$ RNA to be released from the beads. The isolated mRNA is then fragmented by alkaline lysis, heat, or sonication. Random primers as well as oligo-dT primers are used with reverse transcriptase in an RT reaction to create the cDNA, adapters and barcodes are ligated to the cDNA, and the cDNA is replicated by PCR using sequencing primers partially complementary to the adapters to create the library.

The third method (Fig. 18.3, *left panel*) for RNA library construction, which is used to make the RNA-seq libraries described in this chapter, is to remove ribosomal RNAs and capture a larger proportion of the total RNA in the sample, including noncoding RNA, polyA$^+$, and polyA$^-$ RNA including mRNA, tRNA, and some small RNA species (Choudhuri et al., 2020). This method utilizes oligonucleotides specific to rRNA sequences to capture rRNA and remove them by precipitation, a so-called "ribo-depletion" step. RNA is once again fragmented with heat or sonication. Both random primers and oligonucleotide primers are used together with reverse transcriptase to convert the ribo-depleted RNA into cDNA. Using random primers in addition to oligo-dT primers more efficiently converts ribo-depleted RNA into cDNA. Adapters and barcodes are then ligated to the cDNA fragments, and the cDNA is converted to sequencing libraries as described before.

3.3 ChIP-seq and ATAC-seq library preparation

For ChIP-seq and ATAC-seq (Fig. 18.3, *middle and right panels*) to isolate DNA fragments of interest before undergoing similar library construction strategies as RNA-seq to add Illumina sequencing adapters.

To create a ChIP-seq library (Choudhuri et al., 2020), cells are treated with formaldehyde to cross-link proteins bound to the genomic DNA. The cross-linked chromatin is fragmented by physical shearing via ultrasonication. The fragments containing the TF and protein of interest are pulled down with an antibody against the protein of interest, a step known as immunoprecipitation. The precipitated protein–DNA complexes are reverse-cross-linked by heat to purify the protein-bound DNA fragments. Adapters and barcodes are ligated to the remaining fragments, and PCR is performed using sequencing adapters to amplify the fragments and add sequencing primers.

To construct ATAC-seq libraries (Choudhuri et al., 2020), cells are first lysed to isolate nuclei. Tn5 transpose in a complex with two different linkers is introduced to treat nuclei in their native states. The transposase simultaneously cuts open chromatin and adds the two different linkers to each of the ends produced by the cut. DNA is then purified, and the resulting DNA mixture is PCR-amplified using barcode- and linker-containing NGS sequencing primers. Fragments of DNA with different linkers on the two ends will be exponentially amplified, while the fragments with the same linker sequences on both ends will only be amplified linearly. Size analysis and quantification are performed, and the resulting libraries are ready for sequencing.

3.4 Illumina next-generation sequencing

Following careful quantification and quality control by fragment analysis, libraries are loaded onto a flow cell that contains anchored oligonucleotide sequences reverse complementary to the adapters attached to the DNA fragments. The sequencing reaction commences by replicating the DNA fragments on the surface of a flow cell multiple times. Finally, the sequences of each fragment can be captured by imaging the synthesized fluorescent DNA labeled by incorporating fluorescently tagged nucleotides one at a time. The Illumina NGS instruments used to sequence the RNA-, ChIP-, and ATAC-seq libraries described in the following sections are HiSeq 2500 or HiSeq 4000. Sequencing results of the pooled libraries were demultiplexed by nucleotide barcodes corresponding to each sample as unprocessed zipped fastq sequence files

4. Bioinformatic pipelines for analyzing transcriptional regulation

4.1 Data clean-up for next-generation sequencing—based datasets

To begin processing the raw reads, it is necessary to remove adapter contamination and ensure the reads used for analysis are high quality. Adapter removal is an essential step, and many command line tools developed to trim reads will map NGS reads to a database of known Illumina adapters. However, it may be necessary to specify sequencer-specific adapter sequences to be removed when trimming reads. Following are command-line tools used to process the data prior to genome assembly mapping. These steps are used across RNA-seq, ATAC-seq, and ChIP-seq NGS datasets (Fig. 18.3).

1. FASTQC is a command line tool that can be used to generate quality information of reads including Q score, GC content, adapter contamination, and sequence length distribution.
2. To remove adapter contamination and maintain high-quality data, the trimming software cutadapt v1.8.1 (https://cutadapt.readthedocs.io/en/stable/guide.html) was utilized (https://doi.org/10.14806/ej. 17.1. 200).
3. Up to 30 base-pairs from the 3′ end of each read if the read is low-quality (-q 30) is trimmed. We routinely trim Illumina Nextera adapters (-a CTGTCTCTTATACACATCT -g AGATGTGTATAAGAGACAG -A CTGTCTCTTATACACATCT -G AGATGTGTATAAGAGACAG) and the Illumina TruSeq adapters (-a GATCGGAAGAGCACACGTCTGAACTCCAGTCAC -A GATCGGAAGAGCGTCGTGTAGGGAAAGAGTGTAGATCTCGGTGGTCGCCGTATCATT). The minimum length for a read to be kept is 50 base pairs (-m 50) with a maximum error rate of 0.02 (-e 0.02). Adapter removal and quality trimming is performed on paired-end reads. Only paired end reads are included for downstream analysis.

Once reads are trimmed and preprocessed, they can be mapped to an assembled genome of choice (Fig. 18.2B). While mapping against the most updated genome assembly is preferred, comparison across historical datasets may require mapping the reads to an older genome assembly to maintain consistency. The genome must be indexed, for which bowtie2-build can be used to generate indices from the whole genome fasta (Langmead & Salzberg, 2012). The steps to develop an indexed assembly and rationale of which assembly to map to are the same for RNA-, ChIP-, and ATAC-sequencing datasets.

4.2 RNA-seq

4.2.1 RNA-sequencing processing and introduction to differential expression

RNA-seq is a widely used method to assess the transcriptional profile of a collection of cells (Stark et al., 2019). Many different workflows for computationally analyzing RNA-seq libraries have emerged, but no unified pipeline for analyzing RNA-seq datasets has been developed and adopted by the field to date. Here, the RNA-seq analysis workflow and the important benchmarks are discussed (*left panels* of Fig. 18.2A—C):

1. To map reads to the genome, the tophat2 package is used (Kim et al., 2013). Reads are filtered by allowing up to two base mismatches and no gaps (-N 2 –read-edit-dist 2). Only gene isoforms with known splicing junctions defined in the annotation .gtf are called, without predicting new ones (–no-novel-juncs).
2. To assemble transcripts and estimate transcript abundances, Cufflinks is employed to specify the location of the annotation .gtf with the -G parameter, and estimate transcript abundances from the .thanno/accepted_hits.bam output of tophat2 (Trapnell et al., 2010). The samtools package is used to sort and index the accepted_hits.bam to obtain the sorted mapped reads (Li et al., 2009). The sorted mapped reads can be used to generate raw counts per transcript per sample with htseq-count (Anders et al., 2015). Additional htseq-count parameters include ones indicating the reads are single stranded (–stranded=no), the input feature file is in the bam format (-f bam), and the annotation .gtf file of the assembly (gene_id/file/path/to/gtf) (Anders et al., 2015).
3. Additionally, the genomecov feature from the bedtools package is deployed to generate bedgraph summaries of the aligned sequences and the regions they represent in the genome (Quinlan & Hall, 2010). The bedgraph is an intermediary file format to generate a bigwig file from, which can be done utilizing the bedGraphtoBigWig command line option with UCSC tools installed (Kent et al., 2010). This step is only necessary for visual inspection and analysis of the mapping of RNA-seq reads to the genome. It is beneficial.

The raw counts generated from htseq-count can be used for downstream differential gene expression analyses with R packages such as DESeq2 (Love et al., 2014) and EdgeR (Robinson et al., 2010). The documentation and usage of differential expression in R is well described; however, differential expression analyses with the cuffdiff and cuffnorm features of cufflinks are commonly performed. With cuffnorm, reads per kilobase million (RPKM) of each sample at each gene or locus can be generated. This is useful to compare across samples with different library preparation or sequencing conditions. Additionally, cuffdiff with the baseline number of transcripts per gene per differential expression condition (-c 3) is a useful method to perform differential expression. Replicates per condition can be considered as well. To consider a gene differentially expressed, 1.5-fold or 2-fold change difference is widely accepted as a current benchmark (Love et al., 2014). The differentially expressed transcripts, whether increased or decreased in relation to the control treated condition, can then be correlated to TF binding profiles and chromatin accessibility measurements determined via ChIP-seq and ATAC-seq, respectively.

4.3 ChIP-seq

4.3.1 ChIP-sequencing processing (right panel of Fig. 18.2B)

ChIP-seq is a technique developed to isolate and sequence genomic regions bound by a protein (Park, 2009). The genome is cross-linked to preserve protein binding at all regions and fragmented by physical shearing, and regions bound by the protein of interest are protein−DNA complexes isolated through protein-specific antibody pulled-down. Once isolated, protein−DNA interactions are de−cross-linked and the DNA is sequenced. In addition, an aliquoted library of sheared genomic DNA is prepped as the "input." Including an input is an important control for this technique and allows for regions that are specifically bound by the protein to be considered statistically significant compared with randomly captured regions. Here, ChIP-seq analysis workflow and the important benchmarks are discussed.

1. To map reads to the genome, deploy the bowtie2 package with the specification that the entire read aligns to the genome without removing base-pairs from either end of the read (−end-to-end) and return the results in a sequence alignment/map format (-S) (Langmead & Salzberg, 2012).
2. A few intermediary steps to reformat the data are necessary prior to identifying regions of the genome enriched and bound by the protein of interest. Convert the sequence alignment/map file to a binary format with the samtools package (view -bS mapped_reads.sam > mapped_reads.bam) (Li et al., 2009). Next, sort and index the mapped_reads.bam with the sort and index parameters of the samtools package (Li et al., 2009). The results are sorted and indexed mapped reads in the binary file format, which will be referred to as mapped_reads.si.bam moving forward.
3. To identify regions of the genome enriched with aligned reads, commonly referred to as peak calling, the macs2 package is deployed (Zhang et al., 2008). The macs2 command-line tool takes multiple factors into consideration to identify genomic regions enriched within the treatment compared with the control. To summarize, it removes redundant reads, randomly selects a handful of regions to calculate a genome background enrichment score, develops a statistical model of background enrichment from the input, and then compares the mapped reads to this statistical model to identify significantly enriched regions (Zhang et al., 2008). Briefly, macs2 first identifies peaks that are enriched in the indexed treatment sample compared with the input (callpeak -t treatment_mapped_reads.si.bam -c input_mapped_reads.si.bam), with additional parameters indicating the input file format (-f BAM), the genomic species, the threshold to call a region as significantly enriched in the sample (-p 1e-9), and the normalization of reads with respect to the library size (−SPMR) (Zhang et al., 2008). Specify the output file format in bedGraph form to best use this pipeline (-B) (Zhang et al., 2008).
4. Sort the bedGraph with the sortBed command from the bedtools2 package (-i called_peaks.bdg > called_peaks.s.bdg) (Quinlan & Hall, 2010).
5. Convert the bedGraph format to bigWig, which is a format accessible by a genome browser and contains the peak coordinates as well as the peak intensity calculated from macs2 using the bedGraphToBigWig package (Kent et al., 2010). It is necessary to indicate a file of the chromosomal coordinates from the assembled genome at this step.

4.4 ATAC-seq

4.4.1 ATAC-sequencing processing

ATAC-seq is a widely used technique to identify open chromatin regions in the genome of a collection of cells (Buenrostro et al., 2015). It is useful to compare the changes in chromatin accessibility between multiple conditions, as particular changes in region accessibility may underlie observed altered gene expression following experimental perturbation. As is the case with transcriptomics, different methods to analyze ATAC-seq libraries have emerged, but a standard pipeline across the field has not been developed and adopted. Here the ATAC-seq analysis workflow and the important benchmarks are discussed.

1. To map reads to the assembled and indexed genome, use the bowtie2 package with parameters to indicate the file path to the genome (-x/file/path/genome), set maximum fragment length for valid paired-end alignments to 2000 (-X) (Langmead & Salzberg, 2012). Indicate the path to the cutadapt trimmed and filtered paired-end reads (-1 /cutadapt/sample/sample_name_R1.trim.fq.gz -2 /cutadapt/sample/sample_name_R2.trim.fq.gz). Output the mapped reads in a sequence alignment/map format (-S mapped_reads.sam).

2. After mapping reads to the genome, intermediary steps are necessary to appropriately format the data to then identify areas in the genome enriched with aligned reads. First is to convert the sequence alignment/map file to a binary format with the samtools package (view -bS mapped_reads.sam > mapped_reads.bam) (Li et al., 2009). Next, sort and index the mapped_reads.bam with the sort and index parameters of the samtools package (Li et al., 2009). The results are sorted and indexed mapped reads in the binary file format, which will be referred to as mapped_reads.si.bam moving forward.

3. To identify regions of the genome enriched with aligned reads, deploy the macs2 package (Zhang et al., 2008). When using macs2, first identify peaks that are enriched in the indexed and sorted mapped reads (callpeak -t mapped_reads.si.bam), and specify the format (-f BAM), genome size (-g), output folder for the results (-n), and the output format of a bedGraph (-B). The peaks are not called in comparison to an input, so there is no need to calculate a background statistical model as is done for ChIP-sequencing (–nomodel), and instead macs2 extends reads in $5' \rightarrow 3'$ direction to fragments of a fixed-size, which is set to 200 base-pairs (–extsize 200), and smoothens the signal by extending all $5'$ ends in both directions (–shift -100) (Zhang et al., 2008). Additional parameters to utilize are the threshold to call a region as significantly enriched in the sample (-q 0.05) and to normalize the reads with respect to the library size (–SPMR) (Zhang et al., 2008). The result is the aggregate unique peak signal of the ATAC-seq sample compared with the input in a bedGraph format (Zhang et al., 2008), referred to as called_peaks.bdg moving forward.

4. Sort the bedGraph with the sortBed command from the bedtools2 package (-i called_peaks.bdg > called_peaks.s.bdg) (Quinlan & Hall, 2010).

5. Convert the bedGraph format to bigWig, which is a format accessible by a genome browser and contains the peak co-ordinates as well as the peak intensity calculated from macs2 using the bedGraphToBigWig package (Kent et al., 2010). It is necessary to indicate a file of the chromosomal coordinates from the assembled genome at this step.

5. Uncovering GATA-bound regulatory regions: GATA1 and GATA2 ChIP-seq data integration reveals stage-specific TF switching during erythroid differentiation

By identifying the genomic regions bound by GATA2 on D0 and GATA1 on D5, loci that are uniquely bound by GATA2 and GATA1 at D0 and D5 were identified, respectively, as well as the regions where GATA2 is replaced by GATA1 during erythroid commitment from D0 to D5 (GATA-switch regions). For this purpose, the called peaks in the .narrowPeak output from macs2 of GATA2 ChIP-seq of CD34 HSPCs on D0 and GATA1 ChIP-seq of CD34 HSPCs on D5 were utilized. The bedtools package with the intersect feature and union parameter (-u) was deployed to identify common peaks (GATA-switch) from the GATA2 D0 ChIP-seq and GATA1 D5 ChIP-seq. With the same tool, the unique (-v) parameter was utilized to define the peaks unique to GATA2 D0. The peaks where the binding of GATA2 at D0 CD34[+] progenitors overlapped with the binding of GATA1 at D5 CD34[+] erythroid were the GATA-switch regions. These three peak sets as "GATA-switch," "GATA2-D0-progenitor-unique" (G2-D0), and "GATA1-D5-erythroid-unique" (G1-D5) were compiled. Altogether, there were 27,740, 5103, and 5779 peaks that fell in G2-D0, G1-D5, and GATA-switch categories, respectively. From the ChIP-seq data there were 549 GATA2-occupies-only genomic regions at D5, and 629 GATA1-occupies-only regions at D0, comparably less than their binding at D0 and D5, respectively. This validates the scarce expression, as well as a minimal enhancer occupancy, of GATA2 at D5 and GATA1 at D0. Hence, for all our downstream analyses described here, GATA2 binding at D5 and GATA1 binding at D0 were not considered. Additionally, GATA2 and GATA1 antibodies may pull down each TF and the associated bound DNA with distinct efficiencies, which could presumably explain the difference in total numbers of called peaks in the GATA2 and GATA1 ChIP-seq datasets.

5.1 Correlating GATA-bound regions with gene expression profile

Correlating changes in gene expression with TF regulation is a powerful way to investigate the regulatory role of TFs during cell differentiation. To approximate TF regulation of genes, the nearest gene was assigned to each GATA-switch, G2-D0, and G1-D5 peak with the homer command-line tool and annotatePeaks.pl parameter. The same genome for mapping the raw reads was used, hg19, and identified the genes in each category from the output (Fig. 18.4A). The gene-annotated peaks with $CD34^+$ D0 and D5 RNA-seq RPKM results by binding the peaks were next correlated with a matrix of the RPKM values for each gene from the D0 and D5 time points using the merge function in R. To identify regions associated with significantly differentially expressed genes, the $\log_2$ fold-change ($\log_2$FC) of D5/D0 RNA-seq RPKM values was calculated, and cutoffs of D5/D0 RPKM ≥ 2 for "expression-increased" genes, D5/D0 RPKM $<= -2$ for "expression-decreased" genes, and $-2 <$ D5/D0 RPKM < 2 for "expression-unchanged" genes were used. This was performed for GATA-switch, G2-D0, and G1-D5 regions and resulted in increased, decreased, and unchanged gene sets associated ("regulated") with each category of identified chromatin accessibility regions.

To identify the gene expression programs associated with increased, reduced, and unchanged differentially regulated gene sets from RNA-seq for GATA-switch, G2-D0, and G1-D5 ChIP-seq peaks, the metascape was deployed to perform a gene ontology-term (GO) enrichment on the gene sets. This analysis is a proxy for understanding what pathways the GATA-switch, G2-D0, and G1-D5 peaks may be regulating. Since GATA-switch plays a central role to promote erythroid commitment, the biological pathways associated with the GATA-switch genes that are either increased or decreased in their expression as the cells differentiated from D0 to D5 became the focus. While the genes with increased expression were significantly associated with erythroid functions (e.g., erythrocyte differentiation, heme-metabolic pathways), genes with decreased expression showed predominance of hematopoietic progenitor/stem cell—related activities (e.g., chemotaxis/cell migration, blood vessel morphogenesis) (compare Fig. 18.4B-*top panel* with Fig. 18.4B-*bottom panel*).

To further investigate the regulatory landscape of regions associated with expression -increased, -decreased, and -unchanged genes for GATA-switch peaks, a genome-wide motif analysis of these regions was performed. The GATA-switch ChIP-seq peaks were specifically utilized to predict the GATA-associated cofactors that coordinate with GATA2 and GATA1 to promote coordinated gene expression programs at D0 and D5, respectively. The motif analysis under GATA-switch peaks that are located nearby genes with increased or decreased gene expression from D0 to D5 was performed using Homer's findMotifsGenome.pl functionality to calculate the probability of motif enrichment at the relevant regions. The whole-genome background parameter (hg19) was specified to find peaks within 200bp of the peak (-size 200), as most motifs are found within 50—75 bp of the peak center. The repeat-masked region (-mask) was used to prevent counting TF binding to repeat sequences from being specified as unique binding occurrences. To identify the highly abundant TF motifs within the GATA-switch peaks, which are associated with increased and decreased gene expression, the top 40 statistically significant enriched TF motifs in each category were enlisted and the percentage occurrence of individual TF-motif hits estimated. This analysis revealed a higher abundance of progenitor TF motifs (FLI1, PU.1) in peaks related to genes with decreased expression, whereas increased genes from D0 to D5 were associated with TF motifs that are relevant to the erythroid lineage (e.g., KLF1, SCL/TAL1) (Fig. 18.4C). Similar appearance of GATA motifs in both categories provides strong evidence that the genomic regions considered for this analysis are indeed GATA-switch regions (Fig. 18.4C). Overall, these results serve as a proof-of-principle for our large-scale data analysis methodologies, largely validating the inherent aspects of human erythroid differentiation.

5.2 Investigating the chromatin accessibility at GATA-bound loci with differentially regulated gene expression

To visualize chromatin accessibility of regions associated with increased, reduced, and unchanged genes for GATA-switch, G2-D0, and G1-D5 peaks, the peak intensity of TF-bound regions from ChIP-seq was plotted at ATAC-seq accessible coordinates. To do so, the computeMatrix function from deepTools (Ramirez et al., 2014) was deployed. The reference-point parameter was utilized when applying bigWig scores from the ATAC-seq results of CD34 progenitors at D5 and D0 (-S). 2 kb up- and downstream of the center of the gene-annotated peak (-b 2000 -a 2000) and averaged the bigWig peak intensity score over 10bp (-bs 10) were plotted (Ramirez et al., 2014). Additionally, peaks with bigWig peak intensity score of zero (–skipZeros) were skipped and peaks without scores were set to zero (–missingDataAsZero). The result is an intensity matrix that can be plotted with the plotHeatMap function from deepTools, specifying the intensity matrix (-m) and the output.pdf visualization (-out) (Ramirez et al., 2014). Regions by peak intensity (–sortRegions no) were not sorted intentionally. Additionally, it is helpful to visualize the aggregate

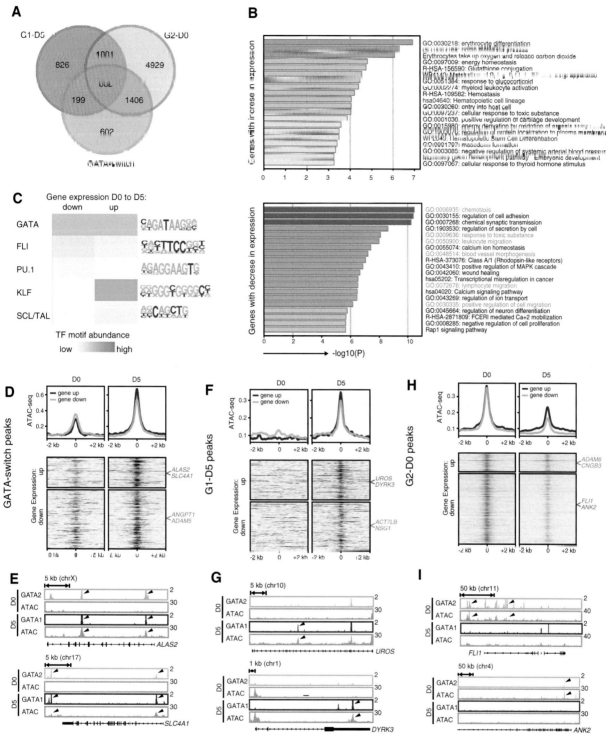

FIGURE 18.4 **Integration of ChIP-seq, ATAC-seq, and RNA-seq data during erythroid differentiation.** (A) Venn diagram representing the number of genes with enhancers that are occupied by GATA2 only at D0, GATA1 only at D5 and show a switch from GATA2 to GATA1 from D0 to D5. (B) *Upper panel*, metascape analysis for genes that show rise in expression upon GATA-switch during erythroid differentiation. Important erythroid GO terms are highlighted in *red*. *Lower panel*, metascape analysis for genes that show decrease in expression upon GATA-switch during erythroid commitment. Important progenitor GO terms are highlighted in *blue*. (C) Heatmaps showing the percentage occurrence of TF motifs under GATA-switch peaks for genes that show increase and decrease in expression during GATA2 to GATA1 switch. Heatmap represents the percentage appearance of motifs related to each TF shown, considering the top 40 motif hits ranked based on the significance value. Representative motifs (position weight matrices, PWMs) for each TF are indicated. Histograms and region heatmaps depicting the change of ATAC-seq signals from D0 to D5 at (D) GATA-switch peaks, (F) GATA1-D5 unique peaks and (H) GATA2-D0 unique peaks. Each plot represents signal intensities centered around $+/-$ 2 kb of the peaks shown in Y-axis, as indicated. Histograms depict ATAC-seq peaks with upregulated genes in magenta, whereas ATAC-seq peaks with decrease in gene expression are shown in *orange*. Region heatmaps are divided in upregulated and downregulated categories, as indicated. Example genes from each category are shown. Genes depicted in *Blue* show the change in ATAC-seq and RNA-seq in the same direction, whereas genes depicted in *red* show ATAC-seq and RNA-seq change in the opposite direction. Gene tracks of representative genes that show change in ATAC-seq and RNA-seq in the same direction in GATA switch (E), GATA1-D5 unique (G) and GATA2-D0 unique (I) peaks. Y-axis represents reads/million. *Black arrowheads* indicate peaks that potentially control the expression of associated genes.

peak intensity distribution across all of the regions in a smoothed histogram using the plotProfile function. Each of the peaks associated with expression -increased, -decreased, and -unchanged genes separately (−perGroup) was plotted, and the peak intensity matrix as the input (-m) was specified (Ramirez et al., 2014). Besides the genome-scale level analysis, using Integrative Genomics Viewer (IGV_2.8.2) (Robinson et al., 2011; Thorvaldsdottir et al., 2013), changes at individual DNA regions from D0 to D5 that contained erythroid differentiation−specific genes and their associated enhancers were shown. Analysis on ATAC-accessible regions associated with increased, reduced, and unchanged gene sets at each time point was used specifically to identify enhancers presumably important for erythroid cell specification.

With this analysis, genes at each stage of differentiation associated with nearby chromatin-accessible enhancers were identifiable. For example, GATA-switch genes include two well-known erythroid genes, *ALSA2* (Sadlon et al., 1999) and SLC4A1 (Salamin et al., 2018), which showed increase in gene expression upon erythroid commitment that may be caused by the increase in accessibility of nearby enhancer regions, presumably due to replacement of GATA2 binding with GATA1 (Fig. 18.4D and E). Similarly, *UROS* and *DYRK3* are two exemplary erythroid genes (Blouin et al., 2017; Zhang et al., 2005) that contain G1-D5 peaks and show gain in enhancer accessibility with increase in gene expression, likely due to gain in GATA1 binding during erythroid transition (Fig. 18.4F and G). Finally, G2-D0 regions include genes with progenitor functions, *FLI1* (Pimanda et al., 2007) and *ANK2* (Peters et al., 1993), which lose GATA2 binding and concomitant enhancer accessibility resulting in decrease in gene expression as cells transition from D0 to D5 (Figs. 18.4H and I).

Besides this mechanism, our analysis also suggested additional modes of transcriptional control. For example, irrespective of the direction of change observed in gene expression, there is an overall increase in the ATAC accessibility at both the GATA-switch and G1-D5 peaks (Fig. 18.4F and H). In contrast, chromatin accessibility is largely decreased at G2-D0 sites regardless of the change in gene expression from D0 to D5 (Fig. 18.4D). Hence, the GATA-switch and G1-D5 categories involve genes that were downregulated upon differentiation, in spite of gain of ATAC accessibility from an increase in GATA1 binding. Presumably, these are potential progenitor genes (e.g., *ANGPT1*, GATA-switch category, Fig. 18.5A *upper panel*; *ACT7LB*, GATA1-D5 unique category, Fig. 18.5B *upper panel*) that are controlled by additional GATA2-bound enhancers (red arrows) that lose openness due to loss of GATA2 during erythroid commitment. Alternatively, there may be genes with majorly progenitor functions (e.g., *ADAM5*, GATA-switch category, Fig. 18.5A *lower panel*; *NSG1*, GATA1-D5 unique category, Fig. 18.5B *lower panel*) that are downregulated by GATA cofactors (KLF1, TAL1) that need additional chromatin accessibility to occupy those enhancers during differentiation. In fact, such GATA-associated TF motifs within these enhancers were identified as shown in Fig. 18.5A and B, *lower panels*). Potential erythroid genes in G2-D0 class were also observed, where gene expression increased substantially irrespective of the decrease of ATAC accessibility of nearby enhancers (Fig. 18.4H). Those genes could be regulated by additional GATA-switch enhancers (*ADAM6*, Fig. 18.5C, *upper panel, red arrows*). Such genes could also be upregulated by erythroid TFs, other than GATA1, which are able to bind those enhancers despite limited accessibility (*CNGB3*, Fig. 18.5C, *lower panel* showing plausible erythroid TF motifs). In this specific context, the gene may not be regulated by the nearest enhancers, but is rather under the control of a distantly located trans-acting enhancer.

6. Conclusion and future directions—a therapeutic approach

Our multiomic approach, described here, identifies multiple modalities of transcriptional control during human erythroid differentiation and opens up different avenues to experimentally test in the future. Using CRISPR-Cas9 mutagenesis, an ideal approach would be to delete entire enhancers, or mutate small DNA-sequences within those enhancers that contain relevant TF motifs, and subsequently monitor for any alteration of gene expression and erythroid differentiation defects. Alternatively, single-nucleotide polymorphisms (SNPs) associated with human hematopoietic disorders may occur naturally within those enhancers to compromise their function and genetics control such conditions. This is an active area of research in the current genome-wide association studies (GWAS), connecting genetic variants to their functions to pinpoint the causality of human hematopoietic disorders (Liggett & Sankaran, 2020; Nandakumar et al., 2020; Schaid et al., 2018; Ulirsch et al., 2019; Vuckovic et al., 2020).

Our approach serves as a thorough bioinformatic data analysis platform that not only validates the activity and regulation of known genes, but also predicts novel gene sets with potential hematopoietic functions. Similar to previously mentioned GATA-switch regions, metascape analysis of gene sets containing G2-D0 and G1-D5 enhancers revealed biologically relevant GO terms that warrant further detailed investigation of those genes in near future. The genes that are bound by GATA2 only at D0 and showed decrease in gene expression from D0 to D5 are associated with "blood vessel development," "regulation of cell adhesion," and "chemotaxis," whereas the genes with increased expression were significantly associated with "erythrocytes take up oxygen and release carbon dioxide" and "ferroptosis" (Fig. 18.5D). G1-D5 genes with a decrease in expression were similarly related to "regulation of cell adhesion"

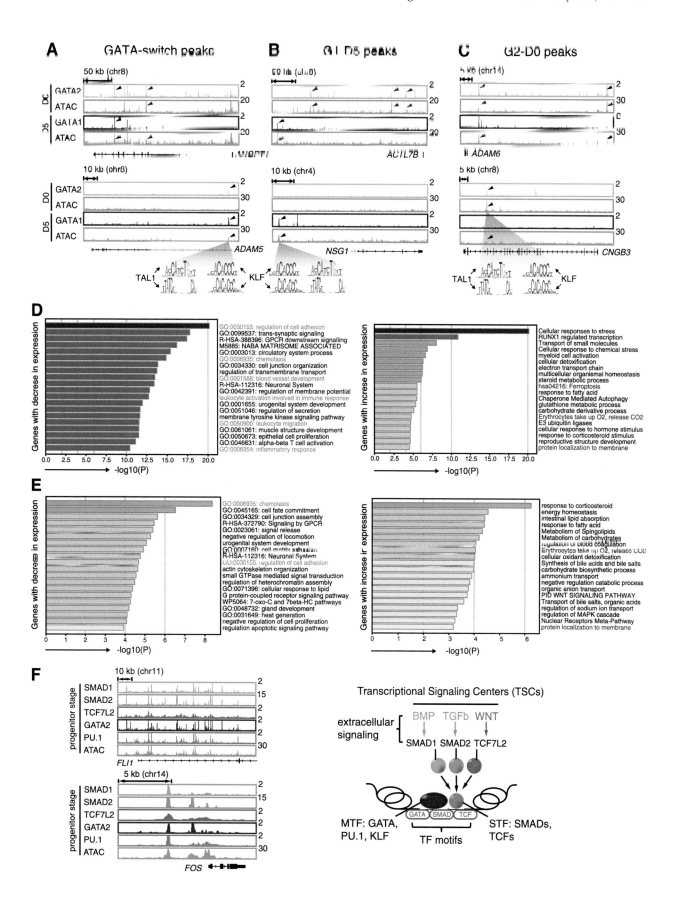

and "chemotaxis," whereas the gene sets with a rise in expression were predominantly associated with "erythrocytes take up oxygen and release carbon dioxide" and "protein localization to plasma membrane" (Fig. 18.5E). Hence, both gene sets that are controlled either by GATA2 at D0 or by GATA1 at D5 contain relevant gene regulatory networks that may play pivotal, yet unknown, roles during human erythropoiesis. Dissection of such locus-specific, as well as genome-wide, mechanisms could pinpoint defects in transcriptional machinery in many human disorders, besides erythroid. This may provide novel therapeutic strategies for clinical benefit. Recent studies have identified crucial genes with GATA-bound enhancers that contain genetic mutations for anemia and fetal hemoglobinopathies (Bauer et al., 2013; Wakabayashi et al., 2016). Remarkably, ongoing gene therapies involving CRISPR-Cas9-mediated modulation of such enhancers of *BCL11A* significantly reduced the severity of sickle cell disease in patients (Canver et al., 2015; Esrick et al., 2021).

Besides GATA2 and GATA1, there are several TFs, e.g., PU.1, RUNX1, KLF1, TAL1, which serve as examples of hematopoietic tissue—specific master transcription factors (MTFs). These MTFs bind hematopoietic cell type—specific enhancers at each differentiation stage to control cell stage—specific transcription (Burda et al., 2010; Orkin, 1995). During erythropoiesis, such MTFs, GATA2 and GATA1, activate critical cell type—specific programs that gradually transition the fate of the progenitor cells to the mature erythrocytes (Kaneko et al., 2010; Katsumura et al., 2017; Suzuki et al., 2013). A critical issue in the field is that the differentiation process involves predetermined changes in cell shape and morphology, which are stage-specific, yet the gene expression or activity of MTFs alone is insufficient to explain all aspects of the differentiation program, especially during regeneration of erythroid system under stress. Other TFs must participate during differentiation. To this end, previous research has shown that while MTFs bind cell type—specific enhancers to establish transcriptional programs associated with specific cell identities, during stress-induced tissue regeneration, growth factors and small molecules activate episodic extracellular signals that activate signaling transcription factors (STFs) to modulate levels of gene expression within these programs (Chen et al., 2008; Cole et al., 2008; Mullen et al., 2011; Trompouki et al., 2011; Verzi et al., 2010; Zhu & Emerson, 2002). Incorporation of ChIP-seq assays for crucial stress-responsive STFs in the presence and absence of appropriate signals, coupled with our holistic approach described here, could resolve the coordinated action of MTFs and STFs, which enables the cellular microenvironment to orchestrate cell type—restricted gene expression. Along this line, as shown previously, a pulse of bone morphogenetic protein (BMP) improves hematopoietic recovery after irradiation and that, upon stimulation, the BMP-induced STF, SMAD1 binds genomic regions that are cooccupied by hematopoietic lineage-specific MTFs, GATA2, and GATA1 in various human hematopoietic cells (Trompouki et al., 2011). Using a similar approach in another recent study (Choudhuri et al., 2020), the set of BMP signaling-associated enhancers that are cooccupied by BMP-responsive STF SMAD1 and hematopoietic MTFs, GATA2, and GATA1 during differentiation of human hematopoietic progenitor cells (CD34$^+$) into proerythroblasts have been identified. These specialized stress-responsive, BMP-target enhancers are a small subset among the overall pool of active enhancers. Importantly, these enhancers operate as "transcriptional signaling centers (TSCs)," since these enhancers serve as docking sites for numerous crucial developmental STFs (e.g., SMAD2 and TCF7L2 upon stimulation with TGFβ and WNT, respectively) and could render nearby genes inducible by growth factors and small molecules under stress (Fig. 18.5F, *left and right panels*). This indicates a crucial role of numerous signaling pathways to mediate transcriptional control during stress hematopoiesis upon cobinding with MTFs to specific stress-inducible enhancers (Fig. 18.5G). Further understanding of the specificity of signaling pathways will allow us to predict mechanisms and gene sets utilized by specific signaling pathways. This may illuminate novel therapeutic insights for many environmentally controlled human disorders, which are potentially caused due to altered response to signaling pathways ("signalopathies"). Modulating enhancer activity with small molecules/drugs for specific signaling pathways may avoid rather challenging and expensive therapeutic alternatives and can pay the way for targeting disease-relevant enhancers with CRISPR-Cas9-mediated gene therapy.

FIGURE 18.5 **Gene tracks of representative GATA-bound gene loci that show the ATAC-seq peaks correlating with RNA-seq in the opposite direction and summary of observations that lay the platform to explore additional outstanding questions in future.** Representative genes with GATA-switch (A), GATA1-D5 unique (B), and GATA2-D0 unique (C) peaks are shown. Y-axis represents reads/million. *Black arrowheads* depict the relevant peaks, whereas the *red arrowheads* indicate potential alternative peaks that may control the expression of associate genes. Important TF motifs that may occupy these enhancers are also indicated. (D) *Left panel*, metascape analysis for genes associated with G2-D0 sites that show decrease in expression during erythroid differentiation. Important erythroid GO-terms are highlighted in *blue*. *Right panel*, metascape analysis for genes associated with G2-D0 sites that show rise in expression during erythroid differentiation. Important progenitor GO-terms are highlighted in *red*. (E) *Left panel*, metascape analysis for genes associated with G1-D5 sites that show decrease in expression during erythroid differentiation. Important erythroid GO-terms are highlighted in *blue*. *Right panel*, metascape analysis for genes associated with G1-D5 sites that show rise in expression during erythroid differentiation. Important progenitor GO-terms are high highlighted in *red*. (F) *Left panel*, example gene tracks depicting peak intensities of SMAD1 (BMP signaling track in *green*), SMAD2 (TGFβ signaling track in magenta), and TCF7L2 (WNT signaling track in *orange*), with the MTFs, GATA2 (*red*), PU.1 (*pink*), and ATAC-seq signals (*gray*) at *FLI1* and *FOS* loci at D0. *Right panel*, schematic representation of "transcriptional signaling centers (TSCs)." TSCs are genomic regions cooccupied by multiple STFs after the induction of the respective signaling pathways. TSCs could be specific responding to signals, which lead to cooccupying a given region by specific combinations of STFs and stage-specific MTFs.

Acknowledgments

All figures are generated using Microsoft PowerPoint, Excel, Adobe Illustrator, and Bio Render
Integration of multiomic data to identify transcriptional targets for human erythroid disorders.

Authors contributions

MP, AC, and YZ are responsible for drafting and editing the manuscript. AC is responsible for performing the described experiments. SY, MP, and AC are responsible for data analysis. MP, AC, EF, and SY are responsible for generating figures and tables; YZ and LZ are responsible for funding and organizing the project.

References

AbuSamra, D. B., Aleisa, F. A., Al-Amoodi, A. S., Jalal Ahmed, H. M., Chin, C. J., Abuelela, A. F., … Merzaban, J. S. (2017). Not just a marker: CD34 on human hematopoietic stem/progenitor cells dominates vascular selectin binding along with CD44. *Blood Advances, 1*(27), 2799–2816. https://doi.org/10.1182/bloodadvances.2017004317

Alekseyev, Y. O., Fazeli, R., Yang, S., Basran, R., Maher, T., Miller, N. S., & Remick, D. (2018). A next-generation sequencing primer-how does it work and what can it do? *Academic Pathology, 5*. https://doi.org/10.1177/2374289518766521, 2374289518766521.

Anders, S., Pyl, P. T., & Huber, W. (2015). HTSeq—A Python framework to work with high-throughput sequencing data. *Bioinformatics, 31*(2), 166–169. https://doi.org/10.1093/bioinformatics/btu638

Andrews, N. C., & Orkin, S. H. (1994). Transcriptional control of erythropoiesis. *Current Opinion in Hematology, 1*(2), 119–124. Retrieved from https://www.ncbi.nlm.nih.gov/pubmed/9371270.

Bauer, D. E., Kamran, S. C., Lessard, S., Xu, J., Fujiwara, Y., Lin, C., … Orkin, S. H. (2013). An erythroid enhancer of BCL11A subject to genetic variation determines fetal hemoglobin level. *Science, 342*(6155), 253–257. https://doi.org/10.1126/science.1242088

Bivens, N. J., & Zhou, M. (2016). RNA-seq library construction methods for transcriptome analysis. *Current Protocols in Plant Biology, 1*(1), 197–215. https://doi.org/10.1002/cppb.20019

Blouin, J. M., Bernardo-Seisdedos, G., Sasso, E., Esteve, J., Ged, C., Lalanne, M., … Richard, E. (2017). Missense UROS mutations causing congenital erythropoietic porphyria reduce UROS homeostasis that can be rescued by proteasome inhibition. *Human Molecular Genetics, 26*(8), 1565–1576. https://doi.org/10.1093/hmg/ddx067

Bresnick, E. H., Lee, H. Y., Fujiwara, T., Johnson, K. D., & Keles, S. (2010). GATA switches as developmental drivers. *Journal of Biological Chemistry, 285*(41), 31087–31093. https://doi.org/10.1074/jbc.R110.159079

Buenrostro, J. D., Wu, B., Chang, H. Y., & Greenleaf, W. J. (2015). ATAC-seq: A method for assaying chromatin accessibility genome-wide. *Current Protocols in Molecular Biology, 109*, 21 29 21–21 29 29. https://doi.org/10.1002/0471142727.mb2129s109

Burda, P., Laslo, P., & Stopka, T. (2010). The role of PU.1 and GATA-1 transcription factors during normal and leukemogenic hematopoiesis. *Leukemia, 24*(7), 1249–1257. https://doi.org/10.1038/leu.2010.104

Cantor, A. B., & Orkin, S. H. (2002). Transcriptional regulation of erythropoiesis: An affair involving multiple partners. *Oncogene, 21*(21), 3368–3376. https://doi.org/10.1038/sj.onc.1205326

Canver, M. C., Smith, E. C., Sher, F., Pinello, L., Sanjana, N. E., Shalem, O., … Bauer, D. E. (2015). BCL11A enhancer dissection by Cas9-mediated in situ saturating mutagenesis. *Nature, 527*(7577), 192–197. https://doi.org/10.1038/nature15521

Chen, X., Xu, H., Yuan, P., Fang, F., Huss, M., Vega, V. B., … Ng, H. H. (2008). Integration of external signaling pathways with the core transcriptional network in embryonic stem cells. *Cell, 133*(6), 1106–1117. https://doi.org/10.1016/j.cell.2008.04.043

Choudhuri, A., Trompouki, E., Abraham, B. J., Colli, L. M., Kock, K. H., Mallard, W., … Zon, L. I. (2020). Common variants in signaling transcription-factor-binding sites drive phenotypic variability in red blood cell traits. *Nature Genetics, 52*(12), 1333–1345. https://doi.org/10.1038/s41588-020-00738-2

Cole, M. F., Johnstone, S. E., Newman, J. J., Kagey, M. H., & Young, R. A. (2008). Tcf3 is an integral component of the core regulatory circuitry of embryonic stem cells. *Genes and Development, 22*(6), 746–755. https://doi.org/10.1101/gad.1642408

Dzierzak, E., & Philipsen, S. (2013). Erythropoiesis: Development and differentiation. *Cold Spring Harbor Perspectives in Medicine, 3*(4), a011601. https://doi.org/10.1101/cshperspect.a011601

Esrick, E. B., Lehmann, L. E., Biffi, A., Achebe, M., Brendel, C., Ciuculescu, M. F., … Williams, D. A. (2021). Post-transcriptional genetic silencing of BCL11A to treat sickle cell disease. *New England Journal of Medicine, 384*(3), 205–215. https://doi.org/10.1056/NEJMoa2029392

Fried, M. G. (1989). Measurement of protein-DNA interaction parameters by electrophoresis mobility shift assay. *Electrophoresis, 10*(5–6), 366–376. https://doi.org/10.1002/elps.1150100515

Goodwin, S., McPherson, J. D., & McCombie, W. R. (2016). Coming of age: Ten years of next-generation sequencing technologies. *Nature Reviews Genetics, 17*(6), 333–351. https://doi.org/10.1038/nrg.2016.49

Hnisz, D., Shrinivas, K., Young, R. A., Chakraborty, A. K., & Sharp, P. A. (2017). A phase separation model for transcriptional control. *Cell, 169*(1), 13–23. https://doi.org/10.1016/j.cell.2017.02.007

Hollbacher, B., Balazs, K., Heinig, M., & Uhlenhaut, N. H. (2020). Seq-ing answers: Current data integration approaches to uncover mechanisms of transcriptional regulation. *Computational and Structural Biotechnology Journal, 18*, 1330–1341. https://doi.org/10.1016/j.csbj.2020.05.018

Hrdlickova, R., Toloue, M., & Tian, B. (2017). RNA-Seq methods for transcriptome analysis. *Wiley Interdisciplinary Reviews: RNA, 8*(1). https://doi.org/10.1002/wrna.1364

International Human Genome Sequencing, C. (2004). Finishing the euchromatic sequence of the human genome. *Nature, 431*(7011), 931–945. https://doi.org/10.1038/nature03001

Kaneko, H., Shimizu, R., & Yamamoto, M. (2010). GATA factor switching during erythroid differentiation. *Current Opinion in Hematology, 17*(3), 163−168. https://doi.org/10.1097/MOH.0b013e32833800b8

Katsumura, K. R., Bresnick, E. H., & Group, G. F. M. (2017). The GATA factor revolution in hematology. *Blood, 129*(15), 2092−2102. https://doi.org/10.1182/blood-2016-09-687871

Kent, W. J., Zweig, A. S., Barber, G., Hinrichs, A. S., & Karolchik, D. (2010). BigWig and BigBed: Enabling browsing of large distributed datasets. *Bioinformatics, 26*(17), 2204−2207. https://doi.org/10.1093/bioinformatics/btq351

Kim, S. I., & Bresnick, E. H. (2007). Transcriptional control of erythropoiesis: Emerging mechanisms and principles. *Oncogene, 26*(47), 6777−6794. https://doi.org/10.1038/sj.onc.1210761

Kim, D., Pertea, G., Trapnell, C., Pimentel, H., Kelley, R., & Salzberg, S. L. (2013). TopHat2: Accurate alignment of transcriptomes in the presence of insertions, deletions and gene fusions. *Genome Biology, 14*(4), R36. https://doi.org/10.1186/gb-2013-14-4-r36

Kim, Y. C., Wu, Q., Chen, J., Xuan, Z., Jung, Y. C., Zhang, M. Q., ... Wang, S. M. (2009). The transcriptome of human CD34$^+$ hematopoietic stem-progenitor cells. *Proceedings of the National Academy of Sciences of the United States of America, 106*(20), 8278−8283. https://doi.org/10.1073/pnas.0903390106

Kornberg, R. D. (1999). Eukaryotic transcriptional control. *Trends in Cell Biology, 9*(12), M46−M49. Retrieved from https://www.ncbi.nlm.nih.gov/pubmed/10611681.

Krumlauf, R. (1994). Analysis of gene expression by northern blot. *Molecular Biotechnology, 2*(3), 227−242. https://doi.org/10.1007/BF02745879

Lander, E. S., Linton, L. M., Birren, B., Nusbaum, C., Zody, M. C., Baldwin, J., ... International Human Genome Sequencing, C. (2001). Initial sequencing and analysis of the human genome. *Nature, 409*(6822), 860−921. https://doi.org/10.1038/35057062

Lane, D., Prentki, P., & Chandler, M. (1992). Use of gel retardation to analyze protein-nucleic acid interactions. *Microbiological Reviews, 56*(4), 509−528. https://doi.org/10.1128/mr.56.4.509-528.1992

Langmead, B., & Salzberg, S. L. (2012). Fast gapped-read alignment with Bowtie 2. *Nature Methods, 9*(4), 357−359. https://doi.org/10.1038/nmeth.1923

Lee, H. Y., Gao, X., Barrasa, M. I., Li, H., Elmes, R. R., Peters, L. L., & Lodish, H. F. (2015). PPAR-alpha and glucocorticoid receptor synergize to promote erythroid progenitor self-renewal. *Nature, 522*(7557), 474−477. https://doi.org/10.1038/nature14326

Lee, T. I., & Young, R. A. (2013). Transcriptional regulation and its misregulation in disease. *Cell, 152*(6), 1237−1251. https://doi.org/10.1016/j.cell.2013.02.014

Liggett, L. A., & Sankaran, V. G. (2020). Unraveling hematopoiesis through the lens of genomics. *Cell, 182*(6), 1384−1400. https://doi.org/10.1016/j.cell.2020.08.030

Li, H., Handsaker, B., Wysoker, A., Fennell, T., Ruan, J., Homer, N., ... Genome Project Data Processing, S. (2009). The sequence alignment/map format and SAMtools. *Bioinformatics, 25*(16), 2078−2079. https://doi.org/10.1093/bioinformatics/btp352

Love, M. I., Huber, W., & Anders, S. (2014). Moderated estimation of fold change and dispersion for RNA-seq data with DESeq2. *Genome Biology, 15*(12), 550. https://doi.org/10.1186/s13059-014-0550-8

Maxam, A. M., & Gilbert, W. (1977). A new method for sequencing DNA. *Proceedings of the National Academy of Sciences of the United States of America, 74*(2), 560−564. https://doi.org/10.1073/pnas.74.2.560

Mullen, A. C., Orlando, D. A., Newman, J. J., Loven, J., Kumar, R. M., Bilodeau, S., ... Young, R. A. (2011). Master transcription factors determine cell-type-specific responses to TGF-beta signaling. *Cell, 147*(3), 565−576. https://doi.org/10.1016/j.cell.2011.08.050

Nandakumar, S. K., Liao, X., & Sankaran, V. G. (2020). The blood: Connecting variant to function in human hematopoiesis. *Trends in Genetics, 36*(8), 563−576. https://doi.org/10.1016/j.tig.2020.05.006

Orkin, S. H. (1995). Transcription factors and hematopoietic development. *Journal of Biological Chemistry, 270*(10), 4955−4958. https://doi.org/10.1074/jbc.270.10.4955

Pariset, L., Chillemi, G., Bongiorni, S., Romano Spica, V., & Valentini, A. (2009). Microarrays and high-throughput transcriptomic analysis in species with incomplete availability of genomic sequences. *Nature Biotechnology, 25*(5), 272−279. https://doi.org/10.1016/j.nbt.2009.03.013

Park, P. J. (2009). ChIP-seq: Advantages and challenges of a maturing technology. *Nature Reviews Genetics, 10*(10), 669−680. https://doi.org/10.1038/nrg2641

Peters, L. L., Turtzo, L. C., Birkenmeier, C. S., & Barker, J. E. (1993). Distinct fetal Ank-1 and Ank-2 related proteins and mRNAs in normal and nb/nb mice. *Blood, 81*(8), 2144−2149. Retrieved from https://www.ncbi.nlm.nih.gov/pubmed/8471772.

Pevny, L., Simon, M. C., Robertson, E., Klein, W. H., Tsai, S. F., D'Agati, V., ... Costantini, F. (1991). Erythroid differentiation in chimaeric mice blocked by a targeted mutation in the gene for transcription factor GATA-1. *Nature, 349*(6306), 257−260. https://doi.org/10.1038/349257a0

Pimanda, J. E., Ottersbach, K., Knezevic, K., Kinston, S., Chan, W. Y., Wilson, N. K., ... Gottgens, B. (2007). Gata2, Fli1, and Scl form a recursively wired gene-regulatory circuit during early hematopoietic development. *Proceedings of the National Academy of Sciences of the United States of America, 104*(45), 17692−17697. https://doi.org/10.1073/pnas.0707045104

Quinlan, A. R., & Hall, I. M. (2010). BEDTools: A flexible suite of utilities for comparing genomic features. *Bioinformatics, 26*(6), 841−842. https://doi.org/10.1093/bioinformatics/btq033

Raghav, P. K., & Gangenahalli, G. (2021). PU.1 mimic synthetic peptides selectively bind with GATA-1 and allow c-Jun PU.1 binding to enhance myelopoiesis. *International Journal of Nanomedicine, 16*, 3833−3859. https://doi.org/10.2147/IJN.S303235

Ramirez, F., Dundar, F., Diehl, S., Gruning, B. A., & Manke, T. (2014). deepTools: a flexible platform for exploring deep-sequencing data. *Nucleic Acids Research, 42*(Web Server Issue), W187—W191. https://doi.org/10.1093/nar/gku365

Robinson, M. D., McCarthy, D. J., & Smyth, G. K. (2010). edgeR: A bioconductor package for differential expression analysis of digital gene expression data. *Bioinformatics, 26*(1), 139—140. https://doi.org/10.1093/bioinformatics/btp616

Robinson, J. T., Thorvaldsdottir, H., Winckler, W., Guttman, M., Lander, E. S., Getz, G., & Mesirov, J. P. (2011). Integrative genomics viewer. *Nature Biotechnology, 29*(1), 24—26. https://doi.org/10.1038/nbt.1754

Ronzoni, L., Bonara, P., Rusconi, D., Frugoni, C., Libani, I., & Cappellini, M. D. (2008). Erythroid differentiation and maturation from peripheral CD34+ cells in liquid culture: Cellular and molecular characterization. *Blood Cells Molecules and Diseases, 40*(2), 148—155. https://doi.org/10.1016/j.bcmd.2007.07.006

Sadlon, T. J., Dell'Oso, T., Surinya, K. H., & May, B. K. (1999). Regulation of erythroid 5-aminolevulinate synthase expression during erythropoiesis. *The International Journal of Biochemistry and Cell Biology, 31*(10), 1153—1167. https://doi.org/10.1016/s1357-2725(99)00073-4

Salamin, O., Mignot, J., Kuuranne, T., Saugy, M., & Leuenberger, N. (2018). Transcriptomic biomarkers of altered erythropoiesis to detect autologous blood transfusion. *Drug Testing and Analysis, 10*(3), 604—608. https://doi.org/10.1002/dta.2240

Sanger, F., & Coulson, A. R. (1975). A rapid method for determining sequences in DNA by primed synthesis with DNA polymerase. *Journal of Molecular Biology, 94*(3), 441—448. https://doi.org/10.1016/0022-2836(75)90213-2

Sanger, F., Nicklen, S., & Coulson, A. R. (1977). DNA sequencing with chain-terminating inhibitors. *Proceedings of the National Academy of Sciences of the United States of America, 74*(12), 5463—5467. https://doi.org/10.1073/pnas.74.12.5463

Sankaran, V. G., Orkin, S. H., & Walkley, C. R. (2008). Rb intrinsically promotes erythropoiesis by coupling cell cycle exit with mitochondrial biogenesis. *Genes and Development, 22*(4), 463—475. https://doi.org/10.1101/gad.1627208

Schaid, D. J., Chen, W., & Larson, N. B. (2018). From genome-wide associations to candidate causal variants by statistical fine-mapping. *Nature Reviews Genetics, 19*(8), 491—504. https://doi.org/10.1038/s41576-018-0016-z

Shields, C. W.t., Reyes, C. D., & Lopez, G. P. (2015). Microfluidic cell sorting: A review of the advances in the separation of cells from debulking to rare cell isolation. *Lab on a Chip, 15*(5), 1230—1249. https://doi.org/10.1039/c4lc01246a

Simon, M. C., Pevny, L., Wiles, M. V., Keller, G., Costantini, F., & Orkin, S. H. (1992). Rescue of erythroid development in gene targeted GATA-1-mouse embryonic stem cells. *Nature Genetics, 1*(2), 92—98. https://doi.org/10.1038/ng0592-92

Spitz, F., & Furlong, E. E. (2012). Transcription factors: From enhancer binding to developmental control. *Nature Reviews Genetics, 13*(9), 613—626. https://doi.org/10.1038/nrg3207

Stark, R., Grzelak, M., & Hadfield, J. (2019). RNA sequencing: The teenage years. *Nature Reviews Genetics, 20*(11), 631—656. https://doi.org/10.1038/s41576-019-0150-2

Suzuki, M., Kobayashi-Osaki, M., Tsutsumi, S., Pan, X., Ohmori, S., Takai, J., … Yamamoto, M. (2013). GATA factor switching from GATA2 to GATA1 contributes to erythroid differentiation. *Genes to Cells, 18*(11), 921—933. https://doi.org/10.1111/gtc.12086

Thorvaldsdottir, H., Robinson, J. T., & Mesirov, J. P. (2013). Integrative genomics viewer (IGV): High-performance genomics data visualization and exploration. *Briefings in Bioinformatics, 14*(2), 178—192. https://doi.org/10.1093/bib/bbs017

Trapnell, C., Williams, B. A., Pertea, G., Mortazavi, A., Kwan, G., van Baren, M. J., … Pachter, L. (2010). Transcript assembly and quantification by RNA-Seq reveals unannotated transcripts and isoform switching during cell differentiation. *Nature Biotechnology, 28*(5), 511—515. https://doi.org/10.1038/nbt.1621

Trompouki, E., Bowman, T. V., Lawton, L. N., Fan, Z. P., Wu, D. C., DiBiase, A., … Zon, L. I. (2011). Lineage regulators direct BMP and Wnt pathways to cell-specific programs during differentiation and regeneration. *Cell, 147*(3), 577—589. https://doi.org/10.1016/j.cell.2011.09.044

Tsai, F. Y., Keller, G., Kuo, F. C., Weiss, M., Chen, J., Rosenblatt, M., … Orkin, S. H. (1994). An early haematopoietic defect in mice lacking the transcription factor GATA-2. *Nature, 371*(6494), 221—226. https://doi.org/10.1038/371221a0

Tsai, F. Y., & Orkin, S. H. (1997). Transcription factor GATA-2 is required for proliferation/survival of early hematopoietic cells and mast cell formation, but not for erythroid and myeloid terminal differentiation. *Blood, 89*(10), 3636—3643. Retrieved from https://www.ncbi.nlm.nih.gov/pubmed/9160668.

Ulirsch, J. C., Lareau, C. A., Bao, E. L., Ludwig, L. S., Guo, M. H., Benner, C., … Sankaran, V. G. (2019). Interrogation of human hematopoiesis at single-cell and single-variant resolution. *Nature Genetics, 51*(4), 683—693. https://doi.org/10.1038/s41588-019-0362-6

Venter, J. C., Adams, M. D., Myers, E. W., Li, P. W., Mural, R. J., Sutton, G. G., … Zhu, X. (2001). The sequence of the human genome. *Science, 291*(5507), 1304—1351. https://doi.org/10.1126/science.1058040

Verzi, M. P., Hatzis, P., Sulahian, R., Philips, J., Schuijers, J., Shin, H., … Shivdasani, R. A. (2010). TCF4 and CDX2, major transcription factors for intestinal function, converge on the same cis-regulatory regions. *Proceedings of the National Academy of Sciences of the United States of America, 107*(34), 15157—15162. https://doi.org/10.1073/pnas.1003822107

Vuckovic, D., Bao, E. L., Akbari, P., Lareau, C. A., Mousas, A., Jiang, T., … Soranzo, N. (2020). The polygenic and monogenic basis of blood traits and diseases. *Cell, 182*(5), 1214—1231 e1211. https://doi.org/10.1016/j.cell.2020.08.008

Wakabayashi, A., Ulirsch, J. C., Ludwig, L. S., Fiorini, C., Yasuda, M., Choudhuri, A., ... Sankaran, V. G. (2016). Insight into GATA1 transcriptional activity through interrogation of cis elements disrupted in human erythroid disorders. *Proceedings of the National Academy of Sciences of the United States of America, 113*(16), 4434–4439. https://doi.org/10.1073/pnas.1521754113

Zhang, D., Li, K., Erickson-Miller, C. L., Weiss, M., & Wojchowski, D. M. (2005). DYRK gene structure and erythroid-restricted features of DYRK3 gene expression. *Genomics, 85*(1), 117–130. https://doi.org/10.1016/j.ygeno.2004.08.021

Zhang, Y., Liu, T., Meyer, C. A., Eeckhoute, J., Johnson, D. S., Bernstein, B. E., ... Liu, X. S. (2008). Model-based analysis of ChIP-seq (MACS). *Genome Biology, 9*(9), R137. https://doi.org/10.1186/gb-2008-9-9-r137

Zhu, J., & Emerson, S. G. (2002). Hematopoietic cytokines, transcription factors and lineage commitment. *Oncogene, 21*(21), 3295–3313. https://doi.org/10.1038/sj.onc.1205318

Chapter 19

Computational approaches to determine stem cell fate

Aiindrila Dhara[1], Sangramjit Mondal[1], Ayushi Gupta[2], Princy Choudhary[2], Sangeeta Singh[2], Pritish Kumar Varadwaj[2] and Nirmalya Sen[1]

[1]*Molecular Oncology Laboratory, Department of Biological Sciences, Bose Institute, Kolkata, West Bengal, India;* [2]*Department of Bioinformatics & Applied Sciences, Indian Institute of Information Technology-Allahabad, Allahabad, Uttar Pradesh, India*

1. Introduction

The general consensus regarding the existence of somatic stem cells came from the pioneering works of Ernest McCulloch and James Till on hematopoietic stem cells (HSCs) in the 1960s (Becker et al., 1963; Siminovitch et al., 1963). Post-discovery of HSCs, which had low potency, work on embryonic origin tumors led to the introduction of pluripotent embryonic carcinoma (EC) cells. Later, EC cells became successful genetic models for stem cell research (Andrews, 2002; Illmensee & Mintz, 1976). Owing to their cancerous origin, clinical applications of EC cells were restricted. Based on EC culture experience, Kaufman and colleagues isolated the first mouse embryonic stem (ES) cells in 1981 from the pre-implantation blastocyst (Evans & Kaufman, 1981; Martin, 1981). The first human ES cell line was derived in 1998 by Thompson and coworker (Thomson et al., 1998). With the advent of gene knockout, knock-in technologies (Bradley et al., 1984; Thomas & Capecchi, 1987), ES cells improved the understanding of mammalian development (Thomson et al., 1996, 1998). However, due to certain ethical concerns, the usage of human embryonic stem cells became controversial and restricted (de Wert & Mummery, 2003; Mendiola et al., 1999). Apart from ESC, adult stem cells (ASCs) are broadly divided into parenchymal and mesenchymal stem cells as per lineage (Alvarez et al., 2012). Mesenchymal stem cells have been used for various medical applications owing to their easy isolation, fewer ethical challenges, and varied developmental potencies (Lim & Khoo, 2021). Yamanaka and colleagues developed induced pluripotent stem cells (iPSCs) from either mouse embryonic or adult fibroblasts by introducing four factors (Oct3/4, Sox2, c-Myc, and Klf4) in ES culture conditions (Takahashi & Yamanaka, 2006). This was probably the first successful attempt to identify a regulatory network of transcription factors (TFs) that could modify cellular differentiation and fate (Rodolfa & Eggan, 2006). Building on extensive knowledge of culture conditions, inducing factors, and regulatory networks, stem cell research has followed a progressive roadmap for successful therapeutic applications (Fig. 19.1) (Zakrzewski et al., 2019). Various biochemical and biophysical means such as STR profiling, surface marker—based detection, and imaging techniques used for cell identification (Nguyen et al., 2010) are now amalgamated with various computational tools and models. With the advent of computational tools, stem cell fates during differentiation are closely correlated with gene regulatory networks (GRNs) and transcriptional landscapes providing a more accurate view of global events during differentiation program (Zhu et al., 2021).

Stem cell researchers around the globe are trying to acquire the perfect differentiation techniques for regeneration of organs such as heart, pancreas, nervous system, and skin that can be used for therapeutic purposes such as regenerative medicine, injury repair, and curing genetic diseases (Bolli et al., 2018; Hoang et al., 2022). However, the success rate regarding stem cell—based clinical trials is low owing to low survival, loss during delivery, failed differentiation in vivo, immune rejection, etc. of stem cells (Deinsberger et al., 2020; Trounson & McDonald, 2015). Currently, computer-based algorithms are being applied, taking into account various intrinsic and extrinsic parameters that can predict the fate of a stem cell more accurately and hence increase the chances of a positive translational outcome. This chapter provides an overview regarding development of computational tools that utilize metadata, mathematical models, and experimental outcomes to predict regulatory circuits, which are currently utilized for stem cell research.

Computational Biology for Stem Cell Research. https://doi.org/10.1016/B978-0-443-13222-3.00017-4
Copyright © 2024 Elsevier Inc. All rights reserved, including those for text and data mining, AI training, and similar technologies.

FIGURE 19.1 Roadmap of seminal discoveries in stem cell biology.

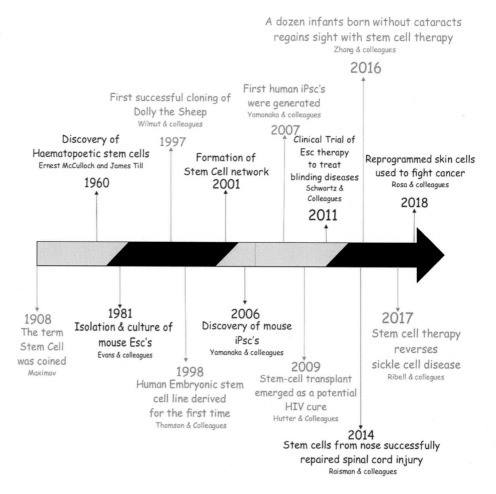

2. Understanding stem cell fates during differentiation: new technologies—software and models

Based on evidence, undifferentiated cells with a potential for prolonged self-renewal and ability to act as precursors of at least one type of mature, differentiated cells can be operationally considered as a stem cell (Weissman, 2000). To alter its fate, stem cells need to differentiate, which requires interplay between molecular signals, transcriptional changes, metabolic reprogramming, and cell communication (Ito & Ito, 2016; Tatapudy et al., 2017). For example, development of central nervous system (CNS) begins from neural stem cells (NSCs), which acquire molecular signals from Wnt signaling pathways and relays them to downstream TFs such as TCFs/LEFs (T cell/lymphoid enhancer factors), HES (hairy and enhancer of split paralogs), Myc, NeuroD, and SOX, resulting in neuronal and astroglial differentiation, specification, and migration (Wen et al., 2009). Interestingly, oligodendrocyte differentiation and neuronal commitment are inhibited by Wnt-Notch-BMP signaling pathways. Again, metabolic triggers such as reactive oxygen species promote proliferation and differentiation of neural lineage cells (Le Belle et al., 2011; Tatapudy et al., 2017). Metabolic genes such as PGC1, SIRT1, and FoxO3 are known to promote mitochondrial biogenesis and homeostasis during neural differentiation (Hisahara et al., 2008; Renault et al., 2009). Similarly, development of ectodermal organs requires epithelial—mesenchymal interplay involving BMP, Notch, Wnt signaling, and transcriptional circuits (Jimenez-Rojo et al., 2012). Owing to its stochastic nature and spatial interactions, studying stem cell differentiation and lineage commitments requires high-throughput experiments, rigorous computational analysis, mathematical modeling, and a defined systems biology approach Fig. 19.2 (Jangid et al., 2021; Khorasani et al., 2020; Macarthur et al., 2009). In the past two decades, microarray technology followed by sequencing and proteomics has enormously improved the understanding of stem cells and their fates postdifferentiation (Ahn et al., 2010; Chang et al.,

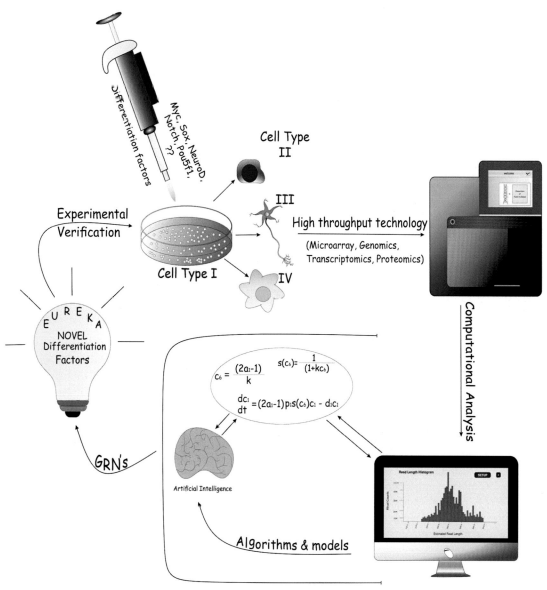

FIGURE 19.2 Workflow of stem cell differentiation research using high-throughput sequencing, computation and mathematical models, and systems biology.

2006; Ramalho-Santos et al., 2002). Deep sequencing—based mapping of genome-wide chromatin interactions revealed chromatin architecture alterations during lineage commitment and differentiation programs (Gorkin et al., 2014). Approximately 36% alterations were observed between inactive and active chromatin compartments with allelic bias for linked promoters and distal enhancers providing information on spatial regulation of the genome during differentiation of stem cells (Dixon et al., 2015). Alternately, functional genomics coupled with sequencing revealed key genes in pathways involving stem cell behaviour, renewal, neural development, and neural and ES cell differentiation (Naxerova et al., 2021; Wang et al., 2019). Recently, advanced single-cell transcriptomics has become a powerful tool for understanding cellular heterogeneity, tissue-specific stem cell niche, and altered GRNs during various stages of development (Chen et al., 2020). New stem cell identification in various tissues and diseases such as lungs, uterus, brain, and cancers became possible via single-cell sequencing techniques (Bandler et al., 2022; Fan et al., 2020; Hou et al., 2016; Treutlein et al., 2014; Wu et al., 2017). Moreover, single-cell sequencing of differentiating iPSCs, ES cells and ASCs revealed that cellular environment, epigenetic modifications, and GRNs modulate genetic effects on expression (Cuomo et al., 2020; Jang et al., 2017). Generation of large data from sequencing techniques imputes a computational burden, which necessities the era for big data—based analysis, machine learning using neural networks, and artificial intelligence

(Ashraf et al., 2021; Del Solet al., 2017; Kusumoto & Yuasa, 2019). Various computational tools used till date for stem cell biology and therapeutics are listed in Table 19.1.

Various mathematical models have been developed for understanding stem cell biology. For example, the Shea—Ackers formalism regarding gene expression dependency on transcriptional machinery architecture was one of the pioneering mathematical models that was applied to study transcriptional bias during stem cell differentiation (Duff et al., 2012; Shea & Ackers, 1985). Various mathematical models based on stem cell differentiation, renewal, proliferation, and cell fate came into existence with advancement in the field (Thomas & Anna, 2011). The Marciniak-Czochra model for stem cell renewal and differentiation, which involves the stochastic stability parameter, is more efficient than older models in calculating extrinsic and intrinsic noise during differentiation (Marciniak-Czochra et al., 2009). Currently, varied nature of stem cells and their differentiation has given rise to newer models accounting for epigenetic changes, cell homing, migration, and oncogenic dedifferentiation (Jilkine & Gutenkunst, 2014; Maric et al., 2022; Situ & Lei, 2017). Broadening our views on big data interpretation, mathematical model simulations, and systems biology has connected (1) pathways and networks with topological states during differentiation, (2) low dimensionality to higher dimensionality (handful of genes to a genome-wide regulatory networks of thousands of genes), and (3) homogeneity to nongenetic heterogeneity (Cahan et al., 2021). More recent approaches in stem cell engineering using systems biology pipelines have identified stem cell niche determinants and novel GRNs with high application potential (Kinney et al., 2019; Ravichandran et al., 2016).

3. Rise of gene regulatory networks: nodes, motifs, gene batteries, and mathematical models

With the seminal work of Yamanaka regarding core transcription factors and iPSCs, one thing became very clear that regulatory networks of genes are the mastermind behind all forms of differentiation. Classically, GRNs are defined as "network that has been inferred from gene expression data which provides a casual map of molecular interactions" (Davidson & Levin, 2005; Emmert-Streib et al., 2014). In the past decade, the concept of GRNs in stem cell differentiation has grown from its infancy into a subject connecting big data analysis, prediction-based AI, and precision medicine (Emmert-Streib et al., 2014; Gerard et al., 2017; Peter & Davidson, 2016). GRNs are a combination of nodes and edges that allow cells to analyze and process commands from the microenvironment (Tewary et al., 2018). Nodes represent the genes and their cis-regulatory control systems, whereas edges represent the interactive relationships (Wolf et al., 2021). To delineate a GRN, the first step is to identify the nodes. For a gene, it refers to the DNA sequences such as promoters, enhancers, etc. that control gene expression. For the regulators, nodes may consist of regulatory TFs, RNA-binding proteins, and additional effector molecules (Fig. 19.3A) (Walhout, 2011). Input signals that trigger GRNs often involve epigenetic alterations such as histone modification and DNA methylation, which can fine-tune the cellular GRN structure by altering active transcription sites via promoter—enhancer looping (Reyes-Palomarcs ct al., 2020). Genetic alterations caused by gene duplication and mutational events often reprogram GRN structure. For example, gene duplication or point mutations of promoters/enhancers often result in strong transcriptional activation of downstream genes resulting in prolonged signals that might affect cellular fates. However, GRN-modulated cell fates and functional states are tightly regulated and adapt to environmental stress signals involving nutrients, salinity, radiations, and temperature so that final outcome of the developmental pathway remains unaltered (Wolf et al., 2021).

Logically, larger GRNs regulating cell fates are based on simpler and distinct network motifs involving a few regulatory nodes and edges. Network motifs represent nonrandom patterns of interconnections and are broadly classified into the following: (1) autoregulatory/feedback loop motif where a gene/node is capable of regulating its expression, (2) multicomponent/circular feedback loop motif where regulators act upon each other via feedback loop, (3) single-input/coregulatory motifs where master regulator controls multiple nodes, (4) multiinput/copointing motif represents regulators that share a group of common targets, (5) Dense Overlapping region/Complex motif is built out of functionally associated genes that might interact physically to achieve regulatory functions, and (6) regulatory chain/feedforward motif consists of only hierarchical regulatory nodes acting upon specific signals (Fig. 19.3B) (Defoort et al., 2018). The construction of a potential GRN comprising of various motif modules in context of cell differentiation is shown in Fig. 19.3C. Topology of these network motifs governs the functionality of a GRN as per their favorable or incompatible states and is often called subcircuits. In fact, Davidson and colleagues described seven canonical subcircuit topologies/logic gates made up of network motif, namely, (1) double-negative gate/signal-mediated toggle switch, (2) inductive signaling, (3) reciprocal repression, (4) spatial repression, (5) negative feedback, (6) intercellular feedback on the ligand gene, and (7) AND logic operation, employed by various GRNs to achieve differentiation during

TABLE 19.1 List of computational tools used in stem cell biology research.

Classifier type	Problems in experimental approach	Tools	Purpose	PMID
Bulk/population classifiers	Improvement of Direct differentiation/cell conversion technology	D'Alessio	Suggests transcription factors for cell differentiation	26603904
		Mogrify	Predicts putative transcription factors from ~300 cell and tissue type. (www.mogrify.net)	26780608
		Heinaniemi	Creates 3D epigenetic landscapes or heatmap to provide a better understanding of transcriptional regulation. (http://trel.systems.biology.net)	23603899
		Lang	Identifies transcription factors from 63 cell type and ~1337 transcription factors of mouse	25122086
		Crespo	Hierarchically arranges predefined GRNs to predict core transcription factors for cell conversion	23873656
		Davis	Recognizes transcription factors from human and mouse databases	23650565
	Measurement of molecular similarity	Pluritest	Checks pluripotency of cells based on gene expression profiles (http://www.pluritest.org)	21378979
		CellNet	Quantifies similarity between engineered population and the target cell type (http://cellnet.hms.harvard.edu)	25126793
		TeratoScore	Predicts pluripotency based on gene expression on teratomas (http://benvenisty.huji.ac.il/teratoscore.php)	26070610
		KeyGenes	Predicts similarity of engineered cells to its in vivo equivalents (www.keygenes.nl/)	26028532
	Linage bias problem in fate determination	Scorecard	Predicts the differential tendency of pluripotent cell lines (http://scorecard.computational-epigenetics.org/)	21295703
Single-cell classifiers	Comparative analysis with their in vivo counterparts	SCmap	Helps project scRNA-seq data onto other cell types from different experiments	29608555
		SingleCellNet	Helps generate quantitative classification of scRNA-seq data in different species	31377170
		RaceID	Identifies cell type from single-cell RNA sequencing data	26287467
	Trajectory identification problem in cell fate prediction	Wanderlust	Aligns single cells in trajectory according to their intermediate course	24766814
		Monocle	Using scRNA-seq data taken at multiple time points, it increases the temporal resolution of single cell transcriptome dynamics	26430159
		Moana	Creates classifiers from heterogenous scRNA-seq databases	DOI: 10.1101/456129
		scClassify	Identifies and classifies cell type based on published scRNA-seq databases	DOI: 10.1101/776948
		scID	Recognizes transcriptionally similar cell type among datasets	32151972
		FateID	Derives lineage biasness to predict different cell fate choices in a cell population	29630061
		StemID	Infers the trajectory of a cell from mass spectrometry or single-cell RNA sequencing data	27345837
		Cellassign	Automatically assign cells in a highly scalable manner throughout databases	31501550
		Capybara	Reveals cell identity and identifies hybrid states of cells to enhance efficiency of cell fate engineering	35354062

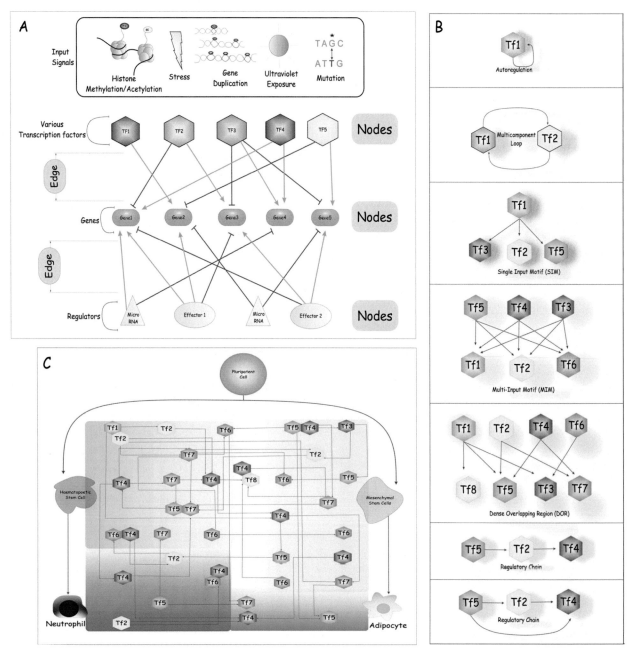

FIGURE 19.3 (A) Various components of a GRN (gene regulatory network) used in cellular signaling, (B) classes of network motifs, (C) network motifs in construction of a GRN for cell differentiation.

development (Peter & Davidson, 2009). Subcircuits in differentiation often regulate a battery of genes in a coordinated manner (Davidson, 2010). Gene batteries are composed of a series of transcriptional switches that are regulated by driver genes and specific GRNs in a signal-dependent fashion. A structured gene battery is often governed by topology of regulatory relationships depending upon the differential expression of genes/nodes as per the tissue specificity (Fig. 19.4).

Understanding mathematical modelling and associated variables that affect the GRN-based prediction models of differentiation is important for better cell conversion technology. Nonetheless, stem cell mathematical models for differentiation are broadly categorized as deterministic, stochastic, or hybrid models (Wu et al., 2013). The deterministic model is based on stimuli-driven transitions of stem cells where fluctuations in cell fates are generally attributed to the culture environment and nonsynchronous nature of the cell population. Mostly ordinary differential equations (ODEs) or partial

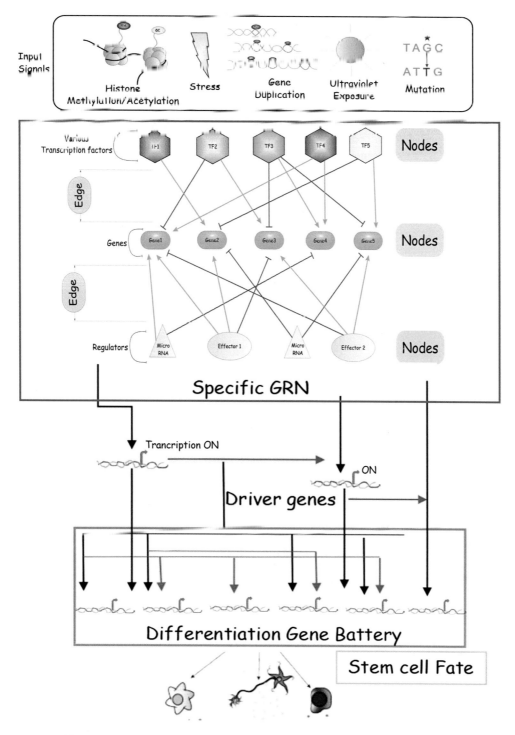

FIGURE 19.4   Structure of a gene battery driven by signal-driven GRNs during differentiation.

differential equations (PDEs) are used to fit the trajectory of the cells to a particular fate in the deterministic model. While the stochastic model uses probability functions and includes various factors that may influence the differentiation process without directly being part of the GRNs, a more advanced hybrid model uses various mathematical, experimental states and machine learning approaches to understand differentiation-associated GRNs at various states of their pluripotency. Again multiscale computational models use four different classes, namely, descriptive model, logical model, continuous model, and probabilistic model (Del Sol & Jung, 2021). The descriptive models are based on cell—cell interaction and prioritize functional modules based on validation experiments. Directed graph-based descriptive models are appropriate for identification of cellular communication that maintains cell states during differentiation or conversion. Logical models use Boolean networks, which are based on coexpression modules and can accurately predict GRN nodes (TFs) required for cell conversions. They usually involve regulatory interactions between many TFs derived from experiments and accounts for TF cooperativity independent of kinetic parameters. An improved version called probabilistic Boolean network (PBN) was developed to account for lack of experimental evidence or multifunctional regulation by an entity (Shmulevich et al., 2002). The PBN generates a sequence of global states that constitute a Markov chain series, thus enabling us to match the predictions with previous data. Continuous models use differential equation suitable for prediction of dynamic behavior of a differentiating cell. The probabilistic model predicts TF combinations for reprogramming experiments and can estimate reprogramming efficiency. It uses a combination of dynamic Bayesian networks and PBNs adequate to stimulate noise and include regulatory interactions in cell system.

4. Conclusion: future of stem cell engineering applications

Currently, GRN-based predictions are widely used for stem cell engineering and stem cell therapeutic purposes (Kinney et al., 2019). Knowledge of master regulatory genes for constructing perfect GRNs will increase our perfection on engineered stem cells. For example, a GRN pattern elucidating the mesoendodermal and ectodermal designation of embryonic stem cells (mESCs) in mice was identified (Shu et al., 2013). Here, the balanced expression of POU5F1, i.e., mesoendodermal cell-fate determinant, and SOX2, an ectodermal cell-fate determinant, retains the pluripotent state. While perturbations in either of them prompt differentiation into the corresponding lineage, replacement of POU5F1 by other mesoendodermal cell-fate determinant promoted pluripotency in fibroblasts (Montserrat et al., 2013; Shu et al., 2013). Applying GRNs and computational tools to predict cell conversions for regenerative medicine and transplantation remains a challenge. The shortcomings of machine learning when compared to mechanistic models include prediction failures due to complexities of biological systems. Moreover, vast data for statistical significance prevents it from understanding complex biological processes. Traditional strategies for improving donor stem cells and autologous iPSCs with better immune tolerance often do not achieve high cell conversion efficiency. Currently, GRN-based predictions with organized tissue-specific TF repertoires are enabling optimization of conversion factors for tissue engineering and regenerative medicines. For example, regeneration of functional epidermis using programmed keratinocytes in patients with junctional epidermolysis bullosa revealed successful skin regeneration (Hirsch et al., 2017). Stem cell rejuvenation strategies that can cure degenerative diseases caused by loss of their renewal property have recently been modified with computational approaches. GRNs regulating stem cell−niche interactions are being used to identify niche-specific events, which can be reintroduced to activate stem cell renewal process (Kalamakis et al., 2019). Importantly, GRN-specific cues to rejuvenate neural stem cells in diseases such as Alzheimer's are currently under trial (Karvelas et al., 2022; Zhang et al., 2022). We hope a closer collaboration between computational and experimental stem cell researchers will result in more fruitful therapeutics.

Acknowledgments

Apologies to those whose related publications were not cited due to space limitations. The authors acknowledge Bose Institute, Kolkata, and the Department of Bioinformatics and Applied Sciences, Indian Institute of Information Technology, Allahabad, for providing opportunity for the work. CorelDRAW and Adobe Photoshop software packages were used for creating the figures. N.S. was funded by DBT−Ramalingaswami Fellowship (Award sanction No: BT/RLF/Reentry/27/2016) from the Department of Biotechnology, India, and SERB-DST (Award sanction No: ECR/2018/000595) from the Department of Science and Technology, India.

Author's contribution

A.D., S.M., A.G., and P.C. drafted the manuscript. N.S., PKV, and S.S. were responsible for conceptualization, methodology, resources, supervision, writing, and editing. S.M., A.D., and N.S. contributed the tables and figures. All authors contributed in writing and editing the chapter.

References

Ahn, S. M., Simpson, R., & Lee, B. (2010). Genomics and proteomics in stem cell research: The road ahead. *Analytical Cell Biology, 43*(1), 1−14. https://doi.org/10.5115/acb.2010.43.1.1

Alvarez, C. V., Garcia-Lavandeira, M., Garcia-Rendueles, M. E., Diaz-Rodriguez, E., Garcia-Rendueles, A. R., Perez-Romero, S., ... Bravo, S. B. (2012). Defining stem cell types: Understanding the therapeutic potential of ESCs, ASCs, and iPS cells. *Journal of Molecular Endocrinology, 49*(2), R89−R111. https://doi.org/10.1530/JME-12-0072

Andrews, P. W. (2002). From teratocarcinomas to embryonic stem cells. *Philosophical Transactions of the Royal Society of London B Biological Sciences, 357*(1420), 405—417. https://doi.org/10.1098/rstb.2002.1058

Ashraf, M., Khalilitousl, M., & Laksman, Z. (2021). Applying machine learning to stem cell culture and differentiation. *Current Protocol, 1*(9), e261 https://doi.org/10.1002/cpz1.261

Dandler, R. C., Vitali, I., Delgado, R. N., Ho, M. C., Dvoretskova, E., Iharra Molinas, J. S., … Mayer, C. (2022). Single-cell delineation of lineage and genetic identity in the mouse brain. *Nature, 601*(7893), 404—409. https://doi.org/10.1038/s41586-021-04237-0

Bcckcr, A. J., Mc, C. E., & Till, J. E. (1963). Cytological demonstration of the clonal nature of spleen colonies derived from transplanted mouse marrow cells. *Nature, 197*, 452—454. https://doi.org/10.1038/197452a0

Bolli, R., Haro, J. M., March, K. L., Pepine, C. J., Willerson, J. T., Perin, E. C., .,, Cardiovascular Cell Therapy Research, N. (2018). Rationale and design of the CONCERT-HF trial (combination of mesenchymal and c-kit(+) cardiac stem cells as regenerative therapy for heart failure). *Circulation Research, 122*(12), 1703—1715. https://doi.org/10.1161/CIRCRESAHA.118.312978

Bradley, A., Evans, M., Kaufman, M. H., & Robertson, E. (1984). Formation of germ-line chimaeras from embryo-derived teratocarcinoma cell lines. *Nature, 309*(5965), 255—256. https://doi.org/10.1038/309255a0

Cahan, P., Cacchiarelli, D., Dunn, S. J., Hemberg, M., de Sousa Lopes, S. M. C., Morris, S. A., … Wells, C. A. (2021). Computational stem cell biology: Open questions and guiding principles. *Cell Stem Cell, 28*(1), 20—32. https://doi.org/10.1016/j.stem.2020.12.012

Chang, H. Y., Thomson, J. A., & Chen, X. (2006). Microarray analysis of stem cells and differentiation. *Methods in Enzymology, 420*, 225—254. https://doi.org/10.1016/S0076-6879(06)20010-7

Chen, T., Li, J., Jia, Y., Wang, J., Sang, R., Zhang, Y., & Rong, R. (2020). Single-cell sequencing in the field of stem cells. *Current Genomics, 21*(8), 576—584. https://doi.org/10.2174/1389202921999200624154445

Cuomo, A. S. E., Seaton, D. D., McCarthy, D. J., Martinez, I., Bonder, M. J., Garcia-Bernardo, J., … Stegle, O. (2020). Single-cell RNA-sequencing of differentiating iPS cells reveals dynamic genetic effects on gene expression. *Nature Communications, 11*(1), 810. https://doi.org/10.1038/s41467-020-14457-z

Davidson, E. H. (2010). Emerging properties of animal gene regulatory networks. *Nature, 468*(7326), 911—920. https://doi.org/10.1038/nature09645

Davidson, E., & Levin, M. (2005). Gene regulatory networks. *Proceedings of the National Academy of Sciences of the United States of America, 102*(14), 4935. https://doi.org/10.1073/pnas.0502024102

Defoort, J., Van de Peer, Y., & Vermeirssen, V. (2018). Function, dynamics and evolution of network motif modules in integrated gene regulatory networks of worm and plant. *Nucleic Acids Research, 46*(13), 6480—6503. https://doi.org/10.1093/nar/gky468

Deinsberger, J., Reisinger, D., & Weber, B. (2020). Global trends in clinical trials involving pluripotent stem cells: A systematic multi-database analysis. *NPJ Regenerative Medicine, 5*, 15. https://doi.org/10.1038/s41536-020-00100-4

Del Sol, A., & Jung, S. (2021). The importance of computational modeling in stem cell research. *Trends in Biotechnology, 39*(2), 126—136. https://doi.org/10.1016/j.tibtech.2020.07.006

Del Sol, A., Thiesen, H. J., Imitola, J., & Carazo Salas, R. E. (2017). Big-data-driven stem cell science and tissue engineering: Vision and unique opportunities. *Cell Stem Cell, 20*(2), 157—160. https://doi.org/10.1016/j.stem.2017.01.006

Dixon, J. R., Jung, I., Selvaraj, S., Shen, Y., Antosiewicz-Bourget, J. E., Lee, A. Y., … Ren, B. (2015). Chromatin architecture reorganization during stem cell differentiation. *Nature, 518*(7539), 331—336. https://doi.org/10.1038/nature14222

Duff, C., Smith-Miles, K., Lopes, I.,, & Tian, T. (2012). Mathematical modelling of stem cell differentiation: The PU.1-GATA-1 interaction. *Journal of Mathematical Biology, 64*(3), 449—468. https://doi.org/10.1007/s00285-011-0419-3

Emmert-Streib, F., Dehmer, M., & Haibe-Kains, B. (2014). Gene regulatory networks and their applications: Understanding biological and medical problems in terms of networks. *Frontiers in Cell and Developmental Biology, 2*, 38. https://doi.org/10.3389/fcell.2014.00038

Evans, M. J., & Kaufman, M. H. (1981). Establishment in culture of pluripotential cells from mouse embryos. *Nature, 292*(5819), 154—156. https://doi.org/10.1038/292154a0

Fan, X., Fu, Y., Zhou, X., Sun, L., Yang, M., Wang, M., … Tang, F. (2020). Single-cell transcriptome analysis reveals cell lineage specification in temporal-spatial patterns in human cortical development. *Science Advances, 6*(34), eaaz2978. https://doi.org/10.1126/sciadv.aaz2978

Gerard, C., Tys, J., & Lemaigre, F. P. (2017). Gene regulatory networks in differentiation and direct reprogramming of hepatic cells. *Seminars in Cell and Developmental Biology, 66*, 43—50. https://doi.org/10.1016/j.semcdb.2016.12.003

Gorkin, D. U., Leung, D., & Ren, B. (2014). The 3D genome in transcriptional regulation and pluripotency. *Cell Stem Cell, 14*(6), 762—775. https://doi.org/10.1016/j.stem.2014.05.017

Hirsch, T., Rothoeft, T., Teig, N., Bauer, J. W., Pellegrini, G., De Rosa, L., … De Luca, M. (2017). Regeneration of the entire human epidermis using transgenic stem cells. *Nature, 551*(7680), 327—332. https://doi.org/10.1038/nature24487

Hisahara, S., Chiba, S., Matsumoto, H., Tanno, M., Yagi, H., Shimohama, S., … Horio, Y. (2008). Histone deacetylase SIRT1 modulates neuronal differentiation by its nuclear translocation. *Proceedings of the National Academy of Sciences of the United States of America, 105*(40), 15599—15604. https://doi.org/10.1073/pnas.0800612105

Hoang, D. M., Pham, P. T., Bach, T. Q., Ngo, A. T. L., Nguyen, Q. T., Phan, T. T. K., … Nguyen, L. T. (2022). Stem cell-based therapy for human diseases. *Signal Transduction and Targeted Therapy, 7*(1), 272. https://doi.org/10.1038/s41392-022-01134-4

Hou, Y., Guo, H., Cao, C., Li, X., Hu, B., Zhu, P., … Peng, J. (2016). Single-cell triple omics sequencing reveals genetic, epigenetic, and transcriptomic heterogeneity in hepatocellular carcinomas. *Cell Research, 26*(3), 304—319. https://doi.org/10.1038/cr.2016.23

Illmensee, K., & Mintz, B. (1976). Totipotency and normal differentiation of single teratocarcinoma cells cloned by injection into blastocysts. *Proceedings of the National Academy of Sciences of the United States of America, 73*(2), 549—553. https://doi.org/10.1073/pnas.73.2.549

Ito, K., & Ito, K. (2016). Metabolism and the control of cell fate decisions and stem cell renewal. *Annual Review of Cell and Developmental Biology, 32,* 399–409. https://doi.org/10.1146/annurev-cellbio-111315-125134

Jang, S., Choubey, S., Furchtgott, L., Zou, L. N., Doyle, A., Menon, V., … Ramanathan, S. (2017). Dynamics of embryonic stem cell differentiation inferred from single-cell transcriptomics show a series of transitions through discrete cell states. *Elife, 6.* https://doi.org/10.7554/eLife.20487

Jangid, A., Selvarajan, S., & Ramaswamy, R. (2021). A stochastic model of homeostasis: The roles of noise and nuclear positioning in deciding cell fate. *iScience, 24*(10), 103199. https://doi.org/10.1016/j.isci.2021.103199

Jilkine, A., & Gutenkunst, R. N. (2014). Effect of dedifferentiation on time to mutation acquisition in stem cell-driven cancers. *PLoS Computational Biology, 10*(3), e1003481. https://doi.org/10.1371/journal.pcbi.1003481

Jimenez-Rojo, L., Granchi, Z., Graf, D., & Mitsiadis, T. A. (2012). Stem cell fate determination during development and regeneration of ectodermal organs. *Frontiers in Physiology, 3,* 107. https://doi.org/10.3389/fphys.2012.00107

Kalamakis, G., Brune, D., Ravichandran, S., Bolz, J., Fan, W., Ziebell, F., … Martin-Villalba, A. (2019). Quiescence modulates stem cell maintenance and regenerative capacity in the aging brain. *Cell, 176*(6), 1407–1419. https://doi.org/10.1016/j.cell.2019.01.040. e1414.

Karvelas, N., Bennett, S., Politis, G., Kouris, N. I., & Kole, C. (2022). Advances in stem cell therapy in alzheimer's disease: A comprehensive clinical trial review. *Stem Cell Investigation, 9,* 2. https://doi.org/10.21037/sci-2021-063

Khorasani, N., Sadeghi, M., & Nowzari-Dalini, A. (2020). A computational model of stem cell molecular mechanism to maintain tissue homeostasis. *PLoS One, 15*(7), e0236519. https://doi.org/10.1371/journal.pone.0236519

Kinney, M. A., Vo, L. T., Frame, J. M., Barragan, J., Conway, A. J., Li, S., … Daley, G. Q. (2019). A systems biology pipeline identifies regulatory networks for stem cell engineering. *Nature Biotechnology, 37*(7), 810–818. https://doi.org/10.1038/s41587-019-0159-2

Kusumoto, D., & Yuasa, S. (2019). The application of convolutional neural network to stem cell biology. *Inflammation and Regeneration, 39,* 14. https://doi.org/10.1186/s41232-019-0103-3

Le Belle, J. E., Orozco, N. M., Paucar, A. A., Saxe, J. P., Mottahedeh, J., Pyle, A. D., … Kornblum, H. I. (2011). Proliferative neural stem cells have high endogenous ROS levels that regulate self-renewal and neurogenesis in a PI3K/Akt-dependant manner. *Cell Stem Cell, 8*(1), 59–71. https://doi.org/10.1016/j.stem.2010.11.028

Lim, S. K., & Khoo, B. Y. (2021). An overview of mesenchymal stem cells and their potential therapeutic benefits in cancer therapy. *Oncology Letters, 22*(5), 785. https://doi.org/10.3892/ol.2021.13046

Macarthur, B. D., Ma'ayan, A., & Lemischka, I. R. (2009). Systems biology of stem cell fate and cellular reprogramming. *Nature Reviews Molecular Cell Biology, 10*(10), 672–681. https://doi.org/10.1038/nrm2766

Marciniak-Czochra, A., Stiehl, T., Ho, A. D., Jager, W., & Wagner, W. (2009). Modeling of asymmetric cell division in hematopoietic stem cells–regulation of self-renewal is essential for efficient repopulation. *Stem Cells and Development, 18*(3), 377–385. https://doi.org/10.1089/scd.2008.0143

Maric, D. M., Velikic, G., Maric, D. L., Supic, G., Vojvodic, D., Petric, V., & Abazovic, D. (2022). Stem cell homing in intrathecal applications and inspirations for improvement paths. *International Journal of Molecular Sciences, 23*(8). https://doi.org/10.3390/ijms23084290

Martin, G. R. (1981). Isolation of a pluripotent cell line from early mouse embryos cultured in medium conditioned by teratocarcinoma stem cells. *Proceedings of the National Academy of Sciences of the United States of America, 78*(12), 7634–7638. https://doi.org/10.1073/pnas.78.12.7634

Mendiola, M. M., Peters, T., Young, E. W., & Zoloth-Dorfman, L. (1999). Research with human embryonic stem cells: Ethical considerations. By geron ethics advisory board. *Hastings Center Report, 29*(2), 31–36. Retrieved from: https://www.ncbi.nlm.nih.gov/pubmed/10321340.

Montserrat, N., Nivet, E., Sancho-Martinez, I., Hishida, T., Kumar, S., Miquel, L., … Izpisua Belmonte, J. C. (2013). Reprogramming of human fibroblasts to pluripotency with lineage specifiers. *Cell Stem Cell, 13*(3), 341–350. https://doi.org/10.1016/j.stem.2013.06.019

Naxerova, K., Di Stefano, B., Makofske, J. L., Watson, E. V., de Kort, M. A., Martin, T. D., … Elledge, S. J. (2021). Integrated loss- and gain-of-function screens define a core network governing human embryonic stem cell behavior. *Genes and Development, 35*(21–22), 1527–1547. https://doi.org/10.1101/gad.349048.121

Nguyen, P. K., Nag, D., & Wu, J. C. (2010). Methods to assess stem cell lineage, fate and function. *Advanced Drug Delivery Reviews, 62*(12), 1175–1186. https://doi.org/10.1016/j.addr.2010.08.008

Peter, I. S., & Davidson, E. H. (2009). Modularity and design principles in the sea urchin embryo gene regulatory network. *FEBS Letters, 583*(24), 3948–3958. https://doi.org/10.1016/j.febslet.2009.11.060

Peter, I. S., & Davidson, E. H. (2016). Implications of developmental gene regulatory networks inside and outside developmental biology. *Current Topics in Developmental Biology, 117,* 237–251. https://doi.org/10.1016/bs.ctdb.2015.12.014

Ramalho-Santos, M., Yoon, S., Matsuzaki, Y., Mulligan, R. C., & Melton, D. A. (2002). Stemness: Transcriptional profiling of embryonic and adult stem cells. *Science, 298*(5593), 597–600. https://doi.org/10.1126/science.1072530

Ravichandran, S., Okawa, S., Martinez Arbas, S., & Del Sol, A. (2016). A systems biology approach to identify niche determinants of cellular phenotypes. *Stem Cell Research, 17*(2), 406–412. https://doi.org/10.1016/j.scr.2016.09.006

Renault, V. M., Rafalski, V. A., Morgan, A. A., Salih, D. A., Brett, J. O., Webb, A. E., … Brunet, A. (2009). FoxO3 regulates neural stem cell homeostasis. *Cell Stem Cell, 5*(5), 527–539. https://doi.org/10.1016/j.stem.2009.09.014

Reyes-Palomares, A., Gu, M., Grubert, F., Berest, I., Sa, S., Kasowski, M., … Zaugg, J. B. (2020). Remodeling of active endothelial enhancers is associated with aberrant gene-regulatory networks in pulmonary arterial hypertension. *Nature Communications, 11*(1), 1673. https://doi.org/10.1038/s41467-020-15463-x

Rodolfa, K. T., & Eggan, K. (2006). A transcriptional logic for nuclear reprogramming. *Cell, 126*(4), 652–655. https://doi.org/10.1016/j.cell.2006.08.009

Shea, M. A., & Ackers, G. K. (1985). The OR control system of bacteriophage lambda. A physical-chemical model for gene regulation. *Journal of Molecular Biology, 181*(2), 211–230. https://doi.org/10.1016/0022-2836(85)90086-5

Shmulevich, I., Dougherty, E. R., Kim, S., & Zhang, W. (2002). Probabilistic boolean networks: A rule based uncertainty model for gene regulatory networks. *Bioinformatics, 18*(2), 261−274. https://doi.org/10.1093/bioinformatics/18.2.261

Shu, J., Wu, C., Wu, Y., Li, Z., Shao, S., Zhao, W., ... Deng, H. (2013). Induction of pluripotency in mouse somatic cells with lineage specifiers. *Cell, 153*(5), 963−975. https://doi.org/10.1016/j.cell.2013.05.001

Siminovitch, L., McCulloch, E. A., & Till, J. E. (1963). The distribution of colony-forming cells among spleen colonies. *Journal of Cellular and Comparative Physiology, 62*, 327−336. https://doi.org/10.1002/jcp.1030620313

Situ, Q., & Lei, J. (2017). A mathematical model of stem cell regeneration with epigenetic state transitions. *Mathematical Biosciences and Engineering, 14*(5−6), 1379−1397. https://doi.org/10.3934/mbe.2017071

Takahashi, K., & Yamanaka, S. (2006). Induction of pluripotent stem cells from mouse embryonic and adult fibroblast cultures by defined factors. *Cell, 126*(4), 663−676. https://doi.org/10.1016/j.cell.2006.07.024

Tatapudy, S., Aloisio, F., Barber, D., & Nystul, T. (2017). Cell fate decisions: Emerging roles for metabolic signals and cell morphology. *EMBO Reports, 18*(12), 2105−2118. https://doi.org/10.15252/embr.201744816

Tewary, M., Shakiba, N., & Zandstra, P. W. (2018). Stem cell bioengineering: Building from stem cell biology. *Nature Reviews Genetics, 19*(10), 595−614. https://doi.org/10.1038/s41576-018-0040-z

Thomas, S., & Anna, M.-C. (2011). Characterization of stem cells using mathematical models of multistage cell lineages. *Mathematical and Computer Modelling, 53*(7), 1505−1517. https://doi.org/10.1016/j.mcm.2010.03.057

Thomas, K. R., & Capecchi, M. R. (1987). Site-directed mutagenesis by gene targeting in mouse embryo-derived stem cells. *Cell, 51*(3), 503−512. https://doi.org/10.1016/0092-8674(87)90646-5

Thomson, J. A., Itskovitz-Eldor, J., Shapiro, S. S., Waknitz, M. A., Swiergiel, J. J., Marshall, V. S., & Jones, J. M. (1998). Embryonic stem cell lines derived from human blastocysts. *Science, 282*(5391), 1145−1147. https://doi.org/10.1126/science.282.5391.1145

Thomson, J. A., Kalishman, J., Golos, T. G., Durning, M., Harris, C. P., & Hearn, J. P. (1996). Pluripotent cell lines derived from common marmoset (*Callithrix jacchus*) blastocysts. *Biology of Reproduction, 55*(2), 254−259. https://doi.org/10.1095/biolreprod55.2.254

Treutlein, B., Brownfield, D. G., Wu, A. R., Neff, N. F., Mantalas, G. L., Espinoza, F. H., ... Quake, S. R. (2014). Reconstructing lineage hierarchies of the distal lung epithelium using single-cell RNA-seq. *Nature, 509*(7500), 371−375. https://doi.org/10.1038/nature13173

Trounson, A., & McDonald, C. (2015). Stem cell therapies in clinical trials: Progress and challenges. *Cell Stem Cell, 17*(1), 11−22. https://doi.org/10.1016/j.stem.2015.06.007

Walhout, A. J. (2011). Gene-centered regulatory network mapping. *Methods in Cell Biology, 106*, 271−288. https://doi.org/10.1016/B978-0-12-544172-8.00010-4

Wang, Z., Zhang, Y., Lee, Y. W., & Ivanova, N. B. (2019). Combining CRISPR/Cas9-mediated knockout with genetic complementation for in-depth mechanistic studies in human ES cells. *Biotechniques, 66*(1), 23−27. https://doi.org/10.2144/btn-2018-0115

Weissman, I. L. (2000). Stem cells: Units of development, units of regeneration, and units in evolution. *Cell, 100*(1), 157−168. https://doi.org/10.1016/s0092-8674(00)81692-x

Wen, S., Li, H., & Liu, J. (2009). Dynamic signaling for neural stem cell fate determination. *Cell Adhesion and Migration, 3*(1), 107−117. https://doi.org/10.4161/cam.3.1.7602

de Wert, G., & Mummery, C. (2003). Human embryonic stem cells: Research, ethics and policy. *Human Reproduction, 18*(4), 672−682. https://doi.org/10.1093/humrep/deg143

Wolf, I. R., Simoes, R. P., & Valente, G. T. (2021). Three topological features of regulatory networks control life-essential and specialized subsystems. *Scientific Reports, 11*(1), 24209. https://doi.org/10.1038/s41598-021-03625-w

Wu, B., An, C., Li, Y., Yin, Z., Gong, L., Li, Z., ... Zou, X. (2017). Reconstructing lineage hierarchies of mouse uterus epithelial development using single-cell analysis. *Stem Cell Reports, 9*(1), 381−396. https://doi.org/10.1016/j.stemcr.2017.05.022

Wu, J., Rostami, M. R., & Tzanakakis, E. S. (2013). Stem cell modeling: From gene networks to cell populations. *Current Opinion in Chemical Engineering, 2*(1), 17−25. https://doi.org/10.1016/j.coche.2013.01.001

Zakrzewski, W., Dobrzynski, M., Szymonowicz, M., & Rybak, Z. (2019). Stem cells: Past, present, and future. *Stem Cell Research and Therapy, 10*(1), 68. https://doi.org/10.1186/s13287-019-1165-5

Zhang, Q., Liu, J., Chen, L., & Zhang, M. (2022). Promoting endogenous neurogenesis as a treatment for alzheimer's disease. *Molecular Neurobiology.* https://doi.org/10.1007/s12035-022-03145-2

Zhu, Y., Huang, R., Wu, Z., Song, S., Cheng, L., & Zhu, R. (2021). Deep learning-based predictive identification of neural stem cell differentiation. *Nature Communications, 12*(1), 2614. https://doi.org/10.1038/s41467-021-22758-0

Chapter 20

Stem cell databases and tools: challenges and opportunities of computational biology

Basudha Banerjee[1,a], Pawan Kumar Raghav[1,a], Rajni Chadha[1], Aditya Raghav[1], Anugya Sengar[2] and Manisha Sengar[3]

[1]BioExIn, Delhi, India; [2]Army Hospital Research and Referral (R&R), Delhi, India; [3]Department of Zoology, Deshbandhu College, University of Delhi, Delhi, India

1. Introduction

The first bone marrow transplantation successfully carried out in 1956 proved the remarkable potential of stem cells in curing tissues/organs (Ezzone, 2009). Another breakthrough study revolutionized stem cell research in 2006 with the foundation of induced pluripotent stem cells (iPSCs) by Takahashi and Shinya Yamanaka using transcription factors (TFs) Oct4, Sox2, c-Myc, and Klf4. These TFs subsequently came to be known as the Yamanaka Factors (Takahashi et al., 2007). As time progressed, cell-based therapeutics such as Holocar and Alofisel were introduced to cure ocular burns and perianal fistulas in Crohn's disease (Abou-El-Enein et al., 2016; Verstockt et al., 2018). Since then, the utility of stem cells was never disregarded, and there has been continuous progress in stem cell research that has been fostering science in various perceptive ways by providing new potential approaches to cure a plethora of diseases.

Stem cells possess two remarkable attributes, namely, self-renewal and differentiation. Self-renewal has the potential to sustain the hematopoietic population by dividing and recreating itself, whereas differentiation generates multipotent progenitors capable of specializing into a more differentiated cell type. The self-renewal and differentiation characteristics of stem cells allow researchers to experiment for drug discovery (Liang et al., 2013), disease modeling (Ebert et al., 2009; Eiges et al., 2007), and regenerative science (Zaret & Grompe, 2008). Several genes and proteins such as c-Kit, SHP-1/2, PU.1, and GATA-1 are known to play an essential role in regulating these properties of stem cells and cancer stem cells (CSCs) (Kumar et al., 2020; Nandan et al., 2021; Raghav, 2022; Raghav & Gangenahalli, 2018; Raghav & Mann, 2021; Raghav et al., 2018a,b; Raghav, Mann, Ahlawat, et al., 2021; Raghav, Mann, Krishnakumar, et al., 2021; Raghav, Mann, Pandey, et al., 2021; Rawat et al., 2021; Sengar et al., 2022; Raghav et al., 2022). Multiple in silico studies have contributed toward delineating the role of essential genes and proteins in stem cells and CSCs (Alisha et al., 2012; Raghav & Gangenahalli, 2021; Raghav et al., 2011, 2012, 2018a,b; Raghav, Kumar, & Raghava, 2019; Raghav, Kumar, Kumar, et al., 2019). Stem cells can be classified as totipotent, pluripotent, multipotent, and unipotent. Transitioning from totipotency to unipotency, the differentiation potential gradually diminishes (Zakrzewski et al., 2019). Pluripotent cells can be sourced in the form of embryonic stem cells (ESCs) and iPSCs, while multipotent cells include a variety of adult stem cells (ASCs) (Zhou & Sears, 2018).

ESCs are acquired from the undifferentiated inner cell mass of blastocyst with high self-renewal capacity due to high telomerase expression (Flores et al., 2006). They have an indefinite life span with higher potency compared with ASCs, but their usage has always been controversial. ASCs are multipotent and have a broad yet restricted differential potential. The absence of ethical concerns makes them attractive candidates for research. Mesenchymal stem cells (MSCs), hematopoietic stem cells (HSCs), and neural stem cells (NSCs) are the commonly used ASCs. HSCs are multipotent stem cells that regulate the process of hematopoiesis by self-renewing and giving rise to myeloid-lineage and lymphoid-lineage

---

[a]Equal contribution.

Computational biology for Stem Cell Research. https://doi.org/10.1016/B978-0-443-13222-3.00032-0
Copyright © 2024 Elsevier Inc. All rights reserved, including those for text and data mining, AI training, and similar technologies.

committed progenitor cells. These progenitor cells become more differentiated and specialized to form mature blood cells. MSCs are stromal cells that exhibit remarkable ability to transform into various cell types within the mesenchymal lineage including adipocytes, osteocytes and chondrocytes. On account of rapid proliferation, simplicity of acquisition, easy procurement, and ease of transplantation, they became the initial choice of stem cells to be employed in clinical regenerative medicine (Javan et al., 2019). NSCs are also an invaluable source for cell-based therapies, especially for neurodegenerative disorders.

This chapter is focused on discussing the significance of computational biology in stem cell research. It emphasizes on the applications such as cell typing, lineage tracing, trajectory inference, and regulatory networks followed by machine learning (ML) algorithms and statistical methods. Further, it highlights the need for a harmonized data repository, advanced stem cell databases, and analysis tools.

Fig. 20.1 provides an overview of the applications of computational biology for stem cell research that are being used to analyze cell typing, the fidelity of cell fate engineering, lineage tracing, and stage cell states via trajectory inference; construct and infer regulatory networks to understand developmental trajectories; develop ML algorithms; and assess the significance through statistical methods such as gene regulatory networks. Fig. 20.2 is a representation of stem cell−rich databases, knowledge portal, and various bioinformatic tools. "Transcriptomic" refers to databases that host transcriptomic data; "genomic/transcriptomic" hosts next-generation sequencing (NGS) data, whole-genome sequencing (WGS), microarray, and RNA-Seq data; "epigenomic/transcriptomic" host ChIP-sequencing, miRNA and DNA methylation analysis data, and other gene expression pattern; "multiomic" refers to data extracted using multiple omic technologies that span proteomics, phosphoproteomics, or metabolomics in addition to the previously mentioned genomics, epigenomics, or transcriptomics data.

2. Applications of computational biology for stem cells research

Since 1952 when Alan Turing devised the first computer program to understand the effect of morphogens on embryonic patterning (Turing, 1990), computational tools have been coupled with developmental and stem cell biology. The

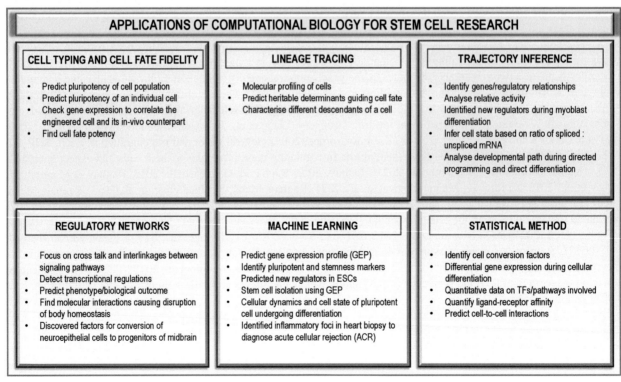

FIGURE 20.1 Applications of computational biology for stem cell research. Cell typing, cell fate fidelity, lineage tracing, trajectory inference, regulatory networks, machine learning, and statistical methods form the backbone of computational science in stem cell biology. Collectively, these fundamental components unveil the intricacies of cellular behavior, differentiation, and development, propelling progress in stem cell research and the exploration of therapeutic opportunities.

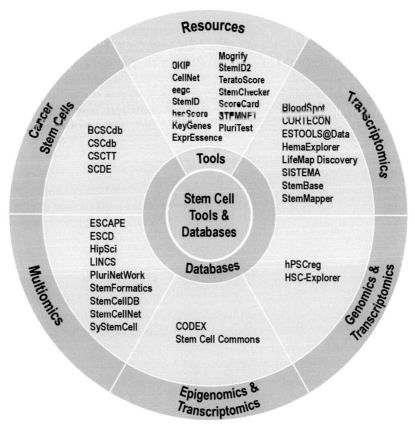

FIGURE 20.2 Representation of databases and tools based on the omic technology used to extract the data hosted. The figure has been classified as resources within the tools category, while transcriptomic, genomic/transcriptomic, epigenomic/transcriptomic, and multiomic within the databases category. Transcriptomics implies the databases focusing on gene expression data, which is essential for understanding gene activation and deactivation. Genomics and transcriptomics refer to databases that hold abundant genetic and gene expression data, playing a pivotal role in comprehending the genetic characteristics of organisms and how genes are expressed in various cellular conditions. Epigenomics and transcriptomics indicate databases that contain data concerning both the epigenctic modifications of genes and their expression profiles. Multiomics encompasses a broader range of data extracted especially using proteomics, or phosphoproteomics in addition to genomics epigenomics, or transcriptomics.

emergence of computational tools has revolutionized the biological field of study parallel to high-throughput technologies. Apart from designing experimental dynamics, it has also progressed to analyze big datasets. Decoding interactive networks, molecular functioning, and understanding multiple omic technologies have determined the various TFs involved in maintaining pluripotency. Algorithms used in computational biology use the data from omics. Subsequent technological advancements help capture gene expression at the cellular and molecular levels and provide a comprehensive understanding of cell morphology and behavior (Cahan et al., 2021).

Researchers have faced significant challenges in obtaining stem cell data and gene expression profiles (GEPs) across diverse lineages and organisms. These difficulties arise from (1) the sheer volume of data, (2) the lack of user-friendly data analysis tools, and (3) the heterogeneity of the available datasets (Pinto et al., 2018). An integrative approach that combines the available data and makes it concise and flexible to simplify the procurement of cell data is very crucial. This advances stem cell research by facilitating the identification and characterization of stem cells and their lineages (Pinto et al., 2018). Computational models allow easy analysis of statistical data and interrelated processes and provide new insights into research.

Complexity can be witnessed at the tiers of cell, tissue, organ, and organ system in all organisms. At the cellular level, cell differentiation and conversion are crucial factors regulated by various signaling molecules, growth factors, and TFs. At tissue levels, cell-to-cell communication is essential to understand tissue homeostasis (Del Sol & Jung, 2021). Computational studies further simplify the biological complexity at the cellular, tissue, and organ levels and find hidden variables responsible for biological processes. Fig. 20.1 represents an overview of the applications of computational stem cell biology. The following sub-sections highlight the applications of computational stem cell biology and ML in medical biotechnology.

2.1 Cell typing and cell fate fidelity

Initially, first-generation computational stem cell biology (CSCB) techniques, including Pluritest (Müller et al., 2011), CellNet (Cahan et al., 2014), KeyGenes (Roost et al., 2015), and ScoreCard (Bock et al., 2011), trained on bulk data predicted the pluripotency of an entire cell population. However, this bulk sequencing is only favorable sometimes since a cell population might comprise both pluripotent and differentiated cells, and profiling such groups of cells via first-generation techniques may lead to inaccurate results and low pluripotency score. Thus, more accurate second-generation methods based on single-cell RNA sequencing (scRNA-Seq) were introduced. scRNA-Seq judges a cell individually and categorizes it as pluripotent, multipotent, or differentiated cell. Single-cell data can be used to predict lineage relationships, which can further deduce cell fate. CellAssign, singleCellNet, and scPred are few cell typing methods available. Single cell/bulk sequencing approaches have also been used to predict the resemblance of and engineered cell to the target cell, i.e., the gene expression of the query sample can be compared to a trained dataset to find cell fate potency, but the extent to which synthetic cells will mimic the natural counterparts is challenging to understand (Neagu et al., 2020). This helps to determine the cell state and function; however, a better prediction of cell state and potency can be recovered by a detailed understanding of cell ancestry. Therefore, lineage tracing becomes essential for comprehending cell states and predicting the regulation of stem cell fate.

2.2 Lineage tracing

The goal of lineage tracing is to identify the individual single cell that initiated the lineage and understand the process through which cells differentiate, leading to the formation of distinct cell types. It also defines cell's fate and helps to determine the differentiation outcomes. Traditional lineage tracing tools provided restricted access to molecular profiling of cells. However, the advent of single-cell transcriptomics has enabled better prediction of the heritable determinants guiding cell fate and overcoming the limitations of traditional approaches. A lineage that fails to convert to a specific cell type is often because the regulatory molecule or TF could not effectively control the desired cell type. This could be due to the target genes being locked in heterochromatin, making them inaccessible for the regulatory molecules (Kamimoto et al., 2020; Soufi et al., 2012). This indicates that chromatin accessibility is crucial for cellular conversion and differentiation. Single-cell lineage tracing approaches having access to chromatin allow for finding hidden variables and a detailed understanding of cellular conversion factors, TFs, regulatory and physical interactions. Heritable marks thus generated help in the construction of phylogenetic trees (Cahan et al., 2021). Tracing tools offer valuable insights into cell fate potential and the differentiation of cell descendants. However, they fall short in providing a comprehensive view of the real-time dynamics within a biological process, including the specific cell states encountered during the process. Furthermore, they are limited in their ability to pinpoint intermediate stages in intricate biological processes like differentiation or reprogramming.

2.3 Trajectory inference

Cellular dynamics are stochastic in a biological system due to various external and internal cues. Trajectory inference (TI) gives an idea of the order of cells in a biological process and identifies genes or regulatory relationships involved during any phase of conversion, differentiation, or programming by pseudotemporal arrangement of cells. It considers the lineage relationships (Cahan et al., 2021). scRNA-Seq can be used by TI tools to analyze the relative activity of a dynamic biological process. Cells or groups of cells are placed on a topologically inferred trajectory, and the starting point, end point, and transient states are well depicted. Intermediate stages during lineage classification and subsequent branch points are predicted in a trajectory. Monocle, one of the earliest TI methods, has been used to identify new regulators during myoblast differentiation and analyze the similarity between the developmental path to form motor neurons, followed during directed programming and direct differentiation (Briggs et al., 2017). RNA velocity has been used to predict spliced and unspliced mRNA and infer the cell state based on the ratio of spliced and unspliced mRNA (Bergen et al., 2020; La Manno et al., 2018). The regulation and mechanism behind trajectories can be described by studying regulatory networks.

2.4 Regulatory networks

The answer to "how" cell states are changed and how multiple regulatory factors and mediators drive the cellular processes and decide cell fate is synchronized by the understanding of regulatory networks. Networks constitute genes, RNA, proteins, macromolecules, and other regulatory elements necessary for a living system. Transcriptional regulation is determined by an interactive link between the gene and its specific TF (Cahan et al., 2021). It focuses on the cross-talk of molecules and interlinkage between multiple pathways. Gene regulatory network (GRN) is a less expensive, commonly used network that improves knowledge about cellular factors. In GRNs, genes are represented by nodes, while the transcriptional regulations detected using chromatin immunoprecipitation (ChIP) assays are represented by edges (Erwin et al., 2009). Single-cell data analysis using Perturb-Seq has also been initiated. Multiple regulatory networks from various physiological processes comparing normal and diseased individuals make it possible to discover the significant molecular interactions responsible for the dysregulation of body homeostasis. Network expression and dynamics can be used to predict the phenotype or biological outcome. A single-cell approach integrated with GRN has discovered the synergistic factors responsible for converting hindbrain neuroepithelial cells into midbrain progenitors (Okawa et al., 2018).

2.5 Machine learning

ML has been extensively used to study the living system, including developmental and stem cell biology. Understanding regulatory interactions, identifying TFs, and developing 3D organoids using microscopic images can be done by ML (Chen et al., 2018).

ML has been broadly classified into supervised and unsupervised approaches depending upon the use of annotated datasets to predict the outcome of an input signal. Supervised learning algorithm uses labeled training data, while unsupervised learning data is neither classified nor labeled (Glass et al., 2022). Data labeling or annotation involves data identification and labeling to make it more specific, which becomes the basis of ML models. A labeled data is tagged with an output variable such that supervised learning strategy is trained to learn the relationship between input features and output variable. Labeled data ensures quality assurance and more precise outcomes with improved data usability but is time-consuming and expensive. However, unlabeled data does not possess any labels such that unsupervised learning predicts relationships without any prior training. Supervised learning is an easier and more accurate method that optimizes performance based on previous experience, for instance, in forecast events.

Supervised learning has been used to predict cell state by taking into account the differential gene expression during cellular differentiation using training data of many GEP (Stumpf et al., 2017). Pixel-by-pixel-based supervised learning has been used to identify inflammatory foci during heart biopsy to diagnose acute cellular rejection in cardiac allografts (Glass et al., 2022). Support vector machine (SVM), a type of supervised learning, is a pattern recognition algorithm that predicts the GEP. Depending upon the hyperplane, the "unknown" is categorized into one of the gene expression profiles/classes (Cortes & Vapnik et al., 1995). It has been used to identify pluripotent and stemness markers (Scheubert et al., 2011; Xu et al., 2010). Bayesian network (BN) is a graph-based model dependent on the conditional probability of distribution between the variables (Jensen, 1996) that has been used to predict new regulators in ESCs (Woolf et al., 2005). Artificial neural network (ANN) is a supervised regression algorithm where signals are transmitted between nodes. One node receives a signal, processes it, and transfers it to the next node, forming an artificial synapse (Makhoul, 1990). ANN facilitates stem cell isolation using GEP (Bidaut & Stoeckert, 2009).

The advantage of unsupervised learning over supervised learning is assessing high-dimensional data while evaluating small query data in case of rare diseases (Lafata et al., 2021). Hierarchical clustering (HCL) is an algorithm that builds a hierarchy of clusters in the form of a dendrogram. Pairwise distances between entities are calculated, and a corresponding hierarchy group is created (Jain et al., 1999). HCL was utilized to compare human embryonic stem cells (hESCs), human induced pluripotent stem cells (hiPSCs), and adult somatic cells, with a conclusion that iPSCs mostly resemble ESCs and are least matched with somatic cells (Chin et al., 2010). In the k-means clustering algorithm, data is randomly assigned to clusters, and then the entity is allocated to the nearest cluster by calculating the mean profiles (Jain et al., 1999). The principal component analysis (PCA) is an unsupervised clustering algorithm used to combat high-dimensional multiomics data. It is a dimension-reduction technique, which finds principal components that increase the variation in data (Bian & Cahan, 2016; Pearson, 1901). PCA and k-clustering were used to characterize ESCs and ASCs to determine the molecular profile of multipotent adult progenitor cells (MAPCs) (Ulloa-Montoya et al., 2007). Single-cell topological data analysis (TDA) has been used to predict the cellular dynamics and cell state of a pluripotent cell undergoing differentiation over time (Rizvi et al., 2017).

2.6 Statistical methods

GRNs are biological—molecular networks with interacting genes. Interactions indicate the control of one gene over another, along with regulatory genes forming a cascade that determines cellular functions (Erwin & Davidson, 2009). Such interactions improve the understanding of cellular differentiation, developmental, and reprogramming biology (Pimanda &

Göttgens, 2010; Zhou et al., 2007). GRN computational models are opted depending upon the research motive. The Boolean network model of GRN consists of an extensive network of regulatory interactions involving TFs that command cellular conversion. It assumes that gene expression levels are either fully activated or deactivated without considering any quantitative changes in expression. As a result, they facilitate the identification of cell conversion variables. A continuous model of GRN predicts differential gene expression during cellular differentiation and provides a quantitative idea about the specific TF or signaling pathways involved. Dynamic Bayesian network/probabilistic Boolean network is used to analyze interactions in the biological network and determine the appropriate TF or combination of TFs necessary for a particular biological process. Directed graphs quantify the ligand—receptor affinity. This model is appropriate for predicting the cell-to-cell interactions that can hamper the tissue regeneration process. To produce host-compatible cells, single-cell computational GRN can be used to predict TFs, conversion factors, and gene interactions necessary to give rise to a specific cell type. Such a cell can later be used in transplantation (Del Sol & Jung, 2021).

Thus, computer algorithms aim to analyze query data and provide information that may benefit early diagnosis and treatment and reduce associated risks. Computational science has remarkable potential to improve medicinal science, but these benefits are not without limitations of privacy and ethical considerations. The amount of stem cell data available is beyond measure, but the data is limited to specific research units and not accessible to all. Developing a stem cell data aggregate could provide a common link to access data generated from multiple experiments. This brings us to the need for stem cell databases and tools.

3. Stem cell databases

The past several years have witnessed the harmonic emergence of computational biology and stem cell biology that orchestrate the systems-level modeling aspects of stem cells with high-throughput molecular data. An open centralized source, freely accessible data portal containing detailed information on cell characteristics, origin, the organism of procurement, and patient details, along with data analysis tools, provide a better understanding of cell potential. Examining the correlation between the attributes of stem cells utilized in the treatment of a specific disease and the resulting therapeutic outcomes is anticipated to facilitate the assessment of clinical effectiveness, guide future decision-making, and identify areas for improvement (Kurtz et al., 2019). It will also facilitate comparing features of different cell lineages obtained from a primary cell type. Databases aim to provide a unified storehouse of stem cell experimental data that stores, curates, and filters data as needed by the user (Finkelstein et al., 2020). Stem cell databases are specialized databases dedicated to stem cell and its properties including stemness, self-renewal, and differentiation. These databases use analytical tools to identify specific stem cell pluripotent markers. Knowledge procured from stem cell—rich database can be used in the field of regenerative medicine, tissue engineering, transplantation, disease modeling, and drug discovery. Fig. 20.2 depicts the stem cell databases categorized on the basis of type of data hosted and omic technology used to extract data. The stem cell databases and tools used frequently for biomedical research are listed in Table 20.1.

3.1 Transcriptomics

Transcriptomics involves transcriptome profiling and the study of RNA transcripts. Apart from the conventional mRNA, tRNA, and rRNA, regulatory RNAs such as miRNA, siRNA, lncRNA, and piRNA are known to regulate gene expression significantly. RNA-Seq, transcriptome microarrays, and whole-transcriptome analysis allow the differential and coexpression study of genes (Kong et al., 2013), cell sorting by visualizing gene expression (Bagger et al., 2013), and identification of housekeeping genes (Wells et al., 2013). Bulk transcriptomics analyze the overall gene expression of population of cells, while single-cell transcriptomics study gene expression and biological complexity at single-cell level. Single-cell sequencing refers to sequencing of RNA molecules at an individual basis to study gene expression at single-cell level and identify cell type.

3.1.1 BloodSpot

BloodSpot is a freely available, more interactive version of HemaExplorer focusing on hematopoietic cells and uses datasets from HemaExplorer. It investigates the expression pattern of a gene at different maturational stages of a cell (Bagger et al., 2016). Using Kaplan—Meier analysis and HCL, it analyzes the data available and portrays association among cell types in the form of a hierarchical tree. It has improved functionalities such as default plot, survival plot, tree plot, etc. Other databases only provide the idea of gene expression of a single gene and do not address the regulation of gene/pathway and the influence of genes on patient.

TABLE 20.1 Bioinformatic databases and tools for stem cell research.

Databases/tools	Description	PMID
	Transcriptomics	
BloodSpot	An enhanced version of HemaExplorer that explores gene expression in malignant hematopoietic cells and normal counterparts. http://www.bloodspot.eu	26507857
CORTECON	Assess the data expression and pathways involved in cortical development. http://cortecon.neuralsci.org/	24991954
ESTOOLSData@Hand	Analyze differential and coexpression of genes and sort samples based on expression similarity. http://estools.cs.tut.fi/	23985726
HemaExplorer	Interface to analyze gene expression in normal and malignant hematopoietic cells. http://servers.binf.ku.dk/hemaexplorer/	23143109
LifeMap Discovery	Explores gene expression and cell differentiation essential for embryonic development to enhance therapeutics. http://discovery.lifemapsc.com	23874394
SISTEMA	Gene expression database for pluripotent stem cell lines (hESCs and hiPSCs) and their derivatives. http://sistema.ens-lyon.fr	34278269
StemBase	Quantify stem cell gene correlation and expression with visualizing aids. http://www.stembase.ca/	19284540
StemMapper	Examines stem cell gene expression of data across stem cell types and lineages. http://stemmapper.sysbiolab.eu	23794736
	Genomics and transcriptomics	
hPSCreg (Human Pluripotent Stem Cell Registry)	Enquire data on stem cell lines (ESCs and iPSCs) and experiments and compare and identify cell lines. https://hpscreg.eu/	35522426
HSC-Explorer	Explores quiescence, differentiation, and self-renewal of HSC populations. http://mips.helmholtz-muenchen.de/HSC/	23936191
	Epigenomics and transcriptomics	
CODEX	Provides NGS data for HSCs and ESCs to understand the transcriptional regulation of cellular identity and fate. http://codex.stemcells.cam.ac.uk/	25270877
Stem Cell Commons	Provides information and allows comparison of various stem cell tissues and types. stemcellcommons.org	24303302
	Multiomics	
ESCAPE	Construct networks, predict lineage specifics, and find existing/new interactions. http://www.maayanlab.net/ESCAPE/	23794736
ESCD (Embryonic Stem Cell Database)	Explore the binding patterns in ESCs and embryonic carcinoma cells. http://biit.cs.ut.ee/escd/	20505756
HipSci (Human iPSC initiative)	Collection of human iPSC lines and multiomic data from healthy and diseased individuals. http://www.hipsci.org/	27733501

Continued

TABLE 20.1 Bioinformatic databases and tools for stem cell research.—cont'd

Databases/tools	Description	PMID
LINCS (Library of Integrated Network-Based Cellular Signatures)	Multiomic data to discover information on perturbation agents (molecules/genetic factors) and metabolic processes. http://lincsportal.ccs.miami.edu/	29140462
PluriNetWork	Analyze pluripotency and association of Oct4, Sox2, and Nanog with other genes/proteins. http://www.ibima.med.unirostock.de/IBIMA/PluriNetWork/	21179244
StemCellDB	Explore experimental data and transcriptional regulation for 21 NIH-approved hESC lines. http://stemcelldb.nih.gov/	23117585
StemCellNet	Generates networks based on regulatory and gene interactions, especially on stemness-specific genes. http://stemcellnet.sysbiolab.eu/	24852251
StemFormatics	Analyzes stem cell gene expression data at genetic, epigenetic, and proteomic levels. https://www.stemformatics.org/	30407577
SyStemCell	Investigates the "stemness" state of stem cells by gene correlation and epigenetic modifications. http://lifecenter.sgst.cn/SyStemCell/	22807998
Cancer stem cells (CSCs)		
BCSCdb (Biomarkers of Cancer Stem Cells)	Compiles information on experimentally validated high-throughput markers and low-throughput markers, i.e., CSC biomarkers, biomarker interactions, and therapeutic targets. http://dibresources.jcbose.ac.in/ssaha4/bcscdb	36169329
CSCdb (Cancer Stem Cells database)	Stem cell portal that analyses CSC genes, markers, and their functional aspects. http://bioinformatics.ustc.edu.cn/cscdb	26989154
CSCTT (Cancer Stem Cells Therapeutic Target)	Focusses on CSC therapeutic target genes, drugs that target CSCs and various therapeutic approaches. http://www.csctt.org/	28191780
Stem Cell Discovery Engine	Compare CSCs' molecular profiles with healthy stem cells to identify molecular targets against CSCs. http://discovery.hsci.harvard.edu.	22121217
Resources		
SKIP (Stem Cell Knowledge and Information Portal)	Facilitates cell/tissue type data and enhances their applications in regenerative medicine. https://skip.stemcellinformatics.org/en/link/	32099411
Tools		
CellNet	Quantifying the analogy between engineered cells and in vivo cells derived from adult tissues using ML and GRN models. http://cellnet.hms.harvard.edu/	25126793
eegc	Predicts engineered cell fidelity and categorizes the dynamics of cell states and status of cell conversion. http://bioconductor.org/packages/release/bioc/html/eegc.htm	28398503
ExprEssence	Identifies the key gene regulators for cell type conversion. http://apps.cytoscape.org/apps/expressence	21118483
hscScore	Identifies rare HSC populations in scRNA-Seq data using HSC transcriptomic profiles in murine bone marrow. https://github.com/fionahamey/hscScore	31513832

TABLE 20.1 Bioinformatic databases and tools for stem cell research, cont'd

Databases/tools	Description	PMID
KeyGenes	Quantifies the analogy between engineered cells and in vivo cells derived from fetal tissue using DE analysis and functional enrichment analysis. http://www.keygenes.nl/key/	26028532
Mogrify	Identifies critical TFs involved in cellular reprogramming facilitating cell conversion. http://www.mogrify.net/	26780608
PluriTest	Assay to determine pluripotency of human cells based on microarray-generated gene expression profiles. https://www.pluritest.org/	21378979
ScoreCard	Quantifies the resemblance of various cell types and determines lineage differentiation of iPSC lines. http://www.medicalepigenomics.org/papers/broad_mirror/scorecard/index.html	21295703
StemChecker	Assesses stemness in query samples by inspecting the stemness signature of gene sets. http://stemchecker.sysbiolab.eu	26007653
StemID	Identifies stem cells in a population using single-cell transcriptomic data. https://github.com/dgrun/StemID	27345837
StemID2	Algorithm used to infer differentiation trajectories and predict stem cell identity using scRNA-Seq data. https://github.com/dgrun/RaceID3_StemID2	31062315
STEMNET	Evaluates stem cell progression using single-cell gene expression for each cell and provides a visualization of the lineage process. https://git.embl.de/velten/STEMNET	28319093
TeratoScore	Quantitative approach to evaluate pluripotency using teratoma gene expression data. http://benvenisty.huji.ac.il/teratoscore.php	26070610

Features of BloodSpot:

- Enhanced visualization of similar GEP.
- Compares GEP in normal and cancerous hematopoietic cells at multiple differentiation stages.
- Determines the therapeutic target based on gene/pathway studies.
- Addresses genetic programs/patient survival.

Limitations of BloodSpot:

- Includes high-quality curated datasets, but only some have been considered for human single cell normal HSCs.

3.1.2 CORTECON

CORTECON is a repository of human ESCs data at the transcriptomic level that proves to be helpful in cases where change in RNA expression and different spliced forms modulate corticogenesis (Van et al., 2014). Gene and lncRNA levels that change their expression levels during corticogenesis were identified using RNA-Seq. The CORTECON dataset assessed that the expression peak of FGF-8 is essential during early corticogenesis, and WNT antagonists DKK1 and FRZB expression vary during specific stages of cortical development. This open-to-browsing database, still available and updated, uses gene expression profiling, Venn diagram, Spearman correlation, and heatmaps for data analysis.

Features of CORTECON:

- Aims to compare regulation of corticogenesis in both human and murine cells

- Splice form analysis focuses on splicing mechanisms and predicts the function of multiple splice forms in cancer and other neural disorders.
- Improves treatment of neurological diseases (autism, schizophrenia, and age-linked memory loss).

Limitations of CORTECON:

- Limited to single-cell type (ESCs).

3.1.3 ESTOOLSData@Hand

ESTOOLSData@Hand hosts human ESCs and iPSCs to look for gene coexpression with emphasis on cell differentiation, dedifferentiation, and pluripotency. This tool had free access for academic purposes but is no longer available (Kong et al., 2013). The clustering tool sorted the sample by expression similarity to generate a dendrogram and heat map. It also used coexpression tool, differential expression tool, and expression profile tool.

Features of ESTOOLS@Data:

- Provide human pluripotent stem cell (hPSC) gene expression data with several analysis tool.
- Integrated with enrichment tool and genomes pathways.

Limitations of ESTOOLS@Data:

- Old datasets from ArrayExpress and Gene Expression Omnibus (GEO) were integrated and needed to be updated.

3.1.4 HemaExplorer

HemaExplorer is a user interface that hosts data on hematopoietic stem and progenitor cells and provides insights into hematological disorders (Bagger et al., 2013). HemaExplorer is the first publicly accessible database with full data integration and comparison of the stem, progenitor, and mature/differentiated hematopoietic cell populations in human and murine models curated using GEP. Entrez Gene, CORUM, and Gene Ontology resources were used to curate information. The data analysis tools used are clustering, heatmap visualization, HCL, and PCA plots.

Features of HemaExplorer:

- Identification of correlated genes and their expression across multiple cell types.
- Sort cell population based on cell characterization with expression.
- Single gene study.
- Study of gene expression at multiple cell maturation stages.

Limitations of HemaExplorer:

- Lacks the necessary signaling pathways involved in hematopoiesis.

3.1.5 LifeMap Discovery

LifeMap Discovery is a valuable interface expanded over ESCs, embryonic progenitor cells, fetal and adult SCs, iPSCs, and primary cells across human and mice (Edgar et al., 2013). LifeMap Discovery brings in a cell nomenclature system and provides each cell with a unique EIndex specific to its organ/tissue, anatomical location, and cell. It consolidates embryology and stem cell biology with an interlink of in vitro and in vivo data. The "ontology tree" briefs organ development and focuses on cell therapy applications and regenerative diseases. It is a freely available database that is currently active and still updated.

Features of LifeMap Discovery:

- Interactive visualization of in vivo anatomical compartment development.
- Data stored belongs to work at research, preclinical, clinical, and marketing stages.
- Detailed information about the model organism, treatment route, and mechanism of action.
- Solid platform for therapeutic success that assists to find potential targets in various clinical applications.

Limitations of LifeMap Discovery:

- Provides mammalian in vitro data and in vivo stem cell differentiation data.

3.1.6 SISTEMA

SISTEMA is a user-friendly web portal that evaluates comparison of human gene expression data from cells in pluripotent state to differentiated states in both normal and pathological complications (Jarrige et al., 2021). It is a further extension of StemCellDB and HipSci database and encompasses hESCs and hiPSCs information. Data resources are mostly retrieved from NCBI and EBI expression atlas. The web portal currently hosts eight monogenic diseases. SISTEMA is used for single-gene as well as multiple-gene analysis, further applied for an enhanced understanding of disease-specific genes, drug-specific genes, coexpressed genes, and cell type biomarkers.

Features of SISTEMA.

- Uses AmpliSeq technology for targeted amplification of human RNA targets.
- Small amplicon size generates smaller raw data compared with the conventional RNA-Seq and is highly advantageous since data generated by RNA-Seq at large scale poses storage problems.

Limitations of SISTEMA:

- Encompass data generated from AmpliSeq only and foresees to integrate scRNA-Seq data and whole-transcriptome analysis.
- Human-specific web portal.

3.1.7 StemBase

StemBase is an open browsing data portal that utilizes sample v/s gene analysis tools and genome viewer to study gene relationships. Expression patterns are indicated by positive/negative correlation according to Spearman's correlation coefficient and their mutual dependence (Sandie et al., 2009). It emphasizes on deriving data from Affymetrix GeneChip DNA microarrays. StemBase uses Genome Viewer to allow graphical representation of expression data on genomic regions.

Features of StemBase:

- Spans, human, mice, and rat data.
- Hosts web-based tools that do not require external software usage.
- Provides an alternative to DNA microarray analysis software.

Limitations of StemBase:

- The database comprises 210 stem cell samples from 60 experiments, as of 2008, and no further updates have been incorporated since then.

3.1.8 StemMapper

StemMapper is a freely accessible database that hosts data on stem cells and progenitor cells (Pinto et al., 2018). Transcriptomic datasets were retrieved from NCBI's GEO. The data points in the PCA plot are alternatively colored as per surface markers, tissue of origin, type of SC, and differentiation. StemMapper lets the users upload their dataset for analysis and compare it with in-house data. It uses heatmap generation, PCA, and Pearson correlation. Genome-wide enhanced circular chromosomal conformation capture (e4C) was used to find 18 genes linked with Gfap in astrocytes. It is currently available.

Features of StemMapper:

- Data study across both human and mice.
- "Preferential Expression Measure" allows gene expression specificity in a cell type.
- "Percentile Expression" investigates the expression of a gene in given sample in comparison with its expression in other samples.

Limitations of StemMapper:

- Only SC expression profiles from Affymetrix sequencing platforms supported.
- Data retrieved from RNA-seq and other transcriptomic platforms are not regarded.
- Lack of single-cell expression platforms to deal with population heterogeneity.

3.2 Genomics and transcriptomics

Genomics and transcriptomics involve the study of the structural and functional aspects of the entire gene content of an organism, along with its transcriptome profiling. Genomics assess whole genome both structurally and functionally using high-throughput methods. NGS for stem cells improves understanding of transcriptomic regulation, cellular identity, and cell fate (Sánchez-Castillo et al., 2015).

3.2.1 Human Pluripotent Stem Cell Registry (hPSCreg)

Human Pluripotent Stem Cell Registry (hPSCreg) stores freely accessible experimental information on human ESCs and iPSCs. It focuses on many diseases, including blood, circulatory, genitourinary, nervous, musculoskeletal, endocrine, cancer types, and COVID-19 (Kurtz et al., 2022). It follows the guidelines of FAIR to make information findable, accessible, interoperable, and reusable. Stem cells are biological twins of donor that can be used to create stem cell twin of a person. In contrast, donor data (lifestyle, gender, ethnicity) can be used to create a digital model of the donor. Now, a biodata hybrid created using a biological twin, i.e., a stem cell and a digital twin, can help to assess a patient's individuality and enhance personalized medicine outcomes.

Features of hPSCreg:

- Collaboration with clinical trial registries (https://trialsearch.who.int/ and http://clinicaltrials.gov/).
- Assess treatment efficacies.

Limitations of hPSCreg:

- Minimal information on studies.
- Standardization of principles to bring therapies into practice needs to be improved.
- Search filter options based on cell type/reprogramming factors need to be updated.

3.2.2 HSC-Explorer

HSC-Explorer is a public database that focusses on mice ESCs and HSCs. Provision of comprehensive detail on experimental details/results such as organism strain/gender and quality/quantity of cells provided by the database improves understanding of bodily functions and respective treatment (Montrone et al., 2013). It focuses on signaling pathways from the most primitive HSCs to differentiated progenitor cells. Other databases, such as Hematopoietic Fingerprints and StemBase, contain hematopoiesis-specific gene expression only. Web resources including Entrez Gene, Gene Ontology, and CORUM were used to obtain structured data.

Features of HSC-Explorer:

- Specific to hematology.
- Analyzes signaling pathways in the HSCs niche (homing, migration, and cell adhesion).
- Stem cell characterization (proliferation state, self-renewal ability, repopulation ability).

Limitations of HSC-Explorer:

- Lack of information regarding other hematopoietic progenitor cells.
- Although manual curation tends to improve data quality, but is error prone.

3.3 Epigenomics and transcriptomics

Epigenomics revolve around modifications of DNA and associated proteins (histones), DNA methylation, and chromatin interactions regulating gene activity.

3.3.1 CODEX

CODEX is a freely accessible data portal hosting high-quality NGS data using a strict processing pipeline. It focusses on ESCs and cells of the hematopoietic system (Sánchez-Castillo et al., 2015). To find new samples, web crawler employs a text-mining approach. The samples are then labeled with CODEX experiment type (histone modification ChIP-Seq, TF ChIP-Seq, RNA-Seq, and DNase-Seq) and sent to the corresponding repository. It uses GEO, ArrayExpress, and DNA Data Bank of Japan to retrieve raw NGS data. CODEX is still available.

Features of CODEX:

- Analyses ESCs and HSCs in both human and mice using gene set control analysis (GSCA) motif.
- Web crawler uniquely searches for any novel sample generated.
- Rapid visualization and comparison of NGS datasets between organisms.

Limitations of CODEX:

- Integration of TF ChIP-Seq and new RNA Seq datasets plugins to compare DNA motif enrichment and TF-binding sites are not available, which are required to understand the transcriptional regulation in different stem cells.

3.3.2 Stem Cell Commons

Stem Cell Commons is an initiative by the Harvard Stem Cell Institute (HSCI) to improve the storage, analysis, and data exchange on stem cells. HSCI investigators have complete access to Stem Cell Commons while limited access to others (Merrill et al., 2013). An ISA tab enables data curation, validation, search, and further submission to repositories. Stem Cell Commons is an integrative and comprehensive approach to stem cell data inspired by Stem Cell Discovery Engine (SCDE) and Blood Genomics databases.

Features of Stem Cell Commons:

- Hosts cancer cell data.
- Match and comparison of stem cell experiments and diseased studies via ISA-Tab.

Limitations of Stem Cell Commons:

- Manual data curation is prone to error and time-consuming.
- Limited access to the data portal.

3.4 Multiomics

Multiomics is an integration of genomics, epigenomics, transcriptomics, and proteomics. Proteomics focuses on protein structure, function, expression, and protein interactions using western blot analysis, mass spectrometry, etc. Multiomics allows for a more comprehensive study combining multiple omic datasets. Techniques such as WGS, SNP genotyping, RNA-Seq, proteomics, and cellular phenotyping analyze TF regulation and the stemness state of a cell. ChIP-Seq, RNA-Seq, and whole-transcriptome analysis were used to detect the binding patterns of Oct4 (Jung et al., 2010) and study the association of Oct4, Sox2, and Nanog with other epigenetic regulators (Som et al., 2010)

3.4.1 Embryonic Stem Cell Atlas from Pluripotency Evidence (ESCAPE)

Embryonic Stem Cell Atlas from Pluripotency Evidence (ESCAPE) includes various data analysis tools, including enrichment analysis, lineage prediction tool, and heatmap generation (Xu et al., 2013). It hosts a huge variety of data type including protein−gene promoter binding, protein−gene transcriptional regulation, protein−protein interactions, miRNA-predicted target interactions, histone modifications, pluripotency genes, ESC-specific proteins, ESC-specific phosphoproteins−phosphosites, gene expression, and miRNA expression. The Enrichr tool analyzes the overlapping of a queried gene with ESCAPE's gene list. The lineage tracing tool identifies a cell's heritable traits and origin. ESCAPE is a freely available database that is still available.

Features of ESCAPE:

- Holds multiomic data across both human and mouse.
- Focusses on cross-talk across multiple regulatory levels (epigenetics, transcriptomic, proteomic, and phosphoproteomic levels).

Limitations of ESCAPE:

- Only one dataset of hESC pluripotent genes from a study is available.

3.4.2 Embryonic Stem Cell Database (ESCD)

Embryonic Stem Cell Database (ESCD) is a freely accessible platform that simplifies the study of published data. OCT4 downstream targets common to both ESCs and embryonic carcinoma cells in inter- and intraspecies have been identified

(Jung et al., 2010). They verified that few genes, viz., NANOG, SOX2, OCT4/POU5F1, PHC1, PHF17, and USP44 are omnipresent for the maintenance of self-renewal in both ESCs and embryonic carcinoma cells. This study's discovery of the cell cycle regulator GADD45G is distinctive. Overexpression of GADD45G upregulates GADD45A involved in cellular differentiation and cell cycle at G2/M phase. The data available has been analyzed using Real-time PCR, western blot, venn diagram, box plot, and STRING analysis. ESCD is currently available.

Features of ESCD:

- Hosts data across both human and mouse.
- Allows comparison of embryonic stem and embryonic carcinoma cells.

Limitations of ESCD:

- Lack of downstream signaling pathways regulating OCT targets in ES and EC cells.

3.4.3 Human induced pluripotent stem cell initiative (HipSci)

Human iPSC initiative (HipSci) is open to browse database that focuses on human iPSCs. The data generated has been assayed through WGS, methylation array, RNA-Seq, proteomic assays, cellular phenotyping, etc. (Streeter et al., 2017). HipSci lines are characterized by cell type, disease state, age/sex of the individual, donor consent (managed/open access), and culture conditions. It is available for academic noncommercial purchase from the European Collection of Authenticated Cell Cultures (ECACC) at https://www.culturecollections.org.uk/collections/ecacc.aspx. However, for commercial units, it can be availed from European Bank of induced Pluripotent Stem Cells (EBiSC) at https://www.culturecollections.org.uk/products/celllines/ebisc-introduction.aspx.

Features of HipSci:

- Open-access detailed information on cell lines obtained from healthy donors and individuals with inherited (rare) genetic diseases.
- HipSci stores information on hypertrophic cardiomyopathy, Alport syndrome, monogenic diabetes, primary immunodeficiency, retinitis pigmentosa, and macular dystrophy.
- RNA sequencing (RNA-Seq), whole-exome sequencing (Exome-seq), DNA methylation profiling (mtarray), proteomics, and cellular phenotyping (cell biol-fn) data are available for iPSCs.

Limitations of HipSci:

- Most of the donors that provided HipSci cell lines are of British heritage. Hence, a population-based study cannot be performed using this dataset.

3.4.4 Library of Integrated Network-Based Cellular Signatures (LINCS)

Library of Integrated Network-Based Cellular Signatures (LINCS) hosts human-specific data generated using over 20 high-throughput assays (Koleti et al., 2018). Entire data has been used from LINCS data registry. It encompasses a diversity of cell types, such as ESCs, iPSCs, differentiated cells, primary cells, and mostly cancer cell lines. The central feature involves Dataset Landing Pages where experimental data is successively designated into different levels, namely level 1 (raw), level 2 (processed), level 3 (normalized), and level 4 (signature). It includes cell lines, cell line biomarkers, microenvironments, protein interaction, and cellular phenotypes focusing on developing therapeutic approaches to treat model diseases. LINCS has free access.

Features of LINCS:

- Comprises multiple cell lines, including CSCs.
- User-friendly portal for the computational and noncomputational background.
- Integrates various assay technologies at the multiomic level.

Limitations of LINCS:

- Evaluations for LINCS datasets and tools are required.

3.4.5 PluriNetWork

PluriNetWork is a network hosting mouse pluripotent stem cell (PSC) data (Som et al., 2010). Relevant literature was sourced from Google Scholar and PubMed. It allows detailed understanding of molecular landscape of pluripotent and reprogramming factors, analyze the addition of stimulators/inhibitor of various genes/proteins, and understand its effect in the initial phases of pluripotency/development.

Features of PluriNetWork:

- Highlights the most differentially altered links between genes/proteins with maximum expression change.
- Analyses partial/full induction of pluripotency by pluripotent genes.
- Manual curation enhances data quality.

    Limitations of PluriNetWork:

- Hosts only mouse data.
- It does not consider a few indirect links, such as histone proteins/genes.
- Manual curation is prone to error and time-consuming.

3.4.6 StemCellDB

StemCellDB is a user-friendly search engine that stands for Stem Cell Database and stores experimental data on hPSC (21 hESCs and 8 hiPSCs derived via retroviral transduction of human fibroblasts) and human adult tissue cell lines (Mallon et al., 2013). It contains information on markers, primers, and antibodies specific to stem cells using covariance PCA and Pearson correlation expression. Pearson correlation studies the expression of the queried gene with other genes in the sample. StemCellDB allows free access.
    Features of StemCellDB:

- In addition to microarray gene expression, it offers data access for SNP genotyping, miRNA, DNA methylation, and array-based comparative genomic hybridization.
- Enhance knowledge of pluripotency by analyzing transcriptional-level data.

    Limitations of StemCellDB:

- It holds human data only.
- Specific to PSCs.

3.4.7 StemCellNet

StemCellNet is a freely accessible data portal that uses the Cytoscape Web application for graphical networking and allows enhanced visualization of molecular interactions of genes/proteins (Pinto et al., 2014). Screening for stemness signatures helps to identify genes associated with stem cell function and enrichment of stemness signatures in a network. It helps to find novel candidate genes linked with stemness. The feature of identifying physical and transcriptional regulatory interactions and tracing the interactions to the original case study help better analyze the results.
    Features of StemCellNet:

- Hosts both human and murine cells.
- Hosts data across multiple cell types (ESCs, iPSCs, NSCs, HSCs).
- Rapid identification of genes regulating stem cell function.
- Useful for regenerative diseases and cancer-requiring stem cell—based therapy.

    Limitations of StemCellNet:

- Lack of updated processing pipelines (query format).
- Less number of central node analysis.
- Unable to analyze more genes (>500) in a short time.

3.4.8 StemFormatics

StemFormatics is a freely accessible database that houses data across different cell types, including stem cells, hematopoietic cells, MSCs, ESCs, etc. (Wells et al., 2013). It is a highly curated data portal that stores data on pluripotency routes and allows scRNA-Seq. It permits various data analysis tools, including GEP, clustering and heatmap generation, HCL, functional enrichment analysis, PCA plots, Glimma report, and YuGene interactive graphs.
    Features of StemFormatics:

- Hosts data over multiple cell/tissue types across human and mice.
- Identification of housekeeping genes.
- Assesses both the quantity and quality of data.

- Interactive YuGene graph that ranks samples according to their expression levels.

Limitations of StemFormatics:

- Data processing failure (due to faulty experimental design, bad data quality, annotation issues in samples, accession hacking).

3.4.9 SyStemCell

SyStemCell is a database that hosts data on stem cells, i.e., ESCs, iPSCs, and NSCs (Yu et al., 2012). Other databases concentrate on limited stem cell types and single-level experimental data. Thus, there was a need to develop the first database to construct stem cell research information on multilevel experiments. The seven levels include histone modification, transcript products, protein products, phosphorylation protein, TF regulation, and DNA CpG 5-hydroxymethylcytosine/methylation. Entrez Gene, UniGene, GeneBank, NCBI, UniProt, and Ensembl were used to derive gene accession numbers, while Gene Ontology was used to extract gene annotations. It is no longer available.

Features of SyStemCell:

- Hosted data across four species (human, mice, rats, and rhesus macaque).
- Hosted data across multiple cell types.
- Stem cell differentiation—associated regulatory mechanisms at multiple levels.
- Helped to understand the evolutionary conserved regulatory patterns (epigenetic modifications) across different species.

Limitations of SyStemCell:

- Data browsing was not easy.
- Lack of update of datasets.

4. Cancer stem cell databases

Efforts to combat cancer have frequently encountered difficulties primarily because of the presence of CSCs. The potential increase in CSCs following conventional treatments is a significant factor contributing to the initiation of tumorigenesis, metastasis, the development of multidrug resistance, and eventual cancer relapse. This is primarily attributed to the CSCs' capability to enter a state of arrest in the G0 phase and maintain quiescence. Therefore, the identification and isolation of CSCs hold great significance in our quest to enhance therapeutic strategies. Surface markers including CD44 and CD133 have been recognized as potential therapeutic targets to mitigate CSCs resistance (Menke-van der Houven et al., 2016; Ning et al., 2016). Omic technologies are used to explore the molecular profiles of CSCs and study various CSC-specific genes, markers, and pathways involved. Such characterization is fundamental to ensure better treatment. Thus, there is an urgent need for CSC-specific data repositories that provide a wholesome information on experimentally validated CSC markers, druggable targets, and therapeutic strategies that selectively kill CSCs. Several databases specific to CSCs have been covered in the following sub-sections.

4.1 Biomarkers of Cancer Stem Cell database (BCSCdb)

Biomarkers of Cancer Stem Cell database (BCSCdb) is a user friendly and freely accessible database that hosts CSC biomarkers, biomarker interactions, and CSCs-therapeutic target data across 10 cancer types in human (Firdous et al., 2022). It comprises experimentally validated high-throughput markers (HTMs) and low-throughput markers (LTMs) and hosts three different tables, viz., CSCs biomarker table that reports biomarker name, type, expression levels, and score. CSCs biomarker interaction table comprises experimentally validated molecular interactions and CSCs therapeutic table that reports therapeutic gene name, biotherapy, and effect of therapy on multiple cancer types. About 445 target genes and 383 drugs responsible for inhibiting CSCs behavior including self-renewal and metastasis have been noted. Based on high confidence and low global score, researchers demonstrated that DCKL1 seems to be a potential CSCs biomarker for colon cancer. Data was collected from PubMed literature.

Features of BCSCdb:

- Contains preclinical and clinical trial biomarkers.
- Differs from CSCdb and CSCTT in terms of information regarding cell line, CSC detection methods, confidence and global score, CSCs enrichment drug, and the effect of drug on CSCs behavior.

- Confidence score used for experimental validation is used to identify new biomarkers.
- Global score gives an idea of the CSC biomarker uniqueness and predicts the frequency of the queried CSC biomarker among the 10 different cancer types.

Limitations of BCSCdb:

- Skips CSC-specific gene—gene interactions for network construction.
- Since global score is dependent on the number of entries, with every version update, the global score also fluctuates; however, the change is small enough
- In situations where combination of drugs has been used to target CSC, only single PubChem ID is designated for the unique drug.

4.2 Cancer Stem Cell database (CSCdb)

Cancer Stem Cell database (CSCdb) is a stem cell portal that allows search for marker genes. It provides data about identified CSCs, marker genes, CSC genes/microRNA, and their functional annotations (Shen et al., 2016). It hosts 74 marker genes of more than 25 tissue type, 1769 CSC-related genes, and 9475 functional annotations. The "functional annotations" tab provides information about the genes function. The "marker" tab and "related gene" tab offer list of all the CSCs-related genes/microRNA. "Related gene" refers to both functionally related and expression-related gene markers. The literature was collected from PubMed, Web of Science, NCBI, and Uniprot with further manual curation. This database is freely available and does not require any registration. Moreover, the data can be easily downloaded for further review.
  Features of CSCdb:

- Differs from other databases (CancerDriver, Brain Tumor Medical Database, CaGe, and GeneCards) as CSCdb highlights CSC-specific genes and markers.
- Easy search for marker genes in specific tissue using the "tissue type" tab.

Limitations of CSCdb:

- No means to predict the confidence value of the marker or its uniqueness.
- Manual curation is prone to error and time-consuming.

4.3 Cancer Stem Cells Therapeutic Target database

Cancer Stem Cells Therapeutic Target database (CSCTT) is an open-to-browse and freely available database that focuses on biological details of multiple targets and therapies (Hu et al., 2017). It comprises about 135 proteins/miRNA as druggable targets that have high binding affinity to its corresponding drug with an ability to differentiate or eliminate the CSC with significant tumor mitigation. About 213 therapeutic strategies including 118 compounds, 75 drug combinations, and 20 biotherapies such as CAR-T therapy, oncolytic virotherapy, immunotherapy, and peptide-based vaccines are reported. Each molecule has been designated with a name, structure, PubChem ID, and functional inhibitory activities against CSCs. Information was collected from literature and web resources such as PubChem, KEGG, UniProt, and Protein Data Bank.
  Features of CSCTT:

- Provides information about each reported molecule.
- Spans over multiple CSC targets (name, gene name, function, related disease, and pathway) and therapeutic targets (known targets and structure).
- Deals with pathways involved in CSCs regulation.

Limitations of CSCTT:

- The standardization of principles to bring therapies into practice needs to be improved.

4.4 Stem Cell Discovery Engine

SCDE is a platform integrating experimental CSCs data on human, mouse, and rat stem cells (Ho Sui et al., 2012). It allows gene comparisons based on molecular profiles and associated pathways. CSCs experiments are particularly blood, intestine, and brain-specific. Manual data curation sorts samples according to surface markers, cell/tissue type, and disease states. The List Match Tool allows us to find similarities between the user's differentially expressed genes with genes from

SCDE experiments. Apart from user-submitted experiments, public stem cell data from GEO and ArrayExpress have been used to curate standardized data. SCDE provides free access and is still available, merged with Harvard's larger Stem cell commons project.

Features of SCDE:

- Analyzes CSCs data.
- Bio Investigation Index ensures uniform, standardized, and comparable stem cell data.
- Data categorized on the type of profiling (transcriptional level) and technology (ChIP-Seq, RT-PCR) used.

Limitations of SCDE:

- Lack of integrative data.
- Lack of various signaling pathways in cancer metabolism and different experimental models.

5. Resources

An informative resource that offers stem cell-specific data, thereby contributing to the improvement of therapeutic strategies.

5.1 Stem Cell Knowledge and Information Portal

Stem Cell Knowledge and Information Portal (SKIP) provides information specific to stem cells (Kurtz et al., 2019). Searches made by specific keywords/diseases help to extract information regarding cell type, cell lineage, cell morphology, source, donor details, diseases or normal condition of the donor, etc. It provides free access and registration for data submission and is still available.

Features of SKIP:

- Enhanced therapeutic applications in various diseases and regenerative medicine.

Limitations of SKIP:

- Only iPSCs and ESCs data are available.
- No information is provided for HSCs and other normal healthy stem cells.

6. Stem cell tools

Significant research efforts are underway in the field of cell engineering across the entire stem cell community. This research holds the promise of advancing various fields such as developmental biology, drug screening, regenerative medicine, and disease treatment. It aims to enhance our understanding of biological processes and ultimately improve therapeutic results. In silico tools are to have significantly contributed to stem cell studies (Agrawal et al., 2018, 2019; Kulshrestha et al., 2021; Kumar et al., 2021; Raghav, Mann, Ahlawat, et al., 2021; Raghav, Mann, Krishnakumar, et al., 2021; Raghav, Mann, Pandey, et al., 2021). The lack of adequate information to determine the growth conditions that induce appropriate cellular differentiation and regulators involved in a biological process are few drawbacks backed by the research. Another complication is to understand how far the engineered cell replicates their in vivo counterpart (Zhou & Sears et al., 2018). Thus, integrative tools are required to ease data extraction. These tools help to understand GEP of cells, monitor the different developmental stages, assess cell engineering efficiency, predict TFs, and check stemness. Primary processes, including differentiation, transdifferentiation, and reprogramming, are also detected (Zhou et al., 2018). Primary bioinformatic tools used in the stem cell community are listed in Table 20.1.

6.1 CellNet

CellNet is based on the principle of GRN that looks for cell conversion expression and cell fidelity in response to biological/environmental cues. It helps to analyze the GEP of cells engineered via reprogramming to iPSCs, directed differentiation, and direct conversion. Cells obtained from directed differentiation are more likely to exhibit the target cell/tissue GRN than those obtained from direct conversion. CellNet indicates that iPSCs and ESCs transcriptomics are virtually not distinguishable. To include more data on different cell types and avoid issues of tissue heterogeneity, CellNet aims to use scRNA-Seq (Cahan et al., 2014).

6.2 eegc

It is a bioinformatics tool that analyzes the precision of cell engineering. It is based on the principle that cell type conversion is accompanied by a differential expression, which also states that if a cell is in the middle of its conversion, it will have an intermediate expression. Based on this, the cell state progression has been classified as "inactive" (the engineered cell has high similarity with the original cell), "insufficient" (intermediate progression of conversion), and "successful activation" (the engineered cell has high similarity with the target cell). Further classification involves "successful/complete reprogramming" and "insufficient/incomplete reprogramming" (Zhou et al., 2017).

6.3 ExprEssence

It is a bioinformatics tool that constructs gene networks based on interactions. It works by highlighting the "most differentially altered" links between genes/proteins with maximum expression change. A link score is generated for every interaction/link taking place (Warsow et al., 2010).

6.4 hscScore

hscScore identifies HSCs population in single-cell transcriptomic data in murine bone marrow (Hamey & Göttgens, 2019). It is a robust and simple method to locate rare HSCs subset in a population. "hscScore" generated is dependent on the gene expression levels of genes/markers upregulated in HSCs. So, hscScore deployed on ML algorithms scores the single-cell transcriptomic data based on its alikeness with the GEP of HSCs that have been prevalidated.

6.5 KeyGenes

Based on gene signatures, it monitors the efficiency of tissue differentiation in human fetal tissues and designates the developmental stage of the sample of interest. It can predict the derivatives of stem cells and tissue of origin with assurance. It uses functional enrichment analysis and DE analysis. KeyGenes is similar to CellNet except that KeyGenes studies fetal tissues at different developmental stages, while CellNet is concerned with adult cells/tissues (Roost et al., 2015).

6.6 Mogrify

Mogrify uses already-stated interactive networks to understand the change in the expression of TFs that might have occurred during cell conversion. Using this strategy, dermal fibroblast to keratinocytes and then to microvascular endothelial cell transdifferentiation was predicted (Rackham et al., 2016).

6.7 PluriTest

PluriTest analyzes the GEP of undifferentiated cells and compares them with pluripotent cells to derive similarities. It uses a pluripotency score and novelty score to report the degree to which pluripotent signature is expressed in the sample and intensity of the signal measured that can be compared with normal PSC lines, respectively. Integration of both pluripotency and novelty score allows better assessment of pluripotent features in the sample. It can also distinguish between a fully and partially reprogrammed iPSC lines and those with non-pluripotent cells as well (Müller et al., 2011). However, it does not take into account the fact that mutations in tissue regulators have the ability to disturb cell potency.

6.8 ScoreCard

ScoreCard quantifies the similarity between engineered cells and target cells, offering predictions about the lineage of PSCs in vitro and assessing the differentiation propensity of PSCs lines. It was demonstrated that the divergence of specific iPSCs lines is similar to ESCs in gene expression. The ScoreCard helps researchers choose the ideal PSC line for their study (Bock et al., 2011).

6.9 StemChecker

StemChecker is a systematic tool that senses "stemness" by evaluating the gene sets predominant in stem cells. It works by input of genes of a query sample to be evaluated for its stemness. Evaluation is done on a hypergeometric test that analyzes the query sample and reference gene list and generates a P value based on the resemblance between the two (Srinivasan et al., 2021).

6.10 StemID

StemID helps to identify abundant as well as rare cells to derive a lineage tree (Grün et al., 2016). The number of branches and homogeneity in transcriptome in the lineage tree is a representation of the degree of pluripotency, thus revealing the position of the cell in the tree. When used to screen human pancreatic cells for multipotent cell populations, it was found that ductal cells expressing CEACAM6 differentiate into α and δ cells while population marked with high ferritin complexes gave rise to β cells and acinar cells. StemID has been used in collaboration with RaceID2, an advanced version of RaceID to improve clustering dynamics and predict multipotent cells.

6.11 StemID2

StemID2 is an algorithm used to construct lineage tree and predict stem cell identity. It is also be used in pseudo-temporal ordering, i.e., visualize and analyze changes in pseudo-temporal gene expression. Trajectory created can thus be useful to arrange the cells from early progenitor stage to more differentiated stage and understand the entire process of differentiation (Grün, 2019). StemID2 has been used in association with RaceID3 and FateID for stem cell characterization using scRNA-Seq. Here, RaceID3 recovered all the cell types and performed clustering, followed by StemID2 used to construct a lineage tree and finally used FateID to quantify the degree of multipotency.

6.12 STEMNET

STEMNET is a reduced algorithm that evaluates cell progression in appropriate lineages by taking into account single-cell gene expression and ending up with a visualized lineage process. STEMNET has been used to obtain insights into transition of stem cells in HSC landscape (Velten et al., 2017). Results reveal that hematopoiesis does not comprise of distinct progenitor cell types; rather, it is in a dynamic and transitory state, described as cellular Continuum of Low-primed UnDifferentiated (CLOUD) hematopoietic stem and progenitor cells unlike the regular aspect of differentiation that states that HSCs pass through distinct intermediate progenitor cell stages.

6.13 TeratoScore

Teratoma formation is currently the most widely accepted qualitative assay to evaluate human PSCs potency to differentiate into three germ layers, but it lacks a quantitative perspective. Thus, TeratoScore was built to estimate tissue expression and lineage distribution and determine pluripotency (Avior et al., 2015). It estimates the differentiation capacity of a tumor initiating cell and has the ability to distinguish PSCs-derived teratomas from malignant tumor.

7. Conclusion and future prospects

This chapter highlights the significant impact of computational biology on the advancement of stem cell research. By integrating technological tools with stem cell biology, researchers can access a unified, centralized resource that provides information on stem cells at the multiomic level. Stem cell databases, such as CSCdb, LifeMap Discovery, StemFormatics, and StemMapper, are among the most enriched and feasible databases that allow researchers to access data on various types of stem cells in specific organisms.

These databases offer several built-in features that facilitate comparison of gene expression and molecular profiling at intra and interspecies levels, linking in vitro and in vivo studies and initiating rapid data visualization. However, while these databases are valuable resources, they are handicapped in terms of data accuracy, privacy, technical limitations, regulatory, and ethical concerns.

This chapter emphasizes the need for databases with enhanced processing pipelines and new plugin tools to handle large datasets, as well as the importance of addressing issues related to data accuracy, privacy, and ethical concerns. To ensure that the scientific community has access to the best possible resources, databases need to be regularly updated with the most current datasets and provide unrestricted access to researchers.

Though current databases offer valuable resources, there is a need for ongoing improvements and updates to ensure that they remain accurate, secure, and in compliance with regulatory and ethical standards.

Acknowledgments

The authors are most grateful to NIH/NLM (U.S. National Institute of Health's National Library of Medicine) to access free full text scientific publications on PubMed Central (www.ncbi.nlm.nih.gov/pmc/), which was integral for the successful completion of this work. PKR designed and conceptualized the research; BB and PKR performed the analysis; BB and PKR wrote the original draft; BB prepared the original figures. BB, PKR, RC, AR, AS, and MS wrote, reviewed, edited, and proofread the manuscript.

References

Abou-El-Enein, M., Elsanhoury, A., & Reinke, P. (2016). Overcoming challenges facing advanced therapies in the EU market. *Cell Stem Cell, 19*(3), 293—297.

Agrawal, P., Raghav, P. K., Bhalla, S., Sharma, N., & Raghava, G. P. (2018). Overview of free software developed for designing drugs based on protein-small molecules interaction. *Current Topics in Medicinal Chemistry, 18*(13), 1146—1167.

Agrawal, P., Patiyal, S., Kumar, R., Kumar, V., Singh, H., Raghav, P. K., & Raghava, G. P. (2019). ccPDB 2.0: an updated version of datasets created and compiled from Protein Data Bank. *Database: The Journal of Biological Databases and Curation, 2019*(bay142).

Alisha, P., Raghav, P. K., & Gangenahalli, G. U. (2012). Computational network model predicts the drug effects on SHP-1 mediated intracellular signaling through c-Kit. *Journal of Proteins and Proteomics, 3*(2), 9.

Avior, Y., Biancotti, J. C., & Benvenisty, N. (2015). TeratoScore: Assessing the differentiation potential of human pluripotent stem cells by quantitative expression analysis of teratomas. *Stem Cell Reports, 4*(6), 967—974.

Bagger, F. O., Rapin, N., Theilgaard-Mönch, K., Kaczkowski, B., Thoren, L. A., Jendholm, J., Winther, O., & Porse, B. T. (2013). HemaExplorer: A database of mRNA expression profiles in normal and malignant haematopoiesis. *Nucleic Acids Research, 41*(D1), D1034—D1039.

Bagger, F. O., Sasivarevic, D., Sohi, S. H., Laursen, L. G., Pundhir, S., Sønderby, C. K., Winther, O., Rapin, N., & Porse, B. T. (2016). BloodSpot: A database of gene expression profiles and transcriptional programs for healthy and malignant haematopoiesis. *Nucleic Acids Research, 44*(D1), D917—D924.

Bergen, V., Lange, M., Peidli, S., Wolf, F. A., & Theis, F. J. (2020). Generalizing RNA velocity to transient cell states through dynamical modeling. *Nature Biotechnology, 38*(12), 1408—1414.

Bian, Q., & Cahan, P. (2016). Computational tools for stem cell biology. *Trends in Biotechnology, 34*(12), 993—1009.

Bidaut, G., & Stoeckert, C. J., Jr. (2009). Characterization of unknown adult stem cell samples by large scale data integration and artificial neural networks. *Biocomputing, 2009*, 356—367.

Bock, C., Kiskinis, E., Verstappen, G., Gu, H., Boulting, G., Smith, Z. D., Ziller, M., Croft, G. F., Amoroso, M. W., Oakley, D. H., Gnirke, A., Eggan, K., & Meissner, A. (2011). Reference Maps of human ES and iPS cell variation enable high-throughput characterization of pluripotent cell lines. *Cell, 144*(3), 439—452.

Briggs, J. A., Li, V. C., Lee, S., Woolf, C. J., Klein, A., & Kirschner, M. W. (2017). Mouse embryonic stem cells can differentiate via multiple paths to the same state. *Elife, 6*, e26945.

Cahan, P., Li, H., Morris, S. A., Lummertz da Rocha, E., Daley, G. Q., & Collins, J. J. (2014). CellNet: Network biology applied to stem cell engineering. *Cell, 158*(4), 903—915.

Cahan, P., Cacchiarelli, D., Dunn, S. J., Hemberg, M., de Sousa Lopes, S. M. C., Morris, S. A., Rackham, O. J. L., Del Sol, A., & Wells, C. A. (2021). Computational stem cell biology: Open questions and guiding principles. *Cell Stem Cell, 28*(1), 20—32.

Chen, J., Ding, L., Viana, M. P., Lee, H., Sluezwski, M. F., Morris, B., et al. (2018). The allen cell and structure segmenter: A new open source toolkit for segmenting 3D intracellular structures in fluorescence microscopy images. *bioRxiv, 491035*.

Chin, M. H., Pellegrini, M., Plath, K., & Lowry, W. E. (2010). Molecular analyses of human induced pluripotent stem cells and embryonic stem cells. *Cell Stem Cell, 7*(2), 263—269.

Cortes, C., & Vapnik, V. (1995). Support-vector networks. *Machine Learning, 20*, 273—297.

Del Sol, A., & Jung, S. (2021). The importance of computational modeling in stem cell research. *Trends in Biotechnology, 39*(2), 126—136.

Ebert, A. D., Yu, J., Rose, F. F., Jr., Mattis, V. B., Lorson, C. L., Thomson, J. A., & Svendsen, C. N. (2009). Induced pluripotent stem cells from a spinal muscular atrophy patient. *Nature, 457*(7227), 277—280.

Edgar, R., Mazor, Y., Rinon, A., Blumenthal, J., Golan, Y., Buzhor, E., Livnat, I., Ben-Ari, S., Lieder, I., Shitrit, A., Gilboa, Y., Ben-Yehudah, A., Edri, O., Shraga, N., Bogoch, Y., Leshansky, L., Aharoni, S., West, M. D., Warshawsky, D., & Shtrichman, R. (2013). LifeMap Discovery™: The embryonic development, stem cells, and regenerative medicine research portal. *PLoS One, 8*(7), e66629.

Eiges, R., Urbach, A., Malcov, M., Frumkin, T., Schwartz, T., Amit, A., Yaron, Y., Eden, A., Yanuka, O., Benvenisty, N., & Ben-Yosef, D. (2007). Developmental study of fragile X syndrome using human embryonic stem cells derived from preimplantation genetically diagnosed embryos. *Cell Stem Cell, 1*(5), 568—577.

Erwin, D. H., & Davidson, E. H. (2009). The evolution of hierarchical gene regulatory networks. *Nature Reviews Genetics, 10*(2), 141—148.

Ezzone, S. A. (May 2009). History of hematopoietic stem cell transplantation. *Seminars in Oncology Nursing, 25*(No. 2), 95—99.

Finkelstein, J., Parvanova, I., & Zhang, F. (2020). Informatics approaches for harmonized intelligent integration of stem cell research. *Stem Cells and Cloning: Advances and Applications*, 1—20.

Firdous, S., Ghosh, A., & Saha, S. (2022). BCSCdb: A database of biomarkers of cancer stem cells. *Database: The Journal of Biological Database and Curation, 2022*(baac082).

Flores, I., Benetti, R., & Blasco, M. A. (2006). Telomerase regulation and stem cell behaviour. *Current Opinion in Cell Biology, 18*(3), 254–260.

Grün, D., Muraro, M. J., Boisset, J. C., Wiebrands, K., Lyubimova, A., Dharmadhikari, G., van den Born, M., van Es, J., Jansen, E., Clevers, H., de Koning, E. J. P., & van Oudenaarden, A. (2016). De novo prediction of stem cell identity using single-cell transcriptome data. *Cell Stem Cell, 19*(2), 266–277.

Glass, C., Lafata, K. J., Jeck, W., Horstmeyer, R., Cooke, C., Everitt, J., et al. (2022). The role of machine learning in cardiovascular pathology. *Canadian Journal of Cardiology, 38*(2), 234–245.

Grün, D., & Grün, D. (2019). Lineage inference and stem cell identity prediction using single-cell RNA-sequencing data. *Computational Stem Cell Biology: Methods and Protocols,* 277–301.

Hamey, F. K., & Göttgens, B. (2019). Machine learning predicts putative hematopoietic stem cells within large single-cell transcriptomics data sets. *Experimental Hematology, 78,* 11–20.

Ho Sui, S. J., Begley, K., Reilly, D., Chapman, B., McGovern, R., Rocca-Sera, P., Maguire, E., Altschuler, G. M., Hansen, T. A. A., Sompallae, R., Krivtsov, A., Shivdasani, R. A., Armstrong, S. A., Culhane, A. C., Correll, M., Sansone, S. A., Hofmann, O., & Hide, W. (2012). The stem cell discovery engine: An integrated repository and analysis system for cancer stem cell comparisons. *Nucleic Acids Research, 40*(D1), D984–D991.

Hu, X., Cong, Y., Luo, H., Wu, S., Zhao, L., Liu, Q., & Yang, Y. (2017). Cancer stem cells therapeutic target database: The first comprehensive database for therapeutic targets of cancer stem cells. *Stem Cells Translational Medicine, 6*(2), 331–334.

Jain, A. K., Murty, M. N., & Flynn, P. J. (1999). Data clustering: A review. *ACM Computing Surveys, 31*(3), 264–323.

Jarrige, M., Polvèche, H., Carteron, A., Janczarski, S., Peschanski, M., Auboeuf, D., & Martinat, C. (2021). Sistema: A large and standardized collection of transcriptome data sets for human pluripotent stem cell research. *iScience, 24*(7), 102767.

Javan, M. R., Khosrojerdi, A., & Moazzeni, S. M. (2019). New insights into implementation of mesenchymal stem cells in cancer therapy: Prospects for anti-angiogenesis treatment. *Frontiers in Oncology,* 840.

Jensen, F. V. (1996). *An introduction to Bayesian networks* (Vol. 210, pp. 1–178). London: UCL press.

Jung, M., Peterson, H., Chavez, L., Kahlem, P., Lehrach, H., Vilo, J., & Adjaye, J. (2010). A data integration approach to mapping OCT4 gene regulatory networks operative in embryonic stem cells and embryonal carcinoma cells. *PLoS One, 5*(5), e10709.

Kamimoto, K., Hoffmann, C. M., & Morris, S. A. (2020). CellOracle: Dissecting cell identity via network inference and in silico gene perturbation. *bioRxiv,* 2020-02.

Koleti, A., Terryn, R., Stathias, V., Chung, C., Cooper, D. J., Turner, J. P., Vidovic, D., Forlin, M., Kelley, T. T., D'Urso, A., Allen, B. K., Torre, D., Jagodnik, K. M., Wang, L., Jenkins, S. L., Mader, C., Niu, W., Fazel, M., Mahi, N., … Schürer, S. C. (2018). Data portal for the library of integrated network-based cellular signatures (LINCS) program: Integrated access to diverse large-scale cellular perturbation response data. *Nucleic Acids Research, 46*(D1), D558–D566.

Kong, L., Aho, K. L., Granberg, K., Lund, R., Järvenpää, L., Seppälä, J., Gokhale, P., Leinonen, K., Hahne, L., Mäkelä, J., Laurila, K., Pukkila, H., Närvä, E., Yli-Harja, O., Andrews, P. W., Nykter, M., Lahesmaa, R., Roos, C., & Autio, R. (2013). ESTOOLS data@Hand: Human stem cell gene expression resource. *Nature Methods, 10*(9), 814–815.

Kulshrestha, S., Arora, T., Sengar, M., Sharma, N., Chawla, R., Bajaj, S., & Raghav, P. K. (2021). Advanced approaches and in silico tools of chemoinformatics in drug designing. In *Chemoinformatics and bioinformatics in the pharmaceutical sciences* (pp. 173–206). Academic Press.

Kumar, R., Lathwal, A., Kumar, V., Patiyal, S., Raghav, P. K., & Raghava, G. P. (2020). CancerEnD: A database of cancer associated enhancers. *Genomics, 112*(5), 3696–3702.

Kumar, R., Lathwal, A., Nagpal, G., Kumar, V., & Raghav, P. K. (2021). Impact of chemoinformatics approaches and tools on current chemical research. In *Chemoinformatics and bioinformatics in the pharmaceutical sciences* (pp. 1–26). Academic Press.

Kurtz, A., Elsallab, M., Sanzenbacher, R., & Abou-El-Enein, M. (2019). Linking scattered stem cell-based data to advance therapeutic development. *Trends in Molecular Medicine, 25*(1), 8–19.

Kurtz, A., Mah, N., Chen, Y., Fuhr, A., Kobold, S., Seltmann, S., & Müller, S. C. (2022). Human pluripotent stem cell registry: Operations, role and current directions. *Cell Proliferation, 55*(8), e13238.

La Manno, G., Soldatov, R., Zeisel, A., Braun, E., Hochgerner, H., Petukhov, V., Lidschreiber, K., Kastriti, M. E., Lönnerberg, P., Furlan, A., Fan, J., Borm, L. E., Liu, Z., van Bruggen, D., Guo, J., He, X., Barker, R., Sundström, E., Castelo-Branco, G., … Kharchenko, P. V. (2018). RNA velocity of single cells. *Nature, 560*(7719), 494–498.

Lafata, K. J., Chang, Y., Wang, C., Mowery, Y. M., Vergalasova, I., Niedzwiecki, D., Yoo, D. S., Liu, J. G., Brizel, D. M., & Yin, F. F. (2021). Intrinsic radiomic expression patterns after 20 Gy demonstrate early metabolic response of oropharyngeal cancers. *Medical Physics, 48*(7), 3767–3777.

Liang, P., Lan, F., Lee, A. S., Gong, T., Sanchez-Freire, V., Wang, Y., Diecke, S., Sallam, K., Knowles, J. W., Wang, P. J., Nguyen, P. K., Bers, D. M., Robbins, R. C., & Wu, J. C. (2013). Drug screening using a library of human induced pluripotent stem cell–derived cardiomyocytes reveals disease-specific patterns of cardiotoxicity. *Circulation, 127*(16), 1677–1691.

Makhoul, J. (1990). Artificial neural networks. *Investigative Radiology, 25*(6), 748–750.

Mallon, B. S., Chenoweth, J. G., Johnson, K. R., Hamilton, R. S., Tesar, P. J., Yavatkar, A. S., Tyson, L. J., Park, K., Chen, K. G., Fann, Y. C., & McKay, R. D. G. (2013). StemCellDB: The human pluripotent stem cell database at the National Institutes of Health. *Stem Cell Research, 10*(1), 57–66.

Menke-van der Houven van Oordt, C. W., Gomez-Roca, C., van Herpen, C., Coveler, A. L., Mahalingam, D., Verheul, H. M. W., van der Graaf, W. T. A., Christen, R., Rüttinger, D., Weigand, S., Cannarile, M. A., Heil, F., Brewster, M., Walz, A. C., Nayak, T. K., Guarin, E., Meresse, V., & Le Tourneau, C. (2016). First-in-human phase I clinical trial of RG7356, an anti-CD44 humanized antibody, in patients with advanced, CD44-expressing solid tumors. *Oncotarget, 7*(48), 80046.

Merrill, E., Gehlenborg, N., Haseley, P., Sytchev, I., Park, R., Rocca-Serra, P., Corlosquet, S., Gonzalez-Beltran, A., Maguire, E., Hofmann, O., & Park, P. (2013). The Stem Cell Commons: an exemplar for data integration in the biomedical domain driven by the ISA framework. *AMIA Joint Summits on Translational Science Proceedings. AMIA Joint Summits on Translational Science,* 70–70.

Montrone, C., Kokkaliaris, K. D., Loeffler, D., Lechner, M., Kastenmüller, G., Schroeder, T., & Ruepp, A. (2013). HSC-Explorer: A curated database for hematopoietic stem cells. *PLoS One, 8*(7), e70348.

Müller, F. J., Schuldt, B. M., Williams, R., Mason, D., Altun, G., Papapetrou, E. P., Danner, S., Goldmann, J. E., Herbst, A. Schmidt, N. O., Aldenhoff, J. B., Laurent, L. C., & Loring, J. F. (2011). A bioinformatic assay for pluripotency in human cells. *Nature Methods, 8*(4), 315–317.

Nandan, A., Raghav, P. K., Srivastava, A., Tiwari, S. K., Shukla, A. K., & Sharma, V. (2021). Current insights to therapeutic targets of ROS induced gastric cancer stem cells. In *Handbook of oxidative stress in cancer: Therapeutic aspects* (pp. 1–13). Singapore: Springer Singapore.

Neagu, A., van Genderen, E., Escudero, I., Verwegen, L., Kurek, D., Lehmann, J., Stel, J., Dirks, R. A. M., van Mierlo, G., Maas, A., Eleveld, C., Ge, Y., den Dekker, A. T., Brouwer, R. W. W., van IJcken, W. F. J., Modic, M., Drukker, M., Jansen, J. H., Rivron, N. C., … Ten Berge, D. (2020). In vitro capture and characterization of embryonic rosette-stage pluripotency between naive and primed states. *Nature Cell Biology, 22*(5), 534–545.

Ning, S. T., Lee, S. Y., Wei, M. F., Peng, C. L., Lin, S. Y. F., Tsai, M. H., Lee, P. C., Shih, Y. H., Lin, C. Y., Luo, T. Y., & Shieh, M. J. (2016). Targeting colorectal cancer stem-like cells with anti-CD133 antibody-conjugated SN-38 nanoparticles. *ACS Applied Materials & Interfaces, 8*(28), 17793–17804.

Okawa, S., Saltó, C., Ravichandran, S., Yang, S., Toledo, E. M., Arenas, E., & Del Sol, A. (2018). Transcriptional synergy as an emergent property defining cell subpopulation identity enables population shift. *Nature Communications, 9*(1), 2595.

Pearson, K. (1901). LIII. On lines and planes of closest fit to systems of points in space. *The London, Edinburgh and Dublin Philosophical Magazine and Journal of Science, 2*(11), 559–572.

Pimanda, J. E., & Göttgens, B. (2010). Gene regulatory networks governing haematopoietic stem cell development and identity. *International Journal of Developmental Biology, 54*(6–7), 1201–1211.

Pinto, J. P., Reddy Kalathur, R. K., Machado, R. S., Xavier, J. M., Bragança, J., & Futschik, M. E. (2014). StemCellNet: An interactive platform for network-oriented investigations in stem cell biology. *Nucleic Acids Research, 42*(W1), W154–W160.

Pinto, J. P., Machado, R. S. R., Magno, R., Oliveira, D. V., Machado, S., Andrade, R. P., Bragança, J., Duarte, I., & Futschik, M. E. (2018). StemMapper: A curated gene expression database for stem cell lineage analysis. *Nucleic Acids Research, 46*(D1), D788–D793.

Rackham, O. J., Firas, J., Fang, H., Oates, M. E., Holmes, M. L., Knaupp, A. S., FANTOM Consortium, Suzuki, H., Nefzger, C. M., Daub, C. O., Shin, J. W., Petretto, E., Forrest, A. R. R., Hayashizaki, Y., Polo, J. M., & Gough, J. (2016). A predictive computational framework for direct reprogramming between human cell types. *Nature Genetics, 48*(3), 331–335.

Raghav, P. K., & Gangenahalli, G. (2018). Hematopoietic stem cell molecular targets and factors essential for hematopoiesis. *Journal of Stem Cell Research & Therapy, 8*(441), 2.

Raghav, P. K., & Gangenahalli, G. (2021). PU.1 mimic synthetic peptides selectively bind with GATA-1 and allow c-Jun PU.1 binding to enhance myelopoiesis. *International Journal of Nanomedicine, 16*, 3833.

Raghav, P. K., & Mann, Z. (2021). Cancer stem cells targets and combined therapies to prevent cancer recurrence. *Life Sciences, 277*, 119465.

Raghav, P. K., Verma, Y. K., & Gangenahalli, G. U. (2011). Silico analysis of flexible loop domain's conformational changes affecting BH3 cleft of Bcl-2 protein. *Journal of Natural Science, Biology and Medicine, 2*(3), 56.

Raghav, P. K., Verma, Y. K., & Gangenahalli, G. U. (2012). Molecular dynamics simulations of the Bcl-2 protein to predict the structure of its unordered flexible loop domain. *Journal of Molecular Modeling, 18*, 1885–1906.

Raghav, P. K., Singh, A. K., & Gangenahalli, G. (2018a). A change in structural integrity of c-Kit mutant D816V causes constitutive signaling. *Mutation Research, Fundamental and Molecular Mechanisms of Mutagenesis, 808*, 28–38.

Raghav, P. K., Singh, A. K., & Gangenahalli, G. (2018b). Stem cell factor and NSC87877 combine to enhance c-Kit mediated proliferation of human megakaryoblastic cells. *PLoS One, 13*(11), e0206364.

Raghav, P. K., Kumar, R., & Raghava, G. P. (2019). Machine learning based identification of stem cell genes involved in stemness. *Journal of Cell Science & Therapy, 10*, 40. https://doi.org/10.4172/2157-7013-C1-049

Raghav, P. K., Kumar, R., Kumar, V., & Raghava, G. P. (2019). Docking-based approach for identification of mutations that disrupt binding between Bcl-2 and Bax proteins: Inducing apoptosis in cancer cells. *Molecular Genetics & Genomic Medicine, 7*(11), e910.

Raghav, P. K., Mann, Z., Ahlawat, S., & Mohanty, S. (2021). Mesenchymal stem cell-based nanoparticles and scaffolds in regenerative medicine. *European Journal of Pharmacology*, 174657.

Raghav, P. K., Mann, Z., Krishnakumar, V., & Mohanty, S. (2021). Therapeutic effect of natural compounds in targeting ROS-induced cancer. In *Handbook of oxidative stress in cancer: Mechanistic aspects* (pp. 1–47). Singapore: Springer Singapore.

Raghav, P. K., Mann, Z., Pandey, P. K., & Mohanty, S. (2021). Systems biology resources and their applications to understand the cancer. In *Handbook of oxidative stress in cancer: Mechanistic aspects* (pp. 1–35). Singapore: Springer Singapore.

Raghav, P. K., Mann, Z., & Mohanty, S. (2022). Therapeutic potential of chemical compounds in targeting cancer stem cells. In S. Chakraborti (Ed.), *Handbook of oxidative stress in cancer: Therapeutic aspects*. Singapore: Springer. https://doi.org/10.1007/978-981-16-5422-0_87

Raghav, P. K. (2022). Hematopoietic stem cell factors: Their functional role in self-renewal and clinical aspects. *Frontiers in Cell and Developmental Biology, 453*.

Rawat, S., Jain, K. G., Gupta, D., Raghav, P. K., Chaudhuri, R., Pinky, Shakeel, A., Arora, V., Sharma, H., Debnath, D., Kalluri, A., Agrawal, A. K., Jassal, M., Dinda, A. K., Patra, P., & Mohanty, S. (2021). Graphene nanofiber composites for enhanced neuronal differentiation of human mesenchymal stem cells. *Nanomedicine, 16*(22), 1963–1982.

Rizvi, A. H., Camara, P. G., Kandror, E. K., Roberts, T. J., Schieren, I., Maniatis, T., & Rabadan, R. (2017). Single-cell topological RNA-seq analysis reveals insights into cellular differentiation and development. *Nature Biotechnology, 35*(6), 551−560.

Roost, M. S., Van Iperen, L., Ariyurek, Y., Buermans, H. P., Arindrarto, W., Devalla, H. D., Passier, R., Mummery, C. L., Carlotti, F., de Koning, E. J. P., van Zwet, E. W., Goeman, J. J., & Chuva de Sousa Lopes, S. M. (2015). KeyGenes, a tool to probe tissue differentiation using a human fetal transcriptional atlas. *Stem Cell Reports, 4*(6), 1112−1124.

Sánchez-Castillo, M., Ruau, D., Wilkinson, A. C., Ng, F. S., Hannah, R., Diamanti, E., Lombard, P., Wilson, N. K., & Gottgens, B. (2015). CODEX: A next-generation sequencing experiment database for the haematopoietic and embryonic stem cell communities. *Nucleic Acids Research, 43*(D1), D1117−D1123.

Sandie, R., Palidwor, G. A., Huska, M. R., Porter, C. J., Krzyzanowski, P. M., Muro, E. M., Perez-Iratxeta, C., & Andrade-Navarro, M. A. (2009). Recent developments in StemBase: A tool to study gene expression in human and murine stem cells. *BMC Research Notes, 2*(1), 1−6.

Scheubert, L., Schmidt, R., Repsilber, D., Luštrek, M. I. T. J. A., & Fuellen, G. (2011). Learning biomarkers of pluripotent stem cells in mouse. *DNA Research, 18*(4), 233−251.

Sengar, A., Sengar, M., Mann, Z., & Raghav, P. K. (2022). Clinical approaches in targeting ROS-induced cancer. In S. Chakraborti (Ed.), *Handbook of oxidative stress in cancer: Therapeutic aspects.* Singapore: Springer. https://doi.org/10.1007/978-981-16-1247-3_256-1

Shen, Y., Yao, H., Li, A., & Wang, M. (2016). CSCdb: A cancer stem cells portal for markers, related genes and functional information. *Database, 2016*(baw023).

Som, A., Harder, C., Greber, B., Siatkowski, M., Paudel, Y., Warsow, G., Cap, C., Schöler, H., & Fuellen, G. (2010). The PluriNetWork: An electronic representation of the network underlying pluripotency in mouse, and its applications. *PLoS One, 5*(12), e15165.

Soufi, A., Donahue, G., & Zaret, K. S. (2012). Facilitators and impediments of the pluripotency reprogramming factors' initial engagement with the genome. *Cell, 151*(5), 994−1004.

Srinivasan, M., Thangaraj, S. R., Ramasubramanian, K., Thangaraj, P. P., & Ramasubramanian, K. V. (2021). Exploring the current trends of artificial intelligence in stem cell therapy: A systematic review. *Cureus, 13*(12).

Streeter, I., Harrison, P. W., Faulconbridge, A., The HipSci Consortium, Flicek, P., Parkinson, H., & Clarke, L. (2017). The human-induced pluripotent stem cell initiative—data resources for cellular genetics. *Nucleic Acids Research, 45*(D1), D691−D697.

Stumpf, P. S., Smith, R. C., Lenz, M., Schuppert, A., Müller, F. J., Babtie, A., Chan, T. E., Stumpf, M. P. H., Please, C. P., Howison, S. D., Arai, F., & MacArthur, B. D. (2017). Stem cell differentiation as a non-Markov stochastic process. *Cell Systems, 5*(3), 268−282.

Takahashi, K., Okita, K., Nakagawa, M., & Yamanaka, S. (2007). Induction of pluripotent stem cells from fibroblast cultures. *Nature Protocols, 2*(12), 3081−3089.

Turing, A. M. (1990). The chemical basis of morphogenesis. *Bulletin of Mathematical Biology, 52*(1−2), 153−197.

Ulloa-Montoya, F., Kidder, B. L., Pauwelyn, K. A., Chase, L. G., Luttun, A., Crabbe, A., Geraerts, M., Sharov, A. A., Piao, Y., Ko, M. S. H., Hu, W. S., & Verfaillie, C. M. (2007). Comparative transcriptome analysis of embryonic and adult stem cells with extended and limited differentiation capacity. *Genome Biology, 8*(8), 1−20.

Van De Leemput, J., Boles, N. C., Kiehl, T. R., Corneo, B., Lederman, P., Menon, V., Lee, C., Martinez, R. A., Levi, B. P., Thompson, C. L., Yao, S., Kaykas, A., Temple, S., & Fasano, C. A. (2014). CORTECON: A temporal transcriptome analysis of in vitro human cerebral cortex development from human embryonic stem cells. *Neuron, 83*(1), 51−68.

Velten, L., Haas, S. F., Raffel, S., Blaszkiewicz, S., Islam, S., Hennig, B. P., Hirche, C., Lutz, C., Buss, E. C., Nowak, D., Boch, T., Hofmann, W. K., Ho, A. D., Huber, W., Trumpp, A., Essers, M. A. G., & Steinmetz, L. M. (2017). Human haematopoietic stem cell lineage commitment is a continuous process. *Nature Cell Biology, 19*(4), 271−281.

Verstockt, B., Ferrante, M., Vermeire, S., & Van Assche, G. (2018). New treatment options for inflammatory bowel diseases. *Journal of Gastroenterology, 53*, 585−590.

Warsow, G., Greber, B., Falk, S. S., Harder, C., Siatkowski, M., Schordan, S., et al. (2010). ExprEssence-revealing the essence of differential experimental data in the context of an interaction/regulation network. *BMC Systems Biology, 4*, 1−18.

Wells, C. A., Mosbergen, R., Korn, O., Choi, J., Seidenman, N., Matigian, N. A., et al. (2013). Stemformatics: Visualisation and sharing of stem cell gene expression. *Stem Cell Research, 10*(3), 387−395.

Woolf, P. J., Prudhomme, W., Daheron, L., Daley, G. Q., & Lauffenburger, D. A. (2005). Bayesian analysis of signaling networks governing embryonic stem cell fate decisions. *Bioinformatics, 21*(6), 741−753.

Xu, H., Lemischka, I. R., & Ma'ayan, A. (2010). SVM classifier to predict genes important for self-renewal and pluripotency of mouse embryonic stem cells. *BMC Systems Biology, 4*, 1−10.

Xu, H., Baroukh, C., Dannenfelser, R., Chen, E. Y., Tan, C. M., Kou, Y., Kim, Y. E., Lemischka, I. R., & Ma'ayan, A. (2013). ESCAPE: Database for integrating high-content published data collected from human and mouse embryonic stem cells. *Database, 2013*(bat045).

Yu, J., Xing, X., Zeng, L., Sun, J., Li, W., Sun, H., He, Y., Li, J., Zhang, G., Wang, C., Li, Y., & Xie, L. (2012). SyStemCell: A database populated with multiple levels of experimental data from stem cell differentiation research. *PLoS One, 7*(7), e35230.

Zakrzewski, W., Dobrzyński, M., Szymonowicz, M., & Rybak, Z. (2019). Stem cells: Past, present, and future. *Stem Cell Research & Therapy, 10*, 1−22.

Zaret, K. S., & Grompe, M. (2008). Generation and regeneration of cells of the liver and pancreas. *Science, 322*(5907), 1490−1494.

Zhou, J., Zhu, F., Li, J., & Wang, Y. (2018). Bioinformatics approaches to stem cell research. *Current Pharmacology Reports, 4*, 314−325.

Zhou, Q., Chipperfield, H., Melton, D. A., & Wong, W. H. (2007). A gene regulatory network in mouse embryonic stem cells. *Proceedings of the National Academy of Sciences, 104*(42), 16438−16443.

Zhou, X., Meng, G., Nardini, C., & Mei, H. (2017). Systemic evaluation of cellular reprogramming processes exploiting a novel R-tool: Eegc. *Bioinformatics, 33*(16), 2532−2538.

Chapter 21

Deciphering the complexities of stem cells through network biology approaches for their application in regenerative medicine

Priyanka Narad[1] and Simran Tandon[2]

[1]Amity Institute of Biotechnology, Amity University, Noida, Uttar Pradesh, India; [2]Amity School of Health Sciences, Amity University, Mohali, Punjab, India

1. Introduction

A human cell contains a repertoire of pathways that interact and cross-talk with each other, thereby leading to cell-specific responses. On an average, thousands of genes and proteins are expressed, which carry out diverse cellular activities leading to specific responses. To understand the role of the plethora of molecular players that are present at the cellular level, it is essential to have a system that can bring together the vast ocean of "omic" data that are generated at the level of either the gene, RNA, protein, or secretome. The advent of bioinformatics-based tools has led to the dawn of a new era wherein the data generated at various levels are applied to a network system. This network system operates at either the gene or protein level or at the level of the highly complex molecular signaling circuits. To reach this level of understanding, it is essential to start with a basic understanding of network-based approaches. In layman's language, a network is a series of connections or interactions. If we apply this network to the intracellular milieu, then this network system would include interactions and outcomes between genes, their transcription factors, promoters, repressors, enzymes, etc. This level of interaction would also be prevalent at the level of RNA and at the protein—protein level. Furthermore, interaction at the metabolic level would also be required so that the outcome be attributed to the sum of all these molecular events. This leads to a mind-boggling amount of data that needs to be mined to decipher what processes are operating at the cellular level, which leads to the proper functioning of the organism.

Stem cells have attracted wide-ranging interest in the scientific community due to their promising potential. However, this potential can act as a double-edged sword, which has to be wielded with caution. Embryonic stem cells (ESCs) are undifferentiated and act through complex signaling pathways. These pathways are responsible for their unique cellular activities of self-renewal, wherein these cells can continue to divide and renew repeatedly. These cells can differentiate into the panoply of specialized cells that make up the organism, which is termed as pluripotency. The stem cells can give rise to identical daughter cells through symmetric division, thereby maintaining the stem cell pool. Another option, depending on the requirement of the body, is for the stem cells to divide in an asymmetric manner to give rise to one stem cell and a daughter cell. This daughter cell then progresses down the differentiation pathway (Melton, 2009). Stem cells have been utilized as model systems in variable settings to assess their unlimited division potential and the extent to which they can differentiate into various cell types of the three lineages, i.e., ectoderm, mesoderm, and endoderm. This unique ability to divide indefinitely and differentiate has been and still is a hot topic in stem cell research (Blau et al., 2001; Weissman et al., 2001).

1.1 Stem cell hierarchy

The hierarchy of stem cells, which is based on their ability to differentiate into various cell types, can be summarized as either totipotent, pluripotent, multipotent, or unipotent (Zakrzewski et al., 2019). The cells with totipotent potential have

Computational Biology for Stem Cell Research. https://doi.org/10.1016/B978-0-443-13222-3.00016-2
Copyright © 2024 Elsevier Inc. All rights are reserved, including those for text and data mining, AI training, and similar technologies.

the ability to divide and differentiate into the 200 cell types present in the human body and hence are capable of generating the entire organism. Hence, these totipotent stem cells can generate both the embryonic and extraembryonic components. A fertilized egg with up to eight cells is an example of a totipotent cell (Larijani et al., 2012). The ESCs that are found in the inner cell mass of a preimplanted embryo are examples of pluripotent stem cells. Although pluripotent stem cells cannot generate extraembryonic structures such as the placenta, they can technically generate the three germ layers, namely, ectoderm, mesoderm, and endoderm and are said to be pluripotent. For research purposes, ESC lines can be derived from spare or discarded embryos available from in vitro fertility (IVF) clinics, after taking the necessary permissions. Experimentally, somatic cells, such as those isolated from the skin, can be manipulated through the introduction of a cocktail of transcription factors or small molecules. These cells then recapitulate the transcriptional machinery of pluripotent stem cells. Japanese scientists successfully created cells, which they named as induced pluripotent stem cells or iPSCs. This pathbreaking work by Shinya Yamanaka's group led to a revolution in stem cell research (Takahashi et al., 2007). Over the years these cells have become an invaluable and crucial tool for scientists to better understand developmental aspects as well as the biology for the onset and progression of various diseases. Furthermore, iPSCs can serve as functional in vitro and in vivo systems, which can be exploited for therapeutic approaches, drug discovery, and regenerative medicine. The tissue-specific, multipotent stem cells have a comparatively limited potential to differentiate and hence give rise to specialized cell lineages, which are needed in a particular tissue. An example of multipotent stem cell is the hematopoietic stem cell (HSC), which differentiates into the myeloid and lymphoid lineages, which give rise to the various blood cells (Dzierzak, 2002). The HSCs normally cannot differentiate into cells of other tissues but, under specific experimental conditions, can be induced to transdifferentiate into either the endodermal or ectodermal lineage. The major function of adult multipotent stem cells is to replace old or damaged cells to maintain homeostasis. In the hierarchy of stem cells, unipotent stem cells are comparatively limited in their differentiation capacity; however, they can undergo repeated division. As these unipotent stem cells can give rise to one specific cell type, this property makes it invaluable to regenerative medicine (Mahla, 2016). One striking difference between the pluripotent ES and multipotent/unipotent adult stem cells is that the adult stem cells lack the ability to be isolated in large numbers owing to their low number in the tissues and stem cell niches, coupled with the want of robust markers (Elkabetz & Studer, 2008).

1.2 Basics of network systems

The sea of experimental data derived through various experimental interventions and conditions in the biological system may appear to be a disconnected circuit to a biologist. The reason for this is the apparent complexity and cross-talk that is occurring under a dynamic umbrella. To make sense of all these data would have been an uphill task a few decades ago; however, with the rapid advancement of computers in the field of biology, this herculean task seems to have miraculously transformed the landscape of cell biology. Using specific types of graphs, the data generated through experimental studies can be visualized, wherein each discrete entity or element, i.e., gene, RNA, protein, or metabolite, is represented graphically as a node. These molecular components or nodes can communicate or interact directly or indirectly with each other and are depicted as links. Hence, biological interactions of the various components can be envisioned graphically through networks as discrete systems of elements that interact through links (Junker & Schreiber, 2011). The graphs that depict the various biological interactions are varied. Some examples of the types of graphs include Bayesian and Boolean networks and other interesting names such as trees and forests.

To be able to build or create networks, the researcher needs to first define what type of network is to be built. This information would need to be conveyed visually through a specific type of graph. The graph in turn would contain nodes and links, depicting the possible direct or indirect interactions. The requirement of a specific method or software program would help to build a network that could convey the information. Various free software packages for noncommercial use are available to researchers and include Cytoscape (Shannon et al., 2003), Pajek (Mrvar & Batagelj, 2016), Genes2Networks (Berger et al., 2007), FANMOD (Wernicke & Rasche, 2006), etc. There are other segments of software that are licensed and require a subscription. The working principles of the very popular free-to-use software Cytoscape and its applications would be discussed later in this chapter.

1.3 Applications of stem cells

To successfully treat and manage various conditions, in-depth knowledge of the pathways influencing disease progression is crucial. Many researchers (Mahla, 2016) have stated that diseases could be better studied if suitable and reliable disease models were available. For a model system to be successful, it needs to mimic the human cellular milieu and be able to perform the various metabolic activities required. Animal models of specific disease conditions have been created for research studies and include animals such as rats, mice, dogs, monkeys, etc. However, their use is limited due to ethical, legal, and social issues. In addition, variations in the genetic makeup of these animal models could lead to differential interspecies variations, thereby making it difficult to extrapolate the research findings. Stem cells offer a very attractive and reproducible model system wherein using network analysis the possible interactions and their biological consequences on disease progression can be studied. Therefore, the various pathways influencing stem cell biology need to be addressed so as to understand the highly complex and interconnected networks governing stem cell fate.

2. Key genes and pathways regulating pluripotent stem cell fate

ESCs through their differentiating ability induce the formation of the three germ layers that culminate in normal development in the organism. Several key pathways are operative within the ESCs and allow them to self-renew and give rise to differentiated cell types of specific lineages. The critical genes within these pathways are subject to transcriptional activation and deactivation, thereby giving rise to distinct tissue patterns and the survival of the organism through cellular turnover. These genes have also been exploited to experimentally induce a pluripotent state in somatic cells, leading to the acquisition of stem cell—like properties.

2.1 A Krüppel-like family of transcription factors

Krüppel-like factors or SP1/KLF are a family of over 19 transcription factors that contain conserved zinc finger motifs that bind to GC-rich or CACCC regions in the promoters and enhancers of various genes (McConnell & Yang, 2010). These TFs play a pivotal role in various biological processes that contribute to embryonic development. KLF4 TFs are transcriptional activators that are highly expressed in the pluripotent stem cells. However, upon differentiation, their levels dramatically drop suggesting their key contribution in the maintenance of pluripotency and lineage differentiation (Ralston & Rossant, 2010). According to Lomberk and Urrutia (2005), the highly conserved C terminal zinc finger domains of the KLF4 TFs are responsible for modulating gene activity. They carry out this function by acting as transcriptional activators or inhibitors. The amino terminal domain of the protein is responsible for protein—protein and protein—DNA binding. Studies have revealed that mutations that lead to the loss of KLF4 function have a bearing on ectodermal and endodermal lineage differentiation, but are nonlethal. The role of various TFs including KLF4 POU5F1 and sex-determining region Y-box 2 (SOX2) in the initiation of reprogramming cannot be undermined (Nakagawa et al., 2008). Genome-wide analysis studies (GWAS) have thrown light on the various targets modulated by KLF4 binding, which include genes involved in pluripotency, namely, POU5F1, SOX2, and Nanog. KLF4 also modulates its activity by binding to its own promoter. Studies have reported that Klf4 inhibits ESC differentiation by binding to the Nanog promoter (Zhang et al., 2010).

2.2 Sex-determining region Y-box 2

The potential of the stem cells to self-renew can be attributed to certain TFs, which directly or indirectly influence genes controlling self-renewal. Sox2 by modulating Oct4 expression in turn influences Nanog expression. This interplay between various TFs provides strong evidence that Sox2 is a major contributor to the maintenance of stemness (Liu et al., 2013). This property of Sox2 to influence pluripotency has been exploited for somatic cell reprogramming for iPSC generation (Yu et al., 2007). However, there is a darker side to this ability of Sox2 to drive stemness as this could influence the development of the cancer-initiating cells or cancer stem cells (CSCs). The aberrant overexpression of Sox2 has been linked with tumorigenesis, and its role in the maintenance of CSC is now being highlighted (Moncho-Amor et al., 2021).

2.3 Octamer-binding transcription factor 4

By far, the most well-studied TF governing stem cell fate is Oct4. Numerous positive and negative regulators bind to the Oct4 gene and thereby control its expression (Goradel et al., 2018). The influence of Oct4 on differentiation at a very early stage of development is, however, debatable (Han et al., 2014). Oct4 plays a critical role in controlling proliferation by modulating the activity of proteins involved in the mammalian cell cycle. By downregulating the expression of the negative regulator of the cell cycle PP1 (protein phosphatase 1) during early and middle phases of G1, Oct4 leads to an increase in CDK4/6-cyclin D activity. This subsequently results in the decreased levels of hypophosphorylated form of the tumor suppressor protein RB (retinoblastoma), which is a major negative regulator of the cell cycle. The consequence is increased proliferation of the stem cells. The complex of CDK1-cyclin B results in the enhanced interaction of Oct4 to its promoter, thereby decreasing the expression of the differentiation marker CDX2 in the G2/M phase of the cell cycle, reiterating its self-renewal characteristics.

The transcription of various genes is controlled by the chromatin organization. Chromatin that is tightly wound is inaccessible to the various enzymes and TFs and results in repression; however, a more open, loosely packed form would be transcriptionally active. Studies report that these chromatin complexes interact with Oct4. Interestingly, proteins that can remodel the chromatin are also targets of Oct4 (Avilion et al., 2003). Growing evidence points to the critical role of Oct4 signaling in CSCs, and Oct4 has been shown to be highly expressed in a wide variety of cancers. The ability of Oct 4 to promote proliferation, and invasion through the epithelial-mesenchymal transitions and chemo- and radioresistance underscores its role in tumorigenesis, leading to a poor prognosis (Zhao et al., 2018).

2.4 c-MYC

MYC is a basal helix−loop−helix/leucine zipper TF that plays a crucial role in a wide number of biological activities, namely, cellular growth, proliferation, and differentiation, which interestingly are key hallmarks of stem cells. The ability of myc to influence cell cycle and apoptosis suggests that it is a critical player in tumorigenesis (Dang, 2013). The aberrant expression of c-myc has been implicated in various cancers including neuroblastomas and in the development of CSCs (Chandrasekaran et al., 2017).

2.5 Nanog

Along with the various TFs that have been discussed till now, another important TF involved in the self-renewal ability of ESC is Nanog. By regulating critical downstream targets, Nanog influences the properties of undifferentiated ESCs and hence the fate of the organism (Chen et al., 2008). Normal ESC subpopulations appear to express varying levels of Nanog. It has been seen that cells with low Nanog expression tend to undergo a greater degree of differentiation and display reduced self-renewal. By acting as a molecular switch, changes in the Nanog levels enable ESCs to quickly decide their fate by undergoing either self-renewal or differentiation (Hadjimichael et al., 2015). Nanog displays a multifaceted role as it not only influences the expression and activity of various TFs involved in the maintenance of pluripotency, but it also modulates its own levels (Torres-Padilla & Chambers, 2014).

Nanog along with the other two highly conserved TFs, namely, Oct4 and Sox2 co-occupy the promoters of various target genes and work as crucial transcriptional regulatory circuits in the maintenance of pluripotency and hence stem cell fate (Mossahebi-Mohammadi et al., 2020).

2.6 FGF signaling pathway

The fibroblast growth factor (FGF) family of mitogenic growth factors contain 22 members having similar structures, functions, and biochemical properties. FGF ligands act by binding and activating their cognate transmembrane tyrosine kinase receptors. These receptors have been designated as FGFR1, FGFR2, FGFR3, and FGFR4. FGF signaling acts through various intracellular effectors, namely, RAS/MAPK, PI3K/AKT, and PLC. Various FGF ligands affect homeostasis and embryonic development, and it has been seen that except for a few FGF ligands (FGF11, 12, 13, and 14), most of the FGF signaling molecules bind to their specific receptors and lead to various mitotic and nonmitotic responses. Kunath et al. (2007) demonstrated the significance of the activation of the FGF4/ERK signaling pathway in mESC polyclonal differentiation.

MAPK signaling is integral to the maintenance of stemness, whereas upregulated PI3K/AKT signaling is seen during differentiation. Modulation of various signals is therefore necessary for stem cells to progress through specific lineage-limiting phases (Kunath et al., 2007).

2.7 TGF-β signaling pathway

There are 33 genes that code for structurally similar polypeptides that are precursors to ligands that make up the TGF family (Sakaki-Yumoto et al., 2013). Signaling through TGF ligands such as nodal and activin is mediated by cell surface and transmembrane kinase type I and type II receptors. There are a large number of players in this pathway, which ultimately act by activating the intracellular Smad TFs (Sakaki-Yumoto et al., 2013). The variable expression of these players present in either the stem cell niche or intracellularly in the stem cells throws light on of the varied responses of TGF-family signaling molecules (Dahle et al., 2010). Smads form complexes with other TFs, thereby leading to the modulation of gene transcription through coactivators and corepressors of this pathway. Smad4 which is a member of the Smad family, serves as a coactivator of the Smad-mediated transcriptional regulation. The Smad TFs play a significant role in governing stem cell responses by regulating the transcriptional activation or repression of crucial genes involved in modulating stem/progenitor cell fate. Not only is this a crucial pathway for the maintenance of pluripotency, but it also plays a role in differentiation (Li et al., 2016).

The Smad-independent signaling pathways, which are usually regarded as significant receptor tyrosine kinase effector pathways, are stimulated by TGF-β as well. TGF-β stimulates the activation of numerous non-Smad pathways by interacting with signaling mediators and type I or type II receptors, either directly or indirectly via the contribution of adaptor proteins. The phosphorylated tyrosine on ShcA serves as a docking site after the recruitment of signaling member proteins Grb2 and Sos, and this complex initiates the Ras activation process, which causes the Erk-MAPK signaling cascade. It further needs to be mentioned that the TGF-induced Smad-independent signaling also plays a pivotal role in stem cells (Li et al., 2016).

2.8 PI3K-Akt signaling pathway

The role of the PI3K pathway in controlling stemness was gathered from elegant studies in which constitutive activation of the pathway by knocking out the PTEN gene led to uncontrolled proliferation of murine and human embryonic stem cell (hESCs) along with the increased expression of Nanog and Oct 4 (Alva et al., 2011; Di Cristofano et al., 1998). Reports have shown that inhibition of this pathway using small molecules results in loss in the undifferentiated state as well as decreased self-renewal in hESCs (Yu et al., 2015).

This pathway can be activated by extracellular ligands and intracellularly through a cross-talk with Ras (Chen et al., 2013) and various other pathways. According to a report (Singh, Bechard, et al., 2012; Singh, Reynolds, et al., 2012), PI3K can inhibit GSK3β and influence the canonical or classical Wnt pathway by its association with MAPK/ERK downstream effectors. Interestingly, results have shown that PI3K has a critical role in signaling involved in reprograming as it promotes glycolysis by upregulating the expression of key genes of the glycolysis pathway (Park et al., 2014).

A number of ligands can activate PI3K/AKT signaling, and studies have shown that insulin and insulin-like growth factor (IGF) can influence stem cell fate. A dual role of FGF2 reveals that while low levels of FGF2 may reduce ERK activity, higher levels appear to activate PI3K/AKT signaling (Dalton, 2013). In addition, there appears to be a cross-talk between the TGF-β and PI3K pathway, which strives to maintain pluripotency through SMAD2,3 signaling. Although the precise molecular mechanism by which PI3K/AKT affects pluripotency is largely unknown, studies suggest that it maintains the activity of mTOR, which has been shown to be a modulator of TGF-β, and may play a significant role in controlling protein synthesis. This shows that mTOR and AKT in hPSCs have a feedback mechanism. C-myc, as discussed earlier, is a crucial regulator of PSC maintenance and is stabilized as a result of the inhibition of GSK3-β (Singh, Bechard, et al., 2012; Singh, Reynolds, et al., 2012).

The outcome of these various pathways acting within the pluripotent and multipotent stem cells leads to metabolic changes, which in turn influence those networks whose role is to control the stem cell fate.

3. Bioinformatics in stem cell research

The molecular structure of stem cells is a result of the complex interaction of a number of cellular processes. Some of these can be listed as (1) pathways, (2) regulation of transcription, (3) epigenetic factors, and (4) miRNA, and their intracellular interaction helps the stem cells to maintain the pluripotent state. Several cross-talks between signaling proteins and the transcriptional regulators make certain for the stem cells to maintain their state of potency. With the start of the post-genomic age, data generation has become a boon for researchers working with stem cells. The problem is an enhanced quantity of data for applications in regenerative medicine and its understanding across the different formats and resources. The high-throughput technologies RNA-seq, Single Seq, ChIP Seq, etc. are facilitating data points to be created at an exceptional rate that is noncurated. As of November 2022, a text-based mining through the NCBI database with the term "Stem Cell Bioinformatics" generates almost 17,030 PubMed articles, 1261 nucleotide sequences, 502 BioProjects, and 566 GEO datasets. The objective of the researchers is to capture the complete picture of the biological process, which is possible through a network-based representation.

3.1 Biological network representation and analysis

Complex biological systems can be represented through network representations. Biological networks are structured to reduce the complexities of a process and improve understanding of the state. The objective of network analysis is to characterize network properties, identify important hub genes, classify, quantify, and correlate biological data, and clustering of data according to similarities. Network visualization is an excellent way to achieve the aforementioned objectives, which help us to explore data overlays and interpret complex mechanisms. Network representations are essentially combined with two modes: nodes and edges. Nodes represent genes/proteins/RNA, and an edge represents the interactions between biomolecular entities. Node degree is an estimator of single node connectivity, and a motif is a congregation of three or more nodes in a network (Junker & Schreiber, 2017). Modules are defined as larger groups of nodes or structural elements together and are known as communities and hierarchies.

There are different types of biological networks in general. Some of the most common are metabolic networks, gene regulatory networks, signaling networks, and protein—protein interaction networks. In the case of stem cells, all these representations have been explored to provide an overview of the cellular processes. There are two kinds of network representation: directed and undirected networks. Directed networks contain only directed edges. In this representation, the network includes only directed edges, and three different interpretations of the network's edge directions are offered by the network. To continue processing the network, the user must choose one of the interpretations. Only undirected edges are present in undirected network representations. Both directed and undirected edges are present in the network. Undirected edges cannot be clearly transformed into directed ones in this situation. Networks having mixed edges are therefore treated as undirected ones. A number of network parameters are useful for analysis of biological networks. These include number of connected components, shortest paths, and neighborhood connections for simple networks, whereas for complex networks, these include clustering coefficient, neighborhood connectivity, and degree distributions. The number of connected components is defined as the node connections in a biological network—a lower value is suggestive of a stronger network. Network diameter is outlined as the longest distance between the two nodes, and the average shortest path is the expected distance between the two connected edges. Another related factor is network centralization. A star topology within a network is representative of network centralization of near 1. On the other hand, a decentralized network is defined as one where the value is near 0. Hub nodes in a network are representative of heterogeneity in a network. Suppose a node is defined by n, then its neighborhood connectivity is calculated as the defined average of all the neighbors (Wuchty et al., 2006).

3.2 Integrative bioinformatics

The Human Genome Project led to the expansion of molecular data and is generating huge amounts of knowledge (Watson, 1990). Notable data are available on the metabolites and reconstructions, cellular pathways, association studies data, and other omics-based approaches such as proteomics and metabolomics for stem cells that could be useful in principle for the analysis of the data. With the availability of open-access data resources and stem cell repositories, there have been tremendous papers and data available for analysis. The objective for us is to congregate individual components of the stem cell system into a comprehensive picture by integrating data from different resources. An emerging field of study known as "integrative bioinformatics" applies electronic representations and in silico science methods to the study of biological processes (Chen et al., 2014). The stem cell bioinformatics approach uses network representations as a key component to understand complicated patterns and incorporate network data. Numerous programs, including well-known ones such as Cytoscape, VisANT (Hu et al., 2004), Pathway Studio (Nikitin et al., 2003), and Cell-Designer (Baek et al., 1992), are available for graphically investigating biological networks. A well-known network visualization program called Cytoscape offers a basic set of capabilities such as standard and adjustable network display styles, the ability to import a wide range of interaction files, and zoomable network views. Currently, there are numerous plugins available for activities such as analyzing networks, creating networks from literature searches, and importing networks from other data formats.

3.3 Related pipelines and work

In 2008, pioneer work was undertaken on hESCs by Muller and his coworkers (Mueller & Tuan, 2008). They reported a robust, reproducible approach for the classification of human cell lines. The group created a database called "STEM CELL

MATRIX " Furthermore, the group extended their work using their understanding of bioinformatics by building a protein—protein interaction network which they named "PluriNet" (Müller et al., 2008). The PluriNet was observed to have common characteristics of the pluripotent stem cells. The prime idea behind the work was to develop a robust classification system that would be able to identify putative stem cell preparations. It had a collection of 299 genes showing common characteristics of pluripotency. In another study done in 2010, the expression patterns across hESC and human-induced pluripotent stem cells (hiPSCs) (Newman & Cooper, 2010) were analyzed. In 2010, another study was conducted on mouse pluripotency using literature curation. This was the first network based on literature-derived data. They manually extracted the binary interactions from over 286 publications reported in the literature, which had a special focus on mouse pluripotency (Som et al., 2010). This was the first knowledge-based network that would be beneficial for the purpose of data integration. Further, in 2018, the reconstruction of human pluripotency was created using Cytoscape (Narad et al., 2018).

3.4 Tools for visualization of networks

Whole-genome profiling has been carried out at the endpoints, and occasionally, in the processes, to better understand reprogramming, direct transformation, self-renewal, and other stem cell biology processes. More than a dozen specialized online resources have been developed using a variety of high-speed data that have been gathered. Different methods can be used to illustrate the connections between significant genes. The use of ESC or somatic cell nuclear transfer (SCNT) technologies a few years ago was fraught with ethical concerns. Recently, iPSC technology has advanced to circumvent these problems. iPSCs can be created by reprogramming fibroblasts and other terminally differentiated cell types using deterministic factors. Tissue-specific inducers can help iPSCs further differentiate into various tissues. Moreover, differentiated cells can be transformed directly into numerous varieties of differentiated cells (Wei et al., 2015). Table 21.1 describes the list of available databases and tools for stem cell—related data analysis.

Case study: Network analysis using Cytoscape for human pluripotent stem cells

In this study, 80 genes were acquired from a publication where the authors worked through the literature-curated network of human pluripotency (Narad et al., 2018). To determine the highly connected nodes that can be potentially employed as candidate biomarkers, network analysis and visualization were performed. The network analysis was performed using Cytoscape. The genes were imported from the string database using the stringApp plug-in. stringApp plug-in imports the protein—protein instructions from the String database into the Cytoscape platform. The confidence threshold while importing the network was set at 0.7, and other parameters were set to default values. The constructed network contained 61 nodes and 193 edges.

Once the network was constructed, the layout was changed to circular to better understand the connections within nodes. To change the layout, yFiles layout plug-in was utilized. The disconnected nodes were removed from the network. The network was then subjected to Cytohubba plug-in for the identification of highly connected nodes. CytoHubba provides 11 descriptors based on topology for shortest paths calculation parameters. In Cytohubba, the top nodes parameter was set at 15, and the method was set as the degree. Other parameters were set as default. The edge color and width represent the interaction score. The greater the width and the darker the color, the better the interaction between nodes and genes. Fig. 21.1 represents the graphical representation of each of the steps mentioned before.

To have a better understanding of the top 15 genes, Jepetto plug-in was used for integrated gene set analysis. In Jepetto plug-in, method of analysis was set as "enrichment," and the reference database was set as "KEGG." The top 15 genes were provided for the pathway analysis. Top 10 pathways in which these genes were involved were endometrial cancer, glioma, colorectal cancer, melanoma, pancreatic cancer, mTOR signaling pathway, aldosterone-regulated sodium reabsorption, prostate cancer, non—small-cell lung cancer, and chronic myeloid leukemia. Table 21.2 represents the pathways and their associated q-values.

For further information, the biological process, molecular function, and cellular component were also analyzed for these genes. For biological process, molecular function, and cellular component, the reference database was set to GO biological process, GO molecular function, and GO cellular component, respectively. The biological process, molecular function, and cellular component results are represented in Table 21.3.

TABLE 21.1 List of databases and tools for in silico analysis of stem cell data.

Database name	Characteristics of database	Link to the server	References
CellNet	1. Cell type classification 2. Analysis on mouse and human cell lines 3. Contrasting analysis from standard analyses 4. Gene regulatory network analyses and status 5. Network influence score	http://cellnet.hms.harvard.edu/	Cahan et al. (2014)
LifeMap	1. Stem cell research studies 2. Regenerative medicine purposes 3. Scientific literature and high-throughput data resources 4. Stem cell differentiation and embryonic development database 5. Tree development and visualization aids 6. Network and pathway analysis	https://discovery.lifemapsc.com	Edgar et al. (2013)
Escape	1. Network building 2. Precise lineage prediction 3. ESCAPE is constructed from *Homo sapiens* and *Mus musculus* ESCs data 4. Representation of the network is as per the input gene list, chromatin immunoprecipitation, protein–protein interaction, and failure of and acquiring of function data 5. RNAi and miRNA experimentation assays	http://www.maayanlab.net/ESCAPE/	Finkelstein et al. (2020)
StemCellNet	1. Network visualization 2. Physical interaction and stemness assessed 3. Node visualization and manual commands for visualization 4. Gene expression profiling is the basis 5. 500 genes can be analyzed in one go	http://stemcellnet.sysbiolab.eu	Pinto et al. (2014)
SCDE	1. SCDE is concentrating on sources for cancer stem cells 2. Samples are blood and brain of almost all mouse and humans 3. Contrasted with molecular signatures in GeneSigB and WikiPathway 4. Enrichment analysis 5. Code sharing with Stem Cell Commons	https://dataverse.harvard.edu/dataverse/stemcellcommons	Sui et al. (2012)
Medusa	1. Java application and available as applet 2. Fruchterman–Reingold algorithm 3. 2D representation and multiedge connections 4. Analyze big datasets in one go 5. Thickness of the line imparts significance 6. Protein–protein, protein–chemical, and chemical–chemical interactions can be assessed	http://combo.dbe.unifi.it/medusa	Hooper and Bork (2005)
Cytoscape	1. Java application 2. 2D representation and large dataset analysis 3. User can change nodes and visualization effects 4. Offers different layout options such as cyclic	https://cytoscape.org	Shannon et al. (2003)

Tool	Features	URL	Reference
	5. mRNA expression profiling, functional gene annotations, gene ontology, and KEGG analysis 6. Organize multiple networks 7. Cluster many connected networks and analyze		
Pajek	1. Stand-alone application and can run only on Windows OS 2. 2D representation and false 3D representation 3. Layout options such as permutation based on algorithms 4. Direct forcing and energy free layout algorithms 5. Hierarchical relationships are showcased 6. Various algorithms help to widen horizon to analyze	http://mrvar.fdv.uni-lj.si/pajek/	Batagelj and Mrvar (1998)
BioLayout express 3D	1. Java language 2. Higher graphic range card is required to run this software 3. 2D and 3D representation 4. Pairwise relationship annotation 5. 2D and 3D graph position and presentation of the network as output 6. Markov clustering algorithm for clustering analysis	http://biolayout.org	Freeman et al. 2007)
proViz	1. Stand-alone application 2. 2D and false 3D representation 3. GEM force algorithm to showcase key elements of the network 4. Protein—protein interaction is focused by this tool 5. System biology uses this tool preferably	http://slim.icr.ac.uk/proviz/	Iragne et al. (2005)
PATIKA	1. 2D representation 2. Supports the split grid of states and transitions 3. Illustrate a variety of edges for analysis 4. Accepts data from variable sources 5. Analysis of cellular and transition states are done effectively	http://www.cs.bilkent.edu.tr/~patikaweb/	Dogrusoz et al. (2006)

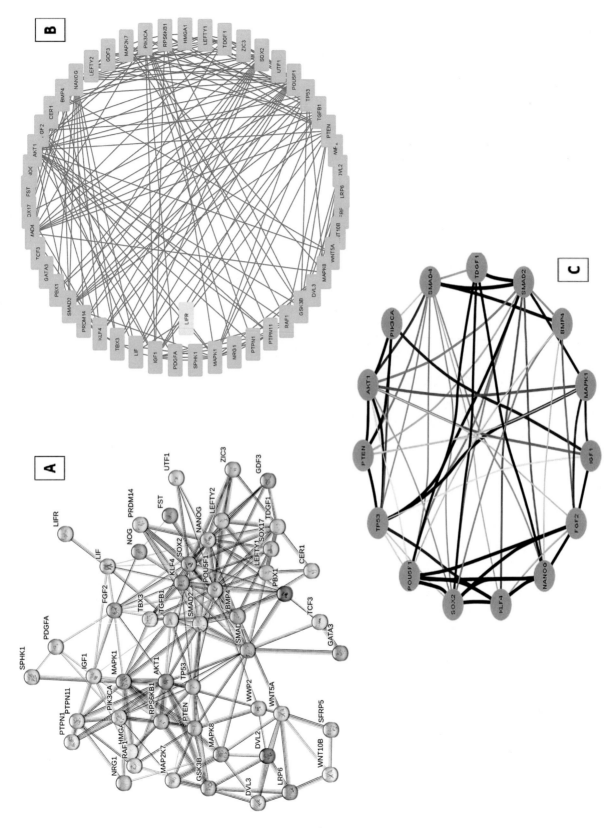

FIGURE 21.1 (A) Network construction using stringApp plug-in; (B) network visualization using yFiles circular layout plug-in; (C) top 15 genes ranked on the basis of degree topology.

TABLE 21.2 Pathways analysis performed using jepetto plug-in.

Pathway or process	XD-score	q-value
Endometrial cancer	0.33482	0
Glioma	0.33482	0
Colorectal cancer	0.32826	0
Melanoma	0.32191	0
Pancreatic cancer	0.27768	0
mTOR signaling pathway	0.26135	0.00011
Aldosterone-regulated sodium reabsorption	0.25061	0.00128
Prostate cancer	0.22053	0
Non−small-cell lung cancer	0.17011	0.00274
Chronic myeloid leukemia	0.1667	0.00039

TABLE 21.3 Gene ontology information for the top 10 biological processes, molecular processes, and cellular components.

Top 10 biological process and their GO annotation for the genes	Top 10 molecular process and their GO annotation for the genes	Top 10 cellular components and their GO annotation
Cell aging—GO:0007569 (12)	MAP kinase activity—GO:0004707 (13)	I band—GO:0031674 (19)
Regulation of protein catabolic process—GO:0042176 (16)	NF-kappa B binding—GO:0051059 (15)	Z disc—GO:0030018 (17)
Striated muscle cell differentiation—GO:0051146 (18)	Phorbol ester receptor activity—GO:0001565 (15)	PML body—GO:0016605 (13)
Protein import into nucleus, translocation—GO:0000060 (22)	Protein kinase C activity—GO:0004697 (15)	PcG protein complex—GO:0031519 (11)
Patterning of blood vessels—GO:0001569 (11)	Protein C-terminus binding—GO:0008022 (70)	Growth cone—GO:0030426 (17)
Negative regulation of neuron apoptosis—GO:0043524 (12)	MAP kinase kinase kinase activity—GO:0004709 (15)	Site of polarized growth—GO:0030427 (18)
Developmental growth—GO:0048589 (27)	Protein kinase inhibitor activity—GO:0004860 (31)	Filopodium—GO:0030175 (12)
Aging—GO:0007568 (19)	Diacylglycerol binding—GO:0019992 (43)	Presynaptic membrane—GO:0042734 (13)
Embryonic pattern specification—GO:0009880 (27)	Double-stranded DNA binding—GO:0003690 (42)	Lamellipodium—GO:0030027 (37)
Neuromuscular junction development—GO:0007528 (12)	Protein domain—specific binding—GO:0019904 (92)	Spindle—GO:0005819 (60)

Acknowledgments

We would like to express our gratitude to Amity School of Health Sciences, Amity University Punjab, Mohali India, and Centre for Computational Biology and Bioinformatics, Amity Institute of Biotechnology for having provided us with the resource base we needed. We would also like to acknowledge Cytoscape software for the generation of the figure represented in the case study.

Author contributions

PN and ST conceived the context of the chapter, ST designed the outline of the chapter, PN performed the case study analysis, and PN and ST authored and reviewed drafts of the chapter.

References

Alva, J. A., Lee, G. M., Escobar, E. E., & Pyle, A. D. (2011). Phosphatase and tensin homolog regulates the pluripotent state and lineage fate choice in human embryonic stem cells. *Stem Cells, 29*(12), 1952–1962. https://doi.org/10.1002/stem.748

Avilion, A. A., Nicolis, S. K., Pevny, L. H., Pérez, L., Vivian, N., & Lovell-Badge, R. (2003). Multipotent cell lineages in early mouse development depend on SOX2 function. *Genes and Development, 17*(1), 126–140. https://doi.org/10.1101/gad.224503

Baek, Y., Cheon, B. Y., Kim, K. H., Lee, H., & Lee, C. (1992). Cell designer: An automatic placement and routing tool for the mixed design of macro and standard cells. *IEICE - Transactions on Fundamentals of Electronics, Communications and Computer Sciences, 2*, 224–232. https://search.ieice.org/bin/summary.php?id=e75-a_2_224&auth=1.

Batagelj, V., & Mrvar, A. (1998). Pajek-program for large network analysis. *Connections, 21*(2), 47–57. https://doi.org/10.1186/s40294-016-0017-8

Berger, S. I., Posner, J. M., & Ma'ayan, A. (2007). Genes2Networks: Connecting lists of gene symbols using mammalian protein interactions databases. *BMC Bioinformatics, 8*(1). https://doi.org/10.1186/1471-2105-8-372

Blau, H. M., Brazelton, T. R., & Weimann, J. M. (2001). The evolving concept of a stem cell. *Cell, 105*(7), 829–841. https://doi.org/10.1016/s0092-8674(01)00409-3

Cahan, P., Li, H., Morris, S. A., Da Rocha, E. L., Daley, G. Q., & Collins, J. J. (2014). CellNet: Network biology applied to stem cell engineering. *Cell, 158*(4), 903–915. https://doi.org/10.1016/j.cell.2014.07.020

Chandrasekaran, S., Zhang, J. Z., Sun, Z., Zhang, L., Ross, C. A., Huang, Y., Asara, J. M., Li, H., Daley, G. Q., & Collins, J. J. (2017). Comprehensive mapping of pluripotent stem cell metabolism using dynamic genome-scale network modeling. *Cell Reports, 21*(10), 2965–2977. https://doi.org/10.1016/j.celrep.2017.07.048

Chen, D., Chen, M., Altmann, T., & Klukas, C. (2014). *Bridging genomics and phenomics* (pp. 299–333). Springer eBooks. https://doi.org/10.1007/978-3-642-41281-3_11

Chen, J., Crawford, R., Chen, C., & Xiao, Y. (2013). The key regulatory roles of the PI3K/Akt signaling pathway in the functionalities of mesenchymal stem cells and applications in tissue regeneration. *Tissue Engineering Part B-Reviews, 19*(6), 516–528. https://doi.org/10.1089/ten.teb.2012.0672

Chen, X. R., Vega, V. B., & Ng, H. H. (2008). Transcriptional regulatory networks in embryonic stem cells. *Cold Spring Harbor Symposia on Quantitative Biology, 73*(0), 203–209. https://doi.org/10.1101/sqb.2008.73.026

Dahle, Ø., Kumar, A., & Kuehn, M. R. (2010). Nodal signaling recruits the histone demethylase Jmjd3 to counteract polycomb-mediated repression at target genes. *Science Signaling, 3*(127). https://doi.org/10.1126/scisignal.2000841

Dalton, S. (2013). Signaling networks in human pluripotent stem cells. *Current Opinion in Cell Biology, 25*(2), 241–246. https://doi.org/10.1016/j.ceb.2012.09.005

Dang, C. V. (2013). MYC, metabolism, cell growth, and tumorigenesis. *Cold Spring Harbor Perspectives in Medicine, 3*(8), a014217. https://doi.org/10.1101/cshperspect.a014217

Di Cristofano, A., Pesce, B., Cordon-Cardo, C., & Pandolfi, P. P. (1998). Pten is essential for embryonic development and tumour suppression. *Nature Genetics, 19*(4), 348–355. https://doi.org/10.1038/1235

Dogrusoz, U., Erson, E. Z., Giral, E., Demir, E., Babur, Ö., Cetintas, A., & Colak, R. (2006). Patika web: A web interface for analyzing biological pathways through advanced querying and visualization. *Bioinformatics, 22*(3), 374–375. https://doi.org/10.1093/bioinformatics/bti776

Dzierzak, E. (2002). Hematopoietic stem cells and their precursors: Developmental diversity and lineage relationships. *Immunological Reviews, 187*(1), 126–138. https://doi.org/10.1034/j.1600-065x.2002.18711.x

Edgar, R., Mazor, Y., Rinon, A., Blumenthal, J. B., Golan, Y., Buzhor, E., Livnat, I., Ben-Ari, S., Lieder, I., Shitrit, A., Gilboa, Y., Ben-Yehudah, A., Edri, O., Shraga, N., Bogoch, Y., Leshansky, L., Aharoni, S., West, M., Warshawsky, D., & Shtrichman, R. (2013). LifeMap DiscoveryTM: The embryonic development, stem cells, and regenerative medicine research portal. *PLoS One, 8*(7), e66629. https://doi.org/10.1371/journal.pone.0066629

Elkabetz, Y., & Studer, L. (2008). Human ESC-derived neural rosettes and neural stem cell progression. *Cold Spring Harbor Symposia on Quantitative Biology, 73*(0), 377–387. https://doi.org/10.1101/sqb.2008.73.052

Finkelstein, J., Parvanova, I., & Zhang, F. (2020). Informatics approaches for harmonized intelligent integration of stem cell research. *Stem Cells and Cloning.* https://doi.org/10.2147/sccaa.s237361

Freeman, T. C., Goldovsky, L., Brosch, M., Van Dongen, S., Mazière, P., Grocock, R. J., Freilich, S., Thornton, J. M., & Enright, A. J. (2007). Construction, visualisation, and clustering of transcription networks from microarray expression data. *PLoS Computational Biology.* https://doi.org/10.1371/journal.pcbi.0030206

Goradel, N. H., Hour, F. Q., Negahdari, B., Malekshahi, Z. V., Hashemzehi, M., Masoudifar, A., & Mirzaei, H. (2018). Stem cell therapy: A new therapeutic option for cardiovascular diseases. *Journal of Cellular Biochemistry, 119*(1), 95–104. https://doi.org/10.1002/jcb.26169

Hadjimichael, C., Chanoumidou, K., Papadopoulou, N., Arampatzi, P., Papamatheakis, J., & Kretsovali, A. (2015). Common stemness regulators of embryonic and cancer stem cells. *World Journal of Stem Cells, 7*(9), 1150–1184. https://doi.org/10.4252/wjsc.v7.i9.1150

Han, S., Han, S., Coh, Y. R., Jang, G., Ra, J. C., Kang, S. K., Lee, H. C., & Youn, H. (2014). Enhanced proliferation and differentiation of Oct4- and Sox2-overexpressing human adipose tissue mesenchymal stem cells. *Experimental and Molecular Medicine, 46*(6), e101. https://doi.org/10.1038/emm.2014.28

Hooper, S. D., & Bork, P. (2005). Medusa: A simple tool for interaction graph analysis. *Bioinformatics, 21*(24), 4432–4433. https://doi.org/10.1093/bioinformatics/bti696

Hu, Z., Mellor, J., Wu, J., & DeLisi, C. (2004). VisANT: An online visualization and analysis tool for biological interaction data. *BMC Bioinformatics, 5*(1), 17. https://doi.org/10.1186/1471-2105-5-17

Iragne, F., Nikolski, M., Mathieu, B., Auber, D., & Sherman, D. H. (2005). ProViz: Protein interaction visualization and exploration. *Bioinformatics, 21*(2), 272–274. https://doi.org/10.1093/bioinformatics/bth494

Junker, B. H., & Schreiber, F. (2011). *Analysis of biological networks*. John Wiley and Sons.

Junker, B. H., & Schreiber, F. (2017). *Analysis of biological networks* (pp. 215–243). John Wiley and Sons, Inc. eBooks. https://doi.org/10.1002/9781119165057.ch10

Kunath, T., Saba-El-Leil, M. K., Almousailleakh, M., Wray, J., Meloche, S., & Smith, A. (2007). FGF stimulation of the Erk1/2 signalling cascade triggers transition of pluripotent embryonic stem cells from self-renewal to lineage commitment. *Development, 134*(16), 2895–2902. https://doi.org/10.1242/dev.02880

Larijani, B., Esfahani, E. N., Amini, P., Nikbin, B., Alimoghaddam, K., Soleimani-Amiri, S., Malekzadeh, R., Yazdi, N. M., Ghodsi, M., Dowlati, Y., Sahraian, M. A., & Ghavamzadeh, A. (2012). Stem cell therapy in treatment of different diseases. *Acta Medica Iranica, 50*(2), 79–96.

Liu, K., Lin, B., Zhao, M., Yang, X., Chen, M., Anding, G., Liu, F., Que, J., & Lan, X. (2013). The multiple roles for Sox2 in stem cell maintenance and tumorigenesis. *Cellular Signalling, 25*(5), 1264–1271. https://doi.org/10.1016/j.cellsig.2013.02.013

Li, W., Wei, W., & Ding, S. (2016). *TGF-B signaling in stem cell regulation* (pp. 137–145). Springer eBooks. https://doi.org/10.1007/978-1-4939-2966-5_8

Lomberk, G., & Urrutia, R. (2005). The family feud: Turning off sp1 by sp1-like KLF proteins. *Biochemical Journal, 392*(1), 1–11. https://doi.org/10.1042/bj20051234

Mahla, R. S. (2016). Stem cells applications in regenerative medicine and disease therapeutics. *International Journal of Cell Biology*, 1–24. https://doi.org/10.1155/2016/6940283, 2016.

McConnell, B. B., & Yang, V. W. (2010). Mammalian Krüppel-like factors in Health and diseases. *Physiological Reviews, 90*(4), 1337–1381. https://doi.org/10.1152/physrev.00058.2009

Melton, D. A. (2009). *"Stemness": Definitions, criteria, and standards* (pp. xxiii–xxix). Elsevier eBooks. https://doi.org/10.1016/b978-0-12-374729-7.00083-4

Moncho-Amor, V., Chakravarty, P., Galichet, C., Matheu, A., Lovell-Badge, R., & Rizzoti, K. (2021). SOX2 is required independently in both stem and differentiated cells for pituitary tumorigenesis in p27 -null mice. *Proceedings of the National Academy of Sciences of the United States of America, 118*(7). https://doi.org/10.1073/pnas.2017115118

Mossahebi-Mohammadi, M., Quan, M., Zhang, J. Z., & Li, X. (2020). FGF signaling pathway: A key regulator of stem cell pluripotency. *Frontiers in Cell and Developmental Biology, 8*. https://doi.org/10.3389/fcell.2020.00079

Mrvar, A., & Batagelj, V. (2016). Analysis and visualization of large networks with program package Pajek. *Complex Adaptive Systems Modeling, 4*(1). https://doi.org/10.1186/s40294-016-0017-8

Mueller, M. D., & Tuan, R. S. (2008). Functional characterization of hypertrophy in chondrogenesis of human mesenchymal stem cells. *Arthritis and Rheumatism, 58*(5), 1377–1388. https://doi.org/10.1002/art.23370

Müller, F., Laurent, L. C., Kostka, D., Ulitsky, I., Williams, R., Lu, C., Park, I., Rao, M. S., Shamir, R., Schwartz, P. H., Schmidt, N. O., & Loring, J. F. (2008). Regulatory networks define phenotypic classes of human stem cell lines. *Nature, 455*(7211), 401–405. https://doi.org/10.1038/nature07213

Nakagawa, M., Koyanagi, M., Tanabe, K., Takahashi, K., Ichisaka, T., Aoi, T., Okita, K., Mochiduki, Y., Takizawa, N., & Yamanaka, S. (2008). Generation of induced pluripotent stem cells without Myc from mouse and human fibroblasts. *Nature Biotechnology, 26*(1), 101–106. https://doi.org/10.1038/nbt1374

Narad, P., Anand, L., Gupta, R., & Sengupta, A. (2018). Construction of discrete model of human pluripotency in predicting lineage-specific outcomes and targeted Knockdowns of essential genes. *Scientific Reports, 8*(1). https://doi.org/10.1038/s41598-018-29480-w

Newman, A. M., & Cooper, J. F. (2010). Lab-specific gene expression signatures in pluripotent stem cells. *Cell Stem Cell, 7*(2), 258–262. https://doi.org/10.1016/j.stem.2010.06.016

Nikitin, A. Y., Egorov, S. A., Daraselia, N., & Mazo, I. (2003). Pathway studio–the analysis and navigation of molecular networks. *Bioinformatics, 19*(16), 2155–2157. https://doi.org/10.1093/bioinformatics/btg290

Park, S., Yeo, H. C., Kang, N., Kim, H., Lin, J., Ha, H., Vendrell, M., Lee, J., Chandran, Y., Lee, D., Yun, S., & Chang, Y. (2014). Mechanistic elements and critical factors of cellular reprogramming revealed by stepwise global gene expression analyses. *Stem Cell Research, 12*(3), 730–741. https://doi.org/10.1016/j.scr.2014.03.002

Pinto, J. C., Kalathur, R. K. R., Machado, R., Xavier, J. R., Bernardino, L., & Futschik, M. E. (2014). StemCellNet: An interactive platform for network-oriented investigations in stem cell biology. *Nucleic Acids Research, 42*(W1), W154–W160. https://doi.org/10.1093/nar/gku455

Ralston, A., & Rossant, J. (2010). The genetics of induced pluripotency. *Reproduction, 139*(1), 35–44. https://doi.org/10.1530/rep-09-0024

Sakaki-Yumoto, M., Katsuno, Y., & Derynck, R. (2013). TGF-β family signaling in stem cells. *Biochimica Et Biophysica Acta - General Subjects, 1830*(2), 2280–2296. https://doi.org/10.1016/j.bbagen.2012.08.008

Shannon, P., Markiel, A., Ozier, O., Baliga, N. S., Wang, J. M., Ramage, D., Amin, N., Schwikowski, B., & Ideker, T. (2003). Cytoscape: A software environment for integrated models of biomolecular interaction networks. *Genome Research, 13*(11), 2498–2504. https://doi.org/10.1101/gr.1239303

Singh, A. B., Bechard, M. E., Smith, K. N., & Dalton, S. (2012). Reconciling the different roles of Gsk3β in "naïve" and "primed" pluripotent stem cells. *Cell Cycle, 11*(16), 2991–2996. https://doi.org/10.4161/cc.21110

Singh, A. B., Reynolds, D., Cliff, T. S., Ohtsuka, S., Mattheyses, A. L., Sun, Y., Menendez, L., Kulik, M., & Dalton, S. (2012). Signaling network crosstalk in human pluripotent cells: A smad2/3-regulated switch that controls the balance between self-renewal and differentiation. *Cell Stem Cell, 10*(3), 312–326. https://doi.org/10.1016/j.stem.2012.01.014

Som, A., Harder, C., Greber, B., Siatkowski, M., Paudel, Y., Warsow, G., Cap, C. H., Schöler, H. R., & Fuellen, G. (2010). The PluriNetWork: An electronic representation of the network underlying pluripotency in mouse, and its applications. *PLoS One, 5*(12), e15165. https://doi.org/10.1371/journal.pone.0015165

Sui, S. J. H., Begley, K., Reilly, D. S., Chapman, B., McGovern, R., Rocca-Sera, P., Maguire, E., Altschuler, G., Hansen, T.a. A., Sompallae, R., Krivtsov, A. V., Shivdasani, R. A., Armstrong, S. A., Culhane, A. C., Correll, M., Sansone, S., Hofmann, O. T., & Hide, W. (2012). The stem cell discovery engine: An integrated repository and analysis system for cancer stem cell comparisons. *Nucleic Acids Research, 40*(D1), D984–D991. https://doi.org/10.1093/nar/gkr1051

Takahashi, K., Okita, K., Nakagawa, M., & Yamanaka, S. (2007). Induction of pluripotent stem cells from fibroblast cultures. *Nature Protocols, 2*(12), 3081–3089. https://doi.org/10.1038/nprot.2007.418

Torres-Padilla, M., & Chambers, I. (2014). Transcription factor heterogeneity in pluripotent stem cells: A stochastic advantage. *Development, 141*(11), 2173–2181. https://doi.org/10.1242/dev.102624

Watson, J. E. M. (1990). The human genome project: Past, present, and future. *Science, 248*(4951), 44–49. https://doi.org/10.1126/science.2181665

Wei, T., Peng, X., Ye, L., Wang, J., Song, F., Bai, Z., Han, G., Ji, F., & Lei, H. (2015). Web resources for stem cell research. *Genomics, Proteomics and Bioinformatics*. https://doi.org/10.1016/j.gpb.2015.01.001

Weissman, I. L., Anderson, D. E., & Gage, F. H. (2001). Stem and progenitor cells: Origins, phenotypes, lineage commitments, and transdifferentiations. *Annual Review of Cell and Developmental Biology, 17*(1), 387–403. https://doi.org/10.1146/annurev.cellbio.17.1.387

Wernicke, S., & Rasche, F. (2006). Fanmod: A tool for fast network motif detection. *Bioinformatics, 22*(9), 1152–1153. https://doi.org/10.1093/bioinformatics/btl038

Wuchty, S., Ravasz, E., & Barabási, A. (2006). *The architecture of biological networks* (pp. 165–181). Springer eBooks. https://doi.org/10.1007/978-0-387-33532-2_5

Yu, J. W., Ramasamy, T. S., Murphy, N., Holt, M. K., Czapiewski, R., Wei, S., & Cui, W. (2015). PI3K/mTORC2 regulates TGF-β/Activin signalling by modulating Smad2/3 activity via linker phosphorylation. *Nature Communications, 6*(1). https://doi.org/10.1038/ncomms8212

Yu, J., Vodyanik, M. A., Smuga-Otto, K., Antosiewicz-Bourget, J., Frane, J. L., Tian, S., Nie, J., Jonsdottir, G., Ruotti, V., Stewart, R., Slukvin, I. I., & Thomson, J. A. (2007). Induced pluripotent stem cell lines derived from human somatic cells. *Science, 318*(5858), 1917–1920. https://doi.org/10.1126/science.1151526

Zakrzewski, W. J., Dobrzyński, M., Szymonowicz, M., & Rybak, Z. (2019). Stem cells: Past, present, and future. *Stem Cell Research and Therapy, 10*(1). https://doi.org/10.1186/s13287-019-1165-5

Zhang, P., Andrianakos, R., Yang, Y., Liu, C., & Lu, W. (2010). Kruppel-like factor 4 (Klf4) prevents embryonic stem (ES) cell differentiation by regulating Nanog gene expression. *Journal of Biological Chemistry, 285*(12), 9180–9189. https://doi.org/10.1074/jbc.m109.077958

Zhao, Y., Li, C., Huang, L., Niu, S., Lu, Q., Gong, D., Huang, S., Yuan, Y., & Chen, H. (2018). Prognostic value of association of OCT4 with LEF1 expression in esophageal squamous cell carcinoma and their impact on epithelial-mesenchymal transition, invasion, and migration. *Cancer Medicine, 7*(8), 3977–3987. https://doi.org/10.1002/cam4.1641

Chapter 22

Bioinformatics approaches to the understanding of Notch signaling in the biology of stem cells

Achala Anand[1], N.S. Amanda Thilakarathna[1], B. Suresh Pakala[2], Ahalya N.[3], Prashanthi Karyala[1], Vivek Kumar[4,6] and B.S. Dwarakanath[5]

[1]Department of Biotechnology, Faculty of Life and Allied Health Sciences, M. S. Ramaiah University of Applied Sciences, Bangalore, Karnataka, India; [2]Department of Biochemistry, School of Life Sciences, University of Hyderabad, Hyderabad, Telangana, India; [3]Department of Biotechnology, Ramaiah Institute of Technology, Bangalore, Karnataka, India; [4]R&D Dept, Shanghai Proton and Heavy Ion Centre (SPHIC), Shanghai Key Laboratory of Radiation Oncology and Shanghai Engineering Research Centre of Proton and Heavy Ion Radiation Therapy, Shanghai, China; [5]Indian Academy Degree College-Autonomous, Bangalore, Karnataka, India; [6]Department of Radiation Oncology, Albert Einstein College of Medicine, Bronx, New York, United States

1. Introduction

Notch signaling is a fundamental signaling pathway required for normal embryonic development, stem cell (SC) maintenance, and cell fate specification. Several human diseases are linked to dysregulated Notch signaling, which can also lead to the growth of tumor by altering the developmental state of the cells and thereby keeping the cells in proliferative or undifferentiated state (Siebel & Lendahl, 2017). Thus, Notch signaling plays a crucial role in the growth of tumors by forcing the cells to adopt a proliferative cell fate (Aster et al., 2017). Moreover, its role in maintenance of SC and in cancer stem cell (CSCs) has been well documented (Meisel et al., 2020). In the recent past, the expansion and development of high-throughput techniques that are accurate and reliable has greatly enhanced our understanding of the molecular features of Notch signaling pathway. However, these technological advancements and the accumulation of data have also made it more challenging to accurately analyze and interpret these constantly growing data. Fortunately, the emergence of bioinformatics has enabled the creation of a variety of biological databases. These databases have offered a comprehensive resource of topic-specific and organized information that is easy to retrieve and use. Moreover, the development of high-resolution computational models from the multiomics datasets, which can capture the synergistic behavior of genes at the molecular and cellular levels, has been made possible by bioinformatics. Indeed, these computational models can make predictions that can be validated experimentally, thereby providing new insights into biological mechanisms (Del Sol & Jung, 2021). This has provided an ideal framework for answering important questions in the field of Notch signaling. Particularly, systems biology models at various levels of complexity can be developed to address pertinent questions in Notch signaling research. This could eventually result in the development of novel therapeutics that target Notch signaling in various diseases and cancers. In this chapter, we provide an overview of various computational and bioinformatics methods that have aided in our comprehension of Notch signaling in SC biology. We also give a brief insight of how this knowledge has influenced the development of therapeutic compounds that target Notch signaling in malignancies as well as diagnostic tools.

2. Notch signaling

In the 1910s, Morgan and colleagues noticed that X-linked dominant mutations resulted in uneven Notches in the wing margins of Drosophila. This observation ultimately led to the discovery of Notch gene in Drosophila (Poulson, 1937).

Computational Biology for Stem Cell Research. https://doi.org/10.1016/B978-0-443-13222-3.00014-9
Copyright © 2024 Elsevier Inc. All rights reserved, including those for text and data mining, AI training, and similar technologies.

Subsequently, several studies demonstrated the embryonic fatal phenotype induced by total loss of Notch function in Drosophila (Welshons & Von Halle, 1962), as well as its intricate genetic interactions and allelic series (Kidd et al., 1986). Notch is generally regarded as an evolutionary conserved local signaling system that is present all through the animal kingdom. Notch is involved in a variety of biological processes that are common to all species, such as organ formation, tissue repair and function etc, and abnormal Notch signaling leads to pathological consequences (Ntziachristos et al., 2014).

Notch receptors and ligands are type I transmembrane protein, and in mammals, there are four Notch receptors (1—4) and five Notch ligands (Delta-like-1, -3, and -4 and Jagged-1, -2). In the endoplasmic reticulum (ER), the Notch receptors are translated as a single precursor protein, which then gets trafficked into the Golgi apparatus where the Notch precursor undergoes S1 cleavage by a furin-like protease prior to translocation to the cell surface. The S1 cleavage results in a mature receptor composed of a Notch extracellular domain (NECD) that is noncovalently attached to Notch intracellular domain (NICD). The extracellular domain of the Notch receptors contains epidermal growth factor (EGF) like repeats, followed by the negative regulatory region (NRR) and a heterodimerization region between NECD and NICD. The NICD contains the RBPJ-associated module (RAM) that is involved in protein—protein interactions, seven ankyrin repeat domain (ANK), two nuclear localization signals on the both sides of the ANK domains, a transcriptional activation domain (TAD), and a PEST domain (proline/glutamic acid/serine/threonine) that regulates protein stability. The structure of Notch ligands is partially similar to Notch receptors and has the following domains: extracellular domain (containing EGF-like repeats), DSL domain, and cysteine-rich region in Serrate (Lobry et al., 2014). The interaction of Notch receptors with the ligands on the nearby cells facilitates short-range intercellular communications. This ligand interaction induces conformational change in the receptors, which causes an S2 cleavage by ADAM10 and ADAM17, resulting in the shedding of the extracellular part. Following this, the S3 site is then cleaved by γ-secretase in the transmembrane region, which leads to the releasing of the NICD from the plasma membrane and allowing it to travel to the nucleus to regulate the transcription of various genes. In the nucleus, it interacts with RBPJK/CBF1/suppressor of hairless/Lag-1 (RBPJK/CSL) to convert the repressor complex into a transcription coactivator complex, thus facilitating the transcription of the target genes (Zhou et al., 2022) (Fig. 22.1).

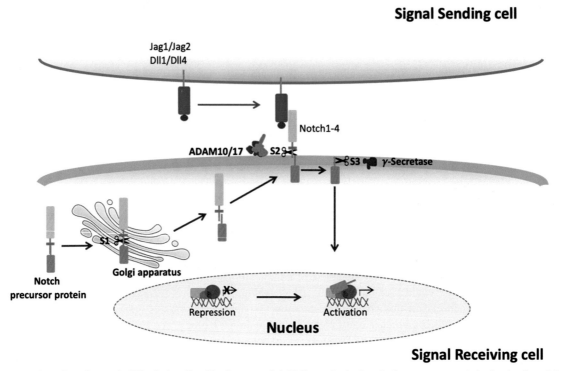

FIGURE 22.1 Overview of canonical Notch signaling. Notch receptor is initially synthesized as single precursor protein in the signal receiving cells, which is then translocated to the Golgi apparatus where it undergoes S1 cleavage by a furin-like convertase to form a heterodimer, which finally gets transported to the cell membrane. The interaction of the Notch ligand on the signal-sending cells with the Notch receptor on the signal-receiving cells causes a conformational change in the Notch receptor, resulting in two successive proteolytic cleavages. The cleavage by ADAMs at S2 site produces the substrate for the S3 cleavage by the γ-secretase complex. This results in the release of the NICD, which moves into the nucleus, where it forms a transcriptional activation complex with coactivators (CoA) and CSL to activate the transcription of target genes.

2.1 Notch signaling in stem cells

SCs are defined as cells found in embryonic and adult stages of an organism with the ability to self-renew and create all cell types found in a certain organ. Throughout development, they give rise to cells of all lineages in a specific tissue, but in adults, they are in charge of maintaining tissue homeostasis and can be activated in response to injury to repair damaged tissues (Batlle & Clevers, 2017). The balance between differentiation and self-renewal in SCs is tightly regulated to promote proper development and prevent unchecked expansion, which in its most extreme form can result in cancer. Numerous studies on somatic SCs have shown that a limited number of evolutionary conserved signaling cascades, including the Notch signaling system, govern these processes (Yang et al., 2020). Indeed, the effectiveness of induced pluripotent SC formation, stemness maintenance, and specific lineage differentiation are all impacted by the modification of Notch signaling. Although the function of Notch signaling has been implicated in numerous tissues, we focus here mainly on the hematopoietic system, muscle, nervous system, and intestine where the Notch signaling has been demonstrated to be essential for the SCs formation, maintenance, and function.

The function of Notch signaling in deciding cell's fate in neurogenesis has been largely preserved throughout evolution (Bohl et al., 2022). In embryonic and adult neurogenesis, Notch regulates the asymmetric division and polarization of neural SCs after differential NUMB segregation (Egger et al., 2010). While overexpression of Notch favors the proliferation of undifferentiated progenitors, the disruption of Notch signaling can cause the development of a neuronal fate (Wen et al., 2009). In adult brain, Notch retains its antineuronal properties, favoring the undifferentiated precursors found in the subgranular and subventricular zone to become glial rather than neuronal cells (Ables et al., 2011).

During hematopoiesis, the SCs arise from the aorta-gonad mesonephros (AGM) region (Taoudi & Medvinsky, 2007). This process is induced and directed by Notch signaling pathway in the hemogenic endothelium via specific receptor—ligand interactions (Gama-Norton et al., 2015). Moreover, the interaction of Notch1 with either JAG1 or DLL4 ligand mediates the development of the earliest blood SCs, which grow and migrate into the hematopoietic organs for the lifelong maintenance of the hematopoietic system (Porcheri et al., 2020). Healthy hematopoietic stem cells (HSCs) can be found in either the perivascular or endosteal niches of adult bone marrow. The NOTCH1 receptor in HSCs and Notch ligands such as DLL1, DLL4, and JAG1 in the adjacent cells regulates the Notch signaling in HSCs, which suggests the significance of Notch signaling in the maintenance of HSCs (Gama-Norton et al., 2015).

Intestinal stem cells (ISCs), which are found at the base of the intestinal crypts, are responsible for maintaining the homeostasis of the intestinal epithelium (Umar, 2010). Notch signaling maintains ISCs in its undifferentiated state and prevents premature differentiation. Inhibition of Notch signaling is known to result in premature differentiation of ISCs and loss of the ISC pool (Pellegrinet et al., 2011). Furthermore, the Notch target gene Hes1 has been shown to promote stemness in ISCs by upregulating CD133 expression and stemness-related genes such as ALDH1, ABCG2, and Nanog. This shows that Notch signaling via Hes1 is important in controlling the stemness of ISCs and maintaining the ISC pool (Pan et al., 2017).

Satellite cells are SCs that reside in skeletal muscle tissue and play a crucial role in muscle regeneration and repair (Relaix et al., 2021). Satellite cells express Notch receptors, in particular NOTCH1 and NOTCH2, and disruption of Notch signaling can lead to escape from quiescence and depletion of the satellite cell population. In addition, age-related decline in muscle regeneration has been linked to insufficient Notch signaling (Conboy et al., 2005; Liu et al., 2018). Studies have shown that inhibition of Notch signaling can impair muscle regeneration in young mice, while forced activation of Notch signaling can restore regenerative potential in aged mice (Luo et al., 2005). This suggests that Notch signaling is a crucial factor in determining the ability of muscle tissue to regenerate, and its dysregulation may cause the age-related loss in muscular function.

2.2 Notch signaling in cancer stem cells

CSCs are a small fraction of cells within the tumor that have SC-like characteristics, and are thought to drive tumor growth, progression, recurrence, and resistance to numerous anticancer therapies. As Notch signaling is necessary for SC maintenance, its dysregulation has been frequently identified in CSCs in a variety of malignancies (Batlle & Clevers, 2017).

In intestinal crypts, the Notch signaling has been shown to affect cell fate, as blocking Notch signaling caused transit amplifying cells to completely transform into differentiated goblet cells. Studies have shown that CSCs in intestinal adenomas with adenomatous polyposis coli (APC) mutations display increased Notch pathway activity compared with non-CSCs. Inhibiting Notch signaling either using genetic approaches or small molecule inhibitors caused reduction in tumor growth as well as reduction in the CSC self-renewal capacity, thus suggesting Notch signaling activation is necessary to maintain CSC in its undifferentiated state (van Es et al., 2005).

The function of Notch signaling in the maintenance of CSC stemness has also been demonstrated in pancreatic cancer. Notch signaling genes were found to be overexpressed in patient-derived pancreatic CSCs (Abel et al., 2014). Moreover, Notch signaling was also found to be upregulated in CSCs, and treatment with gamma-secretase inhibitor reduced the CSC subpopulation as well as the frequency of tumor-sphere formation. Similarly, Hes1 deletion or treatment with a Notch receptor agonist reduced or accelerated the growth of CSC tumor-sphere, respectively, demonstrating that the activity of Notch signaling is essential for CSC stemness in pancreatic cancer (Abel et al., 2014). Moreover, treatment of CSCs with gamma-secretase inhibitors disrupted the cell cycle progression and enhanced apoptosis (Abel et al., 2014), suggesting that Notch signaling may aid in the survival of CSCs.

Notch1 activating mutations were discovered to be prevalent in both the mouse model of T cell acute lymphoblastic leukemia (T-ALL) and in human T-ALL. Furthermore, gain-of-function studies in an animal model of T-ALL showed the significance of Notch1 on leukemia-initiating cells (LICs) (Armstrong et al., 2009), and it was later demonstrated that the inhibition of Notch1 reduced or eliminated LICs, extending animal survival in a transplanting test. Furthermore, gamma-secretase inhibitor treatment of leukemic cells abolished the function of LIC in T-ALL cells. These findings support the theory that Notch1 activating mutations in LIC are needed for clonal proliferation during T-ALL formation (Piya et al., 2022).

Breast cancer stem cells (BCSCs) are a fraction of cells within breast tumors that express ALDH1A1a and CD44 (Hartwig et al., 2014) and display SC-like properties. These cells are purported to drive tumor growth and recurrence. The maintenance and self-renewal of BCSCs are controlled by several signaling pathways, which include the Notch, Wnt, and Hedgehog pathways. Notch signaling, in particular, is important in mammary gland development and has also been implicated in the growth and progression of breast cancer. Several studies have shown that BCSCs have high levels of Notch signaling activity, and inhibition of Notch signaling using small molecule inhibitors or genetic approaches reduces BCSC self-renewal as well as tumor growth in preclinical models of breast cancer (Dontu et al., 2004). In addition, in vitro studies using 3-D cultures of mammospheres have shown that Notch pathway activation can expand the mammary stem/progenitor cell population, indicating a critical role for Notch signaling in BCSCs proliferation (Dontu et al., 2004).

In colon cancer, Hes1 and Notch1 are generally elevated and are shown to be associated with CSCs self-renewal and chemoresistance (Meng et al., 2009). This appears to be the reason why inhibition of Notch signaling by gamma-secreatase inhibitors prevented colorectal cancer cells to form tumors (Ghaleb et al., 2008). Likewise in head and neck cancer models, Notch1 inhibition diminished the CSC proportion both in vitro and in vivo indicating the function for Notch signaling in the maintenance of the CSC phenotype (Zhao et al., 2016). Similar to Notch1, Notch3 was also demonstrated to have a role in growth and maintenance of CSCs, since the inhibition of Notch3 decreased both the cell's capacity to proliferate and form spheroid (Man et al., 2012). Additionally, Notch3 blockage reduced the chemoresistance and tumor volume of the xenograft (Man et al., 2012).

Notch signaling pathway has been found to play an important role in glioma SC generation and maintenance (Hitoshi et al., 2002). Notch signaling can directly upregulate the expression of nestin, a marker of neural stem/progenitor cells. Moreover, Notch signaling can also cooperate with the K-ras oncogene to promote the expansion of CSCs within the subventricular zone in the brain that contains neural stem/progenitor cells (Shih & Holland, 2006). It has been demonstrated that activating the Notch pathway promotes the formation of neurosphere-like colonies, a prominent hallmark of CSCs, while inhibition of Notch signaling can reduce the fraction of CSCs as reflected by a reduction in the CD133-positive and side population, two other hallmarks of CSCs (Fan et al., 2006). These findings suggest that CSCs are sensitive to Notch signaling suppression, making this pathway a prospective therapeutic target for the treatment of gliomas.

3. Applications of bioinformatics in signaling pathways

Bioinformatics can be defined as an interdisciplinary field, which combines biology, mathematics, and computer science. The beginning of bioinformatics dates back to the 1960s with the development of the first software by Margaret Dayhoff to solve the problem of assembling a protein sequence from hundreds of small peptide sequences derived from Edman sequencing reactions (Gauthier et al., 2019). Following which, Dayhoff and Eck developed the first database of 60 protein sequences called the Atlas of Protein Sequence and Structure, and they hypothesized that protein sequence was reflective of the evolutionary relationship between the organisms to which these protein sequences belong (Gauthier et al., 2019). Therefore, the first computational tools for protein analysis constituted de novo protein sequence assembly, protein sequence databases, and amino acid substitution models (Gauthier et al., 2019).

In the 2000s, major improvements, advancements, and reduced cost in sequencing technologies led to generation and exponential increase in sequence information. From the whole-genome sequence, one could predict and annotate all the proteins expressed by the organism through various bioinformatics methods such as (1) sequence analysis to identify homologs with sequence from different organisms, and (2) pattern matching to find regulatory elements such as promoters, transcription factor sites, poly A signals, and ribosome binding sites as well as finding open reading frames in the DNA sequence (Gauthier et al., 2019). The ability to automate proteomics, mass spectrometry and high-throughput workflows has also resulted in generation of large-scale proteomics data (Gauthier et al., 2019). Therefore, molecular biology is greatly impacted by bioinformatics as it plays a crucial role in gene and protein structure, function, and expression analysis.

Current biological research has adopted holistic systems level approaches by integrating information generated by high-throughput technologies such as genomics, transcriptomics, proteomics, and metabolomics (Gauthier et al., 2019). The data obtained from independent investigation of the genome to metabolome are being integrated to build computational models of an organism in entirety along with its environment. Many years of research of metabolism, biochemical pathways have led to compilation of metabolic pathway/network databases such as Kyoto Encyclopedia of Genes and Genomes (KEGG), Reactome and MetaCyc (Gu et al., 2019). The information associated with signaling pathways is available in databases such as TRANSPATH, Netpath, and Signal Link 2.0. Genome-scale metabolic models (GEMs) integrate metabolic reaction, metabolites and gene expression-protein-reaction associations for the entire metabolic genes and reactions in an organism (Gu et al., 2019). These models can be subjected to systems-level metabolic modeling, and simulations to predict metabolic fluxes under different conditions. GEMs have been reconstructed for more than 6000 prokaryotic and eukaryotic organisms (Gu et al., 2019).

Yet another interesting and powerful approach is the use of computational models to simulate cellular signaling pathways and behavior in response to external stimulus. These models encompass biochemical mechanisms that include protein–protein interactions, transcriptional responses, posttranscriptional as well as posttranslational modifications. Quantitative models analyzing a single pathway such as Notch signaling pathway or single-cell models to use heuristic multiscale models to describe cell behavior in a cell population or tissue (Rangamani & Iyengar, 2008). These modeling approaches have a lot of applications in obtaining insights into molecular mechanisms of diseases, basic immunological research, and drug action/response mechanisms involving pharmacokinetic–pharmacodynamic simulations.

4. Bioinformatics approach in the understanding of Notch signaling in the biology of stem cells

4.1 Gene and protein expression databases and their role in deciphering stem cell and Notch pathway function

Elucidation of the human genome sequence has facilitated the measurement of the mRNA levels at a transcriptome-wide scale, using microarray technologies and RNA-sequencing technologies (RNA-seq). Sequencing technologies are superior to other techniques due to its lower cost, greater sensitivity, better reproducibility, improved quantification of the mRNA levels, identification of mutations and novel transcripts, and the ability to quantify splice variants (Marioni et al., 2008). The transcriptome-wide gene expression studies from many studies are being generated and are currently made available for exploration or reuse/reanalysis through public databases such as ArrayExpress (Brazma et al., 2003) and Gene Expression Omnibus (GEO) (Edgar et al., 2002).

The GEO is an internationally recognized public database established in 2000, supported by NCBI and NLM. GEO stores and distributes high-throughput data from gene expression data, chromatin modifications (methylations and acetylation of histones), and DNA–protein interactions. It offers data in multiple formats such as raw data, processed normalized data, and experimental metadata. In addition to giving users access to data from various studies, the database offers several online tools that let them find data pertinent to their individual interests as well as view and analyze the data. Query search for the term "stem cell" in the title field, "Expression profiling by array" and "Expression profiling by high throughput sequencing" in the study type, and "human" in the organism field in the GEO database on February 19, 2023 yielded 31 datasets summarized in Table 22.1. As can be seen from the data and the publications associated with the data that these studies helped in elucidating molecular mechanisms and SC function in normal conditions as well as various diseases such as genetic disorders and CSCs (Lopez-Ayllon et al., 2014; Poon et al., 2013).

Another way to identify the function of any gene or a pathway is by performing inhibition studies or by adding, exchanging, and removing the gene sequences in genomes of model organisms or cell lines. Recently, there has been an international project with an aim to generate mutants of all the genes in the mouse genome (www.knockoutmouse.org) and systematically analyze these knockout mice to identify functions of the genes (White et al., 2013). As majority of the genes of the Notch pathway exhibit embryonic lethality (Gridley, 2010), generation and use of conditional/inducible mutants of loss or gain of function of Notch pathway genes to study their role in different processes of adult mice are being exploited (Murtaugh et al., 2003; Varadkar et al., 2009). Most studies of these studies use phenotypic studies to understand the

TABLE 22.1 List of gene expression data in different stem cell studies as extracted from GEO database on February 19, 2023.

S. no.	Type of stem cell studies	GSE ids	Number of studies
1.	Embryonic stem cell studies	GSE54186, GSE50704, GSE40593, GSE25417, GSE14503, GSE13834, GSE2248	7
2.	Adult and peripheral stem cell studies	GSE42589, GSE43805, GSE33622, GSE35561, GSE30792, GSE25673, GSE24487, GSE19664, GSE3005, GSE4858, GSE3419, GSE2666, GSE1493, GSE1801, GSE1470	15
3.	Cancer stem cell studies	GSE54712, GSE44561, GSE42937, GSE25976, GSE24460, GSE29750, GSE14407, GSE7181, GSE4290	9
	Total number of studies		31

function of protein in Notch signaling pathways. Yet another method of understanding the function of genes is to over-express/inhibit genes and analyze the genome-wide gene expression through microarray and RNA-seq technologies. These data are deposited and available in the GEO database. When we query for the terms "Notch" in all fields with the filters: "Expression profiling by array" and "Expression profiling by high throughput sequencing," we obtain relevant results that are given in Table 22.2. These studies have been extremely useful in understanding the functions of the Notch pathway in different conditions.

The development of various high-throughput omics technologies: genomics, transcriptomics, mass spectrometry (MS)—based proteomics, and antibody-based tissue microarray profiling has resulted in development of a Swedish-based project called the Human Protein Atlas (HPA) (Uhlen et al., 2005). HPA integrates information from these multiple data types and maps the localization and expression of all the human proteins in normal and diseased cells, tissues, and organs, thereby facilitating tissue-based prognosis, diagnosis, and research. The data for each protein can be accessed in 12 different sections such as cell type, tissue type, cell line, pathology, etc. One example of the utility of HPA has been shown in Fig. 22.2, which presents the expression and localization of Notch1 signaling status in colon as extracted from the HPA.

4.2 Bioinformatics of three-dimensional structure of proteins involved in Notch signaling pathway

Proteins are polymers of 20 amino acids and have a defined amino acid sequence. The amino acid sequence is important because it determines the protein structure, and structure dictates the biochemical and enzymatic function. Therefore, knowledge of the protein structure gives us the ability to understand its function and its role in disease pathogenesis. There are several experimental techniques employed to determine the three-dimensional structure of proteins, which include X-

TABLE 22.2 List of datasets in the GEO database which have analyzed gene expression in mammalian cell lines after Notch inhibition/activation under different conditions.

S. No.	Series id	Study type	Organism
1.	GSE57417	Gene expression studies of T helper 1 cells upon exposure to the ligand Dll4	*Mus musculus*
2.	GSE39223	Gene expression profiling of human glioblastoma xenograft tumors treated with anti-Notch and anti-VEGF agents	*Homo sapiens*
3.	GSE29959	Gene expression analysis of human T-ALL cell lines treated with gamma-secretase inhibitor	*H. sapiens*
4.	GSE29544	Gene expression studies of CUTLL1 cells treated with gamma-secretase inhibitor	*H. sapiens*
5.	GSE18198	The effect of treatment of Notch antagonist SAHM1 on gene expression of T-ALL cell lines	*H. sapiens*
6.	GSE6495	Gene expression of MOLT4 cell line treated with gamma-secretase inhibitor	*H. sapiens*

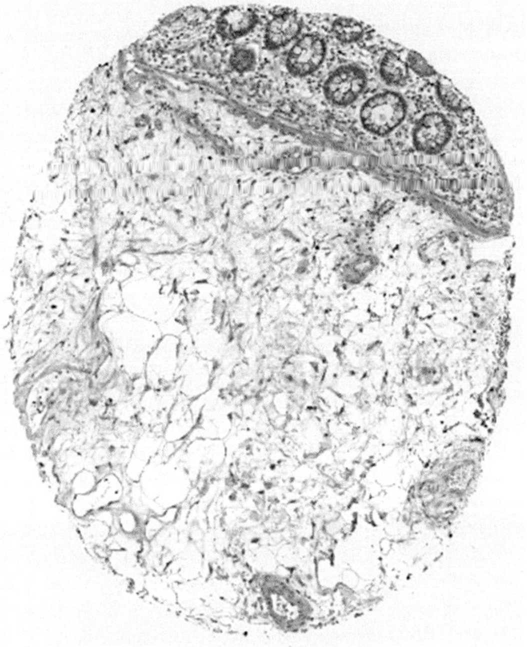

FIGURE 22.2 The expression of Notch1 in colon extracted from the human protein atlas.

ray crystallography, NMR spectroscopy, and cryo-electron microscopy. Since 2003, the 3D structural data of biological molecules and its complexes have been managed by worldwide Protein Data Bank (wwPDB), an international consortium (Laskowski, 2010). Out of the 59 proteins that belong to the Notch signaling pathway (extracted from KEGG database), 38 proteins had 3D structures determined to full length as well as deletions of the proteins (Table 22.3).

Evolutionarily related proteins, in the same or distantly related organisms, share similar sequence, structure, and function. For proteins whose structure is unknown, it is common to predict structure by building homology models based on the 3D structure of a closely related protein. One can build a homology model using online tools such as SWISS-MODEL, which takes a protein sequence and returns a predicted 3D model, if 3D structure of a closely related protein is available (Waterhouse et al., 2018). There are automated servers, such as 3D-JIGSAW and ESyPred3D, that perform homology modeling automatically (Laskowski, 2010). These structures are regularly validated by the EVA, which calculates the accuracy of various servers statistically and uses that information to rank the servers (Eyrich et al., 2001). Recently, there has been a breakthrough in the computational method, in the form of AlphaFold, for the prediction of 3D structure of proteins with or without closely related proteins greatly outperforming other known methods in prediction of structure (Jumper et al., 2021). AlphaFold uses a novel neural network—based approach, which integrates physical and biological aspects of the protein structure and multiple sequence alignments, to predict protein structure. The predicted 3D structures of 21 proteins of the Notch signaling pathway, which have not been experimentally determined, are available in the UniProt database.

4.3 Prediction and experimental validation of small ligands against Notch signaling pathway

An important application of knowledge of a protein's 3D structure is to understand its biological activity and function and its role in disease. This can provide us opportunities to modify and control proteins responsible for disease mechanisms by designing drugs against proteins involved in disease. Prior to the development of techniques for determination of protein 3D structure, drugs with specific biological activity were identified through trial-and-error random screening methods (Kuntz, 1992). Medicinal chemists used computational methods to study the quantitative structure—activity relationship (QSAR) of small molecules with known biological activity and suggested modifications in the molecule to improve biological activity. Significant advances simultaneously in structural biology and affordable and powerful desktop computers led to development of molecular modeling and docking algorithms. Many programs have been written to perform molecular docking to predict small molecules that bind to protein, which operate in this manner. Many ligand conformations are generated and are arranged in the active site in multiple orientations as a pose using sampling procedure such

TABLE 22.3 List of proteins of the Notch signaling pathway whose structure is known extracted from bioDBnet.

Gene symbol	PDB ID	Number of structures (method)
HES1	2MH3	1 (NMR)
DLL4	5MVX	1 (X-ray)
DVL3	6V7O	1 (X-ray)
CIR1	6ZYM	1 (EM)
HEY1	2DB7	1 (X-ray)
NUMBL	3F0W	1 (X-ray)
PTCRA	3OF6	1 (X-ray)
DTX3L	3PG6	1 (X-ray)
DLL1	4XBM	1 (X-ray)
NUMB	5NJJ; 5NJK	2 (X-ray)
DTX1	6Y5P; 6Y5N	2 (X-ray)
NOTCH2	2OO4; 5MWB	2 (X-ray)
CTBP2	4LCJ; 2OME; 6WKW	3 (X-ray, EM)
JAG2	5MW7; 5MW5; 5MWF	3 (X-ray)
DVL1	6TTK; 6LCB; 6LCA	3 (X-ray)
NOTCH3	4ZLP; 5CZV; 5CZX	3 (X-ray)
RBPJ	3NBN; 6PY8; 3V79; 2F8X	4 (X-ray)
MAML1	3V79; 2F8X; 6SMV; 3NBN	4 (X-ray)
DTX2	6Y2X; 6Y22; 6Y3J; 6IR0	4 (NMR, X-ray)
CTBP1	4U6S; 4LCE; 4U6Q; 6CDF; 6CDR; 1MX3	6 (X-ray)
TLE1	4OM2; 1GXR; 2CE8; 5MWJ; 2CE9; 4OM3	6 (X-ray)
JAG1	2KB9; 4XI7; 5BO1; 4CC0; 2VJ2; 4CBZ; 4CC1	7 (NMR, X-ray)
PSENEN	5FN5; 6IYC; 5A63; 5FN2; 5FN4; 5FN3; 6IDF	7 (EM)
APH1A	6IYC; 5A63; 5FN2; 5FN4; 6IDF; 5FN3; 5FN5	7 (EM)
ATXN1	2M41; 4AQP; 4J2J; 4J2L; 1OA8; 6QIU; 4APT	7 (NMR, X-ray)
HDAC1	7AOA; 1TYI; 7AO9; 5ICN; 4BKX; 7AO8; 6Z2K; 6Z2J	8 (X-ray, EM)
PSEN1	5A63; 5FN3; 6IDF; 4UIS; 5FN5; 2KR6; 5FN2; 5FN4; 6IYC	9 (NMR, EM)
KAT2A	5MLJ; 1F68; 5H86; 5H84; 5TRL; 1Z4R; 3D7C; 5TRM; 6J3P	9 (NMR, X-ray)
NCSTN	5FN3; 5FN2; 5FN5; 6IYC; 2N7R; 6IDF; 2N7Q; 4UIS; 5A63; 5FN4	10 (NMR, EM)
DVL2	3CBY; 2REY; 6IW3; 6JCK; 3CBX; 5LNP; 5SUY; 4WIP; 5SUZ; 3CBZ; 3CC0	11 (X-ray)
HDAC2	6XEB; 3MAX; 6WBW; 6XDM; 6XEC; 5IX0; 7KBG; 4LY1; 5IWG; 6WBZ; 7KBH; 4LXZ; 6G3O	13 (X-ray)
SNW1	5YZG; 5Z58; 5MQF; 6ICZ; 6ID0; 6QDV; 7A5P; 7ABF; 7ABG; 5XJC; 6ZYM; 7AAV; 6ID1; 6FF4; 6FF7	15 (EM)
NCOR2	2GPV; 2LTP; 4OAR; 5X8Q; 5X8X; 6IVX; 1KKQ; 6PDZ; 2L5G; 1XC5; 4A69; 5ZOP; 2ODD; 2RT5; 3R29; 3R2A; 1R2B; 6A22; 5ZOO	19 (NMR, X-ray)
ADAM17	3L0V; 3LEA; 3E8R; 3EWJ; 1ZXC; 3CKI; 3EDZ; 3LE9; 3O64; 2DDF; 2I47; 3B92; 2A8H; 2FV5; 2M2F; 3G42; 1BKC; 2FV9; 2OI0; 3KME; 3LGP; 3KMC; 3L0T	23 (NMR, X-ray)
NOTCH1	1TOZ; 1YYH; 2HE0; 3V79; 4D0E; 6IDF; 5FM9; 5KZO; 4CUF; 5FMA; 2F8X; 3ETO; 3L95; 4CUE; 4D0F; 6PY8; 2F8Y; 2VJ3; 4CUD; 5L0R; 5UB5; 1PB5; 3I08; 3NBN	24 (NMR, X-ray, EM)

TABLE 22.3 List of proteins of the Notch signaling pathway whose structure is known extracted from bioDBnet.—cont'd

Gene symbol	PDB ID	Number of structures (method)
KAT2B	1IM4; 1N72; 3GG3; 5FF1; 5MKX; 2RNW; 5FE3; 5FE6; 4NSQ; 5FE0; 5FD7; 5FE2; 5FE7; 5FE8; 1CM0; 1WUM; 1WUG; 2RNX; 5FE9; 5LVR; 6J3U; 1Z5S; 5FF4; 5FE5; 5LVQ	25 (NMR, X-ray)
EP300	3I3J; 4PZT; 5LKT; 5LKX; 5NU5; 6K4N; 6V8B; 2K8F; 3P57; 5LPK; 4BHW; 5XZC; 6FGS; 6V8K; 1L3E; 3IO2; 5KJ2; 5LPM; 6DS6; 6FGN; 5BT3; 5LKU; 5LKZ; 6GYT; 6V8N; 2MH0; 3BIY; 4PZR; 6PF1; 6V90; 1P4Q; 2MZD; 3T92; 4PZS; 6GYR; 6PGU	36 (NMR, X-ray, EM)
CREBBP	2RNY; 4A9K; 4WHU; 5ENG; 5H85; 5JEM; 5MME; 5MPK; 5MQG; 5W0E; 5XXH; 6AY3; 6YIL; 4NR4; 4NR7; 5LPL; 5MQE; 5MQK; 5NLK; 5NRW; 5OWK; 5W0Q; 6ES6; 7CO1; 1RDT; 1WO3; 1ZOQ; 2LXS; 3P1C; 3P1E; 4N3W; 4NR5; 4NR6; 4YK0; 5GH9; 5MMG; 5MPZ; 5NU3; 6FRF; 6SXX; 6YIK; 7JFL; 2L85; 3P1D; 4N4F; 5I83; 5KTW; 5MPN; 6SQM; 6YIJ; 2L84; 2N1A; 3SVH; 4NYX; 4OUF; 5EP7; 5KTU; 6ALB; 6ALC; 1LIQ; 1WO4; 3P1F; 4NYW; 5CGP; 5I86; 5I89; 5I8B; 5LPJ; 5SVH; 5TB6; 5W0L; 6DMK; 6FQO; 6FQT; 6FR0; 6SQE; 1JSP; 1WO5; 1WO6; 1WO7; 2LXT; 4NYV; 4TQN; 4TS8; 5EIC; 5I8G; 5KTX; 5W0F; 6AXQ; 6AY5; 6YIM; 2D82; 2KJE; 2KWF; 3DWY; 5DBM; 5J0D; 6ES5; 6ES7; 6FQU; 6QST; 6SQF; 7JFM	103 (NMR, X-ray, EM)

as Monte Carlo simulation, genetic algorithms, distance geometry, or simulated annealing (Kuntz, 1992; Verdonk et al., 2003). The different poses are scored and ranked to identify the best overall pose (Verdonk et al., 2003). Many studies have performed molecular docking studies against proteins of the Notch signaling pathway proteins such as Notch1 (Gill et al., 2019; Platonova et al., 2017) and gamma-secretase (Kushwaha et al., 2019; Singh et al., 2022). Currently, there are a few drugs against proteins of the Notch signaling pathway in the experimental stage (Table 22.4).

4.4 Protein—protein interactions and signaling pathway database in Notch signaling pathway

Proteins in organisms do not work in isolation. They interact with each other and go from one location to another to perform their functions and contribute to cellular and organism phenotypes. Several low- and high-throughput techniques,

TABLE 22.4 Drugs in the experimental stage to Notch signaling pathway proteins extracted from the DrugBank.

Gene symbol	DrugBank drug name	DrugBank drug ID
HDAC2	Fluvastatin, Pravastatin, Atorvastatin, Theophylline, Beleodaq, Simvastatin, Aminophylline, Oxtriphylline, Lovastatin, Pracinostat, Romidepsin, Vorinostat, Tixocortol, Valproic acid	DB01095, DB00175, DB01076, DB00277, DB05015, DB00641, DB01223, DB01303, DB00227, DB05223, DB06176, DB02546, DB09091, DB00313
CTBP1	Formic acid	DB01942
APH1A	E-2012	DB05171
PSENEN	E-2012	DB05171
HDAC1	Panobinostat, Arsenic trioxide, Fingolimod, Romidepsin, Vorinostat, Belinostat	DB06603, DB01169, DB08868, DB06176, DB02546, DB05015
ADAM17	Benzenesulfonamides, benzenesulfonyl compounds, phenylquinolines, phenylpyrrolidines, quinoline derivatives, indolyl carboxylic acid derivatives	DB07964, DB07079, DB07189, DB07121, DB07145, DB06943, DB07147, DB07233
KAT2B	Nitrobenzenes	DB08291, DB08186
CREBBP	Carbazoles	DB08655

such as CoIP-MS (coimmunoprecipitation-mass spectrometry), affinity chromatography, yeast two hybrid, genome-wide association studies, have been utilized to analyze genetic, biochemical and physical protein—protein interactions (PPIs) generating extensive amount of data (Shoemaker & Panchenko, 2007a). Bioinformaticians have also computationally predicted previously unknown PPIs (Shoemaker & Panchenko, 2007b). Over the past few years, this useful information has been faithfully mined, stored, and easily accessible in different public databases (Klingstrom & Plewczynski, 2011). The PPI databases provide tools to inspect and visualize the interactions, map them to well-known pathways, and generate information required for modeling and simulations. Some of the major protein interaction databases are the IntAct molecular interaction database (IntAct), Biological General Repository for Interaction Datasets (BioGRID), Search Tool for the Retrieval of Interacting Genes/Proteins (STRING), and GeneMania. Fig. 22.3 shows protein interaction network for Notch1 protein from STRING database and GeneMania. Another important resource are pathway databases that meticulously extract and curate information about proteins and their functions and connect them into series of reactions or networks in a cell, leading to formation of products, signaling in a cell, etc. Examples of pathway databases include Signal Link 2.0, BioCyc, KEGG, Reactome, and many others. Fig. 22.3C represents the human Notch signaling pathway extracted from KEGG pathway database.

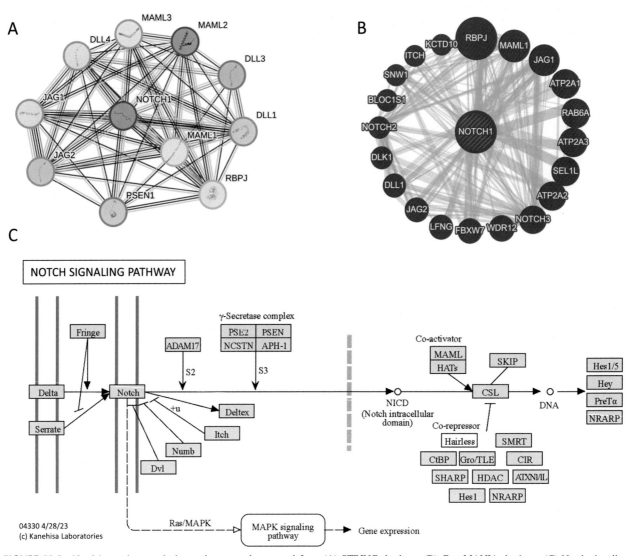

FIGURE 22.3 Notch1 protein—protein interaction network extracted from (A) STRING database, (B) GeneMANIA database, (C) Notch signaling pathway extracted from KEGG database.

4.5 Database of human diseases and Notch signaling pathway

Deciphering the human genome sequence has improved our comprehension of the basic genetic makeup and its variations that lead to disease. The use of novel high-throughput technologies has led to increased pace in molecular dissection of human genetic diseases and revolutionizing biomedical research and treatment of disease. The association of the genes with different genetic disorders has been cataloged in various databases. One of the first comprehensive databases on various genetic disorders in humans is the Online Mendelian Inheritance in Man database (OMIM) maintained by National Center for Biotechnology Information (NCBI) and compiled by John Hopkins University. eDGAR is another database that collates and provides information on gene/disease associations as mined from different databases OMIM, ClinVar, and Humsavar. Yet another disease database is DisGeNET that encompasses information from animal disease models and association of genes with various Mendelian and complex diseases from various curated databases and data extracted from scientific literature through different text mining approaches.

Throughout the development of invertebrate and vertebrate species, the Notch signaling pathway and its genes are crucial in influencing cell fate decisions. Therefore, mutations in Notch pathway genes would result in various genetic disorders. Search in DisGeNET for few important Notch signaling pathway genes showed 5840 gene–disease associations (Fig. 22.4). Abnormal Notch1, Notch3, and Jagged1 gene function has been linked to many disorders including a neoplasia (T cell acute lymphoblastic leukemia/lymphoma), a developmental disorder (Alagille syndrome), and a late-onset neurological condition (CADASIL) (Li et al., 1997), respectively, highlighting the wide range of Notch pathway activity in humans. Notch1 protein mutation can cause a range of developmental aortic valve anomalies, severe valve calcification, and cardiovascular malformation (Garg et al., 2005). Therefore, the Notch pathway proteins can be used as therapeutic targets for designing drugs to cure various diseases.

4.6 In-silico modeling of Notch signaling pathway

Cellular signaling pathways are complex, dynamic, and nonlinear and can be integrated into comprehensive models to facilitate mechanistic understanding of the complex cellular processes. These modeling simulations make predictions that can be experimentally validated and are therefore of interest to theoretical and experimental researchers. There are different types of modeling tools such as Boolean, stochastic, and mathematical modeling using ordinary and partial differential equations (Rangamani & Iyengar, 2008). Notch signaling pathway plays an important role in cell fate determination during embryonic development. Modeling techniques have been used to understand the mechanism of Notch pathway in development and CSCs processes.

Mathematical modeling has been used to find the role of cis signaling, which include cis-inhibition and -activation, on fine-grained cell patterning and cell fate decisions (Formosa-Jordan & Ibanes, 2014). A specific theoretical framework using ordinary differential equations incorporating both ligands, Delta and Jagged, in the Notch signaling circuit was used to determine the role of Jagged in cell fate determination (Boareto et al., 2015). The role of stochasticity of the Notch-Delta signaling pathway in mediating alternate cell fates and spatiotemporal patterning in tissues was modeled using a

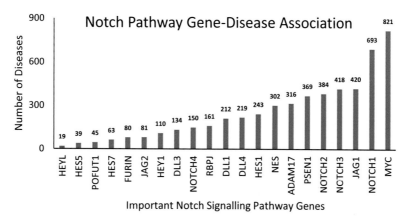

FIGURE 22.4 Notch pathway gene–disease association. The number of gene–disease association of 21 Notch pathway genes extracted from DisGeNET extracted on February 12, 2023.

deterministic multicell model and revealed insights of how Notch-Delta signaling promotes spatiotemporal patterning in a noisy physiological environment (Galbraith et al., 2022). A mathematical framework was used to examine the interplay between Notch signaling pathway, epithelial—mesenchymal transition (EMT) process, and CSCs—three key axes known to contribute to metastases and therapeutic resistance (Bocci et al., 2018). This model predicted a strong correlation between augmented Notch-Jagged signaling and a hybrid E/M phenotype with CSC properties, and not necessarily a completely mesenchymal M state is responsible for cancer aggressiveness (Bocci et al., 2018).

5. Clinical studies

Although the impact of bioinformatics in the diagnosis and design of therapy for various human diseases is still emerging, it has been gainfully employed in understanding the importance of Notch signaling in few of the human diseases, including cancer. These studies not only have unraveled potential of Notch signaling pathway regulators as biomarkers but have also helped in prognosis and identification of targets for the development of effective therapies.

The clinical relevance of Notch1 activation, its heterogeneity among different types of brain tumors, and the status of Notch signaling as a potential biomarker for gliomas as well as a druggable target, besides serving as a prognostic indicator, have been extensively investigated using various bioinformatics tools by employing glioma stem-like cell (GSC) models and clinical cohorts with specific mutations combined with blocking antibodies and the well-known gamma-secretase inhibitor MRK003 (Herrera-Rios et al., 2020). Elevated Notch 1 is generally associated with a larger fraction of the SCs and poor prognosis in glioblastoma multiforme and significantly elevated in tumors with IDH1 mutant without 1p and 19q codeletion (Herrera-Rios et al., 2020). Heterogeneity of gene expression is one of the most important prognostic factors for tumor recurrence and refractoriness to therapy involving temozolomide in GBM. Analysis of genomic data from TCGA and GEO data as well as PPIs to identify differential gene expression, over expression of 14 genes related to developmental pathways including Notch signaling have been identified, which is expected to help in identifying and evaluating TMZ resistance in GBM. These genes are expected to serve as biomarkers as well as potential targets for developing precision molecular targeted therapy (Hsu et al., 2021). Bioinformatics studies of the gene expression and protein levels have revealed the activation of Notch signaling pathway by the spindle and kinetochore-associated complex subunit 3 (SKA3)—promoting SC-like properties leading to tumorigenic abilities and disease dissemination potential (migration and invasion), and resistance to chemotherapy of hepatocellular carcinoma cells (HCCs) suggesting SKA3 to be a potential drug target for HCC (Bai et al., 2021).

Micro-RNAs (miRNAs) play an important role in endometrial cancer. Bioinformatics analysis of miRNAs related to metastasis has shown that miR-1247 and miR-3200 are associated with metastasis and clinical stage of the tumor. Functional enrichment analysis has implicated four miRNAs involved in multiple signaling pathways, including the Wnt, Notch, and TGF-β pathways involved in the regulation of pluripotency of SCs (Zhu et al., 2018). These studies have also suggested the four miRNAs may be explored as biomarkers of endometrial cancer. The molecular mechanisms underlying the brain metastasis from breast cancer have been recently elucidated using a variety of bioinformatics approaches such as GEO database to identify the differentially expressed genes (DEGs), and function enrichment analysis, STRING to analyze PPI and Cytoscape to screen hub genes, with the Oncomine database (Zeng et al., 2022). Further studies are required to establish the various regulators of Notch signaling as potential biomarkers for precision medicine in oncology.

Comprehensive bioinformatics analyses of microarray data on the differentially expressed miRNAs (DEMs) of bone marrow mesenchymal SCs from healthy subjects and adolescent idiopathic scoliosis (AIS) patients have been carried out to provide an insight into the role of miRNAs in the pathogenesis of AIS and the accompanying generalized osteopenia (Hui et al., 2019). These analyses coupled with gene ontology and pathway illustrations have shown that more than 50 miRNAs involved in multiple biological processes including Notch signaling pathway that play an important role in regulating the osteogenic or adipogenic differentiation of MSCs are differentially expressed in AIS (Hui et al., 2019). Interaction network analysis revealed that 7 out of these 50 DEMs play essential roles in AIS pathogenesis and accompanied osteopenia (Hui et al., 2019). Age-associated osteoporosis arises from the disruption of osteogenic SC populations and their functioning. The pathogenic antiosteogenic role for miR-29b-1-5p and other miRNAs in osteogenesis and bone regeneration has been illustrated by the target prediction and Kyoto Encyclopedia of Genes and Genomes (KEGG) pathway analyses and microarray analysis of CD271 positive (+) bone marrow—derived SCs from human surgical BM aspirates. These studies have shown the enrichment of upregulated miRNA targeting genes, involved in the bone development pathways (Eisa et al., 2021). Bioinformatics comprising silico modeling coupled with functional studies in zebrafish embryos have shown that human mindbomb homolog 1 (MIB1), which regulates Notch ligands Delta and Jagged, is involved in left ventricular noncompaction (LVNC) (Luxan et al., 2013). Mutations in MIB1 impair myocardium development and compromises cardiac systolic function.

6. Conclusion

Mounting evidence supports the role of Notch signaling in both SCs and CSCs. Moreover, in recent years, we have witnessed rapid progress in many fields of Notch research. With increasing amount of available multiomics dataset, we are now in a position to develop various in silico and computation models that are expected to accelerate the field of diagnostics and drug discovery, including diseases related to altered function of the notch signaling pathway. The benefit of bioinformatics approach and in silico modeling in gaining insight into the role of gene expression databases in deciphering SC and Notch pathway, prediction of protein structure, and experimental validation using small ligands against Notch signaling in human diseases are already being realized. Exploring the existing and emerging multiomics dataset further using a large array of genome and epigenome, transcriptome, proteome, and metabolome analysis platforms bioinformatics will provide an important insight into understanding of the Notch signaling in SC research and will benefit in developing novel drug and small molecules targeting Notch signaling in CSCs.

Author contributions

VK, PK, and BSD conceptualized the chapter; VK contributed to Sections 1 and 2; AA, NSAT, SBP, AN, and PK contributed to Sections 3 and 4; BSD contributed to Sections 5 and 6.

References

Abel, E. V., Kim, E. J., Wu, J., Hynes, M., Bednar, F., & Proctor, E. (2014). The Notch pathway is important in maintaining the cancer stem cell population in pancreatic cancer. *PLoS One, 9*(3), e91983. https://doi.org/10.1371/journal.pone.0091983

Ables, J. L., Breunig, J. J., Eisch, A. J., & Rakic, P. (2011). Not(ch) just development: Notch signalling in the adult brain. *Nature Reviews Neuroscience, 12*(5), 269−283. https://doi.org/10.1038/nrn3024

Armstrong, F., Brunet de la Grange, P., Gerby, B., Rouyez, M. C., Calvo, J., & Fontenay, M. (2009). Notch is a key regulator of human T-cell acute leukemia initiating cell activity. *Blood, 113*(8), 1730−1740. https://doi.org/10.1182/blood-2008-02-138172

Aster, J. C., Pear, W. S., & Blacklow, S. C. (2017). The varied roles of notch in cancer. *Annual Review of Pathology: Mechanisms of Disease, 12*, 245−275. https://doi.org/10.1146/annurev-pathol-052016-100127

Bai, S., Chen, W., Zheng, M., Wang, X., Peng, W., & Zhao, Y. (2021). Spindle and kinetochore-associated complex subunit 3 (SKA3) promotes stem cell-like properties of hepatocellular carcinoma cells through activating Notch signaling pathway. *Annals of Translational Medicine, 9*(17), 1361. https://doi.org/10.21037/atm-21-1572

Batlle, E., & Clevers, H. (2017). Cancer stem cells revisited. *Nature Medicine, 23*(10), 1124−1134. https://doi.org/10.1038/nm.4409

Boareto, M., Jolly, M. K., Lu, M., Onuchic, J. N., Clementi, C., & Ben-Jacob, E. (2015). Jagged-Delta asymmetry in Notch signaling can give rise to a Sender/Receiver hybrid phenotype. *Proceedings of the National Academy of Sciences of the United States of America, 112*(5), E402−E409. https://doi.org/10.1073/pnas.1416287112

Bocci, F., Jolly, M. K., George, J. T., Levine, H., & Onuchic, J. N. (2018). A mechanism-based computational model to capture the interconnections among epithelial-mesenchymal transition, cancer stem cells and Notch-Jagged signaling. *Oncotarget, 9*(52), 29906−29920. https://doi.org/10.18632/oncotarget.25692

Bohl, B., Jabali, A., Ladewig, J., & Koch, P. (2022). Asymmetric Notch activity by differential inheritance of lysosomes in human neural stem cells. *Science Advances, 8*(6), eabl5792. https://doi.org/10.1126/sciadv.abl5792

Brazma, A., Parkinson, H., Sarkans, U., Shojatalab, M., Vilo, J., & Abeygunawardena, N. (2003). ArrayExpress–a public repository for microarray gene expression data at the EBI. *Nucleic Acids Research, 31*(1), 68−71. https://doi.org/10.1093/nar/gkg091

Conboy, I. M., Conboy, M. J., Wagers, A. J., Girma, E. R., Weissman, I. L., & Rando, T. A. (2005). Rejuvenation of aged progenitor cells by exposure to a young systemic environment. *Nature, 433*(7027), 760−764. https://doi.org/10.1038/nature03260

Del Sol, A., & Jung, S. (2021). The importance of computational modeling in stem cell research. *Trends in Biotechnology, 39*(2), 126−136. https://doi.org/10.1016/j.tibtech.2020.07.006

Dontu, G., Jackson, K. W., McNicholas, E., Kawamura, M. J., Abdallah, W. M., & Wicha, M. S. (2004). Role of Notch signaling in cell-fate determination of human mammary stem/progenitor cells. *Breast Cancer Research, 6*(6), R605−R615. https://doi.org/10.1186/bcr920

Edgar, R., Domrachev, M., & Lash, A. E. (2002). Gene expression Omnibus: NCBI gene expression and hybridization array data repository. *Nucleic Acids Research, 30*(1), 207−210. https://doi.org/10.1093/nar/30.1.207

Egger, B., Gold, K. S., & Brand, A. H. (2010). Notch regulates the switch from symmetric to asymmetric neural stem cell division in the *Drosophila* optic lobe. *Development, 137*(18), 2981−2987. https://doi.org/10.1242/dev.051250

Eisa, N. H., Sudharsan, P. T., Herrero, S. M., Herberg, S. A., Volkman, B. F., & Aguilar-Perez, A. (2021). Age-associated changes in microRNAs affect the differentiation potential of human mesenchymal stem cells: Novel role of miR-29b-1-5p expression. *Bone, 153*, 116154. https://doi.org/10.1016/j.bone.2021.116154

van Es, J. H., van Gijn, M. E., Riccio, O., van den Born, M., Vooijs, M., & Begthel, H. (2005). Notch/gamma-secretase inhibition turns proliferative cells in intestinal crypts and adenomas into goblet cells. *Nature, 435*(7044), 959−963. https://doi.org/10.1038/nature03659

Eyrich, V. A., Marti-Renom, M. A., Przybylski, D., Madhusudhan, M. S., Fiser, A., & Pazos, F. (2001). EVA: Continuous automatic evaluation of protein structure prediction servers. *Bioinformatics, 17*(12), 1242−1243. https://doi.org/10.1093/bioinformatics/17.12.1242

Fan, X., Matsui, W., Khaki, L., Stearns, D., Chun, J., & Li, Y. M. (2006). Notch pathway inhibition depletes stem-like cells and blocks engraftment in embryonal brain tumors. *Cancer Research, 66*(15), 7445−7452. https://doi.org/10.1158/0008-5472.CAN-06-0858

Formosa-Jordan, P., & Ibanes, M. (2014). Competition in Notch signaling with cis enriches cell fate decisions. *PLoS One, 9*(4), e95744. https://doi.org/10.1371/journal.pone.0095744

Galbraith, M., Bocci, F., & Onuchic, J. N. (2022). Stochastic fluctuations promote ordered pattern formation of cells in the Notch-Delta signaling pathway. *PLoS Computational Biology, 18*(7), e1010306. https://doi.org/10.1371/journal.pcbi.1010306

Gama-Norton, L., Ferrando, E., Ruiz-Herguido, C., Liu, Z., Guiu, J., & Islam, A. B. (2015). Notch signal strength controls cell fate in the haemogenic endothelium. *Nature Communications, 6*, 8510. https://doi.org/10.1038/ncomms9510

Garg, V., Muth, A. N., Ransom, J. F., Schluterman, M. K., Barnes, R., & King, I. N. (2005). Mutations in Notch1 cause aortic valve disease. *Nature, 437*(7056), 270−274. https://doi.org/10.1038/nature03940

Gauthier, J., Vincent, A. T., Charette, S. J., & Derome, N. (2019). A brief history of bioinformatics. *Briefings in Bioinformatics, 20*(6), 1981−1996. https://doi.org/10.1093/bib/bby063

Ghaleb, A. M., Aggarwal, G., Bialkowska, A. B., Nandan, M. O., & Yang, V. W. (2008). Notch inhibits expression of the Kruppel-like factor 4 tumor suppressor in the intestinal epithelium. *Molecular Cancer Research, 6*(12), 1920−1927. https://doi.org/10.1158/1541-7786.MCR-08-0224

Gill, B. S., Navgeet, & Kumar, S. (2019). Antioxidant potential of ganoderic acid in Notch-1 protein in neuroblastoma. *Molecular and Cellular Biochemistry, 456*(1−2), 1−14. https://doi.org/10.1007/s11010-018-3485-7

Gridley, T. (2010). Notch signaling in the vasculature. *Current Topics in Developmental Biology, 92*, 277−309. https://doi.org/10.1016/S0070-2153(10)92009-7

Gu, C., Kim, G. B., Kim, W. J., Kim, H. U., & Lee, S. Y. (2019). Current status and applications of genome-scale metabolic models. *Genome Biology, 20*(1), 121. https://doi.org/10.1186/s13059-019-1730-3

Hartwig, F. P., Nedel, F., Collares, T., Tarquinio, S. B., Nor, J. E., & Demarco, F. F. (2014). Oncogenic somatic events in tissue-specific stem cells: A role in cancer recurrence? *Ageing Research Reviews, 13*, 100−106. https://doi.org/10.1016/j.arr.2013.12.004

Herrera-Rios, D., Li, G., Khan, D., Tsiampali, J., Nickel, A. C., & Aretz, P. (2020). A computational guided, functional validation of a novel therapeutic antibody proposes Notch signaling as a clinical relevant and druggable target in glioma. *Scientific Reports, 10*(1), 16218. https://doi.org/10.1038/s41598-020-72480-y

Hitoshi, S., Alexson, T., Tropepe, V., Donoviel, D., Elia, A. J., & Nye, J. S. (2002). Notch pathway molecules are essential for the maintenance, but not the generation, of mammalian neural stem cells. *Genes and Development, 16*(7), 846−858. https://doi.org/10.1101/gad.975202

Hsu, J. B., Lee, T. Y., Cheng, S. J., Lee, G. A., Chen, Y. C., & Le, N. Q. K. (2021). Identification of differentially expressed genes in different glioblastoma regions and their association with cancer stem cell development and temozolomide response. *Journal of Personalized Medicine, 11*(11). https://doi.org/10.3390/jpm11111047

Hui, S., Yang, Y., Li, J., Li, N., Xu, P., & Li, H. (2019). Differential miRNAs profile and bioinformatics analyses in bone marrow mesenchymal stem cells from adolescent idiopathic scoliosis patients. *The Spine Journal, 19*(9), 1584−1596. https://doi.org/10.1016/j.spinee.2019.05.003

Jumper, J., Evans, R., Pritzel, A., Green, T., Figurnov, M., & Ronneberger, O. (2021). Highly accurate protein structure prediction with AlphaFold. *Nature, 596*(7873), 583−589. https://doi.org/10.1038/s41586-021-03819-2

Kidd, S., Kelley, M. R., & Young, M. W. (1986). Sequence of the notch locus of *Drosophila melanogaster*: Relationship of the encoded protein to mammalian clotting and growth factors. *Molecular and Cellular Biology, 6*(9), 3094−3108. https://doi.org/10.1128/mcb.6.9.3094-3108.1986

Klingstrom, T., & Plewczynski, D. (2011). Protein-protein interaction and pathway databases, a graphical review. *Briefings in Bioinformatics, 12*(6), 702−713. https://doi.org/10.1093/bib/bbq064

Kuntz, I. D. (1992). Structure-based strategies for drug design and discovery. *Science, 257*(5073), 1078−1082. https://doi.org/10.1126/science.257.5073.1078

Kushwaha, P. P., Vardhan, P. S., Kapewangolo, P., Shuaib, M., Prajapati, S. K., & Singh, A. K. (2019). Bulbine frutescens phytochemical inhibits notch signaling pathway and induces apoptosis in triple negative and luminal breast cancer cells. *Life Sciences, 234*, 116783. https://doi.org/10.1016/j.lfs.2019.116783

Laskowski, R. A. (2010). Protein structure databases. *Methods in Molecular Biology, 609*, 59−82. https://doi.org/10.1007/978-1-60327-241-4_4

Li, L., Krantz, I. D., Deng, Y., Genin, A., Banta, A. B., & Collins, C. C. (1997). Alagille syndrome is caused by mutations in human Jagged1, which encodes a ligand for Notch1. *Nature Genetics, 16*(3), 243−251. https://doi.org/10.1038/ng0797-243

Liu, L., Charville, G. W., Cheung, T. H., Yoo, B., Santos, P. J., & Schroeder, M. (2018). Impaired notch signaling leads to a decrease in p53 activity and mitotic catastrophe in aged muscle stem cells. *Cell Stem Cell, 23*(4), 544−556. https://doi.org/10.1016/j.stem.2018.08.019. e544.

Lobry, C., Oh, P., Mansour, M. R., Look, A. T., & Aifantis, I. (2014). Notch signaling: Switching an oncogene to a tumor suppressor. *Blood, 123*(16), 2451−2459. https://doi.org/10.1182/blood-2013-08-355818

Lopez-Ayllon, B. D., Moncho-Amor, V., Abarrategi, A., Ibanez de Caceres, I., Castro-Carpeno, J., & Belda-Iniesta, C. (2014). Cancer stem cells and cisplatin-resistant cells isolated from non-small-lung cancer cell lines constitute related cell populations. *Cancer Medicine, 3*(5), 1099−1111. https://doi.org/10.1002/cam4.291

Luo, D., Renault, V. M., & Rando, T. A. (2005). The regulation of Notch signaling in muscle stem cell activation and postnatal myogenesis. *Seminars in Cell and Developmental Biology, 16*(4−5), 612−622. https://doi.org/10.1016/j.semcdb.2005.07.002

Luxan, G., Casanova, J. C., Martinez-Poveda, B., Prados, B., D'Amato, G., & MacGrogan, D. (2013). Mutations in the Notch pathway regulator MID1 cause left ventricular noncompaction cardiomyopathy. *Nature Medicine, 19*(2), 193−201. https://doi.org/10.1038/nm.3046

Man, C. H., Wei-Man Lun, S., Wai-Ying Hui, J., To, K. F. Choy, K. W., & Wing-Hung Chan, A. (2012). Inhibition of Notch3 signalling significantly enhances sensitivity to cisplatin in EBV-associated nasopharyngeal carcinoma. *The Journal of Pathology, 226*(3), 471−481. https://doi.org/10.1002/path.2997

Marioni, J. C., Mason, C. E., Mane, S. M., Stephens, M., & Gilad, Y. (2008). RNA-Seq: An assessment of technical reproducibility and comparison with gene expression arrays. *Genome Research, 18*(9), 1509−1517. https://doi.org/10.1101/gr.079558.108

Meisel, C. T., Porcheri, C., & Mitsiadis, T. A. (2020). Cancer stem cells. Quo Vadis? The notch signaling pathway in tumor initiation and progression. *Cells, 9*(8). https://doi.org/10.3390/cells9081879

Meng, R. D., Shelton, C. C., Li, Y. M., Qin, L. X., Notterman, D., & Paty, P. B. (2009). Gamma-secretase inhibitors abrogate oxaliplatin-induced activation of the Notch-1 signaling pathway in colon cancer cells resulting in enhanced chemosensitivity. *Cancer Research, 69*(2), 573−582. https://doi.org/10.1158/0008-5472.CAN-08-2088

Murtaugh, L. C., Stanger, B. Z., Kwan, K. M., & Melton, D. A. (2003). Notch signaling controls multiple steps of pancreatic differentiation. *Proceedings of the National Academy of Sciences of the United States of America, 100*(25), 14920−14925. https://doi.org/10.1073/pnas.2436557100

Ntziachristos, P., Lim, J. S., Sage, J., & Aifantis, I. (2014). From fly wings to targeted cancer therapies: A centennial for notch signaling. *Cancer Cell, 25*(3), 318−334. https://doi.org/10.1016/j.ccr.2014.02.018

Pan, T., Xu, J., & Zhu, Y. (2017). Self-renewal molecular mechanisms of colorectal cancer stem cells. *International Journal of Molecular Medicine, 39*(1), 9−20. https://doi.org/10.3892/ijmm.2016.2815

Pellegrinet, L., Rodilla, V., Liu, Z., Chen, S., Koch, U., & Espinosa, L. (2011). Dll1- and dll4-mediated notch signaling are required for homeostasis of intestinal stem cells. *Gastroenterology, 140*(4), 1230−1240. https://doi.org/10.1053/j.gastro.2011.01.005. e1231−1237.

Piya, S., Yang, Y., Bhattacharya, S., Sharma, P., Ma, H., & Mu, H. (2022). Targeting the NOTCH1-MYC-CD44 axis in leukemia-initiating cells in T-ALL. *Leukemia, 36*(5), 1261−1273. https://doi.org/10.1038/s41375-022-01516-1

Platonova, N., Parravicini, C., Sensi, C., Paoli, A., Colombo, M., & Neri, A. (2017). Identification of small molecules uncoupling the Notch::Jagged interaction through an integrated high-throughput screening. *PLoS One, 12*(11), e0182640. https://doi.org/10.1371/journal.pone.0182640

Poon, E., Yan, B., Zhang, S., Rushing, S., Keung, W., & Ren, L. (2013). Transcriptome-guided functional analyses reveal novel biological properties and regulatory hierarchy of human embryonic stem cell-derived ventricular cardiomyocytes crucial for maturation. *PLoS One, 8*(10), e77784. https://doi.org/10.1371/journal.pone.0077784

Porcheri, C., Golan, O., Calero-Nieto, F. J., Thambyrajah, R., Ruiz-Herguido, C., & Wang, X. (2020). Notch ligand Dll4 impairs cell recruitment to aortic clusters and limits blood stem cell generation. *EMBO Journal, 39*(8), e104270. https://doi.org/10.15252/embj.2019104270

Poulson, D. F. (1937). Chromosomal deficiencies and the embryonic development of Drosophila melanogaster. *Proceedings of the National Academy of Sciences of the United States of America, 23*(3), 133−137. https://doi.org/10.1073/pnas.23.3.133

Rangamani, P., & Iyengar, R. (2008). Modelling cellular signalling systems. *Essays in Biochemistry, 45*, 83−94. https://doi.org/10.1042/BSE0450083

Relaix, F., Bencze, M., Borok, M. J., Der Vartanian, A., Gattazzo, F., & Mademtzoglou, D. (2021). Perspectives on skeletal muscle stem cells. *Nature Communications, 12*(1), 692. https://doi.org/10.1038/s41467-020-20760-6

Shih, A. H., & Holland, E. C. (2006). Notch signaling enhances nestin expression in gliomas. *Neoplasia, 8*(12), 1072−1082. https://doi.org/10.1593/neo.06526

Shoemaker, B. A., & Panchenko, A. R. (2007a). Deciphering protein-protein interactions. Part I. Experimental techniques and databases. *PLoS Computational Biology, 3*(3), e42. https://doi.org/10.1371/journal.pcbi.0030042

Shoemaker, B. A., & Panchenko, A. R. (2007b). Deciphering protein-protein interactions. Part II. Computational methods to predict protein and domain interaction partners. *PLoS Computational Biology, 3*(4), e43. https://doi.org/10.1371/journal.pcbi.0030043

Siebel, C., & Lendahl, U. (2017). Notch signaling in development, tissue homeostasis, and disease. *Physiological Reviews, 97*(4), 1235−1294. https://doi.org/10.1152/physrev.00005.2017

Singh, A. K., Shuaib, M., Prajapati, K. S., & Kumar, S. (2022). Rutin potentially binds the gamma secretase catalytic site, down regulates the Notch signaling pathway and reduces sphere formation in colonospheres. *Metabolites, 12*(10). https://doi.org/10.3390/metabo12100926

Taoudi, S., & Medvinsky, A. (2007). Functional identification of the hematopoietic stem cell niche in the ventral domain of the embryonic dorsal aorta. *Proceedings of the National Academy of Sciences of the United States of America, 104*(22), 9399−9403. https://doi.org/10.1073/pnas.0700984104

Uhlen, M., Bjorling, E., Agaton, C., Szigyarto, C. A., Amini, B., & Andersen, E. (2005). A human protein atlas for normal and cancer tissues based on antibody proteomics. *Molecular & Cellular Proteomics, 4*(12), 1920−1932. https://doi.org/10.1074/mcp.M500279-MCP200

Umar, S. (2010). Intestinal stem cells. *Current Gastroenterology Reports, 12*(5), 340−348. https://doi.org/10.1007/s11894-010-0130-3

Varadkar, P. A., Kraman, M., & McCright, B. (2009). Generation of mice that conditionally express the activation domain of Notch2. *Genesis, 47*(8), 573−578. https://doi.org/10.1002/dvg.20537

Verdonk, M. L., Cole, J. C., Hartshorn, M. J., Murray, C. W., & Taylor, R. D. (2003). Improved protein-ligand docking using GOLD. *Proteins, 52*(4), 609−623. https://doi.org/10.1002/prot.10465

Waterhouse, A., Bertoni, M., Bienert, S., Studer, G., Tauriello, G., & Gumienny, R. (2018). SWISS-MODEL: Homology modelling of protein structures and complexes. *Nucleic Acids Research, 46*(W1), W296−W303. https://doi.org/10.1093/nar/gky427

Welshons, W. J., & Von Halle, E. S. (1962). Pseudoallelism at the notch locus in drosophila. *Genetics, 47*(6), 743−759. https://doi.org/10.1093/genetics/47.6.743

Wen, S., Li, H., & Liu, J. (2009). Dynamic signaling for neural stem cell fate determination. *Cell Adhesion and Migration, 3*(1), 107−117. https://doi.org/10.4161/cam.3.1.7602

White, J. K., Gerdin, A. K., Karp, N. A., Ryder, E., Buljan, M., & Bussell, J. N. (2013). Genome-wide generation and systematic phenotyping of knockout mice reveals new roles for many genes. *Cell, 154*(2), 452−464. https://doi.org/10.1016/j.cell.2013.06.022

Yang, L., Shi, P., Zhao, G., Xu, J., Peng, W., & Zhang, J. (2020). Targeting cancer stem cell pathways for cancer therapy. *Signal Transduction and Targeted Therapy, 5*(1), 8. https://doi.org/10.1038/s41392-020-0110-5

Zeng, C., Lin, M., Jin, Y., & Zhang, J. (2022). Identification of key genes associated with brain metastasis from breast cancer: A bioinformatics analysis. *Medical Science Monitor, 28*, e935071. https://doi.org/10.12659/MSM.935071

Zhao, Z. L., Zhang, L., Huang, C. F., Ma, S. R., Bu, L. L., & Liu, J. F. (2016). Notch1 inhibition enhances the efficacy of conventional chemotherapeutic agents by targeting head neck cancer stem cell. *Scientific Reports, 6*, 24704. https://doi.org/10.1038/srep24704

Zhou, B., Lin, W., Long, Y., Yang, Y., Zhang, H., & Wu, K. (2022). Notch signaling pathway: Architecture, disease, and therapeutics. *Signal Transduction and Targeted Therapy, 7*(1), 95. https://doi.org/10.1038/s41392-022-00934-y

Zhu, L., Shu, Z., & Sun, X. (2018). Bioinformatic analysis of four miRNAs relevant to metastasis-regulated processes in endometrial carcinoma. *Cancer Management and Research, 10*, 2337−2346. https://doi.org/10.2147/CMAR.S168594

Chapter 23

In silico approaches for the analysis of developmental fate of stem cells

Vinay Bhatt

Amity Institute of Biotechnology, Amity University, Noida, Uttar Pradesh, India

1. Introduction

Stem cell activity relies on interactions with the niche for tissue homeostasis maintenance and wound healing (Brunet et al., 2022). Stem cells are supported throughout their whole lives by stem cell niches, which are made up of changing small-scale environments (Hicks & Pyle, 2022). The primary function of the developing niche, which sets it apart from the adult niche, is to nourish the progenitors that will eventually grow into organ systems (Hicks & Pyle, 2022). With aging, stem cells lose their ability to generate new tissue and differentiate into distinct cells within the tissue, which is associated with a decline in the integrity of tissue and the state of health (Brunet et al., 2022). The structure and function of multicellular organisms are maintained by the development system. Two key mechanisms of cell proliferation are proliferation by the single cell to divide and produce two daughter cells and the cell development mechanism to adopt a specialized cell type (Robert, 2004; Setty, 2014). Individual population cell development is regulated by molecular processes, for example, activation of receptor, expression of gene, and protein breakdown. All of these systems work together to form an intricate network that involves both cell-intrinsic and cell-extrinsic pathways. Existing experimental procedures, however, still have some challenges to overcome when trying to collect data for numerous stages of growth. Modern time-point approaches can typically only capture a few cell divisions at a time, and long-term experimental research that could produce time dependent data is currently impractical (Jensen, 2004). The most important factors are how cells coordinate connections across time, how chemical interactions affect the formation of structures and tissue patterning, and how signaling pathway interactions affect the system. The system-level view reveals knowledge gaps and promotes the development of hypotheses to fill them through experimentation. In the long term, these types of studies may aid in understanding the factors that contribute to diseases such as cancer and tissue deterioration (Ebben et al., 2010; Zhang et al., 2009). The three essential elements will be combined into a database called stem cell lineage database that will be searchable and able to store data "about cell type gene expression, cell lineage maps, differentiation of stem cell and its procedures for both human and mouse stem cells, and endogenous regulatory factors."

Numerous biological systems have been the subject of in silico modeling. The models have been built for a variety of phenomena, which comprises the heart's electrical activity (Noble, 2002), the early *Drosophila* embryo's diffusion (Sample & Shvartsman, 2010), pattern development in zebra fish (Caicedo-Carvajal & Shinbrot, 2008), and investigations involving dual perturbations (Jamshidi & Palsson, 2009). The modeling technique can be improved by focusing on specific models created to describe a certain system in biology, which are usually cell-intrinsic models. Several others have explored a broader approach, they include Cellerator and the Virtual Cell (Shapiro et al., 2003; Slepchenko et al., 2003), which use a software environment based on differential equations to describe cell biology. Understanding cell population dynamics was the goal of several interrelated modeling projects. Expression patterns in the development of stem cell systems were identified by mathematical models at the level of gene and protein (Meier-Schellersheim et al., 2006; Yener et al., 2008).

The epithelial stem cells quantity in experimental systems was quantified using a platform known as STORM (Wang et al., 2010). Additionally, mathematical models are used to infer stem cell division patterns from clone size distributions and stem cell proliferation in the intestinal crypt (Buske et al., 2011). Various mathematical techniques are employed by

Copyright © 2024 Elsevier Inc. All rights are reserved, including those for text and data mining, AI training, and similar technologies.

other in silico models to simulate cell-regulating subsystems. For example, chemotaxis and cell motility models, the circadian cycle, cellular development, and the cell cycle process. This work offers a framework for studying and analyzing the specific system in question and aids in illuminating new information (Setty, 2014). Most of these modeling efforts simply take into account one scale of the biological system's growth, ignoring the influence of the associated scale (Cohen & Harel, 2007). An illustration would be the modeling of single cell behavior without taking into account the numerous interactions on a tissue scale. In silico models are necessary to accurately comprehend the entirety of cellular growth and to incorporate the control of individual subsystems into a thorough multifaceted dynamic model in the tissue (Setty, 2014). Techniques and comparable methods can be used to understand stem cell regulation. Some of these techniques and methods are sequencing, microarray profiling, ChiP-Seq, CLIP-Seq, RIP-Seq, Chip-Seq of histone modification, Bisulfite-Seq, proteomics, and Y2H and DNase hypersensitivity mapping. The fluorescent in situ RNA sequencing (FISSEQ) method affects stem cells' in situ state and facilitates the measurement of quantitative gene expression variation. The parameters at the molecular and cellular levels that control stem cell activity and development are examined by researchers in this field using a number of computational tools, including mathematical modeling, bioinformatics, and high-throughput data analysis. There are a number of techniques used in stem cell biology and development biology, including ScoreCard, Pluritest, TeratoScore, KeyGenes, CellNet, and Mogrify. These tools help to create databases and reduce work load. The most sensitive factors for stem cells can be identified with the aid of mathematical models (Stiehl & Marciniak-Czochra, 2011). The utilization of mathematical modeling, which is a powerful technique, has had a significant positive impact on a number of complex biological systems, including the hematopoietic system (Whichard et al., 2010). Machine learning strategies and recommended methodologies are aimed at stem cell research with a particular emphasis on the potential application of stem cell biosafety and bioefficacy evaluation. Building models that could ultimately help medical professionals make more informed decisions about stem cell therapy (Zaman et al., 2021).

2. Computational techniques in stem and development biology

Understanding how stem cells choose to differentiate into distinct cell types and how these differentiated cells develop into an organism with numerous tissues and organs, it is a major focus of computational stem and development biology. Using a variety of computational techniques such as mathematical modeling, bioinformatics, and high-throughput data analysis, researchers in this discipline investigate the molecular and cellular variables that govern stem cell activity and development. Another focus of computational stem and development biology is to use computational tools to generate predictive models of development and disease; this is helpful in identifying novel therapy targets and to test the effects of drugs and other interventions on stem cell behavior and development.

Computational techniques in developmental biology and stem cell biology have been incredibly influential. Its primary goal is to better understand how theoretical morphogenesis processes work, including modeling the dynamics of adult stem cell self-renewal and the production of positional information during embryogenesis (Lander, 2007; Till et al., 1964). Massive molecular datasets have replaced modeling as the primary use of computer tools in stem cell and developmental biology, starting around 23 years ago with the major genome sequencing projects. With system-level models, which are based on stem cell function and behavior, large-scale molecular data can be effectively generated. In that model, methodologies of system as well as network biology have started to be used. Second, modern technology now allows for molecular profiling of a single cell genome. Using Chip-Chip, its predecessor, the transcriptional regulatory network of essential pluripotency transcription factors (TFs) in human embryonic stem cells (hESCs) was replicated (Table 23.1). This process revealed a self-regulatory cycle that protects the pluripotent state prevents a brief downregulation of any one pluripotency TF (Boyer et al., 2005). Overall, computational stem cell and development biology play a crucial role in understanding the underlying development mechanisms of organisms, and it has the potential to lead to novel therapies for a variety of disorders and illnesses (such as type 1 diabetes, spinal cord injuries, Parkinson's disease, Alzheimer's disease, cancer and osteoarthritis).

3. Stem cell biology with OMICs

OMICS methods have been used a crucial function in understanding the complex regulation of pluripotency. The transcriptional characteristics of embryonic stem cell (ESC) lines and the molecular basis of pluripotency were determined, along with how it differs from that of adult stem cells. The exploration of posttranscriptional regulation in pluripotency was made possible by a variety of novel OMICs approaches, many of which were based on next-generation sequencing, including Ribo-Seq, RIP-Seq, and CLIP-Seq (Table 23.1) (Fig. 23.1) (Cho et al., 2012; Ingolia et al., 2011; Wang et al., 2013). As a demonstration, the mouse and human pluripotency networks MeRIP-Seq were used to

TABLE 23.1 Techniques used for the stem cell biology and developmental biology.

Approach	Description	References
Sequencing	Figuring out the precise arrangement of bases, or nucleotides, in a DNA/RNA molecule.	Gore et al. (2011)
Microarray profiling	Connecting a group of a few hundred to a few million distinguishable nucleic acid fragments to a "chip," or surface that is solid. Then, a study sample (such as cells or tissue) and pure RNA or DNA are applied to the chip.	Josephson et al. (2006)
ChiP-Seq	Finding transcription factors' and other proteins' DNA-binding locations across the entire genome.	Mikkelsen et al. (2007)
Chip-chip	ChIP is used in conjunction with DNA microarray-based profiling to describe how proteins interact with genomic DNA.	Boyer et al. (2005)
CLIP-Seq	Protein–RNA interaction	Kwon et al. (2013)
RIP-Seq	Protein–RNA interaction	Van Wynsberghe et al. (2011)
Chip-Seq of histone modification	Epigenetic	Lienert et al. (2011)
Bisulfite-Seq	DNA methylation	Lister et al. (2011)
Proteomics	Posttranslational modifications, immunoassay, mass spectrometry, and protein profiling	Ingolia et al. (2009)
Y2H	Protein–protein interaction	Zheng et al. (2013)
DNase hypersensitivity mapping	Gene regulatory regions	Vierstra et al. (2014)

characterize the m6A methylome of ESCs. Using this method, it was found that the zinc finger protein 217 (ZFP217) connected with chromatin and ZFP217 interacted via epigenetic networks to govern hESC pluripotency (Bian & Cahan, 2016).

4. Epigenetic memory and induced pluripotent stem cells

It is crucial to comprehend how induced pluripotent stem cell (iPSC) and ESC differ from one another, as well as how reprogramming itself may alter in vitro lineage bias, for iPSC to be utilized in these circumstances. According to a theory that has emerged, epigenetic residual marks that are still present in iPSCs and are transcriptionally suppressed in the pluripotent state and emerge evident in targeted differentiation create lineage bias. Different genomic regions with methylation in iPSC derived from various starting cell types were identified using "comprehensive high-throughput arrays for relative methylation" (CHARM) to investigate this hypothesis. In mice and human iPSCs, these differentially methylated genomic regions were found to be concentrated in TF promoters that determine lineages different from the beginning cell type (Irizarry et al., 2008; Takahashi et al., 2007; Takahashi & Yamanaka, 2006). A unique method based on targeted bisulfite sequencing suggests that the persistence of partially reprogrammed cells may be related to residually constrictive DNA methylation at pluripotency TF promoters (Table 23.1) (Mikkelsen et al., 2008). Genome-wide profiling is combined with sophisticated and frequently customized algorithms to investigate the molecular basis of pluripotency and the processes involved in resetting to pluripotency.

5. Stem cell biology through computation

Technologies that regulate cell fate, such as direct transition of guided differentiation of pluripotent stem cells (PSC) among somatic cell types (one example is transforming fibroblasts into cardiomyocytes by ectopic expression of genes such as Tbx5, Mef2c, and Gata4 (Qian et al., 2013)). It is practiced in thousands of labs across the world to model diseases, to investigate inaccessible developmental times, pharmacological evaluations, and development of regenerative

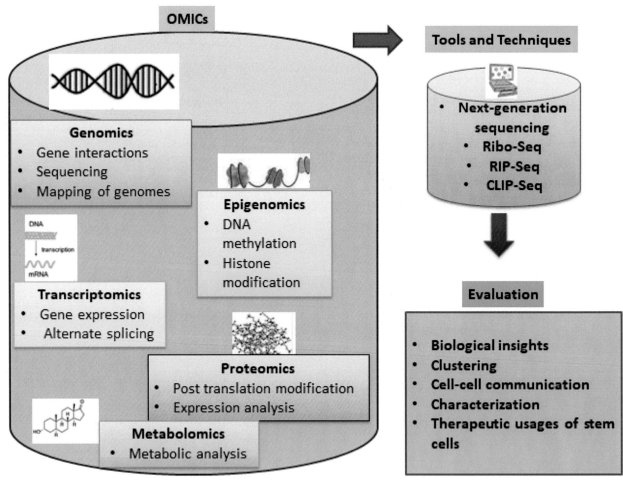

FIGURE 23.1 OMICS and its approaches.

medicine treatments or therapies. However, there are three significant challenges that cell fate engineering must overcome if it is to completely realize its potential to fundamentally transform the biomedical sector. In the first class of barriers is caused by the absence of rational, verified, and hypothesis-driven system to select the conditions that will guide directed differentiation or to select the components that will be employed for direct conversion. The directed differentiation strategies that were motivated by an understanding of signaling cues and forces in mouse development, are constrained due to their inability to study highly transient embryonic stages. As opposed to that, methods for choosing the concept of "master controllers" for usage during immediate conversion are based on a "kernel" gene regulatory network (GRN) made up of a few TFs that positively control the transcription of genes related to particular cell types as well as alternative lineages while negatively repressing other lineages. GRNs analyze gene expression information to understand how a biological process works. While reprogramming back to pluripotency was effective using this method of locating and utilizing TFs of a kernel GRN, it is unknown if it can be applied to additional cell types (Bian & Cahan, 2016). Another challenge is the inability to measure the extent to which changed cells resemble their in vivo counterparts. This relates to the lineage bias problem or the persistent variation among lines in the PSC effectiveness, and they can be suggested to choose their lineages wisely and properly. There has been a lot of focus on the molecular causes of this variation as well as the involvement of both epigenetic and genetic variables (Cahan & Daley, 2013; Osafune et al., 2008). The differentiation tendency of the PSC line can be accurately predicted using the ScoreCard (Table 23.2). With the assumption that either improper expression or methylation of the DNA of lineage-specific controllers would prohibit embryonic conversion to particular lineages, it was initially trained on the data on ESC line expression and methylation of DNA. To develop a reference database that might be used to choose iPSCs for certain applications, iPSCs and ESCs were compared for changes in gene expression and DNA methylation at key genes for lineage development. Bock et al. (2011) additionally choose a collection of 500 genes that identify all three layers of

TABLE 23.2 Tools for computational stem cell biology.

Tool	Description	References
Lang	• Gene expression profile Choose potential TFs for cell type conversion • microRNAs, histone modification information	Lang et al. (2014), Bian and Cahan (2016)
D'Alessio	• Expression of TFs in 233 types of tissue and cells • A list of potential TFs that can change fate	D'Alessio et al. (2015), Bian and Cahan (2016)
Heinanemi	• A gene or kind of cell • Heatmaps for the 3D epigenetic landscape for the data that was chosen feature (gene or type of cell) • Information on the expression of genes in 2602 TFs and 166 cell types • Investigating and comprehending how cells regulate their transcription • RNA-Seq Web link: http://trel.systemsbiology.net/	Crespo and Del Sol (2013)
Mogrify	• Beginning and ending cell types • Gene expression information for around 300 different cell and tissue types • Predict the TFs involved in cell differentiation and conversion Web link: http://www.mogrify.net	Rackham et al. (2016)
CellNet	• Profile of gene expression profile (microarray data from Affymetrix and Illumina) • 20 different cell and tissue types' gene expression profiles • Evaluates the degree by which a cell type specific GRN is created, predicts TFs that can permit cell destiny modification, and evaluates similarity to 20 cell and tissue types • Single-cell and bulk RNA-Seq Web application: http://cellnet.hms.harvard.edu Code: https://github.com/pcahan1/CellNet	Cahan et al. (2014)
KeyGene	• Data from 21 fetal organs' next-generation sequencing (NGS) and gene expression profiles (Affymetrix, Illumina) • Estimates a developmental stage, the identification of stem cell descendants, and the tissue of origin Web link: http://www.keygenes.nl/	Roost et al. (2015)
TeratoScore	• 12 cell line, 26 types of cells and tissues of gene expression profiles • Affymetrix Human Genome U133 Plus 2.0 Array gene expression data (.CEL file) • Contribution of the germ layer to teratomas provides a quantitative evaluation of pluripotency • RNA-Seq and qPCR extensions are theoretically feasible Web link: http://benvenisty.huji.ac.il/teratoscore.php	Avior et al. (2015)
Pluritest	• Gene expression information from the HT12v3 and HT12v4 illumina microarrays (.idat) • 233 ES 41 iPS cell line gene expression profiles • Estimate the query sample's pluripotency and divergence from pluripotency (whether it is a PSC or not). Extending to global DNA methylation and RNA-Seq is theoretically feasible Web link: http://www.pluritest.org	Müller et al. (2011)
ScoreCard	• Data on DNA methylation and nanostring and qPCR gene expression • Genome-wide reference maps for 20 ES cell lines and 12 iPS cell lines for DNA methylation and gene expression • Scorecard for each lineage that summarizes the propensities for differentiation in each cell line based on the expression of the three germ layers as well as the hematopoietic and neural	Tsankov et al. (2015)

Continued

TABLE 23.2 Tools for computational stem cell biology.—cont'd

Tool	Description	References
	lineages. DNA methylation and gene expression deviation scorecard and reference corridor • It is theoretically conceivable to include new cell types in the lineage scorecard prediction Web link: http://scorecard.computational-epigenetics.org/	
Hierarchical clusterin	• Create a cluster hierarchy • The closest pairings of entities are combined first, and then another distance is determined between the merged entity and all other entities • The operation is carried out again and again until all entities have been integrated	Bian and Cahan (2016)
Clustering of k-means	• Cluster information into a predetermined digit (k) of clusters by randomly allocating items to clusters • Computing each cluster's average profile, calculating the distances between clusters, allocating to the nearest cluster of things, and recalculating the mean profiles • Either till the entities' cluster membership remains the same, or until a predetermined number of times, this process is repeated	Bian and Cahan (2016)
Principal component analysiss	• A dimension-reduction technique is to find orthogonal axes or orientations that correspond to a linear mix of variables that optimize the total range in the information set	Bian and Cahan (2016)
Differential analysis	• Aims to find distinctively expressed genes among various subsets by applying methods that take into account the frequently high number of statistical tests being run	Bian and Cahan (2016)
Enrichment analysis	• Checking if certain expression patterns have a tendency to cluster toward the top or bottom of an ordered each gene list, gene set analysis employs tools	Subramanian et al. (2005)
Mutation calling	• The discovery of variations in genes across samples (such as person's germ line or a tumor) and a genomic reference sequence	Bian and Cahan (2016)
Peak comparison	• To identify DNA regions where sequencing readings made using ChIP-Seq or DNase-Seq have been found to be enriched	Bian and Cahan (2016)

germs as well as neuronal and hematological lineages to make the future scoring of new ESC and iPSC easier (Bock et al., 2011). They showed that there is a chance to evaluate PSCs' capacity for differentiation in a neutral manner by tracking how these genes are expressed throughout uncontrolled embryoid body (EB) development. For instance, they verified that the H1 as well as H9 lines showed a significant tendency for brain lineage development, while the endoderm lines were favored by the HUES8 line. This platform has since been enhanced by adding support for qPCR, an expression method that is more commonly available. Notably, no other technique in use today makes an effort to forecast the in vitro lineage bias of PSC lines (Bock et al., 2011; Tsankov et al., 2015). Pluritest may evaluate a cell's pluripotency based on its expression profile of gene (Table 23.2). For this work, the researchers assembled hundreds of publicly available gene expression data from hiPSC (human-induced pluripotent stem cell) and hESC lines to construct a database related to pluripotency expression of gene. Negative matrix factorization (NMF), which gauges the degree of departure from the state of pluripotency, was used to classify the data after logistic regression to discriminate between the likelihood score of pluripotency and nonpluripotency. Pluritest, which has been shown to be useful in assessing PSCs made from various sources, such as chemically stimulated iPSCs as well as those obtained from amniotic fluid of human, can accurately and specifically predict pluripotency (Müller et al., 2011). The TeratoScore was developed to evaluate the potential for differentiation of hPSCs quantitatively in relation to the teratomas' design of gene expression (Table 23.2). The theoretical foundation for this algorithm was that teratoma development is one of the benchmarks for assessing the effectiveness of hPSCs (the ability to distinguish between descendants of the three different germ layers), and

TeratoScore additionally categorizes whether a tumor develops from pluripotent cells or a particular tissue (Avior et al., 2015). KeyGenes is a platform for assessing the effectiveness of tissue differentiation based on gene signatures (Table 23.2). RNA-Seq or microarray analysis of 21 different human fetal tissues produced at multiple levels of development. KeyGenes was applied to recently collected, publicly available data that included human embryonic and adult organs, tissue organoids, and hPSCs that had differentiated into three germ lines. KeyGenes managed to determine the origin of the tissue in the samples and effectively detected the stem cell offspring. Additionally, KeyGenes allows you to assign distinct hPSC descendants from early stages of development (Roost et al., 2015). CellNet was developed with the goal of evaluating and enhancing stem cell engineering paradigms with GRNs with tissue- and cell-specific bases (Table 23.2). CellNet was useful for the analysis of expression of gene profiles for GRN unique to cells and predicted the TFs that can enable cell fate change. CellNet is only capable of microarray data, and because it is dependent on reconstructed GRNs, its predictive power is limited. Bulk tissue is mainly used rather than from homogenous groups of cells (Cahan et al., 2014). The Mogrify system was specifically created by Rackham et al. (2016) to forecast various TF pairings that allow goals transformations among 173 different types of human cells and tissues (Table 23.2). Mogrify estimates the global expression changes in each TF that could result from ectopically expressing them in a particular starting cell type using previously documented regulatory and interaction networks. A group of TFs that will most efficiently upregulate the target cell type expression program can be found by searching all TFs (Rackham et al., 2016). To identify TFs that might be crucial when cell identity modifications are induced, Heinaniemi's theory made use of the idea of TF intersect-repression in determining cell fate. By contrasting their predictions to three cases of trans-differentiation, from myeloid cells to erythroid cells, hepatocytes to fibroblasts, and from T-helper type Th2 to type Th1 lymphocytes, this unique method in silico was tested. The technique was effective in identifying experimentally shown fate-altering TFs (Crespo & Del Sol, 2013).

Despite the rapidly growing single-cell analysis interest for many uses including heterogeneity in cancer and discovery of new sorts of cells, molecular method reviews have subsequently risen. After exposure to the appropriate signaling events, stem cells handle transcription differently to make lineage division easier to access. However, the computational element of maintaining this data correctly has gotten very little attention (Bian & Cahan, 2016). Single-cell OMICs are very useful in stem cell research and are utilized in the subpopulation with stem cell qualities. Based on single-cell transcriptional patterns, a homogenous number of parents of regenerating planarians might be more divided into subgroups with distinct regeneration potential (Van Wolfswinkel et al., 2014). Through single-cell analysis, it is feasible to derive regulatory linkages among genes which were hidden in mass data. Numerous fundamental problems with reconstructed gene controlling networks, such as the confounding effect on the population underneath and the dependence on nonphysiological changes to induce associated changes in gene expression, may be resolved with the potential to thousands individual cell samples (Bian & Cahan, 2016).

6. Single cell genomics with spatial resolution

The FISSEQ method can be used to quantify RNA in thousands numbers, fixed the cells in situ, tissue slices, and entire-mount embryos. FISSEQ can still detect RNA even though it, compared with single-cell RNA-Seq, is less-delicate and has strongly expressed transcripts with functional relevance (Fig. 23.2). The adoption of techniques such as FISSEQ would be

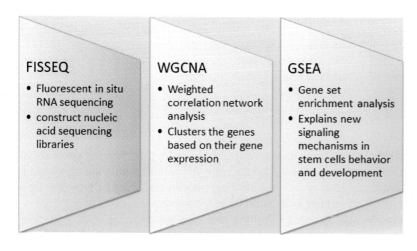

FIGURE 23.2 Functions of FISSEQ, WGCNA, and GSEA.

FISSEQ
- Fluorescent in situ RNA sequencing
- construct nucleic acid sequencing libraries

WGCNA
- Weighted correlation network analysis
- Clusters the genes based on their gene expression

GSEA
- Gene set enrichment analysis
- Explains new signaling mechanisms in stem cells behavior and development

beneficial for researching the types of cells in their surroundings as well as how a niche impacts the in situ status of stem cells (Lee et al., 2015).

7. Mathematical modeling

To understand the fate mechanism of the cells, different mathematical modeling is used in which mathematical equations are used for simulating and predicting the activity of stem cells and developmental processes. It may be used to investigate gene regulatory networks, cell–cell interactions, and signaling cascades.

8. Weighted correlation network analysis

Weighted correlation network analysis (WGCNA) is a computer tool for studying the interactions of genes and other macromolecules in stem cells (Fig. 23.2). WGCNA identifies genes groupings with similar pattern of expression using a measure of resemblance, such as Pearson correlation. These modules can be used to discover pathways and biological processes relevant to stem cell activity and development (Langfelder et al., 2008). WGCNA has the benefit of identifying modules of coexpressed genes even when the underlying biology is complicated and poorly understood. It may also be used to discover transcripts that are active distinctly in various several types of stem cells or circumstances, as well as prospective therapeutic targets.

9. Gene set enrichment analysis

Gene set enrichment analysis (GSEA) is a computer tool for identifying pathways and biological processes that are overrepresented within a set of differentially expressed genes in stem cells (Fig. 23.2). A powerful technique for analyzing a high-throughput data of gene expression is GSEA, and it is frequently used in stem cell research to discover pathways and processes involved in stem cell behavior and development (Subramanian et al., 2005).

10. Machine learning classifiers

Machine learning classifiers are a useful computational tool for analyzing stem cell data and predicting cell fate, phenotypes, and other properties. These classifiers evaluate large-scale data sets with algorithms to uncover patterns that can be used to forecast stem cell activity or differentiate stem cells into distinct subtypes (Ashraf et al., 2021).

Machine learning classifiers may be applied in stem cell studies in a variety of ways, including:

1. Supervised learning classifiers: These classifiers are trained on a labeled data set with a known outcome, such as cell type, and then used to predict the outcome for fresh, unknown data. Trees that make decisions, support vector machines (SVMs), random forests and k-nearest neighbors (k-NN) are among examples.
2. Unsupervised learning classifiers: These classifiers are trained on unlabeled data sets, and the system detects patterns or clusters in the data automatically. k-means, hierarchical clustering (Table 23.2), and t-SNE are a few examples (t-distributed stochastic neighbor embedding).
3. Deep learning classifiers: These classifiers examine and predict stem cell activity using neural networks. They may be trained to identify patterns in large-scale datasets, such as photographs, and can be used to analyze large-scale datasets. Examples include recurrent neural networks (RNNs) and convolutional neural networks (CNNs).

Machine learning classifiers are a useful computational tool for analyzing stem cell data and cell fate prediction, phenotypes, and other properties. They may be used to examine a wide range of data types, including transcriptomics, proteomics, epigenomics, and imaging data, and they can help identify new treatment targets for stem cell–related disorders.

11. Conclusion

The development of "computational stem cell biology" is comparable with the shift from biology of cancer to genomics of cancer, where "Big Data," particularly entire genome sequences, necessitated the computation hiring and training lists with attention on biology of cancer. This pattern is having profound effects on how cancer begins and spreads, as well as how new therapies are being created as a result. The study of stem cells is currently at a comparable turning point. Computational stem cell biology will produce data that can be used to define identity of cell type and characterize the molecular sense of tradition commitment and development, laying the groundwork for understanding how inherited and the epigenetic process flaws interfere with normal growth and development and lead to illnesses (Bian & Cahan, 2016).

Acknowledgments
I thank Dr. Rajesh Kumar for the discussion and suggestions.

Author contributions
VB developed the research topic and carried out the analytical and theoretical experiments. VB wrote this manuscript.

References

Ashraf, M., Khalilitousi, M., & Laksman, Z. (2021). Applying machine learning to stem cell culture and differentiation. *Current Protocols, 1*(9), e261.

Avior, Y., Biancotti, J. C., & Benvenisty, N. (2015). TeratoScore: Assessing the differentiation potential of human pluripotent stem cells by quantitative expression analysis of teratomas. *Stem Cell Reports, 4*(6), 967−974.

Bian, Q., & Cahan, P. (2016). Computational tools for stem cell biology. *Trends in Biotechnology, 34*(12), 993−1009.

Bock, C., Kiskinis, E., Verstappen, G., Gu, H., Boulting, G., Smith, Z. D., et al. (2011). Reference maps of human ES and iPS cell variation enable high-throughput characterization of pluripotent cell lines. *Cell, 144*(3), 439−452.

Boyer, L. A., Lee, T. I., Cole, M. F., Johnstone, S. E., Levine, S. S., Zucker, J. P., et al. (2005). Core transcriptional regulatory circuitry in human embryonic stem cells. *Cell, 122*(6), 947−956.

Brunet, A., Goodell, M. A., & Rando, T. A. (2022). Ageing and rejuvenation of tissue stem cells and their niches. *Nature Reviews Molecular Cell Biology, 24*, 1−18.

Buske, P., Galle, J., Barker, N., Aust, G., Clevers, H., & Loeffler, M. (2011). A comprehensive model of the spatio-temporal stem cell and tissue organisation in the intestinal crypt. *PLoS Computational Biology, 7*(1), e1001045.

Cahan, P., & Daley, G. Q. (2013). Origins and implications of pluripotent stem cell variability and heterogeneity. *Nature Reviews Molecular Cell Biology, 14*(6), 357−368.

Cahan, P., Li, H., Morris, S. A., Da Rocha, E. L., Daley, G. Q., & Collins, J. J. (2014). CellNet: Network biology applied to stem cell engineering. *Cell, 158*(4), 903−915.

Caicedo-Carvajal, C. E., & Shinbrot, T. (2008). *In-silico* zebrafish pattern formation. *Developmental Biology, 315*(2), 397−403.

Cho, J., Chang, H., Kwon, S. C., Kim, B., Kim, Y., Choe, J., et al. (2012). LIN28A is a suppressor of ER-associated translation in embryonic stem cells. *Cell, 151*(4), 765−777.

Cohen, I. R., & Harel, D. (2007). Explaining a complex living system: Dynamics, multi-scaling and emergence. *Journal of The Royal Society Interface, 4*(13), 175−182.

Crespo, I., & Del Sol, A. (2013). A general strategy for cellular reprogramming: The importance of transcription factor cross-repression. *Stem Cells, 31*(10), 2127−2135.

D'Alessio, A. C., Fan, Z. P., Wert, K. J., Baranov, P., Cohen, M. A., Saini, J. S., et al. (2015). A systematic approach to identify candidate transcription factors that control cell identity. *Stem Cell Reports, 5*(5), 763−775.

Ebben, J. D., Treisman, D. M., Zorniak, M., Kutty, R. G., Clark, P. A., & Kuo, J. S. (2010). The cancer stem cell paradigm: A new understanding of tumor development and treatment. *Expert Opinion on Therapeutic Targets, 14*(6), 621−632.

Gore, A., Li, Z., Fung, H. L., Young, J. E., Agarwal, S., Antosiewicz-Bourget, J., et al. (2011). Somatic coding mutations in human induced pluripotent stem cells. *Nature, 471*(7336), 63−67.

Hicks, M. R., & Pyle, A. D. (2022). The emergence of the stem cell niche. *Trends in Cell Biology, 33*.

Ingolia, N. T., Ghaemmaghami, S., Newman, J. R., & Weissman, J. S. (2009). Genome-wide analysis *in vivo* of translation with nucleotide resolution using ribosome profiling. *Science, 324*(5924), 218−223.

Ingolia, N. T., Lareau, L. F., & Weissman, J. S. (2011). Ribosome profiling of mouse embryonic stem cells reveals the complexity and dynamics of mammalian proteomes. *Cell, 147*(4), 789−802.

Irizarry, R. A., Ladd-Acosta, C., Carvalho, B., Wu, H., Brandenburg, S. A., Jeddeloh, J. A., et al. (2008). Comprehensive high-throughput arrays for relative methylation (CHARM). *Genome Research, 18*(5), 780−790.

Jamshidi, N., & Palsson, B. O. (2009). Using *in-silico* models to simulate dual perturbation experiments: Procedure development and interpretation of outcomes. *BMC Systems Biology, 3*, 1−11.

Jensen, J. (2004). Gene regulatory factors in pancreatic development. *Developmental Dynamics, 229*(1), 176−200.

Josephson, R., Sykes, G., Liu, Y., Ording, C., Xu, W., Zeng, X., Shin, S., Loring, J., Maitra, A., Rao, M. S., & Auerbach, J. M. (2006). A molecular scheme for improved characterization of human embryonic stem cell lines. *BMC Biology, 4*(1), 1−13.

Kwon, S. C., Yi, H., Eichelbaum, K., Föhr, S., Fischer, B., You, K. T., Castello, A., Krijgsveld, J., Hentze, M. W., & Kim, V. N. (2013). The RNA-binding protein repertoire of embryonic stem cells. *Nature Structural & Molecular Biology, 20*(9), 1122−1130.

Lander, A. D. (2007). Morpheus unbound: Reimagining the morphogen gradient. *Cell, 128*(2), 245−256.

Langfelder, P., & Horvath, S. (2008). WGCNA: an R package for weighted correlation network analysis. *BMC Bioinformatics, 9*(1), 1−13.

Lang, A. H., Li, H., Collins, J. J., & Mehta, P. (2014). Epigenetic landscapes explain partially reprogrammed cells and identify key reprogramming genes. *PLoS Computational Biology, 10*(8), e1003734.

Lee, J. H., Daugharthy, E. R., Scheiman, J., Kalhor, R., Ferrante, T. C., Terry, R., et al. (2015). Fluorescent in situ sequencing (FISSEQ) of RNA for gene expression profiling in intact cells and tissues. *Nature Protocols, 10*(3), 442–458.

Lienert, F., Mohn, F., Tiwari, V. K., Baubec, T., Roloff, T. C., Gaidatzis, D., et al. (2011). Genomic prevalence of heterochromatic H3K9me2 and transcription do not discriminate pluripotent from terminally differentiated cells. *PLoS Genetics, 7*(6), e1002090.

Lister, R., Pelizzola, M., Kida, Y. S., Hawkins, R. D., Nery, J. R., Hon, G., et al. (2011). Hotspots of aberrant epigenomic reprogramming in human induced pluripotent stem cells. *Nature, 471*(7336), 68–73.

Meier-Schellersheim, M., Xu, X., Angermann, B., Kunkel, E. J., Jin, T., & Germain, R. N. (2006). Key role of local regulation in chemosensing revealed by a new molecular interaction-based modeling method. *PLoS Computational Biology, 2*(7), e82.

Mikkelsen, T. S., Hanna, J., Zhang, X., Ku, M., Wernig, M., Schorderet, P., et al. (2008). Dissecting direct reprogramming through integrative genomic analysis. *Nature, 454*(7200), 49–55.

Mikkelsen, T. S., Ku, M., Jaffe, D. B., Issac, B., Lieberman, E., Giannoukos, G., et al. (2007). Genome-wide maps of chromatin state in pluripotent and lineage-committed cells. *Nature, 448*(7153), 553–560.

Müller, F. J., Schuldt, B. M., Williams, R., Mason, D., Altun, G., Papapetrou, E. P., et al. (2011). A bioinformatic assay for pluripotency in human cells. *Nature Methods, 8*(4), 315–317.

Noble, D. (2002). Modeling the heart from genes to cells to the whole organ. *Science, 295*(5560), 1678–1682.

Osafune, K., Caron, L., Borowiak, M., Martinez, R. J., Fitz-Gerald, C. S., Sato, Y., Cowan, C. A., Chien, K. R., & Melton, D. A. (2008). Marked differences in differentiation propensity among human embryonic stem cell lines. *Nature Biotechnology, 26*(3), 313–315.

Qian, L., Berry, E. C., Fu, J. D., Ieda, M., & Srivastava, D. (2013). Reprogramming of mouse fibroblasts into cardiomyocyte-like cells in vitro. *Nature Protocols, 8*(6), 1204–1215.

Rackham, O. J., Firas, J., Fang, H., Oates, M. E., Holmes, M. L., Knaupp, A. S., Consortium, T. F., Suzuki, H., Nefzger, C. M., Daub, C. O., Shin, J. W., Petretto, E., Forrest, A. R. R., Hayashizaki, Y., Polo, J. M., & Gough, J. (2016). A predictive computational framework for direct reprogramming between human cell types. *Nature Genetics, 48*(3), 331–335.

Robert, J. S. (2004). Model systems in stem cell biology. *BioEssays, 26*(9), 1005–1012.

Roost, M. S., Van Iperen, L., Ariyurek, Y., Buermans, H. P., Arindrarto, W., Devalla, H. D., Passier, R., Mummery, C. L., Carlotti, F., de Koning, E. J. P., van Zwet, E. W., Goeman, J. J., & de Sousa Lopes, S. M. C. (2015). KeyGenes, a tool to probe tissue differentiation using a human fetal transcriptional atlas. *Stem Cell Reports, 4*(6), 1112–1124.

Sample, C., & Shvartsman, S. Y. (2010). Multiscale modeling of diffusion in the early *Drosophila* embryo. *Proceedings of the National Academy of Sciences, 107*(22), 10092–10096.

Setty, Y. (2014). *In silico* models of stem cell and developmental systems. *Theoretical Biology and Medical Modelling, 11*(1), 1–12.

Shapiro, B. E., Levchenko, A., Meyerowitz, E. M., Wold, B. J., & Mjolsness, E. D. (2003). Cellerator: Extending a computer algebra system to include biochemical arrows for signal transduction simulations. *Bioinformatics, 19*(5), 677–678.

Slepchenko, B. M., Schaff, J. C., Macara, I., & Loew, L. M. (2003). Quantitative cell biology with the virtual cell. *Trends in Cell Biology, 13*(11), 570–576.

Stiehl, T., & Marciniak-Czochra, A. (2011). Characterization of stem cells using mathematical models of multistage cell lineages. *Mathematical and Computer Modelling, 53*(7–8), 1505–1517.

Subramanian, A., Tamayo, P., Mootha, V. K., Mukherjee, S., Ebert, B. L., Gillette, M. A., Paulovich, A., Pomeroy, S. L., Golub, T. R., Lander, E. S., & Mesirov, J. P. (2005). Gene set enrichment analysis: A knowledge-based approach for interpreting genome-wide expression profiles. *Proceedings of the National Academy of Sciences, 102*(43), 15545–15550.

Takahashi, K., Tanabe, K., Ohnuki, M., Narita, M., Ichisaka, T., Tomoda, K., & Yamanaka, S. (2007). Induction of pluripotent stem cells from adult human fibroblasts by defined factors. *Cell, 131*(5), 861–872.

Takahashi, K., & Yamanaka, S. (2006). Induction of pluripotent stem cells from mouse embryonic and adult fibroblast cultures by defined factors. *Cell, 126*(4), 663–676.

Till, J. E., McCulloch, E. A., & Siminovitch, L. (1964). A stochastic model of stem cell proliferation, based on the growth of spleen colony-forming cells. *Proceedings of the National Academy of Sciences, 51*(1), 29–36.

Tsankov, A. M., Akopian, V., Pop, R., Chetty, S., Gifford, C. A., Daheron, L., Tsankova, N. M., & Meissner, A. (2015). A qPCR ScoreCard quantifies the differentiation potential of human pluripotent stem cells. *Nature Biotechnology, 33*(11), 1182–1192.

Van Wolfswinkel, J. C., Wagner, D. E., & Reddien, P. W. (2014). Single-cell analysis reveals functionally distinct classes within the planarian stem cell compartment. *Cell Stem Cell, 15*(3), 326–339.

Van Wynsberghe, P. M., Kai, Z. S., Massirer, K. B., Burton, V. H., Yeo, G. W., & Pasquinelli, A. E. (2011). LIN-28 co-transcriptionally binds primary let-7 to regulate miRNA maturation in *Caenorhabditis elegans*. *Nature Structural & Molecular Biology, 18*(3), 302–308.

Vierstra, J., Rynes, E., Sandstrom, R., Zhang, M., Canfield, T., Hansen, R. S., et al. (2014). Mouse regulatory DNA landscapes reveal global principles of cis-regulatory evolution. *Science, 346*(6212), 1007–1012.

Wang, Z., Matsudaira, P., & Gong, Z. (2010). STORM: A general model to determine the number and adaptive changes of epithelial stem cells in teleost, murine and human intestinal tracts. *PLoS One, 5*(11), e14063.

Wang, L., Miao, Y. L., Zheng, X., Lackford, B., Zhou, B., Han, L., et al. (2013). The THO complex regulates pluripotency gene mRNA export and controls embryonic stem cell self-renewal and somatic cell reprogramming. *Cell Stem Cell, 13*(6), 676–690.

Whitchard, Z. L., Sarkar, C. A., Kimmel, M., & Corey, S. J. (2010). Hematopoiesis and its disorders: A systems biology approach. *Blood, The Journal of the American Society of Hematology, 115*(12), 2339−2347.

Yener, B., Acar, E., Aguis, P., Bennett, K., Vandenberg, S. L., & Plopper, G. E. (2008). Multiway modeling and analysis in stem cell systems biology. *BMC Systems Biology, 2*, 1−17.

Zaman, W. S. W. K., Karman, S. B., Ramlan, E. I., Tukimin, S. N. B., & Ahmad, M. Y. B. (2021). Machine learning in stem cells research: Application for biosafety and bioefficacy assessment. *IEEE Access, 9*, 25926−25945.

Zhang, X. Z., Li, X. J., Ji, H. F., & Zhang, H. Y. (2009). Impact of drug discovery on stem cell biology. *Biochemical and Biophysical Research Communications, 383*(3), 275−279.

Zheng, Y., Tan, X., Pyczek, J., Nolte, J., Pantakani, D. K., & Engel, W. (2013). Generation and characterization of yeast two-hybrid cDNA libraries derived from two distinct mouse pluripotent cell types. *Molecular Biotechnology, 54*, 228−237.

Chapter 24

Computational approaches for hematopoietic stem cells: Advancing regenerative therapeutics

Pawan Kumar Raghav[1], Basudha Banerjee[1], Rajesh Kumar[1], Aditya Raghav[1], Anjali Lathwal[2] and Rajni Chadha[1]

[1]BioExIn, Delhi, India; [2]Department of Computational Biology, Indraprastha Institute of Information Technology (IIIT), Delhi, India

1. Introduction

The rejuvenation of hematopoietic stem cells (HSCs) offers a promising approach to restore impaired HSCs function and enhance tissue repair in age-linked disorders and degenerative diseases. Aging or diseased conditions can disrupt signaling pathways that regulate HSCs' functions, leading to their dysfunction (Zhang et al., 2020). Consequently, the targeted disruption of these signals holds promise in mitigating the adverse effects of aging and other diseases. Downstream signaling pathways can be modulated, either activated or inhibited, by targeting transcription factors (TFs) to preserve the quiescent state of HSCs (Nakagawa et al., 2018). Integration of transcriptional regulatory networks, including single-cell RNA sequencing (scRNA-Seq) data, phosphoproteomics data, and signaling pathways, enables the identification of the molecules responsible for maintaining the phenotype associated with age-linked diseases.

Utilizing computational models facilitates the comprehension of interactions within the stem cell niche and enables the prediction of potential dysregulation in stem cell-specific niches and signaling pathways under conditions related to aging or disease (Del Sol & Jung, 2021; Khorasani et al., 2020). In summary, the integration of computational models with experimental techniques emerges as a promising strategy for devising therapeutic interventions aimed at rejuvenating HSCs and enhancing tissue repair in age-related disorders and degenerative diseases (Sason & Shamay, 2020; Stillman et al., 2020). In silico approaches have the capability to generate transgenic cells for transplantation through the identification of disease-related cell fate determinants (Del Sol & Jung, 2021). Numerous in silico tools have proven to be highly significant in the realm of stem cell studies (Agrawal et al., 2018; Kulshrestha et al., 2021; Kumar et al., 2021; Raghav et al., 2021). Although these in silico methods open the door to revolutionary insights into the behavior of HSCs, it is of utmost importance to harmonize this advancement with ethical considerations advocated by regulatory bodies.

Initially, the United States had a history of imposing restrictions on the utilization of stem cells, resulting in a setback in stem cell research. However, these restrictions were later overturned (Shen et al., 2018). The development of any cellular biological product with marketable therapeutic value faces significant challenges within the pharmaceutical regulatory framework (Abou-El-Enein et al., 2017). The success of stem cell research hinges on adherence to specific guidelines and standardized protocols, given the regulatory and ethical norms governing the production of biological therapeutics. One notable effort in this regard is the Minimum Information About a Cellular Assay for Regenerative Medicine (MIACARM) recommendations, which seek to integrate data related to stem cell products (Sakurai et al., 2016). Legislative bodies such as FAIR, ISSCR, ISCBI, and NHLBI, comprising experts from the scientific community, and industrial organizations, recognize the necessity of establishing equitable guidelines for an effective data ecosystem. The FAIR guiding principles

Computational Biology for Stem Cell Research. https://doi.org/10.1016/B978-0-443-13222-3.00013-7
Copyright © 2024 Elsevier Inc. All rights reserved, including those for text and data mining, AI training, and similar technologies.

are designed to simplify data findability, accessibility, interoperability, and reuse (FAIR) (Wilkinson et al., 2016). The International Society for Stem Cell Research (ISSCR) has established the Guidelines for Stem Cell Research and Clinical Translation, which outline a set of regulations for stem cell biology (Daley et al., 2016). Likewise, the International Stem Cell Banking Initiative (ISCBI) provides guidance on the culture, storage, distribution, translation, and characterization of stem cells and their related products (Crook et al., 2014). ISCBI has defined key features that best describe an iPSC (induced pluripotent stem cell) and characteristics to consider when selecting a clinical-grade iPSC line (Sullivan et al., 2018).

A successful bone marrow transplantation typically requires 2×10^6 CD34$^+$ cells/kg of the recipient's body weight (Raghav & Gangenahalli, 2018). The Indian Council of Medical Research (ICMR) has released the National Guidelines for Hematopoietic Cell Transplantation (HCT) (ICMR, 2021). These guidelines emphasize the significant role of human leukocyte antigen (HLA) typing in facilitating HLA compatibility between a donor, whether related or unrelated, and the recipient. In cases where a related donor is unavailable, options include a mismatched unrelated donor, haploidentical donor, or umbilical cord blood donor. Both HLA and ABO incompatibility can pose significant health risks. Plasma depletion is employed to address minor ABO incompatibility, while red cell depletion is used for major ABO incompatibility issues. The volume of cells used for engraftment varies depending on factors such as the cell source, disease, conditioning regimen, HLA incompatibility, and ABO incompatibility. Depending on the degree of HLA disparity leading to graft versus host disease (GvHD), patients may be administered immunosuppressants. For allogeneic transplants, cyclosporine with or without methotrexate serves as a standard immunosuppressant, acting as a calcineurin inhibitor. The process of HCT involves a rigorous pretransplant and posttransplant workup that encompasses quality control, disease-specific investigations, and close monitoring of therapeutic drug levels.

The applications of HSCs co-culture with mesenchymal stem cells (MSCs) in regenerative medicine, as well as bioinformatics approaches in stem cell research, are depicted in Fig. 24.1. Additionally, the design of in vivo techniques to promote gene correction, such as CRISPR/Cas9, can be achieved through computational predictions of dysfunctional cell fate determinants for congenital disorders (Beck et al., 2013; Schütte et al., 2016). Fig. 24.2 provides an overview of sample sequencing based on a barcoded pattern using multiplexing technologies. This method allows for the identification of a sample's origin based on the combination of single-nucleotide polymorphism (SNP) and barcoding, using computational tools. Fig. 24.3 illustrates the significant challenges faced by computational approaches in stem cell biology. These challenges include data and database maintenance difficulties due to limited portal access, vast datasets, restricted data sharing, and outdated analytical tools. The lack of donor-to-patient details and an incomplete understanding of disease pathology and average tissue/data quality further complicate data analysis. Additionally, concerns related to biosafety, bioprivacy, and bioefficacy add to the complexity of the field.

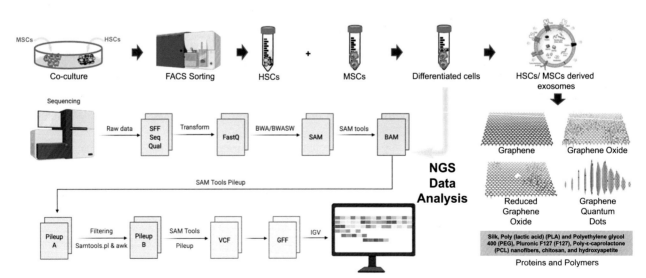

FIGURE 24.1 Application of co-culture of FACS sorted HSCs with MSCs identified by RNA-Seq analysis. Interactions and dynamics of graphene and its derivatives, silk protein, and polymers used in scaffold fabrication considered potential dressing material in wound healing can be analyzed using bioinformatics studies. *FACS*, fluorescence-activated cell sorting; *NGS*, next-generation sequencing.

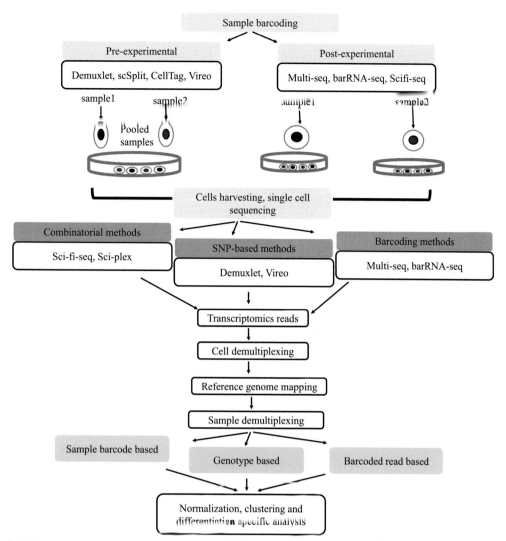

FIGURE 24.2 Workflow of computational tools to analyze stem cell differentiation at the molecular level. Multiplexing technologies enable samples to be sequenced together or individually based on a barcoded pattern. Following sequencing, computational tools, including combination-based, SNP-based, and barcoding-based methods, facilitate the identification of sample origins within the multiplexed dataset.

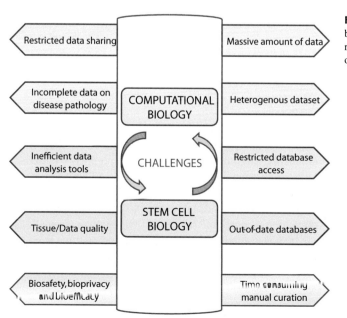

FIGURE 24.3 Challenges of computational biology in stem cell biology. Significant drawbacks include database and data maintenance, restricted sharing access, lack of donor-to-patient data, insufficient data on disease pathology, bioefficacy and bioprivacy issues.

2. Big data analysis of HSCs

The self-renewal and differentiation of HSCs are studied using microarrays and next-generation sequencing (NGS). These technologies have been utilized to profile gene expression patterns in HSCs, identify proliferation-associated TFs, and determine the genetic and epigenetic landscape of HSCs (Chambers et al., 2007; Sharma et al., 2021). In the wake of these discoveries, the potential for developing personalized medicine strategies that guide the selection of HSC therapy has emerged. Given the dynamic and intricate nature of HSCs, there is an increasing demand to profile gene expression across a panel of self-renewal genes rather than focusing solely on individual genes. Consequently, there is a suggestion to investigate the impact of employing Next-Generation Sequencing (NGS) in genomics for personalized medicine. For this purpose, NGS has been harnessed to pinpoint "signature genes" associated with HSC self-renewal. Signature genes, which are genes uniquely expressed in cells, serve as molecular markers for the identification and characterization of specific cell types. HSCs possess a distinctive gene expression profile characterized by the upregulation of CD34, CD38, CD44, CXCR4, and GATA2. These markers set HSCs apart from other cell types found in the bone marrow. Consequently, based on these "signature genes," tailored HSC-based therapies can be employed for the treatment of hematologic malignancies and HSC transplantation. Furthermore, identifying mutated genes in patients responsible for abnormal proliferation and self-renewal is essential for refining therapeutic approaches. For instance, the identification of genomic-level mutations, such as the c-Kit mutation (D816V), which is a "cancer hotspot" mutation associated with a proliferation marker, can guide therapeutic selection (Arock et al., 2018). Multiple in silico studies have delineated the role of essential genes and proteins in stem cells and cancer stem cells (CSCs) (Raghav et al., 2011; Praveen et al., 2012; Raghav, Verma, & Gangenahalli, 2012a, 2012b; Raghav, Singh, & Gangenahalli, 2018a, 2018b, 2018c; Raghav, Kumar, & Raghava, 2019b; Raghav, Kumar, & Raghava, 2019). c-Kit and SHP-1/2 have been identified as regulators of self-renewal and proliferation, while PU.1, c-Jun, and GATA-1 have been implicated in the regulation of differentiation. Beyond these, numerous cytokines and transcription factors (TFs) play pivotal roles in maintaining the self-renewal and proliferation of HSCs (Mann et al., 2022).

Further, the identified "cancer hotspots" from the mixed mutation can also be triggered to differentiate into myeloid and lymphoid cells to eradicate CSCs. Thus, the main objective of this section is to discuss the available tools required to sequence the entire genome of cancer patients to identify "cancer hotspot" somatic mutations in HSCs and their expression profiles.

Clinicians are actively engaged in preserving the self-renewal capacity of progenitor stem cells by utilizing gene therapy in conjunction with cytokines and chemical compounds. This innovative approach aims to supplant traditional cryopreservation and biobanking methods. Achieving this phenomenon necessitates a deep understanding of the specific microenvironment, or niche, in which stem cells reside, and replicating these conditions during their expansion. Numerous genes and proteins play a pivotal role in regulating these unique properties of both stem cells and cancer stem cells (CSCs) (Kumar et al., 2020, 2022; Nandan et al., 2021; Raghav, 2022; Raghav et al., 2018, 2021, 2021, 2022; Raghav & Gangenahalli, 2018; Raghav & Mann, 2021; Rawat et al., 2021; Sengar et al., 2022; Raghav et al., 2022). The identified genes that are hotspots to maintain the stemness of stem cells have been discussed. Using combination therapy, these genes regulate and transplant the treated stem cells, which migrate toward the recipient's bone marrow. They are eventually incorporated and can be differentiated into myeloid or lymphoid lineages. This process will reduce the maintenance cost of stem cell banking and can be directly used as personalized and regenerative medicine. The formulated transplanted stem cells will maintain their stem cell phenotype and self-renew by maintaining the stemness of the genes attained by pluripotent stem cells.

3. Metabolic cross-talk

Metabolic and physiological processes in humans and other organisms are tightly regulated through intricate molecular mechanisms involving complex networks of genes, TFs, and miRNAs. Disruptions of regulatory control often result in various disorders, including hematologic malignancies (Dai et al., 2019). Alterations in metabolic and regulatory networks have been found in HSCs and cancers (Yamashita et al., 2020). Lung adenocarcinoma rewires the STAT3-Socs3 inflammatory signaling axis that causes dysregulation of lipid metabolism occurring in liver and also hinders hepatic insulin signaling, indicating cross-talk between regulatory and metabolic networks in distal tissues (Paul, 2020). Emerging mechanisms of miRNA-mediated dynamic posttranscriptional regulation of genes/mRNAs in various biological functions, such as cell differentiation, development, oncogenes, and immune responses, indicate miRNAs to be crucial links in cross-talk between metabolic and regulatory circuits. Therefore, to understand complex human diseases such as different forms of HSC cancers, it is necessary to develop systems-level genome-scale integrative networks of metabolic and transcriptional processes. The availability of high-throughput data from functional genomics, transcriptomics, proteomics, and metabolomics studies and parallel developments in systems-level mathematical modeling has opened up opportunities for the genome-scale reconstruction of integrated metabolic and regulatory networks. In this section, the integration of functional genomics and transcriptomics data with the genome-scale metabolic model's discussion is required to study cross-talk between metabolic and regulatory networks.

Optimizing the culture conditions and their evaluation using cell based assays is essential to enhance the migration. Standard CD34+ cell expansion involves the use of freshly or cryopreserved hematopoietic progenitor cells for cell plating incubated at 37°C and 5% CO_2 for 4 days. Sufficient cell expansion requires partial medium change. The cells can then be harvested to obtain a single-cell suspension and used further for experimentation. It is essential to select an appropriate growth medium with other supplements, including growth and differentiation factors, to allow a systematic equilibrium between the self-renewal and differentiation properties of HSCs. Recombinant proteins, including stem cell factor (SCF), erythropoietin, thrombopoietin, and IL-3, regulate the fate of HSC culture (Zhang & Lodish, 2008). DMEM is one of the most widely used media, which is usually supplemented with 10% FBS and valproic acid to culture HSCs obtained from umbilical cord blood. RPMI and IMDM have also been used as HSC culture media (Yadav et al., 2020). HSCs differentiation has been maintained using X-VIVO expansion medium supplemented with SCF, IL-3, thrombopoietin, and 4% fetal calf serum (Stec et al., 2007). Methylcellulose is another HSCs culture medium that generates erythroid and myeloid progenitors (Yadav et al., 2020). During transplantation, the collection of peripheral blood stem cells is accompanied by growth factors such as GM-CSF, G-CSF, or CXCR4 inhibitor to improve the migration of HSCs from bone marrow to peripheral blood. These factors can increase the concentration of HSCs in peripheral blood by 30−1000 fold (Micallef et al., 2018). Hence, it is possible to optimize the concentration of individual cytokines or combinations thereof to create a computational stem cell model capable of identifying novel stem cell lineages.

The identification of molecule combinations will lead to the evolution of therapeutic strategies. In silico approach will expose molecules' function in stem cell signaling. Through the systems biology approach, the identified molecules will synthesize lead molecules to evaluate their efficacy and adverse effects. These research areas will lead us one step ahead toward stem cell drug discovery. Using drug discovery strategies, the new molecules can behave as possible drug contenders against the target. Different combinations can be made using cell-based assays, which are likely helpful in formulations for effective cell delivery. Small PU.1-mimetic synthetic peptides have been designed to prime stem cells into myeloid cells both in vitro and in vivo (Raghav & Gangenahalli, 2021). The combination of SCF and NSC87877 enhanced the proliferation and expansion of human megakaryoblastic and erythroid cells mediated by increased c-Kit expression (Raghav et al., 2018a, 2018b, 2018c). Application of molecular docking is used to estimate molecular interactions and analyze the variability of the molecules competing for binding at the target protein's active site. It predicts the precision and efficacy of ligand and receptor interaction in signaling pathways. Thus, it is essential to critically analyze the metabolic processes by molecular docking to understand the dysregulation of body homeostasis. Drugs namely mitoketocins, andrographolide, emetine, cortistatin, solasonine, tylophorine, CIN-RM, 89Zr@CS-MLPs, ZINC000085569211, ZINC000085569178, and ZINC000085569190 have been analyzed by computational techniques and molecular docking to understand the mode of interaction and target pathways (Hongwiangchan et al., 2021; Jaitak, 2016; Liu et al., 2014; Madhunapantula et al., 2011; Mohamed et al., 2022; Ozsvari et al., 2017; Wanandi et al., 2020; Yang et al., 2020). Mechanism of role of molecules and drugs identified through computational approaches to enhance cell differentiation into specific lineages or target cancer stem cells (CSCs) have been cataloged in Table 24.1.

4. Computational approaches for stem cell research

In the last two decennium, computational techniques have been significant in translating high-throughput data into knowledge. Computational tools have been developed to predict cell fate engineering, inspect inaccessible developmental time points, expand regenerative medicine therapies, and screen drugs. Numerous tools can be used to perform stem cell engineering based on the analysis of expression and interaction data derived from bulk samples (Table 24.2). These methods fail when the primary goal is identifying homogenous stem cells at the cellular level for stem cell engineering and their differentiation for organ culture/tissue replacement. In this aspect, using single-cell technologies can enable the identification of pure and homogenous single stem cells. Several reports indicate that molecular techniques are available for stem cells, but the computational side of analyzing and managing single-cell data is relatively less. A hypothetical pipeline is provided for simplicity and practical usability of the available resources that can be used to discover the mechanism of stem cell differentiation at the single-cell level (Fig. 24.2). CODEX, HscExplorer, and hscScore host HSCs-specific data (Sánchez-Castillo et al., 2015; Montrone et al., 2013; Hamey et al., 2019). CODEX retrieves raw NGS data and uses peak

TABLE 24.1 Mechanism of molecules/drugs in drug discovery identified using computational biology.

Molecule name	Description	References
Synthetic PU.1 β3/β4 mimic peptide analogs	These molecules bind to GATA-1 and disrupt its interaction with PU.1, resulting in increased expression of myeloid cell markers such as CD33, CD11b, CD116, and CD114.	Raghav and Gangenahalli (2021)
NSC87877	NSC87877 functions as an inhibitor of SHP-1 and SHP-2, thereby exerting a positive regulatory effect on c-Kit-mediated proliferation of human cells.	Raghav et al. (2018a), (2018b), (2018c)
Mitoketoscins	This compound targets tumor promoters OXCT1 and ACAT1, leading to the inhibition of ketone body recycling, ATP production, and oxidative mitochondrial pathways.	Ozsvari et al. (2017)
Andrographolide	This substance interacts with the survivin protein, inducing apoptosis in human breast CSCs.	Wanandi et al. (2020)
Emetine	Bind to Shh, Smo, and Gli protein and targets CSCs.	Jaitak (2016)
Cortistatin		
Solasonine	Affect Gli proteins and alter the hedgehog pathway.	Liu et al. (2014)
Tylophorine		
CIN-RM	This inhibits Akt pathway, downregulation of CSC markers, and inhibition of the mTOR pathway in lung cancer. It results in decreased expression levels of c-Myc, Nanog, Sox2, and Oct4.	Hongwiangchan et al. (2021), Madhunapantula et al. (2011)
89Zr@CS-MLPs	Target and interact with overexpressed CD44 surface markers in TNBC tissues	Yang et al. (2020)
ZINC000085569211 ZINC000085569178 ZINC000085569190	Form hydrogen bonds to interact with MET126, GLU92, and GLU160 in the MNK-2 active site and with ASP124, ASP127, GLU167, and GLU117 in the PIM-2 binding pocket; improve AML treatment.	Mohamed et al. (2022)

AML, acute myeloid leukemia; *CIN-RM*, hydroquinone 5-O cinnamoyl ester of renieramycin; *CSC*, cancer stem cells; *SCF*, stem cell factor; *TNBC*, triple-negative breast cancer; *89Zr@CS-MLPs*, zirconium-89 (89Zr)-labeled, chitosan (CS)-decorated multifunctional liposomes (MLPs).

profile correlation and motif discovery analyses (Gottgens, 2015). Computational methods have been used to predict the hematopoietic landscape via NGS at bulk and single-cell level (Monga et al., 2022). The transcriptional and post-transcriptional diversity of HSCs during differentiation due to CLK3 and HMGA2 have been delineated using high-throughput genomic strategies (Cesana et al., 2018). RNA-Seq and methylated RNA immunoprecipitation sequencing (MeRIP-Seq) were used to evaluate downstream genes mediated by ALKBH5 in regulating its role in hepatocellular carcinoma (HCC) radiosensitivity (Chen et al., 2023). Identification of SNPs by NGS using the Illumina platform has significantly contributed toward the HLA typing and comprehending the effect of HLA disparities post hematopoietic stem cell transplantation (HSCT).

Computational stem cell biology has evolved as a field that visualizes cellular mechanisms and gains comprehensive knowledge of self-renewal, cellular differentiation, reprogramming, and other biological processes. Biomedical data integration is one of the most critical factors serving the research community, which needs the constant support of databases and analytical tools. Databases and associated resources, including data analysis tools, serve as convenient repositories for data retrieval and provide an interactive platform for exploring molecular interactions across diverse cell types in various organisms. They play a crucial role in facilitating research and seamless integration into bioinformatics workflows. To ensure a harmonized integration of stem cell information, it is essential to consider the following key aspects for robust and insightful computational analyses.

TABLE 24.2 Stem cell-based tools and databases.

Methods	Applications	References
	Bulk sequencing-based computational methods	
ScoreCard	Calculate cell line-specific differentiation propensities, hematopoietic, and neural lineages.	Bock et al. (2011)
PluriTest	Predict pluripotency	Müller et al. (2011)
CellNet	Calculate similarity to 20 cell and tissue types	Cahan et al. (2014)
DAlesio	Predict putative fate-altering TFs	D'Alessio et al. (2015)
KeyGenes	Predict tissue of origin, identify stem cell derivative, and estimate developmental stage.	Roost et al. (2015)
StemCellDB	Database of 21 hESC line	Mallon et al. (2013)
ESCAPE	Embryonic stem cell atlas of pluripotency evidence	Xu et al. (2013)
PSCRIdb	A database of regulatory interaction and network of pluripotency cell lines	Banerjee et al. (2020)
	Single cell-based computational methods	
Sci-fi-Seq	Tool for single-cell combinatorial indexing	Datlinger et al. (2021)
Sci-plex	Method for ssDNA oligonucleotide labeling	Srivatsan et al. (2020)
Demuxlet	Tool for demultiplexing genetically distinct samples	Kang et al. (2018)
Vireo	Bayesian interference-based demultiplexing tools	Huang et al. (2019)
Multi-Seq	Cell demultiplexing tool for lipid-tagged barcode	McGinnis et al. (2019)
barRNA-Seq	Tool for demultiplexing of single cell based on short RNA barcode	Yeo et al. (2020)

4.1 Cell quality is dependent on cell type information

Cell quality can be judged by interrogating the characteristics of the specific cell type and its primary/native cell source. Comparing the two's transcriptomic, proteomic, and epigenomic profiles helps to analyze the functional relativeness and the resulting phenotype (Finkelstein et al., 2020).

4.2 Therapeutic follow-ups are dependent on correct cell characterization

Stem cells are employed in regenerative medicine to replace damaged tissues or restore components of tissues. However, the success of therapy relies on the availability of adequate information regarding the specific cell or tissue requiring replacement and its active regeneration or retirement processes. Unfortunately, there needs to be more reliable data on the functional and structural aspects of cell/tissue types that need replacement (Warren et al., 2010). For instance, stem cells have distinct properties based on the tissue type they are isolated from, donor specifics, disease conditions, cell-specific markers, self-renewal, and differentiation capacity. Moreover, general traits are required to judge the behavior of the stem cells in various culture conditions or disease states since they might behave differently in each condition in the presence/absence of specific growth factors/supplements to show differential proliferation rates (Horgan et al., 2011).

4.3 Donor-to-recipient

Characteristic details regarding the donor from whom the cell was obtained and the recipient who undergoes treatment need to be updated along with its therapeutic outcome in the patient to improve the further course of action in a similar mode of treatment. Linking the cell line details with the patient details and its outcome will allow a relative understanding of the molecular characterization of the cell and therapeutic outcome (Revilla et al., 2016; Park et al., 2009).

4.4 Providing quality assurance

Registries must allow only clinical-grade cell lines that ensure safety and high efficacy for therapeutic purposes so that researchers can link their studies. The European Commission only funds pluripotent SC and ensures the quality of these cells is well maintained (Rubin et al., 2008).

4.5 Data exchange with other databases

The data portal should allow data exchange and comparison with all other databanks to promote unification across multiple experimental studies. Data analysis tools that evaluate all cell lines in different cell banks must be promoted to improve cell type selection and study outcomes. There are multiple data analysis tools: Venn diagram plotting, DAVID enrichment analysis, coexpression and differential expression analysis, heatmap analysis, hierarchical analysis, principal component analysis, and LINCS analytics (Finkelstein et al., 2020).

5. Computational biology in HSCT

The existence of stem cells with remarkable attributes of self-renewal, differentiation, tumor tropism, and migration has allowed us to expand their handling in immunotherapy, regenerative medicine, anticancer drug screening, and delivery vehicles. In today's era, where the focus has shifted to regenerative medicine, stem cells have also been engineered to improve therapeutic effectiveness (Srinivasan et al., 2021). HSCT has been successfully established therapy for treating hematological malignancies (Henig & Zuckerman, 2014). They have also been used to treat neurological, immunological, orthopedic, and traumatic disorders (Srinivasan et al., 2021; El-Sadik, 2010; Shroff, 2018). Gradually, stem cell success has radiated for their use in chronic disease treatments (Preethy et al., 2013; Watt & Driskell et al., 2010). Stem cells with biomaterials can be used and play a pivotal role in combatting COVID-19 (Raghav et al., 2021, 2022; Raghav & Mohanty, 2020).

5.1 Hematological complications

HSCs have been primarily used for autologous and allogenic transplants to treat malignant and nonmalignant immune complications. The three cancerous hematological disorders are leukemia, lymphoma, and myeloma. Different disease indications are provided with a specific standard of care in both pediatrics and adults. Allogenic HSCT is currently the only treatment for myeloproliferative conditions such as primary myelofibrosis. Patients suffering from severe aplastic anemia are also advised for first-line HLA identical sibling HSCT; a haploidentical transplant may opt if unavailable. Auto-HSCT stands out as the standard of care for Hodgkin and follicular lymphoma patients in first or subsequent relapse. It is also given as first-line treatment to patients under 60−65 years suffering from peripheral T cell lymphoma and mantle cell lymphoma. A few other malignant conditions include acute lymphoblastic leukemia (ALL), chronic myeloid leukemia (CML), multiple myeloma (MM), Burkitt's lymphoma, anaplastic large cell lymphoma, and solid tumors such as Ewing's sarcoma. Apart from this, HSCT has been prominent in curing multiple nonmalignant diseases such as Fanconi's anemia, hemophilia, sickle cell disease, thalassemia, primary immunodeficiencies, severe combined immunodeficiency disease (SCID), autoimmune disorders, and metabolic complications (Galgano & Hutt, 2018).

5.2 Regenerative medicine

Stem cells find applications in drug testing, rejuvenation, and regenerative medicine. The integration of artificial intelligence (AI) with medical science represents a revolutionary concept, contingent upon the fulfillment of ethical considerations, including privacy, biosafety, and the efficacy of both cells and patients.

5.2.1 Cell transplantation

Cell transplantation replaces the damaged/defective cells/tissues in the body with healthy functioning cells. Transplantation can have immune reactions and get rejected due to host immunosurveillance, or in vitro cells produced for transplantation might not achieve the specific phenotype or achieve it with low assurance and efficiency (Xu et al., 2015). Overcoming this challenge can be achieved through computational modeling and single-cell RNA sequencing (scRNA-Seq), which enable us to gain insights into the gene expression profiles and other cellular characteristics of both the donor and host cells or tissues. Utilizing a single-cell computational gene regulatory network (GRN), we can predict the transcription factors (TFs), conversion factors, and gene interactions required to generate host-compatible cells of a specific type. These cells can then be used effectively in transplantation procedures (Berneman-Zeitouni et al., 2014; Xu et al., 2015).

5.2.2 Stem cell rejuvenation

Multiple disease conditions, aging factors, or sudden trauma might hinder the stem cell—specific signaling pathways leading to dysregulation and impairment of stem cell functionality. To regain stem cell functionality, stem cell rejuvenation has been strategized. Stem cell—niche interactions can be computationally modeled to analyze the niche signals that maintain a particular cell specificity (Browaeys et al., 2020). An effective alternative approach is to adopt an integrative model that combines scRNA-Seq data with phosphoproteomics data. This integrated analysis provides a more comprehensive understanding of signaling and transcription networks, allowing for the prediction of interactions among signaling molecules that contribute to the dysregulated gene expression (Qin et al., 2020).

5.3 Case studies of AI success

AI has made significant contributions to cancer biology. It has demonstrated its capability to differentiate between benign and metastatic cancer (Osareh & Shadgar et al., 2010). Additionally, AI has been instrumental in studying the staging and increased aggressiveness of prostate cancer, analyzing the localization of immune cells in breast cancer, and detecting relapse in oropharyngeal cancer by examining the features of the tumor microenvironment (Osareh & Shadgar, 2010; Lee et al., 2014; Heindl et al., 2018; Ali et al., 2013). International Business Management Corporation reported a new algorithm using machine and deep learning to detect early-stage breast cancer. Trained mammographic images, clinical trials, and cellular features helped them detect breast cancer in 87% of cases and thus helped improve diagnosis (Akselrod-Ballin et al., 2019). Apart from cancer, AI has succeeded in nephrotic syndrome, hepatic fibrosis, and renal transplants (Barisoni et al., 2020; Hermsen et al., 2019; Wong et al., 2021).

5.3.1 Deep neural network

Deep neural network (DNN) was used with GooLeNet to identify viable or nonviable cells to treat AMD (age-related macular degeneration). An automated, noninvasive technique like this progresses treatment by reducing errors and saving time (Schaub et al., 2020). A study demonstrated bright field imaging analysis to observe a time lapse of the morphological changes that occur during mesenchymal to epithelial transition (MET) before iPSCs colony formation. Machine learning (ML) algorithms have been developed to automatically track the earliest cellular changes leading to iPSCs formation (Fan et al., 2017). A mathematical model is capable of analyzing and determining the optimal phase for iPSCs selection. Importantly, this modeling approach is applicable to both mouse and human systems (Fan et al., 2017). convolutional neural network (CNN) has conventionally been used for two dimensional (2D) image analysis. However, a vector-based convolutional neural network (V-CNN) is a specialized neural network designed to consider the third dimension of volume, capture complex patterns, and process three dimensional (3D) data. Tailored for 3D data analysis, it applies to medical imaging, drug discovery, and disease classification. A new model of V-CNN identified the quality of iPSCs colonies (Kavitha et al., 2017). Appropriate healthy characteristics integrated with the V-CNN model showed that the deep V-CNN model detected the quality of the colony with 95.5% accuracy, which turned out to be more efficient and accurate than support vector machine (SVM) (75.2%). Thus, V-CNN is a robust and cost-effective method (Kavitha et al., 2017). AI was trained using Nanog-GFP images expressed in CSCs in phase contrast images. AI visualized the CSCs, but the results were manageable (Aida et al., 2020).

Deep learning convolutional networks could detect and analyze progressive cardiomyopathy (PCM) in rodent heart sections (Tokarz et al., 2021). CNN trained using brightfield images to identify hiPSCs derived from cardiomyocytes showed remarkable results and detected hiPSCs-CM with high accuracy (Orita et al., 2019). Confocal images of sarcolemma stained with a fluorescent dye trained a densely connected convolutional network (DenseNet) and identified the various developmental stages using ML (Škrabánek & Zahradnikova, 2019). DenseNet is a variant of CNN comprising dense connectivity patterns between layers. The connection of each layer with all the previous layers allows information propagation of previous layers to the new layers, thus improving gradient flow. It has been a popular choice over others due to its feature reusability, improved gradient flow, and diminished gradient vanishing. The histological material from heart explants of patients was used in the algorithm to detect clinical heart failure, which resulted in 99% sensitivity and 94% specificity (Nirschl et al., 2018).

5.3.2 AI in blood-based biomarkers

A novel use of AI is in developing blood-based biomarkers that could improve patient survival outcomes by forecasting rejection during transplantation. Blood-based markers have been used as "liquid biopsy," a noninvasive analysis (Giarraputo et al., 2021). Peripheral blood possessing clinically significant acute cellular rejection was profiled for its gene expression (Deng et al., 2006). The GEP thus obtained was devised into an algorithm that helps in rejection analysis during transplantation. It is currently being marketed as AlloMap (Crespo-Leiro et al., 2016). Flow cytometry was used to quantify the specific proteins present on the surface of EVs that were used as biomarkers to distinguish acute cell-mediated rejection (ACR) from antibody-mediated rejection (AMR) as well as grades of ACR (Castellani et al., 2020).

5.3.3 miRNA analysis using bioinformatics tools

R script was employed to identify common microRNA (miRNA) dysregulation associated with stemness, EMT, and drug resistance in gastric cancer. Further, bioinformatics tools ONCO.IO and KEGG were used to identify miRNA targets and associated signaling pathways. Results revealed seven miRNAs that lead to stemness, EMT, and drug resistance in gastric cancer, of which four miRNAs (miR-23a, miR-30a, miR-34a, and miR-100) are significant in the process. Downregulating these miRNA or their respective targets will help to strategize treatment accordingly and prevent cancer metastasis and relapse (Azimi et al., 2022). Multiple bioinformatic tools such as KEGG software DIANA tools were used to investigate up/downregulated miRNA during NSCs differentiation in humans. The Fisher's Exact Test with a P-value of 0.05 was used to observe the regulated miRNA and union pathways. Seven significant union pathways and the most frequently observed miRNA were miR-125a-5p, miR-320, and miR-423-5p (Silveira et al., 2020).

6. Major concerns and drawbacks

Significant advancements have been achieved in computational biology, yet they come with inherent limitations that demand attention. This discussion delves into the challenges confronted by computational biologists working within the stem cell research community (Fig. 24.3).

6.1 Database maintenance

The unavailability of several databases and the necessity for regular updates pose significant challenges for computational biologists and stem cell researchers. This issue underscores the importance of securing adequate financial support for maintaining data infrastructure and incentivizing data repository contributors (Finkelstein et al., 2020).

Manually curated data, while ensuring accuracy, can be susceptible to human errors. The process of collating data based on multiple factors is time-consuming. It is essential to filter out data with faulty experimental designs and implement rigorous data processing pipelines to maintain data quality (Wells et al., 2013). The integration of ontologies for cells, tissues, and organs can enhance data structuring. However, databases are not always open and accessible to everyone, which can hinder unified knowledge sharing and workflow. Additionally, data sharing is complicated, as many countries have regulations promoting raw file sharing (Kurtz et al., 2022).

6.2 Lack of adequate information

Databases like HemaExplorer and BloodSpot, which contain data of both normal and malignant hematopoietic cells, greatly benefit from comprehensive information about the disease and the affected patient (Kurtz et al., 2019). Insufficient or restricted data on the disease's pathology can hinder the translational and therapeutic potential of the cell, particularly in terms of its applicability. Having a thorough understanding of the disease's characteristics facilitates data analysis and subsequent treatment strategies.

6.3 OMIC technologies

High-throughput technologies have improved data analysis, but their data need to be better standardized and curated (Hasin et al., 2017). Being artificially generated, they need the inherent property to decode the data obtained from a biological perspective. Data analysis tools must be upgraded to make it easier for computational and stem cell biologists to interpret (Finkelstein et al., 2020).

6.4 Disease pathology

Medical imaging is crucial to disease pathology, which is used to train machines. Imaging science is crucial to AI algorithms. Image quality is highly dependent on the tissue's quality, technicians' expertise, and specimen amount, which can be improved by histological standardization, tissue processing, and enhancement of image digitalization techniques (Glass et al., 2021).

6.5 Biosafety and bioprivacy

It is important to consider the bioefficacy of the treatment both before and after the experiment. Updating any data should not be done without the explicit consent of the donor or recipient, taking into account ethical considerations (Srinivasan et al., 2021).

7. Conclusion and future perspectives

In conclusion, the fusion of computational models with experimental techniques represents a promising avenue for rejuvenating HSCs and advancing the field of regenerative medicine. Ethical considerations and regulatory oversight are essential to guide these advancements responsibly. Notably, computational tools like NGS are instrumental in identifying signature genes that underpin personalized therapies, opening new horizons in precision medicine. Clinicians are actively exploring innovative approaches such as gene therapy and combination treatments to preserve self-renewal in progenitor stem cells, providing alternatives to conventional preservation methods. This interdisciplinary approach holds the potential to revolutionize HSC research and therapy, particularly for age-related disorders and degenerative diseases. This highlights the critical role of metabolic cross-talk and its implications in diseases like hematologic malignancies, shedding light on the intricate molecular mechanisms involving genes, TFs, and miRNAs. It emphasizes the necessity of comprehensive genome-scale integrative networks to unravel the complexities of these diseases. Furthermore, it is important to optimize the culture conditions for HSCs, emphasizing the selection of appropriate growth media and factors for their maintenance and differentiation. It also highlights the application of computational approaches in various aspects of stem cell research, from predicting cell fate to drug screening, and the utilization of single-cell technologies. Effective data integration, cell characterization, and information exchange are pivotal for robust and insightful stem cell research. Overall, computational science has ushered in a new era in medical research, harnessing the potential of stem cells for immunotherapy, regenerative medicine, and drug discovery. From hematological malignancies to a wide range of non-malignant conditions, computational modeling, single-cell RNA sequencing, and AI have become indispensable tools. While challenges such as database maintenance, data quality assurance, and ethical considerations persist, the transformative impact of computational science on medical advancements continues to be profound.

Authors' contributions

PKR designed and conceived the study. PKR, BB, RK, AR and RC wrote the original manuscript draft. PKR, BB, RK, AR, AL, and RC contributed to the review collection and analysis and revised the manuscript. All the authors read and approved the manuscript.

References

Abou-El-Enein, M., Duda, G. N., Gruskin, E. A., & Grainger, D. W. (2017). Strategies for derisking translational processes for biomedical technologies. *Trends in Biotechnology, 35*(2), 100−108.

Agrawal, P., Raghav, P. K., Bhalla, S., Sharma, N., & Raghava, G. P. (2018). Overview of free software developed for designing drugs based on protein-small molecules interaction. *Current Topics in Medicinal Chemistry, 18*(13), 1146−1167.

Aida, S., Okugawa, J., Fujisaka, S., Kasai, T., Kameda, H., & Sugiyama, T. (2020). Deep learning of cancer stem cell morphology using conditional generative adversarial networks. *Biomolecules, 10*(6), 931.

Akselrod-Ballin, A., Chorev, M., Shoshan, Y., Spiro, A., Hazan, A., Melamed, R., Barkan, E., Herzel, E., Naor, S., Karavani, E., Koren, G., Goldschmidt, Y., Shalev, V., Rosen-Zvi, M., & Guindy, M. (2019). Predicting breast cancer by applying deep learning to linked health records and mammograms. *Radiology, 292*(2), 331−342.

Ali, S., Lewis, J., & Madabhushi, A. (2013). Spatially aware cell cluster (spaccl) graphs: Predicting outcome in oropharyngeal p16+ tumors. In *Medical image computing and computer-assisted intervention−MICCAI 2013: 16th international conference* (pp. 412−419). Nagoya, Japan: Springer Berlin Heidelberg. September 22-26, 2013, Proceedings, Part I 16.

Arock, M., Wedeh, G., Hoermann, G., Bibi, S., Akin, C., Peter, B., Gleixner, K. V., Hartmann, K., Butterfield, J. H., Metcalfe, D. D., & Valent, P. (2018). Preclinical human models and emerging therapeutics for advanced systemic mastocytosis. *Haematologica, 103*(11), 1760.

Azimi, M., Totonchi, M., & Ebrahimi, M. (2022). Determining the role of MicroRNAs in self-renewal, metastasis and resistance to drugs in human gastric cancer based on data mining approaches: A systematic review. *Cell Journal (Yakhteh), 24*(1), 1.

Banerjee, K., Jana, T., Ghosh, Z., & Saha, S. (2020). PSCRIdb: A database of regulatory interactions and networks of pluripotent stem cell lines. *Journal of Biosciences, 45*, 1–8.

Barisoni, L., Lafata, K. J., Hewitt, S. M., Madabhushi, A., & Balis, U. G. (2020). Digital pathology and computational image analysis in nephropathology. *Nature Reviews Nephrology, 16*(11), 669–685.

Beck, D., Thoms, J. A., Perera, D., Schütte, J., Unnikrishnan, A., Knezevic, K., Kinston, S. J., Wilson, N. K., O'Brien, T. A., Göttgens, B., Wong, J. W., & Pimanda, J. E. (2013). Genome-wide analysis of transcriptional regulators in human HSPCs reveals a densely interconnected network of coding and noncoding genes. *Blood, The Journal of the American Society of Hematology, 122*(14), e12–e22.

Berneman-Zeitouni, D., Molakandov, K., Elgart, M., Mor, E., Fornoni, A., Domínguez, M. R., Kerr-Conte, J., Ott, M., Meivar-Levy, I., & Ferber, S. (2014). The temporal and hierarchical control of transcription factors-induced liver to pancreas transdifferentiation. *PLoS One, 9*(2), e87812.

Bock, C., Kiskinis, E., Verstappen, G., Gu, H., Boulting, G., Smith, Z. D., Ziller, M., Croft, G. F., Amoroso, M. W., Oakley, D. H., Gnirke, A., Eggan, K., & Meissner, A. (2011). Reference Maps of human ES and iPS cell variation enable high-throughput characterization of pluripotent cell lines. *Cell, 144*(3), 439–452.

Browaeys, R., Saelens, W., & Saeys, Y. (2020). NicheNet: Modeling intercellular communication by linking ligands to target genes. *Nature Methods, 17*(2), 159–162.

Cahan, P., Li, H., Morris, S. A., Da Rocha, E. L., Daley, G. Q., & Collins, J. J. (2014). CellNet: Network biology applied to stem cell engineering. *Cell, 158*(4), 903–915.

Castellani, C., Burrello, J., Fedrigo, M., Burrello, A., Bolis, S., Di Silvestre, D., Tona, F., Bottio, T., Biemmi, V., Toscano, G., Gerosa, G., Thiene, G., Basso, C., Longnus, S. L., Vassalli, G., Angelini, A., & Barile, L. (2020). Circulating extracellular vesicles as non-invasive biomarker of rejection in heart transplant. *The Journal of Heart and Lung Transplantation, 39*(10), 1136–1148.

Cesana, M., Guo, M. H., Cacchiarelli, D., Wahlster, L., Barragan, J., Doulatov, S., Vo, L. T., Salvatori, B., Trapnell, C., Clement, K., Cahan, P., Tsanov, K. M., Sousa, P. M., Tazon-Vega, B., Bolondi, A., Giorgi, F. M., Califano, A., Rinn, J. L., Meissner, A., … Daley, G. Q. (2018). A CLK3-HMGA2 alternative splicing axis impacts human hematopoietic stem cell molecular identity throughout development. *Cell Stem Cell, 22*(4), 575–588.

Chambers, S. M., Boles, N. C., Lin, K. Y. K., Tierney, M. P., Bowman, T. V., Bradfute, S. B., Chen, A. J., Merchant, A. A., Sirin, O., Weksberg, D. C., Merchant, M. G., Fisk, C. J., Shaw, C. A., & Goodell, M. A. (2007). Hematopoietic fingerprints: An expression database of stem cells and their progeny. *Cell Stem Cell, 1*(5), 578–591.

Chen, Y., Zhou, P., Deng, Y., Cai, X., Sun, M., Sun, Y., & Wu, D. (2023). ALKBH5-mediated m6A demethylation of TIRAP mRNA promotes radiation-induced liver fibrosis and decreases radiosensitivity of hepatocellular carcinoma. *Clinical and Translational Medicine, 13*(2), e1198.

Crespo-Leiro, M. G., Stypmann, J., Schulz, U., Zuckermann, A., Mohacsi, P., Bara, C., oss, H., Parameshwar, J., Zakliczyński, M., Fiocchi, R., Hoefer, D., Colvin, M., Deng, M. C., Leprince, P., Elashoff, B., Yee, J. P., & Vanhaecke, J. (2016). Clinical usefulness of gene-expression profile to rule out acute rejection after heart transplantation: Cargo II. *European Heart Journal, 37*(33), 2591–2601.

Crook, J. M., & Stacey, G. N. (2014). Setting quality standards for stem cell banking, research and translation: The international stem cell banking initiative. *Stem Cell Banking*, 3–9.

Dai, Y., Jin, F., Wu, W., & Kumar, S. K. (2019). Cell cycle regulation and hematologic malignancies. *Blood Science, 1*(01), 34–43.

D'Alessio, A. C., Fan, Z. P., Wert, K. J., Baranov, P., Cohen, M. A., Saini, J. S., Cohick, E., Charniga, C., Dadon, D., Hannett, N. M., Young, M. J., Temple, S., Jaenisch, R., Lee, T. I., & Young, R. A. (2015). A systematic approach to identify candidate transcription factors that control cell identity. *Stem Cell Reports, 5*(5), 763–775.

Daley, G. Q., Hyun, I., Apperley, J. F., Barker, R. A., Benvenisty, N., Bredenoord, A. L., Breuer, C. K., Caulfield, T., Cedars, M. I., Frey-Vasconcells, J., Heslop, H. E., Jin, Y., Lee, R. T., McCabe, C., Munsie, M., Murry, C. E., Piantadosi, S., Rao, M., Rooke, H. M., … Kimmelman, J. (2016). Setting global standards for stem cell research and clinical translation: The 2016 ISSCR guidelines. *Stem Cell Reports, 6*(6), 787–797.

Datlinger, P., Rendeiro, A. F., Boenke, T., Senekowitsch, M., Krausgruber, T., Barreca, D., & Bock, C. (2021). Ultra-high-throughput single-cell RNA sequencing and perturbation screening with combinatorial fluidic indexing. *Nature Methods, 18*(6), 635–642.

Del Sol, A., & Jung, S. (2021). The importance of computational modeling in stem cell research. *Trends in Biotechnology, 39*(2), 126–136.

Deng, M. C., Eisen, H. J., Mehra, M. R., Billingham, M., Marboe, C. C., Berry, G., Kobashigawa, J., Johnson, F. L., Starling, R. C., Murali, S., Pauly, D. F., Baron, H., Wohlgemuth, J. G., Woodward, R. N., Klingler, T. M., Walther, D., Lal, P. G., Rosenberg, S., Hunt, S., CARGO, Investigators, C. A. R. G. O. (2006). Noninvasive discrimination of rejection in cardiac allograft recipients using gene expression profiling. *American Journal of Transplantation, 6*(1), 150–160.

El-Sadik, A. O. (2010). Potential sources of stem cells as a regenerative therapy for Parkinson's disease. *Stem Cells and Cloning: Advances and Applications*, 183–191.

Fan, K., Zhang, S., Zhang, Y., Lu, J., Holcombe, M., & Zhang, X. (2017). A machine learning assisted, label-free, non-invasive approach for somatic reprogramming in induced pluripotent stem cell colony formation detection and prediction. *Scientific Reports, 7*(1), 13496.

Finkelstein, J., Parvanova, I., & Zhang, F. (2020). Informatics approaches for harmonized intelligent integration of stem cell research. *Stem Cells and Cloning: Advances and Applications*, 1–20.

Galgano, L., & Hunt, D. (2018). *Hsct: How does it work?*. The European blood and marrow transplantation textbook for nurses: Under the auspices of EBMT.

Giarraputo, A., Barison, I., Fedrigo, M., Burrello, J., Castellani, C., Tona, F., Bottio, T., Gerosa, G., Barile, L., & Angelini, A. (2021). A changing paradigm in heart transplantation: An integrative approach for invasive and non-invasive allograft rejection monitoring. *Biomolecules, 11*(2), 201.

Glass, C., Lafata, K. J., Jeck, W., Horstmeyer, R., Cooke, C., Everitt, J., Glass, M., Dov, D., & Seidman, M. A. (2021). The role of machine learning in cardiovascular pathology. *Canadian Journal of Cardiology, 38*(2), 234–245.

Gottgens, B. (2015). CODEX: A next-generation sequencing experiment database for the hematopoietic and embryonic stem cell communities. *Nucleic Acids Research, 43*(D1), D1117–D1123.

Hamey, F. K., & Gottgens, B. (2019). Machine learning predicts putative hematopoietic stem cells within large single-cell transcriptomics data sets. *Experimental Hematology, 78*, 11–20.

Hasin, Y., Seldin, M., & Lusis, A. (2017). Multi-omics approaches to disease. *Genome Biology, 18*(1), 1–15.

Heindl, A., Sestak, I., Naidoo, K., Cuzick, J., Dowsett, M., & Yuan, Y. (2018). Relevance of spatial heterogeneity of immune infiltration for predicting risk of recurrence after endocrine therapy of ER+ breast cancer. *Journal of the National Cancer Institute: Journal of the National Cancer Institute, 110*(2), 166–175.

Henig, I., & Zuckerman, T. (2014). Hematopoietic stem cell transplantation—50 years of evolution and future perspectives. *Rambam Maimonides Medical Journal, 5*(4).

Hermsen, M., de Bel, T., Den Boer, M., Steenbergen, E. J., Kers, J., Florquin, S., Roelofs, J. J. T. H., Stegall, M. D., Alexander, M. P., Smith, B. H., Smeets, B., Hilbrands, L. B., & van der Laak, J. A. (2019). Deep learning–based histopathologic assessment of kidney tissue. *Journal of the American Society of Nephrology, 30*(10), 1968–1979.

Hongwiangchan, N., Sriratanasak, N., Wichadakul, D., Aksorn, N., Chamni, S., & Chanvorachote, P. (2021). Hydroquinone 5-O-Cinnamoyl ester of renieramycin M suppresses lung cancer stem cells by targeting Akt and destabilizes c-Myc. *Pharmaceuticals, 14*(11), 1112.

Horgan, R. P., & Kenny, L. C. (2011). 'Omic'technologies: Genomics, transcriptomics, proteomics and metabolomics. *The Obstetrician and Gynaecologist, 13*(3), 189–195.

Huang, Y., McCarthy, D. J., & Stegle, O. (2019). Vireo: Bayesian demultiplexing of pooled single-cell RNA-seq data without genotype reference. *Genome Biology, 20*, 1–12.

ICMR. (2021). *National Guidelines for Hematopoietic Stem Cell Transplantation*. https://main.icmr.nic.in/sites/default/files/upload_documents/Nat_Guide_HCT.pdf.

Jaitak, V. (2016). Molecular docking study of natural alkaloids as multi-targeted hedgehog pathway inhibitors in cancer stem cell therapy. *Computational Biology and Chemistry, 62*, 145–154.

Kang, H. M., Subramaniam, M., Targ, S., Nguyen, M., Maliskova, L., McCarthy, E., Wan, E., Wong, S., Byrnes, L., Lanata, C. M., Gate, R. E., Mostafavi, S., Marson, A., Zaitlen, N., Criswell, L. A., & Ye, C. J. (2018). Multiplexed droplet single-cell RNA-sequencing using natural genetic variation. *Nature Biotechnology, 36*(1), 89–94.

Kavitha, M. S., Kurita, T., Park, S. Y., Chien, S. I., Bae, J. S., & Ahn, B. C. (2017). Deep vector-based convolutional neural network approach for automatic recognition of colonies of induced pluripotent stem cells. *PLoS One, 12*(12), e0189974.

Khorasani, N., Sadeghi, M., & Nowzari-Dalini, A. (2020). A computational model of stem cell molecular mechanism to maintain tissue homeostasis. *PLoS One, 15*(7), e0236519

Kulshrestha, S., Arora, T., Sengar, M., Sharma, N., Chawla, R., Bajaj, S., & Raghav, P. K. (2021). Advanced approaches and in silico tools of chemoinformatics in drug designing. In *Chemoinformatics and bioinformatics in the pharmaceutical sciences* (pp. 173–206). Academic Press.

Kumar, D., Ishaque, M., & Raghav, P. K. (2022). Regenerative medicines of ROS-induced cancers treatment. In S. Chakraborti (Ed.), *Handbook of oxidative stress in cancer: Therapeutic aspects*. Singapore: Springer.

Kumar, R., Lathwal, A., Kumar, V., Patiyal, S., Raghav, P. K., & Raghava, G. P. (2020). CancerEnD: A database of cancer associated enhancers. *Genomics, 112*(5), 3696–3702.

Kumar, R., Lathwal, A., Nagpal, G., Kumar, V., & Raghav, P. K. (2021). Impact of chemoinformatics approaches and tools on current chemical research. In *Chemoinformatics and bioinformatics in the pharmaceutical sciences* (pp. 1–26). Academic Press.

Kurtz, A., Elsallab, M., Sanzenbacher, R., & Abou-El-Enein, M. (2019). Linking scattered stem cell-based data to advance therapeutic development. *Trends in Molecular Medicine, 25*(1), 8–19.

Kurtz, A., Mah, N., Chen, Y., Fuhr, A., Kobold, S., Seltmann, S., & Müller, S. C. (2022). Human pluripotent stem cell registry: Operations, role and current directions. *Cell Proliferation, 55*(8), e13238.

Lee, G., Sparks, R., Ali, S., Shih, N. N., Feldman, M. D., Spangler, E., Rebbeck, T., Tomaszewski, J. E., & Madabhushi, A. (2014). Co-occurring gland angularity in localized subgraphs: Predicting biochemical recurrence in intermediate-risk prostate cancer patients. *PLoS One, 9*(5), e97954.

Liu, Y., Liu, X., Chen, L. C., Du, W. Z., Cui, Y. Q., Piao, X. Y., Li, Y., & Jiang, C. L. (2014). Targeting glioma stem cells via the Hedgehog signaling pathway. *Neuroimmunology and Neuroinflammation, 1*, 51–59.

Madhunapantula, S. V., Mosca, P. J., & Robertson, G. P. (2011). The Akt signaling pathway: An emerging therapeutic target in malignant melanoma. *Cancer Biology & Therapy, 12*(12), 1032–1049.

Mallon, B. S., Chenoweth, J. G., Johnson, K. R., Hamilton, R. S., Tesar, P. J., Yavatkar, A. S., Tyson, L. J., Park, K., Chen, K. G., Fann, Y. C., & McKay, R. D. (2013). StemCellDB: The human pluripotent stem cell database at the national institutes of health. *Stem Cell Research, 10*(1), 57–66.

Mann, Z., Sengar, M., Verma, Y. K., Rajalingam, R., & Raghav, P. K. (2022). Hematopoietic stem cell factors: Their functional role in self-renewal and clinical aspects. *Frontiers in Cell and Developmental Biology, 10*, 453.

McGinnis, C. S., Patterson, D. M., Winkler, J., Conrad, D. N., Hein, M. Y., Srivastava, V., Hu, J. L., Murrow, L. M., Weissman, J. S., Werb, Z., Chow, E. D., & Gartner, Z. J. (2019). MULTI-Seq: Sample multiplexing for single-cell RNA sequencing using lipid-tagged indices. *Nature Methods, 16*(7), 619−626.

Micallef, I. N., Stiff, P. J., Nademanee, A. P., Maziarz, R. T., Horwitz, M. E., Stadtmauer, E. A., Kaufman, J. L., McCarty, J. M., Vargo, R., Cheverton, P. D., Struijs, M., Bolwell, B., & DiPersio, J. F. (2018). Plerixafor plus granulocyte colony-stimulating factor for patients with non-Hodgkin lymphoma and multiple myeloma: Long-term follow-up report. *Biology of Blood and Marrow Transplantation, 24*(6), 1187−1195.

Mohamed, L. M., Eltigani, M. M., Abdallah, M. H., Ghaboosh, H., Jardan, Y. A. B., Yusuf, O., Elsaman, T., Mohamed, M. A., & Alzain, A. A. (2022). Discovery of novel natural products as dual MNK/PIM inhibitors for acute myeloid leukemia treatment: Pharmacophore modeling, molecular docking, and molecular dynamics studies. *Frontiers in Chemistry, 10*.

Monga, I., Kaur, K., & Dhanda, S. K. (2022). Revisiting hematopoiesis: Applications of the bulk and single-cell transcriptomics dissecting transcriptional heterogeneity in hematopoietic stem cells. *Briefings in Functional Genomics, 21*(3), 159−176.

Montrone, C., Kokkaliaris, K. D., Loeffler, D., Lechner, M., Kastenmüller, G., Schroeder, T., & Ruepp, A. (2013). HSC-Explorer: A curated database for hematopoietic stem cells. *PLoS One, 8*(7), e70348.

Müller, F. J., Schuldt, B. M., Williams, R., Mason, D., Altun, G., Papapetrou, E. P., Danner, S., Goldmann, J. E., Herbst, A., Schmidt, N. O., Aldenhoff, J. B., Laurent, L. C., & Loring, J. F. (2011). A bioinformatic assay for pluripotency in human cells. *Nature Methods, 8*(4), 315−317.

Nakagawa, M. M., Chen, H., & Rathinam, C. V. (2018). Constitutive activation of NF-κB pathway in hematopoietic stem cells causes loss of quiescence and deregulated transcription factor networks. *Frontiers in Cell and Developmental Biology, 6*, 143.

Nandan, A., Raghav, P. K., Srivastava, A., Tiwari, S. K., Shukla, A. K., & Sharma, V. (2021). Current insights to therapeutic targets of ROS induced gastric cancer stem cells. In *Handbook of oxidative stress in cancer: Therapeutic aspects* (pp. 1−13). Singapore: Springer Singapore.

Nirschl, J. J., Janowczyk, A., Peyster, E. G., Frank, R., Margulies, K. B., Feldman, M. D., & Madabhushi, A. (2018). A deep-learning classifier identifies patients with clinical heart failure using whole-slide images of H&E tissue. *PLoS One, 13*(4), e0192726.

Orita, K., Sawada, K., Koyama, R., & Ikegaya, Y. (2019). Deep learning-based quality control of cultured human-induced pluripotent stem cell-derived cardiomyocytes. *Journal of Pharmacological Sciences, 140*(4), 313−316.

Osareh, A., & Shadgar, B. (2010). Machine learning techniques to diagnose breast cancer. In *2010 5th international symposium on health informatics and bioinformatics* (pp. 114−120). IEEE.

Ozsvari, B., Sotgia, F., Simmons, K., Trowbridge, R., Foster, R., & Lisanti, M. P. (2017). Mitoketoscins: Novel mitochondrial inhibitors for targeting ketone metabolism in cancer stem cells (CSCs). *Oncotarget, 8*(45), 78340−78350.

Park, J. Y., Kim, S. K., Woo, D. H., Lee, E. J., Kim, J. H., & Lee, S. H. (2009). Differentiation of neural progenitor cells in a microfluidic chip-generated cytokine gradient. *Stem cells, 27*(11), 2646−2654.

Paul, D. (2020). The systemic hallmarks of cancer. *Journal of Cancer Metastasis and Treatment, 6*, 29.

Praveen, A., Raghav, P. K., & Gangenahalli, G. (2012). Computational network model predicts the drug effects on SHP-1 mediated intracellular signaling through c-Kit. *Journal of Proteins and Proteomics, 3*(2), 9.

Preethy, S., John, S., Ganesh, J. S., Srinivasan, T., Terunuma, H., Iwasaki, M., & Abraham, S. J. (2013). Age-old wisdom concerning cell-based therapies with added knowledge in the stem cell era: Our perspectives. *Stem Cells and Cloning: Advances and Applications*, 13−18.

Qin, X., Sufi, J., Vlckova, P., Kyriakidou, P., Acton, S. E., Li, V. S., Nitz, M., & Tape, C. J. (2020). Cell-type-specific signaling networks in heterocellular organoids. *Nature Methods, 17*(3), 335−342.

Raghav, P. K., & Gangenahalli, G. (2018). Hematopoietic stem cell molecular targets and factors essential for hematopoiesis. *Journal of Stem Cell Research & Therapy, 8*(441), 2.

Raghav, P. K., & Gangenahalli, G. (2021). PU. 1 mimic synthetic peptides selectively bind with GATA-1 and allow c-Jun PU. 1 binding to enhance myelopoiesis. *International Journal of Nanomedicine, 16*, 3833.

Raghav, P. K., Kalyanaraman, K., & Kumar, D. (2021). Human cell receptors: Potential drug targets to combat COVID-19. *Amino Acids, 53*(6), 813−842.

Raghav, P. K., Kumar, R., Kumar, V., & Raghava, G. P. (2019). Docking-based approach for identification of mutations that disrupt binding between Bcl-2 and Bax proteins: Inducing apoptosis in cancer cells. *Molecular genetics & genomic medicine, 7*(11), e910.

Raghav, P. K., Kumar, R., & Raghava, G. P. (2019). Machine learning based identification of stem cell genes involved in stemness. *Journal of Cell Science & Therapy, 10*, 40.

Raghav, P. K., & Mann, Z. (2021). Cancer stem cells targets and combined therapies to prevent cancer recurrence. *Life Sciences, 277*, 119465.

Raghav, P. K., Mann, Z., Ahlawat, S., & Mohanty, S. (2022). Mesenchymal stem cell-based nanoparticles and scaffolds in regenerative medicine. *European Journal of Pharmacology, 918*, 174657.

Raghav, P. K., Mann, Z., Krishnakumar, V., & Mohanty, S. (2021). Therapeutic effect of natural compounds in targeting ROS-induced cancer. In *Handbook of oxidative stress in cancer: Mechanistic aspects* (pp. 1−47). Singapore: Springer Singapore.

Raghav, P. K., Mann, Z., & Mohanty, S. (2022). Therapeutic potential of chemical compounds in targeting cancer stem cells. In S. Chakraborti (Ed.), *Handbook of oxidative stress in cancer. Therapeutic aspects*. Singapore: Springer.

Raghav, P. K., Mann, Z., Pandey, P. K., & Mohanty, S. (2021). Systems biology resources and their applications to understand the cancer. In *Handbook of oxidative stress in cancer: Mechanistic aspects* (pp. 1−35). Singapore: Springer Singapore.

Raghav, P., Maruthamuthu, S., Lee, N., Kong, D., Rajalingam, R., Kaur, N., Butte, A., Reyes, K., Chiu, C., & Ng, D. (2022). Next-generation sequencing revealed linked alleles of different Hla loci associated with susceptibility and protection to covid-19. *Human Immunology*, 127−128.

Raghav, P. K., & Mohanty, S. (2020). Are graphene and graphene-derived products capable of preventing COVID-19 infection? *Medical Hypotheses, 144*, 110031.

Raghav, P. K., Singh, A. K., & Gangenahalli, G. (2018a). A change in structural integrity of c-Kit mutant D816V causes constitutive signaling. *Mutation Research/Fundamental and Molecular Mechanisms of Mutagenesis, 808*, 28−38.

Raghav, P. K., Singh, A. K., & Gangenahalli, G. (2018b). Stem cell factor and NSC87877 combine to enhance c-Kit mediated proliferation of human megakaryoblastic cells. *PLoS One, 13*(11), e0206364.

Raghav, P. K., Singh, A. K., & Gangenahalli, G. (2018c). Stem cell factor and NSC87877 synergism enhances c-Kit mediated proliferation of human erythroid cells. *Life Sciences, 214*, 84−97.

Raghav, P., Verma, Y., & Gangenahalli, G. (2011). In Silico analysis of flexible loop domain's conformational changes affecting BH3 cleft of Bcl 2 protein. *Journal of Natural Science, Biology and Medicine, 2*(3), 56.

Raghav, P. K., Verma, Y. K., & Gangenahalli, G. U. (2012a). Molecular dynamics simulations of the Bcl-2 protein to predict the structure of its unordered flexible loop domain. *Journal of Molecular Modeling, 18*, 1885−1906.

Raghav, P. K., Verma, Y. K., & Gangenahalli, G. U. (2012b). Peptide screening to knockdown Bcl-2's anti-apoptotic activity: Implications in cancer treatment. *International Journal of Biological Macromolecules, 50*(3), 796−814.

Rawat, S., Jain, K. G., Gupta, D., Raghav, P. K., Chaudhuri, R., Pinky, ., Shakeel, A., Arora, V., Sharma, H., Debnath, D., Kalluri, A., Agrawal, A. K., Jassal, M., Dinda, A. K., Patra, P., & Mohanty, S. (2021). Graphene nanofiber composites for enhanced neuronal differentiation of human mesenchymal stem cells. *Nanomedicine, 16*(22), 1963−1982.

Revilla, A., González, C., Iriondo, A., Fernández, B., Prieto, C., Marín, C., & Liste, I. (2016). Current advances in the generation of human iPS cells: Implications in cell-based regenerative medicine. *Journal of tissue engineering and regenerative medicine, 10*(11), 893−907.

Roost, M. S., Van Iperen, L., Ariyurek, Y., Buermans, H. P., Arindrarto, W., Devalla, H. D., Passier, R., Mummery, C. L., Carlotti, F., de Koning, E. J., van Zwet, E. W., Goeman, J. J., & de Sousa Lopes, S. M. C. (2015). KeyGenes, a tool to probe tissue differentiation using a human fetal transcriptional atlas. *Stem Cell Reports, 4*(6), 1112−1124.

Rubin, D. L., Shah, N. H., & Noy, N. F. (2008). Biomedical ontologies: A functional perspective. *Briefings in Bioinformatics, 9*(1), 75−90.

Sakurai, K., Kurtz, A., Stacey, G., Sheldon, M., & Fujibuchi, W. (2016). First proposal of minimum information about a cellular assay for regenerative medicine. *Stem cells translational medicine, 5*(10), 1345−1361.

Sánchez-Castillo, M., Ruau, D., Wilkinson, A. C., Ng, F. S., Hannah, R., Diamanti, E., & 2015.

Sason, H., & Shamay, Y. (2020). Nanoinformatics in drug delivery. *Israel Journal of Chemistry, 60*(12), 1108−1117.

Schaub, N. J., Hotaling, N. A., Manescu, P., Padi, S., Wan, Q., Sharma, R., George, A., Chalfoun, J., Simon, M., Ouladi, M., Simon, C. G., Jr., Bajcsy, P., & Bharti, K. (2020). Deep learning predicts function of live retinal pigment epithelium from quantitative microscopy. *The Journal of Clinical Investigation, 130*(2), 1010−1023.

Schütte, J., Wang, H., Antoniou, S., Jarratt, A., Wilson, N. K., Riepsaame, J., Calero-Nieto, F. J., Moignard, V., Basilico, S., Kinston, S. J., Hannah, R. L., Chan, M. C., Nürnberg, S. T., Ouwehand, W. H., Bonzanni, N., de Bruijn, M. F., & Göttgens, B. (2016). An experimentally validated network of nine haematopoietic transcription factors reveals mechanisms of cell state stability. *Elife, 5*, e11469.

Sengar, A., Sengar, M., Mann, Z., & Raghav, P. K. (2022). Clinical approaches in targeting ROS-induced cancer. In S. Chakraborti (Ed.), *Handbook of oxidative stress in cancer: Therapeutic aspects.* Singapore: Springer.

Sharma, R., Dever, D. P., Lee, C. M., Azizi, A., Pan, Y., Camarena, J., Köhnke, T., Bao, G., Porteus, M. H., & Majeti, R. (2021). The TRACE-Seq method tracks recombination alleles and identifies clonal reconstitution dynamics of gene targeted human hematopoietic stem cells. *Nature Communications, 12*(1), 472.

Shen, C., Quan, Q., Yang, C., Wen, Y., & Li, H. (2018). Histone demethylase JMJD6 regulates cellular migration and proliferation in adipose-derived mesenchymal stem cells. *Stem Cell Research & Therapy, 9*, 1−13.

Shroff, G. (2018). A review on stem cell therapy for multiple sclerosis: Special focus on human embryonic stem cells. *Stem Cells and Cloning: Advances and Applications*, 1−11.

Silveira, R. G., Ferrua, C. P., do Amaral, C. C., Garcia, T. F., de Souza, K. B., & Nedel, F. (2020). MicroRNAs expressed in neuronal differentiation and their associated pathways: Systematic review and bioinformatics analysis. *Brain Research Bulletin, 157*, 140−148.

Škrabánek, P., & Zahradnikova, A., Jr. (2019). Automatic assessment of the cardiomyocyte development stages from confocal microscopy images using deep convolutional networks. *PLoS One, 14*(5), e0216720.

Srinivasan, M., Thangaraj, S. R., Ramasubramanian, K., Thangaraj, P. P., & Ramasubramanian, K. V. (2021). Exploring the current trends of artificial intelligence in stem cell therapy: A systematic review. *Cureus, 13*(12).

Srivatsan, S. R., McFaline-Figueroa, J. L., Ramani, V., Saunders, L., Cao, J., Packer, J., Pliner, H. A., Jackson, D. L., Daza, R. M., Christiansen, L., & Zhang, F. (2020). Massively multiplex chemical transcriptomics at single-cell resolution. *Science, 367*(6473), 45−51.

Stec, M., Weglarczyk, K., Baran, J., Zuba, E., Mytar, B., Pryjma, J., & Zembala, M. (2007). Expansion and differentiation of CD14+ CD16− and CD14++ CD16+ human monocyte subsets from cord blood CD34+ hematopoietic progenitors. *Journal of Leukocyte Biology, 82*(3), 594−602.

Stillman, N. R., Kovacevic, M., Balaz, I., & Hauert, S. (2020). In silico modelling of cancer nanomedicine, across scales and transport barriers. *Npj Computational Materials, 6*(1), 92.

Sullivan, S., Stacey, G. N., Akazawa, C., Aoyama, N., Baptista, R., Bedford, P., Shockley, K. R., Singletary, F., Cesta, M. F., Thomas, H. C., Chen, V. S., Hobbie, K., & Song, J. (2018). Quality control guidelines for clinical-grade human induced pluripotent stem cell lines. *Regenerative Medicine, 13*(7), 859−866.

Tokarz, D. A., Steinbach, T. J., Lokhande, A., Srivastava, G., Ugalmugle, R., Co, C. A., Shockley, K. R., Singletary, E., Cesta, M. F., Thomas, H. C., & Chen, V. S. (2021). Using artificial intelligence to detect, classify, and objectively score severity of rodent cardiomyopathy. *Toxicologic Pathology, 49*(4), 888−896.

Wanandi, S. I., Limanto, A., Yunita, E., Syahrani, R. A., Louisa, M., Wibowo, A. E., & Arumsari, S. (2020). In silico and in vitro studies on the anticancer activity of andrographolide targeting survivin in human breast cancer stem cells. *PLoS One, 15*(11), e0240020.

Warren, L., Manos, P. D., Ahfeldt, T., Loh, Y. H., Li, H., Lau, F., Ebina, W., Mandal, P. K., Smith, Z. D., Meissner, A., & Daley, G. Q. (2010). Highly efficient reprogramming to pluripotency and directed differentiation of human cells with synthetic modified mRNA. *Cell Stem Cell, 7*(5), 618−630.

Watt, F. M., & Driskell, R. R. (2010). The therapeutic potential of stem cells. *Philosophical Transactions of the Royal Society B: Biological Sciences, 365*(1537), 155−163.

Wells, C. A., Mosbergen, R., Korn, O., Choi, J., Seidenman, N., Matigian, N. A., Vitale, A. M., & Shepherd, J. (2013). Stemformatics: Visualisation and sharing of stem cell gene expression. *Stem Cell Research, 10*(3), 387−395.

Wilkinson, M. D., Dumontier, M., Aalbersberg, I. J., Appleton, G., Axton, M., Baak, A., Blomberg, N., Boiten, J. W., da Silva Santos, L. B., Bourne, P. E., Bouwman, & Finkers, R. (2016). Comment: The FAIR Guiding Principles for scientific data management and stewardship. *Scientific Data, 3*(1), 1−9.

Wong, G. L. H., Yuen, P. C., Ma, A. J., Chan, A. W. H., Leung, H. H. W., & Wong, V. W. S. (2021). Artificial intelligence in prediction of non-alcoholic fatty liver disease and fibrosis. *Journal of Gastroenterology and Hepatology, 36*(3), 543−550.

Xu, H., Baroukh, C., Dannenfelser, R., Chen, E. Y., Tan, C. M., Kou, Y., Kim, Y. E., Lemischka, I. R., & Ma'ayan, A. (2013). *Escape: Database for integrating high-content published data collected from human and mouse embryonic stem cells.* Database.

Xu, J., Du, Y., & Deng, H. (2015). Direct lineage reprogramming: Strategies, mechanisms, and applications. *Cell Stem Cell, 16*(2), 119−134.

Yadav, P., Vats, R., Bano, A., & Bhardwaj, R. (2020). Hematopoietic stem cells culture, expansion and differentiation: An insight into variable and available media. *International Journal of Stem Cells, 13*(3), 326−334.

Yamashita, M., Dellorusso, P. V., Olson, O. C., & Passegué, E. (2020). Dysregulated haematopoietic stem cell behaviour in myeloid leukaemogenesis. *Nature Reviews Cancer, 20*(7), 365−382.

Yang, R., Lu, M., Ming, L., Chen, Y., Cheng, K., Zhou, J., Jiang, S., Lin, Z., & Chen, D. (2020). 89Zr-Labeled multifunctional liposomes conjugate chitosan for PET-trackable triple-negative breast cancer stem cell targeted therapy. *International Journal of Nanomedicine*, 9061−9074.

Yeo, G. H. T., Lin, L., Qi, C. Y., Cha, M., Gifford, D. K., & Sherwood, R. I. (2020). A multiplexed barcodelet single-cell RNA-seq approach elucidates combinatorial signaling pathways that drive ESC differentiation. *Cell Stem Cell, 26*(6), 938−950.

Zhang, C. C., & Lodish, H. F. (2008). Cytokines regulating hematopoietic stem cell function. *Current Opinion in Hematology, 15*(4), 307−311.

Zhang, L., Mack, R., Breslin, P., & Zhang, J. (2020). Molecular and cellular mechanisms of aging in hematopoietic stem cells and their niches. *Journal of Hematology & Oncology, 13*, 1−22.

Chapter 25

Approaches to construct and analyze stem cells regulatory networks

Vinay Randhawa[1] and Shivalika Pathania[2]

[1]Division of Cardiovascular Medicine, Brigham and Women's Hospital, Harvard Medical School, Boston, MA, United States; [2]Department of Biotechnology, Panjab University, Chandigarh, Punjab, India

1. Introduction

Molecular interactions occur between molecules of different biochemical pathways in almost every biological domain, and a cellular system can be referred to as forming molecular interaction networks, also called interactomes. Therefore, the knowledge of molecular interactions is very important for obtaining deeper mechanistic and biological insights. For example, genes and proteins regulate most of the biological mechanisms and processes by regulating each other at the molecular level. Also, regulations among biological molecules (e.g., genes, proteins, miRNAs, etc.) exist as complex regulatory networks and not in a linear one-to-one process. In contrast to gene coexpression networks (GCNs), which examine a group of coexpressed genes and make predictions about the likelihood of regulatory interactions based on the strength of the correlation, protein-protein interaction (PPI) networks give information about all physical interactions among proteins. Gene regulatory networks (GRNs), on the other hand, are also the networks inferred from gene expression data, but they have information about the directionality of gene regulation and the type of coexpression relationship, which is not determined in GCNs. For example, a directed edge in GRN connecting two genes represents a biochemical process such as a reaction, activation/inhibition, interaction, etc. The molecular processes and cell-intrinsic effects contributing to a particular phenotype are under the joint molecular regulation of genes, transcription factors (TFs) (Kashyap et al., 2009), and miRNAs (Shivdasani, 2006). Therefore, recently, there has been a deep interest in analyzing genes, miRNAs, and TFs together to solve the underlying biological problems. It has become crucial to identify these molecular regulators to well understand the various molecular mechanisms of stem cells. In this respect, various studies have been done to predict regulatory miRNAs and TFs in various biological systems. In stem cell research, the roles of molecular regulators and the complex interplay among molecular regulators to regulate biological processes/mechanisms are also well-known. For instance, miRNAs are established cell-intrinsic regulators of various processes (Heinrich & Dimmeler, 2012). These regulate stem cells either by targeting the 3′ untranslated region of pluripotency factors (Xu et al., 2009) or by targeting the coding regions of TFs for modulating stem cell differentiation (Tay et al., 2008). For example, in embryonic stem cells (ESCs), various miRNAs have been discovered that control the pluripotency, self-renewal (e.g., members of the miR-290 cluster (miR-291 to 3p), miR-294, miR-295; members of the miR-302 cluster), and differentiation (e.g., miR-296; miR-134, miR-470, miR-200c, miR-203, miR-183, miR-145) of ESCs. Besides, miRNAs also control the reprogramming in ESCs (e.g., miR-291-3p, miR-294, miR-295, miR-294, miR-290 cluster, miR-302, miR-369 cluster, miR-17-92 cluster, miR-106b-25 cluster, miR-106a-363 cluster) (Heinrich & Dimmeler, 2012). Several methodologies have been developed to predict regulatory interactions and infer respective interactomes by including information from experimental as well as computational methods. While various studies are done to understand molecular mechanisms or processes in various domains, these integrated studies are somehow limited in stem cell research. Therefore, the main aim of this chapter is to provide information on various approaches to infer and analyze the regulatory interaction networks in stem cells.

We begin this chapter by giving a brief introduction to some of the key biological networks. Next, we start focusing on detailed information on various machine learning (ML)-based methods for stem cell regulatory network prediction.

Computational Biology for Stem Cell Research. https://doi.org/10.1016/B978-0-443-13222-3.00029-0
Copyright © 2024 Elsevier Inc. All rights reserved, including those for text and data mining, AI training, and similar technologies.

Following, a detailed second section covers multiomics data integration methods for stem cell network prediction. In this section, we discussed strategies for data integration and network-based data integration methods. Additionally, in the next section, we discussed some of the most recent studies that have used bioinformatics and network-based approaches in stem cell research. We finished the section by providing the reader with a practical example where the regulatory network motifs were identified and integrated to establish a gene-TF-miRNA regulatory network in HSCs as a case study. In the conclusion and future implications section, we highlight the possibilities and challenges of inferring and understanding regulatory networks in stem cell research.

2. Machine learning methods for stem cell regulatory network prediction

ML methods have been implemented to infer GRNs, though the number of studies in this domain is still limited. ML techniques are implemented in two aspects: (i) Network inference and (ii) assessment of network characteristics. To infer regulatory networks from stem cells, ML-based approaches utilize a varying set of definitions and features. According to Patel and Wang (2015), GRN prediction methods can be categorized primarily into (i) supervised and (ii) unsupervised.

(i) Supervised algorithms: The methods developed on supervised algorithms that predict networks by considering +ve and −ve training examples. The +ve dataset comprises real interactions existing among pairs of genes, proteins, or any other biological molecule. On the other hand, the −ve dataset comprises interactions that are not happening among biological molecules. There are many supervised ML-based methods available for network prediction. One of the methods using the supervised approach for gene network prediction is multifeature relatedness (MFR). MMR is a supervised ML-based algorithm that determines how closely related two genes are by taking into account the similarities in their expression profiles and similarity based on already existing information (Wang, Yang, et al., 2019; Wang, Zhang, et al., 2019). MFR was implemented to predict interactions among genes by using interactions data gathered from the databases GeneFriends (https://www.genefriends.org/) and Database of Interacting Proteins; https://dip.doe-mbi.ucla.edu/dip/Main.cgi) for additional verification as well as to identify gene interactions taken from various databases by implementing the n-fold cross-validation strategy. The findings demonstrated that MFR was performing well (with maximum area under the curve values). The workflow of MFR comprises: (i) Gathering gene pair samples from the source databases; (ii) extracting gene features from published databases; (iii) calculating similarity-based gene pair features; (iv) building a classification model from the training dataset; and (v) using the prediction model to find gene interactions in the testing dataset. Joint density-based nonparametric differential interaction network analysis and classification is another example of a supervised method for gene network prediction. It is a neighborhood-based approach for the prediction of GRN that computes the closeness among genes based on common neighbors in genetic interaction networks (Chen et al., 2019). While these methods have been implemented for gene network prediction in general, to the best of our knowledge, they have not been implemented for inferring networks in stem cells.

(ii) Unsupervised algorithms: Unsupervised methods predict gene/protein networks based on input data only and do not require any training dataset. Unsupervised algorithms have been used in stem cell research. For example, Kusumoto and Yuasa (2019) developed a deep learning-based method that led to an automation approach that could identify cell types from phase contrast microscope images only. Also, Kusumoto et al. (2018) developed an automated method to identify endothelial cells derived from induced pluripotent stem cells (iPSCs) using a deep learning technique. To measure the conditional relatedness among genes, Wang, Feng, and Li (2019) developed a convolutional neural network using the gene coexpression profiles and existing knowledge. The developed model gained higher accuracy in predicting gene-gene interactions collected from other databases. The model was then used to establish cancer gene networks. Unsupervised learning methods are also implemented for predicting molecular regulatory networks underlying stem cells. For example, Stumpf and MacArthur (2019) used a method to establish the regulatory network patterns within individual cells from the single-cell expression data; these analyses led to the identification of (i) three distinct network configurations in cultured mouse ESCs and (ii) changes in regulatory network dynamics leading to variability in stem cells.

3. Toward multiomics data integration for stem cell networks

While each omics data type is distinctly useful in elucidating the functions of a particular cellular domain, together they compose layers of biological complexity. However, multiomics data-based networks are well-capable in dissecting various underlined mechanisms that are hard to infer from single omics data. For example, an ensemble-based network inference approach was developed to combine GRNs inferred in different studies and produce comparatively more accurate networks (Zhong et al., 2014). Iwasaki et al. (2022) Iwasaki et al. (2022) compared the global protein and mRNA levels among iPSCs/ESCs and differentiated cells to predict posttranscriptionally regulated 228 essential genes entirely in iPSCs/ESCs, which presented specific biological processes for RNA and nucleic acid binding. Zou et al. (2022) analyzed the biological effects of UVB irradiation on the HaCaT cells by analyzing their transcriptomics and proteomics profiles; their analyses revealed that UVB irradiation could affect many biological functions of the cells. In another study, a humaniPSC-based in vitro model of cardiac hypertrophy was developed by Johansson et al. (2022); in this study, the integrated tran scriptome and secretome based analyses led to the discovery of multimodal biomarkers that could help in observing disease progression.

3.1 Strategies for data integration

Data stemming from various individual omics platforms tends to be produced separately. Hence, simultaneous integration and analysis of these individual omics datasets is a challenge. Simple correlation analyses are the most common methods to predict the relationships between omics datasets and then data integration. Apart from commonly implemented general data integration methods, these methods have been well categorized from different aspects. In the continuous section, we have discussed and summarized the recent approaches in this field. Lin and Lane (2017) have summarized ML-based data integration into three major frameworks: (i) Model-based integration, (ii) concatenation-based integration, and (iii) transformation-based integration. Based on the type of data integrated, Gligorijević et al. (2015) classified data integration methods into three categories: (i) Early (or full) data integration, (ii) late (or decision) data integration, and (iii) intermediate (or partial) data integration. Details of each of these methods are summarized in a review by Randhawa and Pathania (2020), where the authors have profiled the progress achieved in the field of multi-omics integration and discussed some comprehensive tools and methods available.

3.2 Network-based data integration methods

Regulatory networks inferred by the integration of multi-omics data provide a powerful approach for identifying links among and across various layers of biological complexity. Combining multiomics data between molecular information layers (e.g., genomics, proteomics, metabolomics, etc.) generally comprises the development of network models from single and/or multiple omics data combinations. This state-of-the-art network model considers molecular interaction, network topology, and other statistical parameters to provide a more robust representation of the biological system under question. To provide more comprehensive information regarding regulatory interactions among molecules across multiple omics layers, methods for reconstructing molecular networks by linking multiple omics data, referred to as "trans-omics", have been developed; these have been reviewed in detail by Yugu et al. (2016), who classify trans-omics network reconstruction approaches into two aspects: (i) Having prior knowledge of an established empirical molecular network and (ii) not having prior knowledge of an established empirical molecular network. The former approach comprises associating various omics data layers utilizing existing information on molecular networks; in this case, the inferred network can predict kinetic models directly from reconstructed networks that help in analyzing its static and dynamic nature. Because multilayer omics data is available for only a few model organisms, this approach has limited usefulness. The latter approach computes associations between molecules across multiomics data using simple correlations. It could be applied to understand regulatory relationships among omics data for a wide range of organisms; however, since a computationally inferred network does not present a biochemical network, such networks could not be considered for the analysis of static or dynamic signal transmission across the inferred network and require experimental validation. Among the proposed "prior knowledge-based methods/classes" for connecting omics layers, the "metabolic regulation class" of methods has been used in trans-omic studies to reveal regulatory interactions among transcriptome and metabolome datasets in various species.

4. Constructing the stem cell gene coexpression networks

The best method for using transcriptomics data to produce testable hypotheses is to build a network graph based on the coexpression profiles of the genes. Finding genes with a similar expression pattern is the initial aim of expression profile analysis. For instance, genes that are up- or down-regulated in response to a certain factor, such as genotype, a human disease, or a medication. Studying the coexpression measurements is of primary biological relevance because two genes whose expression levels fluctuate together across samples may be controlled by a single regulatory mechanism. Gene expression profiling techniques concurrently monitor the levels of a large number of genes' expression, but they may not reveal much about how those genes interact with one another. coGCN, a graph where genes and edges represent nodes and the extent of coexpression relationships, respectively, is inferred to provide a global outline of transcriptional relationships among genes and provide functional and regulatory connections among genes. Weighted and unweighted GCNs and signed and unsigned GCNs are two subcategories of GCNs. The databases hosting gene expression datasets for stem cells offer a wide range of opportunities for gene coexpression-based investigations, including (i) obtaining the gene expression data and (ii) creating the GCNs. The databases Gene Expression Omnibus (GEO) (https://www.ncbi.nlm.nih.gov/geo/) and ArrayExpress (https://www.ebi.ac.uk/biostudies/arrayexpress) are two of the largest databases that host exponential datasets for biological research. Users can search these databases to fetch the gene expression profile datasets for stem cells by using specific keywords and logical operators. Additionally, stem cell-specific databases have also been established (Table 25.1). There have been several ways used to establish GCN in stem cells. However, in theory, each of these methods uses a two-step process that includes (i) calculating the coexpression and (ii) choosing a threshold before inferring the network of genes that control their expression. The first stage involves selecting a coexpression metric and computing a similarity score for each pair of genes. Edge weights in a weighted GCN indicate how strongly genes are expressed together. One of the most frequently used techniques is the Pearson correlation coefficient (PCC), which assesses the strength of the linear link between variables with normally distributed distributions. It is computed between two genes X and Y, as follows:

$$PCC_{XY} = \frac{\sum (X_i - \overline{X})(Y_i - \overline{Y})}{\sqrt{\sum (X_i - \overline{X})^2 \sum (Y_i - \overline{Y})^2}}$$

where X_i and Y_i represent the observations of the gene X and Y on the i-th sample, respectively. n is the number of samples, $\overline{X}$ and $\overline{Y}$ are the mean observations on the gene X and Y in all samples.

The second stage involves establishing a threshold and connecting gene pairs in the network via an edge if their similarity scores are greater than the chosen threshold and are thought to indicate substantial coexpression associations. Before using any correlation method, it is typically advised to identify pertinent thresholds. The PCC threshold,

TABLE 25.1 List of omics databases for stem cell research.

Database	Website link	References
StemCellDB	http://stemcelldb.nih.gov	(Mallon et al., 2013)
BCSCdb (biomarker of cancer stem cell database)	http://dibresources.jcbose.ac.in/ssaha4/bcscdb	(Firdous et al., 2022)
UESC (The urologic epithelial stem cell database)	http://scgap.systemsbiology.net/	(Pascal et al., 2007)
ESCAPE (embryonic stem cell atlas from pluripotency evidence)	http://www.maayanlab.net/ESCAPE/	(Xu et al., 2013)
HSC (hematopoietic stem cells)-Explorer	http://mips.helmholtz-muenchen.de/HSC/	(Montrone et al., 2013)
StemCellCKB(chemogenomics knowledge base for stem cell (SC) research)	https://www.cbligand.org/StemCellCKB/	(Zhang et al., 2016)
SCDevDB (single-cell developmental databbase)	https://scdevdb.deepomics.org	(Wang, Yang, et al., 2019; Wang, Zhang, et al., 2019)
StemMapper	http://stemmapper.sysbiolab.eu/	(Pinto et al., 2018)
SIStemA	http://sistema.ens-lyon.fr/	(Jarrige et al., 2021)
ScRNASeqDB (single-cell RNA-Seq database)	https://bioinfo.uth.edu/scrnaseqdb/	(Cao et al., 2017)
PSCRIdb (pluripotent stem cell regulatory interactions database)	http://bicresources.jcbose.ac.in/ssaha4/pscridb	(Banerjee et al., 2020)
Embryonic stem cell database	https://biit.cs.ut.ee/escd/	(Jung et al., 2010)
FunGenES (functional genomics in embryonic stem cells consortium)	http://www.fungenes.org	(Schulz et al., 2009)
StemBase	http://www.stembase.ca	(Porter et al., 2007)

for instance, might be chosen using previous knowledge of biological processes or by statistical comparison with randomized expression datasets. Although PCC is frequently used to compute the correlations between gene pairs, it is still advised to assess the biological correlations among genes as well. Examining the variations in the density of the gene correlation network as a function of correlation is one method for identifying biologically important correlation thresholds. Examining the variations in gene correlation network density as a function of correlation coefficient cut-off values is one method for figuring out biologically relevant correlation thresholds. For the prediction of large-scale GCN networks, several additional approaches and frameworks are available in addition to PCC. In the absence of linear correlations, measures such as Spearman's correlation coefficient, Kendall correlation coefficient, Gini correlation coefficient, Spearman's rank correlation, and biweight midcorrelation may also be taken into consideration. Bayesian networks, mutual information (MI), and probabilistic graphical models are other techniques for GCN inference. It is important to remember that MI and correlation-based approaches are frequently taken into account when building large-scale GCNs that contain more than 10,000 genes (Krouk et al., 2013). Weighted coGCN analysis (WGCNA) (Langfelder & Horvath, 2008) is one of the most widely used methods for building GCNs and has been used in numerous applications, including their implementation for establishing stem cell networks. For example, Mason et al. (2009) examined the use of unsigned and signed WGCNA to find genes in ESCs relevant to pluripotency and differentiation. Yang et al. (2019) investigated the main genes that control the development of osteogenesis in several mesenchymal stem cell (MSC) lines using WGCNA. A study by Liu et al. (2021) discovered the essential genes, biological processes, and control mechanisms for each stage of BMSC adipogenic and osteogenic development using WGCNA.

5. Constructing the stem cell regulatory networks

The ability of TFs and miRNAs to regulate one another results in the establishment of the network motif known as the feedforward loop (FFL) when both of them regulate a common target gene. According to Bo et al. (2021), FFLs are crucial regulatory units that serve as the foundation for regulatory networks and could contain 3/4/5 nodes. The 3-node FFL (composed of TF, miRNA, and their mutually regulated genes) is among the most common types of motifs in transcriptional networks. This motif type could be extended to a 4-node FFL by including a coexpressed gene as a shared molecular target of miRNA and TF. A miRNA-miRNA interaction could be added to the 4-node FFL to build a 5-node FFL. Based on the regulatory relationships among genes, miRNAs, and TFs, all FFLs can be categorized into three main categories: (i) TF-FFL, (ii) miRNA-FFL, and (iii) composite FFL. FFLs are therefore significantly important network motifs in the genome, and various of these have been identified in biological systems. For example, MYC-miR-26a-EZH2 FFL was discovered to be associated with lymphoma progression in aggressive B cell lymphomas (Zhao et al., 2013). The role of the NF-κB/STAT5/miR-155 motif in FLT3 ITD-driven acute myeloid leukemia is also well studied (Gerloff et al., 2015). FFLs have also been found to be critically important in stem cell systems. For example, miR-34a is found to be directly suppressing the protein NUMB (NUMB endocytic adapter protein) in early-stage colon cancer stem cells by establishing a FFL that targets the Notch pathway (Bu et al., 2016). In another study, it was discovered that a niche-dependent feedback loop, which involves thickveins (a type I receptor of Dpp) and serine/threonine-protein kinase fused, generates a steep gradient of bone morphogenetic protein activity and decides the fate of Drosophila ovarian germline stem cells (Xia et al., 2012). In another example, it was discovered that there is a FFL between dietary raffinose and *Lactobacillus* associated with the renewal of intestinal stem cells and epithelial repair during stress (Hou et al., 2021).

5.1 Establishing the regulatory network for hematopoietic stem cells aging: a case study

In the continuous section, we present a summary of work by Randhawa and Kumar (2021) where authors have used an integrated approach to establish HSC as an aging-related regulatory network by taking into account the interactions among genes, miRNAs, and TFs. miRNAs control lineage specification, self-renewal, and age-related diseases in HSCs. It was discovered that HSC aging could be modulated by regulating miR-125b expression (Yalcin et al., 2014). While miRNA-126 is a key regulator of HSC self-renewal potential and outcomes, miRNA-22 is an inducer of HSC maintenance and self-renewal (Mayani, 2016). Additionally, miR-125a expression improves HSCs' capacity for self-renewal and pluripotency. In addition, miR-155 may target genes that govern HSCs development into other progenitor cells (Ferreira et al., 2018). In addition, TFs also regulate stem cell aging at the cellular level. For instance, SOX2 is known to be crucial for preserving the ability of undifferentiated ESCs to self-renew (Feng & Wen, 2015), but PLZF limits enhancer activity as hematopoietic progenitors age (Poplineau et al., 2019). HSC proliferation and differentiation are crucially regulated by the transcriptional network. TF networks control several HSC functions, including cell growth (Ciau-Uitz et al., 2013), hierarchical determination of hematopoietic lineages (Iwasaki et al., 2006), and specification, differentiation, and reprogramming (Nakajima, 2011). In humans, several TFs have been discovered that control HSC states (such as SOX8, SOX18, and NFIB), HSC differentiation into MPPs (such as MYC and IKZF1), and HSC multilineage potential expansion (such as BMI1) (Mayani, 2016). Fetal HSC formation depends on BRPF1 (You et al., 2016), while Zfp90 (Liu et al., 2018) and Lhx2 (Kitajima et al., 2013) regulate differentiation in HSC and HSC-like cells, respectively. While Gfi1, Egr1, FOXOs, and PBX1 regulate HSC quiescence (Nakagawa et al., 2018), Notch is implicated in HSC emergence (Guiu et al., 2014). The age-related reduction in functions connected to modifications in gene expression also has an impact on HSCs. For instance, gene expression analyses of young and old HSCs suggested that, during aging, a total of 1600 genes were up-regulated (which were involved in pathways such as nitric oxide-mediated signaling, protein folding, and inflammatory responses) and 1500 were down-regulated (which were involved in pathways such as chromatin remodeling and repair) (Chambers et al., 2007).

5.2 Database resources for gene, miRNA, and TF interactions

Following are some of the widely used databases that host information on interactions among genes, miRNAs, and TFs.

(i) DIANA-TarBase: DIANA-TarBase (http://diana.imis.athena-innovation.gr/DianaTools) is a reference database that comprises a human-curated dataset for experimentally tested miRNA targets and information on interactions for various organisms such as humans, mice, fruit flies, worms, and zebrafish.
(ii) ChIPBase: ChIPBase (https://rna.sysu.edu.cn/chipbase/) is a resource having transcriptional regulation information for RNAs such as ncRNAs, lincRNAs, microRNAs, ncRNAs, and genes.
(iii) miRTarBase: miRTarBase (https://mirtarbase.cuhk.edu.cn) is a database of miRNA-target interactions. The miRNA-target interactions are manually collected from the literature after data mining.

5.3 Establishing the network

Once the information of candidate genes, miRNAs, and TFs is known, the information of regulatory interactions among these candidates can be retrieved from public databases to establish the global interaction regulatory network. The overall approach is to initially identify the FFLs, followed by creating the miRNA-TF-gene (MTG) regulatory network by integrating the identified FFLs. Various studies have used FFLs to infer transcriptional regulatory networks (Wang et al., 2017; Wang et al., 2013). The MTG regulatory network was predicted as follows (Fig. 25.1):

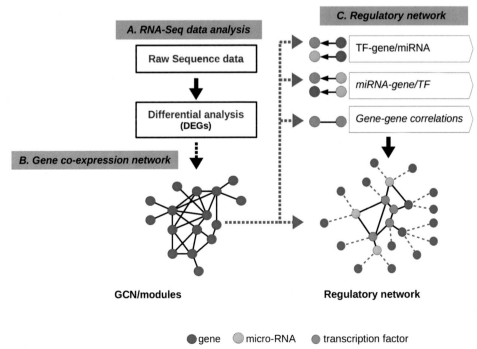

FIGURE 25.1 The pipeline implemented for establishing a regulatory network in HSCs.

(i) Identifying candidate aging-associated genes and modules: Candidate mouse HSC-aging-related genes were obtained by differential expression studies of HSC-aging-related transcriptomics studies. Next, the human homologs (orthologs) of DEGs from mice were predicted.

(ii) Identifying coGCN modules: The RNA-seq dataset of HCS aging was analyzed to obtain a gene expression matrix that was first used to establish the HSC aging-related GCN. Afterward, the large GCN was fragmented into a large number of small-sized significant gene modules (clusters). The WGCNA tool was used for these analyses.

(iii) Finding potential TFs and miRNAs and the related regulatory connections: The significant gene modules were next enriched to include information on regulatory miRNAs and TFs. From the DIANA-TarBase database, information on regulatory miRNAs and their related gene-miRNA interactions was obtained. Also, from the ChIPBase v2.0 database, information on TFs that regulate both module genes and miRNAs (from DIANA-TarBase) as well as their connections (TF-gene/miRNA) was found. The combined study of DIANA-TarBase and miRTarBase v7 predicted miRNA-gene/TF regulatory connections that have been empirically validated.

(iv) Finding significant FFLs: Once all the interaction information was retrieved, all interaction pairs obtained were integrated to predict FFLs, the network motif in which miRNA and TF jointly regulate a target gene (Zhang et al., 2015). To evaluate the significance of FFLs, a random permutation approach was used to simultaneously permutate TFs and miRNAs (Lin et al., 2015). The nonsignificant network loops (p-value <0.05) were removed.

(v) Regulatory network inference: By integrating significant composite FFLs, a comprehensive MTG regulatory network was predicted for HSCs aging. The statistical significance of the MTG network was evaluated by the network randomization approach. Important TFs and miRNAs were prioritized by computing their shortest path lengths with important genes.

6. Studies using integrated bioinformatics and network-based approaches in stem cells research

Shi et al. (2018) implemented an integrated bioinformatics approach to understand hypoxia's effect on neural stem cells (NSCs). In their approach, gene expression profiles of hypoxic and normoxic NSCs were compared to identify DEGs, and by associating the data with PPIs, important genes, pathways, and molecular mechanisms were discovered. In another study, He et al. (2021) analyzed the gene expression of long noncoding RNAs and genes of olfactory mucosa MSCs (OM-MSCs) in hypoxic and normoxic states by bioinformatics and interaction network analysis; this analysis identified that the latter condition encouraged the inhibition of apoptosis in OM-MSCs. Khodabandehloo et al. (2020) analyzed the gene expression profiles of adult bone marrow-derived MSCs to identify key hub genes and linked these hub genes with the Wnt pathways enriched from the PPIs; the findings from their analysis provided valuable information on their association with osteogenesis and also their roles in the PI3K/AKT and Wnt pathways. Zou et al. (2009) combined information from PPI information and overexpressed genes from human ESCs (hESCs) to establish a protein interaction network; this analysis shed new light on how hESC-enriched proteins might interact with other proteins in hESCs. The analysis discovered NF−Y as an important TF that regulates many highly connected proteins in hESCs. Xiao et al. (2020) performed a gene coexpression analysis study to predict gene modules related to bone marrow MSC differentiation; this integrated study identified TFs ZNF740 and FOS and two hub genes, FOXQ1 and SGK1, as potential novel biomarkers.

7. Conclusion and future perspectives

Recently, there has been an increased trend towards implementing ML and multiomics data integration for understanding biological systems. While many studies have been done to understand molecular mechanisms/processes in various domains using interaction networks, these integrated studies are somehow limited in stem cell research. Over the last few years, there has been a great advancement in stem cell research, which has led to the availability of vast omics datasets. This provides an opportunity to integrate and analyze these datasets from an interactomics perspective. This chapter is focused on enhancing the current knowledge about various approaches to first inferring and then analyzing the regulatory interaction networks in stem cells. We gave a succinct summary of the most recent methods used for regulatory network functional analyses and inference in stem cell research, including briefly discussing a pipeline to infer the regulatory network of stem cells starting from gene expression profiles. The current area is constantly developing new tools and techniques to comprehend the complexity of GRNs. However, due to the regular evolution of data and thus the methods to analyze the complex data, this book chapter presents a whistle-stop tour of regulatory stem cell network biology. As a whole, this chapter provides a thorough introduction to the approaches used in the contemporary postgenomic age to infer and analyze stem cell regulatory networks.

Acknowledgments
SP acknowledges the support of University Grants Commission (UGC), New Delhi, India, for Dr. D.S. Kothari postdoctoral fellowship (UGC Award Letter No. F.4-2/2006(BSR)/BL/18-19/0277) and Department of Biotechnology, Panjab University, Chandigarh for providing computational infrastructure.

Authors' contributions
All authors contributed equally to this manuscript.

References

Banerjee, K., Jana, T., Ghosh, Z., & Saha, S. (2020). PSCRIdb: A database of regulatory interactions and networks of pluripotent stem cell lines. *Journal of Biosciences, 45*(1). https://doi.org/10.1007/S12038-020-00027-4

Bo, C., Zhang, H., Cao, Y., Lu, X., Zhang, C., Li, S., Kong, X., Zhang, X., Bai, M., Tian, K., Saitgareeva, A., Lyaysan, G., Wang, J., Ning, S., & Wang, L. (2021). Construction of a TF–miRNA–gene feed-forward loop network predicts biomarkers and potential drugs for myasthenia gravis. *Scientific Reports, 11*(1), 1–15. https://doi.org/10.1038/s41598-021-81962-6

Bu, P., Wang, L., Chen, K. Y., Srinivasan, T., Murthy, P. K. L., Tung, K. L., Varanko, A. K., Chen, H. J., Ai, Y., King, S., Lipkin, S. M., & Shen, X. (2016). A miR-34a-Numb feed-forward loop triggered by inflammation regulates asymmetric stem cell division in intestine and colon cancer. *Cell Stem Cell, 18*(2), 189. https://doi.org/10.1016/J.STEM.2016.01.006

Cao, Y., Zhu, J., Jia, P., & Zhao, Z. (2017). scRNASeqDB: A database for RNA-Seq based gene expression profiles in human single cells. *Genes, 8*(12). https://doi.org/10.3390/GENES8120368

Chambers, S. M., Shaw, C. A., Gatza, C., Fisk, C. J., Donehower, L. A., & Goodell, M. A. (2007). Aging hematopoietic stem cells decline in function and exhibit epigenetic dysregulation. *PLoS Biology, 5*(8), e201. https://doi.org/10.1371/journal.pbio.0050201

Chen, H., He, Y., Ji, J., & Shi, Y. (2019). A machine learning method for identifying critical interactions between gene pairs in Alzheimer's disease prediction. *Frontiers in Neurology, 10.* https://doi.org/10.3389/FNEUR.2019.01162

Ciau-Uitz, A., Wang, L., Patient, R., & Liu, F. (2013). ETS transcription factors in hematopoietic stem cell development. *Blood Cells, Molecules, and Diseases, 51*(4), 248–255. https://doi.org/10.1016/j.bcmd.2013.07.010

Feng, R., & Wen, J. (2015). Overview of the roles of Sox2 in stem cell and development. *Biological Chemistry, 396*(8), 883–891. https://doi.org/10.1515/hsz-2014-0317

Ferreira, A. F., Calin, G. A., Picanço-Castro, V., Kashima, S., Covas, D. T., & de Castro, F. A. (2018). Hematopoietic stem cells from induced pluripotent stem cells – considering the role of microRNA as a cell differentiation regulator. *Journal of Cell Science, 131*(4), jcs203018. https://doi.org/10.1242/jcs.203018

Firdous, S., Ghosh, A., & Saha, S. (2022). BCSCdb: A database of biomarkers of cancer stem cells. *Database,* 1–8. https://doi.org/10.1093/DATABASE/BAAC082, 2022.

Gerloff, D., Grundler, R., Wurm, A. A., Bräuer-Hartmann, D., Katzerke, C., Hartmann, J.-U., Madan, V., Müller-Tidow, C., Duyster, J., Tenen, D. G., Niederwieser, D., & Behre, G. (2015). NF-κB/STAT5/miR-155 network targets PU.1 in FLT3-ITD-driven acute myeloid leukemia. *Leukemia, 29*(3), 535–547. https://doi.org/10.1038/leu.2014.231

Gligorijević, V., & Pržulj, N. (2015). Methods for biological data integration: Perspectives and challenges. *Journal of the Royal Society Interface, 12*(112), 20150571. https://doi.org/10.1098/rsif.2015.0571

Guiu, J., Bergen, D. J. M., De Pater, E., Islam, A. B. M. M. K., Ayllón, V., Gama-Norton, L., Ruiz-Herguido, C., González, J., López-Bigas, N., Menendez, P., Dzierzak, E., Espinosa, L., & Bigas, A. (2014). Identification of Cdca7 as a novel Notch transcriptional target involved in hematopoietic stem cell emergence. *Journal of Experimental Medicine, 211*(12), 2411–2423. https://doi.org/10.1084/jem.20131857

He, J., Huang, Y., Liu, J., Ge, L., Tang, X., Lu, M., & Hu, Z. (2021). Hypoxic conditioned promotes the proliferation of human olfactory mucosa mesenchymal stem cells and relevant lncRNA and mRNA analysis. *Life Sciences, 265*, 118861. https://doi.org/10.1016/J.LFS.2020.118861

Heinrich, E.-M., & Dimmeler, S. (2012). MicroRNAs and stem cells. *Circulation Research, 110*(7), 1014–1022. https://doi.org/10.1161/CIRCRESAHA.111.243394

Hou, Y., Wei, W., Guan, X., Liu, Y., Bian, G., He, D., Fan, Q., Cai, X., Zhang, Y., Wang, G., Zheng, X., & Hao, H. (2021). A diet-microbial metabolism feedforward loop modulates intestinal stem cell renewal in the stressed gut. *Nature Communications, 12*(1), 1–15. https://doi.org/10.1038/s41467-020-20673-4

Iwasaki, M., Kawahara, Y., Okubo, C., Yamakawa, T., Nakamura, M., Tabata, T., Nishi, Y., Narita, M., Ohta, A., Saito, H., Yamamoto, T., Nakagawa, M., Yamanaka, S., & Takahashi, K. (2022). Multi-omics approach reveals posttranscriptionally regulated genes are essential for human pluripotent stem cells. *iScience, 25*(5). https://doi.org/10.1016/J.ISCI.2022.104289

Iwasaki, H., Mizuno, S., Arinobu, Y., Ozawa, H., Mori, Y., Shigematsu, H., Takatsu, K., Tenen, D. G., & Akashi, K. (2006). The order of expression of transcription factors directs hierarchical specification of hematopoietic lineages. *Genes and Development, 20*(21), 3010–3021. https://doi.org/10.1101/gad.1493506

Jarrige, M., Polvèche, H., Carteron, A., Janczarski, S., Peschanski, M., Auboeuf, D., & Martinat, C. (2021). Sistema: A large and standardized collection of transcriptome data sets for human pluripotent stem cell research. *iScience, 24*(7), 102767. https://doi.org/10.1016/J.ISCI.2021.102767

Johansson, M., Ulfenborg, B., Andersson, C. X., Heydarkhan-Hagvall, S., Jeppsson, A., Sartipy, P., & Synnergren, J. (2022). Multi-omics characterization of a human stem cell-based model of cardiac hypertrophy. *Life, 12*(2), 293. https://doi.org/10.3390/LIFE12020293

Jung, M., Peterson, H., Chavez, L., Kahlem, P., Lehrach, H., Vilo, J., & Adjaye, J. (2010). A data integration approach to mapping OCT4 gene regulatory networks operative in embryonic stem cells and embryonal carcinoma cells. *PLoS One, 5*(5). https://doi.org/10.1371/JOURNAL.PONE.0010709

Kashyap, V., Rezende, N. C., Scotland, K. B., Shaffer, S. M., Persson, J. L., Gudas, L. J., & Mongan, N. P. (2009). Regulation of stem cell pluripotency and differentiation involves a mutual regulatory circuit of the NANOG, OCT4, and SOX2 pluripotency transcription factors with polycomb repressive complexes and stem cell microRNAs. *Stem Cells and Development, 18*(7), 1093–1108. https://doi.org/10.1089/scd.2009.0113

Khodabandehloo, F., Taleahmad, S., Aflatoonian, R., Rajaei, F., Zandieh, Z., Nassiri-Asl, M., & Eslaminejad, M. B. (2020). Microarray analysis identification of key pathways and interaction network of differential gene expressions during osteogenic differentiation. *Human Genomics, 14*(1), 1–13. https://doi.org/10.1186/S40246-020-00293-1/TABLES/2

Kitajima, K., Kawaguchi, M., Iacovino, M., Kyba, M., & Hara, T. (2013). Molecular functions of the LIM-homeobox transcription factor Lhx2 in hematopoietic progenitor cells derived from mouse embryonic stem cells. *Stem Cells, 31*(12), 2680–2689. https://doi.org/10.1002/stem.1500

Krouk, G., Lingeman, J., Colon, A. M., Coruzzi, G., & Shasha, D. (2013). Gene regulatory networks in plants: Learning causality from time and perturbation. *Genome Biology, 14*(6), 123. https://doi.org/10.1186/gb-2013-14-6-123

Kusumoto, D., Lachmann, M., Kunihiro, T., Yuasa, S., Kishino, Y., Kimura, M., Katsuki, T., Itoh, S., Seki, T., & Fukuda, K. (2018). Automated deep learning-based system to identify endothelial cells derived from induced pluripotent stem cells. *Stem Cell Reports, 10*(6), 1687–1695. https://doi.org/10.1016/J.STEMCR.2018.04.007

Kusumoto, D., & Yuasa, S. (2019). The application of convolutional neural network to stem cell biology. *Inflammation and Regeneration, 39*(1). https://doi.org/10.1186/S41232-019-0103-3

Langfelder, P., & Horvath, S. (2008). WGCNA: An R package for weighted correlation network analysis. *BMC Bioinformatics, 9*, 559. https://doi.org/10.1186/1471-2105-9-559

Lin, E., & Lane, H.-Y. (2017). Machine learning and systems genomics approaches for multi-omics data. *Biomarker Research, 5*, 2. https://doi.org/10.1186/s40364-017-0082-y

Lin, Y., Zhang, Q., Zhang, H.-M., Liu, W., Liu, C.-J., Li, Q., & Guo, A.-Y. (2015). Transcription factor and miRNA co-regulatory network reveals shared and specific regulators in the development of B cell and T cell. *Scientific Reports, 5*(1), 15215. https://doi.org/10.1038/srep15215

Liu, T., Kong, W., Tang, X., Xu, M., Wang, Q., Zhang, B., Hu, L., & Chen, H. (2018). The transcription factor Zfp90 regulates the self-renewal and differentiation of hematopoietic stem cells. *Cell Death and Disease, 9*(6), 677. https://doi.org/10.1038/s41419-018-0721-8

Liu, Y., Tingart, M., Lecouturier, S., Li, J., & Eschweiler, J. (2021). Identification of co-expression network correlated with different periods of adipogenic and osteogenic differentiation of BMSCs by weighted gene co-expression network analysis (WGCNA). *BMC Genomics, 22*(1). https://doi.org/10.1186/S12864-021-07584-4

Mallon, B. S., Chenoweth, J. G., Johnson, K. R., Hamilton, R. S., Tesar, P. J., Yavatkar, A. S., Tyson, L. J., Park, K., Chen, K. G., Fann, Y. C., & McKay, R. D. G. (2013). StemCellDB: The human pluripotent stem cell database at the national Institutes of Health. *Stem Cell Research, 10*(1), 57–66. https://doi.org/10.1016/J.SCR.2012.09.002

Mason, M. J., Fan, G., Plath, K., Zhou, Q., & Horvath, S. (2009). Signed weighted gene co-expression network analysis of transcriptional regulation in murine embryonic stem cells. *BMC Genomics, 10*. https://doi.org/10.1186/1471-2164-10-327

Mayani, H. (2016). The regulation of hematopoietic stem cell populations. *F1000Research, 5*, F1000. https://doi.org/10.12688/f1000research.8532.1. Faculty Rev-1524.

Montrone, C., Kokkaliaris, K. D., Loeffler, D., Lechner, M., Kastenmüller, G., Schroeder, T., & Ruepp, A. (2013). HSC-explorer: A curated database for hematopoietic stem cells. *PLoS One, 8*(7), e70348. https://doi.org/10.1371/JOURNAL.PONE.0070348

Nakagawa, M. M., Chen, H., & Rathinam, C. V. (2018). Constitutive activation of NF-κB pathway in hematopoietic stem cells causes loss of quiescence and deregulated transcription factor networks. *Frontiers in Cell and Developmental Biology, 6*, 143. https://doi.org/10.3389/fcell.2018.00143

Nakajima, H. (2011). Role of transcription factors in differentiation and reprogramming of hematopoietic cells. *Keio Journal of Medicine, 60*(2), 47–55. http://www.ncbi.nlm.nih.gov/pubmed/21720200.

Pascal, L. E., Deutsch, E. W., Campbell, D. S., Korb, M., True, L. D., & Liu, A. Y. (2007). The urologic epithelial stem cell database (UESC) - a web tool for cell type-specific gene expression and immunohistochemistry images of the prostate and bladder. *BMC Urology, 7*. https://doi.org/10.1186/1471-2490-7-19

Patel, N., & Wang, J. T. L. (2015). Semi-supervised prediction of gene regulatory networks using machine learning algorithms. *Journal of Biosciences, 40*(4), 731–740. https://doi.org/10.1007/S12038-015-9558-9/METRICS

Pinto, J. P., Machado, R. S. R., Magno, R., Oliveira, D. V., Machado, S., Andrade, R. P., Bragança, J., Duarte, I., & Futschik, M. E. (2018). Stem-Mapper: A curated gene expression database for stem cell lineage analysis. *Nucleic Acids Research, 46*(D1), D788–D793. https://doi.org/10.1093/NAR/GKX921

Poplineau, M., Vernerey, J., Platet, N., N'guyen, L., Hérault, L., Esposito, M., Saurin, A. J., Guilouf, C., Iwama, A., & Duprez, E. (2019). PLZF limits enhancer activity during hematopoietic progenitor aging. *Nucleic Acids Research, 47*(9), 4509–4520. https://doi.org/10.1093/nar/gkz174

Porter, C. J., Palidwor, G. A., Sandie, R., Krzyzanowski, P. M., Muro, E. M., Perez-Iratxeta, C., & Andrade-Navarro, M. A. (2007). StemBase: A resource for the analysis of stem cell gene expression data. *Methods in Molecular Biology, 407*, 137–148. https://doi.org/10.1007/978-1-59745-536-7_11/ COVER

Randhawa, V., & Kumar, M. (2021). An integrated network analysis approach to identify potential key genes, transcription factors, and microRNAs regulating human hematopoietic stem cell aging. *Molecular Omics, 17*(6), 967–984. https://doi.org/10.1039/D1MO00199J

Randhawa, V., & Pathania, S. (2020). Advancing from protein interactomes and gene co-expression networks towards multi-omics-based composite networks: Approaches for predicting and extracting biological knowledge. *Briefings in Functional Genomics, 19*(5–6), 364–376. https://doi.org/ 10.1093/bfgp/elaa015

Schulz, H., Kolde, R., Adler, P., Aksoy, I., Anastassiadis, K., Bader, M., Billon, N., Boeuf, H., Bourillot, P. Y., Buchholz, F., Dani, C., Doss, M. X., Forrester, L., Gitton, M., Henrique, D., Hescheler, J., Himmelbauer, H., Hübner, N., Karantzali, E., … Hatzopoulos, A. K. (2009). The FunGenES database: A genomics resource for mouse embryonic stem cell differentiation. *PLoS One, 4*(9). https://doi.org/10.1371/JOURNAL.PONE.0006804

Shivdasani, R. A. (2006). MicroRNAs: Regulators of gene expression and cell differentiation. *Blood, 108*(12), 3646–3653. https://doi.org/10.1182/blood-2006-01-030015

Shi, Z., Wei, Z., Li, J., Yuan, S., Pan, B., Cao, F., Zhou, H., Zhang, Y., Wang, Y., Sun, S., Kong, X., & Feng, S. (2018). Identification and verification of candidate genes regulating neural stem cells behavior under hypoxia. *Cellular Physiology and Biochemistry, 47*(1), 212–222. https://doi.org/10.1159/ 000489799

Stumpf, P. S., & MacArthur, B. D. (2019). Machine learning of stem cell identities from single-cell expression data via regulatory network archetypes. *Frontiers in Genetics, 10*. https://doi.org/10.3389/FGENE.2019.00002

Tay, Y., Zhang, J., Thomson, A. M., Lim, B., & Rigoutsos, I. (2008). MicroRNAs to Nanog, Oct4 and Sox2 coding regions modulate embryonic stem cell differentiation. *Nature, 455*(7216), 1124–1128. https://doi.org/10.1038/nature07299

Wang, Z., Feng, X., & Li, S. C. (2019). SCDevDB: A database for insights into single-cell gene expression profiles during human developmental processes. *Frontiers in Genetics, 10*. https://doi.org/10.3389/FGENE.2019.00903

Wang, D., Haley, J. D., & Thompson, P. (2017). Comparative gene co-expression network analysis of epithelial to mesenchymal transition reveals lung cancer progression stages. *BMC Cancer*, 1–12. https://doi.org/10.1186/S12885-017-3832-1

Wang, J., Peng, X., Li, M., & Pan, Y. (2013). Construction and application of dynamic protein interaction network based on time course gene expression data. *Proteomics, 13*(2), 301–312. https://doi.org/10.1002/pmic.201200277

Wang, Y., Yang, S., Zhao, J., Du, W., Liang, Y., Wang, C., Zhou, F., Tian, Y., & Ma, Q. (2019). Using machine learning to measure relatedness between genes: A multi-features model. *Scientific Reports, 9*(1), 1–15. https://doi.org/10.1038/s41598-019-40780-7

Wang, Y., Zhang, S., Yang, L., Yang, S., Tian, Y., & Ma, Q. (2019). Measurement of conditional relatedness between genes using fully convolutional neural network. *Frontiers in Genetics, 10*. https://doi.org/10.3389/FGENE.2019.01009/FULL

Xiao, B., Wang, G., & Li, W. (2020). Weighted gene correlation network analysis reveals novel biomarkers associated with mesenchymal stromal cell differentiation in early phase. *PeerJ, 2020*(4), e8907. https://doi.org/10.7717/PEERJ.8907/SUPP-10

Xia, L., Zheng, X., Zheng, W., Zhang, G., Wang, H., Tao, Y., & Chen, D. (2012). The niche-dependent feedback loop generates a BMP activity gradient to determine the germline stem cell fate. *Current Biology, 22*(6), 515–521. https://doi.org/10.1016/J.CUB.2012.01.056

Xu, H., Baroukh, C., Dannenfelser, R., Chen, E. Y., Tan, C. M., Kou, Y., Kim, Y. E., Lemischka, I. R., & Ma'ayan, A. (2013). *Escape: Database for integrating high-content published data collected from human and mouse embryonic stem cells*. Oxford): Database. https://doi.org/10.1093/ DATABASE/BAT045, 2013.

Xu, N., Papagiannakopoulos, T., Pan, G., Thomson, J. A., & Kosik, K. S. (2009). MicroRNA-145 regulates OCT4, SOX2, and KLF4 and represses pluripotency in human embryonic stem cells. *Cell, 137*(4), 647–658. https://doi.org/10.1016/j.cell.2009.02.038

Yalcin, S., Carty, M., Shin, J. Y., Miller, R. A., Leslie, C., & Park, C. Y. (2014). Microrna mediated regulation of hematopoietic stem cell aging. *Blood, 124*(21). http://www.bloodjournal.org/content/124/21/602?sso-checked=true.

Yang, W., Xia, Y., Qian, X., Wang, M., Zhang, X., Li, Y., & Li, L. (2019). Co-expression network analysis identified key genes in association with mesenchymal stem cell osteogenic differentiation. *Cell and Tissue Research, 378*(3), 513–529. https://doi.org/10.1007/S00441-019-03071-1

You, L., Li, L., Zou, J., Yan, K., Belle, J., Nijnik, A., Wang, E., & Yang, X.-J. (2016). BRPF1 is essential for development of fetal hematopoietic stem cells. *Journal of Clinical Investigation, 126*(9), 3247–3262. https://doi.org/10.1172/JCI80711

Yugi, K., Kubota, H., Hatano, A., & Kuroda, S. (2016). Trans-Omics: How to reconstruct biochemical networks across multiple 'omic' layers. *Trends in Biotechnology, 34*(4), 276–290. https://doi.org/10.1016/j.tibtech.2015.12.013

Zhang, H.-M., Kuang, S., Xiong, X., Gao, T., Liu, C., & Guo, A.-Y. (2015). Transcription factor and microRNA co-regulatory loops: Important regulatory motifs in biological processes and diseases. *Briefings in Bioinformatics, 16*(1), 45–58. https://doi.org/10.1093/bib/bbt085

Zhang, Y., Wang, L., Feng, Z., Cheng, H., McGuire, T. F., Ding, Y., Cheng, T., Gao, Y., & Xie, X. Q. (2016). StemCellCKB: An integrated stem cell-specific chemogenomics KnowledgeBase for target identification and Systems-pharmacology research. *Journal of Chemical Information and Modeling, 56*(10), 1995–2004. https://doi.org/10.1021/ACS.JCIM.5B00748

Zhao, X., Lwin, T., Zhang, X., Huang, A., Wang, J., Marquez, V. E., Chen-Kiang, S., Dalton, W. S., Sotomayor, E., & Tao, J. (2013). Disruption of the MYC-miRNA-EZH2 loop to suppress aggressive B-cell lymphoma survival and clonogenicity. *Leukemia, 27*(12), 2341–2350. https://doi.org/ 10.1038/leu.2013.94

Zhong, R., Allen, J. D., Xiao, G., & Xie, Y. (2014). Ensemble-based network aggregation improves the accuracy of gene network reconstruction. *PLoS One, 9*(11), e106319. https://doi.org/10.1371/journal.pone.0106319

Zou, X., Zou, D., Li, L., Yu, R., Li, X. H., Du, X., Guo, J. P., Wang, K. H., & Liu, W. (2022). Multi-omics analysis of an in vitro photoaging model and protective effect of umbilical cord mesenchymal stem cell-conditioned medium. *Stem Cell Research and Therapy, 13*(1), 435. https://doi.org/10.1186/S13287-022-03137-Y

Zuo, C., Liang, S., Wang, Z., Li, H., Zheng, W., & Ma, W. (2009). Enriching protein-protein and functional interaction networks in human embryonic stem cells. *International Journal of Molecular Medicine, 23*(6), 811−819. https://doi.org/10.3892/IJMM_00000197

Computational approaches for stem cell tissue engineering

Chapter 26

Tissue engineering in chondral defect

Madhan Jeyaraman[1], Arulkumar Nallakumarasamy[1], Naveen Jeyaraman[1] and
Swaminathan Ramasubramanian[2]

[1]Department of Orthopaedics, ACS Medical College and Hospital, Dr. MGR Educational and Research Institute, Chennai, Tamil Nadu, India;
[2]Department of Orthopaedics, Government Medical College, Omandurar Government Estate, Chennai, Tamil Nadu, India

1. Introduction

Articular cartilage is the highly specialized tenacious and load-bearing connective tissue of hyaline type, which requires tedious tissue homeostasis for long-term survivability. The developed cartilage lacks its regenerative capacity since it lacks neochondrogenesis due to its intrinsic avascularity, alymphatic, and the dormancy of the naïve underlying stem cells (Sophia Fox et al., 2009). Nevertheless, these tissues in the joints are unable to completely regenerate on their own when get affected due to trauma, inflammation, or aging. Once the hyaline cartilage is damaged, it gets repaired with fibrous cartilage of poor mechanical stability of suboptimal clinical utility (Eschweiler et al., 2021; Sophia Fox et al., 2009). The intercellular communication and the paracrine signaling of the administered mesenchymal stem cells (MSCs) and biological therapies standardize the self-regulating process of traumatized or degenerating cartilage tissue (Fernández-Francos et al., 2021). Computational cartilage tissue engineering integrates biology, engineering, and computer science to create approaches for restoring damaged cartilage. This field employs computational models and simulations to assist in tasks such as scaffold design, analysis of chondrocyte behavior, and optimization of bioreactors. The process of cartilage tissue engineering entails introducing stem cells into specially designed scaffolds under controlled conditions. By aligning with the complex signaling mechanisms of immature chondrocytes, this approach enables effective interactions between proteins and extracellular matrices, ultimately promoting cartilage regeneration (Jelodari et al., 2022).

2. Cartilage structure and signaling

The undifferentiated MSCs in the lateral plate mesoderm undergo the condensation process and thereafter get differentiated into chondrocytes. The interaction between signaling molecules and chondrocytes leads to the production of extracellular matrices, which are essential for the development of a cartilage environment (Akkiraju & Nohe, 2015; Deng et al., 2021). The formed epiphyseal cartilage niche undergoes endochondral ossification, which delineates the developing joint. The histological layers of the formed cartilage are the superficial zone, intermediate zone, deep layer (basal layer), tidemark, and subchondral bone (Sophia Fox et al., 2009). The structural organization of the extracellular matrix (water—which forms 75% of total weight of cartilage, collagen, and proteoglycans) decreases the friction and further dissipates the stress across the joint. Thus, 75% of the articular cartilage's total weight is water (Gao et al., 2014).

The comprehensive modularity of the dynamic stress load—bearing articular cartilage exhibits a characteristic biomechanical property (Eschweiler et al., 2021; Landínez-Parra et al., 2012). It depends on the structural integrity of the extracellular matrix interposed with the type II collagenous network. Chondroitin and keratan sulfate are the important glycosaminoglycans (GAGs) found anchored in the large glycosylated proteoglycans (mainly aggrecan and hyaluronic acid) and maintain the shear and compressive stress over the articular cartilage (Boschetti et al., 2004). It constitutes 95% of the total ECM; other small core glycosylated protein molecules are decorin, biglycan, and perlecan. The sulfates in the GAGs of the proteoglycans are negatively charged, which maintains the cartilage well hydrated. Further, it attracts the cations such as Ca^{2+}, Na^+, and K^+ and regulates the electrophysiology of the cartilage within the matrix (Poole et al., 2001; Sophia Fox et al., 2009).

Computational Biology for Stem Cell Research. https://doi.org/10.1016/B978-0-443-13222-3.00033-2
Copyright © 2024 Elsevier Inc. All rights are reserved, including those for text and data mining, AI training, and similar technologies.

In the articular cartilage, the basic functional unit is the chondrocyte, which is a nonexcitable cell surrounded by a pericellular matrix. Connexin and pannexin are the transmembrane hemichannel proteins that play a significant role in the stimulus-evoked ATP release from the chondrocytes. It maintains the hyperpolarization state of the endoplasmic reticulum, and it determines the secretion, migration, and proliferation of the cartilage (Cai et al., 2020; Iwamoto et al., 2013). TRPV4 channels (responsible for Ca^{2+} influx) have been highly evaluated for their sensitivity to osmotic modularity and dynamic loading of the cartilage (Phan et al., 2009).

The inherent avascular cartilaginous template, in its hypoxic environment, gets primed by the hypoxia-inducible factors (HIFs) for their survivability. Many biologic molecules and growth factors are involved in intercellular communication and chondrocyte physiology in articular cartilage development. In vitro, various common molecules such as transforming growth factor-β (TGF-β) subfamily members, bone morphogenetic proteins (BMPs), insulin-like growth factors (IGFs), fibroblast growth factors (FGFs), vascular endothelial growth factor (VEGF), perlecan, cartilage-specific β-catenin, Indian hedgehog protein (IHH), and sex-determining region Y (SRY)-box (SOXs) are employed to stimulate chondrogenesis (Cai et al., 2022; Chen et al., 2021; Li & Dong, 2016; Zhao et al., 2020). Extensive research has revealed the role of signaling molecules and their associated signal transduction pathways in engineering cartilage, emphasizing the importance of repurposing cues from cartilage development and advancing our understanding of growth and transcription factors for effective cartilage tissue engineering.

3. Signaling in tissue engineering

Cell sources and biochemical/mechanophysiology stimuli represent pillars of cartilage tissue engineering (Somoza et al., 2014; Tan et al., 2021). Pericytes, perivascular stem cells, bone marrow−derived mesenchymal stem cells (BM-MSCs), adipose-derived MSCs, embryonic stem cells (ESCs), Wharton's jelly stem cells, and induced pluripotent stem cells (iPSCs) are all promising cell sources for articular cartilage repair and regeneration (de Windt et al., 2017; Goldberg et al., 2017; Hanxiang Le, 2020; Zha et al., 2021). A variety of intercellular signaling molecules have been applied in cartilage tissue engineering to trigger chondrogenic differentiation and to stimulate the synthesis of the cartilage-specific matrix (Kwon et al., 2016; Li, Liu, et al., 2020; Li, Gao, et al., 2020). The following subsections review the role of signaling molecules in various cell types used in cartilage tissue engineering. They cover (1) signaling factors for cell expansion and differentiation, as shown in Fig. 26.1, and (2) biochemical molecules and biophysical agents to improve biomechanical properties and ECM components in engineered articular cartilage, depicted in Fig. 26.2.

4. Pathophysiology of chondral defects and degeneration

Softening, microtears, fibrillations, and fissuring of the articular cartilage can occur as a result of acute trauma, repetitive chronic stress with suboptimal impact, altered metabolism, genetic abnormalities, or vascular damage to the affected area (Chubinskaya et al., 2015; Jones et al., 2004). It is a well-known fact that articular cartilage has limited regenerative potential and untreated lesions lead to early degeneration. The impaction force of more than 24 MPa will alter the mechanophysiology of the naïve cartilage leading to posttraumatic arthritis (PTA) (Punzi et al., 2016). Following the impact, there is an upregulation of matrix metalloproteinases (MMPs 1, 3, and 13) with thrombospondin motifs (ADAMTS 4 and 5), aggrecanase (disintegrin), and sulfated glycosaminoglycans (S-GAGs) released by the damaged chondrocytes. This process is accompanied by an increased release of inflammatory mediators such as interleukin (IL)-1, -6, and tumor necrosis factor (TNF)-α, leading to chondrocyte necrosis and apoptosis (Punzi et al., 2016). Nevertheless, the imbalance between the ECM synthesis and the ongoing deterioration decreases the chondrocyte per cubic mm of the articular area paving the way for early cartilage damage and degeneration (Punzi et al., 2016). In this hostile environment, the electrophysiology of the chondrocyte gets altered (defective Na-K ATPase pump activity), and decreased synthesis of proteoglycans leads to a poorly hydrated matrix (Akkiraju & Nohe, 2015; Martin & Buckwalter, 2001; Mobasheri et al., 2012). Thus, the defective matrix architecture with minimal water content fails to withstand the shear and compressive forces acting on it over a period leading to early disease progression (Eschweiler et al., 2021). The lesion can be categorized into (1) partial thickness and (2) full thickness with subchondral involvement.

5. Management of chondral injuries

Untreated chondral defects have been established to be a major risk factor for early degeneration leading to osteoarthritis (OA) (Houck et al., 2018). Currently, no pharmacotherapy or surgical interventions have shown proven efficacy in halting disease progression.

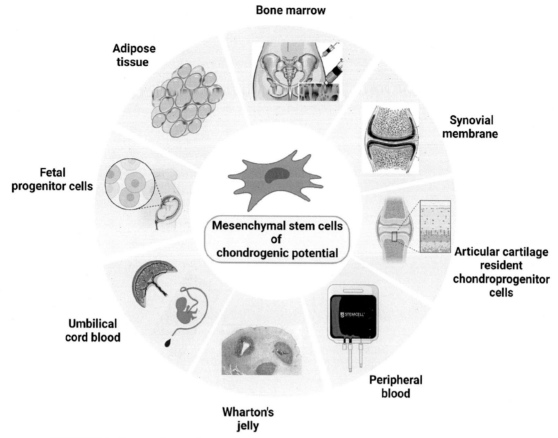

FIGURE 26.1   Sources of mesenchymal stem cells with chondrogenic potential. *Created with Biorender.com.*

6. Diagnosis of chondral defects

Most of the cases are asymptomatic at the early stage of presentation. However, symptomatic patients present with pain, swelling, clicking/popping, limited movements, and locking. Though the symptoms are nonspecific, provocative tests such as Wilson's test can be utilized, which is used to diagnose the osteochondritis dissecans in the medial femoral condyle.

Plain X-rays are seldom useful to diagnose chondral lesions in the early stage. However, it is possible to screen the joint for osteophytes, subchondral cysts, sclerosis, and loose bodies, which are indirect evidence of sequelae of chondral lesions (Dallich et al., 2019).

Computed tomography arthrography (CTA) using iodinated contrast can be performed in patients where MRI is contraindicated. Lee et al. observed that the sensitivity and specificity of MRI and CTA for the detection of acetabular chondral lesions were 36% and 84%, respectively, and 46% and 72%, respectively (Lee et al., 2019). Naraghi and White found that magnetic resonance arthrography (MRA) has a high negative predictive value for diagnosing chondral lesions (Naraghi & White, 2015). T2 CartiGram and MRI-delayed gadolinium-enhanced magnetic resonance imaging of cartilage (dGEMRIC) can be used to detect chondral lesions with improved sensitivity, specificity, and accuracy (Bekkers et al., 2013). Elevated levels of biomarkers such as fibronectin–aggrecan complex (FAC) in the synovial fluid, cartilage oligomeric matrix protein (COMP) in the plasma, may assist in diagnosing the chondral lesions before radiological involvement (El-Arman et al., 2010; Posey et al., 2018; Tseng et al., 2009).

7. Management of chondral defects

Various treatment options for the management of chondral lesions include the following:

(A) **Mechanical:** Chondroplasty, abrasion, drilling, microfracture (Erggelet & Vavken, 2016). Microfracture and subchondral drilling techniques often result in inferior fibrocartilage with suboptimal biomechanical properties.

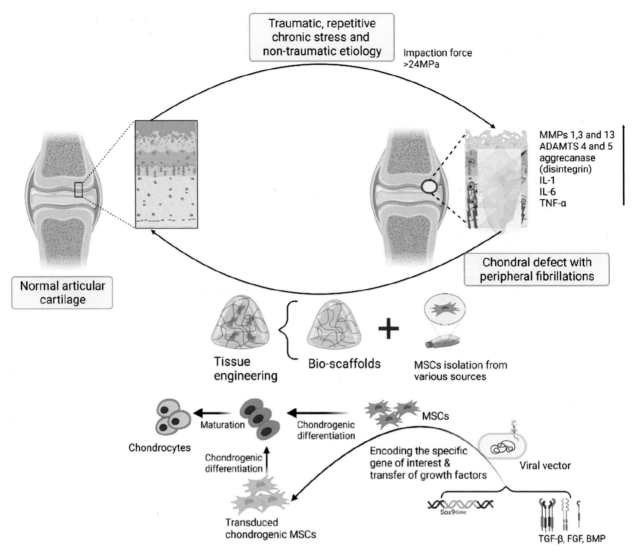

FIGURE 26.2 Schematic representation of computational tissue engineering in chondral defects. *Created with Biorender.com.*

(B) Cartilage transfer procedures: Mosaicplasty, osteochondral autograft transfer system (OATS) (Memon & Quinlan, 2012). The osteochondral autograft/allografting technique has a limitation wherein the size of the cartilage defect restricts its applicability, and the use of autografts leads to donor site morbidity.

(C) Allogenic osteochondral transplantation: Allogenic osteochondral transplantation can be used to manage larger defects without donor site morbidity. However, the demerits include graft rejection, graft resorption, and chondrocyte apoptosis (Dean et al., 2016).

(D) Autologous chondrocyte implantation

a. First generation: Autologous chondrocyte implantation (ACI) consists of two surgical stages: harvesting chondrocytes from a healthy donor area and culturing them under specific conditions, followed by implantation of the cultured and multiplied chondrocytes at the recipient site. The implanted chondrocytes are protected by securing a layer of periosteum around the lesion edges (Brittberg, 2021).

b. Second generation: In second-generation autologous chondrocyte transplantation, a collagen membrane is utilized instead of a periosteum, eliminating the requirement for an additional incision to extract it and avoiding potential periosteal hypertrophy (Brittberg, 2021).

c. Third generation: Third-generation ACI, called matrix-autologous chondrocyte implantation (MACI), uses culture-expanded chondrocytes embedded in a hydrated scaffold. It prevents chondrocyte leakage, treats larger defects, and ensures proper chondrocyte distribution (Niethammer et al., 2021).

d. Fourth generation: Advanced ACI techniques utilize scaffolds composed of collagen, chondroitin sulfate, and natural biopolymers. These scaffolds are infused with chondrogenic MSCs from diverse sources and supplemented with growth factors including TGF-β, IGF, VEGF, and platelet-rich plasma (Kim et al., 2022).

8. Molecular methods of chondral defects management

8.1 Genetic transfer-mediated stem cell therapy

In the past two decades, scientists have made significant progress in exploring the efficacy of MSCs in cartilage repair. Gene therapy can be directly implicated in cartilage tissue repair; however, transduced chondrogenic MSCs deliver more regenerative potential (Steinert et al., 2008). Encoding the specific gene of interest in the MSCs enhances the chondrogenic phenotype expression, trophism, maturation, proliferation, and delivery of growth factors (Lefebvre et al., 2019). Sox9 is a pivotal transcription factor that directly enhances collagen II expression by transcribing the enhancer region of the col2a1 gene, facilitated through interaction with the Barx2 homeobox transcription factor. This regulatory mechanism operates in both differentiating MSCs and chondrocytes (Akiyama et al., 2002; Haseeb et al., 2021; Lefebvre et al., 2019). Further, the induced transcription factors minimize the premature dedifferentiation process and early senescence. It includes (1) identifying the ideal MSCs that have proper chondrogenic differentiation efficiency, (2) identifying the anabolic growth factors (TGF-beta, IGF-I, BMPs, and FGF-2) that stimulate the maturation and proliferation of the differentiated chondrocytes, and (3) identifying the ideal vector for gene transfer with high transduction potential, minimal cytotoxicity, and immunological reactions.

Transfections based on liposome-based systems are highly challenging, because of their cytotoxicity and poor DNA delivery to the target cell (Bardania et al., 2017; Sercombe et al., 2015). Later, Im et al. found the transfection process and the efficacy of the transfected adipose-derived MSCs cells can be enhanced by incorporating the chondrogenic plasmid DNA containing SOX-trio genes into the porous polylactic co-glycolic acid (PLGA) scaffold biopolymer (Im et al., 2011).

In addition, another perspective of efficient transgene delivery (transduction) can be done by using adenovirus, recombinant adeno-associated virus (rAAV), retrovirus, and lentiviral vectors. Madry et al. suggested the usage of rAAV-mediated transduction since it is less immunogenic and is associated with a reduced risk of insertional mutagenesis (Madry et al., 2003).

Gelse et al. observed that the stimulation of perichondrium-derived mesenchymal cells by transfer of growth factor (BMP, IGF-1) cDNA using adenovirus in a partial-thickness defect model allows for satisfactory cartilage restoration by a repair tissue comparable with hyaline articular cartilage (Gelse et al., 2003). Denker et al. observed that the transduction of human BMP-2 cDNA using replication-defective retroviral vector demonstrates improved chondrogenesis of the multipotential murine embryonic C3H10T1/2 mesenchymal cell line in a high-density micromass environment (Denker et al., 1999).

The efficacy of the gene delivery by viral and nonviral vectors to the MSCs can be quantified by using green fluorescence protein (GFP) reporter gene expression (Kumar et al., 2013; Soboleski et al., 2005). As insights into MSC regulation advance, it becomes possible to clinically repair injured articular and growth plate cartilage. This can be achieved through ex vivo—expanded or genetically modified MSC transplantation, or by mobilizing endogenous MSCs from adjacent source tissues such as synovium, bone marrow, or trabecular bone.

8.2 Human adipose tissue—derived mesenchymal stem cells

Human adipose tissue—derived mesenchymal stem cells (hAMSCs) are progenitor cells obtained from adipose tissue, possess multilineage differentiation, and can regenerate various biological tissues. ASCs can potentially be isolated from subcutaneous adipose tissue of various regions of the body such as the lower abdomen, thoracic back thigh, arm, and infrapatellar fat pad (Sharma et al., 2021). Different anatomic locations have varying ASCs characteristics; it has been shown that the thoracic back and lower abdomen are rich in ASCs mainly because the adipose tissue in these areas is rich in fibrous connective tissue (Zwick et al., 2018).

Felimban et al. characterized the chondrogenic potential of the human infrapatellar fat pad (hIPFP)—derived MSCs. They were found to have chondrocyte lineage markers such as SOX-9, aggrecan, and collagen type II alpha 1. The growth factors released (TGF beta-3, and BMP-6) from hIPFP are highly chondrogenic, and they yield a hyaline matrix in micromass cultures (Felimban et al., 2014).

Adipose tissue has a distinct type of cellular morphology—both unilocular large and multiple small lipid vacuoles can perform the physiological functions of both white (storing energy) and brown adipose tissue (thermogenesis) (Richard et al., 2000). Adipose cells have enormous mitochondria and express high levels of UCP-1 (Choe et al., 2016). Adipose tissue has been harvested using a resection technique, syringe, pump-assisted liposuction, tumescent liposuction, and ultrasound-assisted liposuction (Bellini et al., 2017; Francesco et al., 2019). Mojallal et al. observed the impact of power-assisted liposuction with negative pressure of about −350 mmHg on the yield of SVF cells and gave satisfactory results when compared with non−power-assisted liposuction (Mojallal et al., 2008).

After harvesting, the tissue is minced with collagenase type II, which is centrifuged to yield an abundant population of multipotent adipose tissue-derived stem cells (ASCs) called stromal vascular fraction (SVF) (Aronowitz et al., 2015; Oberbauer et al., 2015; Senesi et al., 2019). The nonhomogenous population was assayed for the presence of different stem and progenitor cell types using colony formation assays including fibroblast and alkaline phosphatase-positive colony-forming units. Further, the flow cytometry analyses depict ASCs showing a $CD31^-/CD34^+$ after the serial passage of culture can have the greatest proliferation and differentiation potential (Forghani et al., 2020). Greenwood et al. observed that microfragmented adipose tissue (MFAT) preserves the cellular integrity and morphology within an intact extracellular environment, resulting in superior biological potential potentially over enzymatically released stromal vascular fraction (Greenwood et al., 2022). ASCs secrete multiple growth factors basic fibroblast growth factor (bFGF), vascular endothelial growth factor (VEGF), insulin-like growth factor 1, hepatocyte growth factors (HGF), platelet-derived growth factor (PDGF), epidermal growth factor (EGF), and transforming growth factor (TGF)-β (Danišovič et al., 2012).

These factors facilitate chondrogenic differentiation of the MSCs and maintenance of the differentiated cell in that specific lineage. However, the heterogeneity of the derived cell phenotype may affect their chondrogenic potential. Recently, Kabiri et al. found that the serial culture passage containing FGF in optimal concentrations regulates various signaling cascades (increased mRNA levels of Sox-9 and N-cadherin) of chondrogenic differentiation (Kabiri et al., 2012). Tissue engineering of ADSCs with bioactive factors such as platelet-rich plasma (PRP), L-ascorbic acid (LAA), and biocompatible silk fibroin scaffold provides improved cell adhesion, proliferation, and differentiation within 21 days of culture (Patel et al., 2019).

Pak et al. proposed a clinical utility protocol for the isolation and administration of ADSCs-SVF for cartilage regeneration, which was approved by the Korean Food and Drug Administration (KFDA). It includes the composite containing ADSCs, and SVF with the extracellular matrix is injected into the desired site along with HA and $CaCl_2$. The results showed improved clinic-radiological outcomes with the formation of hyaline cartilage (Pak, Lee, Pak, Pak, et al., 2018; Pak, Lee, Pak, Park, et al., 2018).

Li et al. demonstrated the healing of an artificially created chondral defect in a rabbit model using cocultured ADSC and chondrocytes, which were impregnated on a novel TGF-β1/poly(3-hydroxybutyrate-co-4-hydroxybutyrate) (P3HB4HB) scaffold. It showed better proliferation, migration, and differentiation of the chondrocytes (Li, Fu, et al., 2015; Li, Huang, et al., 2015). In another study by Ba et al. (2019) the addition of SVF to the cocultured ASCs and chondrocytes showed a better proliferation rate. The composite cocultured cells are seeded on a 1:1 ratio over polyhydroxy butyrate (PHB)/poly-hydroxybutyrate-co-hydroxyhexanoate scaffold material, which showed better adhesion and proliferation.

Kabiri et al. demonstrated that the expression levels of SOX-9, collagen type II, and aggrecan were all significantly increased in hADSCs expanded in presence of FGF-2 (Kabiri et al., 2012). Our results suggest that FGF-2 induces hADSCs chondrogenesis in Transwell culture, which may be beneficial in cartilage tissue engineering.

8.3 Bone marrow aspirate concentrate

The functional component of the marrow constitutes hematopoietic precursors, marrow adipocytes, fibroblasts, macrophages, progenitor cells, and stromal cells. The stem cells present in the marrow are quantitatively insufficient for the therapeutic purpose of cartilage regeneration (Brozovich et al., 2021; El-Jawhari et al., 2020).

Common bone marrow aspiration sites are anterior or posterior superior iliac spine/crest regions (Chahla et al., 2017). The MSCs count in the bone marrow aspirate concentrate (BMAC) may vary depending on the site of harvest, aspiration technique, the syringe volume, and the quantity of the aspirate (Jeyaraman et al., 2022). The concentrations of MSCs are higher under ideal conditions using a 10 mL syringe than a 50 mL syringe. The aspirate ranges from 60 to 120 mL and is subjected to a density gradient centrifugation process (soft and hard spin of about 2400 rpm:15 min and 3600 rpm:10 min respectively), which concentrates the buffy coat layer with mononuclear cells and a sufficient amount of MSCs with a proportional increase in chondrogenic progenitor cells (Jeyaraman et al., 2022).

BMAC implantation into chondral defect appears to be an easy and less invasive modality. Centeno et al. documented the increased cartilage thickness and volume of the meniscus after injecting autologous BM-MSCs but failed to document whether the regenerated cartilage is either hyaline or fibrocartilage (Centeno et al., 2008). Wong et al. injected cultured BM-MSCs along with high tibial osteotomy and microfractures for varus knees with cartilage defects and reported statistically significant MOCART scores at the end of 1 year (Wong et al., 2013). However, the injected BM-MSCs bind to the synovial membrane around the knee joint and produce synovial hypertrophy. Hence the delivery of MSCs into desired chondral lesions was directed appropriately with the help of external magnetic beads coupled with an external magnetic force (Ikuta et al., 2015; Kobayashi et al., 2008).

Clinical evidence reported regeneration of hyaline cartilage in chondral defects and osteoarthritis with implantation of BM-MSCs through paracrine effects and chondrocyte differentiation and proliferation of MSCs (Cotter et al., 2018). Intraarticular injection of BM-MSCs of 10^8 cells improves pain and functional outcome in patients with cartilage defects by hyaline-like cartilage regeneration (Jo et al., 2014). Significant improvement has been observed in the treatment of large osteochondral lesions (>4 cm^2) using BM-MSCs compared with MSCs obtained from peripheral blood (Skowroński & Rutka, 2013).

In the goat model, Saw et al. reported excellent functional outcomes with subchondral drilling along with HA or BMAC/HA injection in cartilage repair histologically than subchondral drilling alone. At the end of 2 years follow-up, the animals were sacrificed to assess the quality of the cartilage tissue, which revealed the superior quality of cartilage formed in the BMAC group according to Gill scoring, which demonstrated hyaline-like cartilage structure (Saw et al., 2009). In the equine model, Fortier et al. (2023) compared the functional outcomes of microfracture and microfracture with BMAC augmentation in a 15-mm cartilage defect along the lateral trochlear ridge. In second-look arthroscopy at 3 months, macroscopy, histology, and MRI were performed to analyze the quality of cartilage regenerated. They observed superior quality of hyaline-like cartilage with enhanced type 2 collagen expression and enhanced integration of cartilage-like tissue along the cartilage defect in the BMAC-augmented group.

Uncultured BM-MSCs from BMAC may not be sufficient enough to enhance cartilage regeneration. Various researchers expanded BM-MSCs from BMAC to attain millions of BM-MSCs to obtain the desired cartilage regenerate (Goodrich et al., 2016; Mahmoud et al., 2019). Mahmoud et al. reported a significant better histological appearance of hyaline-like cartilage tissue at 1 and 3 months to follow up in 1.25 and 6.25 million BM-MSCs than 0.125 million BM-MSCs with an enhanced amount of hyaline cartilage content. These findings suggest that higher concentrations of MSCs produce a robust prochondrogenic effect (Mahmoud et al., 2019). In a minipig full-thickness chondral defect model, Jung et al. reported an enhanced GAG and type 2 collagen expression in MACI augmented with culture-expanded BM-MSCs (Jung et al., 2009). Microfracture augmented with BM-MSCs and HA compared with microfracture and HA in 1 cm^2 condylar defects of the stifle joint in the equine model revealed enhanced levels of aggrecan content and tissue integrity and firmness without any clinical or histological differences in the cartilage regenerate at the end of 12 months (McIlwraith et al., 2011).

Combining BM-MSCs with autologous platelet-rich fibrin (PRF) scaffold improved cartilage repair in a 15 mm full-thickness chondral defect equine model. However, after 1 year, no significant differences were found compared with the isolated PRF scaffold in terms of arthroscopy, MR T2 cartilage mapping, histology, regenerated tissue integrity, and firmness. Both treatment groups had lower proteoglycan contents than normal cartilage (Goodrich et al., 2016). Amid contradictory preclinical evidence on cartilage tissue regeneration, it is crucial to explore different scaffolds and tissue engineering methods. The goal is to achieve equivalent biomechanical stability and high-quality regeneration.

8.4 Synovial mesenchymal stem cells

The two different sources of synovial MSCs are synovial membrane (SM) and synovial fluid (SF). The SM is made of the outer fibrous and inner cellular layers. Type A or macrophage-like synoviocytes and type B fibroblast-like (mesenchyme—FLS) synoviocytes are abundant in the inner cellular layer made of thin connective tissue along with blood vessels. Synovial MSCs have higher multidirectional differentiation potential, proliferation rate, low immunogenicity, and high clonogenic tendency (Jeyaraman et al., 2021).

Nakagawa et al. in the experimental study found the synovial MSCs might get delivered from the perivascular channels, originating from the bone marrow, which passes through the synovial vessels and finally reaches the synovial membrane and fluid (Nakagawa et al., 2015). Sakaguchi et al. demonstrated the efficacy of synovial MSCs for chondrogenesis. In this study, the harvested MSCs were subjected to a density gradient and the nucleated cells were plated without interference. They analyzed MSCs yield, expandability, differentiation potential, and epitope profile of synovial origin; in addition, they compared it with other sources such as bone marrow, adipose tissue, periosteum, and skeletal muscle. The progenitor synovial cells are rich in collagen type II and aggrecans. The fact is evident that the metaplasia of synovial tissue forms a loose cartilage body called synovial chondromatosis, which was observed to be hyaline in nature (Sakaguchi et al., 2005).

Sekiya et al. prepared culture-expanded synovial MSCs with zero passage in autologous human serum for 14 days. The production of passage 0 cells is cost-effective, and it reduces the risk of chromosomal aberrations. Synovial MSCs express

CD44 (hyaluronic acid receptor), CD29, CD 73, CD 105, CD106, CD31 (platelet endothelial cell adhesion molecule—PECAM-1), VCAM-1 (surface glycoprotein), STRO1 (osteoblast precursor cells), and reduced CD10 level, which fulfill the International Society for Cell Therapy's (ISCT) criteria for human MSCs (Jeyaraman et al., 2021; Sekiya et al., 2021).

Roelofs et al. stated that the synovial MSCs originated from the embryonic joint interzone (Roelofs et al., 2017). Later Kurth et al. confirmed the fact that synovial tissue has an identical source of origin as that of articular cartilage (Kurth et al., 2011). The implanted synovial MSCs differentiated to form a cartilaginous matrix in the artificially created chondral model (Jeyaraman et al., 2021; Li, Liu, et al., 2020). Sekiya et al. studied the efficacy of the chondrogenic potential of synovial MSCs in 10 patients with chondral defects involving the femoral condyle. The culture-expanded synovial MSC suspension was transplanted into the defect with a syringe under arthroscopic guidance. All the patients were followed up for 3 years and showed satisfactory clinic-radiological outcomes. Second-look arthroscopy was performed on four patients; the biopsy taken from the regenerated tissue has been studied histologically, which showed the formation of mature hyaline cartilage (Sekiya et al., 2021). Since it is less invasive than mosaicplasty and autologous chondrocyte implantation, the repair of chondral defects using synovial MSCs gained considerable interest in regenerative medicine.

8.5 Articular cartilage resident chondroprogenitor cells

A promising unique population of progenitor cells in the superficial zone of the naïve cartilage is called articular cartilage resident chondroprogenitor cells (ACPCs) (Vinod et al., 2018; Xu et al., 2022). The characteristic expression of universal MSC markers with multilineage differentiation, clonogenicity, migration, and antiapoptotic potential makes ACPCs the preferable cell source for cartilage regeneration. The markers used to distinguish ACPCs from MSCs and chondrocytes are COL2, CD146, CD166, RAB3B, FZD7, laminin, integrinα5β1, Jagged, Notch1,2, differential adhesion to fibronectin, and STRO1 (De Luca et al., 2019; Vinod et al., 2018). However, none of the markers were found to be specific. A low-glucose, low-density two-dimensional (2DLL) culture medium is essential for phenotype conversion of the adult chondrocytes, a phenomenon called reversed stemness (differentiated chondrocytes into progenitor-like cells—dedifferentiation) (Jiang et al., 2016).

Numerous studies have demonstrated that the growth factor differentiation factor (GDF5[+]) at the interzone layer and PRG4+ (proteoglycan 4) cells at the superficial zone release ACPCs for cartilage development and growth in the antenatal and postnatal period respectively (Sun et al., 2021; Takahata et al., 2022). Recently, it was found that creb5 was the transcription factor for the expression of TGFβ, IGF1, and EGFR growth factor signaling molecules to stimulate PRG4+ expression on the superficial zone of the articular cartilage (Zhang et al., 2021).

Seol et al. (2014) found that factor such as high mobility group protein B1 (HMGB1), and IGF1, influences the release of ACPCs at the injured site within 7—14 days, whereas Decker et al. (2017) observed that the makers of PRG4+ cells were found in the injured site within 7 days in the mice model. Levato et al. found that cultured ACPCs impregnated on 3D hydrogel scaffolds yield more COLII, PRG4+, and less COL X hypertrophy markers than other sources of MSCs, which facilitates early neochondrogenesis in an equine model (Levato et al., 2017).

Jiang et al. conducted a pilot study on 15 patients with chondral defects (range, 6—13 cm^2) in the distal femur managed with ACPCs-impregnated collagen scaffolds. This study evaluated the effectiveness of cartilage repair using chondroprogenitor cells in vivo, and the cells were harvested from the non—weight-bearing area of the trochlea. The harvested cells were cultured in a suitable medium (100—300 cells per cm^2 in a low glucose medium as a monolayer culture). Further, the cells were quantified and assessed for multilineage differentiation and colony formation. Histological examination of the biopsy tissue taken during second-look arthroscopy showed that the regenerated tissue possesses a hyaline cartilage-like structure and matrix. All patients showed significant improvement in IKDC and Lysholm scores within a year of the index procedure (Jiang et al., 2016).

Wang et al. found that the cartilage resident stem cells migrate into the biopolymers scaffolds such as PLGA, facilitating cartilage repair. It further showed that the scaffolds incorporated with culture-expanded CPSCs can get released from the scaffold into the artificially created chondral defect of the animal model (Wang et al., 2014). The concept of dedifferentiation, the ability to resume their chondrogenic phenotype brings new insights into cartilage biology and the development of chondrocyte-based cellular therapies.

8.6 Human umbilical cord blood and Wharton's jelly—derived mesenchymal stem cells

The human umbilical cord blood—derived MSCs (hUCB-MSCs) are frequently used as a source of stem cells because they are readily available, and easy to collect and store (Marino et al., 2019). Nevertheless, pluripotent stem cells can also be derived from the perivascular mucoid connective tissue called Wharton's jelly. Both the cells of different origins maintain immunological homeostasis, and they do not require a human leukocyte antigen (HLA) matching, which reduces the graft-versus-host reactions (Marino et al., 2019). They are highly compatible and considered the safe mode of allogeneic stem cell therapy for the recipients. They can self-renewal and differentiate into various lineages of somatic cell types, including adipocytes, osteoblasts, chondrocytes, myoblasts, hepatocytes, and neurocytes. Two methods are commonly employed to isolate MSCs from the umbilical cord connective tissue, the explant, and enzymatic digestion methods (Kim et al., 2013; Lee et al., 2022).

They originated from the extraembryonic mesoderm, which suggests they are highly primitive. Further, it shows consistent proliferation even after a series of 30 passages and possesses a higher multiplication rate (faster cell-doubling time) than the MSCs isolated from other sources such as adipose tissue, bone marrow, and so on. The frequency of colony-forming units-fibroblasts (CFU-F) assay measures the differentiation and proliferation ability of the individual cells, which are significantly higher in umbilical-derived stem cells (Kim et al., 2013; Kim et al., 2010).

Jeong et al. reported that hUCB-MSCs release the thrombospondin-2 factor, which exerts its paracrine effect on the endogenous chondroprogenitor cells; further, it hastens the differentiation. The available stimulatory factors in the synovial fluid enhance maturation and proliferation (Jeong et al., 2013). The chondrogenic differentiation potential of these progenitor cells using safranin O staining marks the specific proteoglycans. It was also found that the mRNA expression of collagen II, a higher amount of aggrecan, and GAGs production on 14, 21, and 42 days of the differentiation process. Site-specific injection of hUCB-MSCs increases the local release of IL-10, whereas it downregulates the production of other inflammatory markers TNF-α, IL-1, IL-6, MMP-9, and INF-γ. The immunoregulatory effect creates a salutary environment for chondral regeneration (Russo et al., 2022; Sadlik et al., 2017).

Chung et al. confirmed that there was an improved healing rate of the artificially created chondral defect treated with hUCB-MSCs-based hydrogel scaffolds in rat models (Chung et al., 2014). Kao et al. demonstrated the chondrogenesis of hUCB-MSCs incorporated in a chemically synthesized thermoreversible gelation polymer (TGP) supplemented with ascorbic acid and TGFβ3. The differentiated cells in the TGP scaffold enhance the production of specific mRNA of collagen type II, aggrecan, glucosaminoglycans, and Sox9, which facilitates early cartilage repair (Kao et al., 2008).

South Korean research group successfully conducted a large clinical trial evaluating the regenerative potential of hUCB-MSCs on animal models. Later, they conducted a human clinical trial using culture-expanded hUCB-MSCs incorporated on HA hydrogel scaffolds (Cartistem) on patients with articular degenerative conditions. The results of the study at 1-year follow-up showed satisfactory clinic-radiological improvement, and histological analyses of the regenerative tissue were found to be hyaline-like cartilage with high GAG content (Park et al., 2017).

8.7 Peripheral blood-derived MSCs

Peripheral blood-derived MSCs (PB-MSCs) are the potential alternative source of MSC for cartilage repair, which demonstrated a similar chondrogenic differentiation with BM-MSCs in both in vitro and in vivo studies (Muthu et al., 2021). Mononuclear cells are mobilized to peripheral blood with the help of subcutaneous injection of GM-CSF. With the rise in the leukocyte count, peripheral blood is collected, and mononuclear cells are isolated and subjected to culture expansion; hence, PB-MSCs are obtained, which can be utilized for various clinical applications (ouryazdanpanah et al., 2018). These PB-MSCs possess similar surface markers, and their procurement and isolation appear to be less invasive and safe than other sources of MSCs (Li, Fu, et al., 2015; Li, Huang, et al., 2015). The manipulation of PB-MSCs possesses no risk, no immunological rejection, and no ethical issues. MSCs derived from bone marrow, adipose tissue, or synovium reach the confluence in 7 days (Jin et al., 2016; Shimomura et al., 2017; Zhang et al., 2014), whereas MSCs derived from peripheral blood reach the confluence in 21 days after the primary culture (Chen et al., 2019), hence limiting the usage of PB-MSCs for various clinical applications.

The presence of MSCs in human PB is debatable due to the presence of MSC in human PB at very low frequency and biological variations in terms of age, sex, and disease status of the individual (Moll et al., 2019). Chen et al. reviewed the safety, efficacy, and feasibility of PB-MSCs for cartilage repair and regeneration in vivo by analyzing the functional outcomes and the adverse events associated with the peripheral blood source of MSCs (Chen et al., 2020). The noncultured PB-MSCs comprise hematopoietic stem cells, fibrocytes, a population of MSCs, WBCs, growth factors, cytokines, and a minimal population of RBCs. Hopper et al. reported upregulation of eight genes associated with chondrogenic differentiation, enhanced number of MSCs, and enhanced naïve chondrocyte migration (Hopper, Henson, et al., 2015; Hopper, Wardale, et al., 2015).

Skowroński et al. demonstrated an inferior quality of cartilage regeneration in the BMAC group when compared with the PB-MSC group by inferring that the number of MSC in BMAC appears to be low when compared with PB-MSC (Skowroński & Rutka, 2013). To increase the yield of MSC counts in autologous PB sources, repeated intraarticular injections were given in a few studies (Saw et al., 2011, 2013, 2015; Turajane et al., 2013). Cultured-expanded allogenic PB-MSCs are not recommended due to immune recognition after the second injection of allogenic PB-MSCs (Joswig et al., 2017). The optimal therapeutic dose of PB-MSCs in cartilage regeneration have to be standardized with large-scale blinded controlled clinical trials.

Saw et al. demonstrated a hyaline-like cartilage in second-look arthroscopy in five patients with chondral defects managed with intraarticular injections of scaffold loaded (HA) with PB-MSCs along with arthroscopic subchondral drilling. Saw et al. studied histologic and MRI evaluation of cartilage regeneration in grade 3 and 4 chondral lesions managed by arthroscopic subchondral drilling alone and arthroscopic subchondral drilling along with PB-MSCs + HA augmentation in the postoperative period. Subjective IKDC scores and follow-up MRI scans were performed periodically. Total ICRS II histological scores, MRI morphological scores, and 2-year IKDC scores were superior in arthroscopic subchondral drilling along with PB-MSCs + HA augmentation group (Saw et al., 2013). Saw et al. demonstrated hyaline-like cartilage regeneration followed by arthroscopic subchondral drilling followed by postoperative IA injection of PB-MSCs + HA with concomitant medial wedge HTO in varus deformity of the knee joint (Saw et al., 2015).

Chen et al. (2020) reported that the autologous, nonculture-expanded PB-MSCs produce hyaline-like cartilage in chondral defects. In preclinical studies, autologous, allogenic, and xenogenic forms of PB-MSCs are utilized for assessing cartilage regeneration in chondral defects and osteoarthritis. The utilization of PB-MSCs further warrants the improvisation of technologies used for mobilizing and purifying the progenitor cells or mononuclear cells to the periphery. Though technically PB-MSCs are safe and less invasive, the clinical evidence in human studies is very limited. Hence, large-scale blinded and controlled trials are required to prove the similar or superior results of PB-MSCs in terms of cartilage regeneration when compared with other forms of MSCs.

8.8 Scaffolds in cartilage tissue engineering

Scaffolds form an integral part of tissue engineering. An ideal scaffold is a 3D and porous template that assists tissue regeneration. Scaffold forms a mechanical support and framework where cells survive, thrive, and make up the target tissue. It acts as a building block that helps fill the tissue gaps. In cartilage tissue engineering, the scaffold must be soft, pliable, and flexible like a naïve cartilage (Hutmacher, 2000; Moutos & Guilak, 2008). The types of scaffolds are natural (chitosan, collagen, silk fibroin, hyaluronan, alginate, agarose, fibrin, and gelatin) (Zhao et al., 2021), synthetics (poly-ethylene glycol [PEG], polycaprolactone [PCL], polylactic acid [PLA], polyglycolic acid [PGA], polyurethane, and PLGA) (Wasyłeczko et al., 2020), and composite (growth factors, embedded fibers, and textiles, embedded solid struc-tures, multilayered designs, and woven composite scaffold) (Moutos & Guilak, 2008) varieties. The various forms of scaffolds are hydrogels, nanofiber matrices, and decellularized ECM.

Scaffolds are used to preset MSCs in 3D circumstances. The 3D scaffold provides a surface for cell adhesion, dif-ferentiation, proliferation, and maturation of cartilage. Various in vitro, preclinical, and clinical studies have demonstrated the use of scaffold-based tissue engineering for cartilage regeneration. hWJ-MSCs cultured with chondrogenic differen-tiation medium for 21 days demonstrate an enhanced transcriptional activity of type 2 collagen gene in 2D or 3D PLGA scaffold. Expressions of type 1 and 3 collagen were reduced in 3D. Hence, MSCs in 3D scaffolds play a vital role in hyaline cartilage regeneration (Paduszyński et al., 2016). Yamagata et al. cultured MSCs with cell growth medium in 2D monolayer or 3D PLGA plug scaffold, which demonstrated the upregulation of SOX-9 gene expression in the 3D PLGA plug group, which indicates enhanced chondrogenesis (Yamagata et al., 2018).

Hydrogel loaded with chondrocyte along with type 1 and 2 collagen helps to maintain chondrocyte morphology and secrete chondrocyte-specific ECM (Yuan et al., 2016). Toyokawa et al. (2010) implanted PLGA without MSCs in an osteochondral defect in rabbit model, which displayed an adequate cover of cartilaginous tissue with a satisfactory repair. This infers that endogenous MSC adheres to PLGA scaffold and enhances articular cartilage repair. Sonomoto et al. (2016) reported spontaneous differentiation of MSCs in a PLGA medium. In a rabbit model with cartilage defect, Qi et al. revealed the integration of BM-MSC sheet and BM-MSCs-loaded bilayer PLGA scaffold in the cartilage defect forms a hyaline-like cartilage (Qi et al., 2014).

In an ovine stifle joint model with chronic osteochondral defect, Marquass et al. reported chondrogenic differentiation in vitro when cultured with autologous MSC seeded in collagen 1 hydrogel. At the end of 1 year, pre-MSC hydrogels display better histological, O'Driscoll, and ICRS scores (Marquass et al., 2011). Morille et al. (2016) induced the chon-drogenic differentiation with PLGA-P188-PLGA-based microspheres loaded with MSCs in pathological OA environment. Bouffi et al. demonstrated cartilage formation with MSC loaded with microcarriers releasing TGF-β3 (Bouffi et al., 2010).

A chondromimetic hyaluronan microsphere loaded with TGF-β3 in hMSC medium display an enhanced accumulation of GAGs and proteoglycans (Ansboro et al., 2014). Enhanced chondrogenesis was observed with gelatin-loaded microsphere containing TGF-β3 along with MSC coculture. At the end of 1 month, MSCs were uniformly distributed in the pellet with excellent cellular viability and counts. Engineered cells exhibit a faster proliferation rate along with enhanced production of GAGs, type 2 collagen, and proteoglycans (Fan et al., 2008). Bhang et al. revealed the additive effects of TGF-β3 and HA on the chondrogenic differentiation of hMSCs. They demonstrated an increased expression of chondrogenic mediators (Bhang et al., 2011). The treating physician or surgeon must have a wide knowledge of the usage of appropriate scaffolds for cartilage disorders.

9. Computational tissue engineering

The advent of computational models and three-dimensional (3D) printing has become a critical aspect of scaffold development progression. The complex interplay of stem cells, derived from a variety of sources, demands the incorporation of appropriate scaffolding to engender specific tissue structures, thus augmenting anatomical and functional outcomes. Computational modeling's adoption in the discipline of tissue engineering paves the way to scrutinize cellular reactions to different scaffolds and explore intercellular communication before the application in physical settings. This preliminary assessment not only conserves invaluable time and resources but also yields substantial economic savings. This strategy cultivates an environment that encourages the acceleration of research and efficient use of resources, embodying a revolutionary approach in tissue engineering.

Further emphasizing the significance of computational tissue engineering, it is particularly instrumental in addressing chondral defects. The intricacy of cartilage tissue, characterized by its distinctive biomechanical properties and limited self-repair ability, presents a considerable challenge in tissue engineering. Thus, a method that accurately simulates and predicts chondrocytes' responses to various scaffold designs emerges as a vital research trajectory. Computational models, in this context, can serve as potent instruments for optimizing scaffold properties for cartilage regeneration. These models can project the ideal parameters for aspects such as porosity, mechanical robustness, and degradation rate, which are vital in the scaffold design for cartilage tissue engineering.

Moreover, computational simulations can assist in unraveling the mechanisms of chondrocyte differentiation and maturation, potentially steering the development of more efficacious strategies for chondral defect repair. These simulations may include elements such as cell–cell interaction, the function of growth factors, and the impact of biomechanical stimuli on chondrocyte behavior. Merging these computational models with 3D printing technology opens possibilities for personalized medicine in chondral defects. Computational designs can be directly transposed into patient-specific scaffolds via 3D printing, thus enhancing the clinical outcomes of tissue-engineered treatments for chondral defects.

10. Limitations

Though there are treatment options available for chondral defects, there are certain limitations available that hinder researchers to create an engineered cartilage tissue with equivalent biomechanical strength. The major limitations in tissue engineering are (1) optimal cell source (source of MSC, autologous or allogenic), (2) standard protocol for isolation of cells, (3) GMP laboratory facilities for culture expansion, (4) triggers for ECM production, either biochemical or mechanical, (5) ideal scaffold for cartilage regeneration, (6) regulatory protocols as per geographical distribution, and (7) histology, macroscopy, and radiological documentation of the engineered cartilage tissue. The literature lacks human trials for cartilage tissue engineering. Hence, we recommend large-scale triple-blinded controlled trials on cartilage tissue engineering to develop a consensus for clinical usage.

11. Conclusions

Tissue engineering paves a way for creating an engineered tissue or organ with an equivalent biomechanical strength to withstand stress. Unnoticed chondral defects may progress to global osteoarthritis of the joint involved. Early diagnosis and appropriate management with biological therapy with tissue engineering modality prevents the progression of chondral defect to osteoarthritis of the joint. The interdisciplinary approach involving computational modeling, tissue engineering, and 3D printing technologies harbors potential for catalyzing advancements in the treatment of chondral defects.

References

Ouryazdanpanah, N., Dabiri, S., Derakhshani, A., Vahidi, R., & Farsinejad, A. (2018). Peripheral blood-derived mesenchymal stem cells: Growth factor-free isolation, molecular characterization and differentiation. *Iranian Journal of Pathology, 13*(4), 461−466.

Akiyama, H., Chaboissier, M.-C., Martin, J. F., Schedl, A., & Crombrugghe, B. de (2002). The transcription factor Sox9 has essential roles in successive steps of the chondrocyte differentiation pathway and is required for expression of Sox5 and Sox6. *Genes and Development, 16*(21), 2813−2828. https://doi.org/10.1101/gad.1017802

Akkiraju, H., & Nohe, A. (2015). Role of chondrocytes in cartilage formation, progression of osteoarthritis and cartilage regeneration. *Journal of Developmental Biology, 3*(4), 177−192. https://doi.org/10.3390/jdb3040177

Ansboro, S., Hayes, J. S., Barron, V., Browne, S., Howard, L., Greiser, U., Lalor, P., Shannon, F., Barry, F. P., Pandit, A., & Murphy, J. M. (2014). A chondromimetic microsphere for in situ spatially controlled chondrogenic differentiation of human mesenchymal stem cells. *Journal of Controlled Release: Official Journal of the Controlled Release Society, 179*, 42−51. https://doi.org/10.1016/j.jconrel.2014.01.023

Aronowitz, J. A., Lockhart, R. A., & Hakakian, C. S. (2015). Mechanical versus enzymatic isolation of stromal vascular fraction cells from adipose tissue. *SpringerPlus, 4*, 713. https://doi.org/10.1186/s40064-015-1509-2

Bardania, H., Tarvirdipour, S., & Dorkoosh, F. (2017). Liposome-targeted delivery for highly potent drugs. *Artificial Cells, Nanomedicine, and Biotechnology, 45*(8), 1478−1489. https://doi.org/10.1080/21691401.2017.1290647

Ba, K., Wei, X., Ni, D., Li, N., Du, T., Wang, X., & Pan, W. (2019). Chondrocyte Co-cultures with the stromal vascular fraction of adipose tissue in polyhydroxybutyrate/poly-(hydroxybutyrate-co-hydroxyhexanoate) scaffolds: Evaluation of cartilage repair in rabbit. *Cell Transplantation, 28*(11), 1432−1438. https://doi.org/10.1177/0963689719861275

Bekkers, J. E. J., Bartels, L. W., Benink, R. J., Tsuchida, A. I., Vincken, K. L., Dhert, W. J. A., Creemers, L. B., & Saris, D. B. F. (2013). Delayed gadolinium enhanced MRI of cartilage (dGEMRIC) can be effectively applied for longitudinal cohort evaluation of articular cartilage regeneration. *Osteoarthritis and Cartilage, 21*(7), 943−949. https://doi.org/10.1016/j.joca.2013.03.017

Bellini, E., Grieco, M. P., & Raposio, E. (2017). A journey through liposuction and liposculture: Review. *Annals of Medicine and Surgery, 24*, 53−60. https://doi.org/10.1016/j.amsu.2017.10.024

Bhang, S. H., Jeon, J.-Y., La, W.-G., Seong, J. Y., Hwang, J. W., Ryu, S. E., & Kim, B.-S. (2011). Enhanced chondrogenic marker expression of human mesenchymal stem cells by interaction with both TGF-β3 and hyaluronic acid. *Biotechnology and Applied Biochemistry, 58*(4), 271−276. https://doi.org/10.1002/bab.39

Boschetti, F., Pennati, G., Gervaso, F., Peretti, G. M., & Dubini, G. (2004). Biomechanical properties of human articular cartilage under compressive loads. *Biorheology, 41*(3−4), 159−166.

Bouffi, C., Thomas, O., Bony, C., Giteau, A., Venier-Julienne, M.-C., Jorgensen, C., Montero-Menei, C., & Noël, D. (2010). The role of pharmaco-logically active microcarriers releasing TGF-beta3 in cartilage formation in vivo by mesenchymal stem cells. *Biomaterials, 31*(25), 6485−6493. https://doi.org/10.1016/j.biomaterials.2010.05.013

Brittberg, M. (2021). The illustrative first and second generation autologous chondrocyte implantation (ACI) for cartilage repair. In D. R. Goyal (Ed.), *The illustrative book of cartilage repair* (pp. 137−146). Springer International Publishing. https://doi.org/10.1007/978-3-030-47154-5_13

Brozovich, A., Sinicrope, B. J., Bauza, G., Niclot, F. B., Lintner, D., Taraballi, F., & McCulloch, P. C. (2021). High variability of mesenchymal stem cells obtained via bone marrow aspirate concentrate compared with traditional bone marrow aspiration technique. *Orthopaedic Journal of Sports Medicine, 9*(12). https://doi.org/10.1177/23259671211058459, 23259671211058460.

Cai, L., Liu, W., Cui, Y., Liu, Y., Du, W., Zheng, L., Pi, C., Zhang, D., Xie, J., & Zhou, X. (2020). Biomaterial stiffness guides cross-talk between chondrocytes: Implications for a novel cellular response in cartilage tissue engineering. *ACS Biomaterials Science and Engineering, 6*(8), 4476−4489 https://doi.org/10.1021/acsbiomaterials.0c00367

Cai, L., Pi, C., Guo, D., Li, J., Chen, H., Zhang, D., Zhou, X., & Xie, J. (2022). TGF-β3 enhances cell-to-cell communication in chondrocytes via the ALK5/p-Smad3 axis. *Biochemical and Biophysical Research Communications, 636*, 64−74. https://doi.org/10.1016/j.bbrc.2022.10.069

Centeno, C. J., Busse, D., Kisiday, J., Keohan, C., Freeman, M., & Karli, D. (2008). Regeneration of meniscus cartilage in a knee treated with percu-taneously implanted autologous mesenchymal stem cells. *Medical Hypotheses, 71*(6), 900−908. https://doi.org/10.1016/j.mehy.2008.06.042

Chahla, J., Mannava, S., Cinque, M. E., Geeslin, A. G., Codina, D., & LaPrade, R. F. (2017). Bone marrow aspirate concentrate harvesting and processing technique. *Arthroscopy Techniques, 6*(2), e441−e445. https://doi.org/10.1016/j.eats.2016.10.024

Chen, H., Tan, X.-N., Hu, S., Liu, R.-Q., Peng, L.-H., Li, Y.-M., & Wu, P. (2021). Molecular mechanisms of chondrocyte proliferation and differentiation. *Frontiers in Cell and Developmental Biology, 9*, 664168. https://doi.org/10.3389/fcell.2021.664168

Chen, Y.-R., Yan, X., Yuan, F.-Z., Ye, J., Xu, B.-B., Zhou, Z.-X., Mao, Z.-M., Guan, J., Song, Y.-F., Sun, Z.-W., Wang, X.-J., Chen, Z.-Y., Wang, D.-Y., Fan, B.-S., Yang, M., Song, S.-T., Jiang, D., & Yu, J.-K. (2020). The use of peripheral blood-derived stem cells for cartilage repair and regeneration in vivo: A review. *Frontiers in Pharmacology, 11*, 404. https://doi.org/10.3389/fphar.2020.00404

Chen, Y.-R., Zhou, Z.-X., Zhang, J.-Y., Yuan, F.-Z., Xu, B.-B., Guan, J., Han, C., Jiang, D., Yang, Y.-Y., & Yu, J.-K. (2019). Low-molecular-weight heparin-functionalized chitosan-chondroitin sulfate hydrogels for controlled release of TGF-β3 and in vitro neocartilage formation. *Frontiers in Chemistry, 7*, 745. https://doi.org/10.3389/fchem.2019.00745

Choe, S. S., Huh, J. Y., Hwang, I. J., Kim, J. I., & Kim, J. B. (2016). Adipose tissue remodeling: Its role in energy metabolism and metabolic disorders. *Frontiers in Endocrinology, 7*, 30. https://doi.org/10.3389/fendo.2016.00030

Chubinskaya, S., Haudenschild, D., Gasser, S., Stannard, J., Krettek, C., & Borrelli, J. (2015). Articular cartilage injury and potential remedies. *Journal of Orthopaedic Trauma, 29*(Suppl. 12), S47−S52. https://doi.org/10.1097/BOT.0000000000000462

Chung, J. Y., Song, M., Ha, C. W., Kim, J.-A., Lee, C.-H., & Park, Y.-B. (2014). Comparison of articular cartilage repair with different hydrogel-human umbilical cord blood-derived mesenchymal stem cell composites in a rat model. *Stem Cell Research and Therapy, 5*(2), 39. https://doi.org/10.1186/scrt427

Cotter, E. J., Wang, K. C., Yanke, A. B., & Chubinskaya, S. (2018). Bone marrow aspirate concentrate for cartilage defects of the knee: From bench to bedside evidence. *Cartilage, 9*(2), 161–170. https://doi.org/10.1177/1947603517741169

Dallich, A. A., Rath, E., Atzmon, R., Radparvar, J. R., Fontana, A., Sharfman, Z., & Amar, E. (2019). Chondral lesions in the hip: A review of relevant anatomy, imaging and treatment modalities. *Journal of Hip Preservation Surgery, 6*(1), 3–15. https://doi.org/10.1093/jhps/hnz002

Danišovič, Ĺ., Varga, I., & Polák, Š. (2012). Growth factors and chondrogenic differentiation of mesenchymal stem cells. *Tissue and Cell, 44*(2), 69–73. https://doi.org/10.1016/j.tice.2011.11.005

De Luca, P., Kouroupis, D., Viganò, M., Perucca-Orfei, C., Kaplan, L., Zagra, L., de Girolamo, L., Correa, D., & Colombini, A. (2019). Human diseased articular cartilage contains a mesenchymal stem cell-like population of chondroprogenitors with strong immunomodulatory responses. *Journal of Clinical Medicine, 8*(4), 423. https://doi.org/10.3390/jcm8040423

Dean, C. S., Chahla, J., Serra Cruz, R., & LaPrade, R. F. (2016). Fresh osteochondral allograft transplantation for treatment of articular cartilage defects of the knee. *Arthroscopy Techniques, 5*(1), e157–e161. https://doi.org/10.1016/j.eats.2015.10.015

Decker, R. S., Um, H.-B., Dyment, N. A., Cottingham, N., Usami, Y., Enomoto-Iwamoto, M., Kronenberg, M. S., Maye, P., Rowe, D. W., Koyama, E., & Pacifici, M. (2017). Cell origin, volume and arrangement are drivers of articular cartilage formation, morphogenesis and response to injury in mouse limbs. *Developmental Biology, 426*(1), 56–68. https://doi.org/10.1016/j.ydbio.2017.04.006

Deng, Z., Zhang, Q., Zhao, Z., Li, Y., Chen, X., Lin, Z., Deng, Z., Liu, J., Duan, L., Wang, D., & Li, W. (2021). Crosstalk between immune cells and bone cells or chondrocytes. *International Immunopharmacology, 101*, 108179. https://doi.org/10.1016/j.intimp.2021.108179

Denker, A. E., Haas, A. R., Nicoll, S. B., & Tuan, R. S. (1999). Chondrogenic differentiation of murine C3H10T1/2 multipotential mesenchymal cells: I. Stimulation by bone morphogenetic protein-2 in high-density micromass cultures. *Differentiation, 64*(2), 67–76. https://doi.org/10.1046/j.1432-0436.1999.6420067.x

El-Arman, M. M., El-Fayoumi, G., El-Shal, E., El-Boghdady, I., & El-Ghaweet, A. (2010). Aggrecan and cartilage oligomeric matrix protein in serum and synovial fluid of patients with knee osteoarthritis. *HSS Journal, 6*(2), 171–176. https://doi.org/10.1007/s11420-010-9157-0

El-Jawhari, J. J., Ilas, D. C., Jones, W., Cuthbert, R., Jones, E., & Giannoudis, P. V. (2020). Enrichment and preserved functionality of multipotential stromal cells in bone marrow concentrate processed by vertical centrifugation. *European Cells and Materials, 40*, 58–73. https://doi.org/10.22203/eCM.v040a04

Erggelet, C., & Vavken, P. (2016). Microfracture for the treatment of cartilage defects in the knee joint – a golden standard? *Journal of Clinical Orthopaedics and Trauma, 7*(3), 145–152. https://doi.org/10.1016/j.jcot.2016.06.015

Eschweiler, J., Horn, N., Rath, B., Betsch, M., Baroncini, A., Tingart, M., & Migliorini, F. (2021). The biomechanics of cartilage—an overview. *Life, 11*(4). https://doi.org/10.3390/life11040302. Article 4.

Fan, H., Zhang, C., Li, J., Bi, L., Qin, L., Wu, H., & Hu, Y. (2008). Gelatin microspheres containing TGF-beta3 enhance the chondrogenesis of mesenchymal stem cells in modified pellet culture. *Biomacromolecules, 9*(3), 927–934. https://doi.org/10.1021/bm7013203

Felimban, R., Ye, K., Traianedes, K., Di Bella, C., Crook, J., Wallace, G. G., Quigley, A., Choong, P. F. M., & Myers, D. E. (2014). Differentiation of stem cells from human infrapatellar fat pad: Characterization of cells undergoing chondrogenesis. *Tissue Engineering Part A, 20*(15–16), 2213–2223. https://doi.org/10.1089/ten.tea.2013.0657

Fernández-Francos, S., Eiro, N., Costa, L. A., Escudero-Cernuda, S., Fernández-Sánchez, M. L., & Vizoso, F. J. (2021). Mesenchymal stem cells as a cornerstone in a galaxy of intercellular signals: Basis for a new era of medicine. *International Journal of Molecular Sciences, 22*(7), 3576. https://doi.org/10.3390/ijms22073576

Forghani, A., Koduru, S. V., Chen, C., Leberfinger, A. N., Ravnic, D. J., & Hayes, D. J. (2020). Differentiation of adipose tissue–derived CD34+/CD31– cells into endothelial cells in vitro. *Regenerative Engineering and Translational Medicine, 6*(1), 101–110. https://doi.org/10.1007/s40883-019-00093-7

Fortier, L. M., Knapik, D. M., Dasari, S. P., Polce, E. M., Familiari, F., Gursoy, S., & Chahla, J. (2023). Clinical and magnetic resonance imaging outcomes after microfracture treatment with and without augmentation for focal chondral lesions in the knee: A systematic review and meta-analysis. *American Journal of Sports Medicine*. https://doi.org/10.1177/03635465221087365. Sage publications inc.

Francesco, S., Nicolò, B., Michele, P. G., & Edoardo, R. (2019). From liposuction to adipose-derived stem cells: Indications and technique. *Acta Bio-Medica: Atenei Parmensis, 90*(2), 197–208. https://doi.org/10.23750/abm.v90i2.6619

Gao, Y., Liu, S., Huang, J., Guo, W., Chen, J., Zhang, L., Zhao, B., Peng, J., Wang, A., Wang, Y., Xu, W., Lu, S., Yuan, M., & Guo, Q. (2014). The ECM-cell interaction of cartilage extracellular matrix on chondrocytes. *BioMed Research International*, e648459. https://doi.org/10.1155/2014/648459

Gelse, K., von der Mark, K., Aigner, T., Park, J., & Schneider, H. (2003). Articular cartilage repair by gene therapy using growth factor-producing mesenchymal cells. *Arthritis and Rheumatism, 48*(2), 430–441. https://doi.org/10.1002/art.10759

Goldberg, A., Mitchell, K., Soans, J., Kim, L., & Zaidi, R. (2017). The use of mesenchymal stem cells for cartilage repair and regeneration: A systematic review. *Journal of Orthopaedic Surgery and Research, 12*(1), 39. https://doi.org/10.1186/s13018-017-0534-y

Goodrich, L. R., Chen, A. C., Werpy, N. M., Williams, A. A., Kisiday, J. D., Su, A. W., Cory, E., Morley, P. S., McIlwraith, C. W., Sah, R. L., & Chu, C. R. (2016). Addition of mesenchymal stem cells to autologous platelet-enhanced fibrin scaffolds in chondral defects does it enhance repair? *Journal of Bone and Joint Surgery-American Volume, 98*(1), 23–34. https://doi.org/10.2106/JBJS.O.00407. Lippincott williams and wilkins.

Greenwood, V., Clausen, P., & Matuska, A. M. (2022). Micro-fragmented adipose tissue cellular composition varies by processing device and analytical method. *Scientific Reports, 12*(1), 16107. https://doi.org/10.1038/s41598-022-20581-1

Hanxiang Le, W. X. (2020). Mesenchymal stem cells for cartilage regeneration. *Journal of Tissue Engineering, 11.* https://doi.org/10.1177/2041731420943839

Haseeb, A., Kc, R., Angelozzi, M., de Charleroy, C., Rux, D., Tower, R. J., Yao, L., Pellegrino da Silva, R., Pacifici, M., Qin, L., & Lefebvre, V. (2021). SOX9 keeps growth plates and articular cartilage healthy by inhibiting chondrocyte dedifferentiation/osteoblastic redifferentiation. *Proceedings of the National Academy of Sciences, 118*(8). https://doi.org/10.1073/pnas.2019152118

Hopper, N., Henson, F., Brooks, R., Ali, E., Rushton, N., & Wardale, J. (2015). Peripheral blood derived mononuclear cells enhance osteoarthritic human chondrocyte migration. *Arthritis Research and Therapy, 17*(1), 199. https://doi.org/10.1186/s13075-015-0709-z

Hopper, N., Wardale, J., Howard, D., Brooks, R., Rushton, N., & Henson, F. (2015). Peripheral blood derived mononuclear cells enhance the migration and chondrogenic differentiation of multipotent mesenchymal stromal cells. *Stem Cells International*, e323454. https://doi.org/10.1155/2015/323454

Houck, D. A., Kraeutler, M. J., Belk, J. W., Frank, R. M., McCarty, E. C., & Bravman, J. T. (2018). Do focal chondral defects of the knee increase the risk for progression to osteoarthritis? A review of the literature. *Orthopaedic Journal of Sports Medicine, 6*(10). https://doi.org/10.1177/2325967118801931, 2325967118801931.

Hutmacher, D. W. (2000). Scaffolds in tissue engineering bone and cartilage. *Biomaterials, 21*(24), 2529–2543. https://doi.org/10.1016/S0142-9612(00)00121-6

Ikuta, Y., Kamei, N., Ishikawa, M., Adachi, N., & Ochi, M. (2015). In vivo kinetics of mesenchymal stem cells transplanted into the knee joint in a rat model using a novel magnetic method of localization. *Clinical and Translational Science, 8*(5), 467–474. https://doi.org/10.1111/cts.12284

Im, G.-I., Kim, H.-J., & Lee, J. H. (2011). Chondrogenesis of adipose stem cells in a porous PLGA scaffold impregnated with plasmid DNA containing SOX trio (SOX-5,-6 and -9) genes. *Biomaterials, 32*(19), 4385–4392. https://doi.org/10.1016/j.biomaterials.2011.02.054

Iwamoto, T., Ishikawa, M., Ono, M., Nakamura, T., Fukumoto, S., & Yamada, Y. (2013). Biological roles of gap junction proteins in cartilage and bone development. *Journal of Oral Biosciences, 55*(1), 29–33. https://doi.org/10.1016/j.job.2012.12.001

Jelodari, S., Ebrahimi Sadrabadi, A., Zarei, F., Jahangir, S., Azami, M., Sheykhhasan, M., & Hosseini, S. (2022). New insights into cartilage tissue engineering: Improvement of tissue-scaffold integration to enhance cartilage regeneration. *BioMed Research International, 2022*, e7638245. https://doi.org/10.1155/2022/7638245

Jeong, S. Y., Kim, D. H., Ha, J., Jin, H. J., Kwon, S.-J., Chang, J. W., Choi, S. J., Oh, W., Yang, Y. S., Kim, G., Kim, J. S., Yoon, J.-R., Cho, D. H., & Jeon, H. B. (2013). Thrombospondin-2 secreted by human umbilical cord blood-derived mesenchymal stem cells promotes chondrogenic differentiation. *Stem Cells, 31*(10), 2136–2148. https://doi.org/10.1002/stem.1471

Jeyaraman, M., Bingi, S. K., Muthu, S., Jeyaraman, N., Packkyarathinam, R. P., Ranjan, R., Sharma, S., Jha, S. K., Khanna, M., Rajendran, S. N. S., Rajendran, R. L., & Gangadaran, P. (2022). Impact of the process variables on the yield of mesenchymal stromal cells from bone marrow aspirate concentrate. *Bioengineering, 9*(2). https://doi.org/10.3390/bioengineering9020057. Article 2.

Jeyaraman, M., Muthu, S., Jeyaraman, N., Ranjan, R., Jha, S. K., & Mishra, P. (2021). Synovium derived mesenchymal stromal cells (Sy-MSCs): A promising therapeutic paradigm in the management of knee osteoarthritis. *Indian Journal of Orthopaedics.* https://doi.org/10.1007/s43465-021-00439-w

Jiang, Y., Cai, Y., Zhang, W., Yin, Z., Hu, C., Tong, T., Lu, P., Zhang, S., Neculai, D., Tuan, R. S., & Ouyang, H. W. (2016). Human cartilage-derived progenitor cells from committed chondrocytes for efficient cartilage repair and regeneration. *Stem Cells Translational Medicine, 5*(6), 733–744. https://doi.org/10.5966/sctm.2015-0192

Jin, G.-Z., Park, J.-H., Wall, I., & Kim, H.-W. (2016). Isolation and culture of primary rat adipose derived stem cells using porous biopolymer microcarriers. *Tissue Engineering and Regenerative Medicine, 13*(3), 242–250. https://doi.org/10.1007/s13770-016-0040-z

Jo, C. H., Lee, Y. G., Shin, W. H., Kim, H., Chai, J. W., Jeong, E. C., Kim, J. E., Shim, H., Shin, J. S., Shin, I. S., Ra, J. C., Oh, S., & Yoon, K. S. (2014). Intra-articular injection of mesenchymal stem cells for the treatment of osteoarthritis of the knee: A proof-of-concept clinical trial. *Stem Cells, 32*(5), 1254–1266. https://doi.org/10.1002/stem.1634

Jones, G., Ding, C., Zhai, G., Scott, F., Cooley, H., & Cicuttini, F. (2004). Chondral defects: Genetic contribution and relevance and associations with pain, age, body mass index, joint surface area, cartilage volume and radiographic features of osteoarthritis. *Arthritis Research and Therapy, 6*(Suppl. 3), 54. https://doi.org/10.1186/ar1389

Joswig, A.-J., Mitchell, A., Cummings, K. J., Levine, G. J., Gregory, C. A., Smith, R., & Watts, A. E. (2017). Repeated intra-articular injection of allogeneic mesenchymal stem cells causes an adverse response compared to autologous cells in the equine model. *Stem Cell Research and Therapy, 8*(1), 42. https://doi.org/10.1186/s13287-017-0503-8

Jung, M., Kaszap, B., Redöhl, A., Steck, E., Breusch, S., Richter, W., & Gotterbarm, T. (2009). Enhanced early tissue regeneration after matrix-assisted autologous mesenchymal stem cell transplantation in full thickness chondral defects in a minipig model. *Cell Transplantation, 18*(8), 923–932. https://doi.org/10.3727/096368909X471297

Kabiri, A., Esfandiari, E., Hashemibeni, B., Kazemi, M., Mardani, M., & Esmaeili, A. (2012). Effects of FGF-2 on human adipose tissue derived adult stem cells morphology and chondrogenesis enhancement in Transwell culture. *Biochemical and Biophysical Research Communications, 424*(2), 234–238. https://doi.org/10.1016/j.bbrc.2012.06.082

Kao, I.-T., Yao, C.-L., Chang, Y.-J., Hsieh, T.-B., & Hwang, S.-M. (2008). Chondrogenic differentiation of human mesenchymal stem cells from umbilical cord blood in chemically synthesized thermoreversible polymer. *The Chinese Journal of Physiology, 51*(4), 252–258.

Kim, J.-Y., Jeon, H. B., Yang, Y. S., Oh, W., & Chang, J. W. (2010). Application of human umbilical cord blood-derived mesenchymal stem cells in disease models. *World Journal of Stem Cells, 2*(2), 34–38. https://doi.org/10.4252/wjsc.v2.i2.34

Kim, J., Park, J., Song, S.-Y., & Kim, C. (2022). Advanced Therapy medicinal products for autologous chondrocytes and comparison of regulatory systems in target countries. *Regenerative Therapy, 20*, 126–137. https://doi.org/10.1016/j.reth.2022.04.004

Kim, D.-W., Staples, M., Shinozuka, K., Pantcheva, P., Kang, S.-D., & Borlongan, C. V. (2013). Wharton's jelly-derived mesenchymal stem cells: Phenotypic characterization and optimizing their therapeutic potential for clinical applications. *International Journal of Molecular Sciences, 14*(6). https://doi.org/10.3390/ijms140611692. Article 6.

Kobayashi, T., Ochi, M., Yanada, S., Ishikawa, M., Adachi, N., Deie, M., & Arihiro, K. (2008). A novel cell delivery system using magnetically labeled mesenchymal stem cells and an external magnetic device for clinical cartilage repair. Arthroscopy : the Journal of Arthroscopic and Related Surgery, 24(1), 69–76. https://doi.org/10.1016/j.arthro.2007.08.017. W B Saunders CO-Elsevier INC.

Kumar, M., Yasodha, T., Singh, R. K., Singh, R., Kumar, K., Ranjan, R., Meshram, C. D., Das, B. C., & Bag, S. (2013). Generation of transgenic mesenchymal stem cells expressing green fluorescent protein as reporter gene using no viral vector in caprine. *Indian Journal of Experimental Biology, 51*(7), 502–509.

Kurth, T. B., Dell'accio, F., Crouch, V., Augello, A., Sharpe, P. T., & De Bari, C. (2011). Functional mesenchymal stem cell niches in adult mouse knee joint synovium in vivo. *Arthritis and Rheumatism, 63*(5), 1289–1300. https://doi.org/10.1002/art.30234

Kwon, H., Paschos, N. K., Hu, J. C., & Athanasiou, K. (2016). Articular cartilage tissue engineering: The role of signaling molecules. *Cellular and Molecular Life Sciences, 73*(6), 1173–1194. https://doi.org/10.1007/s00018-015-2115-8

Landínez-Parra, N. S., Garzón-Alvarado, D. A., Vanegas-Acosta, J. C., Landínez-Parra, N. S., Garzón-Alvarado, D. A., & Vanegas-Acosta, J. C. (2012). Mechanical behavior of articular cartilage. In *Injury and skeletal biomechanics*. IntechOpen. https://doi.org/10.5772/48323

Lee, G. Y., Kim, S., Baek, S.-H., Jang, E.-C., & Ha, Y.-C. (2019). Accuracy of magnetic resonance imaging and computed tomography arthrography in diagnosing acetabular labral tears and chondral lesions. *Clinical Orthopaedic Surgery, 11*(1), 21–27. https://doi.org/10.4055/cios.2019.11.1.21

Lee, D. H., Kim, S. A., Song, J.-S., Shetty, A. A., Kim, B.-H., & Kim, S. J. (2022). Cartilage regeneration using human umbilical cord blood derived mesenchymal stem cells: A systematic review and meta-analysis. *Medicina, 58*(12). https://doi.org/10.3390/medicina58121801. Article 12.

Lefebvre, V., Angelozzi, M., & Haseeb, A. (2019). SOX9 in cartilage development and disease. *Current Opinion in Cell Biology, 61*, 39–47. https://doi.org/10.1016/j.ceb.2019.07.008

Levato, R., Webb, W. R., Otto, I. A., Mensinga, A., Zhang, Y., van Rijen, M., van Weeren, R., Khan, I. M., & Malda, J. (2017). The bio in the ink: Cartilage regeneration with bioprintable hydrogels and articular cartilage-derived progenitor cells. *Acta Biomaterialia, 61*, 41–53. https://doi.org/10.1016/j.actbio.2017.08.005

Li, J., & Dong, S. (2016). The signaling pathways involved in chondrocyte differentiation and hypertrophic differentiation. *Stem Cells International*, e2470351. https://doi.org/10.1155/2016/2470351

Li, G., Fu, N., Xie, J., Fu, Y., Deng, S., Cun, X., Wei, X., Peng, Q., Cai, X., & Lin, Y. (2015). Poly(3-hydroxybutyrate-co-4-hydroxybutyrate) based electrospun 3D scaffolds for delivery of autogeneic chondrocytes and adipose-derived stem cells: Evaluation of cartilage defects in rabbit. *Journal of Biomedical Nanotechnology, 11*(1), 105–116. https://doi.org/10.1166/jbn.2015.2053

Li, N., Gao, J., Mi, L., Zhang, G., Zhang, L., Zhang, N., Huo, R., Hu, J., & Xu, K. (2020). Synovial membrane mesenchymal stem cells: Past life, current situation, and application in bone and joint diseases. *Stem Cell Research and Therapy, 11*(1), 381. https://doi.org/10.1186/s13287-020-01885-3

Li, S., Huang, K.-J., Wu, J.-C., Hu, M. S., Sanyal, M., Hu, M., Longaker, M. T., & Lorenz, H. P. (2015). Peripheral blood-derived mesenchymal stem cells: Candidate cells responsible for healing critical-sized calvarial bone defects. *Stem Cells Translational Medicine, 4*(4), 359–368. https://doi.org/10.5966/sctm.2014-0150

Li, T., Liu, D., Chen, K., Lou, Y., Jiang, Y., & Zhang, D. (2020). Small molecule compounds promote the proliferation of chondrocytes and chondrogenic differentiation of stem cells in cartilage tissue engineering. *Biomedicine and Pharmacotherapy, 131*, 110652. https://doi.org/10.1016/j.biopha.2020.110652

Madry, H., Cucchiarini, M., Terwilliger, E. F., & Trippell, S. B. (2003). Recombinant adeno-associated virus vectors efficiently and persistently transduce chondrocytes in normal and osteoarthritic human articular cartilage. *Human Gene Therapy, 14*(4), 393–402. https://doi.org/10.1089/104303403321208998

Mahmoud, E. E., Kamei, N., Kamei, G., Nakasa, T., Shimizu, R., Harada, Y., Adachi, N., Misk, N. A., & Ochi, M. (2019). Role of mesenchymal stem cells densities when injected as suspension in joints with osteochondral defects. *Cartilage, 10*(1), 61–69. https://doi.org/10.1177/1947603517708333

Marino, L., Castaldi, M. A., Rosamilio, R., Ragni, E., Vitolo, R., Fulgione, C., Castaldi, S. G., Serio, B., Bianco, R., Guida, M., & Selleri, C. (2019). Mesenchymal stem cells from the Wharton's jelly of the human umbilical cord: Biological properties and therapeutic potential. *International Journal of Stem Cells, 12*(2), 218–226. https://doi.org/10.15283/ijsc18034

Marquass, B., Schulz, R., Hepp, P., Zscharnack, M., Aigner, T., Schmidt, S., Stein, F., Richter, R., Osterhoff, G., Aust, G., Josten, C., & Bader, A. (2011). Matrix-associated implantation of predifferentiated mesenchymal stem cells versus articular chondrocytes: In vivo results of cartilage repair after 1 year. *The American Journal of Sports Medicine, 39*(7), 1401–1412. https://doi.org/10.1177/0363546511398646

Martin, J. A., & Buckwalter, J. A. (2001). Roles of articular cartilage aging and chondrocyte senescence in the pathogenesis of osteoarthritis. *The Iowa Orthopaedic Journal, 21*, 1–7.

McIlwraith, C. W., Frisbie, D. D., Rodkey, W. G., Kisiday, J. D., Werpy, N. M., Kawcak, C. E., & Steadman, J. R. (2011). Evaluation of intra-articular mesenchymal stem cells to augment healing of microfractured chondral defects. *Arthroscopy: The Journal of Arthroscopic and Related Surgery: Official Publication of the Arthroscopy Association of North America and the International Arthroscopy Association, 27*(11), 1552–1561. https://doi.org/10.1016/j.arthro.2011.06.002

Memon, A. R., & Quinlan, J. F. (2012). Surgical treatment of articular cartilage defects in the knee: Are we winning? *Advances in Orthopedics*, 528423. https://doi.org/10.1155/2012/528423

Mobasheri, A., Trujillo, E., Arteaga, M.-F., & Martín-Vasallo, P. (2012). Na(+), K(+)-ATPase subunit composition in a human chondrocyte cell line; evidence for the presence of α1, α3, β1, β2 and β3 isoforms. *International Journal of Molecular Sciences, 13*(4), 5019–5034. https://doi.org/10.3390/ijms13045019

Mojallal, A., Auxenfans, C., Lequeux, C., Braye, F., & Damour, O. (2008). Influence of negative pressure when harvesting adipose tissue on cell yield of the stromal-vascular fraction. *Bio-Medical Materials and Engineering, 18*(4–5), 193–197.

Moll, G., Ankrum, J. A., Kamhieh-Milz, J., Bieback, K., Ringdén, O., Volk, H.-D., Geissler, S., & Reinke, P. (2019). Intravascular mesenchymal stromal/stem cell therapy product diversification: Time for new clinical guidelines. *Trends in Molecular Medicine, 25*(2), 149–163. https://doi.org/10.1016/j.molmed.2018.12.006

Morille, M., Toupet, K., Montero-Menei, C. N., Jorgensen, C., & Noël, D. (2016). PLGA-based microcarriers induce mesenchymal stem cell chondrogenesis and stimulate cartilage repair in osteoarthritis. *Biomaterials, 88*, 60–69. https://doi.org/10.1016/j.biomaterials.2016.02.022

Moutos, F. T., & Guilak, F. (2008). Composite scaffolds for cartilage tissue engineering. *Biorheology, 45*(3–4), 501–512.

Muthu, S., Jeyaraman, M., Jain, R., Gulati, A., Jeyaraman, N., Prajwal, G. S., & Mishra, P. C. (2021). Accentuating the sources of mesenchymal stem cells as cellular therapy for osteoarthritis knees—a panoramic review. *Stem Cell Investigation, 8*, 13. https://doi.org/10.21037/sci-2020-055

Nakagawa, Y., Muneta, T., Kondo, S., Mizuno, M., Takakuda, K., Ichinose, S., Tabuchi, T., Koga, H., Tsuji, K., & Sekiya, I. (2015). Synovial mesenchymal stem cells promote healing after meniscal repair in microminipigs. *Osteoarthritis and Cartilage, 23*(6), 1007–1017. https://doi.org/10.1016/j.joca.2015.02.008

Naraghi, A., & White, L. M. (2015). MRI of labral and chondral lesions of the hip. *AJR American Journal of Roentgenology, 205*(3), 479–490. https://doi.org/10.2214/AJR.14.12581

Niethammer, T. R., Altmann, D., Holzgruber, M., Goller, S., Fischer, A., & Müller, P. E. (2021). Third generation autologous chondrocyte implantation is a good treatment option for athletic persons. *Knee Surgery, Sports Traumatology, Arthroscopy: Official Journal of the ESSKA, 29*(4), 1215–1223. https://doi.org/10.1007/s00167-020-06148-5

Oberbauer, E., Steffenhagen, C., Wurzer, C., Gabriel, C., Redl, H., & Wolbank, S. (2015). Enzymatic and non-enzymatic isolation systems for adipose tissue-derived cells: Current state of the art. *Cell Regeneration, 4*(1). https://doi.org/10.1186/s13619-015-0020-0, 4:7.

Paduszyński, P., Aleksander-Konert, E., Zajdel, A., Wilczok, A., Jelonek, K., Witek, A., & Dzierżewicz, Z. (2016). Changes in expression of cartilaginous genes during chondrogenesis of Wharton's jelly mesenchymal stem cells on three-dimensional biodegradable poly(L-lactide-co-glycolide) scaffolds. *Cellular and Molecular Biology Letters, 21*, 14. https://doi.org/10.1186/s11658-016-0012-2

Pak, J., Lee, J. H., Pak, N., Pak, Y., Park, K. S., Jeon, J. H., Jeong, B. C., & Lee, S. H. (2018). Cartilage regeneration in humans with adipose tissue-derived stem cells and adipose stromal vascular fraction cells: Updated status. *International Journal of Molecular Sciences, 19*(7), 2146. https://doi.org/10.3390/ijms19072146

Pak, J., Lee, J. H., Pak, N. J., Park, K. S., Jeon, J. H., Jeong, B. C., & Lee, S. H. (2018). Clinical protocol of producing adipose tissue-derived stromal vascular fraction for potential cartilage regeneration. *Journal of Visualized Experiments, 139*, 58363. https://doi.org/10.3791/58363

Park, Y., Ha, C., Lee, C., Yoon, Y. C., & Park, Y. (2017). Cartilage regeneration in osteoarthritic patients by a composite of allogeneic umbilical cord blood-derived mesenchymal stem cells and hyaluronate hydrogel: Results from a clinical trial for safety and proof-of-concept with 7 Years of extended follow-up. *Stem Cells Translational Medicine, 6*(2), 613–621. https://doi.org/10.5966/sctm.2016-0157

Patel, J. M., Saleh, K. S., Burdick, J. A., & Mauck, R. L. (2019). Bioactive factors for cartilage repair and regeneration: Improving delivery, retention, and activity. *Acta Biomaterialia, 93*, 222–238. https://doi.org/10.1016/j.actbio.2019.01.061

Phan, M. N., Leddy, H. A., Votta, B. J., Kumar, S., Levy, D. S., Lipshutz, D. B., Lee, S., Liedtke, W., & Guilak, F. (2009). Functional characterization of TRPV4 as an osmotically sensitive ion channel in articular chondrocytes. *Arthritis and Rheumatism, 60*(10), 3028–3037. https://doi.org/10.1002/art.24799

Poole, A. R., Kojima, T., Yasuda, T., Mwale, F., Kobayashi, M., & Laverty, S. (2001). Composition and structure of articular cartilage: A template for tissue repair. *Clinical Orthopaedics and Related Research, 391*, S26.

Posey, K. L., Coustry, F., & Hecht, J. T. (2018). Cartilage oligomeric matrix protein: COMPopathies and beyond. *Matrix Biology: Journal of the International Society for Matrix Biology, 71–72*, 161–173. https://doi.org/10.1016/j.matbio.2018.02.023

Punzi, L., Galozzi, P., Luisetto, R., Favero, M., Ramonda, R., Oliviero, F., & Scanu, A. (2016). Post-traumatic arthritis: Overview on pathogenic mechanisms and role of inflammation. *RMD Open, 2*(2), e000279. https://doi.org/10.1136/rmdopen-2016-000279

Qi, Y., Du, Y., Li, W., Dai, X., Zhao, T., & Yan, W. (2014). Cartilage repair using mesenchymal stem cell (MSC) sheet and MSCs-loaded bilayer PLGA scaffold in a rabbit model. In *Knee surgery sports traumatology arthroscopy* (Vol. 22, pp. 1424–1433). https://doi.org/10.1007/s00167-012-2256-3. Issue 6; Springer.

Richard, A. J., White, U., Elks, C. M., & Stephens, J. M. (2000). Adipose tissue: Physiology to metabolic dysfunction. In K. R. Feingold, B. Anawalt, A. Boyce, G. Chrousos, W. W. de Herder, K. Dhatariya, K. Dungan, J. M. Hershman, J. Hofland, S. Kalra, G. Kaltsas, C. Koch, P. Kopp, M. Korbonits, C. S. Kovacs, W. Kuohung, B. Laferrère, M. Levy, E. A. McGee, … D. P. Wilson (Eds.), *Endotext*. MDText.com, Inc. http://www.ncbi.nlm.nih.gov/books/NBK555602/.

Roelofs, A. J., Zupan, J., Riemen, A. H. K., Kania, K., Ansboro, S., White, N., Clark, S. M., & De Bari, C. (2017). Joint morphogenetic cells in the adult mammalian synovium. *Nature Communications, 8*, 15040. https://doi.org/10.1038/ncomms15040

Russo, E., Caprnda, M., Kruzliak, P., Conaldi, P. G., Borlongan, C. V., & La Rocca, G. (2022). Umbilical cord mesenchymal stromal cells for cartilage regeneration applications. *Stem Cells International, 2022*, 2454168. https://doi.org/10.1155/2022/2454168

Sadlik, B., Jaroslawski, G., Gladysz, D., Puszkarz, M., Markowska, M., Pawelec, K., Boruczkowski, D., & Oldak, T. (2017). Knee cartilage regeneration with umbilical cord mesenchymal stem cells embedded in collagen scaffold using dry arthroscopy technique. In M. Pokorski (Ed.), *Clinical research and practice* (Vol. 1020, pp. 113–122). Springer International Publishing AG. https://doi.org/10.1007/5584_2017_9

Sakaguchi, Y., Sekiya, I., Yagishita, K., & Muneta, T. (2005). Comparison of human stem cells derived from various mesenchymal tissues: Superiority of synovium as a cell source. *Arthritis and Rheumatism, 52*(8), 2521—2529. https://doi.org/10.1002/art.21212

Saw, K.-Y., Anz, A., Jee, C. S.-Y., Ng, R. C.-S., Mohtarrudin, N., & Ragavanaidu, K. (2015). High tibial osteotomy in combination with chondrogenesis after stem cell therapy: A histologic report of 8 cases. *Arthroscopy-the Journal of Arthroscopic and Related Surgery, 31*(10), 1909—1920. https://doi.org/10.1016/j.arthro.2015.03.038. W B Saunders Co-Elsevier Inc.

Saw, K.-Y., Anz, A., Merican, S., Tay, Y.-G., Ragavanaidu, K., Jee, C. S. Y., & McGuire, D. A. (2011). Articular cartilage regeneration with autologous peripheral blood progenitor cells and hyaluronic acid after arthroscopic subchondral drilling: A report of 5 cases with histology. *Arthroscopy-the Journal of Arthroscopic and Related Surgery, 27*(1), 493—506. https://doi.org/10.1016/j.arthro.2010.11.054. W B Saunders Co-Elsevier Inc.

Saw, K.-Y., Anz, A., Siew-Yoke Jee, C., Merican, S., Ching-Soong Ng, R., Roohi, S. A., & Ragavanaidu, K. (2013). Articular cartilage regeneration with autologous peripheral blood stem cells versus hyaluronic acid: A randomized controlled trial. *Arthroscopy: The Journal of Arthroscopic and Related Surgery: Official Publication of the Arthroscopy Association of North America and the International Arthroscopy Association, 29*(4), 684—694. https://doi.org/10.1016/j.arthro.2012.12.008

Saw, K.-Y., Hussin, P., Loke, S.-C., Azam, M., Chen, H.-C., Tay, Y.-G., Low, S., Wallin, K.-L., & Ragavanaidu, K. (2009). Articular cartilage regeneration with autologous marrow aspirate and hyaluronic Acid: An experimental study in a goat model. *Arthroscopy: The Journal of Arthroscopic and Related Surgery: Official Publication of the Arthroscopy Association of North America and the International Arthroscopy Association, 25*(12), 1391—1400. https://doi.org/10.1016/j.arthro.2009.07.011

Sekiya, I., Katano, H., & Ozeki, N. (2021). Characteristics of MSCs in synovial fluid and mode of action of intra-articular injections of synovial MSCs in knee osteoarthritis. *International Journal of Molecular Sciences, 22*(6), 2838. https://doi.org/10.3390/ijms22062838

Senesi, L., De Francesco, F., Farinelli, L., Manzotti, S., Gagliardi, G., Papalia, G. F., Riccio, M., & Gigante, A. (2019). Mechanical and enzymatic procedures to isolate the stromal vascular fraction from adipose tissue: Preliminary results. *Frontiers in Cell and Developmental Biology, 7.* https://www.frontiersin.org/articles/10.3389/fcell.2019.00088.

Seol, D., Yu, Y., Choe, H., Jang, K., Brouillette, M. J., Zheng, H., Lim, T.-H., Buckwalter, J. A., & Martin, J. A. (2014). Effect of short-term enzymatic treatment on cell migration and cartilage regeneration: In vitro organ culture of bovine articular cartilage. *Tissue Engineering Part A, 20*(13—14), 1807—1814. https://doi.org/10.1089/ten.TEA.2013.0444

Sercombe, L., Veerati, T., Moheimani, F., Wu, S. Y., Sood, A. K., & Hua, S. (2015). Advances and challenges of liposome assisted drug delivery. *Frontiers in Pharmacology, 6*, 286. https://doi.org/10.3389/fphar.2015.00286

Sharma, S., Muthu, S., Jeyaraman, M., Ranjan, R., & Jha, S. K. (2021). Translational products of adipose tissue-derived mesenchymal stem cells: Bench to bedside applications. *World Journal of Stem Cells, 13*(10), 1360—1381. https://doi.org/10.4252/wjsc.v13.i10.1360

Shimomura, K., Moriguchi, Y., Nansai, R., Fujie, H., Ando, W., Horibe, S., Hart, D. A., Gobbi, A., Yoshikawa, H., & Nakamura, N. (2017). Comparison of 2 different formulations of artificial bone for a hybrid implant with a tissue-engineered construct derived from synovial mesenchymal stem cells: A study using a rabbit osteochondral defect model. *The American Journal of Sports Medicine, 45*(3), 666—675. https://doi.org/10.1177/0363546516668835

Skowroński, J., & Rutka, M. (2013). Osteochondral lesions of the knee reconstructed with mesenchymal stem cells—results. *Ortopedia Traumatologia Rehabilitacja, 15*(3), 195—204. https://doi.org/10.5604/15093492.1058409

Soboleski, M. R., Oaks, J., & Halford, W. P. (2005). Green fluorescent protein is a quantitative reporter of gene expression in individual eukaryotic cells. *The FASEB Journal: Official Publication of the Federation of American Societies for Experimental Biology, 19*(3), 440—442. https://doi.org/10.1096/fj.04-3180fje

Somoza, R. A., Welter, J. F., Correa, D., & Caplan, A. I. (2014). Chondrogenic differentiation of mesenchymal stem cells: Challenges and unfulfilled expectations. *Tissue Engineering Part B Reviews, 20*(6), 596—608. https://doi.org/10.1089/ten.teb.2013.0771

Sonomoto, K., Yamaoka, K., Kaneko, H., Yamagata, K., Sakata, K., Zhang, X., Kondo, M., Zenke, Y., Sabanai, K., Nakayamada, S., Sakai, A., & Tanaka, Y. (2016). Spontaneous differentiation of human mesenchymal stem cells on poly-lactic-Co-glycolic acid nano-fiber scaffold. *PLoS One, 11*(4), e0153231. https://doi.org/10.1371/journal.pone.0153231

Sophia Fox, A. J., Bedi, A., & Rodeo, S. A. (2009). The basic science of articular cartilage. *Sport Health, 1*(6), 461—468. https://doi.org/10.1177/1941738109350438

Steinert, A. F., Nöth, U., & Tuan, R. S. (2008). Concepts in gene therapy for cartilage repair. *Injury, 39*(Suppl. 1), S97—S113. https://doi.org/10.1016/j.injury.2008.01.034

Sun, K., Guo, J., Yao, X., Guo, Z., & Guo, F. (2021). Growth differentiation factor 5 in cartilage and osteoarthritis: A possible therapeutic candidate. *Cell Proliferation, 54*(3), e12998. https://doi.org/10.1111/cpr.12998

Takahata, Y., Hagino, H., Kimura, A., Urushizaki, M., Yamamoto, S., Wakamori, K., Murakami, T., Hata, K., & Nishimura, R. (2022). Regulatory mechanisms of Prg4 and Gdf5 expression in articular cartilage and functions in osteoarthritis. *International Journal of Molecular Sciences, 23*(9), 4672. https://doi.org/10.3390/ijms23094672

Tan, S., Fang, W., Vangsness, C. T., & Han, B. (2021). Influence of cellular microenvironment on human articular chondrocyte cell signaling. *Cartilage, 13*(Suppl. 1), 935S—946S. https://doi.org/10.1177/1947603520941219

Toyokawa, N., Fujioka, H., Kokubu, T., Nagura, I., Inui, A., Sakata, R., Satake, M., Kaneko, H., & Kurosaka, M. (2010). Electrospun synthetic polymer scaffold for cartilage repair without cultured cells in an animal model. *Arthroscopy: The Journal of Arthroscopic and Related Surgery: Official Publication of the Arthroscopy Association of North America and the International Arthroscopy Association, 26*(3), 375—383. https://doi.org/10.1016/j.arthro.2009.08.006

Tseng, S., Reddi, A. H., & Di Cesare, P. E. (2009). Cartilage oligomeric matrix protein (COMP): A biomarker of arthritis. *Biomarker Insights, 4*, 33—44

Turajane, T., Chaweewannakorn, U., Larbpaiboonpong, V., Aojanepong, J., Thitiset, T., Honsawek, S., Fongsarun, J., & Papadopoulos, K. I. (2013). Combination of intra-articular autologous activated peripheral blood stem cells with growth factor addition/preservation and hyaluronic acid in conjunction with arthroscopic microdrilling mesenchymal cell stimulation improves quality of life and regenerates articular cartilage in early osteoarthritic knee disease. *Journal of the Medical Association of Thailand, 96*(5), 580−588.

Vinod, E., Boopalan, P. R. J. V. C., & Sathishkumar, S. (2018). Reserve or resident progenitors in cartilage? Comparative analysis of chondrocytes versus chondroprogenitors and their role in cartilage repair. *Cartilage, 9*(2), 171−182. https://doi.org/10.1177/1947603517736108

Wang, J., Cui, X., Zhou, Y., & Xiang, Q. (2014). Core-shell PLGA/collagen nanofibers loaded with recombinant FN/CDHs as bone tissue engineering scaffolds. *Connective Tissue Research, 55*(4), 292−298. https://doi.org/10.3109/03008207.2014.918112

Wasyłeczko, M., Sikorska, W., & Chwojnowski, A. (2020). Review of synthetic and hybrid scaffolds in cartilage tissue engineering. *Membranes, 10*(11), E348. https://doi.org/10.3390/membranes10110348

de Windt, T. S., Vonk, L. A., Slaper-Cortenbach, I. C. M., van den Broek, M. P. H., Nizak, R., van Rijen, M. H. P., de Weger, R. A., Dhert, W. J. A., & Saris, D. B. F. (2017). Allogeneic mesenchymal stem cells stimulate cartilage regeneration and are safe for single-stage cartilage repair in humans upon mixture with recycled autologous chondrons. *Stem Cells, 35*(1), 256−264. https://doi.org/10.1002/stem.2475

Wong, K. L., Lee, K. B. L., Tai, B. C., Law, P., Lee, E. H., & Hui, J. H. P. (2013). Injectable cultured bone marrow-derived mesenchymal stem cells in varus knees with cartilage defects undergoing high tibial osteotomy: A prospective, randomized controlled clinical trial with 2 years' follow-up. *Arthroscopy: The Journal of Arthroscopic & Related Surgery: Official Publication of the Arthroscopy Association of North America and the International Arthroscopy Association, 29*(12), 2020−2028. https://doi.org/10.1016/j.arthro.2013.09.074

Xu, W., Wang, W., Liu, D., & Liao, D. (2022). Roles of cartilage-resident stem/progenitor cells in cartilage physiology, development, repair and osteoarthritis. *Cells, 11*(15), 2305. https://doi.org/10.3390/cells11152305

Yamagata, K., Nakayamada, S., & Tanaka, Y. (2018). Use of mesenchymal stem cells seeded on the scaffold in articular cartilage repair. *Inflammation and Regeneration, 38*(1), 4. https://doi.org/10.1186/s41232-018-0061-1

Yuan, L., Li, B., Yang, J., Ni, Y., Teng, Y., Guo, L., Fan, H., Fan, Y., & Zhang, X. (2016). Effects of composition and mechanical property of injectable collagen I/II composite hydrogels on chondrocyte behaviors. *Tissue Engineering Part A, 22*(11−12), 899−906. https://doi.org/10.1089/ten.TEA.2015.0513

Zha, K., Li, X., Yang, Z., Tian, G., Sun, Z., Sui, X., Dai, Y., Liu, S., & Guo, Q. (2021). Heterogeneity of mesenchymal stem cells in cartilage regeneration: From characterization to application. *NPJ Regenerative Medicine, 6*(1). https://doi.org/10.1038/s41536-021-00122-6. Article 1.

Zhang, C.-H., Gao, Y., Jadhav, U., Hung, H.-H., Holton, K. M., Grodzinsky, A. J., Shivdasani, R. A., & Lassar, A. B. (2021). Creb5 establishes the competence for Prg4 expression in articular cartilage. *Communications Biology, 4*(1), 332. https://doi.org/10.1038/s42003-021-01857-0

Zhang, W., Zhang, F., Shi, H., Tan, R., Han, S., Ye, G., Pan, S., Sun, F., & Liu, X. (2014). Comparisons of rabbit bone marrow mesenchymal stem cell isolation and culture methods in vitro. *PLoS One, 9*(2), e88794. https://doi.org/10.1371/journal.pone.0088794

Zhao, X., Hu, D. A., Wu, D., He, F., Wang, H., Huang, L., Shi, D., Liu, Q., Ni, N., Pakvasa, M., Zhang, Y., Fu, K., Qin, K. H., Li, A. J., Hagag, O., Wang, E. J., Sabharwal, M., Wagstaff, W., Reid, R. R., ... Athiviraham, A. (2021). Applications of biocompatible scaffold materials in stem cell-based cartilage tissue engineering. *Frontiers in Bioengineering and Biotechnology, 9*, 603444. https://doi.org/10.3389/fbioe.2021.603444

Zhao, Z., Li, Y., Wang, M., Zhao, S., Zhao, Z., & Fang, J. (2020). Mechanotransduction pathways in the regulation of cartilage chondrocyte homoeostasis. *Journal of Cellular and Molecular Medicine, 24*(10), 5408−5419. https://doi.org/10.1111/jcmm.15204

Zwick, R. K., Guerrero-Juarez, C. F., Horsley, V., & Plikus, M. V. (2018). Anatomical, physiological and functional diversity of adipose tissue. *Cell Metabolism, 27*(1), 68−83. https://doi.org/10.1016/j.cmet.2017.12.002

Chapter 27

Recent advances in computational modeling: An appraisal of stem cell and tissue engineering research

Pinky, Neha and Suhel Parvez

Department of Medical Elementology and Toxicology, School of Chemical and Life Sciences, Jamia Hamdard, New Delhi, India

1. Introduction

The two major qualities of stem cells make them the most desirable cell types for the application of tissue engineering and a highly promising platform for therapeutics at cellular and molecular levels. i.e., the potential of self-renewal activities and their capacity for multilineage differentiation. Stem cells could proliferate in vitro, and various biomaterials can be used to stimulate or maintain the differentiation of cells into numerous cell types (Neuss et al., 2008). Stem cell research produced huge omics data and possibly processed large datasets as well as constructed computer-based models. The potential of computational modeling to direct their experimental research appears to be underutilized by the stem cell community (Bian & Cahan, 2016).

Tissue engineering is a broad category under stem cell research that applies the principles of life sciences and engineering to the creation of biomimetic substitutes, including the maintenance, restoration, and improvement of tissue function. Mechanical replacement and organ transplantation are examples of traditional procedures, although both have restrictions on availability and bodily acceptance. They might also provide insight into how cells function as a whole or how to efficiently use growth factors to regulate gene expression. While these methods are useful, their utility is not always obvious to tissue engineers.

The highest advancement in terms of tissue engineering is the design of "computational models" that are a potential guide in bioprinting and biomanufacturing technologies (Mozafari et al., 2019). Through this book chapter, an effort was made to discuss the advancement of computational biology in stem cell modeling and tissue engineering. As the importance of computation in stem cell biology continues to rise, we employ these applications to explicate underlying principles.

2. The node of computational stem cell biology

Our understanding of stem cells and developmental biology has benefited from computational methods. Alan Turing has been researching reaction-diffusion as a theory for embryonic patterning and how morphogen concentration may impact embryo formation in an in silico setting since 1952 (Dawes, 2016). In quite a few past decades, computational systems have been fundamentally tied to the field of stem cell and developmental biology research due to the emergence of high-throughput technologies, such as nucleic acid sequencing, and the growing frequency of modeling and simulation. Before the era of OMIC developments, the high use of computational tools in stem cell biology was to explore the mechanisms of morphogenesis (Lander, 2007; Till et al., 1964). Moreover, single-cell technologies offer the ideal approach for computational modeling at multiple levels to report problems in stem cell biology for treating multirange diseases. "Tissue engineering" is the heart of biotechnologists to approach organ transplantation, and the most noticeable organ skin has the long ability to self-renewal. Several appropriate gene therapies have been isolated from the stem cells of various tissues. The single-cell-based mechanistic approach must be highly accurate for the characterization of various transcription factors (TFs) or the biological processes/interactions that are responsible for tissue engineering.

Computational Biology for Stem Cell Research. https://doi.org/10.1016/B978-0-443-13222-3.00006-X
Copyright © 2024 Elsevier Inc. All rights are reserved, including those for text and data mining, AI training, and similar technologies.

3. A brief introduction to common modeling approaches

It is commonly impossible to understand the entire multiscale biological system by evaluating experimental and clinical facts and figures using exclusively verbal thinking and linear interpretations. Highly expressed theoretical or in silico-*based* models in mathematically sound language can present fresh and more insightful information. The key elements used to construct a mathematical model include model construction, calibration, prediction, validation, and refinement (Del Sol & Jung, 2021). A theoretical model is the base of numerous models, and it is used to depict various biological and molecular processes. It may also be phenomenological or mechanistic at multiple steps including a single-cell/a tissue/an organ/or an organ system.

Integrating a mathematical model with experimental data is a crucial step in creating predictive mathematical models that are physiologically plausible, but the most inverse problem arises with the sets of parameters and which parameters are valued the most to produce the observational data. Similarly, the assessment of different models among the various competing hypotheses is more likely to produce precise data, and complexity-based models are highly required to analyze the area of sample/data collection. A thorough comparison of mathematical model predictions and observational evidence is then used to validate theoretical models after they have been calibrated (Waters et al., 2021). Any mathematical modeling techniques require the use of hypotheses, whether they are founded on already-existing but concealed data or on predictions that will be assessed by recently created data. Predictive test refine cycles are essential to the development of all models (Fig. 27.1). The advancement of the model can therefore be augmented by taking benefit of the inconsistencies developed between the predictions of the model and the experimental evidence. This section provides a concise and comprehensive classification of commonly used mathematical and computational modeling techniques in biology and medicine.

The questions that arise with biological processes and their quantitive analysis of experimental data are highly responsible for making decisions on what type of modeling approach is beneficial to combating the problems. Along with this, the book chapter provides more information on concise and general categories of mathematical/computational modeling that are highly used in biology and medicine.

3.1 Mechanistic models

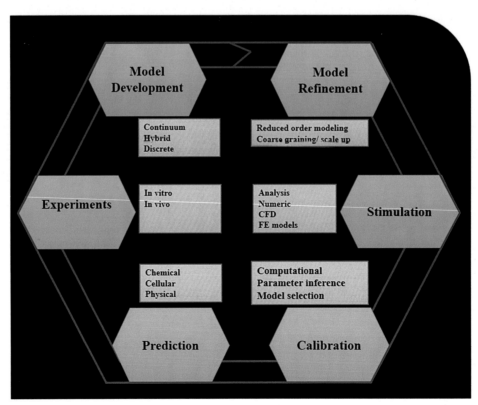

FIGURE 27.1 A schematic representation of predicted test-refine cycles necessary for model development.

The elements of a hypothesis are cell-cell interactions and the functional interactions of the biomolecules in the behavioral sciences, quantitatively represented by the mathematical models. The advance of mechanistic models is regularly driven by the examination of experimental data, allowing the formulation of hypotheses underlying the causal mechanisms of a biological system (Pearson et al., 2014). When debating the development of mechanosensitive tissues, including bone in a bioreactor, produced the response of mechanosensitive cells to imposed mechanical loads (like fluid shear-based stress and hydrostatic pressure, etc.). The identification of the dependent factors of the system includes cell number, fluid velocity, and substrate density. These variables are highly dependent upon space, time, geography, and temporal scales.

Coarse-graining methods with varied spatial and temporal proportions are critical for empowering rigorous mathematical examination and classifying models according to their projected developing behaviors (O'Dea et al., 2015; Davit et al., 2013). The highly precison-based computational tools used for the simulation of mathematical models are essential to facilitate behavioral-based alternative models' parameter sensitivity and calibration (Deisboeck et al., 2011). Instead, a reductionist or molecular biological system is more phenomenological (Price et al., 2020). The phenomenological models have a great advantage where we can directly replicate the experimental results without the terms in the model equations and directly relate to the cellular and molecular levels (O'Dea et al., 2015). Additionally, phenomenological models reveal how a system works and why different types of regulatory interactions are necessary. In such circumstances, numerous molecular paths may produce phenomenological models that are identical. All models, meanwhile, are phenomenological because they radically simplify the underlying physics and chemistry; phenomenological models are created using a similar process to other mechanistic models.

4. Types of stem cell-based modeling

The five different types of primary mechanistic modeling that have been used to combat the problems, Table 27.1 shows that graphs, Boolean networks, regression-based models, ordinary differential equations, and Bayesian networks are utilized in stem cell research.

a. **Graphs** consist of vertices and edges that describe the coexpression between genes or cell interactions. The graph may be directed or undirected, depending on the nature of the interactions. Prominently, graphs do not implement modeling assumptions; they only display the data.

TABLE 27.1 ScRNA-seq-based mechanistic models.

Framework for modeling	Type of models	Application	Biological relevance
Descriptive model	Undirected Graph	Information theory is used to infer gene coexpression networks from time-series data	Organization of functional units by importance
	Directed graph	Using the expression of ligand-receptor pairs, networks of cell-cell communication can be inferred	Predicting the interactions between tissue cells
Logical model	Boolean network	Coexpression modules and predicted TF binding sites are used to infer a Boolean model of GRNs	Cell conversion factors are predicted
		Granger causality analysis of time-series data is used to infer a Boolean model of GRNs	Prediction of the regulatory connections controlling the course of cells
		Using GRNs to infer Boolean logic-rules	Cooperative TF regulation identification
Continuous model	Regression-based	Inference of a quantitative model of GRNs using ridge regression and partial correlation	Prediction of the kinetics of gene expression during differentiation
	Linear ordinary Differential equations	A quantitative model of GRNs can be determined using differentiation trajectories.	
		A quantitative model of GRNs is inferred based on fluctuations in the cell cycle and cell state.	
Probabilistic model	Dynamic Bayesian network	Using knowledge from the past, a probabilistic GRN model is inferred	Forecasting the effectiveness of reprogramming

b. **Boolean networks** are graph-based models where each vertex has a Boolean value describing its state. Principally, Boolean networks undertake that the action of vertices can be considered active or inactive. This hypothesis is supported by gene regulatory networks (GRNs) because the transcription rate is monitored by a Hill function, which is analogous to a Boolean arrangement (Hollister et al., 2000). The regulation of a gene by multiple TFs is specified by Boolean functions. Characteristically, time-series or perturbation data are required for identifying Boolean functions. Though the increasing quantity of single-cell RNA sequencing (scRNA-seq) data belonging to distinct cell subpopulations can be used to originate Boolean functions.

c. **Continuous models** in contrast to Boolean networks, the levels of continuous states are associated with vertices, and regression is a very familiar technique for evaluating the relationship between vertices. Additionally, linear ordinary differential equations are used as a regular modeling approach that forms a linkage between the function of a linear combination and its byproducts, including the relationship between gene expression and the rate of transcription. Notwithstanding being commonly acceptable in the context of GRNs, the conventions of linearity restrict the examination of nonadditive effects, like competing TF regulation. For the interpretation of continuous models, time-series data demonstrating biological variation over time are necessary. In addition to replicating stochastic processes over time, dynamic Bayesian networks can also replicate stochastic processes.

d. **Probabilistic interactions** represent the conditional relationships between a DBN's vertices. It is believable that an event in the framework of gene regulation can induce another event in the forthcoming but not in the past. For continuous models, time-series measurements are also essential for DBN inference.

5. Single-cell isolation-related tools and software

Single-cell isolation refers to the process of isolating high-quality individual cells from a tissue, thereby extracting precise genetic and biochemical information and facilitating the study of unique genetic and molecular mechanisms. Traditional techniques such as the transcriptome, epigenome, and proteome from bulk RNA/DNA samples can only capture the total level of tissue/organ signals, failing to differentiate between individual cell variations. For the isolation of a single cell, various capture procedures are used based on the organism, tissue, or cell properties. It is possible to isolate cells by isolating whole cells, cell-specific nuclei or organelles, or by separating the desired cells expressing specific marker proteins. Limiting dilutions, fluorescence-activated cell sorting, magnetic-activated cell sorting, microfluidic systems, and laser microdissection are among the most common techniques for isolating and capturing single cells. Every single cell is collected in an isolated reaction mixture, and after being converted into complementary DNAs, all transcripts from a single cell are barcoded uniquely (Ding et al., 2021). Numerous computational techniques and software have been created to analyze the number of profiled cells for various anticipated distributions (Table 27.2).

6. Multiscale computational modeling: from intracellular networks to cell-cell interactions in the tissue

Stem cell research requires experimental investigations that interact with several biological levels, such as the behavior of cell populations and their cellular relationships. Multi-scale computational modeling aims to characterize living systems and depict the dimensions of numerous spatial and temporal systems. Then a scientific problem is discussed, and the data supplied defines what kind of model should be built. Computational approaches and the study of intracellular GRNs are advantageous for modeling cell conversion, which defines the optimal parameter range. The proper inference of cell-specific GRNs is necessary for network-based techniques but not always possible, particularly for currently discovered cell subtypes. scRNA-seq is the best tool for detecting exact gene interactions in individual cells. In a single experiment, scRNA-seq captures the expression of genes in multiple cells, providing many data points that enable the extraction of information on gene expression heterogeneity (Berneman-Zeitouni et al., 2014; de Soysa et al., 2019; Qiu et al., 2020) (Table 27.1). The identification of cell TFs through complementary gene expression computational approaches reduces the inherent complexity of GRN inference (Kamimoto et al., 2023; Stumpf & MacArthur, 2019). The methods do not provide information on how genes are regulated.

7. Frameworks for mechanistic modeling in stem cell research: an overview

Mechanistic-based models define a biological system based on a standard relationship between its components, like genes, where the nature of the relationship is well-defined in terms of the biological process that formed the experimental

TABLE 27.2 Tools for single-cell data processing, analysis, and modeling.

S.No.	Name of the software	Availability	Outcomes	Software link	References
1	Sceb	Sceb (Python)	(a) Cell profiling (b) Reads the depth of the sequence	https://github.com/martinjzhang/single_cell_eb	Zhang et al. (2020)
2	TPS	TPS (Python)	(a) Appropriate time to take profiling	https://github.com/topics/tps	Kleyman et al. (2017)
3	Seurat	Seurat	(a) Screening and (b) Trajectory graph	https://github.com/satijalab/seurat	Stuart et al. (2019)
4	SCNN	SCNN (Python)	(a) Annotations in cell types	https://github.com/XingangPan/SCNN	Lin et al. (2017)
5	CellNet	CellNet (R)	(a) Identification of cell type (b) Establishment of GRN for various cell types	https://github.com/pcahan1/CellNet	Morris et al. (2014)
6	Leiden	Leiden (Python)	(a) Segmentation	https://github.com/vtraag/leidenalg	Traag et al. (2019)
7	Local Inverse Simpson's Index (LISI)	LISI (R)	(a) LISI score per cell valued at a real rate	https://github.com/immunogenomics/LISI	Korsunsky et al. (2019)

observations. The components of the living system include physical, biological, and chemical components under the mechanistic model defining their characteristics. Additionally, logical and continuous models have been established recently to describe a biological or living system at multiple levels. The selection of a framework directly impacts the ability to reproduce system dynamics and more accurate biological data.

The basis of logical models, which provide the qualitative framework for enlightening the interaction of genes, governs the behavior system. The entity of each model indicates the regulators, which can be in "active" or "inactive" states. Noteworthy problems with stem cell research have previously been effectively addressed through logical models. For example, the GRN model of embryonic blood formation was inferred using scRNA-seq data to predict the rate of TFs regulating the differentiation of progenitor cells of mesoderm into endothelium and blood cells (Stumpf & MacArthur, 2019). Based on the literature, the GRN model of hematopoietic differentiation was developed and used to estimate the elements that cause transdifferentiation and reprogramming directly (D'Alessio et al., 2015). Additionally, to GRNs, logical models integrate the regulation of genes, intracellular signaling, and interactions with ligand-receptors utilized to depict the molecules responsible for maintaining the conditions of tissues (Okawa et al., 2018).

Due to the growing accessibility of single-cell data, scientists are now able to mimic biological systems using continuous models with greater accuracy. This framework, as different from logical models, can give a measurable description of biological systems without designating components as "active". As a result, the system's temporal dynamics are depicted on a continuous time scale, allowing for a direct comparison of experimental results with the model's state. For example, a stem cell-based interaction model was used to identify the mechanical factors that affect cellular development (Sun et al., 2004). By foreseeing the length of the culture and its stiffness, this model was used to optimize cellular differentiation techniques.

Comparison between the advantages and disadvantages of machine learning and mechanistic modeling.

Mechanistic modeling

➤ Develops a mechanical connection between inputs and results
➤ Accurately integrating data from various space and time scales is challenging
➤ Equipped to handle tiny datasets
➤ Once validated, it can be used to perform difficult or costly experiments

Machine learning

➤ Establishes statistical connections and correlations between the inputs and the outputs
➤ May solve issues involving various time and space scales
➤ Involve large datasets
➤ Only make predictions within data supplied related to patterns

In conjunction with single-cell approaches, VIPER (virtual inference of protein activity by enriched regulon analysis)/ARACNE-based TF modeling can be utilized to designate tissue-specific differentiation programs, which can subsequently be employed for tissue engineering or mimicking. The VIPER (virtual inference of protein activity on a per-sample basis using gene expression profile data) uses the expression of genes that are most directly controlled by a specific protein, such as the targets of a TF, as a reliable indicator of its activity. VIPER is provided in this package in two flavors: a multiple sample version (msVIPER) designed for gene expression signatures based on multiple samples or expression proles, and a single sample version (VIPER) that estimates relative protein activity on a sample-by-sample basis, allowing the transformation of a typical gene expression matrix (i.e., multiple mRNA proles across multiple samples) into a protein activity matrix, representing the relative activity of each gene. In conjunction with single-cell approaches, VIPER/ARACNE-based TF modeling may be used to identify tissue-specific differentiation programs, which can then be utilized for tissue engineering or mimicking (Wan et al., 2022).

8. Computational modeling associated with stem cells in the era of single-cell big data production

During the past 20 years, the discovery of induced pluripotent stem cells (iPSC) has revolutionized stem cell research. This has offered new challenges and opportunities for the examination of human disease and the regeneration of tissues, including cell transplantation. The speedy development of single-cell techniques has characterized cellular phenotypes in various tissue conditions during development and reprogramming. The development of new techniques that produce single-cell phenotypic omics data like epigenomes, proteosomes, and spatial information and the resolution of scRNA-seq permit systematic integration and data analysis, which led to a more detailed characterization of types of cells and their interactions (Del Sol & Jung, 2021). Despite technical limitations like gene dropouts and slow capture rates, single-cell data analysis yields high statistical power by more individual samples and permits the identification of subpopulations. A perfect basis for addressing important questions in stem cell research in the field is made possible by the enormous generation of multiomics single-cell-based data. These high-resolution computational tools can capture the behavior of genes at the molecular or tissue level. Computer models can directly generate original hypotheses and offer possible insights into biological mechanisms. In addition, cellular-level models like GRN-based ones can improve our knowledge of cellular differentiation and conversion, which depict the key TFs and signaling molecules for regulation (Kime et al., 2019).

In conclusion, the statistical data analysis by computational tools addressed several biological processes and established highly novel hypotheses that are directly linked with the experimental study in the field of stem cell research and regenerative medicines. In the fields of stem cell research and regenerative medicine, there are huge biological problems that might be resolved via computer models. It is high time for the development of computational tools due to ongoing advancements in technology. In several areas of research, we can produce or create models based on biological processes more accurately at different levels of complexity. These models have the possibility of significant advancement in regenerative medicine, in vivo reprogramming, and stem cell transplantation by overcoming the limitations of experimental methods.

9. Current obstacles and future possibilities

Cell typing, lineage tracing, trajectory inference (TI), and regulatory networks at the intersections of stem cell and computational biology.

9.1 Cell typing and validation of cell fate engineering

Three specific objectives of computational system cell biology are related to the notion of cell identity and can be attained by employing analogous computational approaches that operate on molecular profiles. These include determining:

(1) Whether or not a population of cells is pluripotent,
(2) Regardless of whether a single cell is multipotent or pluripotent, and
(3) The fidelity of synthetic cell types compared to their in vivo counterparts.

Although some potential experiments, e.g., blastocyst complementation, are perhaps the gold standard for identifying whether a cell group is pluripotent (Darling & Sun, 2004), they are not always possible or desirable for all species.

Molecular surrogates of pluripotency are restricted to a handful of markers, such as Pou5f1/Oct4 in pluripotent stem cells (Price et al., 2020) and Lgr5 in intestinal stem cells (Feng et al., 2021). To answer this query, computer strategies utilizing genome-wide data have evolved. The "first-generation" methods were trained on aggregate data, which is known to have limitations compared to single-cell data. If some cells stay pluripotent while others develop, profiling clusters of cells may result in a poor pluripotency score (Müller et al., 2010). The majority of the second generation of computational pluripotent predictions are based on scRNA-seq and other single-cell genome-wide data. The ability to differentiate into lineages is a defining feature of stem cells. If single-cell data can be used to deduce lineage links, this information can be included in a probable fate metric. Similar to the first generation, the second generation of methods for predicting the potential of a cell's fate depends on large datasets. Several "cell typing" methods already exist, including scamp, scID, single-celled, scPred, scClassify, and CellAssign (Avior et al., 2015; Barker et al., 2007; Cahan et al., 2021).

The purpose of this section is to determine whether cell fate engineering (CFE) products match their "natural" counterparts. Multiple methods for predicting cell type and viability follow standard practices for scientific computational tools. Instructions on how to optimize parameters and evaluate the output's reliability are two areas that require further development. While it would be optimal if these outputs were paired with performance metrics such as sensitivity and precision, they are still sufficient for hypothesis generation. Various current cell-typing techniques were predominantly developed and evaluated for terminally differentiated cell types, whereas stem progenitor cells are, by definition, uncommitted. It is observed that because cells frequently exhibit a continuum of expression states as they differentiate, supervised machine-learning approaches to predict cell development will require exceptionally large training datasets. In addition to scATACseq, emerging approaches include dual profiling of RNA and open chromatin, RNA and protein surface indicators, and RNA and DNA methylation. Methods that account for underlying genetic variability and the function of variations in regulating cell destiny may reveal cellular mechanisms and point the way for future research (Alquicira-Hernandez et al., 2019).

Thirdly, we must establish methods and nomenclature for dealing with instances in which CFE generates cell types that do not exist or have not been detected. Imperfectly differentiated or reprogrammed cells occur in hybrid and frequently transitory stages and the scope of artificial cell types has not been well examined (Okawa et al., 2018). To perform more precise cell typing, we will need to utilize in situ technologies such as cell morphology, structural characteristics, and neighboring cell behavior. In situ, data will be used to characterize the fidelity of tissue and organ engineering.

9.2 Lineage tracing

In contrast to cell types, the primary goal of lineage tracing is to identify all descendants deriving from a single cell. Even though numerous lineage-tracing methods can achieve high spatial resolution, they typically capture few molecular characteristics of cellular components. To combat this, single-cell transcriptomics has assumed a central role in recent years. Multiple paradigms for stem cell differentiation and reprogramming have utilized coding for clonal analysis and lineage tracing (Alquicira-Hernandez et al., 2019). The accessibility and malleability of cells in these systems make them amenable to multiple rounds of viral transduction, a more practical approach for cell barcoding than genome editing. In addition, cells can be sampled via differentiation or reprogramming, allowing for the association of progenitor states with ultimate fate. Using this strategy and two computational methods (multilayer perceptron neuronal network and logistic regression), it was determined how well gene expression state in progenitors predicts eventual cell fate during hematopoiesis (Zhang et al., 2022). Sister cells at later stages of development were utilized to predict the dominant differentiation outcome, with gene expression at later stages being more predictive than the progenitor expression state. The outcomes demonstrate the existence of heritable traits governing fate determination that were not detected by scRNA-seq alone. During the direct reprogramming of mouse embryonic fibroblasts to induced endoderm progenitors (iEPs) by TFs (Alemany et al., 2018), similar observations were made. In contrast, an in vivo state-fate system identified distinct transcriptional states that predicted differentiation potential in HSCs (Berneman-Zeitouni et al., 2014). This apparent discrepancy may be explained by the fact that ex vivo and in vitro progenitors have less well-defined transcriptional and epigenomic molecular states than their in vivo counterparts. To identify these heritable characteristics and resolve these apparent discrepancies, additional information from chromatic accessibility assays, such as ATAC-seq and single-cell ATAC-seq, will be necessary. Specifically, machine learning and GRN analysis of gene expression and chromatin accessibility data revealed that lineages that failed to transition to iEPs did so due to the function and identity of iEPs (de Soysa et al., 2019). The target genes required for pluripotency are "locked" in heterochromatin and therefore inaccessible to reprogramming factors (Weinreb et al., 2020), as indicated by these findings. This combination of lineage tracing, scRNA-seq, and computational investigation suggests that chromatic accessibility determining factors are plausible heritable properties that affect the outcome of cell differentiation and reprogramming. The computational analysis of single-cell lineage tracing is in its infancy, and there are currently numerous unmet needs. Complex experimental platforms are being overcome by innovative computational techniques. Diverse data types will be required to identify the inherited characteristics that determine cell fate.

9.3 Trajectory inference

TI is the computational task of detecting solitary cells involved in temporally regulated biological processes. TI enables the identification of gene clusters whose temporal expression is correlated, thereby facilitating the "guilt-by-association" determination of the function of unannotated genes. TI enables the inference of causal regulatory relationships by arranging cells in pseudotemporal order. This facilitates the identification of differentiated regulators and the subsequent transcriptional cascade. In practice, TI has been employed most often to investigate differentiation and, like lineage tracing, lineage relationships. Wanderlust and Monocle were among the first TI techniques designed for mass cytometry and scRNA-seq data (Biddy et al., 2018; Soufi et al., 2012). Their application led to the discovery of new regulators of branch point decisions. Determining whether directed differentiation and direct programming of motor neurons follow the same developmental pathway was another innovative application. ScRNA-seq was initially devised to predict cell differentiation pathways based on scRNA sequences. Since then, dozens of TI methods that extend beyond the fundamental TI pipeline have been developed. Waddington's optimal transport is robust to the underlying topology of the developmental trajectory. There are many significant unanswered questions and development opportunities in TI (Bendall et al., 2014). The assumption that transcriptional resemblance equals lineage relationship might not be valid in all developmental contexts. In vitro, data continue to significantly rely on temporal ordering as described in the original publications. The inclusion of fate-state data in the expanding TI gold standards is essential.

9.4 Extrapolation and application of regulatory networks

Multiple macromolecules (e.g., genes, proteins, metabolites, and noncoding RNA) interact in diverse ways to generate the extraordinary diversity of emergent behavior observed in each of our cell types. Computational representations of regulatory networks enable system modeling of cell behavior. To investigate cellular decision-making in stem cell biology, transcriptional networks are utilized. Genes are represented by nodes in GRNs, whereas transcriptional regulation is represented by edges. Using ChIP-ChIP or Chip-seq edges for a single TF can be inferred (Karimi et al., 2020; Trapnell et al., 2014). GRN has traditionally been applied to gene-by-gene analysis, which can assess the effect of multiple TFs alone or in combination. Using methods such as Perturb-seq and Reprogram-seq, this strategy has been recently extended to the single-cell level (Briggs et al., 2017). Identifying the role of SOX and HOX TFs in the emergence of hematopoietic lineages from mesoderm has been made possible by reconstructing GRNs from single-cell data (Boyer et al., 2005). Determining how to optimally incorporate pseudotemporal data to infer causal relationships is the logical next step. Approaches based on GRN that account for variations in regular activity will detect TFs that influence the outcome. One of these techniques identified a combination of small molecules that enhanced the efficacy of reprogramming. Another application of GRNs used information theory concepts to identify synergistic factors that promoted the transformation of hindbrain neuroepithelial cells into medial floor plate midbrain progenitor cells (Feng et al., 2021).

10. SWOT analysis of computational approaches

SWOT analysis analyzes the strengths, weaknesses, opportunities, and threats in multiple areas of research with the advanced tools or techniques of the computational model. The primary objective of stem cell tissue engineering is to develop structures that heal, preserve, and enhance damaged tissues or organs. Synthetic skin and cartilage are examples of artificial tissues that have been authorized by the U.S. Food and Drug Administration. Regenerative medicines make considerable use of tissue engineering, and the two concepts have become largely interchangeable. Reports on SWOT analyses of lists of multiple data points can act as a systematic controller for transplant organizations, authorities, and patients (Copes et al., 2019; Mohsin et al., 2011). The SWOT analysis of stem cell tissue engineering is as follows:

10.1 Strengths

In association with tissue engineering, a variety of adult stem cell types that might be used to treat numerous illnesses have been found. Also, multiple clinical investigations have proven the functional improvement of scaffolds derived from tissue engineering. Using a multidisciplinary approach to comprehend cellular treatment may result in improved overall performance as well as the long-term healing and regeneration of functioning tissue. Other added advantages include:

❖ Natural polymer or FDA-approved synthetic polymer
❖ Weak antigenicity
❖ Biocompatibility
❖ High porosity
❖ Promotes vascular and cellular adhesion
❖ Biodegradable

10.2 Weakness

Even though computational models of biological organisms have inherent limitations, this is not cause for despair. Nothing in biology is outside the domain of chemistry and physics, so it should eventually be possible to overcome current obstacles and create much more accurate simulations. This will necessitate the accumulation of an unprecedented quantity and level of information. Additionally, tissue engineering has several flaws that might be viewed as opportunities, including:

❖ Low mechanical property
❖ Lack of understanding regarding the optimal patient, treatment, and timing of treatment
❖ Lack of appraisal of mechanistic approaches in clinical trials
❖ High cost of the material and scaffolds
❖ Difficulty in scaling up

10.3 Opportunity

In the coming years, technological advances in tissue engineering and the creation of in vitro implants utilizing three-dimensional (3D) printers are anticipated to contribute to the expansion of the market. The increasing number of reconstructive and replacement procedures is predicted to contribute to the growth of the market. In addition, the broad scope of growth and substantial research in tissue engineering are anticipated to have a beneficial effect on the market size. To achieve this objective, cross-disciplinary, multiscale, multisource hybrid strategies are required:

❖ Need for understanding cellular mechanisms influencing functional improvement
❖ Functional improvement is modest and variable depending upon a variety of factors, including patient selection, cell selection, and long-term clinical results
❖ Long-term planning of GMP facilities under careful regulation by the FDA

10.4 Threats

Due to the use of disruptive technologies, the current economic climate, and the small scale of the market, the cell therapy market is highly volatile. In such a market, industries and academic research institutes need instruments to advance their knowledge while simultaneously reducing their R&D expenses, increasing product quality and productivity, and shortening the time to market. The regulatory process must be followed for stem cell-based tissue engineering therapeutic products, which is challenging and should be based on quality by design principles. As tissue engineering encounters various concerns, the following obstacles must be addressed to reduce potential dangers:

❖ Pathogenic contamination
❖ Source-dependent variability
❖ Interbatch variability
❖ Current effect of autologous versus allogeneic cell therapy with each type of stem cell
❖ A cost-effective and safe modality may fall into disuse

11. Overview of cell modeling: from basic 2D to 3D innovative system

Cell cultures have proven to be the primary controlling system in the field of cell biology, from fundamental research to progressively advanced translational methods. Standard two-dimensional (2D) culture characteristically comprises a definite type of cell (primary or continuous cultures) as adherent monolayers of cells or freely formed suspension cultures, containing media with necessary nutrients and growth factors. In-vitro 2D culture has produced a high number of populations, allowing the development of metabolic and molecular studies (Duan et al., 2019). They are not appropriate for disease modeling due to their inability to replicate the complexity of the in-vivo biotic environment. The advancement of 3D-cell culture (Moignard et al., 2015) has overcome the limitations of 2D culture. The validation of 3D-cell cultures is to produce cells in microstructures that bear a resemblance to tissues or organs, thus permitting better cell-to-cell/cell-environment interactions and signaling pathway cross-talk, which is beneficial for the proper advancement of tissue function. This 3D-cell culture approach consists of numerous methodologies and procedures that can be categorized as scaffold-based or nonscaffold-based platforms, each with distinct advantages and disadvantages that contribute to their translational efficacy. In silico modeling of multicellular systems reinforces 3D cell culture practices, which necessitate the development of techniques in systems biology (Calzone et al., 2006). These computational tools offer the simulation of a multiplicity of cellular events in space and time, both deterministically and stochastically (Table 27.3).

12. Computer-aided tissue engineering and regenerative medicines

The applications of tissue engineering associated with computer-aided tissue engineering (CATE) apply cutting-edge systems from modern design and manufacturing, biomedical engineering, information technology, and biology. Computer-aided design (CAD), medical image processing, computer-aided manufacturing, and solid freeform fabrication are highly used in the simulation, design, and construction of tissues or organs (D'Alessio et al., 2015).

TABLE 27.3 Computational tools: software and their application.

Software	Application	Links	References
BIOCHAM4	Software for modeling biochemical system	https://gitlab.inria.fr/lifeware/biocham	Duan et al. (2019)
BioSPICE	Stimulation of biological software used for intra and intercellular examination, modeling, and the simulation of spatio-temporal processes	https://github.com/cnastoski/Salem-Website/blob/master/biospice.htm	Jensen and Teng (2020)
CellML	This stores and shares computer-based mathematical models	https://github.com/topics/cellml	Machado et al. (2011)
COPASI Biochemical system stimulator	Stimulation and analysis software for evaluating the biochemical networks and their dynamics	https://github.com/copasi/COPASI	Kumar and Feidler (2003)
E-cell system	Better platform for modeling, simulation, and analysis of cell which is highly complex, heterogeneous, and multiscale systems	https://github.com/ecell/ecell3	Cuellar et al. (2015)
EPISIM	Graphical multiscale modeling platform and stimulation of the multicellular system	https://github.com/StatCan/EpiSim	Bergmann et al. (2017)
MCell	Utilizes spatially accurate 3D cellular models and specific Monte Carlo algorithms to simulate molecular interactions between cells.	https://github.com/mcellteam/mcell	Arjunan (2013), Sütterlin et al. (2013)
Morpheus	Perform modeling and simulation to study the multiscale and multicellular system	https://github.com/nv-morpheus/Morpheus	Tapia et al. (2019)
Tissue Stimulation Toolkit	2D library for the cellular Potts model (CPM) is a computational tool used to study cells and tissue	https://github.com/mpicbg-scicomp/tissue_miner	Kerr et al. (2008)
VCell	The computational context for addressing electrophysiological and physiochemical procedures in living cells	https://github.com/virtualcell/vcell	Starruß et al. (2014)

The significant tissue engineering applications involved in CATE are:

a. 3D anatomic visualization and reconstruction and bio-physical modeling for surgical planning and simulation are the key elements of computer-aided tissue modeling.
b. Computer-assisted tissue classification and application for tissue-level identification and its characterization at various levels of hierarchical organization, biomimetic design, and multiscale modeling.
c. The regeneration of tissues and organs through the process of bio-manufacturing includes tissue constructs, 3D cell culture, and organ printing via bio-blueprint modeling.

13. Bio-CAD

The current advancements in computational technologies, including both hardware and software, have stimulated the rate of CAD applications instead of conventional design and its analysis. Various modern applications of CAD include biomedical engineering (Shevidi et al., 2017), architecture, manufacturing designs, and the advancement of imaging technologies such as computed tomography (CT), magnetic resonance imaging, microCT, and optical microscopy. These technologies' generated data has been used to simulate human joints and produce scaffolds and implants, a process known as bioCAD modeling (Resasco et al., 2012). The ideal characteristics features of bio-scaffolds include imitating structure, dynamics, and biocompatibility. Here, two types of software, namely "scaffold informatics" and "bio-scaffold informatics," are used to describe problems as a whole (Resasco et al., 2012), and the core design of CATE provides information on morphology, anatomy, and the organization of tissue.

14. Biomimetic designing

The advancement in the formation of scaffolds along with a regular and controlled fashion of designing offer a challenge in the field of tissue engineering. There are several scaffolding production methods that enable the internal structure without the involvement of fundamental properties like size, shape, and pores, which then lead to unsuitable and irregular structures for the applications of tissue engineering. There are few boundaries or limitations on the complexity that may be accomplished, and the advent of CAD and manufacturing procedures permits a better way to control the elements of scaffolds. The creation of multifaceted pores with interrelated channel systems in CAD is a tedious and time-taking process that requires a wide range of geometrical elements with pores. The automated design has been developed to combat the problem by integrating CAD along with an exclusive software algorithm that allows the design of a number of geographical elements automatically with highly proficient properties. For instance, this technology is used to design five cubical-shaped scaffolds with interconnected channels associated with pores reaching a diameter of 200–800 mm (D'Alessio et al., 2015). In addition, one of the craniofacial implants has been published and reported in a clinical trial (Cho et al., 2017).

15. Constriction of stem cell-based tissue-engineered patient-specific implants

This novel therapy regimen employs medical imaging, computational modeling, and a rapid prototyping-fabricated bio-resorbable scaffold. A flowchart depicts the design and manufacture of scaffolds customized for each patient (Fig. 27.2).

16. Application of computational modeling to regenerative medicine

To replace damaged or old cells with healthy, functional cells, cell transplantation is one of the primary methods utilized in regenerative medicine. Nevertheless, several obstacles must be surmounted for this strategy to attain its full potential. Therapeutic applications of autologous iPSC-derived cell transplantation have been launched and are ongoing (Mondy et al., 2009; Taguchi & Chida, 2003); For example, cells transplanted from a distinct tissue or grown in vitro. Not all research is effectively integrated with the target tissue (Kikuchi et al., 2017). In order to obtain the correct gene expression identity of host tissue cells, it necessitates the advancement of in vitro manufacturing of donor tissue cells. Indeed, current in vitro experimental techniques frequently suffer from low conversion efficiencies, necessitating the expansion of substantial resources in order to extract sufficient target cells for future functional studies or clinical applications. Moreover, by failing to generate cells with the required phenotypes and functions, in vitro cell conversion frequently produces immature, nonfunctional variations of target cells. In this regard, computer modeling can help overcome these limitations. As a consequence of advancements in scRNA-seq technologies and the development of computational models, researchers can now characterize functionally distinct cell subtypes with subtle gene expression and identify conversion factors for the

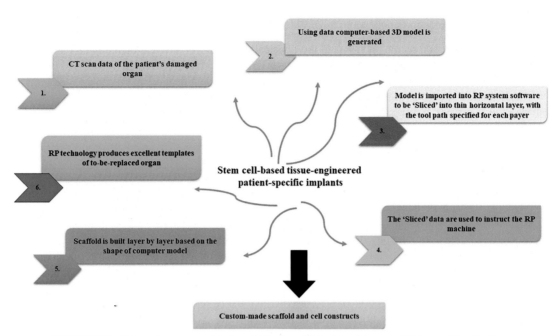

FIGURE 27.2 Schematic representation for solid freeform fabrication of tissue scaffold using RP technology.

in vitro generation of such cell subtypes. In addition, network-based models that incorporate the hierarchical organization of cell TFs with GRNs could predict optimal combinations of conversion factors, such as cell type and subtype-specific TFs. Overexpression of anticipated conversion factors, such as cell type and subtype-specific identity TFs, can facilitate the in vitro generation of functionally mature cell subtypes (Kikuchi et al., 2017; Xu et al., 2015). Combining computational methods with existing or emerging experimental techniques would facilitate the creation of new protocols for treating congenital disease and promoting regeneration. Transplantation of transgenic keratinocyte cultures, for instance, has been demonstrated to regenerate a fully functional epidermis in a patient with junctional epidermolysis bullosa (Hirsch et al., 2017), an inherited disorder characterized by a loss of epidermal stem cells and inadequate keratinocyte differentiation. Nevertheless, the genetic etiology of congenital disorders that affect cell differentiation is frequently unknown, and computational methods can be used to identify disease-related cell fate factors in order to generate transplantable transgenic cells. Computational biology and stem cell tissue engineering are two of the most swiftly expanding academic disciplines. Computational models and simulations can supplement the resource-intensive in vivo and in vitro investigations required for cartilage tissue engineering. Bioengineering human intestinal mucosal grafts employing organoids, fibroblasts, and scaffolds produced from the patients, efficacious production of functional pancreatic cells from dental-derived stem cells using laminin-induced differentiation, CRISPR-based advances in dermatological diseases, and gene editing with "pencils" rather than "scissors" in human pluripotent stem cells to enhance experimental design. Recent advancements in computational stem cell rejuvenation of aging and damaged tissues (Meran et al., 2023; Park et al., 2023; Pearce et al., 2020) indicate that stem cell utility will increase with each advancement in stem cell isolation, cell differentiation, reprogramming technology, and CRISPR/Cas9 genome engineering. This special issue presents cutting-edge research on several types of stem cell computational biology, such as adult stem cells (iPSCs), embryonic stem cells, and cancer stem cells, as well as organoids derived from stem cells.

The use of scRNA-seq in biomedical research has improved our understanding of the pathogenesis of disease and provided valuable insights for the development of novel diagnostic and therapeutic methods. Through the expansion of capacity for high-throughput scRNA-seq, including clinical samples, the analysis of these massive quantities of data has become intimidating for newcomers to this field. Similarly, stem cell rejuvenation is a promising technique for restoring diminished stem cell function and optimizing tissue repair in degenerative diseases and age-related disorders. The negative impact of an ill or aging microenvironment on the endogenous stem cell signaling pathway is one of the causes of stem cell failure. Continuous activation or inhibition of critical signaling pathways that target functionally significant TFs (Kalamakis et al., 2019) maintains the quiescent stem cell phenotype. Computational models of stem-cell niche interactions can aid in identifying niche-specific cues and affected stem-cell signaling pathways, thus casting light on the mechanism of stem-cell functional dysregulation in aging and disease conditions.

17. Concluding remarks

The combination of biomaterials and computer technologies/tools, as well as their translational value to tissue engineering, offers new opportunities to surmount the obstacles faced by current scaffold-building technologies. With the expansion of multiscale computational models, a variety of stem cell research obstacles can be surmounted. In this circumstance, as the quantity of obtainable single-cell data, predominantly scRNA-seq data, increases, we are now able to hypothesize computational models with varying degrees of complexity, including intracellular and cell-cell communication network-based models. In addition, the various types of single-cell omics data will provide a more comprehensive explanation of the biological system and enhance the precision of multiscale models. Through a combination of scRNA-seq and scTAC-seq, we are able to better describe unique cellular subtypes/phenotypes with their gene regulation, permitting the application of cellular differentiation to create specific cell subtypes. Computational software programs and machines have transformed the controlled manufacturing of multifaceted artificial tissue scaffolds. Due to developments in bioprinting technology, 3-dimensional tissue printing has long drawn out to organ printing in recent years. 3D organ printing technology is still being advanced to confront the foremost hindrances in tissue engineering, such as reproducing tissue difficulties and providing vascularization. In the upcoming years, there may be a better development in patient wounds by direct replacement of wound-specific regions by putting in vitro cells/tissues using the endoscopic technique. It is probable that the speedy development of this high-tech research field will reform traditional tissue engineering's therapeutic potential.

18. Future prospects

Through this book chapter, efforts are made to enlighten the close association between experimental and computational models to advance the area of stem cell research along with the clinical applications for regenerative medicine. Scientists should employ their research efforts from the outset to postulate the best-fitting computational model to combat particular biological problems and verify the experimental setup's precision.

Acknowledgments

The author would like to thank the funding agencies, ICMR Senior Research Fellowship (File No. 45/14/2020/PHA/BMS) awarded to Ms. Pinky. Ms. Neha received the Senior Research Fellowship from DST under the INSPIRE program (File No. DST/INSPIRE/03/2019/001769. IVR No. 201900031405]. Prof. Suhel Parvez appreciates the PURSE (SR/PURSE Phase 2/39[C]) and FIST (SR/FST/LS-I/2017/05[C]) grant of DST for financial support.

References

Alemany, A., Florescu, M., Baron, C. S., Peterson-Maduro, J., & van Oudenaarden, A. (2018). Whole-organism clone tracing using single-cell sequencing. *Nature, 556*(7699), 108–112. https://doi.org/10.1038/nature25969

Alquicira-Hernandez, J., Sathe, A., Ji, H. P., Nguyen, Q., & Powell, J. E. (2019). scPred: Accurate supervised method for cell-type classification from single-cell RNA-seq data. *Genome Biology, 20*(1), 264. https://doi.org/10.1186/s13059-019-1862-5

Arjunan, S. N. V. (2013). A guide to modeling reaction-diffusion of molecules with the E-cell system. In *E-cell system. Molecular biology intelligence unit*. New York, NY: Springer. https://doi.org/10.1007/978-1-4614-6157-9_4

Avior, Y., Biancotti, J. C., & Benvenisty, N. (2015). TeratoScore: Assessing the differentiation potential of human pluripotent stem cells by quantitative expression analysis of teratomas. *Stem Cell Reports, 4*(6), 967–974. https://doi.org/10.1016/j.stemcr.2015.05.006

Barker, N., van Es, J. H., Kuipers, J., Kujala, P., van den Born, M., Cozijnsen, M., Haegebarth, A., Korving, J., Begthel, H., Peters, P. J., & Clevers, H. (2007). Identification of stem cells in small intestine and colon by marker gene Lgr5. *Nature, 449*(7165), 1003–1007. https://doi.org/10.1038/nature06196

Bendall, S. C., Davis, K. L., Amir, E.-A. D., Tadmor, M. D., Simonds, E. F., Chen, T. J., Shenfeld, D. K., Nolan, G. P., & Pe'er, D. (2014). Single-cell trajectory detection uncovers progression and regulatory coordination in human B cell development. *Cell, 157*(3), 714–725. https://doi.org/10.1016/j.cell.2014.04.005

Bergmann, F. T., Hoops, S., Klahn, B., Kummer, U., Mendes, P., Pahle, J., & Sahle, S. (2017). COPASI and its applications in biotechnology. *Journal of Biotechnology, 261*, 215–220. https://doi.org/10.1016/j.jbiotec.2017.06.1200

Berneman-Zeitouni, D., Molakandov, K., Elgart, M., Mor, E., Fornoni, A., Domínguez, M. R., Kerr-Conte, J., Ott, M., Meivar-Levy, I., & Ferber, S. (2014). The temporal and hierarchical control of transcription factors-induced liver to pancreas transdifferentiation. *PLoS One, 9*(2), e87812. https://doi.org/10.1371/journal.pone.0087812

Bian, Q., & Cahan, P. (2016). Computational tools for stem cell biology. *Trends in Biotechnology, 34*(12), 993–1009. https://doi.org/10.1016/j.tibtech.2016.05.010

Biddy, B. A., Kong, W., Kamimoto, K., Guo, C., Waye, S. E., Sun, T., & Morris, S. A. (2018). Single-cell mapping of lineage and identity in direct reprogramming. *Nature, 564*(7735), 219−224. https://doi.org/10.1038/s41586-018-0744-4

Boyer, L. A., Lee, T. I., Cole, M. F., Johnstone, S. E., Levine, S. S., Zucker, J. P., Guenther, M. G., Kumar, R. M., Murray, H. L., Jenner, R. G., Gifford, D. K., Melton, D. A., Jaenisch, R., & Young, R. A. (2005). Core transcriptional regulatory circuitry in human embryonic stem cells. *Cell, 122*(6), 947−956. https://doi.org/10.1016/j.cell.2005.08.020

Briggs, J. A., Li, V. C., Lee, S., Woolf, C. J., Klein, A., & Kirschner, M. W. (2017). Mouse embryonic stem cells can differentiate via multiple paths to the same state. *Elife, 6*, e26945. https://doi.org/10.7554/eLife.26945

Cahan, P., Cacchiarelli, D., Dunn, S. J., Hemberg, M., de Sousa Lopes, S. M. C., Morris, S. A., Rackham, O. J. L., Del Sol, A., & Wells, C. A. (2021). Computational stem cell biology: Open questions and guiding principles. *Cell Stem Cell, 28*(1), 20−32. https://doi.org/10.1016/j.stem.2020.12.012

Calzone, L., Fages, F., & Soliman, S. (2006). Biocham: An environment for modeling biological systems and formalizing experimental knowledge. *Bioinformatics, 22*(14), 1805−1807. https://doi.org/10.1093/bioinformatics/btl172

Cho, Y. S., Hong, M. W., Jeong, H. J., Lee, S. J., Kim, Y. Y., & Cho, Y. S. (2017). The fabrication of well-interconnected polycaprolactone/hydroxyapatite composite scaffolds, enhancing the exposure of hydroxyapatite using the wire-network molding technique. *Journal of Biomedical Materials Research, Part B: Applied Biomaterials, 105*(8), 2315−2325. https://doi.org/10.1002/jbm.b.33769

Copes, F., Pien, N., Van Vlierberghe, S., Boccafoschi, F., & Mantovani, D. (2019). Collagen-based tissue engineering strategies for vascular medicine. *Frontiers in Bioengineering and Biotechnology, 7*, 166. https://doi.org/10.3389/fbioe.2019.00166

Cuellar, A., Hedley, W., Nelson, M., Lloyd, C., Halstead, M., Bullivant, D., Nickerson, D., Hunter, P., & Nielsen, P. (2015). The CellML 1.1 specification. *Journal of Integrative Bioinformatics, 12*(2), 259. https://doi.org/10.2390/biecoll-jib-2015-259

D'Alessio, A. C., Fan, Z. P., Wert, K. J., Baranov, P., Cohen, M. A., Saini, J. S., Cohick, E., Charniga, C., Dadon, D., Hannett, N. M., Young, M. J., Temple, S., Jaenisch, R., Lee, T. I., & Young, R. A. (2015). A systematic approach to identify candidate transcription factors that control cell identity. *Stem Cell Reports, 5*(5), 763−775. https://doi.org/10.1016/j.stemcr.2015.09.016

Darling, A. L., & Sun, W. (2004). 3D microtomographic characterization of precision extruded poly-epsilon-caprolactone scaffolds. *Journal of Biomedical Materials Research, Part B: Applied Biomaterials, 70*(2), 311−317. https://doi.org/10.1002/jbm.b.30050

Davit, Y., Bell, C. G., Byrne, H. M., Chapman, L. A., Kimpton, L. S., Lang, G. E., Leonard, K. H., Oliver, J. M., Pearson, N. C., Shipley, R. J., & Waters, S. L. (2013). Homogenization via formal multiscale asymptotics and volume averaging: How do the two techniques compare? *Advances in Water Resources, 62*, 178−206. https://doi.org/10.1016/j.advwatres.2013.09.006

Dawes, J. H. (2016). After 1952: The later development of Alan Turing's ideas on the mathematics of pattern formation. *Historia Mathematica, 43*(1), 49−64. https://doi.org/10.1016/j.hm.2015.03.003

Deisboeck, T. S., Wang, Z., Macklin, P., & Cristini, V. (2011). Multiscale cancer modeling. *Annual Review of Biomedical Engineering, 15*(13), 127−155. https://doi.org/10.1146/annurev-bioeng-071910-124729

Del Sol, A., & Jung, S. (2021). The importance of computational modeling in stem cell research. *Trends in Biotechnology, 39*(2), 126−136. https://doi.org/10.1016/j.tibtech.2020.07.006

Ding, J., Alavi, A., Ebrahimkhani, M. R., & Bar-Joseph, Z. (2021). Computational tools for analyzing single-cell data in pluripotent cell differentiation studies. *Cell Reports Methods, 1*(6), 100087. https://doi.org/10.1016/j.crmeth.2021.100087

Duan, J., Li, B., Bhakta, M., Xie, S., Zhou, P., Munshi, N. V., & Hon, G. C. (2019). Rational reprogramming of cellular states by combinatorial perturbation. *Cell Reports, 27*(12), 3486−3499. https://doi.org/10.1016/j.celrep.2019.05.079

Feng, J., Dewitt, W. S., McKenna, A., Simon, N., Willis, A. D., & Matsen, F. A. (2021). Estimation of cell lineage trees by maximum-likelihood phylogenetics. *Annals of Applied Statistics, 15*(1), 343−362. https://doi.org/10.1214/20-aoas1400

Hirsch, T., Rothoeft, T., Teig, N., Bauer, J. W., Pellegrini, G., De Rosa, L., Scaglione, D., Reichelt, J., Klausegger, A., Kneisz, D., Romano, O., Secone Seconetti, A., Contin, R., Enzo, E., Jurman, I., Carulli, S., Jacobsen, F., Luecke, T., Lehnhardt, M., Fischer, M., … De Luca, M. (2017). Regeneration of the entire human epidermis using transgenic stem cells. *Nature, 551*(7680), 327−332. https://doi.org/10.1038/nature24487

Hollister, S. J., Levy, R. A., Chu, T. M., Halloran, J. W., & Feinberg, S. E. (2000). An image-based approach for designing and manufacturing craniofacial scaffolds. *International Journal of Oral and Maxillofacial Surgery, 29*(1), 67−71. https://doi.org/10.1034/j.1399-0020.2000.290115.x

Jensen, C., & Teng, Y. (2020). Is it time to start transitioning from 2D to 3D cell culture? *Frontiers in Molecular Biosciences, 7*, 33. https://doi.org/10.3389/fmolb.2020.00033

Kalamakis, G., Brüne, D., Ravichandran, S., Bolz, J., Fan, W., Ziebell, F., Stiehl, T., Catalá-Martinez, F., Kupke, J., Zhao, S., Llorens-Bobadilla, E., Bauer, K., Limpert, S., Berger, B., Christen, U., Schmezer, P., Mallm, J. P., Berninger, B., Anders, S., Del Sol, A., … Martin-Villalba, A. (2019). Quiescence modulates stem cell maintenance and regenerative capacity in the aging brain. *Cell, 176*(6), 1407−1419. https://doi.org/10.1016/j.cell.2019.01.040

Kamimoto, K., Stringa, B., Hoffmann, C. M., Jindal, K., Solnica-Krezel, L., & Morris, S. A. (2023). Dissecting cell identity via network inference and in silico gene perturbation. *Nature, 614*(7949), 742−751. https://doi.org/10.1038/s41586-022-05688-9

Karimi, D., Dou, H., Warfield, S. K., & Gholipour, A. (2020). Deep learning with noisy labels: Exploring techniques and remedies in medical image analysis. *Medical Image Analysis, 65*, 101759. https://doi.org/10.1016/j.media.2020.101759

Kerr, R. A., Bartol, T. M., Kaminsky, B., Dittrich, M., Chang, J. C., Baden, S. B., Sejnowski, T. J., & Stiles, J. R. (2008). Fast Monte Carlo simulation methods for biological reaction-diffusion systems in solution and on surfaces. *SIAM Journal on Scientific Computing: A Publication of the Society for Industrial and Applied Mathematics, 30*(6), 3126. https://doi.org/10.1137/070692017

Kikuchi, T., Morizane, A., Doi, D., Magotani, H., Onoe, H., Hayashi, T., Mizuma, H., Takara, S., Takahashi, R., Inoue, H., Morita, S., Yamamoto, M., Okita, K., Nakagawa, M., Parmar, M., & Takahashi, J. (2017). Human iPS cell-derived dopaminergic neurons function in a primate Parkinson's disease model. *Nature, 548*(7669), 592−596. https://doi.org/10.1038/nature23664

Kime, C., Kiyonari, H., Ohtsuka, S., Kohbayashi, E., Asahi, M., Yamanaka, S., Takahashi, M., & Tomoda, K. (2019). Induced 2C expression and implantation-competent blastocyst-like cysts from primed pluripotent stem cells. *Stem Cell Reports, 13*(3), 485−498. https://doi.org/10.1016/j.stemcr.2019.07.011

Kleyman, M., Sefer, E., Nicola, T., Espinoza, C., Chhabra, D., Hagood, J. S., Kaminski, N., Ambalavanan, N., & Bar-Joseph, Z. (2017). Selecting the most appropriate time points to profile in high-throughput studies. *Elife, 6*, e18541. https://doi.org/10.7554/eLife.18541

Korsunsky, I., Millard, N., Fan, J., Slowikowski, K., Zhang, F., Wei, K., Baglaenko, Y., Brenner, M., Loh, P. R., & Raychaudhuri, S. (2019). Fast, sensitive and accurate integration of single-cell data with Harmony. *Nature Methods, 16*(12), 1289−1296. https://doi.org/10.1038/s41592-019-0619-0

Kumar, S. P., & Feidler, J. C. (2003). BioSPICE: A computational infrastructure for integrative biology. *OMICS: A Journal of Integrative Biology, 7*(3), 225. https://doi.org/10.1089/153623103322452350

Lander, A. D. (2007). Morpheus unbound: Reimagining the morphogen gradient. *Cell, 128*(2), 245−256. https://doi.org/10.1016/j.cell.2007.01.004

Lin, C., Jain, S., Kim, H., & Bar-Joseph, Z. (2017). Using neural networks for reducing the dimensions of single-cell RNA-Seq data. *Nucleic Acids Research, 45*(17), e156. https://doi.org/10.1093/nar/gkx681

Machado, D., Costa, R. S., Rocha, M., Ferreira, E. C., Tidor, B., & Rocha, I. (2011). Modeling formalisms in systems biology. *AMB Express, 1*, 45. https://doi.org/10.1186/2191-0855-1-45

Meran, L., Tullie, L., Eaton, S., De Coppi, P., & Li, V. S. W. (2023). Bioengineering human intestinal mucosal grafts using patient-derived organoids, fibroblasts and scaffolds. *Nature Protocols, 18*(1), 108−135. https://doi.org/10.1038/s41596-022-00751-1

Mohsin, S., Siddiqi, S., Collins, B., & Sussman, M. A. (2011). Empowering adult stem cells for myocardial regeneration. *Circulation Research, 109*(12), 1415−1428. https://doi.org/10.1161/CIRCRESAHA.111.243071

Moignard, V., Woodhouse, S., Haghverdi, L., Lilly, A. J., Tanaka, Y., Wilkinson, A. C., Buettner, F., Macaulay, I. C., Jawaid, W., Diamanti, E., Nishikawa, S. I., Piterman, N., Kouskoff, V., Theis, F. J., Fisher, J., & Göttgens, B. (2015). Decoding the regulatory network of early blood development from single-cell gene expression measurements. *Nature Biotechnology, 33*(3), 269−276. https://doi.org/10.1038/nbt.3154

Mondy, W. L., Cameron, D., Timmermans, J. P., De Clerck, N., Sasov, A., Casteleyn, C., & Piegl, L. A. (2009). Computer-aided design of micro-vasculature systems for use in vascular scaffold production. *Biofabrication, 1*(3), 035002. https://doi.org/10.1088/1758-5082/1/3/035002

Morris, S. A., Cahan, P., Li, H., Zhao, A. M., San Roman, A. K., Shivdasani, R. A., Collins, J. J., & Daley, G. Q. (2014). Dissecting engineered cell types and enhancing cell fate conversion via CellNet. *Cell, 158*(4), 889−902. https://doi.org/10.1016/j.cell.2014.07.021

Mozafari, M., Sefat, F., & Atala, A. (2019). Computational design of tissue engineering scaffolds. *Handbook of tissue engineering scaffolds: Volume one* (pp. 73−92). https://doi.org/10.1016/B978-0-08-102563-5.00004-6

Müller, F. J., Goldmann, J., Löser, P., & Loring, J. F. (2010). A call to standardize teratoma assays used to define human pluripotent cell lines. *Cell Stem Cell, 6*(5), 412−414. https://doi.org/10.1016/j.stem.2010.04.009

Neuss, S., Ape, C., Buttler, P., Denecke, B., Dhanasingh, A., Ding, X., Grafahrend, D., Groger, A., Hemmrich, K., Herr, A., Jahnen-Dechent, W., Mastitskaya, S., Perez-Bouza, A., Rosewick, S., Salber, J., Wöltje, M., & Zenke, M. (2008). Assessment of stem cell/biomaterial combinations for stem cell-based tissue engineering. *Biomaterials, 29*(3), 302−313. https://doi.org/10.1016/j.biomaterials.2007.09.022

O'Dea, R. D., Nelson, M. R., El Haj, A. J., Waters, S. L., & Byrne, H. M. (2015). A multiscale analysis of nutrient transport and biological tissue growth in vitro. *Mathematical Medicine and Biology: A Journal of the IMA, 32*(3), 345−366. https://doi.org/10.1093/imammb/dqu015

Okawa, S., Saltó, C., Ravichandran, S., Yang, S., Toledo, E. M., Arenas, E., & Del Sol, A. (2018). Transcriptional synergy as an emergent property defining cell subpopulation identity enables population shift. *Nature Communications, 9*(1), 2595. https://doi.org/10.1038/s41467-018-05016-8

Park, J. C., Park, M. J., Lee, S. Y., Kim, D., Kim, K. T., Jang, H. K., & Cha, H. J. (2023). Gene editing with 'pencil' rather than 'scissors' in human pluripotent stem cells. *Stem Cell Research & Therapy, 14*(1), 164. https://doi.org/10.1186/s13287-023-03394-5

Pearce, D., Fischer, S., Huda, F., & Vahdati, A. (2020). Applications of computer modeling and simulation in cartilage tissue engineering. *Tissue Engineering and Regenerative Medicine, 17*(1), 1−13. https://doi.org/10.1007/s13770-019-00216-9

Pearson, N. C., Shipley, R. J., Waters, S. L., & Oliver, J. M. (2014). Multiphase modelling of the influence of fluid flow and chemical concentration on tissue growth in a hollow fibre membrane bioreactor. *Mathematical Medicine and Biology: A Journal of the IMA, 31*(4), 393−430. https://doi.org/10.1093/imammb/dqt015

Price, J. C., Krause, A. L., Waters, S. L., & El Haj, A. J. (2020). Predicting bone formation in mesenchymal stromal cell-seeded hydrogels using experiment-based mathematical modeling. *Tissue Engineering Part A, 26*(17−18), 1014−1023. https://doi.org/10.1089/ten.TEA.2020.0027

Qiu, X., Rahimzamani, A., Wang, L., Ren, B., Mao, Q., Durham, T., McFaline-Figueroa, J. L., Saunders, L., Trapnell, C., & Kannan, S. (2020). Inferring causal gene regulatory networks from coupled single-cell expression dynamics using scribe. *Cell Systems, 10*(3), 265−274. https://doi.org/10.1016/j.cels.2020.02.003

Resasco, D. C., Gao, F., Morgan, F., Novak, I. L., Schaff, J. C., & Slepchenko, B. M. (2012). Virtual cell: Computational tools for modeling in cell biology. *Wiley Interdisciplinary Reviews. Systems Biology and Medicine, 4*(2), 129−140. https://doi.org/10.1002/wsbm.165

Shevidi, S., Uchida, A., Schudrowitz, N., Wessel, G. M., & Yajima, M. (2017). Single nucleotide editing without DNA cleavage using CRISPR/Cas9-deaminase in the sea urchin embryo. *Developmental Dynamics: An Official Publication of the American Association of Anatomists, 246*(12), 1036−1046. https://doi.org/10.1002/dvdy.24586

Soufi, A., Donahue, G., & Zaret, K. S. (2012). Facilitators and impediments of the pluripotency reprogramming factors' initial engagement with the genome. *Cell, 151*(5), 994−1004. https://doi.org/10.1016/j.cell.2012.09.045

de Soysa, T. Y., Ranade, S. S., Okawa, S., Ravichandran, S., Huang, Y., Salunga, H. T., Schricker, A., Del Sol, A., Gifford, C. A., & Srivastava, D. (2019). Single-cell analysis of cardiogenesis reveals basis for organ-level developmental defects. *Nature, 572*(7767), 120−124. https://doi.org/10.1038/s41586-019-1414-x

Starruß, J., de Back, W., Brusch, L., & Deutsch, A. (2014). Morpheus: A user-friendly modeling environment for multiscale and multicellular systems biology. *Bioinformatics, 30*(9), 1331−1332. https://doi.org/10.1093/bioinformatics/btt772

Stuart, T., Butler, A., Hoffman, P., Hafemeister, C., Papalexi, E., Mauck, W. M., 3rd, Hao, Y., Stoeckius, M., Smibert, P., & Satija, R. (2019). Comprehensive integration of single-cell data. *Cell, 177*(7), 1888−1902. https://doi.org/10.1016/j.cell.2019.05.031

Stumpf, P. S., & MacArthur, B. D. (2019). Machine learning of stem cell identities from single-cell expression data via regulatory network archetypes. *Frontiers in Genetics, 10*, 2. https://doi.org/10.3389/fgene.2019.00002

Sun, W., Darling, A., Starly, B., & Nam, J. (2004). Computer-aided tissue engineering: Overview, scope and challenges. *Biotechnology and Applied Biochemistry, 39*(Pt1), 29−47. https://doi.org/10.1042/BA20030108

Sütterlin, T., Kolb, C., Dickhaus, H., Jäger, D., & Grabe, N. (2013). Bridging the scales: Semantic integration of quantitative SBML in graphical multi-cellular models and simulations with EPISIM and COPASI. *Bioinformatics, 29*(2), 223−229. https://doi.org/10.1093/bioinformatics/bts659

Taguchi, M., & Chida, K. (2003). Computer reconstruction of the three-dimensional structure of mouse cerebral ventricles. *Brain Research Protocols, 12*(1), 10−15. https://doi.org/10.1016/s1385-299x(03)00055-2

Tapia, J. J., Saglam, A. S., Czech, J., Kuczewski, R., Bartol, T. M., Sejnowski, T. J., & Faeder, J. R. (2019). MCell-R: A particle-resolution network-free spatial modeling framework. *Methods in Molecular Biology, 203−229*. https://doi.org/10.1007/978-1-4939-9102-0_9

Till, J. E., Mcculloch, E. A., & Siminovitch, L. (1964). A stochastic model of stem cell proliferation, based on the growth of spleen colony-forming cells. *Proceedings of the National Academy of Sciences of the United States of America, 51*(1), 29−36. https://doi.org/10.1073/pnas.51.1.29

Traag, V. A., Waltman, L., & van Eck, N. J. (2019). From Louvain to Leiden: Guaranteeing well-connected communities. *Scientific Reports, 9*(1), 5233. https://doi.org/10.1038/s41598-019-41695-z

Trapnell, C., Cacchiarelli, D., Grimsby, J., Pokharel, P., Li, S., Morse, M., Lennon, N. J., Livak, K. J., Mikkelsen, T. S., & Rinn, J. L. (2014). The dynamics and regulators of cell fate decisions are revealed by pseudotemporal ordering of single cells. *Nature Biotechnology, 32*(4), 381−386. https://doi.org/10.1038/nbt.2859

Wan, H., Gao, W., Zhang, W., Tao, Z., Lu, X., Chen, F., & Qin, J. (2022). Network-based inference of master regulators in epithelial membrane protein 2-treated human RPE cells. *BMC Genomic Data, 23*(1), 52. https://doi.org/10.1186/s12863-022-01047-9

Waters, S. L., Schumacher, L. J., & El Haj, A. J. (2021). Regenerative medicine meets mathematical modelling: Developing symbiotic relationships. *NPJ Regenerative Medicine, 6*(1), 24. https://doi.org/10.1038/s41536-021-00134-2

Weinreb, C., Rodriguez-Fraticelli, A., Camargo, F. D., & Klein, A. M. (2020). Lineage tracing on transcriptional landscapes links state to fate during differentiation. *Science (New York, N.Y.), 367*(6479), eaaw3381. https://doi.org/10.1126/science.aaw3381

Xu, J., Du, Y., & Deng, H. (2015). Direct lineage reprogramming: Strategies, mechanisms, and applications. *Cell Stem Cell, 16*(2), 119−134. https://doi.org/10.1016/j.stem.2015.01.013

Zhang, M. J., Ntranos, V., & Tse, D. (2020). Determining sequencing depth in a single-cell RNA-seq experiment. *Nature Communications, 11*(1), 774. https://doi.org/10.1038/s41467-020-14482-y

Zhang, X., Qiu, H., Zhang, F., & Ding, S. (2022). Advances in single-cell multi-omics and application in cardiovascular research. *Frontiers in Cell and Developmental Biology, 10*, 883861. https://doi.org/10.3389/fcell.2022.883861

Chapter 28

Computational approaches for bioengineering of cornea

Subodh Kumar[1], Shivi Uppal[2], V.S. Vipin[2], Nishant Tyagi[1], Ratnesh Singh Kanwar[2], Reena Wilfred[2], Sweta Singh[1] and Yogesh Kumar Verma[1]

[1]*Stem Cell and Tissue Engineering Research Group, INMAS, DRDO, Delhi, India;* [2]*Division of Clinical Research and Medical Management, Institute of Nuclear Medicine and Allied Sciences (INMAS), DRDO, Delhi, India*

1. Introduction

Corneal disease is the second leading cause of blindness worldwide prompting the urge to develop a bioactive tissue to overcome the deficiencies of transplanted cornea. Due to avascularity, nonimmunogenicity, and basic cellular structure, Bioengineered corneal substitute reconstruction is simpler than that of other tissues. The availability of bioengineered cornea would relieve demand on organ banks, which have limited cadaveric cornea outputs. Time required for transplantation can also be reduced because several synthetic corneas can be created from a single pool of donor cells (Risbud et al., 2001). Ideal biomaterials for corneal bioengineering should have adequate durability, transparency, biodegradability, biocompatibility with cultured cellular elements, and improved clinical compliance. Tissue engineering (TE) may produce a corneal prosthesis that is similar to living cornea. Apart from silk fibroin, alginate, gelatin, collagen, chitosan, cellulose, hyaluronic acid, and decellularized corneas, native matrix proteins (as basis material), and collagen-col-1 (collagen type I present in cornea) have been intensively examined as appropriate possibilities (Karamichos et al., 2015). The creation of a functional cornea from natural and synthetic biomaterials necessitates the use of specialized tools such as as electro spinning, lyophilizers, and 3D bioprinters. We seek to emphasize the benefits of computational techniques such as computer-aided TE (computer-aided design [CAD] and bioprinting), virtual docking, molecular modeling, in silico Swiss target and toxicity prediction, response surface methodology (RSM), and so on in this chapter. These strategies aid in the standardization and optimization of accessible biomaterials, allowing for the reduction of toxicity and the selection of appropriate growth factors. This method will help in the development of a corneal replacement with the requisite transparency and cell density (Giegengack et al., 2013).

2. Ultrastructure of the cornea

The human cornea is made up of five layers: epithelium, Bowman layer, stroma, Descemet's layer, and endothelium (Fig. 28.1). Only epithelium, stroma, and endothelium are cellular in nature.

2.1 Epithelium

This layer is separated into three layers: the core cornea, the limbus with blood vessels, and the conjunctiva. This layer's phenotypic design inhibits tissue necrosis since it comprises tear film (containing lysozyme) and lactoferrin for epithelial healing. The epithelium can rejuvenate by migrating cells from the limbus area. Surface cells and wing cells make up the next level. Limbal tissue is transplanted from this layer, and in the absence of this layer, corneal tissue is produced from allogenic/autogenic stem cells (Hancox et al., 2020). There are two types of stem cells used in limbal tissue transplantation viz. epithelial stem cells and mesenchymal stem cells (MSCs). The epithelial stem cells are present in the cornea's outermost layer. They are responsible for regenerating the surface of the cornea and maintaining its clarity. These cells are

Computational Biology for Stem Cell Research. https://doi.org/10.1016/B978-0-443-13222-3.00012-5
Copyright © 2024 Elsevier Inc. All rights reserved, including those for text and data mining, AI training, and similar technologies.
395

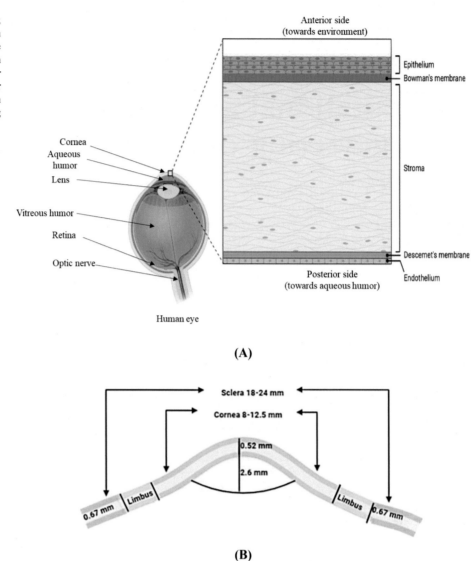

harvested from the healthy limbus of the patient or a donor and transplanted onto the damaged cornea. The MSCs are found in the stromal layer of the cornea, which lies beneath the epithelium and can be used to differentiate into various cell types, including those involved in repairing and regenerating the damaged tissue. MSCs can be harvested from multiple tissues including bone marrow, adipose tissue, umbilical cord tissue, etc.

2.2 Stroma

This largest layer is composed of collagen type I fibrils, which determines the shape, transparency, and tensile strength of the cornea. Keratocytes present in this layer are responsible for providing the stability to the extracellular matrix that is associated with wound healing (Chen et al., 2018).

2.3 Endothelium

The endothelium layer is responsible for maintaining the water balance in the eye. The anterior part of this layer attaches to the Descemet's layer, and the posterior part attaches to the microvilli. This layer is known to maintain osmotic gradient, and if destroyed, it does not regenerate easily due to absence of G1 phase during cell division. Once damaged, endothelial keratoplasty (EK) can be performed based on the biochemical and physical properties of the normal cornea. Another function of this layer is to prevent stromal dehydration and development of corneal edema (a cause of loss of transparency and reduced vision) (Guérin et al., 2021).

3. Corneal diseases

Corneal diseases can result from various causes, including trauma induced by thermal, chemical, or electrical sources (Ren et al., 2010), as well as abrasions from foreign bodies or physical contact. Dystrophies may lead to conditions such as cysts, corneal scarring, and opacity (Whitcher et al., 2001). Dry eyes or keratoconjunctivitis sicca occurs due to vitamin A deficiency or collagen vascular disorders. Corneal edema may arise from damage to the endothelial or outer epithelial layer. Infections are observed following eye surgery, contact lens use, or other eye diseases (Klintworth et al., 2009). Corneal ulcers can develop due to infections (fungal, bacterial, or viral), trauma, contact lens wear, or dryness. Corneal developmental anomalies are hereditary or sporadic in nature, and aging may also contribute to certain conditions (Kaercher et al., 2008). Corneal inflammatory disorders, often associated with autoimmune conditions, would result from trauma or infections. Metabolic disorders, such as lipid disorders, Wilson's disease, and osteogenesis imperfecta, may impact corneal health (Hazlett et al., 2016; Jameson et al., 2021; Yi et al., 2002). Other diseases affect cornea are immunological disorders viz. multiple myeloma and Stevens—Johnson syndrome, as well as corneal curvature defects such as keratoconus, keratoglobus, and corneal ectasia. Various databases have been developed currently to study corneal diseases that also serve as data repository (Table 28.1) in this domain.

4. Corneal transplantation

Different types of corneal transplantations are (a) Penetrating keratoplasty (b) Anterior lamellar keratoplasty (c) Descement stripping automated endothelial keratoplasty, and (d) Descement membrane endothelial keratoplasty (Fig. 28.2). Keratoplasty, keratoprosthesis, and ocular surface reconstruction are the surgical options for the replacement of damaged cornea. Major problems encountered with corneal transplantation are donor cornea shortage, infection risk, and graft rejection. Rejection is an immune-mediated process characterized by cell-mediated response against the transplanted cornea. Prophylactic topical steroids have been demonstrated to be an effective therapy in reducing the rejection of donor cornea. Another therapy, known as amniotic membrane (AM) transplantation, is used for diseases such as pterygium. Gene therapy for corneal endothelial regeneration has shown promising results in few studies (Córdoba et al., 2020).

4.1 Penetrating keratoplasty (PK)

This technique uses femtosecond laser assisted keratoplasty to replace all corneal layers for conditions such as corneal ectasias, infections, and stromal, endothelial, and immunological problems. This surgery has complications such as transplant failure, resulting in endothelial cell loss and astigmatism, choroidal hemorrhage, endophthalmitis, wound leaks, transplant dehiscence, detachment (in 1%—2% of cases), corneal ulceration, keratitis, postsurgical inflammation (scleritis), and neurotrophic cornea (corneal nerves are cut resulting in delayed epithelialization). Retinal detachment, glaucoma, and cataract may appear sometimes following the surgery (Rahman et al., 2009).

TABLE 28.1 Databases for corneal diseases.

Database	URL
National Eye Institute Refractive Error Correction Study (NEI-RECS)	https://www.nei.nih.gov/research/clinical-trials/national-eye-institute-refractive-error-correction-study-nei-recs
The CorneaGen Keratoconus Registry	https://www.corneagen.com/research-and-development/keratoconus-registry/
The Australian Corneal Graft Registry	https://www.acgr.org.au/
The Corneal Transplant Follow-up Study	https://www.ctfstudy.co.uk/
The Singapore Corneal Transplant Study	https://www.singhealth.com.sg/patient-care/clinical-specialties/cornea-transplant-centre/corneal-transplant-study
The Corneal Epithelium Database	https://webeye.ophth.uiowa.edu/ced/

FIGURE 28.2 Types of corneal transplantation.

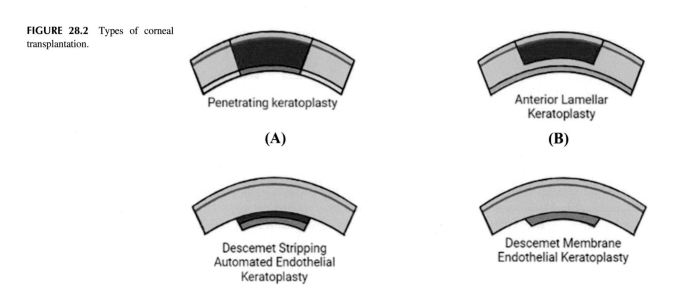

4.2 Anterior lamellar keratoplasty (ALK)

The epithelium and stroma in the anterior lamellar keratoplasty (ALK) procedure are replaced by donor's Bowman's membrane and stroma. This extraocular technique requires less use of topical steroids and reduces transplant rejection. One of the main complication associated with this procedure is astigmatism. Surgical complications such as keratitis, scleritis, epithelial defects, and neurotrophic cornea are also observed in this procedure (Melles et al., 1999; Rahman et al., 2009).

4.3 Endothelial keratoplasty (EK)

This procedure repairs the Descemet's membrane and endothelium in endothelial disorders. EK is better than penetrating keratoplasty (PK) and ALK due to lower risk of astigmatism and better visual outcomes. Complications such as endothelial failure and transplant rejection are very common in EK. Surgical complications of EK posterior include lamellar detachment, severe uveitis associated with mydriasis, corneal ulceration, and neurotrophic cornea. Retinal detachment and cataract due to direct lens trauma are often observed after endothelial keratoplasty (Melles et al., 1999).

4.4 Keratoprosthesis (KP)

This prosthesis replaces all five corneal layers and is used preferentially in late endothelial disorders. Endothelial failure and transplant rejection are less prevalent, although retroprosthetic membrane formation (impeding vision) is a common complication of keratoprosthesis. Other complications associated with keratoprosthesis are glaucoma device dehiscence, fungal keratitis, and retinal detachment (Price et al., 2010; Khan et al., 2001).

4.5 Ocular surface reconstruction

Limbal epithelial and mucosal epithelial transplantation are the other procedures followed for ocular surface reconstruction. The donor limbus is obtained from either conjunctival limbal autograft or allograft from a living/cadaveric donor. Occular surface reconstruction techniques are used for stromal disorder treatment. Transplant rejection, dehiscence, and detachment are commonly observed. Therefore, immunosuppression is indicated post-procedure. Other complications associated with ocular surface reconstruction are microbial keratitis and epithelial defects (Khan et al., 2001; Tan et al., 2012).

5. Biomaterials for corneal bioengineering

Biomaterials are uniquely designed to interact physically and chemically with biological systems. Biomaterials can be obtained from natural materials or through synthetic materials or by modifying natural materials. These biomaterials can be employed in corneal regeneration approaches that rely on scaffold-based corneal regeneration. The selection of biomaterials for corneal bioengineering is based on the ability of the materials or their combined effect on tissue regeneration. A scaffold provides an appropriate milieu for both cellular and extracellular elements. Biomaterials selected for corneal bioengineering must show features such as transparency, biocompatibility, biodegradability, and mechanical strength and allow cells to adhere and proliferate (Ahearne et al., 2020). For corneal bioengineering, both natural and synthetic biomaterials have been studied (Table 28.2). Biocompatibility ensures the physiological

TABLE 28.2 Natural and synthtic biomaterials used in bioengineering of cornea.

	A: Natural biomaterials			
S. No.	Biomaterial	Application	Advantages	Disadvantages
1	Collagen	Substrate for bioengineering of the cornea, epithelium, and endothelium	High biocompatibility	Inferior mechanical properties (Shah et al., 2008)
2	Silk	Corneal stroma equivalent. Substrate for corneal epithelial cells, fibroblasts, and endothelial cells	Optical clarity; adjustable deterioration rate; and mechanical properties	Surface alterations or combinations are required for its effective functionality
3	Chitosan	Substrate for corneal epithelium regeneration, delivery of growth factors and for endothelial cell growth	Easily bio-functionalized Favorable biocompatibility Controllable properties Biodegradability	Combination of other materials required such as gelatin, collagen, and so on
4	Gelatin	Substrate for corneal endothelium cell transplantation and corneal stroma cell growth	Inherent biocompatibility; suitable biodegradability	Inferior mechanical properties (Tarsitano et al., 2022)
5	Silk sericin	Substrate for corneal limbal cells	Highly biocompatible	Mechanical properties are not satisfactory
	B: Synthetic biomaterials			
1	Poly(methyl methacrylate) (PMMA)	Keratoprosthetics	Proper mechanical properties	Lack of cell integration
2	Poly(lactic-co-glycolic acid) (PLGA)	Substrate for corneal endothelium regeneration	Highly transparent; proper mechanical properties	Lack of cell integration, surface modification required
3	Poly(vinyl alcohol) (PVA)	Delivery and regeneration of corneal epithelium and stroma	Biocompatible'sufficient mechanical properties	Low degradation rate, lack of cell integration
				Surface alteration is required, and light transmittance is unsatisfactory (Mahdavi et al., 2020)
4	Polycaprolactone (PCL)	Limbal epithelial cell substrate	Biocompatible; proper mechanical properties	Low degradation rate
5	Poly(ethylene glycol) (PEG)	Substrate for regeneration of corneal endothelium	Biocompatible; have proper mechanical properties	Lack of cell integration
6	Polyethylene (glycol) diacrylate (PEGDA)	Substrate for corneal epithelial cells, epithelial wound healing	Biocompatible; have proper mechanical properties	Surface modification requires, lack of cell integration (Lai et al., 2012)
			Biochemical and topographical stimuli are controlled; the relaxed swelling ratio is minimal	

functioning of bioengineered cornea. The toxic nature of material would bring several other issues such as vision impedement and transplant rejection.

5.1 Natural biomaterials for corneal regeneration

Collagen is a typical biomaterial utilized in corneal bioengineering which is made up of both insoluble fibers and microfibrils. It is a fibrous structural protein comprising of a right-handed bundle of three parallel and left-handed polyproline II-type helices. It is recognized to be the most abundant protein in extracellular matrix. Collagen accounts for approximately 70% of the dry weight of the cornea (Duan et al., 2006), and its variations are regarded as the primary constituents of the cellular scaffold. For manufacturing corneal scaffolds, collagens are produced from animal skin and tendons (bovine or pig) or rat tails. Physical properties of bioengineered scaffolds depend upon the source of collagen. Collagen type I is the major collagen of skin, tendon, and ligament; collagen type II is present in the cartilage; collagen type III is an important component of the blood vessels. The type I collagen is the most prominent collagen in the connective tissue that is extensively studied (Ramshaw et al., 1996). It is comparatively cheaper to isolate (Chen et al., 2018; Antoine et al., 2014). To enhance the functionality of bioengineered cornea, collagen has been used with various other natural and synthetic biomaterials to make composites including collagen and poly(vinyl alcohol) (PVA) composite (Miyashita et al., 2006; Rafat et al., 2008), poly(ethylene glycol) diacrylate (PEGDA), collagen and chitosan IV composite, PEGDA and collagen composite (Kadakia et al., 2008), compressed collagen and poly lactic-co-glycolide (PLGA) composite (Kong et al., 2017), and polycaprolactone (PCL) and collagen composite (Kim et al., 2018). Youngkoo Lee et al. used three-dimensional model for collagen network simulation for their shear and tensile tests. They observed that different networks determine the mechanical properties of collagen. These results helped in collagen optimisation as ECM.

Several in vitro studies have shown that collagen stimulates cell differentiation and proliferation, as well as structural integrity (Jameson et al., 2021). According to C. Somaiah et al., collagen-coated tissue culture plates promote mesenchymal cell adhesion and proliferation under stress (Torricelli et al., 2013). The scaffolds employed in corneal bioengineering encounter a variety of challenges including low mechanical strength and lack of stroma's native organization. Various studies have explored different methods to overcome these challenges by using collagen as their primary material for fabricating corneal scaffolds (Kular et al., 2014).

Fibrin and sericin are the two constitutive proteinaceous molecular complexes formed by silk-producing insects. Because of its biocompatibility, better mechanical properties, and faster degradation,. silk fibroin is widely employed in TE and regenerative medicine. Silk fibroin has become a promising biomaterial for the bioengineering of cornea due to its inherent optical clarity (Harkin et al., 2011). Electrospinning and evaporation techniques are used for the development of silk fibroin—based corneal scaffolds. Laura J. Bray et al. have prepared a protein membrane fibroin, which is isolated from silkworm (*Bombyx mori*) and is known to support the cultivation of human limbal epithelial (HLE) cells and thus shows significant potential as a biomaterial for ocular surface reconstruction (Bray et al., 2013).

Chitosan is known as a naturally occurring polymer that is synthesized by deacetylation of chitin. It is the second-most prevalent polymer with benefits such as biocompatibility, hemostatic activity, biodegradability, and cell proliferation potential for bioengineering applications. Lung-Kun Yeh et al. have reported that primary bovine corneal epithelial cells (BCECs) seeded on pure chitosan membrane showed better proliferation and attached quickly to amniotic membrane (AM). Yeh et al. stated that the artificial chitosan membrane can preserve the phenotype of BCECs in vitro as compared with AM. Incorporating chitosan into the collagen membrane improves the optical properties and mechanical properties of the collagen membrane (Chirila et al., 2013). Gelatin, a molecular derivative of collagen, is produced by the irreversible denaturation of collagen proteins. Gelatin has a similar chemical structure and function to collagen, frequently used to replace collagen in cell and tissue cultures. It is generally isolated from highly collagenous raw materials such as pig skin, cattle bones, and bovine hides (Mistry et al., 2009). Carbodiimide cross-linked porous gelatin is considered as an alternative to collagen for TE of the stromal layer of the cornea (Li et al., 2014).

5.2 Synthetic biomaterials

A large number of synthetic biomaterials have been investigated for the bioengineering of cornea such as poly(2-hydroxyethyl methacrylate) (PHEMA), PVA, PLGA, and PEGDA (Khan et al., 2001). Application of synthetic biomaterials in corneal bioengineering is associated with cytotoxicity and require chemical modification to enhance cell proliferation and adhesion. According to previous studies, it has been observed that the biodegradation of synthetic biomaterials is higher over natural biomaterials (Lai et al., 2012). The uses of natural and synthetic biomaterials has become limitations due to their unknown interaction pattern and manual permutation and combination, which may not

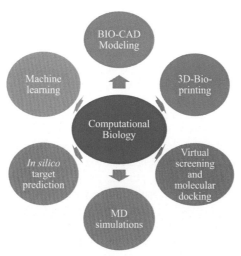

FIGURE 28.3 Computational approaches in TE to fabricate tissues of desired properties for multiple applications. The computational approaches use mathematical algorithms/software to identify growth factors and biomaterials for designing tissue of interest. These softwares include virtual screening and molecular docking, MD simulation, BIO-CAD (biological computer-aided design), in silico target prediction, 3D bioprinting, and artificial intelligence. The image was created using Microsoft PowerPoint.

result in proper binding of biomaterials. The application of computational biology would eliminate the outliers and may help in selecting proper binding of biomaterial(s). This would stabilize the mechanical strength and viability of the synthesized bioengineered cornea.

6. Computer-aided tissue engineering

Computer-aided TE is a modern field that encompasses the application of biology, engineering (Fig. 28.3), computational biology, and manufacturing to make a scaffold that mimics a natural tissue for cell migration, cell proliferation, and tissue repair/regeneration (Chen et al., 2018).

6.1 BIO-CAD modeling

Beyond the conventional design and analysis software, the application of CAD has revolutionized the field of TE, especially in the development of artificial cornea, heart, lung models, etc. The advancement in CAD has taken place due to advances in magnetic resonance imaging (MRI), microcomputed tomography, and optical microscopy (Geris et al., 2013). In CAD modeling, the MRI data are converted into CAD format to make a 3D model of interest using various algorithms. The data retrieved from the CAD software are extensively analyzed for the development of scaffolds that mimic natural tissue. Simulation of bone, muscle, etc. can be done in CAD for better utilization of this technology. The BIO-CAD modeling work includes three steps, acquisition of the noninvasive image, imaging process, and construction of a 3D-based model. The application of BIO-CAD would significantly enhance the properties of the artificial cornea, for instance, cell viability, cell attachment, transparency, and sustainability (Geris et al., 2013).

Ikram et al. used machine learning application in BIO-CAD to include a Grossberg layer based on Runge-Kutta numerical schemes, which improved the stability and accuracy of the standard feedforward neural network. The system demonstrated the ability to detect keratoconus at early stages without relying on a large set of interdependent clinical programs. The study compared the CAD system to other available approaches for keratoconus detection using cross-validation techniques and receiver operating characteristic (ROC) curves. The CAD system achieved high accuracy in distinguishing clinical keratoconus from normal eyes, with accuracy rates of 99.58% for moderate cases and 99.40% for mild cases. Furthermore, the CAD system outperformed other systems in detecting suspect keratoconus, with an accuracy of 96.56%. The applicability of the BIO-CAD has not been demonstrated in the bioengineered cornea optimization, but in TE approaches. Therefore, the uses of BIO-CAD versed with machine learning algorithm may drastically help the researcher to develop functional bioengineered cornea that mimics the natural one (Issarti et al., 2019).

6.2 Three-dimensional printing technology

The three-dimensional printing (3DP) technology is the latest in the field of TE. There have been several approaches to design 3D scaffold models such as CAD-based methods, image-based method, implicit surface, and space-filling curve. A major advantage of 3DP is to make layer-by-layer deposition of biomaterials, rapidly incorporating cells to make artificial constructs. Like other techniques, the 3DP is capable of simultaneously delivering biomaterials, living cells, and growth factors spatially to form living cells/extracellular matrix for in vitro or in vivo application. 3DP uses powerful commercial CAD software packages. The model is designed into two main parts plus a support part. These design parts are exported individually as STeroLithography (STL) files describing the 3D surface as triangular polygons in a Cartesian coordinate system. Further, the slicing algorithm helps in generation of a cross- section of design model. The printing paths are composed of lines between points 1, 2,3, ….n and print head moves starting from point 1. The printing paths are arranged in a zig-zag manner to minimize the total length of the printing path. The final step of the algorithm is the generation of commands to control the four head printing system, based on user defined parameters to control speed and dispensing pressure. The computer algorithm helps in controlling most of the parameters of design construct for their well-suited 3D structure (Jung et al., 2016). Recently, several organ models have been printed to check the therapeutic effects of drugs designed to counter COVID-19 infection. A planning program controls the process of 3DP of tissues and organs. During the printing process, the tool path directs the movement of the printing head(s) to deposit cells as needed to produce a 3D construct. Besides, the toolpath program should include cell composition information to help print heads in depositing the correct cells spatially and temporally.

The 3DP builds the 3D object by first spreading fresh powder on a platform and then depositing the binder solution into the powder bed with an "inkjet" print head. Following the printing of each 2D layer, a fresh layer of powder is laid down, followed by the deposition of binder solution. The process is repeated until the entire structure is finished. The final product is extracted by removing loose powder after the binder has dried in the powder bed, integrating each layer.

Several studies have reported that 3DP constructs show enhanced cell viability as compared with electrospinning and lyophilization techniques. In printed samples, the vitality of numerous cell types was found to be 90%. Despite these excellent technologies and findings, which demonstrate the promising scope of organ printing, significant difficulties remain before full potential can be realized (Shahrubudin et al., 2019). The 3DP technique may facilitate layer-by-layer printing of corneal layers by maintaining the orientation of collagen for improved shape, transparency, and strength of the cornea. Ruiqi Wang et al. printed a 3D bioengineered cornea using CAD software containing ink of gelatin methacrylate modified by hyaluronic acid. Further, the cornea was enriched with primary rabbit-derived corneal stromal cells to secrete and form ECM with a highly arranged architecture.

6.3 Virtual screening and molecular docking

In TE, the selection of a suitable growth factor that promotes the proliferation of cells at the site of injury is followed by their migration throughout the scaffold. The selection of the right growth factor is a tedious task; therefore, the application of computational biology becomes beneficial in TE (Niu et al., 2013). Currently, many softwares have been used for molecular docking (Table 28.3) study, for instance, AutoDock (Huey et al., 2012) docks ligand and protein by Lamarckian genetic algorithm and empirical free energy scoring function; the DOCK software works through geometric matching algorithm; the EADock program functions through evolutionary algorithms. A number of binding poses and flexibility of macromolecules and ligands ensure proper binding of ligands at suitable binding sites. By the application of molecular docking, researchers investigate possible lead compounds/proteins/ligands for their binding pattern (Yang et al., 2011). Ardan et al. investigated the potential of modified polyether ether ketone (PEEK) incorporated with nanohydroxyapatite (HA) for orthodontic mini-implant fabrication. They performed molecular docking simulations to analyze the binding affinity of HA, PEEK, and the HA + PEEK complex with 12 target proteins related to osseointegration. They observed that the HA + PEEK complex exhibited the strongest negative binding affinity with osteogenic markers ALP and IGF-1, indicating its potential for enhanced osseointegration compared to nano-HA or PEEK alone (Ardani et al., 2022). Furthermore, Nihan et al. also used molecular docking to investigate the interaction between Nisin bacteriocin and PVA polymer. Nisin is a natural antimicrobial compound suitable for use in food packaging and drug design, while PVA is a flexible, soft, and water-soluble synthetic polymer with biodegradable and non-toxic properties. The molecular docking method was employed to simulate the binding interactions between Nisin and a single monomer of PVA. Through molecular docking between ligand and protein potential synergistic effects and bonding modes between Nisin and PVA were depicted. The work was employed to develop advance material for food packaging (Nihan et al., 2023).

TABLE 28.3 Different types of docking.

S.No.	Docking type	Advantages	Limitations
1	Ligand-based docking	The ligand-based docking does not require a protein structure and can be used for ligands with unknown targets. It is more efficient over protein—ligand docking due to the high rate of screening of a large number of compounds in a small time span.	The major limitation includes an inaccurate prediction of binding affinity and conformational changes of the proteins.
2	Protein—ligand docking	It provides insights into protein—ligand interactions and can aid in drug design. It can be used to predict the binding site and explore the binding of a range of ligands.	The major limitation of this method includes the accuracy of the docking algorithm and the quality of the protein structure and ligand data. It does not take into account the other variables such as solubility, bioavailability, and toxicity, which are important in drug development.
3	Hybrid docking	Combination of ligand-based and protein—ligand docking methods for improved accuracy. It can account for protein flexibility and conformational changes, which can be useful in predicting ligand binding.	The limitation of this method is computationally intensive and requires extensive parameter optimization.

6.4 Molecular dynamics simulation

Molecular dynamics (MD) simulations are regarded as a useful method for studying the atomic-scale interactions between inorganic fillers and polymer matrices. The MD simulations have been successfully applied to analyze the interactions among different polymers. Mechanical properties and the role of binder molecules can be studied by MD simulations. This method was used to evaluate the nature of polyacrylic acid—hydroxyapatite composites (Wei et al., 2017). The MD simulations assists in the prediction of interactions between biomaterials and drugs and suggests the stability and compatibility of two or more than two biomaterials for in vitro and in vivo applications. Since mimicking the natural properties of bioengineered cornea requires correct orientation of individual polymers and specific interactions, the MD simulations would remove all those polymers, which do not occur in the set parameters (Dodson et al., 2008) (Table 28.4). In TE, the application of MD simulation allows finite element method to explore the response of the materials under the influence of multiple field. By applying the finite element method, researchers can simulate and analyze the macroscopic properties and responses of biomaterials. As computing technology and modeling methods continue to advance, computer simulations are expected to play an increasingly essential role in biomaterial research. The development of more

TABLE 28.4 Softwares used to design corneal substitutes.

S.No.	Software (Names/websites)	Functions
1	Modeling software (AutoCAD)	This tool helps to design 3D model of cornea.
2	Finite element analysis tool (https://www.autodesk.in/solutions/finite-element-analysis)	This computational method is used to analyze the stress and strain of complex structures, such as the cornea. It helps in prediction of mechanical behavior to external loads.
3	Molecular dynamics simulation software (https://www.mat3ra.com; gromacs https://www.gromacs.org/)	It helps in prediction of biological behavioral of molecules at the atomic level (interaction between atoms).
4	Image analysis software (http://fiji.sc/Fiji)	This software works on images of the cornea, such as those obtained using confocal microscopy and optical coherence tomography. The image analysis software can be used to measure corneal thickness, curvature, and other parameters, which can be used to diagnose and monitor corneal diseases.

sophisticated simulation techniques will provide researchers with powerful tools to understand the structure, properties, and behavior of biomaterials, ultimately leading to the design and development of novel biomaterials with tailored functionalities for various applications. Biovia Material studio is a modeling and simulation environment to predict the materials properties. It is used to engineer various materials including polymers, composites, etc. The procedure for material simulation in Material Studio involves a combination of modeling, simulation, and analysis, with a focus on understanding the properties and behavior of materials at the atomic and molecular levels. The Material Studio software works in different steps as follows. Building the model (constructing a model): This step involves the construction of a model of the material using the Material Studio graphical user interface, which helps in writing the material script language (MSL). The constructed model would have all the relevant molecules and atoms. The second step includes the simulation of material constructed in a periodic box having all the necessary parameters such as temperature, forces, pressure, and time step for MD simulations, etc.

The third step includes running of simulation via simulation engines of Material Studio (DMol3 [for density functional theory calculations]). In the fourth step results are analyzed for visualizing the trajectories of individual atoms or molecules, calculating thermodynamic properties, or analyzing electronic structure. The fifth step includes repeating the simulation if desired result is not obtained using the aforementioned four steps. The best-simulated materials may be replicated via 3DP technology. Chidapha Kusinram et al. used Monte Carlo (MC) simulation to generate and equilibrate the amorphous structure of polymers of polyvinyl alcohol (PVA). The PVA chains were coarse-grained by grouping each monomer unit into one bead and then placed onto the high coordination lattice with intra- and inter-molecular interactions. The simulated structure was compared with experimentally obtained data, which showed that most of the parameters were similar to the experimentally obtained data. The PVA material possess high mechanical property, therefore, use of PVA for corneal bioengineered would help mimic the natural properties (Kusinram et al., 2022).

6.5 Machine learning

Application of artificial intelligence (AI) has revolutionized the field of Biology. AI allows us to simulate the environment of a polymer or biomaterial according to ambient conditions. Given that laboratory condition varies with the computational environment, affecting reproducibility of results (Biswal et al., 2007), application of AI ensures to reproduce desired environment to mimic the natural condition. The AI may help in prediction of best corneal model based on previously available data and running different algorithms. A case study in corneal diseases where a group of researchers from National Eye Institute Refractive Error Correction Study (NEI-RECS) used AI/ML-based approach for predicting corneal disease using datasets of corneal topography images and patient information. They used a combination of feature extraction, feature selection, and classification algorithms to develop their model. In feature extraction step, relevant features were extracted from the corneal topography images using the Zernike polynomials method. In correlation analysis and principal component analysis step, they selected the most important features for the model. In the last step, they classified data using support vector machine (SVM) algorithm into healthy or diseased categories of cornea. The application of Machine learning based models helps in prediction of toxicity, aqueous solubility, and permeability of molecules. This model was developed by considering the chemical and structural information of the molecules. The developed tool, called ToxiM, showed high performance in assigning molecules as toxic or nontoxic, as well as predicting aqueous solubility and permeability. The bioengineered cornea also needs to be free from toxicity and has cell viability in nature. Therefore, using AI-based algorithm would drastically reduce time and toxic effects associated with the known compounds (Sharma et al., 2017).

6.6 Response surface methodology

This approach enables an understanding of the relationship between variables (reactants) and the response (product), generally termed as design of experiment (DoE). A DoE approach can overcome the limitations posed by the traditional trial-and-error or one factor at a time method, which involves alteration of single random values to measure the response (Table 28.4) (Politis et al., 2017). Contrarily, DoE deduces the values of the variables that generate the best response. The RSM exploits this approach to provide optimal experimental conditions by employing a regression analysis of the collective data (Bezerrama et al., 2008; Lamidi et al., 2022). This technique was first developed in 1951 by Box and Wilson to provide an impetus to the chemical manufacturing industry. The methodology acts as a statistical tool that functions to estimate a feasible combination of variables to either maximize or minimize or equalize a specific target value of the response (Dean et al., 2017).

6.6.1 Response surface methodology workflow

The RSM scheme involves three broad categories namely, (1) experiment design, (2) modeling technique, and (3) optimization. The experimental design is initiated by defining the input variables and output response (Lamidi et al., 2022). This aids in developing a strategy for the identification of variables and interactions influencing the process. This is followed by the generation of a run experiment based on the modeling method of choice, viz., factorial design, Box–Behnken, and central composite, to express a relationship between dependent and independent variables (Dean et al., 2017). Accordingly, a number of runs are generated, and a corresponding response is measured for each run according to the factorial design and center point. This is followed by the interpretation of the selected model by curvature analysis of the experimental condition and obtaining an estimated model for each response variable.

Finally, the optimal conditions are formulated to maximize or minimize the variables for improving the response and resolving the problem (Fig. 28.4).

6.6.2 Response surface methodology model and design

The models for RSM can be broadly categorized (based on uniform precision) into first- and second-degree models (Lamidi et al., 2022). Designs for fitting first-degree models are called as first-order designs, while those fitting second-degree models are referred to as second-order designs. The most common first-order designs are factorial designs, while central composite design (CCD) and Box–Behnken design (BBD) are classified as second-order designs.

6.6.3 Response surface methodology in tissue engineering

RSM has evolved to find applications in various research fields starting from chemical manufacturing to biological applications. Over the years, a wide range of scaffold materials have emerged to enhance understanding and substantial role played by the substrates in terms of cellular development and function. Recent studies have indicated that choice of a substrate and material properties can individually have a significant role to alter phenotype of the cell, rate of growth, and differentiation. Consequently, the variation in fabrication techniques such as electrospinning and 3D printing has

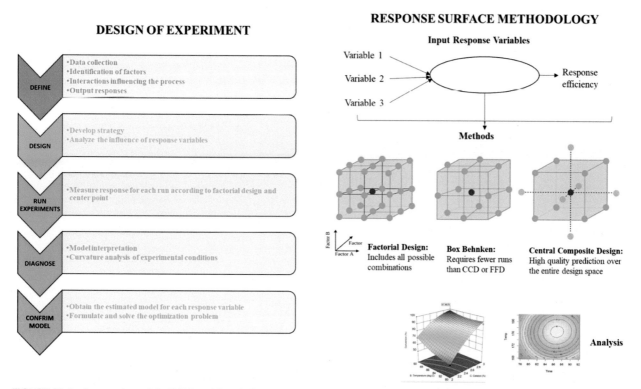

FIGURE 28.4 An overview of the RSM workflow indicating two aspects that are (A) design of experiment (DoE) that allows initial selection of parameters and control variables for the desired response and (B) the modeling technique used to design a space for the selected criteria upon which the significance and the interactions between the parameter and the response would be estimated to ultimately optimize a targeted response using statistics. Image was created using merging of multiple images in PowerPoint.

significantly increased the number of control variables, thus limiting the scope of predictive tissue scaffold models. To counter this, several sophisticated statistical approaches have been developed to address the challenges associated with TE with respect to optimization of stem cell differentiation, biomaterial fabrication, imaging protocols, etc. This has led to the identification of tissue scaffold composition for specific cell populations primarily through a multifactor design of experiment approach. Rezaei et al. have reported the use of Design-Expert software to study various compositions of 3D printing to determine an optimum formulation of polycaprolactone/chitosan (PLC/CS) scaffold that may have applications in lung TE and COVID-19 (Kong et al., 2016). TE has also shifted its focus toward a globally limited and major graft failure transplanted organs such as cornea. Several new materials and strategies have been developed to engineer a full-/partial-thickness corneal graft to regenerate or replace the diseased cornea. However, a standing challenge still remains in the field as to how the materials can be optimized pertaining to the construction of scaffold that stimulates the ECM environment to retain functionality. The RSM as a strategy may present as a useful technique to design a confluent corneal epithelial layer and its integration with functional nerves for vision restoration by tackling complications associated with the fiber structures (isotropic or anisotropic), functionalization (improved mechanical properties and transparency), survival of corneal cells, phenotype maintenance, and formation of corneal tissue (Mista & Kacprzyk, 2008). This methodology allows a multifactor approach to optimize the factors such as biomaterials, cell attachment, movement, proliferation, and differentiation altogether and ultimately enhance the repair mechanism of the tissue (Table 28.5). Elaheh Esmaeili et al. employed RSM for optimization of nanoclay/polyacrylonitrile (PAN) nanofibers using electrospinning. By employing RSM, the optimum values for the electrospinning process were determined (17 kV for the applied voltage, 0.41 mL/h for the flow rate, and a nanoclay/PAN ratio of 19.06%). The resulting scaffold exhibited a fiber diameter of 145 ± 12 nm and a Young's modulus of 267 ± 8.7 MPa, which aligned well with the predicted values determined by RSM. The in vitro result showed that nanoclay/polyacrylonitrile fiber was biocompatible with the cells. These results suggest possible use of RSM in cornea designing. Furthermore, the alignment of high-throughput computational and data ML may enhance the success rate in development of functional bioengineered cornea.

TABLE 28.5 Softwares used in design of experiments.

Tool	Description	Features
Minitab (https://www.minitab.com/en-us/products/minitab/)	This is a statistical software that is used for data analysis and quality improvement.	This software has the following features: full factorial, fractional factorial, response surface, mixture designs, Taguchi designs, Monte Carlo simulations, dynamic DOE, etc.
JMP (https://www.jmp.com/en_in/home.html)	It uses DOE-based visualization and data analysis.	It applies the use of full factorial, fractional factorial, response surface, mixture designs, D-optimal, I-optimal, central composite, and Box–Behnken for data analysis.
Design-Expert (https://www.statease.com/software/design-expert/)	It helps in optimization of independent variables to design a product or process.	This software works by using full factorial, fractional factorial, response surface, mixture designs, D-optimal, I-optimal, central composite, and Box–Behnken to retrieve better model to design a product.
RSM Design (https://develve.net/Response_surface_methodology.html)	It is an online and plugin software with Design-Expert tool for designing response surface methodology experiments to remove undesired responses.	It uses various algorithms including central composite, Box–Behnken, face-centered cubic, full factorial, factorial points.
Easy DOE (https://www.jmp.com/en_dk/events/mastering/topics/easy-doe.html)	It is a cloud-based tool for designing and analyzing DOE experiments.	It works through full factorial or fractional factorial, response surface, mixture designs, Taguchi designs, and robust parameter to design a model.
Stat-Ease (https://www.statease.com/)	This software is dedicated for optimization and process improvement.	Full factorial, fractional factorial, response surface, mixture designs, Taguchi designs, and central composite design.

7. Conclusion

Efficient TE approaches for the development of bioengineered cornea requires computational approaches, which would further enrich the therapeutic uses of biomaterials, growth factors, and stem cells to enhance cell attachment, proliferation, transparency, biodegradation, and suturability. For bioengineered cornea, these parameters are of utmost importance to replace traditional approaches for corneal dysfunction management. The present study highlights computational approaches, modeling software, lyophilizer, 3DP, and electrospinning to fabricate optimized corneal substitutes. Furthermore, the stem cells, nanoparticles, repurposed drugs, and growth factors can be optimized by RSM to develop a functional bioengineered cornea. This approach has the potential to replace traditional graft transplantation and associated limitations. The RSM-based techniques would reduce independent variables in the selection of dependent variables for the desired result. For example, in bioengineering cornea, it would select only those biomaterials (natural and synthetic), which enhance their mechanical strength, transparency, cell attachment, and proliferation, while it rejects the variables (cytokine storm, biomaterial concentration, cell transformation, etc.), which may have a role in the rejection of transplanted corneal substitute.

8. Future prospective

The emergence of computational biology has opened a new field in medicine and life sciences. This field has the ability to provide answer to puzzling questions. In the personalized medicine, computational models can be used to simulate the response of an individual's cornea to various treatments, allowing the development of personalized treatment plans. The computer-based model would help in the prediction of disease progression and allow an early intervention and better outcomes. The optimization of surgical interventions using computational models can increase the success rate of corneal transplantation or keratoplasty. By simulating the surgical procedure and predicting the outcomes, surgeons can have better treatment plans with reduced complications. By utilizing computer-based drug discovery and molecular simulation between drugs and corneal proteins, researchers can identify potential targets for drug development. Overall, the future prospects of computational biology in the corneal bioengineering are promising, and it is likely that these technologies will play an increasingly important role in the diagnosis and treatment of corneal diseases.

Acknowledgments
We thank the director of INMAS for his continuous support.

Authors contribution
The manuscript was written and edited by Subodh Kumar, Shivi Uppal, Vipin V.S, and Nishant Tyagi, edited by Ratnesh Singh Kanwar, Reena Wilfred, and Sweta Singh. The manuscript was conceptualized, designed, and edited by Yogesh Kumar Verma.

References

Ahearne, M., Fernández-Pérez, J., Masterton, S., Madden, P. W., & Bhattacharjee, P. (2020). Designing scaffolds for corneal regeneration. *Advanced Functional Materials, 30*(44), 1908996.

Antoine, E. E., Vlachos, P. P., & Rylander, M. N. (2014). Review of collagen I hydrogels for bioengineered tissue microenvironments: Characterization of mechanics, structure, and transport. *Tissue Engineering, Part B: Reviews, 20*(6), 683–696.

Ardani, I. G. A. W., Nugraha, A. P., Suryani, M. N., Pamungkas, R. H. P., Vitamamy, D. G., Susanto, R. A., ... Putera, A. (2022). Molecular docking of polyether ether ketone and nano-hydroxyapatite as biomaterial candidates for orthodontic mini-implant fabrication. *Journal of Pharmacy and Pharmacognosy Research, 10*(4), 676–686.

Bezerra, M. A., Santelli, R. E., Oliveira, E. P., Villar, L. S., & Escaleira, L. A. (2008). Response surface methodology (RSM) as a tool for optimization in analytical chemistry. *Talanta, 76*(5), 965–977.

Biswal, D., Yadav, P., Nayak, S. K., & Moharana, A. (2007). Artificial intelligence in advancement of regenerative medicine and tissue engineering. *Artificial Intelligence in Medicine, 41*(2), 151–159.

Bray, L. J., Suzuki, S., Harkin, D. G., & Chirila, T. V. (2013). Incorporation of exogenous RGD peptide and inter-species blending as strategies for enhancing human corneal limbal epithelial cell growth on Bombyxmori silk fibroin membranes. *Journal of Functional Biomaterials, 4*(2), 74–88.

Chen, Z., You, J., Liu, X., Cooper, S., Hodge, C., Sutton, G., & Wallace, G. G. (2018). Biomaterials for corneal bioengineering. *Biomedical Materials, 13*(3), 032002.

Chirila, T. V., Suzuki, S., Bray, L. J., Barnett, N. L., & Harkin, D. G. (2013). Evaluation of silk sericin as a biomaterial: *In-vitro* growth of human corneal limbal epithelial cells on Bombyxmorisericin membranes. *Progress in Biomaterials, 2*, 1–10.

Córdoba, A., Mejía, L. F., Mannis, M. J., Navas, A., Madrigal-Bustamante, J. A., & Graue-Hernandez, E. O. (2020). Current global bioethical dilemmas in corneal transplantation. *Cornea, 39*(4), 529−533.

Dean, A., Voss, D., Draguljić, D., Dean, A., Voss, D., & Draguljić, D. (2017). *Response surface methodology. Design and analysis of experiments* (pp. 565−614).

Dodson, G. G., Lane, D. P., & Verma, C. S. (2008). Molecular simulations of protein dynamics: New windows on mechanisms in biology. *EMBO Reports, 9*(2), 144−150.

Duan, X., & Sheardown, H. (2006). Dendrimer crosslinked collagen as a corneal tissue engineering scaffold: Mechanical properties and corneal epithelial cell interactions. *Biomaterials, 27*(26), 4608−4617.

Geris, L. (2013). *Computational modeling in tissue engineering*. Berlin: Springer.

Giegengack, M., & Soker, S. (2013). Constructing the cornea: Hopes and challenges for regenerative medicine. *Expert Review of Ophthalmology, 8*(3), 209−211.

Guérin, L. P., Le-Bel, G., Desjardins, P., Couture, C., Gillard, E., Boisselier, É., & Guérin, S. L. (2021). The human tissue-engineered cornea (hTEC): Recent progress. *International Journal of Molecular Sciences, 22*(3), 1291.

Hancox, Z., Keshel, S. H., Yousaf, S., Saeinasab, M., Shahbazi, M. A., & Sefat, F. (2020). The progress in corneal translational medicine. *Biomaterials Science, 8*(23), 6469−6504.

Harkin, D. G., George, K. A., Madden, P. W., Schwab, I. R., Hutmacher, D. W., & Chirila, T. V. (2011). Silk fibroin in ocular tissue reconstruction. *Biomaterials, 32*(10), 2445−2458.

Hazlett, L., Suvas, S., McClellan, S., & Ekanayaka, S. (2016). Challenges of corneal infections. *Expert Review of Ophthalmology, 11*(4), 285−297.

Huey, R., Morris, G. M., & Forli, S. (2012). Using AutoDock 4 and AutoDockvina with AutoDockTools: A tutorial. *The Scripps Research Institute Molecular Graphics Laboratory, 10550*(92037), 1000.

Issarti, I., Consejo, A., Jiménez-García, M., Hershko, S., Koppen, C., & Rozema, J. J. (2019). Computer aided diagnosis for suspect keratoconus detection. *Computers in Biology and Medicine, 109*, 33−42.

Jameson, J. F., Pacheco, M. O., Nguyen, H. H., Phelps, E. A., & Stoppel, W. L. (2021). Recent advances in natural materials for corneal tissue engineering. *Bioengineering, 8*(11), 161.

Jung, J. W., Lee, J. S., & Cho, D. W. (2016). Computer-aided multiple-head 3D printing system for printing of heterogeneous organ/tissue constructs. *Scientific Reports, 6*(1), 21685.

Kadakia, A., Keskar, V., Titushkin, I. A., Djalilian, A., Gemeinhart, R. A., & Cho, M. (2008). Hybrid superporous scaffolds: An application for cornea tissue engineering. *Critical Reviews in Biomedical Engineering, 36*(5−6).

Kaercher, T., & Bron, A. J. (2008). Classification and diagnosis of dry eye. *Surgery for the Dry Eye, 41*, 36−53.

Karamichos, D. (February 17, 2015). Ocular tissue engineering: Current and future directions. *Journal of Functional Biomaterials, 6*(1), 77−80.

Khan, B., Dudenhoefer, E. J., & Dohlman, C. H. (2001). Keratoprosthesis: An update. *Current Opinion in Ophthalmology, 12*(4), 282−287.

Kim, J. I., Kim, J. Y., & Park, C. H. (2018). Fabrication of transparent hemispherical 3D nanofibrous scaffolds with radially aligned patterns via a novel electrospinning method. *Scientific Reports, 8*(1), 3424.

Klintworth, G. K. (2009). Corneal dystrophies. *Orphanet Journal of Rare Diseases, 4*, 1−38.

Kong, B., & Mi, S. (2016). Electrospun scaffolds for corneal tissue engineering: A review. *Materials, 9*(8), 614.

Kong, B., Sun, W., Chen, G., Tang, S., Li, M., Shao, Z., & Mi, S. (2017). Tissue-engineered cornea constructed with compressed collagen and laser-perforated electrospun mat. *Scientific Reports, 7*(1), 1−13.

Kular, J. K., Basu, S., & Sharma, R. I. (2014). The extracellular matrix: Structure, composition, age-related differences, tools for analysis and applications for tissue engineering. *Journal of Tissue Engineering, 5*, 2041731414557112.

Kusinram, C., & Vao-soongnern, V. (2022). A multiscale simulation of amorphous poly (vinyl alcohol). *Materials Today Communications, 30*, 103029.

Lai, J. Y., Li, Y. T., Cho, C. H., & Yu, T. C. (2012). Nanoscale modification of porous gelatin scaffolds with chondroitin sulfate for corneal stromal tissue engineering. *International Journal of Nanomedicine, 7*, 1101.

Lamidi, S., Olaleye, N., Bankole, Y., Obalola, A., Aribike, E., & Adigun, I. (2022). *Applications of response surface methodology (RSM) in product design, development, and process optimization. Response surface methodology-research Advances and applications.*

Li, W., Long, Y., Liu, Y., Long, K., Liu, S., Wang, Z., Wang, Y., & Ren, L. (November 22, 2014). Fabrication and characterization of chitosan−collagen crosslinked membranes for corneal tissue engineering. *Journal of Biomaterials Science, Polymer Edition, 25*(17), 1962−1972.

Mahdavi, S. S., Abdekhodaie, M. J., Mashayekhan, S., Baradaran-Rafii, A., & Djalilian, A. R. (2020). Bioengineering approaches for corneal regenerative medicine. *Tissue Engineering and Regenerative Medicine, 17*, 567−593.

Melles, G. R., Lander, F., Rietveld, F. J., Remeijer, L., Beekhuis, W. H., & Binder, P. S. (1999). A new surgical technique for deep stromal, anterior lamellar keratoplasty. *British Journal of Ophthalmology, 83*(3), 327−333.

Mista, W., & Kacprzyk, R. (2008). Decomposition of toluene using non-thermal plasma reactor at room temperature. *Catalysis Today, 137*(2−4), 345 349.

Mistry, A. S., Cheng, S. H., Yeh, T., Christenson, E., Jansen, J. A., & Mikos, A. G. (2009). Fabrication and *in-vitro* degradation of porous fumarate-based polymer/alumoxane nanocomposite scaffolds for bone tissue engineering. *Journal of Biomedical Materials Research Part A: An Official Journal of the Society for Biomaterials, 89*(1), 68−79. The Japanese Society for Biomaterials, and the Australian Society for Biomaterials and the Korean Society for Biomaterials.

Miyashita, H., Shimmura, S., Kobayashi, H., Taguchi, T., Asano-Kato, N., Uchino, Y., Kato, M., Shimazaki, J., Tanaka, J., & Tsubota, K. (January 2006). Collagen-immobilized poly (vinyl alcohol) as an artificial cornea scaffold that supports a stratified corneal epithelium. *Journal of Biomedical*

Materials Research Part B: Applied Biomaterials: An Official Journal of the Society for Biomaterials, 76(1), 56–63. The Japanese Society for Biomaterials, and the Australian Society for Biomaterials and the Korean Society for Biomaterials.

Nihan, Ü. N. L.Ü., Özgen, A., & Canbay, C. A. (2023). Molecular docking study on interaction of polyvinyl alcohol (PVA) with group IA bacteriocin. *Turkish Journal of Science and Technology, 18*(1), 177–182.

Niu, M., Dong, F., Tang, S., Fida, G., Qin, J., Qiu, J., & Gu, Y. (2013). Pharmacophore modeling and virtual screening for the discovery of new type 4 cAMP phosphodiesterase (PDE4) inhibitors. *PLoS One, 8*(12), e82360.

Politis, N., Colombo, P., Colombo, G., & M. Rekkas, D. (2017). Design of experiments (DoE) in pharmaceutical development. *Drug Development and Industrial Pharmacy, 43*(6), 889–901.

Price, M. O., & Price, F. W., Jr. (2010). Endothelial keratoplasty—A review. *Clinical and Experimental Ophthalmology, 38*(2), 128–140.

Rafat, M., Li, F., Fagerholm, P., Lagali, N. S., Watsky, M. A., Munger, R., … Griffith, M. (2008). PEG-stabilized carbodiimide crosslinked collagen—chitosan hydrogels for corneal tissue engineering. *Biomaterials, 29*(29), 3960–3972.

Rahman, I., Carley, F., Hillarby, C., Brahma, A., & Tullo, A. B. (2009). Penetrating keratoplasty: Indications, outcomes, and complications. *Eye, 23*(6), 1288–1294.

Ramshaw, J. A., Werkmeister, J. A., & Glattauer, V. (1996). Collagen-based biomaterials. *Biotechnology and Genetic Engineering Reviews, 13*(1), 335–382.

Ren, S., Zhang, F., Li, C., Jia, C., Li, S., Xi, H., Zhang, H., Yang, L., & Wang, Y. (2010). Selection of housekeeping genes for use in quantitative reverse transcription PCR assays on the murine cornea. *Molecular Vision, 16*, 1076.

Risbud, M. (June 2001). Tissue engineering: Implications in the treatment of organ and tissue defects. *Biogerontology, 2*(2), 117–125.

Shah, A., Brugnano, J., Sun, S., Vase, A., & Orwin, E. (2008). The development of a tissue-engineered cornea: Biomaterials and culture methods. *Pediatric Research, 63*(5), 535–544.

Shahrubudin, N., Lee, T. C., & Ramlan, R. J. P. M. (2019). An overview on 3D printing technology: Technological, materials, and applications. *Procedia Manufacturing, 35*, 1286–1296.

Sharma, A. K., Srivastava, G. N., Roy, A., & Sharma, V. K. (2017). ToxiM: A toxicity prediction tool for small molecules developed using machine learning and chemoinformatics approaches. *Frontiers in Pharmacology, 8*, 880.

Tan, D. T., Dart, J. K., Holland, E. J., & Kinoshita, S. (May 5, 2012). Corneal transplantation. *The Lancet, 379*(9827), 1749–1761.

Tarsitano, M., Cristiano, M. C., Fresta, M., Paolino, D., & Rafaniello, C. (2022). Alginate-based composites for corneal regeneration: The optimization of a biomaterial to overcome its limits. *Gels, 8*(7), 431.

Torricelli, A. A., Singh, V., Santhiago, M. R., & Wilson, S. E. (2013). The corneal epithelial basement membrane: Structure, function, and disease. *Investigative Ophthalmology and Visual Science, 54*(9), 6390–6400.

Wei, Q., Wang, Y., Chai, W., Zhang, Y., & Chen, X. (2017). Molecular dynamics simulation and experimental study of the bonding properties of polymer binders in 3D powder printed hydroxyapatite bioceramic bone scaffolds. *Ceramics International, 43*(16), 13702–13709.

Whitcher, J. P., Srinivasan, M., & Upadhyay, M. P. (2001). Corneal blindness: A global perspective. *Bulletin of the World Health Organization, 79*(3), 214–221.

Yang, T., Wu, J. C., Yan, C., Wang, Y., Luo, R., Gonzales, M. B., & Ren, P. (2011). Virtual screening using molecular simulations. *Proteins: Structure, Function, and Bioinformatics, 79*(6), 1940–1951.

Yi, D. H., & Dana, M. R. (2002, January). Corneal edema after cataract surgery: Incidence and etiology. In *Seminars in ophthalmology* (Vol 17, pp. 110–114). Taylor and Francis. No. 3–4.

Chapter 29

Targeting cancer stem cells and harnessing of computational tools offer new strategies for cancer therapy

Kaushala Prasad Mishra

Ex Radiation Biology and Health Sciences Division, Bhabha Atomic Research Center, Mumbai, Maharashtra, India

1. Introduction

Cancer remains a major health problem in the 21st century in the world due mainly to the growing population and increasing aging subpopulation. Most cancer patients undergo surgery, radiotherapy, and chemotherapy modalities, but many limitations are encountered. Physicians are compelled to abandon these treatments due to severe side effects arising from toxicity to nontarget areas, drug resistance, and recurrence of tumor. It is usually difficult to eliminate metastasized cancer cells by employing traditional anticancer therapies, and rapid spread to multiple sites in the body poses a severe problem in clinics. Therefore, intensive research efforts are needed to develop new therapies that cause minimal or no toxicity to normal cells and tissues but help achieve selective and efficient killing of cancer cells. Conventional therapeutic methods for cancer treatment often lead to unacceptable normal tissue damage, resistance to therapy, and tumor recurrence. Medical researchers have proposed to transplant the stem cells, which have ability to self-renew and differentiate, into patients' body to help recover from the unavoidable normal tissue damage caused by chemo- and radiotherapy. It is commonly suggested that stem cells may be utilized in regenerative medicine, as therapeutic carriers, for drug targeting, and activation of immune cells (Chu et al., 2020).

Cancer research has significantly progressed over the years, and it is observed that after the significant shrinkage of tumor by traditional treatment modalities, a tiny mass in the core of tumor lump called cancer stem cells (CSCs) exhibits resistance to therapies and becomes a foci for recurrence. The tumor initiation and progression are believed to be driven by CSCs, which have self-renewal properties. Research focus in cancer therapy has shifted to use stem cell transplant for mitigating the adverse effects and treating cancer patients. It is considered crucial to discover a new class of drugs targeting CSCs and developing strategies to sensitize CSCs to anticancer therapies, including radiotherapy. CSCs are capable to form single cell—derived clonal cell populations and have ability to differentiate into various cell types (Trachootham et al., 2009). The residual stem cell pool plays a vital role in tissue regeneration and cellular homeostasis (Dawood et al., 2014). CSCs can be obtained from embryonic stem cells (ESCs) or adult/somatic stem cells (SSCs) sources. It is observed that CSCs maintain lower levels of reactive oxygen species (ROS) inside the cells, perhaps by raising the antioxidant levels favoring their survival and giving a clue to target cancer cells by raising prooxidant levels by therapeutic agents. At this point of time, the discovery of CSC-targeted drugs is considered a priority research to circumvent the limitations of conventional therapies (Chu et al., 2020). In clinic, it is found that chemo- and radiotherapies significantly kill nonstem cancer cell mass, but remaining small mass of CSCs poses resistance perhaps due partly to lower levels of ROS in their cytosol.

The new hope is growing with the development of the multidimensional potential of stem cell technologies for success in cancer therapy (Dawood et al., 2014). Stem cells are known to migrate to solid tumors and micrometastatic lesions, and this property can be utilized for the site-specific delivery of antitumor drugs and bioactive agents preloaded in them. The engineered stem cells can stably express a variety of new antitumor drugs, overcoming the short half lives of conventional

Computational Biology for Stem Cell Research. https://doi.org/10.1016/B978-0-443-13222-3.00021-6
Copyright © 2024 Elsevier Inc. All rights reserved, including those for text and data mining, AI training, and similar technologies.

chemotherapy. Many limitations, however, do exist, and conquering stem cell therapy—related issues needs further intensive research to understand the underlying mechanisms of interaction between normal stems cells and CSCs. But there remains scientific concern regarding stem cell therapies, especially the possibility of new sites of stem cell—based secondary tumors (tumorigenesis). Cancer therapists are hoping to gain deeper insight on certain properties of stem cells such as resistance to therapies, ability to migrate to areas of tumor, release of encapsulated bioactive agents, and immunogenic suppressors, which may help circumvent obstacles currently impeding anticancer therapies (Kim et al., 2015; Koizumi et al., 2011; Park et al., 2011).

This chapter aims to examine the possibility of using CSCs for improving cancer therapy by delivery of CSC-specific drugs, delineating the role of ROS for more significant tumor toxicity, evaluating the possibility of CSC-based drug delivery, assessing the role of cellular signaling mechanisms, and harnessing the power of computing methods in predicting the treatment outcomes in cancer therapy. Studies on the expression of genes in CSCs for exhibiting stemness display self-renewal and differentiate into other cell types, and mechanisms of resistance to anticancer therapies are considered vital (Kim et al., 2015). Attempts are made to briefly state the role of in silico methods in sequencing of DNA and RNA to identify pathogens and diseases for better predictions of treatment outcomes. The treated tumors are believed to relapse mainly due to proliferation of remaining CSCs. It is believed that targeting CSCs may considerably improve upon the clinical challenges of drug resistance and recurrence of tumors (Park et al., 2011).

2. Targeting cancer stem cells for therapy

CSCs seem to offer a new target for the treatment of cancer patients and for screening and identifying drugs that can cause toxicity to them. The CSCs are multipotent and have high proliferative capacities reflected in aggressive tumor invasion and metastasizing properties. The targeting of CSCs seems promising to enhance therapeutic efficacies and prevention of tumor recurrence (Masahiro & Mohammad Obaidul, 2019; Park et al., 2011). It seems possible to develop therapies based on the fact that CSCs attract normal stem cells, which can be potentially used to target CSCs (Yang et al., 2020). Evidently, engineering CSCs for anticancer drug delivery to improve therapy appears a reasonable and practical approach (Ling et al., 2010).

3. Reactive oxygen species and sources

ROS are oxygen-centered free radicals generated in cellular activities, which are capable of reacting with diverse biomolecules. The radicals can be broadly classified into oxygen-centered and nonfree radicals or molecular species. Primary reactive oxygen radicals consist in superoxide ($O_2^{\bullet-}$), hydroxyl radical ($^{\bullet}OH$), nitric oxide ($NO^{\bullet}$), organic radicals ($R^{\bullet}$), peroxyl radicals ($ROO^{\bullet}$), alkoxyl radicals ($RO^{\bullet}$), thiyl radicals ($RS^{\bullet}$), sulfonyl radicals ($ROS^{\bullet}$), thiyl peroxyl radicals ($RSOO^{\bullet}$), and disulfides ($RSSR$). The main nonradical ROS are hydrogen peroxide (H_2O_2), and organic hydroperoxides ($ROOH$) are known to react with metal ions and generating reactive free radicals such as hydroxyl radicals causing damage to crucial cellular molecules. Among them, superoxide, singlet oxygen, hydrogen peroxide, and hydroxyl radicals play a major role in causing cytotoxicity, and mechanisms of these reactions have been the subject of many studies especially in the processes of redox signaling and oxidative stress (Behrend et al., 2003; Fridovich, 1978; Singh et al., 2014).

Cancer cells possess higher levels of ROS due mainly to increased cellular metabolism, mitochondrial dysfunction, peroxisome activity, altered cellular receptor signaling, oncogenic activities, etc. (Behrend et al., 2003; Koizumi et al., 2011). ROS are produced in mitochondria, as an inevitable by-product due to electron leakage in chain reactions during oxidative phosphorylation. The superoxide is generated by one-electron reduction process in the intermembrane space and mitochondria (Raha & Robinson, 2000).

Apart from mitochondria, ROS are also generated in peroxisomes. Peroxisomal matrix and peroxisomal membranes generate hydrogen peroxide and superoxide radicals due to lipid oxidation and hydrogen peroxide catabolism involving xanthine oxidase and other enzymes (Behrend et al., 2003; Turrens, 2003).).

4. Reactive oxygen species level in cancer stem cells and drug resistance

Normal living cells maintain a highly delicate redox status necessary for cell functions. On the other hand, cancer cells possess an excess level of ROS, and redox status of the cell is altered (Kumari et al., 2018; Mishra, 2022). The oxidative stress conditions of cancer cells suggest a marked target to exploit for therapeutic benefits. Cancer cells exhibit multiple genetic alterations and adapt to high oxidative stress, suggesting the possibility of preferentially eliminating these cells by pharmacological drugs and ionizing radiation through generation of excess ROS inside cells causing toxicity. It is reported that cancer cells upregulate their antioxidant capacity for adapting to intrinsic oxidative stress and confer resistance to anticancer therapies (Ahire et al., 2015, Raghav et al., 2021; Yakes & Van Houten, 1997). Clinicians aim to abrogate the drug-resistant mechanisms in tumor cells by exploiting redox modulation strategies. Most anticancer therapies, including ionizing radiation, rely on the modulation of redox regulatory mechanisms of cancer cells for effectively eliminating tumor cells (Ahire et al., 2015; Tiwari & Mishra, 2022). However, CSCs pose a challenge to radiotherapy as CSCs are more resistant to radiation than differentiated or non-CSCs. Researchers have prompted to consider developing CSC-targeted therapies by sensitizing them to drugs/radiation for improving treatment outcomes (Fridovich, 1978; Mishra, 2022; Singh et al., 2014).

5. Cancer stem cell for secretion and delivery of anticancer drugs

Studies have shown that healthy stem cells can be used in situ as drug stores, secreting antitumor agents for a prolonged period at tumor site, and allow overcoming limitations of various cancer therapies such as drug induced high systematic toxicity and short life time of the drug in nontargeted circulatory system. Mesenchymal stem cells (MSCs) capable to express TNF-α-related apoptosis-inducing ligand (TRAIL) shown to migrate to and home with CSCs and prevent their growth and metastases (Loebinger et al., 2010). The authors reported that MSCs secreting TRAIL protein induce apoptosis and reduce colony formation of CSCs in squamous and lung cell lines. In fact, stem cell transplant therapy works through secretion of cytokines and chemokines, which prevent tumorigenesis and suppress immune system apart from believed damage repair of affected tissues (Perillo et al., 2020). These findings suggest possibility of using stem cell—based delivery of endogenous and loaded anticancer agents to treat delicate tumor such as glioblastoma multiforme (GBM) after surgical debulking (Duiker et al., 2012). Stem cell—based therapy of cancer holds promise for effective treatment of a variety of cancer by virtue of continuous secretion of anticancer, antiinflammatory, and immune-suppressive agents and release of preloaded therapeutic agents at the tumor sites. This approach especially seems useful in treating brain tumors and metastasized circulating tumor cells. Preliminary experiments with mouse model have shown marked retardation of malignancy and invasion of brain tumors resulting in increased survival of animal.

Normal stem cells secrete cytokines, chemokines, and other bioactive agents that inhibit tumor cell growth, cause death of CSCs, and sensitize them to anticancer therapies such as chemotherapy and radiotherapy. Researchers have reported migration of MSCs expressing growth inhibitory proteins and their homing to primary breast tumor sites, and causing suppression of tumor cell growth and reduction in metastases (Ling et al., 2010). It was observed that MSCs secreted IFN-β at elevated levels in the tumor microenvironment but not in the blood circulation. The study also suggested that in situ IFN-β expression in MSCs suppressed and prevented the tumor growth by inactivating signal transduction activating factors (Mi et al., 2011; Turrens, 2003).

6. Issues of secondary cancer

It is suggested that normal stem cells share some characteristics with CSCs, including self-renewal, differentiation, etc. The stem cell transplant therapy may potentially increase in cancer risk, as observed in tumor induction after fetal neural stem cell transplantation in brain-related diseases. It is imperative and warrants further studies on the delineation of controlling factors after transplantation of stem cells. It is important to determine whether stem cells promote the growth of certain tumors or transform themselves into tumors. In this context, it is pertinent to note that the breast cancer cells stimulated secretion of chemokines such as chemotactic cytokine ligand (CCL5) in MSCs, which significantly increased cellular motility, invasive ability, and metastasis aggression. It is significant to note that metastatic breast cancer cells were preventable to nonmetastatic state by the action of signaling molecules involving chemokine receptors. The MSCs in the tumor microenvironment facilitated metastasis by reversibly changing cancer cell phenotypes through signaling mechanisms (Mi et al., 2011).

Pluripotent stem cells that are undifferentiated exhibit potentially tumorigenic properties, and they exhibit specific cell surface biomarkers. But following differentiation of stem cells, biomarkers are altered, and they are predominantly upregulated or downregulated. Stem cells can be identified and separated by using monoclonal antibodies methodology to surface protein molecules such as expression of CD133. It is significant to note that the technology of induced pluripotent stem cell (iPSCs) has enabled detailed studies and monitoring of the progress of differentiation processes (Montero & Jassem, 2011; Zhang et al., 2017).

7. In silico modeling of drug development for cancer treatment

Laboratory research for developing drugs for various diseases is an expensive and time-consuming process. In silico modeling for drug discovery offers an alternative method to enable researchers gain in terms of reduced costs and faster screening of the drug candidates (Trisilowati & Mallet, 2012). Conventional laboratory-based cancer research requires time-consuming and labor-intensive trial-and-error experimentation using human and animal models. The in silico experiments involve the combination of current computing technologies with screening of a large number of drugs and identify most probable drug structures for in vitro testing and enable short-listing the most potent drugs for clinical evaluations. The computational methods turn out speedier to screen in potential drugs based on vast numbers of selection choices for repetitions under multiparametric controls of pharmacological factors for optimization of therapeutic effectiveness (Trisilowati & Mallet, 2012; Vu et al., 2021).

Studies involving in vitro and in vivo animal experiments allow building hypothesis and to justify the need for preclinical and clinical trials. Computational and mathematical biology scientists emphatically recognize the fact that in silico optimizations in combination with preclinical results provide basis for formulation of developing more logical and technically sound hypotheses on molecular structure–activity relations for functions and enable verify the efficacy of protein and ligand interactions. The dynamics of molecular interactions of various conformational structures of protein with the ligands or drugs are better understood to determine the topological hotspot for specific binding and structural stability. In silico and mathematical modeling allow obtaining deeper insight on intermolecular and intramolecular interactions, nature of forces in the interaction mechanism, and influence of solvent and other factors on the strength of binding. In silico methods empower researchers to predict most probable drug structures that could bind at specific conformational spot in the target protein. Computer simulations have revealed the dynamics of protein and drug binding in many studies and have allowed to determine preliminary information on such effects as toxicity, pharmacokinetics, and efficacy, which could guide preclinical and clinical outcomes (Raghav & Mann, 2021; Vu et al., 2021).

In silico modeling permits mapping of protein and determine most favorable spot for binding of the ligand or drug. The modeling methods help determine most stable binding sites for the drug indicating clinical testing in laboratory (Trisilowati & Mallet, 2012). Researchers find computer-based modeling as a powerful tool to identify drug candidates and study the influence of multiple factors, say solvent on the protein–drug interactions and their stability in vivo conditions. Using physiological and pharmacological parameters based on advanced biological knowledge level for molecular functions in cellular environments or using actual research data as inputs into the in silico models, it is possible to carry out computations for personalized patient treatment avoiding unwanted drug exposures and unacceptable side effects. Appropriate use of in silico models allows making predictions on the patient treatment outcomes from stem cell implant and anticancer drug therapy, leading to clinical trials reducing costs and increasing efficiency (Trisilowati & Mallet, 2012; Zhand 2017).

In silico methods and technologies have enormous new potential in the sequencing technology of the DNA or RNA to identify bacteria and other pathogens for diseases including cancer. The labor-intensive and time-consuming laboratory sequencing methods will be replaced by biocomputing methods. Understanding the biology and genetic alterations in cancer cells would revolutionize the diagnosis, treatment, and prediction of cancer by in silico methods. Most importantly, in silico methods an important role in development of precision medicine that concerns development of drugs effective to individual patients.

The COVID-19 pandemic that inflicted people on large scale world over and caused morbidity has pushed up the possibility of drug discovery and repurposing of drugs for the control and prevention of the diseases. Additionally, it is important to note that virtual and computational methods help avoid many bureaucratic and financial hurdles to develop drugs and processes. Moreover, the possibility of virtual trials are much safer than conventional clinical experiments. It is a cheap and secure way to discover new drugs and, also, repurpose the existing drugs.

8. In silico trial methods for anticancer drugs

Computer modeling has accelerated drug discovery research, which is proving acceptable alternative to in vivo animal models. The success of in silico modeling requires closer interactions between cancer researchers and biocomputation specialists. Importantly, research data on biology and behavior of tumor are required for developing the in silico model for target–molecule applications. The next stage involves the abstraction of biological information into a mathematical or computational algorithm, that is, building the model on the basis of updated molecular architectural information and biological responses/requirements. This requires the appropriate modeling mathematical formulations relevant to micro-level cellular functions and cellular signaling controls such as structure–activity characterizations, target–ligand recognition and binding, conformational variations of target protein and ligand positional orientations, etc. The optimization of physicochemical interactions with respect to entropy, solvation, and other controlling factors leads to macrolevel descriptions of intermolecular binding and interactions for designing and testing in the actual experiments in laboratory.

Quantitative structure—activity relations (QSARs) have been subject of studies for many years and have yielded useful guidance for investigating the interactions between a drug or ligand and the target protein. More recent developments have focused research on the known 3D proteins and small molecules as activator or inhibitor of function. In silico models use specifically designed computer programs to mimic the "real" experimental environments and to conduct computation al experiments for maximization of drug action. Since target molecules possess multifunctional sites, it becomes important to determine specific site (cavity) in the protein after binding with the ligand, antigen, or drug molecule. On this account, in silico methods have turned out crucial new tools for suitable model building, target—molecule interactions, saving in both time and investments for drug discovery and development. More importantly, in silico methods allow studies on unknown drug molecule interactions with the surface receptors proteins of unavailable structures. The power of such model development is that the drug candidate under study can be tested for examining the influence of various parameters such as most stable conformation, hotspots for binding, solvent effects, affinity, and entropy properties. The concept of the in silico trial, called virtual screening, can be thought of as akin to clinical trials for molecules with similar structures or even unidentified structures. Just as each patient in a clinical trial has his/her own set of characteristics such as disease status, health condition, height, weight, age, and habit of smoking and alcohol consumption, etc. Clinicians can be empowered to run the software program of an in silico model multiple times with varied required parameters to produce "computational patients" for assessing the treatment outcomes (Chu et al., 2020; Marcu & Marc, 2016), which is a stupendous task without computational techniques.

The computational methods involve rigorous calculations as a function of variations in cellular conditions from the models developed from a large database on protein structures, receptors, enzymes, and other functional molecules. Potential drugs from natural and synthetic sources can be mass-screened at accelerated scales by computational methods. Results are then analyzed and interpreted in the context of cells, tissues, and organs. This involves the use of custom-designed visualization of cellular response to drugs and therapies. The investigators use the outputs of the model to determine what results are already helpful in formulating associated laboratory results and what parts of the in silico model are deficient and require refinement along with a follow-up round of in-depth in silico modeling experiments. The whole process can be repeated, with the objectives of treatment outcome improvements, as often as new observations are found, which are needed to be addressed. In general, the costs of follow-up in silico experimentation substantially decrease as the relevant computational framework has already been developed and optimized. The computational tools such as Score Card, KeyGene, and CellNet are contributing significantly to learn about the molecular cellular factors and gene functions and improve stem cell transplants. It is pertinent to develop and test computer-based and AI-assisted anticancer drugs capable to enhance toxicity to CSCs for improving treatment outcome.

9. Cancer driver genes and computational methods

It is considered important to identify genes that cause tumorigenesis and support progression of tumor. A number of computational tools have been developed to predict cancer driver genes, and efforts are made to apply the new knowledge for designing the treatments in the clinics. Many tools have been developed such as MuSiC, OncodriveFM, and OncodriveCLUST to predict the driver genes in pan-cancer as well as in particular cancer types for gaining knowledge on cancer incident and their regulations. It is seen that the number of driver genes varies from method to method ranging from 150 to 2600 genes. Active research is in progress to predict and evaluate cancer driver genes in cancer types. It is hoped to develop and apply suitable computational tool to discover cancer drivers from available software tools and employ the gene database such as GO (Gene Ontology) for the genes in genome and also from KEGG database through identification of a small cancer driver cohort that is capable of promoting cancer patient survival (Choudhury et al., 2013). In the context of CSCs, they continue to receive attention as a promising target for cancer therapy.

From the studies on computational methods, it seems clear that computational methods may prove complementary to wet lab results. It is undisputed that lab experiments generate inadequate data to identify and predict the gene or genes responsible for the induction of cancer. Therefore, more rigorous methods of computations must be developed to predict gene mutations responsible for cancer induction or expression of genes for functional predictions including treatment trend predictions (Marcu & Marc, 2016). From the critical analysis of results obtained by computational methods, it seems unlikely that these methods will entirely replace the wet laboratory experimental approaches to derive information on therapeutic effectiveness on either already known or new set of potential drugs for study. However, due to the complicated mechanisms of cancer initiation and development processes, further extensive studies on cancer driver genes are highly warranted by in silico methods. It is hoped that future development and refinements of computer methods would help identify the single gene or a group of genes that mutate in the induction of cancer. Undoubtedly, computational methods seem to possess power and capability to push forward the drug discovery, regenerative medicine, and gene science. The computational stem cell biology is expected to help produce quantitative basis to determine the identity of cell types and to define the molecular logic of lineage commitment for providing basis to understand the genetic and epigenetic alterations to control healthy development and eventually contribute to understand and overcome diseases for improving human health (Patrik, 2021).

10. Developing new cancer therapies

Cancer researchers and oncologists have aspired and developed computer science algorithms, data collection, and analysis for developing the personalized and effective cancer treatment. Use of robotics in surgery is a growing specialty allowing physician to achieve perfection and accuracy in patient treatment. A surgeon trained on the device sits at a console and uses controls to guide multiple robotic arms that hold scalpels, employing a high-resolution 3D video camera and other medical devices. In addition, the robotic-assisted tools can obtain biopsies from hard-to-reach tissues deeper in the body such as in patients' lungs. Robotic surgery and earlier lung cancer diagnoses through biopsies can prove highly beneficial for early medical intervention and offer greater confidence of the clinician and more comfort to cancer patients. Computer technology has massive potential for the future of cancer treatment, especially in the evolving fields of artificial intelligence, virtual reality, implantable sensor, and other technologies (Ekins et al., 2007; Zenil et al., 2022).

11. Prospects for anticancer drug screening

The effective treatment of cancer patients continues to be at the core of physician concern, whereas considerable improvements in treatment success have been achieved, but the observed side effects and discomforts of therapeutic procedures pose challenges. Therefore, discovery of new drugs for therapeutic efficiency and reduced normal tissue toxicity are high on cancer research, diagnosis, and therapy objectives. It seems legitimate to conduct research and screen drugs on CSCs in vitro, in animal models, and also employ in silico methods, another procedure of refinements. For example, normal somatic cells from easily available body parts can be converted into cancer cell lines using iPSC method to screen potential anticancer drugs on large scale. Research has progressed to obtain cancer patient's tissue derived by iPSCs that may be biologically more relevant to human tumors than traditionally available drug screening methods, such as use of cancer cell lines and nude mouse tumor models. Moreover, frequently observed drug-induced hepatotoxicity prevents many potential antitumor drugs from further clinical testing; however, they can now be screened using hepatocytes produced from human iPSCs with various genetically induced cellular alterations. The applications of stem cells in cancer therapy are rapidly growing, but pathological factors, patients' health, disease status, and outcome uncertainty pose challenges that are needed to be tackled by advanced clinical research.

12. Conclusion

CSCs manifest an unusual redox status characterized by lower intracellular ROS levels and they exhibit resistance to anticancer therapies, including ionizing radiation. Therefore, targeting CSCs for treatment objectives has acquired significance in clinical settings, especially by sensitizing these cells to therapies. Cancer cells become vulnerable to treatments when ROS levels increase beyond a threshold. In addition, MSCs possesses specific properties, such as their homing with solid or metastatic tumor cells (tumor tropic properties) giving new opportunities to engineer these cells to carry novel drugs or preforms of drugs to deliver them to achieve targeted therapy of patients treated with chemo- and radiotherapy (Zhang et al., 2017). Cell-based drug delivery of endogenous or loaded antiinflammatory cytokines and immune modulatory agents seems possible in near future. Human stem cells have intrinsic tumor tropic properties, which need to be exploited for delivery of encapsulated toxic agents/chemokines and anticancer drugs to CSCs. The properties of stem cells need to be optimized for maximizing the therapeutic benefits to patients. Also, researchers are fervently making efforts to identify genes or groups of genes in DNA responsible for tumorigenesis and targeted therapy of CSCs. In the wet laboratory, it requires elaborate experiments, considerable cost, and huge data collection and analysis for predicting the treatment outcomes in patients. To overcome these tedious problems, harnessing of the immense power of computational modeling has brought innovative solutions to problems (Vu et al., 2021; Zenil et al., 2022). Proposing a hypothesis for achieving specific objectives and designing appropriate models for verification of its validity can be accomplished a faster speed and in shorter periods. It is suggestive that collaborations between experimental biologists, cancer physicians, and computational experts can empower cancer radiotherapy and help fulfill the dream of precision medicine (Masahiro & Mohammad Obaidul, 2019; Yang et al., 2020).

Acknowledgments

The author is grateful to his family members, especially his wife, Usha, for exhibiting utmost patience for his indulgence in developing this chapter.

Author contribution

The author singly conceived and prepared the chapter on the topic under discussion.

References

Ahire, V. R., Mishra, K. P., & Kulkarni, G. R. (2015). Apoptotic radiosensitivity of cervical tumor cells enhanced by ellagic acid. *European Journal of Biotechnology and Bioscience, 3*(4), 56−58.

Behrend, L., Henderson, G., & Zwacka, R. M. (2003). Reactive oxygen species in oncogenic transformation. *Biochemical Society Transactions, 31*, 1441−1444.

Choudhury, Tanupriya, Kumar, Vivek, & Nigam, Darshika (2013). International journal of computer science and mobile computing cancer research through the help of soft computing techniques: A survey. *International Journal of Computer Science and Mobile Computing, 2*(4), 467−477.

Chu, D. T., Nguye, T. T., Tien, N. B., Tran, D. K., Jeong, J. H., Anh, P. G., Thanh, V. V., Truong, D. T., & Dinh, T. C. (2020). Recent progress of stem cell therapy in cancer treatment: Molecular mechanisms and potential applications. *Cells, 9*(3), 563. https://doi.org/10.3390/cells9030563

Dawood, S., Austin, L., & Cristofanilli, M. (2014). Cancer stem cells: Implications for cancer therapy. *Oncology (Williston Park), 28*, 1101−1107.

Duiker, E. W., Dijkers, E. C., Lambers, H. H., de Jong, S., van der Zee, A. G., Jager, P. L., Kosterink, J. G., de Vries, E. G., & Lub-de, H. M. (2012). Development of a radioiodinated apoptosis-inducing ligand, rhTRAIL, and a radiolabelled agonist TRAIL receptor antibody for clinical imaging studies. *British Journal of Pharmacology, 165*, 2203−2212.

Ekins, S., Mestres, J., & Testa, B. (2007). In silico pharmacology for drug discovery: Methods for virtual ligand screening and profiling. *British Journal of Pharmacology, 152*(1), 9−20. https://doi.org/10.1038/sj.bjp.0707305

Fridovich, I. (1978). The biology of oxygen radicals. *Science, 201*, 875−880.

Kim, B. M., Hong, Y., Lee, S., Liu, P., Hong, J. L., Lee, Y. H., Lee, T. H., Chang, K. T., & Hong, Y. (2015). Therapeutic implications of overcoming radiation resistance in cancer therapy. *International Journal of Molecular Sciences, 16*(11), 26880−26913. https://doi.org/10.3390/ijms161125991

Koizumi, S., Gu, C., Amano, S., Yamamoto, S., Ihara, H., Tokuyama, T., & Namba, H. (2011). Migration of mouse-induced pluripotent stem cells to glioma-conditioned medium is mediated by tumor-associated specific growth factors. *Oncology Letters, 2*, 283−288.

Kumari, Seema, Badana, A. K., Murali, M. G., Shailender, G., & Malla, R. R. (2018). Reactive oxygen species: A key constituent in cancer survival. *Biomarker Insights, 13*. https://doi.org/10.1177/1177271918755391, 1177271918755391.

Ling, X., Marni, F., Konopleva, M., Schober, W., Shi, Y., Burks, J., Clise-Dwyer, K., Wang, R. Y., Zhang, W., Yuan, X., Lu, H., Caldwell, L., & Andreeff, M. (2010). Mesenchymal stem cells overexpressing IFN-β inhibit breast cancer growth and metastases through Stat 3 signaling in a syngeneic tumour model. *Cancer Microenvironment, 3*(1), 83−95. https://doi.org/10.1007/s12307-010-0041-8

Loebinger, M., Sage, E., Davies, D., & Janes, S. (2010). TRAIL-expressing mesenchymal stem cells kill the putative cancer stem cell population. *British Journal of Cancer, 103*, 1692−1697. https://doi.org/10.1038/sj.bjc.6605952

Marcu, L. G., & Marc, D. (2016). In silico modelling of a cancer stem cell-targeting agent and its effects on tumour control during radiotherapy. *Scientific Reports, 6*, 32332. https://doi.org/10.1038/srep32332

Masahiro, Shibata, & Mohammad Obaidul, Hoque (2019). Targeting cancer stem cells: A strategy for effective eradication of cancer. *Cancers, 11*(5), 732. https://doi.org/10.3390/cancers11050732

Mishra, K. P. (2022). Radiosensitivity of cancer stem cells holds promise for the outcome in radiotherapy. In Yogesh Verma, et al. (Eds.), *Cancer stem cell and cytokine*. Elsevier Nature (in press).

Mi, Z., Syamal, D. B., Kim, V. M., Hongtao, G., Lindsay, J. T., & Kuo, P. C. (2011). Osteopontin promotes CCL5-mesenchymal stromal cell-mediated breast cancer metastasis. *Carcinogenesis, 32*(4), 477−487. https://doi.org/10.1093/carcin/bgr009

Montero, A. J., & Jassem, J. (2011). Cellular redox pathways as a therapeutic target in the treatment of cancer. *Drugs, 71*(11), 1385−1396.

Park, S. A., Ryu, C. H., Kim, S. M., Lim, J. Y., Park, S. I., Jeong, C. H., Jun, J. A., Oh, J. H., Park, S. H., Oh, W., & Jeun, S. S. (2011). CXCR4-transfected human umbilical cord blood-derived mesenchymal stem cells exhibit enhanced migratory capacity toward gliomas. *International Journal of Oncology, 38*, 97−103.

Perillo, B., Donato, M. D., Pezone, A., Zazzo, E. D., Giovannelli, P., Galasso, G., Castoria, G., & Migliaccio, A. (2020). ROS in cancer therapy: The bright side of the moon. *Experimental and Molecular Medicine, 52*, 192−203.

Raghav, P. K., & Mann, Zoya (2021). Cancer stem cells targets and combined therapies to prevent cancer recurrence. *Life Sciences, 277*, 119465. https://doi.org/10.1016/j.lfs.2021.119465. ISSN 0024-3205.

Raghav, P. K., Mann, Z., Pandey, P. K., & Mohanty, S. (2021). Systems biology resources and their applications to understand the cancer. In *Handbook of oxidative stress in cancer: Mechanistic aspects*. Publisher: Springer Nature LE. https://doi.org/10.1007/978-981-15-4501-6_140-1. Edition: Living.

Raha, S., & Robinson, B. H. (2000). Mitochondria, oxygen free radicals, disease and ageing. *Trends in Biochemical Sciences, 25*, 502−508.

Singh, R., Pandey, A., Dayal, A., & Mishra, K. P. (2014). Reactive oxygen species as mediator of tumor radiosensitivity. *Journal of Cancer Research and Therapeutics, 1014*, 811–816.

Tiwari, Prabha, & Mishra, Kaushala Prasad (2022). Role of plant-derived flavonoids in cancer treatment. *Nutrition and Cancer*. https://doi.org/10.1080/01635581.2022.2135744

Trachootham, D., Alexandre, J., & Huang, P. (2009). Targeting cancer cells by ROS-mediated mechanisms: A radical therapeutic approach? *Nature Reviews Drug Discovery, 8*(7), 579–591. https://doi.org/10.1038/nrd2803

Trisilowati, Trisilowati, & Mallet, D. G. (2012). In silico experimental modeling of cancer treatment. *International Scholarly Research Network. ISRN Oncology, 2*, 828701. https://doi.org/10.5402/2012/828701

Turrens, J. F. (2003). Mitochondrial formation of reactive oxygen species. *Journal of Physiology, 552*, 335–344.

Vu, Viet, Pham, Hoang, Liu, Lin, Cameron, B., Gregory, G., Li, J., & Duy Le, T. (2021). Computational methods for cancer driver discovery: A survey. *Theranostics, 11*(11), 5553–5568. https://doi.org/10.7150/thno.52670

Yakes, F. M., & Van Houten, B. (1997). Mitochondrial, DNA damage is more extensive and persists longer than nuclear DNA damage in human cells following oxidative stress. *Proceedings of the National Academy of Sciences, 94*, 514–519.

Yang, Liqun, Shi, Pengfei, Zhao, Gaichao, & Cui, Hongjuan (2020). Targeting cancer stem cell pathways for cancer therapy. *Signal Transduction and Targeted Therapy, 5*(1), 8. https://doi.org/10.1038/s41392-020-0110-5

Zenil, H., Kiani, N. A., & Tegnér, J. (2022). *Algorithmic information dynamics: A computational approach to causality with applications to living systems*. Cambridge, UK: Cambridge University Press.

Zhang, C. L., Huang, Ting, Wu, Bi-Li, He, Wen-Xi, & Liu, D. (2017). Stem cells in cancer therapy: Opportunities and challenges. *Oncotarget, 8*(43), 75756–75766. https://doi.org/10.18632/oncotarget.20798

Further reading

Buetler, T. M., Krauskopf, A., & Ruegg, U. T. (2004). Role of superoxide as a signaling molecule. *News in Physiological Sciences, 19*, 120–123.

Patrick, C., Davide, C., Sara-Jane, D., Martin, H., Susana, M., Chuva, S. L., Samantha, A. M., Owen, J. L., Rackham, A., & Christine, A. W. (2021). Computational stem cell biology: Open questions and guiding principles cancer stem. *Cell, 28*, 20–32.

Chapter 30

Introduction to machine learning and its applications in stem cell research

Nirbhay Raghav[1], Anil Vishnu G.K.[2], Neha Deshpande[3] and Annapoorni Rangarajan[3]

[1]Department of Interdisciplinary Mathematical Sciences, Indian Institute of Science, Bangalore, Karnataka, India; [2]Centre for BioSystems Science and Engineering, Indian Institute of Science, Bangalore, Karnataka, India; [3]Department of Developmental Biology and Genetics, Indian Institute of Science, Bangalore, Karnataka, India

1. Introduction

Stem cells are undifferentiated, unspecialized cells with the potential to self-renew and differentiate into diverse cell types (Zakrzewski et al., 2019). They exist both in developing embryos and in adult tissues. As stem cells commit to a lineage differentiation program, they lose their potency at each step. Potency is the capacity of stem cells to differentiate into a number of specialized cell types. Depending on their differentiation potential, stem cells are categorized as totipotent, pluripotent, multipotent, oligopotent, and unipotent. Totipotent stem cells are capable of dividing and differentiating into any cell type found within an organism. For example, a zygote (formed after an egg is fertilized by a sperm) is a totipotent stem cell. As embryonic development progresses, the zygotic cell divides and loses its totipotency, first giving rise to pluripotent cells (e.g., embryonic stem cells [ESCs] derived from the inner cell mass) that form the three germ layers. Multipotent stem cells can produce discrete cell types of a specific cell lineage; however, their differentiated potential is narrower than that of pluripotent stem cells (e.g., hematopoietic stem cells). Oligopotent stem cells can differentiate into several cell types, whereas unipotent stem cells, which have the narrowest differentiating capacity, can give rise to only one cell type. Thus, each stem cell can self renew and differentiate into a stem cell with lesser potency, ultimately ending with a completely differentiated mature cell.

Stem cells are relatively abundant in developing embryos, particularly ESCs. However, in adult organisms, a minor population of quiescent stem cells resides in the tissues, where they serve to replenish normal differentiated tissue cells during normal wear-and-tear or after tissue injury. For instance, intestinal stem cells (Lgr5+) are located at the base of the intestinal crypt ("stem cell niche") and are actively involved in maintaining a healthy gut epithelium (Barker et al., 2007). Thus, stem cells might be valuable candidates for regenerative medicine to treat various diseases, including degenerative diseases (such as Alzheimer's, Parkinson's, osteoporosis, and osteoarthritis) and cardiovascular diseases (Lindvall et al., 2004; Segers & Lee, 2008). Stemlike cells are also present in cancers (cancer stem cells [CSCs]), where they function as tumor-initiating cells that contribute to cancer growth and metastasis formation (Deshpande & Rangarajan, 2015). Thus, targeting CSCs might lead to effective cancer treatment. Hence, researchers are focused on understanding basic stem cell biology and developing stem cell-based therapies. However, culturing these cells in vitro outside their biological niche has been challenging.

To date, several specialized culture techniques and growth media have been developed for this purpose (McKee & Chaudhry, 2017). It was in 2006 that a major breakthrough occurred in stem cell research when Kazutoshi Takahashi and Shinya Yamanaka discovered that introducing four transcription factors (Oct-3/4, Sox2, KLF4, and c-Myc) transformed murine multipotent stem cells into pluripotent stem cells (Takahashi & Yamanaka, 2006). This experiment was later replicated with human cells (Takahashi et al., 2007), and the derived cells were named induced pluripotent stem cells (iPSCs). Reprogramming of iPSCs has generated multiple cell types, which might have translational applications, and has rapidly advanced the field of stem cell research. However, our understanding of stem cells remains incomplete, primarily owing to difficulties in identifying and studying this elusive population in their native biological settings. Recent advances

Computational Biology for Stem Cell Research. https://doi.org/10.1016/B978-0-443-13222-3.00025-3
Copyright © 2024 Elsevier Inc. All rights reserved, including those for text and data mining, AI training, and similar technologies.

in high-throughput technologies have been critical in studying the elusive stem cells. Stem cell researchers are actively utilizing machine learning (ML) algorithms to process the vast amounts of data generated by high-throughput techniques to identify stem cells, predict their cell fate, and comprehend complex molecular signaling networks.

2. Introduction to machine learning

ML is a branch of artificial intelligence (AI) that aims to develop algorithms capable of learning from data. The concept has been around since the early days of computer research in the 1940s. One of the first ML algorithms was the "perceptron," developed by McCulloch and Pitts in 1943 to imitate the function of a biological neuron (McCulloch & Pitts, 1943). Frank Rosenblatt later implemented the perceptron algorithm on custom-built hardware called the "Mark 1 perceptron" and trained it to recognize images (Frank, 1957). However, early successes in the field led to unrealistic expectations, and it was later proved that the perceptron was only suitable for solving linearly separable problems. This led to reduced funding and a general lack of interest in the field, known as the "AI winter."

In the late 1970s, funding was further reduced by government agencies, which expected natural language systems and autonomous tank applications. Funding cuts led to a shift from symbolic approaches to probabilistic methods. Following the discovery of the backpropagation algorithm in the 1980s by Rumelhart, ML resurfaced, with the training of larger models made possible (Rumelhart et al., 1986). The combination of graphical processing units (GPUs) and large-scale datasets led to the dominance of larger neural networks in competitions such as ImageNet. These networks have been adopted in various applications, including image, speech, and language processing (Brown et al., 2020; Krizhevsky et al., 2017). With the advent of omics platforms and large-scale datasets, ML is increasingly being applied to biology. This chapter provides an introduction to the theory of ML, neural networks, and deep learning and their applications in stem cell research.

2.1 What is machine learning?

According to Mitchell, "A computer program is said to learn from experience E with respect to some class of tasks T and performance measure P if its performance at tasks in T, as measured by P, improves with experience E" (Mitchell & Mitchell, 1997). Experience "E" is called "training", and tasks "T" can be classification, regression, or clustering. In other words, if a computer program performs better and better with increasing experience, it can be said that the program can "learn". We measure if a program performs better with experience using a performance measure P.

Consider an example dataset with three parameters—radius, texture, and compactness—obtained from digitized images of cytology smears for breast cancer diagnosis. For each image, these three parameters are measured and used to calculate *radius_mean*, *texture_mean*, and *compactness_mean*. The original dataset has 30 parameters; we only consider three for this simple illustration (Dua & Graff, 2019) (https://archive.ics.uci.edu/ml/datasets/breast+cancer+wisconsin+(diagnostic)). The goal is to develop a model that can accurately classify the sample as benign or malignant. This is known as a *classification* task since it involves assigning an observation to a category based on available data. Here, *radius_mean*, *texture_mean*, and *compactness_mean* are *input variables*. They are also known as *features* or *predictors*. The symbol X commonly denotes input variables, and a subscript denotes a specific feature. The output variable is commonly denoted by the symbol Y. Given a dataset with p such *predictors*, say, $X_1, X_2, \ldots, X_p$, and a *response variable* Y, there exists a function

$$Y = f(X) \tag{30.1}$$

which explains the relationship between the input (X) and output (Y). In Eq. (30.1), f maps the list of input variables X to the output Y. However, this exact (true) relationship is unknown in practical cases. For example, if X is a list of parameters collected from patients' serum samples and Y encodes a drug's efficacy, then the exact relationship between these parameters and the drug's efficacy is unknown beforehand. In ML, the goal is to estimate this function f or true relationship using given data and predict the response variable Y for future inputs (Breiman, 2001). Hence, Eq. (30.1) becomes

$$\widehat{Y} = \widehat{f}(X) \tag{30.2}$$

where $\widehat{f}$ (read as "f hat") denotes the learned representation or estimate of f, and $\widehat{Y}$ represents the output predicted by the model. In general, $\widehat{f}$ will not be a flawless estimate of f and will always suffer from errors that can either be reduced, are impossible to reduce, or are both. These errors will collectively reduce the predictive accuracy of our model. In summary, ML refers to the development of algorithms that can learn to perform a task from given data and apply this learning for prediction (disease classification, drug efficacy, etc.) on unseen data (test data).

ML is essential in data analytics and is applied in diverse scientific fields. Its goal is to estimate the function f (unknown relationship; see Eq. 30.1) and use it for prediction, which is useful when direct measurements of the target variable are constrained by cost, time, or other reasons. In molecular biology, computer vision models such as convolutional neural networks (CNNs) have been used to speed up image analysis (Dao et al., 2016; Falk et al., 2019; McQuin et al., 2018). Deep learning models trained on sequence-structure pairs can be used for predicting the structure of novel drug candidates, designing novel enzymes, and drug discovery (Jumper et al., 2021; Sanchez-Lengeling & Aspuru-Guzik, 2018). As ML models become more sophisticated, disciplines such as stem cell biology can leverage ML to tackle complex problems.

2.2 Supervised and unsupervised machine learning models

Most models in ML come under what is known as supervised ML. The examples that we have discussed above also come under supervised ML (see Fig. 30.1). In supervised ML, we are given a dataset $\mathscr{D}$ with n examples, that is, n-pairs of (x_i, y_i). The error between the ground truth (y_i) and the model's prediction is formulated as a loss function. Loss functions capture the error between the actual value and the model's prediction, such that the lower the loss, the better the model's prediction. This loss function is then minimized to train a model. Many ML algorithms, such as logistic regression, decision trees, and neural networks operate under the supervised learning framework. Unsupervised learning problems do not have labels associated with each example. We have x_i for $i = 1, 2, \ldots, n$ but we do not have the y_i associated with each sample. This task is more challenging than supervised learning as we do not have labels to train ML models. In this domain, the focus is on discovering novel groups and patterns in the data. This is called clustering analysis. Clustering analysis aims to discover distinct groups present in the data based on observations x_i for $i = 1, 2, \ldots, n$. Fig. 30.1B shows

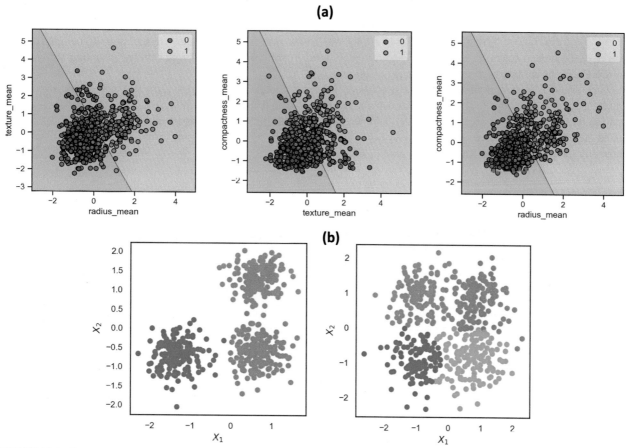

FIGURE 30.1 Supervised and unsupervised learning. (A) The scatter plot displays the distribution of examples colored by diagnosis. In each plot, we also show the decision boundaries of a supervised machine learning model when trained using two features only. (B) Clustering on a simulated dataset for both plots, the groups were created using known distribution; their memberships are known, but in practice, these are unavailable. On the left, a cluster with three distinct groups represented by different colors. The groups can be easily identified in this plot. On the right, a cluster with four distinct groups and represented by different colors. 0- Benign, 1- malignant. *Data from: https://archive.ics.uci.edu/ml/datasets/breast+cancer+wisconsin+(diagnostic).*

a clustering analysis on a simulated dataset. The plot on the left shows the scatter plot based on two variables, making it easier to separate the groups visually. However, in practice, we usually have more than two variables, which may not be possible to plot or cluster visually. The plot on the right shows an example of clustering where it is difficult to visually inspect and assign groups to data points. This difficulty can be due to overlap in the clusters or an unmeasured variable that is not included in the dataset but affects the clustering. However, we cannot visualize the data beyond three variables using a scatter plot. Hence, it is necessary to have algorithms that can automatically assign group membership to data points based on some measure without visual inspection.

ML algorithms have shown great promise in stem cell research, especially in cell classification and differentiation (Ashraf et al., 2021). Supervised learning algorithms such as logistic regression, k-nearest neighbors, and support vector machines have been used to classify stem cells based on various features, including gene expression patterns, cell morphology, and functional characteristics (Dao et al., 2016; Falk et al., 2019; Malta et al., 2018). Unsupervised learning algorithms such as clustering and dimensionality reduction have been used to identify subpopulations of stem cells based on their transcriptomic profiles (Cao et al., 2019; Wagner et al., 2018). These algorithms may help to discover novel markers of stem cell subpopulations. In addition, deep learning algorithms such as CNNs have been applied to the image-based analysis of stem cell colonies, enabling accurate and high-throughput monitoring of stem cell differentiation (Buggenthin et al., 2017; Kusumoto & Yuasa, 2019). In this chapter, we discuss neural networks and deep learning models, which are increasingly being applied for various applications in stem cell research.

2.3 Neural networks and deep learning

Artificial neural networks (ANNs), simply called neural networks, are ML models inspired by the structure and function of the human brain. The first computational version of a biological neuron was created by Rosenblatt in 1957, which laid the foundation for neural network research (Frank, 1957). While the human brain is made up of billions of neurons connected by trillions of synapses, artificial neurons are much simpler in comparison but can still perform complex tasks. However, even the most advanced ANNs today still have a long way to go in terms of completely resembling the complexity and power of the human brain. At present, ANNs are widely used in various real-world applications, including autonomous vehicles and natural language understanding (LeCun et al., 2015).

Each artificial neuron in an ANN can be described by three components: A *weight vector* (w), a scalar *bias* term (b), and an *activation function* ($g(z)$). Fig. 30.2A depicts the general architecture of a multilayer neural network. The weights and the bias term (see Fig. 30.2B) are model parameters. The inputs $x_1, x_2, \ldots, x_n$ can be considered a layer, collectively called the 0th layer of the network. Alternatively, the inputs can also be viewed as outputs from a previous layer. Eq. (30.3) shows the output of a single neuron by combining inputs and weights.

$$a = g(z) = g(w_1 x_1 + \ldots + w_n x_n + bias) \tag{30.3}$$

The activation function $g(z)$ is usually a nonlinear function that outputs a scalar value. Nonlinear activation functions are crucial because the function modeled by a neural network would collapse into a linear model without them. Commonly used activation functions include *rectilinear linear unit* (ReLU), sigmoid, hyperbolic tangent, and leaky ReLU. Although activation functions can introduce some nonlinearity in the modeling, we need more than a single layer to model more complex functions, hence the use of a multilayered architecture. All the weights and biases of a neural network model are parameters of the model, which must be estimated using available data. Fitting any ML model requires a loss function that aids the model in finding optimal parameter values. Several loss functions are available for different tasks; however, the squared-error function is commonly adopted for quantitative response variables and categorical cross-entropy for qualitative response variables. The expression for categorical cross-entropy is shown in Eq. (30.4).

$$-\sum_{i=1}^{n} \sum_{m=0}^{9} y_{im} log(f_m(x_i)) \tag{30.4}$$

Fig. 30.2A shows the general form of a multilayer neural network. Fig. 30.2B shows a computational graph of a multilayer network with computations and activation at each layer. ReLU and SoftMax activation functions are shown in Eqs. (30.5) and (30.6), respectively.

$$g(z) = \begin{cases} 0, z < 0 \\ z, \text{otherwise} \end{cases} \tag{30.5}$$

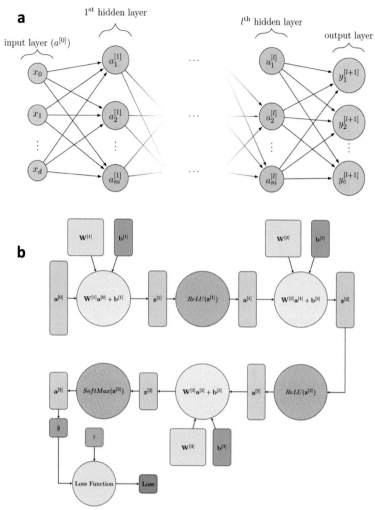

FIGURE 30.2   Multilayer neural network. (A) A multilayer network consisting of l hidden layers or a l-layer neural network. The input layer is the 0th layer with d dimensional input, while the output is the $(l+1)^{th}$ layer with c outputs. The number l is also called the depth of the network. The weights and biases of individual neurons are not shown for the sake of clarity. (B) Computational graph of a multilayer neural network. All scalars are shown as small squares, vectors as vertical bars (rectangles), matrices as big squares, and computations in circles. All input vectors are in pink (vertical bars), nodes that compute output ($z^{[l]}$) in light orange circles, and nodes that compute activations ($a^{[l]}$) in deep orange circles with their activation function shown inside. All weight matrices are in light teal (big squares) and bias vectors are blue (vertical bars). The prediction $\hat{y}$ (gray square) and ground truth y (green square) serve as inputs to the loss function. For a single neuron in the layer, the inputs from the previous layer are combined with weights associated with each input, and then the bias term is added. The output vector $z^{[1]}$ is obtained by multiplying the input vector $a^{[0]}$, which has all the inputs in the 0th layer, by the weight matrix of the first layer (W) and adding bias ($b^{[1]}$). It is passed through the ReLU function to compute the activations/output of the first layer $a^{[1]}$. The output of second layer is computed using the activations $a^{[1]}$ from first layer as input. The final output $\hat{y}$ is a scalar value corresponding to the model's prediction, which is combined with the ground truth value y to compute the loss using a loss function. Please note that, the choice of activation functions was only for illustrative purpose and arbitrarily selected for this example.

$$f_m(Y) = \frac{e^{Y_m}}{\sum\limits_{i=1}^{K} e^{Y_i}} \tag{30.6}$$

Neural networks are trained using a method called *backpropagation*. The details of the backpropagation algorithm lie outside the scope of this chapter. While training, the inputs to a model are passed in batches to estimate the error vectors better. After every batch is passed through the network, the values of all model parameters are updated using backpropagation. An epoch is when a neural network has seen all the examples in the dataset exactly once. A validation set is a subset of the training data used to monitor the model while training. It allows the practitioner to stop the training if the model performance degrades over epochs on the validation dataset.

2.4 Deep learning

Deep learning refers to ML techniques based on multilayer ANNs with more than three hidden layers. However, today when we use the term deep learning, we usually mean networks with 50 or even more than 100 layers (LeCun et al., 2015). CNN developed in the 1980s was one of the earliest deep learning models (Fukushima, 1980). However, the lack of large-scale datasets and powerful processors limited its capabilities and usage. The revival of neural networks that happened around 2010 could be partially attributed to the success of CNNs combined with powerful GPUs and the availability of large datasets such as those from the Canadian Institute for Advanced Research (CIFAR)-10 and CIFAR-100 (LeCun et al., 2015). Table 30.1 summarizes the various deep learning models and their current applications in stem cell biology.

TABLE 30.1 Application of deep learning models in computational biology.

Deep learning models	Description	Applications	References
Convolutional neural network (CNN)	• First proposed in 1998 (Lecun et al., 1998) • Primarily used for image and video processing • Several CNN variants have been proposed (He et al., 2016; Krizhevsky et al., 2017; Szegedy et al., 2014)	• Stem cell lineage prediction • Cell counting and live tracking • Segmentation and phenotyping • Cell motility prediction	• (Buggenthin et al., 2017; Kusumoto et al., 2018) • (Dao et al., 2016; Falk et al., 2019; McQuin et al., 2018) • (Dao et al., 2016; Dürr & Sick, 2016; Godinez et al., 2017; Wang et al., 2022) • (Nishimoto et al., 2019)
Deep autoencoder	• First proposed in 2006 (Hinton & Salakhutdinov, 2006) • Used for rerepresenting the input vector • Size of the input and output layers is equal	• Dimensionality reduction • Stem cell identification	• (Wang & Gu, 2018) • (Song et al., 2017)
Recurrent neural network (RNN)	• First proposed in 1989 by Williams and Zipser (1989) Major contributions by Hochreiter and Schmidhuber (1997) through their work on LSTMs • Primarily used for sequential data such as text	• Cell motility discrimination and prediction • Proarrhythmia prediction (in HPSCs-derived cardiomyocytes)	• (Kimmel et al., 2021) • (Golgooni et al., 2019)
Generative adversarial network (GAN)	• First proposed in 2014 by Goodfellow et al. (2014) • The generator tries to fool the adversarial classifier network by generating fake samples that get more realistic with training • Several variants have been proposed for specific applications (Karras et al., 2018; Zhu et al., 2020)	• Cell motility and trajectory prediction	• (Comes et al., 2020)
Transformers	• First proposed in 2017 (Vaswani et al., 2017) • Mainly used for text and sequential inputs, such as music and speech • Contains attention blocks that can weigh the importance of different parts of a sequence • Ability to handle parallel inputs and variable-length sequences • Several variants have been proposed (Devlin et al., 2019; Raffel et al., 2020)	• Cell-type annotation	• (Yang et al., 2022)

Table shows the types of deep learning models and their applications in computational and stem cell biology.

Table 30.1 additionally shows the applications in computational biology, and these models might also have future applications in stem cell research.

CNNs are based on the architecture of the human visual cortex and leverage the spatial information present in the image to learn valuable features for image classification. A CNN consists of two specialized layers that help build powerful feature maps: *Convolution* and *pooling* layers. Convolution layers comprise convolution filters that perform the convolution operation. The filter slides over the image matrix and extracts features that resemble the filter. Pooling layers condense the feature maps from convolution layers into smaller maps. As the feature maps are condensed, it also reduces the number of parameters fit by the model. Finally, the maps are flattened to vectors at the last layer and passed to a simple classifier network, which outputs the class probabilities for a given input (O'Shea & Nash, 2015). CNNs have been increasingly used in stem cell biology to analyze and interpret complex imaging data. They have been applied to a range of tasks, including the detection and lineage tracking of stem cells, cell motility prediction, and the identification of different subpopulations of stem cells (Buggenthin et al., 2017; Kimmel et al., 2021; Kusumoto et al., 2018). Outside of stem cell biology, they have been extensively used in advancing image analysis and are part of popular analysis tools (Dao et al., 2016; McQuin et al., 2018).

Recurrent neural networks (RNNs) are designed to process sequential data such as text, speech, or time series. Unlike CNNs, RNNs can retain information from previous inputs and use that information to make predictions about future inputs, rendering them suitable for tasks such as speech recognition, machine translation, and sentiment analysis (Lipton et al., 2015). However, RNNs can suffer from the vanishing gradient problem, which limits their ability to capture long-term dependencies in sequential data (Bengio et al., 1994). To address this issue, neural network architectures such as long-short-term memory networks (Sherstinsky, 2020) and transformers have been developed. Although the application of RNNs in stem cell research has been limited, a few studies have utilized RNNs to predict cell motility and proarrhythmia in cardiomyocytes derived from human pluripotent stem cells (Kimmel et al., 2021). RNNs can prove to be very useful in the future, as they may prove useful in modeling the temporal dynamics of gene expression during stem cell differentiation, allowing for the identification of key regulatory genes and pathways.

The Transformer model architecture is based on the concept of self-attention, which allows the model to weigh the importance of different parts of the input sequence when making predictions (Vaswani et al., 2017). Unlike traditional RNNs, which sequentially process input sequences, transformers can process the entire input sequence in parallel. This mechanism allows the model to weigh the importance of different parts of the input sequence when making predictions, which enables the model to capture long-term dependencies in the data. Transformers have achieved state-of-the-art performance on various natural language processing tasks, including language understanding, generation, and machine translation. Some popular transformer-based models include Bidirectional Encoder Representations from Transformers (BERT), GPT-3, and T5, which have been pretrained on large datasets and fine-tuned for specific downstream tasks (Brown et al., 2020; Devlin et al., 2019; Raffel et al., 2020). Transformers have proven to be exceptionally good at inferring meaningful information from long sequences They have been applied for the task of annotating single-cell omics data and inferring gene regulatory networks (Yang et al., 2022). As transformer-based large language models become more and more powerful, computational biologists can greatly benefit from such models. Transformer models such as GPTs and BERTs can help researchers annotate large datasets and can also help discover new cell types (Yang et al., 2022).

Generative adversarial networks (GANs) were proposed by Goodfellow et al. (2014). A GAN consists of two neural networks: a generator and a discriminator. The generator produces fake data similar to real data, while the discriminator tries to distinguish between fake and real data. The two networks are trained in an adversarial way, meaning that the generator tries to fool the discriminator, and the discriminator tries to identify the fake data accurately. GANs have been used for various applications, including image, text, and speech generation (Alqahtani et al., 2021). In stem cell research, GANs have been applied to augment organ-on-chip (OoC) models in predicting cell trajectories in OoC models (Comes et al., 2020). The amount of data generated by time-lapse microscopy in OoC experiments can become a bottleneck in the future as more and more drug discovery pipelines shift toward OoC validation. GANs may also prove crucial in modeling cell trajectories in OoCs to predict the movement of immune cells toward tumors and cell motility in the presence of chemotherapeutic agents (Comes et al., 2020).

2.5 Limitations

One of the major limitations of deep neural networks and ML models is their vulnerability to overfitting, especially when they have more parameters than the number of training examples (Marcus, 2018). Another major limitation of deep learning models is interpretability (Marcus, 2018). In many practical applications, interpretability can play a more important role than performance, especially in clinical settings where simpler models with explainability can be more useful. Deep learning models are also called "data-hungry" models, and for good reason. Typical deep learning applications require large datasets (with a large number of samples), which are often scarce in computational biology. Further, the high dimensionality in computational biology datasets limits the power of deep learning models. In stem cell biology, we often have partial a priori knowledge of the system, which could be beneficial in a prediction task. Deep learning works on the basis of automated feature extraction, which allows minimal inclusion of prior knowledge into the prediction framework (Marcus, 2018). Lastly, deep learning models are not able to distinguish causation from correlation. Correlation does not necessarily imply causation, especially for applications in computational biology. For instance, a model can learn the correlation between increased levels of a marker and the presence (or absence) of a disease, but it cannot ascertain the underlying mechanism. Hence, caution is warranted while interpreting the results of neural network models.

3. Applications of deep learning in stem cell research

In this section, we discuss the applications of deep learning models in stem cell research, specifically for computer vision tasks.

We consider two major applications of deep learning models in stem cell biology.

1. Lineage prediction
2. Automated cell type identification

There is a subtle difference between the two: the first application seeks predictions (based on specific features) about the lineage choice of cells before they have differentiated (lineage prediction), while the second aims to identify different cell types in a *differentiated population* of cells (lineage classification). For lineage prediction, we must provide reliable predictions before conventional methods detect the lineage choice. Otherwise, there is no incentive to use computational methods.

3.1 Stem cell lineage prediction

In stem cell biology, we are often interested in constructing quantitative and reproducible strategies to guide the differentiation of stem cells or their derivatives for various applications, especially therapeutics. The conventional methods include immunocytochemical analysis of fluorescently labeled lineage-specific proteins to detect specific cell types within a lineage. However, protein marker expression changes are frequently preceded by transcriptional, epigenetic, cytoskeletal, morphological, or metabolic changes. Hence, protein marker-based assays might not serve as early predictors of cell fate decisions because lineage markers will be expressed only after a cell has committed to a lineage (i.e., differentiated). Moreover, label-based assays are time-consuming, require fixation (which kills cells), expensive antibodies, and do not *predict* lineage (Kusumoto & Yuasa, 2019). In contrast, certain cellular properties can predict lineage commitment before the cell differentiates or expresses the lineage-specific protein marker. We refer to cellular features that can be quantified without using external reagents as cell-autonomous features. Cell morphology is one such factor that can be easily measured using various high-resolution microscopy techniques. Hence, morphology-dependent lineage prediction techniques prove to be cost-efficient and less time-consuming.

Stem cells are known to undergo various morphological changes during and before lineage commitment (McBeath et al., 2004). There are two clear advantages to using morphology-based features for lineage prediction. First, and perhaps most important, predictions are generated before differentiation occurs, as opposed to using conventional immunostaining assays. Second, the cells used for prediction remain intact and can be directly used for clinical applications.

An early application of ML for lineage prediction using cell-autonomous properties was demonstrated by Treiser et al. (2010). They used human mesenchymal stem cells (hMSCs) to forecast lineage within hours of stimulation. They obtained 43 morphological (descriptive) features from the hMSC cultures of multiple donors. They used an unsupervised learning algorithm to extract the most relevant features for the task, which reduced the number of features to three. They demonstrated that morphological features extracted from confocal microscopy could be used to distinguish clusters of stem cells, establishing that cell morphology contains relevant information for lineage prediction tasks.

Although this work established the importance of morphological features in lineage prediction, the reliance on manual feature extraction from confocal images was a significant bottleneck. This is where the power of deep learning models such as CNN comes into play. CNNs learn these feature maps through training over the dataset rather than utilizing handcrafted features for classification.

Buggenthin et al. used a combination of CNN and RNN to prospectively forecast lineage in hematopoietic stem and progenitor cells (HSPCs) (Buggenthin et al., 2017). They purified (sorted by flow cytometry) HSPCs from the bone marrow extracts of 12−14-week-old mice (PU.1^{eYFP}GATA1^{mCherry}). After sorting, the cells were immediately incubated with CD16/32 antibodies to detect lineage choice (granulocytic/monocytic [GM]). Approximately 6000 cells were imaged from three independent experiments using brightfield and confocal microscopy. For training any ML model, accurate labels are critical,

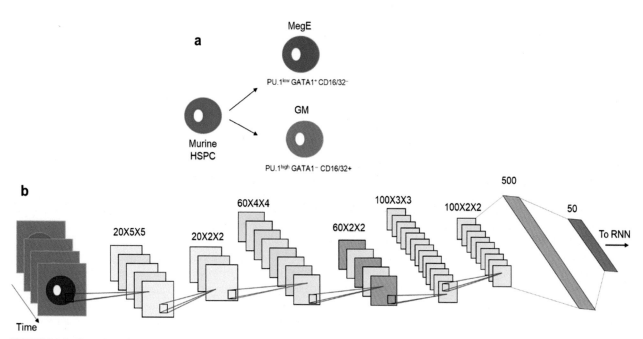

FIGURE 30.3   Deep learning model for lineage prediction. (A) An HSPC cell can differentiate into two lineages, MegE and GM. The conventional markers used for both the lineages are also shown. These markers appear after the cell commits to a lineage. (B) The schematic depicts the modified LeNet model used for lineage prediction. Image patches of size 27 × 27 are fed to the model and the extracted features are passed on to an RNN model which outputs a lineage score. *GM*, granulocytic/monocytic; *HSPC*, Hematopoietic stem cell; *MegE*, Megakaryocytic/erythroid. *Adapted from Buggenthin et al. (2017).*

and for this task, single-cell patches were manually annotated for lineage commitment based on the expression of CD16/32 (GM) and GATA1-mCherry (megakaryocytic/erythroid lineage). Fig. 30.3 shows an overview of the prediction task.

They used a modified version of vanilla LeNet (Lecun et al., 1998). The differentiation of HSPCs is a dynamic process and must be modeled similarly. The authors employed an RNN model to account for this time-varying attribute. The output of the CNN model was a continuous value called the lineage score, which was fed to a bidirectional long short-term memory RNN model. For each cell, the conventional markers are expressed only after a lineage choice is made; hence, predicting lineage before marker expression was a crucial component of this work. The model achieved an area under the curve (AUC) of 0.87 for annotated patches and 0.79 for latent patches. A high AUC value for latent patches suggests two things. first, there were identifiable morphological differences between latent and annotated patches, and second, the model could learn these differences and correctly identify the lineages based on them. The authors further demonstrated that the model could attain an AUC of 0.84 up to three generations before a conventional marker appeared.

Buggenthin et al. successfully demonstrated that cell-autonomous characteristics, like cytoskeleton rearrangement happening prior to lineage commitment, could be used for lineage prediction. A major advantage of deep learning models is that they eliminate the need to design handcrafted features and can independently extract potentially better features directly from the data. This was also demonstrated in another study by Dürr and Sick (2016), in which they compared the performance of a CNN model with conventional ML algorithms that use handcrafted features. The CNN model outperformed the existing state-of-the-art model by reducing the error rate from 8.9% to 6.6% (Dürr & Sick, 2016). The features obtained using CNN models were used as input for an RNN model. Changes in cell morphology occur over time, and using a model that can capture time-varying information is essential. Thus, it is important to choose models that are relevant to the task. While a CNN model may give satisfactory results, applying task-specific models can improve performance and increase the model's generalization capability. Several other studies have also proposed the application of deep learning for lineage prediction (Boldú et al., 2021; Moen et al., 2019; Sugawara et al., 2022; Waisman et al., 2019). Sugawara and colleagues developed a platform called efficient learning using sparse human annotations for nuclear tracking (Sugawara et al., 2022). The platform used an incremental learning approach, starting with a small amount of annotated data and gradually improving tracking performance through successive prediction-validation cycles. They used it to track lineages spanning the entire course of leg regeneration in a crustacean over 1 week, with 504 time points. Developing such semiautomated platforms with human-in-the-loop can help researchers significantly reduce the time and effort required to track cells accurately, particularly in large-scale studies such as those involving long-term tracking of cell lineages.

Buggenthin et al. developed a model using murine hematopoietic lineage stem cells. However, their model was not validated with data from other lineages or human iPSC cell lines. The generalization capabilities of the model could have been tested if it performed similarly on iPSCs. However, working with datasets generated from multiple cell lines sourced from different organisms is often challenging in stem cell research. Nevertheless, the performance of a model must be scrutinized as rigorously as possible, and if available, the model should be tested with multiple datasets before drawing any strong conclusions.

3.2 Automated identification of cell type

Identifying cell types from a differentiated population is essential for various applications, such as toxicological drug testing, the study of prenatal development, and regenerative medicine (Zakrzewski et al., 2019). Immunostaining methods are used as gold standard assays for classifying different cell types. The ideal method to classify cell types or identify subpopulations in a heterogeneous culture should be fast, accurate, and cost-efficient, such that the researchers can use most of their time working on applications of interest rather than manual annotation and lineage identification. The use of cell-autonomous features is a compelling idea that we discussed previously. Using morphological features for cell identification eliminates the need for time-consuming experiments and labor-intensive manual annotation.

Pluripotent stem cells can be extracted from human embryonic tissues or from unique niches in adult tissues such as bone marrow. Alternatively, in vitro reprogramming of somatic cells can generate iPSCs (Takahashi & Yamanaka, 2006). In one of the earliest applications of ML for identifying cell types, used handcrafted features based on morphological information from phase-contrast images (Theriault et al., 2012). However, as discussed earlier, handcrafting morphology-based features is time-consuming. Niioka et al. (2018) applied a CNN model to classify differentiated cells obtained from culturing mouse C2C12 myoblasts. They achieved an accuracy of 91.8% with an image size of 400 pixels. Kusumoto et al. (2018) applied deep learning models to identify endothelial cells within a population of differentiated iPSCs using phase-contrast images (Kusumoto et al., 2018). This work is one of the most cited works in the field, and hence we delve into the details of this study to highlight a few important aspects. The cells were imaged using phase-contrast microscopy and then stained for CD31, a marker of endothelial cells. The acquired immunofluorescent images were then binarized into black and white pixels based on a cut-off value. The binarized images were labeled as "unstained" or "stained" based on the ratio of white to black pixels. In total, 160,000 blocks were generated from 800 images obtained from four experiments, out of which 128,000 were used for training and the rest were used for testing—an 80:20 split. They compared the performance of two CNN models, AlexNet and LeNet. Interestingly, AlexNet had won the ImageNet competition in 2012, revolutionizing the field of neural networks and deep learning (Krizhevsky et al., 2017). This network won the ImageNet competition in 2012 by a huge margin, which inspired further research into deep learning models, especially for computer vision tasks.

Kusumoto et al. trained both models and studied the effect of various parameters on model performance. Varying the number and size of input blocks for training the model increased the performance, but only up to 32,000 blocks, beyond which the improvement was slow. The model yielded maximum performance with 32,000 blocks and a 512×512-pixel input block size. The target block size did not significantly affect performance. Further, they varied the staining threshold to test its effect on the model's performance. They found that a threshold of 0.3 and an input size of 512×512 resulted in the best F1 score of 0.7604. The authors also highlight the effect of model size on performance. AlexNet is a larger network with a greater number of convolution layers. The size of the feature maps (first convolution layer) is also large, which helps with better feature extraction. The final number of feature maps extracted by AlexNet is eight times that of LeNet. Lastly, the number of neurons in the fully connected layer is more than 20 times that of LeNet, which helps model a highly nonlinear function. The performance was higher for AlexNet, suggesting that larger networks can model more complex relationships in data and find optimal feature maps (Fig. 30.4).

A critical aspect in studies such as the one by Kusumoto et al. is the effect of staining, which can arise due to multiple factors (Van Eycke et al., 2017). The authors rebinarized the images manually and evaluated the performances of both models again. The F_1 score for LeNet and AlexNet increased above 0.9 and 0.95, respectively. Finally, they also performed 4-fold cross-validation. When used with the rebinarized blocks for cross-validation, the large network had an F_1 score of 0.93 and an accuracy of 0.91. This result justified the use of a larger network and manual binarization efforts. This establishes that the developed model had reasonable generalization capabilities based on cross-validation results.

4. Discussion

In this chapter, we discussed neural networks and deep learning models and their applications in computational stem cell biology. We also looked at specific models and how they have been applied to solve various problems in computational

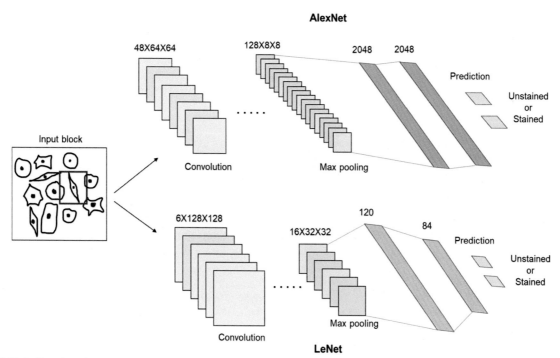

FIGURE 30.4   Deep learning models used to predict cell type. Two deep learning models were trained for the task of identifying endothelial cells in a differentiated iPSC population. *Adapted from Kusumoto et al. (2018).*

biology (see Table 30.1). Finally, we discussed two major applications of deep learning models in detail, highlighting their merits and demerits. Problems in stem cell biology have fascinated not only biologists but also engineers, physicists, and mathematicians, leading to a fusion of ideas from these fields. For example, the problem of embryonic pattern formation in developmental and stem cell research is modeled differently by researchers from different fields. The transition from knowledge-based to data-driven systems has transformed many life science fields. Research in various biology divisions, such as microbiology, genetics, and ecology, uses computational techniques for various purposes. Stem cell research has also benefited from the infusion of computational techniques such as deep learning. However, there are challenges associated with applying deep learning in stem cell research; we discuss a few of them below.

5. Challenges and future directions

Biomedical research is drowning in data but starving for knowledge. High-throughput sequencing technologies such as single-cell omics generate vast amounts of data daily that are incomprehensible for humans to process. As more and more datasets are added each year, experimental data increases exponentially without fully realizing its potential. The combined size and number of existing datasets in all such databases is mind-boggling (Perez-Riverol et al., 2019). Combining these datasets for deep learning is an even more difficult task. Image datasets are structured arrays that are easy to use, simple to manipulate, faster to process, and can be fed to a model with minimal transformations. Unstructured datasets such as gene regulatory networks, protein-protein interaction networks, and textual data from patient files are difficult to extract, transform, and load and require customized processing before modeling. Datasets in stem cell omics and other experiments are varied, each with its own advantages and disadvantages. Datasets generated to train deep learning models must be unbiased and ideally cover diverse genetic backgrounds. We must ensure that our conscious or unconscious biases do not get transferred to the models we develop and datasets we generate (Ouyang et al., 2023). Another challenge is that deep learning models are affected by intrinsic noise in the systems used to generate data, such as sequencing platforms and imaging systems. Deep learning models, specially developed for clinical utility, must account for these noise factors and strive to test prototypes as rigorously as possible.

Stem cell biology is conveniently placed at a juncture where computational methods are an essential component of biological research. All aspects of stem cell research, from generating iPSC cell lines to their characterization and clinical usage, require some computational methods, directly or indirectly. At this juncture also lies the field of deep learning, which has proven to perform excellently in many areas of biomedical science and is a promising technology for various applications. For applications in stem cell biology, such as regenerative medicine, deep learning can accelerate pipelines to screen cell types, predict tumorigenicity, and screen novel biomaterials (Srinivasan et al., 2021). In conclusion, deep learning has the potential to revolutionize stem cell research by providing an automated data analysis platform for researchers. We believe that applications of deep learning in stem cell research are still in their early stages, and the abovementioned challenges must be overcome. However, with continued advancements in deep learning techniques and stem cell biology and collaborations between computer scientists and stem cell researchers, the potential benefits of deep learning in stem cell research are limitless.

Acknowledgments

We acknowledge and confirm that all figures and tables in the chapter are original and made by authors. The figures were created and compiled using LaTex (www.overleaf.com) and Microsoft Power Point. Plots for illustrations in figures were coded in Python 3.7.

Author contributions

Nirbhay Raghav: Writing—original draft; writing—review and editing; **Anil Vishnu:** writing—review and editing; **Neha Deshpande:** writing—review and editing; **Annapoorni Rangarajan:** writing—review and editing; supervision.

References

Alqahtani, H., Kavakli-Thorne, M., & Kumar, G. (2021). Applications of generative adversarial networks (GANs): An updated review. *Archives of Computational Methods in Engineering, 28*(2), 525−552. https://doi.org/10.1007/s11831-019-09388-y

Ashraf, M., Khalilitousi, M., & Laksman, Z. (2021). Applying machine learning to stem cell culture and differentiation. *Current Protocols, 1*(9), e261. https://doi.org/10.1002/cpz1.261

Barker, N., van Es, J. H., Kuipers, J., Kujala, P., van den Born, M., Cozijnsen, M., Haegebarth, A., Korving, J., Begthel, H., Peters, P. J., & Clevers, H. (2007). Identification of stem cells in small intestine and colon by marker gene Lgr5. *Nature, 449*(7165). https://doi.org/10.1038/nature06196. Article 7165.

Bengio, Y., Simard, P., & Frasconi, P. (1994). Learning long-term dependencies with gradient descent is difficult. *IEEE Transactions on Neural Networks, 5*(2), 157−166. https://doi.org/10.1109/72.279181

Boldú, L., Merino, A., Acevedo, A., Molina, A., & Rodellar, J. (2021). A deep learning model (ALNet) for the diagnosis of acute leukaemia lineage using peripheral blood cell images. *Computer Methods and Programs in Biomedicine, 202*, 105999. https://doi.org/10.1016/j.cmpb.2021.105999

Breiman, L. (2001). Statistical modeling: The two cultures (with comments and a rejoinder by the author). *Statistical Science, 16*(3), 199−231. https://doi.org/10.1214/ss/1009213726

Brown, T., Mann, B., Ryder, N., Subbiah, M., Kaplan, J. D., Dhariwal, P., Neelakantan, A., Shyam, P., Sastry, G., Askell, A., Agarwal, S., Herbert-Voss, A., Krueger, G., Henighan, T., Child, R., Ramesh, A., Ziegler, D., Wu, J., Winter, C., … Amodei, D. (2020). Language models are few-shot learners. *Advances in Neural Information Processing Systems, 33*, 1877−1901. https://papers.nips.cc/paper/2020/hash/1457c0d6bfcb4967418bfb8ac142f64a-.

Buggenthin, F., Buettner, F., Hoppe, P. S., Endele, M., Kroiss, M., Strasser, M., Schwarzfischer, M., Loeffler, D., Kokkaliaris, K. D., Hilsenbeck, O., Schroeder, T., Theis, F. J., & Marr, C. (2017). Prospective identification of hematopoietic lineage choice by deep learning. *Nature Methods, 14*(4). https://doi.org/10.1038/nmeth.4182. Article 4.

Cao, J., Spielmann, M., Qiu, X., Huang, X., Ibrahim, D. M., Hill, A. J., Zhang, F., Mundlos, S., Christiansen, L., Steemers, F. J., Trapnell, C., & Shendure, J. (2019). The single-cell transcriptional landscape of mammalian organogenesis. *Nature, 566*(7745). https://doi.org/10.1038/s41586-019-0969-x. Article 7745.

Comes, M. C., Filippi, J., Mencattini, A., Corsi, F., Casti, P., De Ninno, A., Di Giuseppe, D., D'Orazio, M., Ghibelli, L., Mattei, F., Schiavoni, G., Businaro, L., Di Natale, C., & Martinelli, E. (2020). Accelerating the experimental responses on cell behaviors: A long-term prediction of cell trajectories using social generative adversarial network. *Scientific Reports, 10*(1). https://doi.org/10.1038/s41598-020-72605-3. Article 1.

Dao, D., Fraser, A. N., Hung, J., Ljosa, V., Singh, S., & Carpenter, A. E. (2016). CellProfiler Analyst: Interactive data exploration, analysis and classification of large biological image sets. *Bioinformatics, 32*(20), 3210−3212. https://doi.org/10.1093/bioinformatics/btw390

Deshpande, N., & Rangarajan, A. (2015). Cancer stem cells: Formidable allies of cancer. *Indian Journal of Surgical Oncology, 6*(4), 400−414. https://doi.org/10.1007/s13193-015-0451-7

Devlin, J., Chang, M.-W., Lee, K., & Toutanova, K. (2019). *Bert: Pre-Training of deep bidirectional Transformers for language understanding.* https://doi.org/10.48550/arXiv.1810.04805. arXiv:1810.04805). arXiv.

Dua, D., & Graff, C. (2019). *UCI machine learning repository.* Irvine, CA: University of California, School of Information and Computer Science. https://archive.ics.uci.edu/ml/datasets/breast+cancer+wisconsin+(diagnostic.

Dürr, O., & Sick, B. (2016). Single-cell phenotype classification using deep convolutional neural networks. *SLAS Discovery, 21*(9), 998−1003. https://doi.org/10.1177/1087057116631284

Falk, T., Mai, D., Bensch, R., Çiçek, Ö., Abdulkadir, A., Marrakchi, Y., Böhm, A., Deubner, J., Jäckel, Z., Seiwald, K., Dovzhenko, A., Tietz, O., Dal Bosco, C., Walsh, S., Saltukoglu, D., Tay, T. L., Prinz, M., Palme, K., Simons, M., … Ronneberger, O. (2019). U-Net: Deep learning for cell counting, detection, and morphometry. *Nature Methods, 16*(1). https://doi.org/10.1038/s41592-018-0261-2. Article 1.

Frank, R. (1957). *The perceptron, a perceiving and recognizing automaton.* Cornell Aeronautical Laboratory.

Fukushima, K. (1980). Neocognitron: A self-organizing neural network model for a mechanism of pattern recognition unaffected by shift in position. *Biological Cybernetics, 36*(4), 193−202. https://doi.org/10.1007/BF00344251

Godinez, W. J., Hossain, I., Lazic, S. E., Davies, J. W., & Zhang, X. (2017). A multi-scale convolutional neural network for phenotyping high-content cellular images. *Bioinformatics, 33*(13), 2010−2019. https://doi.org/10.1093/bioinformatics/btx069

Golgooni, Z., Mirsadeghi, S., Soleymani Baghshah, M., Ataee, P., Baharvand, H., Pahlavan, S., & Rabiee, H. R. (2019). Deep learning-based proar-rhythmia analysis using field potentials recorded from human pluripotent stem cells derived cardiomyocytes. *IEEE Journal of Translational Engineering in Health and Medicine, 7*, 1−9. https://doi.org/10.1109/JTEHM.2019.2907945

Goodfellow, I. J., Pouget-Abadie, J., Mirza, M., Xu, B., Warde-Farley, D., Ozair, S., Courville, A., & Bengio, Y. (2014). *Generative adversarial networks*. https://doi.org/10.48550/arXiv.1406.2661. arXiv:1406.2661). arXiv.

He, K., Zhang, X., Ren, S., & Sun, J. (2016). Deep residual learning for image recognition. In *2016 IEEE conference on computer vision and pattern recognition (CVPR)* (pp. 770−778). https://doi.org/10.1109/CVPR.2016.90

Hinton, G. E., & Salakhutdinov, R. R. (2006). Reducing the dimensionality of data with neural networks. *Science, 313*(5786), 504−507. https://doi.org/10.1126/science.1127647

Hochreiter, S., & Schmidhuber, J. (1997). Long short-term memory. *Neural Computation, 9*(8), 1735−1780. https://doi.org/10.1162/neco.1997.9.8.1735

Jumper, J., Evans, R., Pritzel, A., Green, T., Figurnov, M., Ronneberger, O., Tunyasuvunakool, K., Bates, R., Žídek, A., Potapenko, A., Bridgland, A., Meyer, C., Kohl, S. A. A., Ballard, A. J., Cowie, A., Romera-Paredes, B., Nikolov, S., Jain, R., Adler, J., … Hassabis, D. (2021). Highly accurate protein structure prediction with AlphaFold. *Nature, 596*(7873). https://doi.org/10.1038/s41586-021-03819-2. Article 7873.

Karras, T., Aila, T., Laine, S., & Lehtinen, J. (2018). *Progressive growing of GANs for improved quality, stability, and variation*. https://doi.org/10.48550/arXiv.1710.10196. arXiv:1710.10196). arXiv.

Kimmel, J. C., Brack, A. S., & Marshall, W. F. (2021). Deep convolutional and recurrent neural networks for cell motility discrimination and prediction. *IEEE/ACM Transactions on Computational Biology and Bioinformatics, 18*(2), 562−574. https://doi.org/10.1109/TCBB.2019.2919307

Krizhevsky, A., Sutskever, I., & Hinton, G. E. (2017). ImageNet classification with deep convolutional neural networks. *Communications of the ACM, 60*(6), 84−90. https://doi.org/10.1145/3065386

Kusumoto, D., Lachmann, M., Kunihiro, T., Yuasa, S., Kishino, Y., Kimura, M., Katsuki, T., Itoh, S., Seki, T., & Fukuda, K. (2018). Automated deep learning-based system to identify endothelial cells derived from induced pluripotent stem cells. *Stem Cell Reports, 10*(6), 1687−1695. https://doi.org/10.1016/j.stemcr.2018.04.007

Kusumoto, D., & Yuasa, S. (2019). The application of convolutional neural network to stem cell biology. *Inflammation and Regeneration, 39*(1), 14. https://doi.org/10.1186/s41232-019-0103-3

LeCun, Y., Bengio, Y., & Hinton, G. (2015). Deep learning. *Nature, 521*(7553). https://doi.org/10.1038/nature14539. Article 7553.

Lecun, Y., Bottou, L., Bengio, Y., & Haffner, P. (1998). Gradient-based learning applied to document recognition. *Proceedings of the IEEE, 86*(11), 2278−2324. https://doi.org/10.1109/5.726791

Lindvall, O., Kokaia, Z., & Martinez-Serrano, A. (2004). Stem cell therapy for human neurodegenerative disorders−how to make it work. *Nature Medicine, 10*(7). https://doi.org/10.1038/nm1064. Article 7.

Lipton, Z. C., Berkowitz, J., & Elkan, C. (2015). *A critical review of recurrent neural networks for sequence learning*. https://doi.org/10.48550/arXiv.1506.00019 (arXiv:1506.00019). arXiv.

Malta, T. M., Sokolov, A., Gentles, A. J., Burzykowski, T., Poisson, L., Weinstein, J. N., Kamińska, D., Huelsken, J., Omberg, L., Gevaert, O., Colaprico, A., Czerwińska, P., Mazurek, S., Mishra, L., Heyn, H., Krasnitz, A., Godwin, A. K., Lazar, A. J., Caesar-Johnson, S. J., … Wiznerowicz, M. (2018). ML identifies stemness features associated with oncogenic dedifferentiation. *Cell, 173*(2), 338−354.e15. https://doi.org/10.1016/j.cell.2018.03.034

Marcus, G. (2018). *Deep learning: A critical appraisal*. https://doi.org/10.48550/arXiv.1801.00631. arXiv:1801.00631). arXiv.

McBeath, R., Pirone, D. M., Nelson, C. M., Bhadriraju, K., & Chen, C. S. (2004). Cell shape, cytoskeletal tension, and RhoA regulate stem cell lineage commitment. *Developmental Cell, 6*(4), 483−495. https://doi.org/10.1016/s1534-5807(04)00075-9

McCulloch, W. S., & Pitts, W. (1943). A logical calculus of the ideas immanent in nervous activity. *Bulletin of Mathematical Biophysics, 5*(4), 115−133. https://doi.org/10.1007/BF02478259

McKee, C., & Chaudhry, G. R. (2017). Advances and challenges in stem cell culture. *Colloids and Surfaces B: Biointerfaces, 159*, 62−77. https://doi.org/10.1016/j.colsurfb.2017.07.051

McQuin, C., Goodman, A., Chernyshev, V., Kamentsky, L., Cimini, B. A., Karhohs, K. W., Doan, M., Ding, L., Rafelski, S. M., Thirstrup, D., Wiegraebe, W., Singh, S., Becker, T., Caicedo, J. C., & Carpenter, A. E. (2018). CellProfiler 3.0: Next-generation image processing for biology. *PLoS Biology, 16*(7), e2005970. https://doi.org/10.1371/journal.pbio.2005970

Mitchell, T. M., & Mitchell, T. M. (1997). *Machine learning* (Vol. 1). New York: McGraw-hill.

Moen, E., Borba, E., Miller, G., Schwartz, M., Bannon, D., Koe, N., Camplisson, I., Kyme, D., Pavelchek, C., Price, T., Kudo, T., Pao, E., Graf, W., & Valen, D. V. (2019). *Accurate cell tracking and lineage construction in live-cell imaging experiments with deep learning* (p. 803205). https://doi.org/10.1101/803205. bioRxiv.

Niioka, H., Asatani, S., Yoshimura, A., Ohigashi, H., Tagawa, S., & Miyake, J. (2018). Classification of C2C12 cells at differentiation by convolutional neural network of deep learning using phase contrast images. *Human Cell, 31*(1), 87−93. https://doi.org/10.1007/s13577-017-0191-9

Nishimoto, S., Tokuoka, Y., Yamada, T. G., Hiroi, N. F., & Funahashi, A. (2019). Predicting the future direction of cell movement with convolutional neural networks. *PLoS One, 14*(9), e0221245. https://doi.org/10.1371/journal.pone.0221245

O'Shea, K., & Nash, R. (2015). *An introduction to convolutional neural networks*. https://doi.org/10.48550/arXiv.1511.08458. arXiv:1511.08458). arXiv.

Ouyang, J. F., Chothani, S., & Rackham, O. J. L. (2023). Deep learning models will shape the future of stem cell research. *Stem Cell Reports, 18*(1), 6–12. https://doi.org/10.1016/j.stemcr.2022.11.007

Perez-Riverol, Y., Zorin, A., Dass, G., Vu, M.-T., Xu, P., Glont, M., Vizcaíno, J. A., Jarnuczak, A. F., Petryszak, R., Ping, P., & Hermjakob, H. (2019). Quantifying the impact of public omics data. *Nature Communications, 10*(1). https://doi.org/10.1038/s41467-019-11461-w. Article 1.

Raffel, C., Shazeer, N., Roberts, A., Lee, K., Narang, S., Matena, M., Zhou, Y., Li, W., & Liu, P. J. (2020). *Exploring the limits of transfer learning with a unified text-to-text transformer.* https://doi.org/10.48550/arXiv.1910.10683. arXiv:1910.10683). arXiv.

Rumelhart, D. E., Hinton, G. E., & Williams, R. J. (1986). Learning representations by back-propagating errors. *Nature, 323*(6088). https://doi.org/10.1038/323533a0. Article 6088.

Sanchez-Lengeling, B., & Aspuru-Guzik, A. (2018). Inverse molecular design using machine learning: Generative models for matter engineering. *Science, 361*(6400), 360–365. https://doi.org/10.1126/science.aat2663

Segers, V. F. M., & Lee, R. T. (2008). Stem-cell therapy for cardiac disease. *Nature, 451*(7181). https://doi.org/10.1038/nature06800. Article 7181.

Sherstinsky, A. (2020). Fundamentals of recurrent neural network (RNN) and long short-term memory (LSTM) network. *Physica D: Nonlinear Phenomena, 404*, 132306. https://doi.org/10.1016/j.physd.2019.132306

Song, T.-H., Sanchez, V., ElDaly, H., & Rajpoot, N. M. (2017). Hybrid deep autoencoder with Curvature Gaussian for detection of various types of cells in bone marrow trephine biopsy images. In *2017 IEEE 14th international symposium on biomedical imaging* (pp. 1040–1043). https://doi.org/10.1109/ISBI.2017.7950694 (ISBI 2017).

Srinivasan, M., Thangaraj, S. R., Ramasubramanian, K., Thangaraj, P. P., Ramasubramanian, K. V., Srinivasan, M., Thangaraj, S. R., Ramasubramanian, K., Thangaraj, P. P., & Ramasubramanian, K. V. (2021). Exploring the current trends of artificial intelligence in stem cell therapy: A systematic review. *Cureus, 13*(12). https://doi.org/10.7759/cureus.20083

Sugawara, K., Çevrim, Ç., & Averof, M. (2022). Tracking cell lineages in 3D by incremental deep learning. *Elife, 11*, e69380. https://doi.org/10.7554/eLife.69380

Szegedy, C., Liu, W., Jia, Y., Sermanet, P., Reed, S., Anguelov, D., Erhan, D., Vanhoucke, V., & Rabinovich, A. (2014). *Going deeper with convolutions.* https://doi.org/10.48550/arXiv.1409.4842. arXiv:1409.4842). arXiv.

Takahashi, K., Tanabe, K., Ohnuki, M., Narita, M., Ichisaka, T., Tomoda, K., & Yamanaka, S. (2007). Induction of pluripotent stem cells from adult human fibroblasts by defined factors. *Cell, 131*(5), 861–872. https://doi.org/10.1016/j.cell.2007.11.019

Takahashi, K., & Yamanaka, S. (2006). Induction of pluripotent stem cells from mouse embryonic and adult fibroblast cultures by defined factors. *Cell, 126*(4), 663–676. https://doi.org/10.1016/j.cell.2006.07.024

Theriault, D. H., Walker, M. L., Wong, J. Y., & Betke, M. (2012). Cell morphology classification and clutter mitigation in phase-contrast microscopy images using machine learning. *Machine Vision and Applications, 23*(4), 659–673. https://doi.org/10.1007/s00138-011-0345-9

Treiser, M. D., Yang, E. H., Gordonov, S., Cohen, D. M., Androulakis, I. P., Kohn, J., Chen, C. S., & Moghe, P. V. (2010). Cytoskeleton-based forecasting of stem cell lineage fates. *Proceedings of the National Academy of Sciences, 107*(2), 610–615. https://doi.org/10.1073/pnas.0909597107

Van Eycke, Y.-R., Allard, J., Salmon, I., Debeir, O., & Decaestecker, C. (2017). Image processing in digital pathology: An opportunity to solve inter-batch variability of immunohistochemical staining. *Scientific Reports, 7*(1). https://doi.org/10.1038/srep42964. Article 1.

Vaswani, A., Shazeer, N., Parmar, N., Uszkoreit, J., Jones, L., Gomez, A. N., Kaiser, L., & Polosukhin, I. (2017). *Attention is all you need.* https://doi.org/10.48550/arXiv.1706.03762. arXiv:1706.03762). arXiv.

Wagner, D. E., Weinreb, C., Collins, Z. M., Briggs, J. A., Megason, S. G., & Klein, A. M. (2018). Single-cell mapping of gene expression landscapes and lineage in the zebrafish embryo. *Science, 360*(6392), 981–987. https://doi.org/10.1126/science.aar4362

Waisman, A., La Greca, A., Möbbs, A. M., Scarafía, M. A., Santín Velazque, N. L., Neiman, G., Moro, L. N., Luzzani, C., Sevlever, G. E., Guberman, A. S., & Miriuka, S. G. (2019). Deep learning neural networks highly predict very early onset of pluripotent stem cell differentiation. *Stem Cell Reports, 12*(4), 845–859. https://doi.org/10.1016/j.stemcr.2019.02.004

Wang, D., & Gu, J. (2018). VASC: Dimension reduction and visualization of single-cell RNA-seq data by deep variational autoencoder. *Genomics, Proteomics and Bioinformatics, 16*(5), 320–331. https://doi.org/10.1016/j.gpb.2018.08.003

Wang, A., Zhang, Q., Han, Y., Megason, S., Hormoz, S., Mosaliganti, K. R., Lam, J. C. K., & Li, V. O. K. (2022). A novel deep learning-based 3D cell segmentation framework for future image-based disease detection. *Scientific Reports, 12*(1). https://doi.org/10.1038/s41598-021-04048-3. Article 1.

Williams, R. J., & Zipser, D. (1989). A learning algorithm for continually running fully recurrent neural networks. *Neural Computation, 1*(2), 270–280. https://doi.org/10.1162/neco.1989.1.2.270

Yang, F., Wang, W., Wang, F., Fang, Y., Tang, D., Huang, J., Lu, H., & Yao, J. (2022). ScBERT as a large-scale pretrained deep language model for cell type annotation of single-cell RNA-seq data. *Nature Machine Intelligence, 4*(10). https://doi.org/10.1038/s42256-022-00534-z. Article 10.

Zakrzewski, W., Dobrzyński, M., Szymonowicz, M., & Rybak, Z. (2019). Stem cells: Past, present, and future. *Stem Cell Research and Therapy, 10*(1), 68. https://doi.org/10.1186/s13287-019-1165-5

Zhu, J.-Y., Park, T., Isola, P., & Efros, A. A. (2020). *Unpaired image-to-image translation using cycle-consistent adversarial networks.* https://doi.org/10.48550/arXiv.1703.10593. arXiv:1703.10593). arXiv.

Chapter 31

Multiscale computational and machine learning models for designing stem cell-based regenerative medicine therapies

Shraddha Pandit[1], Tanya Jamal[1], Anamta Ali[1] and Ramakrishnan Parthasarathi[1,2]

[1]Toxicoinformatics and Industrial Research, CSIR-Indian Institute of Toxicology Research, Vishvigyan Bhawan, Lucknow, Uttar Pradesh, India;
[2]Academy of Scientific and Innovative Research (AcSIR), Ghaziabad, Uttar Pradesh, India

1. Introduction

The discovery and advancement of stem cell therapy, mainly induced pluripotent stem cells (iPSCs), have provided a unique possibility to study human disease, focusing on tissue regeneration, cellular transplantation, and developmental research. Advancement in the single-cell approach has helped in the cellular phenotype's characterization, reconstruction, development, and reprogramming of single-cell trajectories. The generation of phenotypic omics data integration and analysis helps to determine cellular classification, biological function, and cell-cell interaction (Asahara & Kawamoto, 2004; Murry et al., 2005). Due to this, stem cells have become an object of intense scrutiny for their application in regenerative medicine, which can potentially repair and replace damaged tissue and organs due to chronic or acute stress. With this, stem cells used for regenerative therapies can combat severe autoimmune responses and overcome organ and tissue failures (Heidary Rouchi & Mahdavi-Mazdeh, 2015).

Despite these factors, the conventional approach of cell-based therapies and regenerative medicine still stands in a primitive phase in terms of commercialization because of its existing shortcomings. This includes the identification of iPSC-derived cells, their characterization, the generation of a humongous amount of complex data, and quality estimation. The manual approach to evaluate the morphology of a stem cell colony is a tedious and complicated process that is generally not applicable to large-scale cultures. Such complexities and complications further make the process difficult, impacting the decision-making process with the risk of error involved. In this scenario, massive multiomics data can be utilized to develop computational models that can predict the collective behavior of genes from the molecular to the tissue level. Thus, a synergistic approach utilizing the computational models and data obtained from the experimental analysis can help overcome the shortcomings and provide a futuristic approach to address relevant queries associated with stem cell and regenerative medicine research (Fig. 31.1). Once the stem cell starts proliferating, experimental analysis like population & developmental kinetics or single-cell RNA sequencing (scRNA-seq) can obtain data regarding the regeneration, continuity, heterogeneity, and robustness of the stem cell pool. This information is utilized further by multiscale computational models like gene regulatory network (GRN) models, deep learning, machine learning (ML), etc. to provide either prognostic or remedial support. While prognostic support generates predictions regarding stem cell therapy, its associated risks, cellular stage, subcellular compartment identification, etc., remedial support, on the other hand, presents opportunities related to decision-making regarding therapeutic choices and optimization of such treatments. This flow of information from experimental analysis to multiscale computational models works in a bidirectional manner, where each step stands in complement to the other.

Computational Biology for Stem Cell Research. https://doi.org/10.1016/B978-0-443-13222-3.00027-7
Copyright © 2024 Elsevier Inc. All rights reserved, including those for text and data mining, AI training, and similar technologies.

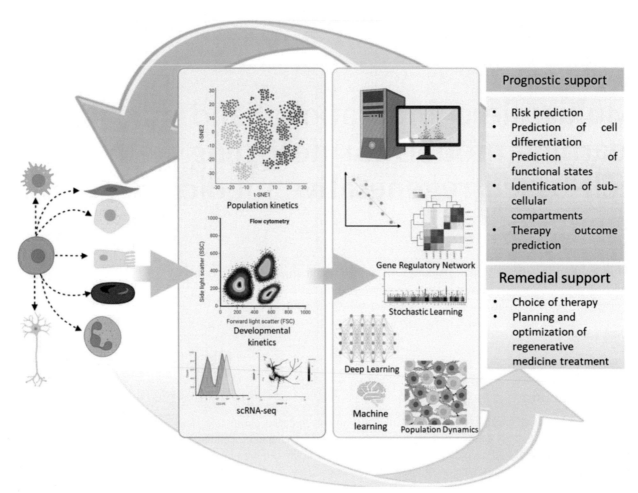

FIGURE 31.1 The current strategies and flow of information applied for generating computational models for providing prognostic and remedial support in regenerative medicine.

2. Attributes of multiscale computational modeling for stem cell research

Multicellular biological systems by utilizing experimental and therapeutic data become too complex to comprehend. The biomedical engineering initiatives primarily focus on integrating data across spatial, temporal, and functional scales. Researchers have become significantly more equipped to analyze enormous amounts of experimental data for a range of applications because of computational modeling. The multiscale models of stem cells may offer mathematical and computational modeling techniques that can potentially clarify the essential dynamics of three different stem cell populations. The main attributes of multiscale model development involve its construction, optimization, prediction, and improvement (Wang et al., 2019). Such multiscale models can then be used to examine biological processes more thoroughly due to the development of robust computing platforms and quantitative data from high-throughput experimental approaches (Walpole et al., 2013). The scales of biological functions and empirical research performed using experimental research and extensive living models can be extrapolated together using computational models. Since they can fill the research gaps between isolated in vitro research and whole-organism in vivo models, computational models are well-positioned to capture the connectivity between these disparate scales of biological function. The potential of ML approaches to analyze and learn from high-dimensional big data has been demonstrated in the areas of computer vision and natural language processing (Bengio & LeCun, 2007). Consequently, they are regarded as a promising method for determining biological data and designing prediction models (Raimundo et al., 2021). These models can automatically analyze and gain knowledge using multiscale statistics in fields like machine vision or expert systems (LeCun et al., 2015). It can be applied to predict the stages of stem cells from their naïve stage to the final differentiated stage. Another approach is provided by mechanistic models based on experimental observations to generate a novel prediction of the behavior of the stem cell in a particular environment. These models contain mathematical concepts of the observed phenomenon and provide a framework for guiding experimental research in cellular conversions. In the following sections, various approaches and computational models are described in terms of their role in biological data analysis and their contribution to enhancing the understanding of stem cell biology. In conclusion, computational frameworks support quantitative analysis of data by offering mechanistic perceptions into biological processes and by predicting in a way that can direct experimental investigation and may be used to solve critical biological problems in the field of stem cell research.

3. Bioinformatics approach in stem cell

Bioinformatics approaches utilize various computational and statistical methods in the field of iPSC containing large data sets that address biological complexities (Müller et al., 2012; Nestor & Noggle, 2013; Onyido et al., 2016; Polouliakh, 2019). A branch of computing algorithms called ML is continually improving and aims to replicate human intelligence by learning from the environment (El Naqa & Murphy, 2015). In the area of stem cell research, including identification of patterns, classification, and prediction-based studies, data owners who seek to train predictive models on their data can access a variety of ML frameworks and services (Song et al., 2017). With this, experimental genomics and proteomics data can also be obtained using different open-access databases that provide information regarding cell regulatory components. For example, the Embryonic Stem Cell Atlas from Pluripotency Evidence database (http://www.maayanlab.net/ESCAPE/index.php) can be utilized to curate and analyze single-cell data related to pluripotency (Xu et al., 2013, 2014). Another tool is StemSight (www.stemsight.org), which predicts the functionally interconnected network of the gene using Bayesian network analysis (Dowell et al., 2013). In stem cell facilities engaged in basic research and the development of regenerative medicines, quality control measures are critically required. For continuous assessment of cell quality and procedures, automated techniques are necessary. One such approach is using the StemCellQC toolkit. It monitors the cellular processes and helps in the qualitative prediction of the pluripotent stem cell colonies (Zahedi et al., 2016). Further, there is a tremendous need to analyze live cell imaging of stem cells to perform relevant toxicological studies. This can be done by applying video bioinformatics software tools to analyze stem cell progression (Talbot et al., 2014). Apart from the discussed approaches, GRN-based models also present an interesting example that provides insights into cellular conversion and differentiation and predicts the signaling molecules forming the foundations of these processes. With this, the multiscale computational models also implement statistical data analysis into the biological processes to generate novel predictions and mechanistic insights, thus aiding experimental research. Such multiscale bioinformatics approaches explore gene expression and changes in cellular behavior spanning different timescales.

4. Artificial intelligence based supervised and unsupervised modeling approach

Data mining, analysis of enormous amounts of data, pattern detection in the test data, and prediction of outcomes to solve biological problems have been made easier through artificial intelligence automation (Kumar et al., 2011; Wang et al., 2019). It provides real-time solutions to aid in the clinical and therapeutic domains of precision medicine (He et al., 2019). The most broadly involved man-made intelligence calculation in the clinical business is AI/ML and deep learning (DL) (Jiang et al., 2017; Kumar et al., 2018; Sharma et al., 2021). Most ML tasks are implemented using supervised learning, intended to generate results in classification and regression by utilizing an expert system and learning a pattern obtained in the test data to perform prediction (Buch et al., 2018). The standard algorithms used are decision tree (DT), logistic regression, and support vector machine (SVM). For the regression analysis, simple linear regression, DT, multivariate regression, and lasso regression algorithms are performed (Salehnasab et al., 2019). In this scenario, a specification presented by the DL approach thriving in computed tomography has demonstrated that convolutional neural networks (CNN) are potent pixel coordinators. Overall, CNNs have the potential to outperform traditional ML approaches in acquiring highly selective visual features related to computer vision and pattern recognition tasks (Cadieu et al., 2014). In the context of stem cell research, CNN can be used as an alternative molecular technique that enables morphology-based identification of multipotent iPSCs. Additionally, DL may use currently accessible techniques like DigitalDLSorter (Torroja & Sanchez-Cabo, 2019), scVI, DeepImpute, DeepMc, autoencoder by deconvolution training, BERMUDA, acAlign, cTP-net-NET, and DESC. These techniques use scRNA-seq data to count and measure tumor-infiltrating colon cells. It employs a deep neural network model allowing it to modify and measure any cell type as specific cell types interpret scRNA-seq data. On the other hand, unsupervised learning detects an unknown pattern present within the data without any predetermined information. These approaches have been employed in the analysis of stem cell experimentation data by recognizing the pattern in the stem cell structure and its classification of various subpopulations with unique potential for stem cell therapy (Chan et al., 2017; Stumpf & MacArthur, 2019).

5. Mechanistic modeling approach

ML models have depicted significant predictiveness; however, their major drawback is that such models can process vast omics and imaging data and can only establish a statistical network between the data input and predictive output. In contrast, mechanistic models can hypothesize a connection between biological entities, such as transcription factors, where this connection has previously been specified using experimental observations. Different mathematical and conceptual frameworks have been developed to represent a biological system at various levels through logical and continuous models.

Logical models represent the relationship between genes as governed by changing biological processes. The model represents the condition of a gene as active or inactive, determined by the state of its regulators. The GRN model is one such logical model that predicts stem cell biological state. For example, a model developed using data obtained from scRNA-seq could predict the transformation of mesodermal progenitors into endothelium and primitive blood cells (Moignard et al., 2015).

On the other hand, the elevated availability of unicellular data and continuous models have provided detailed insights into the stem cell system. As a result, the temporal dynamics are depicted on a continuous time scale, enabling a direct comparison of experimental results with the state of the model. These models can predict the differentiation, trans-differentiation, hematopoietic lineage specification, and niche interaction-determined mechanical properties of the stem cell (Matsumoto et al., 2017). With this, predictions of the culture duration obtained by the continuous models have been used to optimize the cellular differentiation protocols.

6. Dynamical modeling approach

Several modeling approaches are applied to determine the fate of the stem cells. A deterministic process is used as a dynamic system to define the progression of stem cell differentiation over time. The changes in the cellular morphology and expression of specific factors or proteins lead to the characterization of the particular stem cell stage. The biological system is mainly tackled using the nonlinear dynamical system, which considers that the outcome of the system is not impacted by the input change (Goldberger, 2006). It includes the cross-talk between the signal transduction pathway and physiological dynamics. This advancement can be described by the dynamics system using ordinary differential equations (ODEs) (Peltier & Schaffer, 2010). For this purpose, the expression or nonexpression of a particular protein is recognized using the binary approach. Stem cell metabolism is inextricably linked to pluripotency and function. However, the interconnectivity and high complexity of the metabolic networks of stem cells have made it difficult to understand the perplexing metabolic rewiring. Genome-scale metabolic network models are utilized to describe the metabolic activities of cells and tissues by employing transcriptomics data. Nevertheless, these approaches are ideal for modeling steady-state stem cells and not dynamic states. Recently, a genome-scale modeling approach the dynamic flux activity technique, has been created to address this issue (Shen et al., 2019). This technique predicts metabolic flux rewiring using time-course metabolic data. Utilizing the metabolic change data of pluripotent stem cells retrieved from the naïve and primed states of cells, such biological states can be easily differentiated (Chandrasekaran et al., 2017). For example, the differentiated versus undifferentiated embryonic stem cell state was defined using the availability or lack of the transcription factor Oct-4 (Prudhomme et al., 2004). The model generated two linear ODEs, depicting the cell's original or transformed state.

7. Spatiotemporal analysis

The spatiotemporal features are accounted for by the stem cell's time and space-occupying volume. Recent computational models describe these spatiotemporal effects on the kinetics of stem cell differentiation (Peltier & Schaffer, 2010; Van Leeuwen et al., 2009). Various biochemical processes, such as the volume exclusion of biological macromolecules, organelles, and the cytoskeleton impact the spatial mechanisms of a cell due to macromolecular crowding and anomalous diffusion (Mika & Poolman, 2011). Discrete and continuous mathematical expressions of spatially detailed subcellular reaction kinetics have been combined recently using in silico models (Cowan et al., 2012). In addition, the spatiotemporal influence on the kinetics of vital processes is demonstrated. Such spatial mechanistic properties like shape, location, diffusive transport of cells in polycystic renal disease, angiogenesis, and spheroid fusion are applied in multicellular systems in development and disease (Magno et al., 2020). Although with such applicability potential, the deployment of spatial models in defining the biological system remains uncommon due to the complex mathematical equations and expensive computational calculations required to cover entire cellular processes. This highlights the need for the development of more coherent methods in order to model the spatial characteristics of biological systems.

8. Comparing approaches & future perspectives of computational models for stem cell-based regenerative medicine therapies

The description and classification of cellular morphology have advanced considerably due to the digital pathology development and its computational analysis (Madabhushi & Lee, 2016). The high definition and dimensionality of image analysis-generated data, feature extraction algorithms, and the presence of high-throughput multiscale models have enhanced their applicability in the scientific domain of regenerative medicine (Table 31.1).

For a successful intervention to readdress pathologies using high-throughput computational techniques, a pragmatic approach in preclinical science is required. Few such cases have been discussed regarding the utilization of multiscale computational models in the available therapeutic applications.

The rate of heart failure and its frequency among humans today are significant and growing issues with a dismal prognosis. Mesenchymal stem cells (MSCs) have been employed as potential therapeutic agents for cardiovascular regeneration therapy for decades (Eckert et al., 2013). The stem cell diversity, or MSC sources, and potential uses for these cells make it difficult to choose the right cell type for cell therapy (Fathi & Farahzadi, 2018). Although myocardial regeneration is an effective alternative to heart transplantation, the best type of stem cells has not yet been identified due to inconsistent results in clinical trials. In the area of regenerative cardiology, one of the prominent cell forms is human iPSCs (hiPSCs). These cells have poor reprogramming efficiency rates because of the genomic instability of the reprogramming process, despite the fact that they may retain epigenetic memory from the starting tissue and revert to the somatic cell type from which they were created (Poetsch et al., 2022). Since current yields of actual hiPSCs can be as low as 0.001%−0.1% of the beginning cell population, the main goal for clinical researchers in hiPSC is to develop a high efficiency of the derivation of hiPSCs. The effectiveness of establishing pluripotency is still just 1%−5%, even in so-called "secondary" reprogramming systems, when all somatic cells uniformly express the reprogramming components (Narsinh et al., 2011). Cell culture media inconsistency is one factor that can affect the outcome of experimental studies and implant recipients. It may be brought on by using various formulations along with data obtained from image analysis using computational approaches to optimize and monitor environmental conditions for differentiation of hiPSC (McGillicuddy et al., 2018).

Another application of MSC-based therapy for the treatment of acute respiratory distress syndrome relies on experimental data of in vivo research. The potential of MSCs to treat lung disorders and safety results for patients have been obtained in clinical trials. So far, the use of MSCs from different sources, including bone marrow or the umbilical cord, has been evaluated using preclinical studies. However, it is restricted due to multiple parameters such as the origin of MSC, the processing of tissues, or in vitro expansion. Based on the stated upcoming rise in demand for MSC-based treatments and the present restrictions on their in vitro production, new strategies that can boost and raise their in vitro expansion and secretome synthesis from MSCs are required (Fernández-Francos et al., 2021).

Using various ML techniques may help overcome this inconsistency and disparity. To stabilize this inconsistent nature of the cells that led to differentiation into cardiomyocytes (CMs), for example, researchers have identified and maintained continuous environmental features. They also used the classifier model of ML, boasting 90% accuracy using a 90% purity criterion, to predict if the end product of the cell was CM adequate or insufficient. However, the process faced certain potholes, which necessitated the introduction of automated, high-throughput validation for the whole cell manufacturing process to ensure that the screening process is selective and precise (Mehta et al., 2022). To distinguish iPSCs from feeder fibroblasts, a different team of researchers used gradient boosting, one of the supervised ML approaches to analyze cell morphology and motility parameters (nucleus-cytoplasm ratio, area, and displacement) (Haishan Zhang et al., 2019).

Specific morphological characteristics of cells can be selected from colony microscopy with the aid of conventional statistics, and then classifier models can be trained in order to distinguish healthy iPSC colonies from unhealthy ones. Kavitha et al. performed the same by using the CNN classifier model (Kavitha et al., 2017). Similarly, Hwang et al. constructed an SVM classifier to identify, measure, and describe aberrant Ca^{2+} transients in hiPSC-CMs for quality assessment (Hwang et al., 2020). The ML technique can also help in tracking the cell. For example, Sun et al. used the unsupervised ML method to monitor the transplantation of hiPSC pancreatic islet organoids using iron-labeled magnetic particles (Sun et al., 2021).

TABLE 31.1 Summary of the computational approaches and their applications in stem cell-based regenerative medicines.

S. No.	Model	Application	References
1.	Supervised ML (XG Boost) model	• The ML model can predict iPSC progenitor cells utilizing morphology and motion patterns via microscopic image analysis. • Validated with fivefold cross-analysis, hold-out analysis, and independent test experiments.	(Zhang et al., 2019)
2.	Vector-based convolution neural network (V–CNN) classifier	• Model can distinguish between the iPSC healthy or unhealthy colonies based on extracted aspects of morphology and texture of the virtual images of colonies. • Performance was checked with fivefold cross-validation.	(Kavitha et al., 2017)
3.	Support vector machine (SVM) classifier	• Prediction of anomalous Ca^{2+} transients for evaluating cardiomyocyte function	(Hwang et al., 2020)
4.	CNN ,VGG16	• Based on the images obtained from bright-field microscopy, the system to categorize the appropriateness of cardiomyocytes produced from human iPSCs-cardiomyocytes (hiPSC-CMs). Ninefold cross-validation	(Orita et al., 2019)
5.	2D + 3D U-Net ensemble + random forest classifier	• Automatic processing pipeline to predict the pathologic disease. • Time series and domain feature extraction using cardiac magnetic resonance imaging	(Isensee et al., 2017)
6.	Deep learning + CNN's	• The framework identifies the motion pattern using optical flow techniques for diagnosing myocardial infarction.	(Xu et al., 2018)
7.	ML regression models + Boruta algorithm	• The model to anticipate changes in left ventricular diastolic dysfunction using electrocardiographic features. • Fivefold cross-validation	(Kagiyama et al., 2020)
8.	Unsupervised ML + regression methods	• Model-based pattern recognition in velocity distribution and deformation graphs of normal and pressure-overloaded cardiac function.	(Loncaric et al., 2021)
9.	DigitalDLSorter (deep neural networks)	• This approach uses denoised bulk RNASeq data to predict and completely identify various cell types.	(Torroja & Sanchez-Cabo, 2019)
10.	scAlign (unsupervised deep learning method)	• It aids in the detection of rare cellular populations and reliably identifies newer data sets without annotation labels.	(Johansen & Quon, 2019)
11.	CNNC (convolutional neural network method)	• Offers a supervised method for carrying out gene connection inference and function allocation.	(Kusumoto & Yuasa, 2019)

9. Stem cell population modeling in wound healing

The recent development of the computational model in regenerative medicine has also provided insights into the study of the dynamic behavior of stem cell lineages in the complex process of wound healing. The details of the individual cell population and temporal-spatial affiliation have been studied by a computational model based on cellular growth (Cao et al., 2013). In this model, the cell's shape, growth, and division are represented by a realistic geometric model, and feedback loops represent the likelihood of whether a cell will differentiate or proliferate, which is influenced by the secretions from nearby cells within an appropriate diffusion radius. A Monte Carlo sampling procedure modeled the cellular prediction of each cell. The process of division is chosen based on the likelihood of each dividing stem cell and progenitor cell (Zio & Zio, 2013).

10. Insights into stem cell therapy for vision defects

The corneal endothelium is the cornea's innermost layer that lacks the self-renewal capacity. Maintaining corneal dehydration and transparency is critical in order to avoid severe damage to the corneal stroma or endothelium. The drawbacks of the transplant include the paucity of donors, the need for continued medicine to avoid rejection, and the graft's life span of 10−20 years (Kumar et al., 2022). To address these issues, Bhattacharya et al. (2017) conducted a clinical experiment on stem cells produced from bone marrow, which were directly implanted into the patient's eye to heal retinal damage caused by age-related macular degeneration and other blinding disorders. Human therapeutic trials employing pluripotent stem cells and iPSCs are now underway because they have the same structure as photoreceptor cells and retinal pigment epithelial cells. New advancements in this branch of medicine will provide insight into new therapies, timescales, and infrastructural provisions (Bhattacharya et al., 2017).

11. Regeneration of pacemakers using stem cells

Novel stem cell-based regenerative therapies can be designed by direct reprogramming, targeting cardiac myocytes as a basis for the regeneration of pacemaker cells (PCs). To address this, it is required to amalgamate ML models for data classification depending on the statistical information with multiscale models to understand the underlying biological mechanism. Coaxing ESCs into inert biomaterial and propagation under specified culture conditions causes ESCs to transdifferentiate into sinoatrial node PCs. Ex vivo genomic insertion of TBox3 into ESCs results in the production of PCs-like cells, which express activated leukocyte cell adhesion molecules and have gene expression and immunological activities comparable to PCs (Vedantham, 2015). Transplantation of PCs can restore pacemaker functions in the ailing heart. In summary, ESCs are considered a potent source of regenerative medicine that can be transdifferentiated into all types of cells. The trained ML algorithms can recognize the primary electrophysiological specifications for the arrhythmias.

12. Interpretation of scRNA-seq for the usage of MSCs in regenerative engineering

As a potent new paradigm for treating human disease, researchers are currently focusing on harnessing bone marrow-derived MSCs in preclinical and clinical trials to cure injury by increasing endogenous repair strategies. MSCs accessibility and better potency make them excellent candidates for cell therapy (Colter et al., 2000). MSCs were described as plastic adherent cells with the potential to differentiate into osteogenic, adipogenic, and chondrogenic lineages and expressing CD73, CD105, and CD90 (90%) but not the hematopoietic markers CD34, CD45, CD14, CD19, or HLA-DR (2%), according to the International Society for Cellular Therapy (Dominici et al., 2006). Currently, there is scant unanimity on the common traits of MSC differentiation potential. Because of these ambiguities, we are unable to assess the effectiveness and capacity for self-renewal of the isolated MSC, which means that using it will likely result in outcomes that are wholly unrelated to or almost the reverse of what is anticipated for therapeutic applications. DL approaches for scRNA-seq data are getting adopted to overcome such contradictory results and determine and analyze gene expression. These promising approaches demonstrate the way forward for significant integration of multiresolution computation and algorithms for advancing precision stem cell-based therapies.

13. Conclusion

Stem cell-based regenerative medicinal therapies are promising, although several challenges, like a lack of target-specific cellular reprogramming proficiency, tissue regeneration, and homeostasis, demand in-depth learning and precise prediction of the underpinning biological operations. Computational methods for the application and evolution of stem cells in the branch of regenerative medicine are in the nascent stage of development. The approach would be tremendously beneficial in directing advancement toward regenerative medicine using stem cells. It is challenging to enhance the effectiveness of cellular reprogramming and to take control over tissue regeneration; thus, it is important to understand the underlying biological mechanisms. Multiscale modeling and ML techniques have effectively classified and identified patterns in the stem cell population. Such models provide prediction and help enhance the biological understanding required in regenerative medicine and treating congenital disorders. The synergistic usage of the models coupled with the experimentation has the potential to improve the precision required to revolutionize the shift beyond the classical approaches applied in stem cell research.

Acknowledgments

All the authors are thankful to the Council of Scientific and Industrial Research (CSIR), New Delhi, India, for the support under Mission Mode Project HCP-31(WP 5.1) and to CSIR-Indian Institute of Toxicology Research, Lucknow, for providing the computational resources. Fig. 31.1 has been generated using online tool BioRender (www.biorender.com). The institutional manuscript number is IITR/SEC/MS/2022/105.

Author contribution

SP and RP designed the chapter table of contents. SP, TJ, and AA wrote the chapter along with the figures and table formation. RP edited the chapter, and all authors gave final approval for its publication.

References

Asahara, T., & Kawamoto, A. (2004). Endothelial progenitor cells for postnatal vasculogenesis. *American Journal of Physiology: Cell Physiology, 287*(3), C572−C579. https://doi.org/10.1152/ajpcell.00330.2003

Bengio, Y., & LeCun, Y. (2007). Scaling learning algorithms towards AI. *Large-Scale Kernel Machines, 34*(5), 1−41.

Bhattacharya, S., Gangaraju, R., & Chaum, E. (2017). Recent advances in retinal stem cell therapy. *Current Molecular Biology Reports, 3*, 172−182.

Buch, V. H., Ahmed, I., & Maruthappu, M. (2018). Artificial intelligence in medicine: Current trends and future possibilities. *British Journal of General Practice, 68*(668), 143−144.

Cadieu, C. F., Hong, H., Yamins, D. L., Pinto, N., Ardila, D., Solomon, E. A., … DiCarlo, J. J. (2014). Deep neural networks rival the representation of primate IT cortex for core visual object recognition. *PLoS Computational Biology, 10*(12), e1003963.

Cao, Y., Naveed, H., Liang, C., & Liang, J. (2013). Modeling spatial population dynamics of stem cell lineage in wound healing and cancerogenesis. *Annual International Conference of the IEEE Engineering in Medicine and Biology − Proceedings*, 5550−5553. https://doi.org/10.1109/EMBC.2013.6610807

Chandrasekaran, S., Zhang, J., Sun, Z., Zhang, L., Ross, C. A., Huang, Y.-C., … Collins, J. (2017). Comprehensive mapping of pluripotent stem cell metabolism using dynamic genome-scale network modeling. *Cell Reports, 21*(10), 2965−2977.

Chan, T. E., Stumpf, M. P. H., & Babtie, A. C. (2017). Gene regulatory network inference from single-cell data using multivariate information measures. *Cell Systems, 5*(3), 251−267 e253. https://doi.org/10.1016/j.cels.2017.08.014

Colter, D. C., Class, R., DiGirolamo, C. M., & Prockop, D. J. (2000). Rapid expansion of recycling stem cells in cultures of plastic-adherent cells from human bone marrow. *Proceedings of the National Academy of Sciences, 97*(7), 3213−3218.

Cowan, A. E., Moraru, I. I., Schaff, J. C., Slepchenko, B. M., & Loew, L. M. (2012). Spatial modeling of cell signaling networks. In *Methods in cell biology, 110* (pp. 195−221). Elsevier.

Dominici, M., Le Blanc, K., Mueller, I., Slaper-Cortenbach, I., Marini, F., Krause, D., … Horwitz, E. (2006). Minimal criteria for defining multipotent mesenchymal stromal cells. The international society for cellular therapy position statement. *Cytotherapy, 8*(4), 315−317.

Dowell, K. G., Simons, A. K., Wang, Z. Z., Yun, K., & Hibbs, M. A. (2013). Cell-type-specific predictive network yields novel insights into mouse embryonic stem cell self-renewal and cell fate. *PLoS One, 8*(2), e56810. https://doi.org/10.1371/journal.pone.0056810

Eckert, M. A., Vu, Q., Xie, K., Yu, J., Liao, W., Cramer, S. C., & Zhao, W. (2013). Evidence for high translational potential of mesenchymal stromal cell therapy to improve recovery from ischemic stroke. *Journal of Cerebral Blood Flow and Metabolism, 33*(9), 1322−1334.

El Naqa, I., & Murphy, M. J. (2015). *What is machine learning?* Springer.

Fathi, E., & Farahzadi, R. (2018). Zinc sulphate mediates the stimulation of cell proliferation of rat adipose tissue-derived mesenchymal stem cells under high intensity of EMF exposure. *Biological Trace Element Research, 184*(2), 529−535.

Fernández-Francos, S., Eiro, N., González-Galiano, N., & Vizoso, F. (2021). Mesenchymal stem cell-based therapy as an alternative to the treatment of acute respiratory distress syndrome: Current evidence and future perspectives. *International Journal of Molecular Sciences, 22*(15), 7850.

Goldberger, A. L. (2006). Giles f. Filley lecture. Complex systems. *Proceedings of the American Thoracic Society, 3*(6), 467−471. https://doi.org/10.1513/pats.200603-028MS

He, J., Baxter, S. L., Xu, J., Xu, J., Zhou, X., & Zhang, K. (2019). The practical implementation of artificial intelligence technologies in medicine. *Nature Medicine, 25*(1), 30−36.

Heidary Rouchi, A., & Mahdavi-Mazdeh, M. (2015). Regenerative medicine in organ and tissue transplantation: Shortly and practically achievable? *International Journal of Organ Transplantation Medicine, 6*(3), 93−98.

Hwang, H., Liu, R., Maxwell, J. T., Yang, J., & Xu, C. (2020). Machine learning identifies abnormal Ca^{2+} transients in human induced pluripotent stem cell-derived cardiomyocytes. *Scientific Reports, 10*(1), 16977. https://doi.org/10.1038/s41598-020-73801-x

Isensee, F., Jaeger, P. F., Full, P. M., Wolf, I., Engelhardt, S., & Maier-Hein, K. H. (2017). Automatic cardiac disease assessment on cine-MRI via time-series segmentation and domain specific features. In *Paper presented at the international workshop on statistical atlases and computational models of the heart.*

Jiang, F., Jiang, Y., Zhi, H., Dong, Y., Li, H., Ma, S., … Wang, Y. (2017). Artificial intelligence in healthcare: Past, present and future. *Stroke and Vascular Neurology, 2*(4), 230.

Johansen, N., & Quon, G. (2019). scAlign: A tool for alignment, integration, and rare cell identification from scRNA-seq data. *Genome Biology, 20*(1), 1−21.

Kagiyama, N., Piccirilli, M., Yanamala, N., Shrestha, S., Farjo, P. D., Casaclang-Verzosa, G., … Sengupta, P. P. (2020). Machine learning assessment of left ventricular diastolic function based on electrocardiographic features. *Journal of the American College of Cardiology, 76*(8), 930−941. https://doi.org/10.1016/j.jacc.2020.06.061

Kavitha, M. S., Kurita, T., Park, S. Y., Chien, S. I., Bae, J. S., & Ahn, B. C. (2017). Deep vector-based convolutional neural network approach for automatic recognition of colonies of induced pluripotent stem cells. *PLoS One, 12*(12), e0189974. https://doi.org/10.1371/journal.pone.0189974

Kumar, R., Sharma, A., Siddiqui, M. H., & Tiwari, R. K. (2018). Prediction of drug-plasma protein binding using artificial intelligence based algorithms. *Combinatorial Chemistry & High Throughput Screening, 21*(1), 57−64.

Kumar, R., Sharma, A., Varadwaj, P., Ahmad, A., & Ashraf, G. M. (2011). Classification of oral bioavailability of drugs by machine learning approaches: A comparative study. *Journal of Computational Interdisciplinary Science, 2*(9), 1−18.

Kumar, A., Yun, H., Funderburgh, M. L., & Du, Y. (2022). Regenerative therapy for the cornea. *Progress in Retinal and Eye Research, 87*, 101011.

Kusumoto, D., & Yuasa, S. (2019). The application of convolutional neural network to stem cell biology. *Inflammation and Regeneration, 39*(1), 1−7.

LeCun, Y., Bengio, Y., & Hinton, G. (2015). Deep learning. *Nature, 521*(7553), 436−444.

Loncaric, F., Marti Castellote, P. M., Sanchez-Martinez, S., Fabijanovic, D., Nunno, L., Mimbrero, M., … Bijnens, B. (2021). Automated pattern recognition in whole-cardiac cycle echocardiographic data: Capturing functional phenotypes with machine learning. *Journal of the American Society of Echocardiography, 34*(11), 1170−1183. https://doi.org/10.1016/j.echo.2021.06.014

Madabhushi, A., & Lee, G. (2016). Image analysis and machine learning in digital pathology: Challenges and opportunities. *Medical Image Analysis, 33*, 170−175.

Magno, V., Meinhardt, A., & Werner, C. (2020). Polymer hydrogels to guide organotypic and organoid cultures. *Advanced Functional Materials, 30*(48), 2000097.

Matsumoto, H., Kiryu, H., Furusawa, C., Ko, M. S., Ko, S. B., Gouda, N., … Nikaido, I. (2017). SCODE: An efficient regulatory network inference algorithm from single-cell RNA-seq during differentiation. *Bioinformatics, 33*(15), 2314−2321.

McGillicuddy, N., Floris, P., Albrecht, S., & Bones, J. (2018). Examining the sources of variability in cell culture media used for biopharmaceutical production. *Biotechnology Letters, 40*(1), 5−21. https://doi.org/10.1007/s10529-017-2437-8

Mehta, C., Shah, R., Yanamala, N., & Sengupta, P. P. (2022). Cardiovascular imaging databases: Building machine learning algorithms for regenerative medicine. *Current Stem Cell Reports*, 1−10.

Mika, J. T., & Poolman, B. (2011). Macromolecule diffusion and confinement in prokaryotic cells. *Current Opinion in Biotechnology, 22*(1), 117−126.

Moignard, V., Woodhouse, S., Haghverdi, L., Lilly, A. J., Tanaka, Y., Wilkinson, A. C., … Gottgens, B. (2015). Decoding the regulatory network of early blood development from single-cell gene expression measurements. *Nature Biotechnology, 33*(3), 269−276. https://doi.org/10.1038/nbt.3154

Müller, G. A., Tarasov, K. V., Gundry, R. L., & Boheler, K. R. (2012). Human ESC/iPSC-based 'omics' and bioinformatics for translational research. *Drug Discovery Today: Disease Models, 9*(4), e161−e170.

Murry, C. E., Field, L. J., & Menasche, P. (2005). Cell-based cardiac repair: Reflections at the 10-year point. *Circulation, 112*(20), 3174−3183. https://doi.org/10.1161/CIRCULATIONAHA.105.546218

Narsinh, K., Narsinh, K. H., & Wu, J. C. (2011). Derivation of human induced pluripotent stem cells for cardiovascular disease modeling. *Circulation Research, 108*(9), 1146−1156. https://doi.org/10.1161/CIRCRESAHA.111.240374

Nestor, M. W., & Noggle, S. A. (2013). Standardization of human stem cell pluripotency using bioinformatics. *Stem Cell Research & Therapy, 4*(2), 1−7.

Onyido, E. K., Sweeney, E., & Nateri, A. S. (2016). Wnt-signalling pathways and microRNAs network in carcinogenesis: Experimental and bioinformatics approaches. *Molecular Cancer, 15*(1), 1−17.

Orita, K., Sawada, K., Koyama, R., & Ikegaya, Y. (2019). Deep learning-based quality control of cultured human-induced pluripotent stem cell-derived cardiomyocytes. *Journal of Pharmacological Sciences, 140*(4), 313−316. https://doi.org/10.1016/j.jphs.2019.04.008

Peltier, J., & Schaffer, D. V. (2010). Systems biology approaches to understanding stem cell fate choice. *IET Systems Biology, 4*(1), 1−11. https://doi.org/10.1049/iet-syb.2009.0011

Poetsch, M. S., Strano, A., & Guan, K. (2022). Human induced pluripotent stem cells: From cell origin, genomic stability, and epigenetic memory to translational medicine. *Stem Cells, 40*(6), 546−555. https://doi.org/10.1093/stmcls/sxac020

Polouliakh, N. (2019). In silico transcription factor discovery via bioinformatics approach: Application on iPSC reprogramming resistant genes. In *Leveraging biomedical and healthcare data* (pp. 183−193). Elsevier.

Prudhomme, W. A., Duggar, K. H., & Lauffenburger, D. A. (2004). Cell population dynamics model for deconvolution of murine embryonic stem cell self-renewal and differentiation responses to cytokines and extracellular matrix. *Biotechnology and Bioengineering, 88*(3), 264–272. https://doi.org/10.1002/bit.20244

Raimundo, F., Meng-Papaxanthos, L., Vallot, C., & Vert, J.-P. (2021). Machine learning for single-cell genomics data analysis. *Current Opinion in Systems Biology, 26,* 64–71.

Salehnasab, C., Hajifathali, A., Asadi, F., Roshandel, E., Kazemi, A., & Roshanpoor, A. (2019). Machine learning classification algorithms to predict aGvHD following allo-HSCT: A systematic review. *Methods of Information in Medicine, 58*(06), 205–212.

Sharma, A., Kumar, R., Ranjta, S., & Varadwaj, P. K. (2021). SMILES to smell: Decoding the structure–odor relationship of chemical compounds using the deep neural network approach. *Journal of Chemical Information and Modeling, 61*(2), 676–688.

Shen, F., Cheek, C., & Chandrasekaran, S. (2019). Dynamic network modeling of stem cell metabolism. *Computational Stem Cell Biology: Methods and Protocols,* 305–320.

Song, C., Ristenpart, T., & Shmatikov, V. (2017). Machine learning models that remember too much. In *Paper presented at the proceedings of the 2017 ACM SIGSAC conference on computer and communications security.*

Stumpf, P. S., & MacArthur, B. D. (2019). Machine learning of stem cell identities from single-cell expression data via regulatory network archetypes. *Frontiers in Genetics, 10,* 2. https://doi.org/10.3389/fgene.2019.00002

Sun, A., Hayat, H., Liu, S., Tull, E., Bishop, J. O., Dwan, B. F., ... Li, W. (2021). 3D in vivo magnetic particle imaging of human stem cell-derived islet organoid transplantation using a machine learning algorithm. *Frontiers in Cell and Developmental Biology, 9.*

Talbot, P., I. zur Nieden, N., Lin, S., Martinez, I., Guan, B., & Bhanu, B. (2014). Use of video bioinformatics tools in stem cell toxicology. In *Handbook of nanotoxicology, nanomedicine and stem cell use in toxicology* (pp. 379–402).

Torroja, C., & Sanchez-Cabo, F. (2019). Digitaldlsorter: Deep-learning on scRNA-seq to deconvolute gene expression data. *Frontiers in Genetics, 10,* 978.

Van Leeuwen, I. M., Mirams, G., Walter, A., Fletcher, A., Murray, P., Osborne, J., ... Doyle, B. (2009). An integrative computational model for intestinal tissue renewal. *Cell Proliferation, 42*(5), 617–636.

Vedantham, V. (2015). New approaches to biological pacemakers: Links to sinoatrial node development. *Trends in Molecular Medicine, 21*(12), 749–761.

Walpole, J., Papin, J. A., & Peirce, S. M. (2013). Multiscale computational models of complex biological systems. *Annual Review of Biomedical Engineering, 15,* 137.

Wang, L., Zhang, H.-C., & Wang, Q. (2019). On the concepts of artificial intelligence and innovative design in product design. In *Paper presented at the IOP conference series: Materials science and engineering.*

Xu, H., Ang, Y. S., Sevilla, A., Lemischka, I. R., & Ma'ayan, A. (2014). Construction and validation of a regulatory network for pluripotency and self-renewal of mouse embryonic stem cells. *PLoS Computational Biology, 10*(8), e1003777. https://doi.org/10.1371/journal.pcbi.1003777

Xu, H., Baroukh, C., Dannenfelser, R., Chen, E. Y., Tan, C. M., Kou, Y., ... Ma'ayan, A. (2013). Escape: Database for integrating high-content published data collected from human and mouse embryonic stem cells. *Database (Oxford),* bat045. https://doi.org/10.1093/database/bat045

Xu, C., Xu, L., Gao, Z., Zhao, S., Zhang, H., Zhang, Y., ... Li, S. (2018). Direct delineation of myocardial infarction without contrast agents using a joint motion feature learning architecture. *Medical Image Analysis, 50,* 82–94. https://doi.org/10.1016/j.media.2018.09.001

Zahedi, A., On, V., Lin, S. C., Bays, B. C., Omaiye, E., Bhanu, B., & Talbot, P. (2016). Evaluating cell processes, quality, and biomarkers in pluripotent stem cells using video bioinformatics. *PLoS One, 11*(2), e0148642.

Zhang, H., Shao, X., Peng, Y., Teng, Y., Saravanan, K. M., Zhang, H., ... Wei, Y. (2019a). A novel machine learning based approach for iPS progenitor cell identification. *PLoS Computational Biology, 15*(12), e1007351. https://doi.org/10.1371/journal.pcbi.1007351

Zio, E., & Zio, E. (2013). *Monte Carlo simulation: The method.* Springer.

Chapter 32

Computational analysis in epithelial tissue regeneration

Priyanka Chhabra[1] and Khushi Gandhi[2]

[1]Center for Medical Biotechnology, Amity Institute of Biotechnology, Amity University, Noida, Uttar Pradesh, India; [2]Amity Institute of Biotechnology, Amity University, Noida, Uttar Pradesh, India

1. Introduction

Ectoderm, mesoderm, and endoderm are the building blocks of epithelial cells, which explains why epithelia line the cavities present in the body and cover organ surfaces (Kurn & Daly, 2022). One of the fundamental functions of epithelia is to act as physical barriers to defend underlying tissues from external attacks (Torras et al., 2018). As a result, epithelial injuries are prevalent. Trauma, infection, inflammation, and toxic chemicals can all cause epithelial tissue harm on any bodily surface, particularly the skin or urethral mucosa tissue (Yin et al., 2022). Fortunately, due to the tissue's resident stem cells' ability to self-renew, sustain their density during homeostasis, and result in more than one specialized cell types to preserve as well as restore tissue function, it has a wide range of regeneration (Gonzales & Fuchs, 2017). This makes them ideal for accelerating wound healing in clinical settings. However to do so, it is important to understand the regeneration process extensively, by taking aid from computational tools and models (Bian & Cahan, 2016).

2. Epithelial tissue

Epithelial tissues are made up of cells arranged in sheets with strong intercellular linkages that line the cavities of major organs and protect them from physical, chemical, and microbial assaults. Epithelial cells are polarized, indicating that the apical side, which is exposed to the organ lumen, is different from the basolateral side in terms of form and content. The basement membrane serves as a growth support and is selectively permeable. Epithelial cells rest at the basement membrane (Torras et al., 2018).

Epithelial tissues are classified on the basis of the cells present (Table 32.1):

The stratified squamous epithelium that makes up the mature mammalian epidermis is what gives the skin its barrier. It is composed of keratinocytes. The basement membrane separates the keratinocytes from the dermis present below it. The external environment regularly attacks the skin's epidermis and its appendages. Repair and regeneration of the skin's barrier is essential for wound healing. Resident stem cells aid in the same. Stem cell compartments are distributed across the basement membrane in various areas whose cellular constituents and other sources of signaling have a substantial influence on their differentiation and expression of specific genes (Gonzales & Fuchs, 2017).

3. Physiology of epithelial tissue regeneration

Hemostasis, inflammation, proliferation, and remodeling are the four different but overlapping phases that make up the epithelial tissue regeneration process (Broughton et al., 2006).

The mechanisms of tissue healing involve several cell types, enzymes, cytokines, proteins, and hormones (Rajendran et al., 2018). To put it briefly, once an injury occurs, the natural healing process will induce hemostasis to produce blood clots and constrict blood vessels to reduce blood flow. Next, proinflammatory cytokines and growth factors will be secreted (Zhao et al., 2016). Through the recruitment of macrophages, neutrophils, and lymphocytes by epithelial cells, these

Computational Biology for Stem Cell Research. https://doi.org/10.1016/B978-0-443-13222-3.00014-1
Copyright © 2024 Elsevier Inc. All rights are reserved, including those for text and data mining, AI training, and similar technologies.

TABLE 32.1 Types of epithelial tissues on the basis of their cellular structure.

S.No.	Type	Structure	Location	Function
1.	Squamous epithelium	Flattened cells with the nucleus in the center Possess uneven edges	Lining of alveoli in lungs Lining of kidney tubules Lining of blood capillaries	Exchange of oxygen and carbon dioxide Absorption Exchange of materials (Choudhary et al., 2021)
2.	Cuboidal epithelium	Cuboidal cells with a nucleus in the center Cells seem to be polygonal	Pancreas and salivary duct lining Salivary and sweat glands	Absorption Secretion (Torras et al., 2018)
3.	Ciliated epithelium	Cuboidal cells having cilia at free ends	Renal tubule lining	Nephrotic filtrate flow (Nakano, 1998)
4.	Columnar epithelium	Cells resembling long columns, each having a nucleus at the basal end	Intestine and stomach lining	Absorption and secretion (Torras et al., 2018)
5.	Ciliated columnar epithelium	Columnar cells having cilia at free ends	Interior of the trachea	Fluid movement in a certain direction (Nakano, 1998)
6.	Brush-bordered columnar epithelium	Columnar cells with several folds at the free ends that resemble brush bristles	Lining of intestine	Increasing the surface area for absorption (Elaine & Marieb, 1995; Torras et al., 2018)

growth factors in turn promote inflammation. Expansion factors subsequently trigger angiogenesis, where reepithelialization takes place as a result of the growth of fibroblasts and keratinocytes. Later, the fibroblasts would further develop into myofibroblasts, resulting in the deposition of extracellular matrix (ECM) (Xue & Jackson, 2015).

3.1 Hemostasis and inflammation (injury—Day 4)

The initial and most important stage of the healing process is hemostasis. Vasodilation and increased blood vessel permeability are brought on by inflammation. However, the body's first response to a wound is to stop bleeding. The endothelium and adjacent platelets interact with the intrinsic coagulation cascade component as the damaged blood vessel vasoconstricts. Collagen, platelets, thrombin, and fibronectin make up the clot that develops. Additionally, these components discharge cytokines and growth factors, which initiate the inflammatory response. Invading cells including fibroblasts, monocytes, neutrophils, and endothelial cells can utilize the fibrin clot as a scaffold. Furthermore, the complicated cytokines and growth factors are concentrated in the clot (Kelm & Anger, 2022; Kurkinen et al., 1980; Witte & Barbul, 1997).

3.2 Chemotaxis and activation

Cellular signals that induce a neutrophil response are produced as soon as the clot forms. Tumor necrosis factor (TNF-α), interleukin-1 (IL-1), platelet factor 4 (PF4), transforming growth factor (TGF), and bacterial "products" attract neutrophils to the site of injury. These factors also trigger the release of prostaglandins and the vasodilation of neighboring blood vessels to facilitate the elevated cellular traffic. 48—96 h after the injury, monocytes in the surrounding tissue and blood are lured to the region and start the process of transforming into macrophages. An active macrophage will promote fibroplasia, angiogenesis, and the production of nitric oxide (Bevilacqua et al., 1985; Pohlman et al., 1986).

Upon entering the wound site, neutrophils begin to clear away cellular debris and invading bacteria. Caustic proteolytic enzymes are released by neutrophils that disintegrate pathogens and nonviable tissues (Broughton et al., 2006).

Matrix metalloproteinase (MMP), which is produced by macrophages, fibroblasts, monocytes, and keratinocytes in response to TNF-α, additionally breaks down the damaged ECM. MMP reduces inflammatory by-products and speeds up a wound cell's migration through the ECM (Abraham et al., 2000).

3.3 Proliferative phase: Epithelization, angiogenesis, and provisional matrix formation (Day 4—14)

In an effort to rebuild a barrier that would stop fluid loss and future bacterial invasion, epithelial cells close to the skin's border start to multiply and send out projections. Epidermal growth factor (EGF) and TGF-α generated by active platelets and macrophages enhance epithelial proliferation and chemotaxis (Grotendorst et al., 1989; Lawrence & Diegelmann, 1994). Inflammatory cytokines (IL-1 and TNF-α) are the initial factors that trigger epithelization shortly following trauma. Keratinocyte growth factor (KGF)-1, KGF-2, and IL-6 are then produced and secreted by fibroblasts, simulating the migration, proliferation, and differentiation of adjacent keratinocytes in the epidermis (Smola et al., 1993; Xia et al., 1999). It appears that KGF-2 is crucial for controlling this process in humans (Jimenez & Rampy, 1999). Fibroblasts and endothelial cells are the main cells that are proliferating at this stage. Vascular endothelial growth factor (VEGF), which is mostly secreted by keratinocytes at the border of the wound but also by fibroblasts, platelets, macrophages, and other endothelial cells, encourages endothelial cells close to intact venules to start generating new capillaries. Fibroblasts invade the wound from adjacent tissue, become active, begin making collagen, and multiply (Kelm & Anger, 2022; Regan et al., 1991).

3.4 Maturation and remodeling (Day 8—1 Year)

From a clinical perspective, the maturation and remodeling phase is arguably the most important phase. The collagen is deposited in a structured and organized network, which is the phase's defining feature.

There is a pattern of how the wound matrix is constructed. Initially, fibrin and fibronectin make up the majority of the matrix (Witte & Barbul, 1997). The fibroblasts then produce glycosaminoglycans, proteoglycans, and other proteins. This haphazardly arranged group of glycans serves as the new matrix's early framework. A collagen-based matrix that is more robust and well-organized takes the place of this transitory matrix (Kurkinen et al., 1980).

For at least 4—5 weeks following injury, net collagen production will continue. Initial collagen formation occurs parallel to the skin and is weaker than collagen in healthy skin. The original collagen threads are reabsorbed, thickened, and arranged along the stress lines over time. Additionally, a wound with higher tensile strength is observed along with these alterations, demonstrating a direct link between collagen fiber orientation and tensile strength (Broughton et al., 2006; Chhabra & Bhati, 2022; Grubbs & Manna, 2022). However, the collagen in the scar will never be as organized as the collagen in healthy skin. Additionally, the strength of a wound never fully recovers. The strength of the wound is only 3% of its ultimate level after 1 week, 30% after 3 weeks, and about 80% after 3 months (and beyond) (Fig. 32.1) (Fried, 1998).

Furthermore, an organized process of reepithelialization is made possible by various transcription factors, which operate as a crucial coordination point for the transfer of information with the external environment. Importantly, transcription factors associate with several chromatin-modifying complexes to interact as well as affect the chromatin structure surrounding important target genes (Boudra & Ramsey, 2020). Some important transcription factors are summarized in Table 32.2.

4. Computational biology and bioinformatics

Bioinformatics is a multidisciplinary discipline that develops and employs computational methods to examine vast amounts of biological data, such as genetic code, collection of cells, or protein samples, to generate novel ideas or comprehend specialized biological processes. It makes use of analytical methods, mathematical models, and simulation for the same (Computational Biology and Bioinformatics, 2022).

Some applications of computational biology and bioinformatics include the following:

4.1 Genomics and proteomics

Genomics and proteomics are both broad topics that cover several genome and proteome investigations (Nair, 2007). Genomics is concerned with the structure, operation, evolution, mapping, and modification of genomes. It provides an opportunity to define and quantify every gene in an organism, in addition to their interactions and their impact on the organism (Lesk, 2017). Proteomics, a mix of proteome experiments and data analysis, examines protein composition, structure, expression, modification status, and total protein interactions and linkages. It provides information that is complimentary to genomics and transcriptomics (Wilhelm et al., 2014) (Table 32.3).

Epithelial Tissue Regeneration

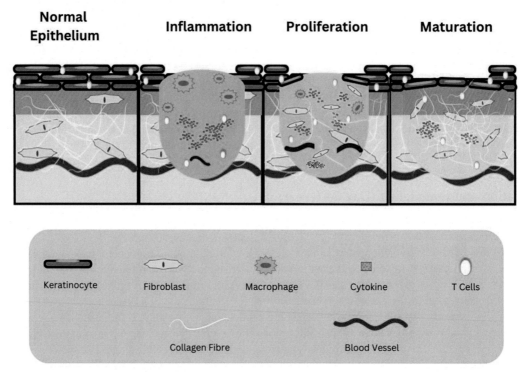

FIGURE 32.1 Schematic representation of the process of epithelial tissue regeneration (Tavakoli et al., 2022).

TABLE 32.2 Significant transcription factors and their role in epithelial tissue regeneration.

S.No	Transcription factor	Function	References
1.	Krüppel-like factor 5 (KLF-5)	Regulation of expression of genes involved in cytokines' activation and cell adhesion.	Bell et al. (2013), Paranjapye et al. (2021)
2.	Signal transducer and activator of transcription 3 (STAT-3)	Activates genes involved in cell migration and remodeling of modeling such as TIAM1, THBS1, LOXL2, and SERPINE2.	Dauer et al. (2005)
3.	GATA-6	Stimulates epidermal cells' migration and regulates expression of genes involved in cellular motility. It has also been identified as a key regulator of the sebaceous duct lineage, leading to dedifferentiation and acquirement of self-renewal properties.	Donati et al. (2017)
4.	Activator protein-1 (AP-1) family	Consists of Jun (JunB, c-Jun, JunD) and Fos (Fos, FosB, Fra1, and Fra2) family proteins. They play a central role in cell migration, focal adhesions, and cell proliferation.	Balli et al. (2019), Boudra and Ramsey (2020)
5.	Suppressor of mothers against decapentaplegic (SMAD) family	Controls expression of genes involved in epidermal–mesenchymal transition (important process preceding cell migration) and cell motility.	Boudra and Ramsey (2020), Miyazawa and Miyazono (2017)
6.	Hypoxia-inducible factor (HIF)	Aids in adaptation to the hypoxic conditions created in the wound environment. Further, HIF α activates VEGFA that promotes angiogenesis. It also promotes migration through activation of SERPINE1, ITGA6, LAMA3, etc.	Boudra and Ramsey (2020), Elson et al. (2001)

TABLE 32.3 Key differences between genomics and proteomics (Willett, 2002).

S.No.	Genomics	Proteomics
1.	Involves the study of a genome's whole gene set	Involves the study of proteins generated through a cell (Aslam, 2016)
2.	Includes genome mapping, sequencing, and analysis	Includes protein function, protein—protein interactions, and 3D protein structure (Witte and Barbul, 1997)
3.	Two types: structural and functional genomics	Three types: structural, functional, and expression proteomics (Willett, 2002)
4.	Genome sequencing initiatives, including the Human Genome Project, are major fields	Major fields include proteome database innovations, such as SWISS-2DPAGE, and software building for computer-aided drug design (Nair, 2007)

Proteomics and genomics have applications in a variety of sectors, including medicine, biotechnology, anthropology, and other social sciences. Genomic modification, drug discovery, and preventive medicine using genetic and biomarkers are a few notable examples (Collins et al., 2006; Sharma et al., 2022; Wilson, 2004).

4.2 Pharmacology

According to its definition, computational pharmacology is "the study of the influence of genetic data to find correlations between certain genotypes and disorders, followed by drug data screening." Pharmacologists were unable to compare biochemical and genetic details related to medication effectiveness using the outdated techniques. To examine these enormous datasets, scientists and researchers develop computational algorithms. This enables an effective comparison of the noteworthy data points and enables the development of more precise drugs (Harmon, 2010).

The use of computational approaches in reducing the search for therapeutic compounds via modeling trials is known as computer-aided drug design (Nair, 2007). In recent years, a number of computational drug design studies have been published. It is an effective strategy for finding potential drug candidates, especially when combined with modern chemical biology screening methods. Due to its capacity to accelerate drug discovery by making use of prior information regarding receptor—ligand communications, energy and morphological optimization, and synthesis, this approach has emerged as a crucial step in the drug design process (Macalino et al., 2015).

4.3 Data and modeling

Mathematical models are used to study the processes that control the structure, development, and behavior of biological systems (Friedman, 2010). Through the use of these mathematical techniques, databases and other techniques for storing, retrieving, and interpreting biological data have been created (Noble, 2002). The main data sources are the DNA databases from the European Molecular Biology Laboratory (EMBL), GenBank (located at the National Center for Biotechnology Information, Bethesda), DNA Data Bank Japan (DDBJ), and the protein databases from SWISS-PROT (the database for protein sequences at the Swiss Institute of Bioinformatics, Geneva), and PDB (3D protein structure databases) (Nair, 2007). Large dataset gathering and analysis have paved the path for developing study areas such as computational biomodeling and data mining (Noble, 2002).

The broad array of approaches and databases available through bioinformatics may be used to analyze, combine, and interpret cancer multiomics data (Jiménez-Santos et al., 2022). Another significant application is computational anatomy, which involves the development of computational mathematics and data analytical tools for modeling and simulating biological systems (Grenander & Miller, 1998). Researchers can utilize this information to ascertain if a system can "preserve their state and functions against internal as well as external disturbances" and can forecast how such systems would respond to various settings (Kitano, 2002).

4.4 Statistics and algorithms

Traditional statistics approaches have a variety of biases and can be time-consuming when dealing with large amounts of data. To process massive data produced by multiomics, new high-throughput methods or machine learning—based

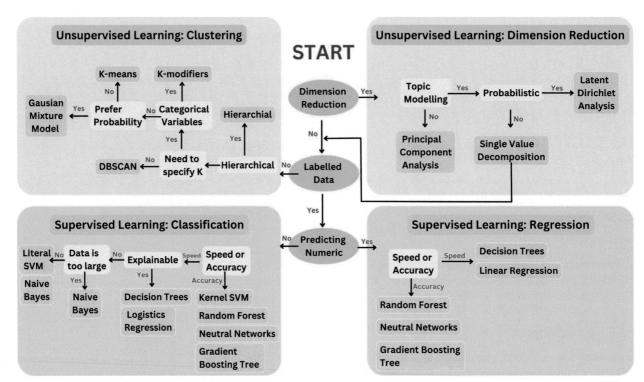

FIGURE 32.2 Outline of algorithms for machine learning. Unsupervised learning and supervised learning are the two primary categories of learning. Depending on the learning objectives, such as clustering, dimension reduction, classification, or regression, the appropriate algorithms for each type of learning methodology are chosen (Nguyen et al., 2022).

algorithms are required. While classic statistical procedures such as normalization and batch effect removal continue to be crucial, artificial intelligence and digital pathology are becoming increasingly significant in research, pathology, and medicine (Baxi et al., 2022; Cui et al., 2022).

Machine learning methods may generally be divided into supervised learning and unsupervised learning strategies. Unsupervised learning is an algorithm that searches for patterns in unlabeled data. One use of this in biology is the 3D mapping of a genome. While supervised learning is a type of algorithm that uses labeled data to learn how to categorize unlabeled data in the future (Fig. 32.2) (Cui et al., 2022; Rashidi et al., 2019; Nguyen et al., 2022).

5. Computational analysis in epithelial tissue regeneration

Epithelial tissue regeneration is dependent on many redundant signals and cross-talk across distinct signaling networks. Even though the biological system benefits greatly from redundancy, it increases the difficulty of understanding the healing process functions as a whole. The components of the signaling network are well understood individually, but when they are coupled, it is exceptionally complicated and it becomes nearly impossible to derive simple conclusions from sparse experimental data. In the near future, computational systems biology enables us to digitally investigate multiple hypotheses and systematically analyze controlled what-if scenarios to better comprehend the healing process as a whole. In the long term, given the wide range of healing reactions across individuals, computational systems biology is a crucial element in determining the direction of personalized medicine (Buganza Tepole & Kuhl, 2013; Kitano, 2002; Vodovotz, 2010).

By offering insights into the principles behind biological processes and developing innovative predictions that may be utilized to direct experimental study, computational models enable statistical analysis of data. They can aid in solving a range of biological issues and difficulties in the realms of regenerative medicine and stem cell research. These models have been successful in guiding experimental work, particularly in the development of methods for cellular conversion (del Sol & Jung, 2021; Maurizi et al., 2021; Ziraldo et al., 2013).

Sherrat and Murray are widely regarded as the fathers of contemporary mathematical modeling of wound healing, as they developed the initial model over two decades ago (Sherratt & Murray, 1990). Since then, several models with varying properties, focuses, and simplifications have been developed.

6. Examples of computational models and tools in epithelial tissue regeneration

6.1 Models in stem cells research

Important issues associated with stem cell research are addressed by building systems biology models at various degrees of complexity, such as cellular, tissue, and even organ levels. Cellular models, such as those based on gene regulatory networks (GRNs), can advance understanding of cell differentiation and conversion by predicting the signaling molecules and critical transcription factors (TFs) directing such processes. The essential concepts underpinning tissue homeostasis and regeneration, however, can be better understood using tissue-level models, such as those based on networks of cell−cell interactions. They can also be utilized to foresee significant cell−cell interactions that promote the ability for tissue regeneration (del Sol & Jung, 2021).

6.1.1 Applications of different modeling types in stem cell research

Computational models can aid in providing critical solutions in the case of stem cells research. However, the modeling framework utilized is highly dependent on the individual study subject. Cell conversion factors may be found using GRN Boolean network models. Despite not including the determination of kinetic parameters, these models typically incorporate regulatory interactions involving several TFs. Additionally, this modeling paradigm enables the determination of the ideal TF combinations controlling cellular conversion by inferring TF cooperativity in regulation.

Ordinary differential equation−based continuous GRN models are ideally designed for predicting the dynamics involved in expression of gene in biological processes such as cellular development. Additionally, by applying these equations to disrupted gene expression, it is possible to statistically anticipate its continuous dynamic behavior.

One of the elements that contributes to the efficiency of cellular reprogramming is the variability in gene expression trends and interactions between regulators in cellular systems. It might be effectively reproduced using probabilistic models such as probabilistic Boolean networks or dynamic Bayesian networks. Therefore, these models may be used to choose the best TF combinations that, when perturbed, effectively trigger cellular reprogramming.

Directed graphs depicting networks of communication between cells may be deduced from the expression of ligand−receptor pairs rather than evaluating receptor−ligand binding affinity. This kind of informative model is efficient for finding crucial structural aspects of cell−cell communication linkages to define the cellular interactions carried out to ensure tissue function during homeostasis. This method may also be used to address significant interactions between cells that might impede tissue regeneration in illness or aging (del Sol & Jung, 2021).

6.1.2 Overview of stem cell research mechanistic modeling frameworks

Mechanistic models define a biological system based on a known or assumed connection among its parts, such as genes, and explain the associated nature in relation to the biological process that produced the experimental data. By explicitly integrating the physical, biological, or chemical components of biological systems, mechanistic models may be distinguished from other types of models. Many theoretical and analytical frameworks, including logical and continuous models, have been created in recent years to depict a biological system at different degrees of complexity. The selected modeling strategy affects how accurately biological reality is depicted, how well system functioning can be replicated, and, most crucially, how well one can predict. Typically, more accurate biological system simulations demand more data but also have more predictive value (del Sol & Jung, 2021).

Logical models are qualitative frameworks for modeling gene interactions that are based on an array of principles that regulate the system's behavior. Each unit in the model can choose to be in either of the two states—"active" or "inactive," which are defined by the conditions of its regulators. Previous studies have demonstrated that using logical models can help solve some of the biggest problems in stem cell research. For example, scRNA-seq data was used to build a GRN model of embryonic blood production and predict TFs controlling the process of differentiation of mesodermal progenitors to endothelium and primary blood cells (Moignard et al., 2015). The elements that lead to directed transdifferentiation and reprogramming were predicted using a GRN model of hematopoietic differentiation that was developed from current research, similar to the prior case (Collombet et al., 2017). To sustain the characteristic condition of tissue colonies, molecules that combine gene control, internal signaling pathways, and external ligand−receptor interactions have also been predicted using logical models (Browaeys et al., 2020).

Researchers are now able to mimic biological systems more precisely using continuous models because of the improved availability of single-cell data. This framework, on contrary to logical models, may provide a quantitative definition of biological models without the need to categorize objects as "active" or "inactive." As a result, it is possible to directly correlate experimental findings with model state since the system's temporal dynamics are shown in continuous time. These continuous models have been used to predict significant transcription factors necessary for hematopoietic lineage determination, transdifferentiation, and stem cell differentiation (Matsumoto et al., 2017; Ocone et al., 2015). However, the applications of continuous models are not simply restricted to TF forecast. For example, mechanical parameters influencing cellular growth were discovered via a stem cell—niche interaction model (Peng et al., 2017). Through the prediction of culture length and stiffness, this model was utilized to improve cellular differentiation processes.

To address particular issues in stem cell research, five fundamental mechanistic modeling types have been used: graphs, Boolean networks, ordinary differential equations, regression-based models, and Bayesian networks.

Graphs are made up of vertices and edges that identify relationships between them, such as gene coexpression or cell connections. Graphs can be either directed or undirected depending on the type of relationship they represent. However, graphs just display the data; they do not impose model assumptions.

Boolean networks are graph-based models that denote the state of each vertex by a Boolean value. The Boolean functions that represent the states of other network vertices determine these states. The fundamental property is that activity of vertices may be categorized as either active or inactive. Given that transcription rate is a Hill function that closely resembles a Boolean function and follows in the context of GRNs, this supposition is validated (Bhaskaran et al., 2015).

Contrary to Boolean networks, continuous models tie vertex positions to continuous states, such as the variable levels of expression of genes. Regression is a common technique for figuring out how the vertices are connected. Ridge regression has been created to assess the dependence between vertices with nearly linear correlations. Linear ordinary differential equations, which link a function to a linear mixture of its derivatives (for example, linking genetic expression to transcription rates), are a popular modeling type in addition to regression. The linearity assumption prevents nonadditive effects, such as competitive TF regulation, from being taken into account, although being frequently justified in the case of GRNs. For the analysis of continuous models, time-series data illustrating biological change through time is frequently needed.

Finally, dynamic Bayesian networks (DBNs) may be used to mimic temporal stochastic processes. The probabilistic interactions of a DBN indicate the conditional reliance between its vertices. DBNs can reasonably assume that one event can lead to another event occurring in the future when discussing gene regulation. Like with continuous model inference, time-series measurements are required for DBN inference (Fig. 32.3) (del Sol & Jung, 2021).

6.1.3 Stem cell bioinformatics in diseases prognosis

Stem cells have a unique ability to predict disease phenotypes because they contain the fundamental molecular baseline that gives birth to genetic content and its derivative transcriptomic, proteomic, and metabolomic strata. Understanding the systems biology of developmental programs through the interrogation of these "-omic" levels is essential to understanding

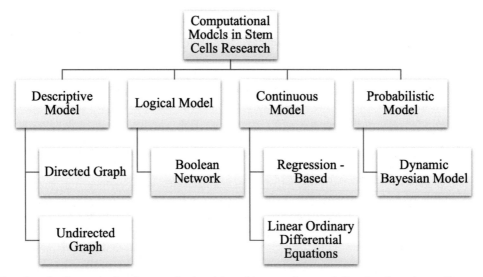

FIGURE 32.3 Overview of various categories of computational models used in stem cells research. Based on the study, specific type of model is chosen such as descriptive, logical, continuous, or probabilistic models (del Sol and Jung, 2021).

the causes of clinical pathologies. In addition to providing a tool for dynamic prognosis to detect alterations concurrent with clinical progression or evaluation of response to clinical treatment, molecular mapping of pluripotent disease-prone backgrounds will promote preventive diagnosis by comparison of healthy templates with pathological states (Arrell & Terzic, 2012; Faustino et al., 2010). While various stem cells have unique clinical benefits, pluripotent stem cells could be the best option for patient implementations (Faustino et al., 2013).

To create a molecular baseline for comparing healthy genomes to pathological genomes, embryonic stem cells (ES cells) can be harvested and analyzed. The creation and development of disease- and patient-specific embryonic stem cell lines has the potential to offer unique experimental cell models, allowing for the performance of tailored experimental design with an emphasis on personalized treatment approaches (Faustino et al., 2013; Silva et al., 2012).

Furthermore, high-throughput technology, advanced approaches in bioinformatics, and recent advancements in cellular reprogramming methodology have created an integration of molecular tools and techniques that allow diseased cells to be reprogrammed to a primordial state, where they can then be examined for characteristics that differentiate them from cells with normal phenotypes. To find new molecular targets for early intervention, comprehensive bioinformatic dissection of phenotypically regressed ES cells can be done at genomic, transcriptomic, proteomic, and metabolomic levels, as well as at any combination of these levels (Faustino et al., 2013; Polo et al., 2012).

6.2 Wound healing assay

Given the importance of cell migration in regeneration processes and the progression of diseases such as cancer angiogenesis and metastasis, it may be utilized as a metric to characterize the health of different cell lines. Furthermore, cell migration quantification may be utilized to assess the impact of medication treatments upon similar cell line. The wound healing assay (WHA) measures mobility by tracking the progress of an injured monolayer of cells. In the experiment, confluent monolayers of cells are grown under certain conditions, wounds or scratches are made on each layer or well, and then the wounds are imaged repeatedly over the course of the experiment. To gauge the rate of migration and subsequently quantify situation-specific cell movement, wound areas from individual images are evaluated. Depending on the conditions, the rate at which wounds close may change. A variety of frames per replication must be recorded and wound areas must be calculated across various copies of the same cell condition to obtain reliable results. This is done to determine whether changes are statistically significant. An automated method must be used to process the massive amounts of data needed for this study and to produce measurements that are unaffected by image artifacts such as uneven lighting, dings or smudges on the well plate lid, as well as some variation in the width and shape of the original wound. One such example is the robust quantitative scratch assay algorithm. It employs MATLAB 7.11.0 to handle the issues provided by huge datasets and unpredictability in picture illuminations, as well as to construct a statistical output for enhanced assessment of cell motility (Vargas et al., 2016).

6.3 Network approach to epithelial tissue regeneration

Network analysis reveals the most distinct genes and proteins that are involved in the healing process. It also aids in the understanding of how they are organized into functional modules and distributed across gene ontology categories for biological processes, molecular activities, and cellular localization. Network analysis may be utilized to gain a better knowledge of the monitoring of critical mediators in wound healing and the regulating microRNAs by using algorithms and software programs (Arodz et al., 2013).

A network approach to epithelial tissue regeneration appears to be a difficult technique, linking numerous cells, cytokines, and various enzymes that engage in normal as well as pathologic regeneration process, assuming that this massive data array is well characterized. Network analysis is a novel approach that has the potential to improve our understanding of the molecular switches that govern normal and pathological tissue regeneration pathways (Wu et al., 2009).

Data-driven techniques and knowledge-based methods are the two primary kinds of methods with a high degree of detail. The primary difference is the function of the network in the analysis. Data-driven approaches are concerned with inference of network, where the outcome of data analysis is the network. The foundation of knowledge-based approaches is a massive network of linkages that represent the most recent biological information gleaned from books and databases. This past knowledge becomes the basis for building a focused network relevant to the results of the experiment being considered (Diegelmann, 2004). Experimental results are used in knowledge-based techniques as well, but they are evaluated in light of the knowledge from publically accessible databases. Knowledge-based techniques seek to identify the subset of connections that make up a subnetwork that differentiates between illness stages, phenotypes, or time periods rather than to create new links.

There are numerous online algorithms, databases, as well as other methods that can be used to create relevance networks, such as the PINA (protein interaction network analysis) (Wu et al., 2009), the Intact—the European Molecular Biology Laboratory—European Bioinformatics Institute (EMBL-EBI) protein interaction database (Kerrien et al., 2012), the Biological General Repository for Interaction Datasets (BioGRID) (Stark et al., 2011), and the Kyoto Encyclopedia of Genetics database (KEGG) (Kanehisa et al., 2012). Some of the top professional network analysis software packages, such as Ingenuity and Pathway Studio, include databases for scanning within their private software (Nikitin et al., 2003). Cytoscape is another popular software program for network analysis that provides a number of extra capabilities (plugins) and is freely accessible (Shannon et al., 2003).

The networks created from experimental or literature data may be further processed as a last step in the study to find common targets and regulators of the proteins, microRNAs, or genes that these nodes represent, or to include the nodes that are closest to the most significant nodes. By using this method, predictions for additional genes related to the disease of interest and its treatment may be generated, and new drug candidates can be discovered. All of this will bring medicine a great deal closer to tailored and effective treatments (Table 32.4) (N Shevtsova, 2018).

6.4 Modeling of angiogenesis

Angiogenesis is a complicated process in which existing microvessels sprout new capillaries capable of providing extra oxygen and nutrients to a developing, wounded, or inflammatory tissue. Angiogenesis is required for normal physiological activities such as tissue regeneration and embryonic development (Peirce, 2008). Assays that can trigger the crucial phases of angiogenesis as well as give a tool for analyzing the efficacy of therapeutic drugs are required for research of angiogenic processes (Rahman et al., 2020). Since the mid-1900s, a number of experimental models for studying both healthy and pathological angiogenesis have been established. Over the past two decades, mathematical and computer models have supplemented experimental techniques and improved our knowledge of this intricate process (N Shevtsova, 2018). Some notable examples include the following:

- Endothelial Cell Migration model developed by Stokes and Lauffenburger—This mathematical model of angiogenesis assumes that the actions of microvessel endothelial cells (MECs) are coordinated. It made use of a structured, probabilistic framework capable of modeling the growth of individual microvessels and the networks that

TABLE 32.4 Examples of popular databases currently employed in network analysis of Epithelial tissue regeneration.

S.No.	Name of databases	Application
1.	Protein interaction network analysis (PINAs)	Incorporates information about protein—protein interactions (PPIs) from six databases and offers tools for building, filtering, analyzing, and visualizing networks (Wu et al., 2009).
2.	Intact—the European Molecular Biology Laboratory—European Bio-informatics Institute (EMBL-EBI)	Houses a collection of open data tools and resources that encompass all molecular biology data types, such as protein sequences, chemical biology, nucleotide sequence data, and literature (Cantelli et al., 2021).
3.	Biological General Repository for Interaction Datasets (BioGRID)	It is a freely available database of genetic and physical interconnections for examining gene and protein activity and examining the characteristics of the entire network. More than 116,000 interactions including *Saccharomyces cerevisiae*, *Drosophila melanogaster*, and *Homo sapiens* have been included in the BioGRID release version 2.0 (Stark, 2006).
4.	Kyoto Encyclopedia of Genetics (KEGG)	It is a knowledge base that connects genomic data with higher level functional data to analyze genes in a systematic way. There are three KEGG databases—the GENES database (collection of gene catalogs), the PATHWAY database (visual representations of biological activities), and the LIGAND database (data on chemical compounds, enzyme molecules, and enzymatic processes) (Kanehisa, 2000).
5.	Cytoscape	It combines high-throughput expression data, various chemical states, and biomolecular interaction networks into a single conceptual framework. It is an open-source software project that provides the users with fundamental tools they need to build and analyze networks, to graphically combine them with phenotypes, expression profiles, and other molecular states, and to connect them to functional annotations' databases (Shannon et al., 2003).

resulted from them. It was created to look at how underlying MEC behaviors cohere during microvascular development. The model investigated how MEC motility properties such as as speed, persistence duration, and chemotactic responsiveness impact the pace of network expansion and the resulting network shape. This allowed for the investigation of a variety of issues about the functions of MEC migration and proliferation behavior (Stokes & Lauffenburger, 1991).

- Merks et al. (2006)—The researchers employed a model of human umbilical vein endothelial cells in in vitro conditions, in Matrigel, to distinguish between the intrinsic morphogenetic capacity of endothelial cells and its control by long-range guidance signals and other cell types. This endothelial cell culture model, which is virtually two-dimensional in nature, was successful in imitating the vasculogenesis that occurs in the yolk sac and other flat parts of the embryo. Using a cell-cantered computer model, they demonstrated that the elongated shape of endothelial cells is necessary for proper spatiotemporal in silico reproduction of stable vascular network expansion (Merks et al., 2006).

- Pettet et al. (1996) developed a technique, comprising of partial differential equations to simulate the development of macrophage-derived chemoattractants, new blood arteries, and capillary-tip endothelial cells during tissue healing. The developed model allowed for traveling wave solutions that display many characteristics of soft tissue wound healing. A dense band of small, tipped blood vessels formed close to the wound-healing unit's outermost edge and higher vessel density was associated with recently healed wounds before vascular remodeling, according to full-model numerical simulations (Fig. 32.4) (Pettet et al., 1996).

6.5 Translational systems biology

Translational systems biology (TSB) is the use of dynamic mathematical modeling and specific engineering concepts to biological systems to integrate mechanism and phenomena and, more crucially, to update clinical practice. TSB may offer a translational link between clinical contexts and data on gene regulation processes. Early tissue wound healing simulation

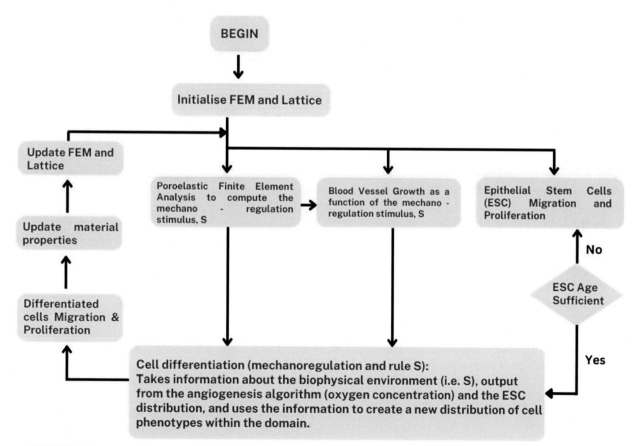

FIGURE 32.4   A schematic representation of how algorithms can be used to model tissue regeneration (Checa et al., 2010, pp. 423–435).

efforts concentrated on epithelial proliferation and migration. These models' scope were then broadened to incorporate the interplay between inflammation and healing. In injuries due to burning, if it is updated to represent the peculiarities of thermal/chemical/electrical harm, these sorts of computational models may give insight into attempts to impact the acute biology of the burn site, as well as project the following repercussions of such actions on future healing and scarring. The regeneration of epithelial tissue may greatly benefit from TSB simulations. Considering how lengthy the "real-world" process takes, being able to test prospective experimental treatments in silico at a much faster rate may greatly speed up our capacity to analyze these relationships (An et al., 2008).

6.6 Modeling for the short- and long-term mechanism guiding the healing of epithelial wounds

For the first two phases of epithelial regeneration, two-dimensional finite element simulations were built and run inside a monolithically totally implicit implementation using a mechanical continuum model. A nonlinear response to force of the tissue was combined with the attraction of actin fibers and myosin motors in the notion for the creation of actomyosin cables. The actomyosin cable's active deformation during the contraction stage was factored into consideration. For dynamic tissue contraction along the course of those fibers, the movement of the myosin motors in relation to the actin filaments was replicated. The quick discharge of calcium into the wound's area upon injury and the subsequent development of the actin ring, exactly mirroring what has been reported in biological literature, were both supported by computational mechanics. The dynamics of wound closure at the actin-ring contraction stage was then connected to the numerical model for actomyosin contraction, which was completely unified with the nonlinear dynamics of the problem. When it comes to simulating the layers of epithelial and embryonic cells, this approach was the first of its kind. Previously, a variety of complex mechanics had been mixed and addressed using engineering computer techniques (Roldán et al., 2019).

6.7 Social genomic models

Human social genomics is the study of the genetic effects of a person's social environment. The study of social genomics in humans and animals can help identify the distinctive genetic and molecular mechanisms that control social behavior, such as the proteins and genes involved in tissue regeneration. Social factors, such as modifications in gene expression and alterations in the local cutaneous environment, obstruct tissue regeneration through a number of different mechanisms. Reduced concentrations of antimicrobial peptides and growth factors, altered macrophage and/or neutrophil phenotype, and altered cytokine production are a few of these (Fayne et al., 2020).

Epigenetic changes such as microRNA expression, DNA methylation, and histone modifications can all be influenced by the social environment. On the other hand, genetic data can affect behavior. For instance, a fascinating clinical experiment examined the efficiency and financial viability of informing people about their individual genetic risk of melanoma to modify sun exposure (Smit et al., 2018). Contrarily, certain environmental settings can regulate a certain gene expression pattern.

One well-known example comes from studies on social isolation, which is one of the strongest epidemiologic risk factors for death and chronic disease, including impairment in tissue regeneration. According to studies, socially isolated people had upregulated gene transcripts for the inflammatory cascade but downregulated gene transcripts for innate antiviral reactions by type I interferon and antibody production (Cole, 2014). High levels of inflammation and poor antibody and antiviral response may increase the risk of slow tissue regeneration and wound infection from the standpoint of wound healing. Another illustration is how the early postnatal environment influences epigenetic changes that regulate neurotrophins such as brain-derived neurotrophic factor (BDNF) (Branchi et al., 2011). Through the activation of extracellular signal—related kinase 1/2 (ERK1/2), BDNF can promote the production of VEGF and affect how an injury is repaired (Fayne et al., 2020; Zhang et al., 2017).

The hypothalamic—pituitary—adrenal and the sympathetic—adrenal—medullary axis are both typically activated in response to psychosocial stress (Gouin & Kiecolt-Glaser, 2012). It is known that increased glucocorticoid and catecholamine synthesis affects the epithelium regeneration process. For instance, using a glucocorticoid receptor antagonist to limit glucocorticoid action eliminates the stress-related delay in tissue regeneration in stressed animals (Detillion et al., 2004). Additionally, adrenalectomy lessens how restraint stress affects tissue regeneration. The stress-related tissue repair delay in mice is reduced when α-adrenergic receptor antagonist is administered (Eijkelkamp et al., 2007; Gouin & Kiecolt-Glaser, 2012). Furthermore, within 24 h following a wound's commencement, women who report feeling more stressed had lower levels of IL-1 and IL-8 near the wound bed (Gouin & Kiecolt-Glaser, 2012). Additionally, following a marital quarrel as opposed to after a social support talk, lower levels of TNF-α IL-1, and IL-6 were discovered (Fayne et al., 2020; Kiecolt-Glaser et al., 2005).

Subsequently, it has been observed that other aspects of the human social environment, linked to genetic composition, such as nutritional status (Arnold & Barbul, 2006; Baltzis et al., 2014) and socioeconomic status (Sen et al., 2009), have an impact on tissue regeneration.

7. Conclusion

Exploring the biology epithelial stem cell and tissue regeneration has significant clinical benefits for human diseases' detection, prevention, and treatment, in addition to for regenerative medicine. This could revolutionize medicine. However, tissue regeneration is an elegant and complex process, comprising of various signaling networks. This creates complications in understanding the basics of individual processes. Also, the overwhelming amount of information generated from research generates a need of a data storage system, wherein data can be added and accessed from any part of the globe. Here, computational biology plays an essential role. It provides with different models and tools, which are specific for the processes and have various desirable feature. It also aids in data globalization for smooth access of information. At present, diverse tools and models have been developed for the same as discussed in this chapter. However, there is a scope of new and efficient models and tools with systematic and logical algorithms for better understanding and analysis of the physiology of epithelial tissue regeneration.

Acknowledgments

We would like to acknowledge Amity Institute of Biotechnology and Amity University, Noida campus, for providing us with facilities and infrastructure of working on this chapter. All the figures were created by the authors using Canva and Microsoft Word SmartArt tool.

Author contribution

Each author has contributed equally to the concept or design of the article; or the acquisition, analysis, or interpretation of data for the article and has approved the final version to be published.

References

Abraham, D. J., Shiwen, X., Black, C. M., Sa, S., Xu, Y., & Leask, A. (2000). Tumor necrosis factor α suppresses the induction of connective tissue growth factor by transforming growth factor-β in normal and scleroderma fibroblasts. *Journal of Biological Chemistry, 275*(20), 15220–15225. https://doi.org/10.1074/jbc.275.20.15220

An, G., Faeder, J., & Vodovotz, Y. (2008). Translational systems biology: Introduction of an engineering approach to the pathophysiology of the burn patient. *Journal of Burn Care and Research, 29*(2), 277–285. https://doi.org/10.1097/BCR.0b013e31816677c8

Arnold, M., & Barbul, A. (2006). Nutrition and wound healing. *Plastic and Reconstructive Surgery, 117*(Suppl. MENT), 42S–58S. https://doi.org/10.1097/01.prs.0000225432.17501.6c

Arodz, T., Bonchev, D., & Diegelmann, R. F. (2013). A network approach to wound healing. *Advances in Wound Care, 2*(9), 499–509. https://doi.org/10.1089/wound.2012.0386

Arrell, D. K., & Terzic, A. (2012). Systems proteomics for translational network medicine. *Circulation: Cardiovascular Genetics, 5*(4), 478. https://doi.org/10.1161/CIRCGENETICS.110.958991

Aslam, B., Basit, M., Nisar, N. M. A., Khurshid, M., & Rasool, M. H. (2016). Proteomics: technologies and their applications. *Journal of Chromatographic Science*, 1–15. https://doi.org/10.1093/chromsci/bmw167

Balli, M., Chui, J. S.-H., Athanasouli, P., Abreu de Oliveira, W. A., El Laithy, Y., Sampaolesi, M., & Lluis, F. (2019). Activator protein-1 transcriptional activity drives soluble micrograft-mediated cell migration and promotes the matrix remodeling machinery. *Stem Cells International, 2019*, 1–19. https://doi.org/10.1155/2019/6461580

Baltzis, D., Eleftheriadou, I., & Veves, A. (2014). Pathogenesis and treatment of impaired wound healing in diabetes mellitus: New insights. *Advances in Therapy, 31*(8), 817–836. https://doi.org/10.1007/s12325-014-0140-x

Baxi, V., Edwards, R., Montalto, M., & Saha, S. (2022). Digital pathology and artificial intelligence in translational medicine and clinical practice. *Modern Pathology, 35*(1), 23–32. https://doi.org/10.1038/s41379-021-00919-2

Bell, S. M., Zhang, L., Xu, Y., Besnard, V., Wert, S. E., Shroyer, N., & Whitsett, J. A. (2013). Kruppel-like factor 5 controls villus formation and initiation of cytodifferentiation in the embryonic intestinal epithelium. *Developmental Biology, 375*(2), 128–139. https://doi.org/10.1016/j.ydbio.2012.12.010

Bevilacqua, M. P., Pober, J. S., Wheeler, M. E., Cotran, R. S., & Gimbrone, M. A. (1985). Interleukin 1 acts on cultured human vascular endothelium to increase the adhesion of polymorphonuclear leukocytes, monocytes, and related leukocyte cell lines. *Journal of Clinical Investigation, 76*(5), 2003–2011. https://doi.org/10.1172/JCI112200

Bhaskaran, S., Umesh, U., & Nair, A. S. (2015). Hill equation in modeling transcriptional regulation. In *Systems and synthetic biology* (pp. 77−92). Springer Netherlands. https://doi.org/10.1007/978-94-017-9514-2_5

Bian, Q., & Cahan, P. (2016). Computational tools for stem cell biology. *Trends in Biotechnology, 34*(12), 993−1009. https://doi.org/10.1016/j.tibtech.2016.05.010

Boudra, R., & Ramsey, M. R. (2020). Understanding transcriptional networks regulating initiation of cutaneous wound healing. *Yale Journal of Biology & Medicine, 93*(1), 161−173.

Branchi, I., Karpova, N. N., D'Andrea, I., Castrén, E., & Alleva, E. (2011). Epigenetic modifications induced by early enrichment are associated with changes in timing of induction of BDNF expression. *Neuroscience Letters, 495*(3), 168−172. https://doi.org/10.1016/j.neulet.2011.03.038

Brandon, S., Cottreau, J., Oviedo, A., & Arnason, T. (2019). Ciliated columnar epithelium in the esophagus and gastroesophageal junction: A different perspective from study of a North American population. *Annals of Diagnostic Pathology, 41*, 90−95. https://doi.org/10.1016/j.anndiagpath.2019.05.008

Broughton, G., Janis, J. E., & Attinger, C. E. (2006). The basic science of wound healing. *Plastic and Reconstructive Surgery, 117*(7 Suppl. l), 12S−34S. https://doi.org/10.1097/01.prs.0000225430.42531.c2

Browaeys, R., Saelens, W., & Saeys, Y. (2020). NicheNet: Modeling intercellular communication by linking ligands to target genes. *Nature Methods, 17*(2), 159−162. https://doi.org/10.1038/s41592-019-0667-5

Buganza Tepole, A., & Kuhl, E. (2013). Systems-based approaches toward wound healing. *Pediatric Research, 73*(2−4), 553−563. https://doi.org/10.1038/pr.2013.3

Cantelli, G., Cochrane, G., Brooksbank, C., McDonagh, E., Flicek, P., McEntyre, J., Birney, E., & Apweiler, R. (2021). The European bioinformatics institute: Empowering cooperation in response to a global health crisis. *Nucleic Acids Research, 49*(D1), D29−D37. https://doi.org/10.1093/nar/gkaa1077

Checa, S., Byrne, D. P., & Prendergast, P. J. (2010). *Predictive modelling in mechanobiology: Combining algorithms for cell activities in response to physical stimuli using a lattice-modelling approach.* https://doi.org/10.1007/978-3-642-05241-5_22

Chhabra, P., & Bhati, K. (2022). Bionanomaterials: Advancements in wound healing and tissue regeneration. In *Recent advances in wound healing.* IntechOpen. https://doi.org/10.5772/intechopen.97298

Choudhary, M., Chhabra, P., & Amit, T. (2021). Scar free healing of full thickness diabetic wounds: A unique combination of silver nanoparticles as antimicrobial agent, calcium alginate nanoparticles as hemostatic agent, fresh blood as nutrient/growth factor supplier and chitosan as base matrix. *International Journal of Biological Macromolecules, 178*, 41−52. https://doi.org/10.1016/j.ijbiomac.2021.02.133

Cole, S. W. (2014). Human social genomics. *PLoS Genetics, 10*(8), e1004601. https://doi.org/10.1371/journal.pgen.1004601

Collins, C. D., Purohit, S., Podolsky, R. H., Zhao, H. S., Schatz, D., Eckenrode, S. E., Yang, P., Hopkins, D., Muir, A., Hoffman, M., McIndoe, R. A., Rewers, M., & She, J. X. (2006). The application of genomic and proteomic technologies in predictive, preventive and personalized medicine. *Vascular Pharmacology, 45*(5), 258−267. https://doi.org/10.1016/j.vph.2006.08.003

Collombet, S., van Oevelen, C., Sardina Ortega, J. L., Abou-Jaoudé, W., di Stefano, B., Thomas-Chollier, M., Graf, T., & Thieffry, D. (2017). Logical modeling of lymphoid and myeloid cell specification and transdifferentiation. *Proceedings of the National Academy of Sciences, 114*(23), 5792−5799. https://doi.org/10.1073/pnas.1610622114

Computational biology and bioinformatics. (2022). Springer Nature Limited.

Cui, M., Cheng, C., & Zhang, L. (2022). High-throughput proteomics: A methodological mini-review. *Laboratory Investigation, 102*(11), 1170−1181. https://doi.org/10.1038/s41374-022-00830-7

Dauer, D. J., Ferraro, B., Song, L., Yu, B., Mora, L., Buettner, R., Enkemann, S., Jove, R., & Haura, E. B. (2005). Stat3 regulates genes common to both wound healing and cancer. *Oncogene, 24*(21), 3397−3408. https://doi.org/10.1038/sj.onc.1208469

del Sol, A., & Jung, S. (2021). The importance of computational modeling in stem cell research. *Trends in Biotechnology, 39*(2), 126−136. https://doi.org/10.1016/j.tibtech.2020.07.006

Detillion, C. E., Craft, T. K. S., Glasper, E. R., Prendergast, B. J., & DeVries, A. C. (2004). Social facilitation of wound healing. *Psychoneuroendocrinology, 29*(8), 1004−1011. https://doi.org/10.1016/j.psyneuen.2003.10.003

Diegelmann, R. F. (2004). Wound healing: An overview of acute, fibrotic and delayed healing. *Frontiers in Bioscience, 9*(1−3), 283. https://doi.org/10.2741/1184

Donati, G., Rognoni, E., Hiratsuka, T., Liakath-Ali, K., Hoste, E., Kar, G., Kayikci, M., Russell, R., Kretzschmar, K., Mulder, K. W., Teichmann, S. A., & Watt, F. M. (2017). Wounding induces dedifferentiation of epidermal Gata6+ cells and acquisition of stem cell properties. *Nature Cell Biology, 19*(6), 603−613. https://doi.org/10.1038/ncb3532

Eijkelkamp, N., Engeland, C. G., Gajendrareddy, P. K., & Marucha, P. T. (2007). Restraint stress impairs early wound healing in mice via α-adrenergic but not β-adrenergic receptors. *Brain, Behavior, and Immunity, 21*(4), 409−412. https://doi.org/10.1016/j.bbi.2006.11.008

Elaine, N., & Marieb, R. N. (1995). Human anatomy and physiology. In *Redwood City, Calif* (3rd ed.). Benjamin/Cummings.

Elson, D. A., Thurston, G., Huang, L. E., Ginzinger, D. G., McDonald, D. M., Johnson, R. S., & Arbeit, J. M. (2001). Induction of hypervascularity without leakage or inflammation in transgenic mice overexpressing hypoxia-inducible factor-1α. *Genes & Development, 15*(19), 2520−2532. https://doi.org/10.1101/gad.914801

Faustino, R. S., Arrell, D. K., Folmes, C. D. L., Terzic, A., & Perez-Terzic, C. (2013). Stem cell systems informatics for advanced clinical biodiagnostics: Tracing molecular signatures from bench to bedside. *Croatian Medical Journal, 54*(4), 319−329. https://doi.org/10.3325/cmj.2013.54.319

Faustino, R. S., Chiriac, A., Niederlander, N. J., Nelson, T. J., Behfar, A., Mishra, P. K., Macura, S., Michalak, M., Terzic, A., & Perez-Terzic, C. (2010). Decoded calreticulin-deficient embryonic stem cell transcriptome resolves latent cardiophenotype. *Stem Cells, 28*(7), 1281–1291. https://doi.org/10.1002/stem.447

Fayne, R. A., Borda, L. J., Egger, A. N., & Tomic-Canic, M. (2020). The potential impact of social genomics on wound healing. *Advances in Wound Care, 9*(6), 325–331. https://doi.org/10.1089/wound.2019.1095

Fried, G. (1998). *Textbook of surgery. The biological basis of modern surgical practice* (15th ed., Vol. 41, p. 3). Philadelphia: W.B. Saunders Company. Harcourt Brace & Co. Canada, Ltd., Toronto.

Friedman, A. (2010). What is mathematical biology and how useful is iT? *Notices of the American Mathematical Society, 57*(7).

Gonzales, K. A. U., & Fuchs, E. (2017). Skin and its regenerative powers: An alliance between stem cells and their niche. *Developmental Cell, 43*(4), 387–401. https://doi.org/10.1016/j.devcel.2017.10.001

Gouin, J.-P., & Kiecolt-Glaser, J. K. (2012). The impact of psychological stress on wound healing. *Critical Care Nursing Clinics of North America, 24*(2), 201–213. https://doi.org/10.1016/j.ccell.2012.03.006

Grenander, U., & Miller, M. I. (1998). Computational anatomy: An emerging discipline. *Quarterly of Applied Mathematics, 56*(4), 617–694. https://doi.org/10.1090/qam/1668732

Grotendorst, G. R., Soma, Y., Takehara, K., & Charette, M. (1989). EGF and TGF-alpha are potent chemoattractants for endothelial cells and EGF-like peptides are present at sites of tissue regeneration. *Journal of Cellular Physiology, 139*(3), 617–623. https://doi.org/10.1002/jcp.1041390323

Grubbs, H., & Manna, B. (2022). *Wound physiology.*

Harmon, K. (2010). *Genome sequencing for the rest of us.* Scientific American, A Division of Springer Nature America, Inc.

Jimenez, P. A., & Rampy, M. A. (1999). Keratinocyte growth factor-2 accelerates wound healing in incisional wounds. *Journal of Surgical Research, 81*(2), 238–242. https://doi.org/10.1006/jsre.1998.5501

Jiménez-Santos, M. J., García-Martín, S., Fustero-Torre, C., di Domenico, T., Gómez-López, G., & Al-Shahrour, F. (2022). Bioinformatics roadmap for therapy selection in cancer genomics. *Molecular Oncology, 16*(21), 3881–3908. https://doi.org/10.1002/1878-0261.13286

Kanehisa, M. (2000). KEGG: Kyoto Encyclopedia of genes and genomes. *Nucleic Acids Research, 28*(1), 27–30. https://doi.org/10.1093/nar/28.1.27

Kanehisa, M., Goto, S., Sato, Y., Furumichi, M., & Tanabe, M. (2012). KEGG for integration and interpretation of large-scale molecular data sets. *Nucleic Acids Research, 40*(D1), D109–D114. https://doi.org/10.1093/nar/gkr988

Kelm, M., & Anger, F. (2022). Mucosa and microbiota – the role of intrinsic parameters on intestinal wound healing. *Frontiers in Surgery, 9.* https://doi.org/10.3389/fsurg.2022.905049

Kerrien, S., Aranda, B., Breuza, L., Bridge, A., Broackes-Carter, F., Chen, C., Duesbury, M., Dumousseau, M., Feuermann, M., Hinz, U., Jandrasits, C., Jimenez, R. C., Khadake, J., Mahadevan, U., Masson, P., Pedruzzi, I., Pfeiffenberger, E., Porras, P., Raghunath, A., ... Hermjakob, H. (2012). The IntAct molecular interaction database in 2012. *Nucleic Acids Research, 40*(D1), D841–D846. https://doi.org/10.1093/nar/gkr1088

Kiecolt-Glaser, J. K., Loving, T. J., Stowell, J. R., Malarkey, W. B., Lemeshow, S., Dickinson, S. L., & Glaser, R. (2005). Hostile marital interactions, proinflammatory cytokine production, and wound healing. *Archives of General Psychiatry, 62*(12), 1377. https://doi.org/10.1001/archpsyc.62.12.1377

Kitano, H. (2002). Computational systems biology. *Nature, 420*(6912), 206–210. https://doi.org/10.1038/nature01254

Kurkinen, M., Vaheri, A., Roberts, P. J., & Stenman, S. (1980). Sequential appearance of fibronectin and collagen in experimental granulation tissue. *Laboratory Investigation; a Journal of Technical Methods and Pathology, 43*(1), 47–51.

Kurn, H., & Daly, D. T. (2022). *Histology, epithelial cell.* https://pubmed.ncbi.nlm.nih.gov/32644489/.

Lawrence, W. T., & Diegelmann, R. F. (1994). Growth factors in wound healing. *Clinics in Dermatology, 12*(1), 157–169. https://doi.org/10.1016/0738-081X(94)90266-6

Lesk, A. M. (2017). *Introduction to genomics* (2nd ed.). Oxford University Press Inc.

Macalino, S. J. Y., Gosu, V., Hong, S., & Choi, S. (2015). Role of computer-aided drug design in modern drug discovery. *Archives of Pharmacal Research, 38*(9), 1686–1701. https://doi.org/10.1007/s12272-015-0640-5

Matsumoto, H., Kiryu, H., Furusawa, C., Ko, M. S. H., Ko, S. B. H., Gouda, N., Hayashi, T., & Nikaido, I. (2017). SCODE: An efficient regulatory network inference algorithm from single-cell RNA-seq during differentiation. *Bioinformatics, 33*(15), 2314–2321. https://doi.org/10.1093/bioinformatics/btx194

Maurizi, E., Adamo, D., Magrelli, F. M., Galaverni, G., Attico, E., Merra, A., Maffezzoni, M. B. R., Losi, L., Genna, V. G., Sceberras, V., & Pellegrini, G. (2021). Regenerative medicine of epithelia: Lessons from the past and future goals. *Frontiers in Bioengineering and Biotechnology, 9.* https://doi.org/10.3389/fbioe.2021.652214

Merks, R. M. H., Brodsky, S.v., Goligorksy, M. S., Newman, S. A., & Glazier, J. A. (2006). Cell elongation is key to in silico replication of in vitro vasculogenesis and subsequent remodeling. *Developmental Biology, 289*(1), 44–54. https://doi.org/10.1016/j.ydbio.2005.10.003

Miyazawa, K., & Miyazono, K. (2017). Regulation of TGF-β family signaling by inhibitory smads. *Cold Spring Harbor Perspectives in Biology, 9*(3), a022095. https://doi.org/10.1101/cshperspect.a022095

Moignard, V., Woodhouse, S., Haghverdi, L., Lilly, A. J., Tanaka, Y., Wilkinson, A. C., Buettner, F., Macaulay, I. C., Jawaid, W., Diamanti, E., Nishikawa, S.-I., Piterman, N., Kouskoff, V., Theis, F. J., Fisher, J., & Göttgens, B. (2015). Decoding the regulatory network of early blood development from single-cell gene expression measurements. *Nature Biotechnology, 33*(3), 269–276. https://doi.org/10.1038/nbt.3154

N Shevtsova, O. (2018). Wound healing management bioinformatics approach. *Biostatistics and Biometrics Open Access Journal, 7*(2). https://doi.org/10.19080/BBOAJ.2018.07.555709

Nair, A. (2007). *Computational biology & bioinformatics: A gentle overview.* Communications of the Computer Society of India.

Nakano, T. (1998). Intermediate epithelium. *Journal of Anatomy (Kaibogaku zasshi), 73*(2), 87−92.

Nguyen, D., Tao, L., & Li, Y. (2022). Integration of machine learning and coarse-grained molecular simulations for polymer materials: Physical understandings and molecular design. *Frontiers in Chemistry, 9.* https://doi.org/10.3389/fchem.2021.820417

Nikitin, A., Egorov, S., Daraselia, N., & Mazo, I. (2003). Pathway studio–the analysis and navigation of molecular networks. *Bioinformatics, 19*(16), 2155−2157. https://doi.org/10.1093/bioinformatics/btg290

Noble, D. (2002). The rise of computational biology. *Nature Reviews Molecular Cell Biology, 3*(6), 459−463. https://doi.org/10.1038/nrm810

Ocone, A., Haghverdi, L., Mueller, N. S., & Theis, F. J. (2015). Reconstructing gene regulatory dynamics from high-dimensional single-cell snapshot data. *Bioinformatics, 31*(12), i89−i96. https://doi.org/10.1093/bioinformatics/btv257

Paranjapye, A., NandyMazumdar, M., Browne, J. A., Leir, S.-H., & Harris, A. (2021). Krüppel-like factor 5 regulates wound repair and the innate immune response in human airway epithelial cells. *Journal of Biological Chemistry, 297*(2), 100932. https://doi.org/10.1016/j.jbc.2021.100932

Peirce, S. M. (2008). Computational and mathematical modeling of angiogenesis. *Microcirculation, 15*(8), 739−751. https://doi.org/10.1080/10739680802220331

Peng, T., Liu, L., MacLean, A. L., Wong, C. W., Zhao, W., & Nie, Q. (2017). A mathematical model of mechanotransduction reveals how mechanical memory regulates mesenchymal stem cell fate decisions. *BMC Systems Biology, 11*(1), 55. https://doi.org/10.1186/s12918-017-0429-x

Pettet, G. J., Byrne, H. M., Mcelwain, D. L. S., & Norbury, J. (1996). A model of wound-healing angiogenesis in soft tissue. *Mathematical Biosciences, 136*(1), 35−63. https://doi.org/10.1016/0025-5564(96)00044-2

Pohlman, T. H., Stanness, K. A., Beatty, P. G., Ochs, H. D., & Harlan, J. M. (1986). An endothelial cell surface factor(s) induced in vitro by lipopolysaccharide, interleukin 1, and tumor necrosis factor-alpha increases neutrophil adherence by a CDw18-dependent mechanism. *Journal of Immunology (Baltimore, Md. : 1950), 136*(12), 4548−4553.

Polo, J. M., Anderssen, E., Walsh, R. M., Schwarz, B. A., Nefzger, C. M., Lim, S. M., Borkent, M., Apostolou, E., Alaei, S., Cloutier, J., Bar-Nur, O., Cheloufi, S., Stadtfeld, M., Figueroa, M. E., Robinton, D., Natesan, S., Melnick, A., Zhu, J., Ramaswamy, S., & Hochedlinger, K. (2012). A molecular roadmap of reprogramming somatic cells into iPS cells. *Cell, 151*(7), 1617−1632. https://doi.org/10.1016/j.cell.2012.11.039

Rahman, H. S., Tan, B. L., Othman, H. H., Chartrand, M. S., Pathak, Y., Mohan, S., Abdullah, R., & Alitheen, N. B. (2020). An overview of in vitro, in vivo, and computational techniques for cancer-associated angiogenesis studies. *BioMed Research International, 2020,* 1−14. https://doi.org/10.1155/2020/8857428

Rajendran, N. K., Kumar, S. S. D., Houreld, N. N., & Abrahamse, H. (2018). A review on nanoparticle based treatment for wound healing. *Journal of Drug Delivery Science and Technology, 44,* 421−430. https://doi.org/10.1016/j.jddst.2018.01.009

Rashidi, H. H., Tran, N. K., Betts, E. V., Howell, L. P., & Green, R. (2019). Artificial intelligence and machine learning in pathology: The present landscape of supervised methods. *Academic Pathology, 6.* https://doi.org/10.1177/2374289519873088

Regan, M. C., Kirk, S. J., Wasserkrug, H. L., & Barbul, A. (1991). The wound environment as a regulator of fibroblast phenotype. *Journal of Surgical Research, 50*(5), 442−448. https://doi.org/10.1016/0022-4804(91)90022-E

Roldán, L., Muñoz, J. J., & Sáez, P. (2019). Computational modeling of epithelial wound healing: Short and long term chemo-mechanical mechanisms. *Computer Methods in Applied Mechanics and Engineering, 350,* 28−56. https://doi.org/10.1016/j.cma.2019.02.018

Sen, C. K., Gordillo, G. M., Roy, S., Kirsner, R., Lambert, L., Hunt, T. K., Gottrup, F., Gurtner, G. C., & Longaker, M. T. (2009). Human skin wounds: A major and snowballing threat to public health and the economy. *Wound Repair and Regeneration, 17*(6), 763−771. https://doi.org/10.1111/j.1524-475X.2009.00543.x

Shannon, P., Markiel, A., Ozier, O., Baliga, N. S., Wang, J. T., Ramage, D., Amin, N., Schwikowski, B., & Ideker, T. (2003a). Cytoscape: A software environment for integrated models of biomolecular interaction networks. *Genome Research, 13*(11), 2498−2504. https://doi.org/10.1101/gr.1239303

Sharma, S., Ray, B., & Mahapatra, S. K. (2022). Genomic and proteomic: Their tools and application. *Asian Journal of Research in Biosciences, 4*(1), 48−61.

Sherratt, J. A., & Murray, J. D. (1990). Models of epidermal wound healing. *Proceedings. Biological Sciences, 241*(1300), 29−36. https://doi.org/10.1098/rspb.1990.0061

Silva, F. de S., Almeida, P. N., Rettore, J. V. P., Maranduba, C. P., Souza, C. M. de, Souza, G. T. de, Zanette, R. de S. S., Miyagi, S. P. H., Santos, M. de O., Marques, M. M., & Maranduba, C. M. da C. (2012). Toward personalized cell therapies by using stem cells: Seven relevant topics for safety and success in stem cell therapy. *Journal of Biomedicine and Biotechnology, 2012,* 1−12. https://doi.org/10.1155/2012/758102

Smit, A. K., Newson, A. J., Morton, R. L., Kimlin, M., Keogh, L., Law, M. H., Kirk, J., Dobbinson, S., Kanetsky, P. A., Fenton, G., Allen, M., Butow, P., Dunlop, K., Trevena, L., Lo, S., Savard, J., Dawkins, H., Wordsworth, S., Jenkins, M., ... Cust, A. E. (2018). The melanoma genomics managing your risk study: A protocol for a randomized controlled trial evaluating the impact of personal genomic risk information on skin cancer prevention behaviors. *Contemporary Clinical Trials, 70,* 106−116. https://doi.org/10.1016/j.cct.2018.05.014

Smola, H., Thiekötter, G., & Fusenig, N. (1993). Mutual induction of growth factor gene expression by epidermal-dermal cell interaction. *Journal of Cell Biology, 122*(2), 417−429. https://doi.org/10.1083/jcb.122.2.417

Stark, C. (2006). BioGRID: A general repository for interaction datasets. *Nucleic Acids Research, 34*(90001), D535−D539. https://doi.org/10.1093/nar/gkj109

Stark, C., Breitkreutz, B.-J., Chatr-aryamontri, A., Boucher, L., Oughtred, R., Livstone, M. S., Nixon, J., van Auken, K., Wang, X., Shi, X., Reguly, T., Rust, J. M., Winter, A., Dolinski, K., & Tyers, M. (2011). The BioGRID interaction database: 2011 update. *Nucleic Acids Research, 39*(Database), D698−D704. https://doi.org/10.1093/nar/gkq1116

Stokes, C. L., & Lauffenburger, D. A. (1991). Analysis of the roles of microvessel endothelial cell random motility and chemotaxis in angiogenesis. *Journal of Theoretical Biology, 152*(3), 377−403. https://doi.org/10.1016/S0022-5193(05)80201-2

Tavakoli, S., Kisiel, M. A., Biedermann, T., & Klar, A. S. (2022). Immunomodulation of skin repair: Cell-based therapeutic strategies for skin replacement (A comprehensive review). *Biomedicines, 10*(1), 118. https://doi.org/10.3390/biomedicines10010118

Torras, N., García-Díaz, M., Fernández-Majada, V., & Martínez, E. (2018). Mimicking epithelial tissues in three-dimensional cell culture models. *Frontiers in Bioengineering and Biotechnology, 6.* https://doi.org/10.3389/fbioe.2018.00197

Vargas, A., Angeli, M., Pastrello, C., McQuaid, R., Li, H., Jurisicova, A., & Jurisica, I. (2016). Robust quantitative scratch assay. *Bioinformatics, 32*(9), 1439−1440. https://doi.org/10.1093/bioinformatics/btv746

Vodovotz, Y. (2010). Translational systems biology of inflammation and healing. *Wound Repair and Regeneration, 18*(1), 3−7. https://doi.org/10.1111/j.1524-475X.2009.00566.x

Wilhelm, M., Schlegl, J., Hahne, H., Gholami, A. M., Lieberenz, M., Savitski, M. M., Ziegler, E., Butzmann, L., Gessulat, S., Marx, H., Mathieson, T., Lemeer, S., Schnatbaum, K., Reimer, U., Wenschuh, H., Mollenhauer, M., Slotta-Huspenina, J., Boese, J.-H., Bantscheff, M., ... Kuster, B. (2014). Mass-spectrometry-based draft of the human proteome. *Nature, 509*(7502), 582−587. https://doi.org/10.1038/nature13319

Willett, J. D. (2002). Genomics, proteomics: what's next? *Pharmacogenomics, 3*(6), 727−728. https://doi.org/10.1517/14622416.3.6.727

Wilson, K. E. (2004). Functional genomics and proteomics: Application in neurosciences. *Journal of Neurology, Neurosurgery & Psychiatry, 75*(4), 529−538. https://doi.org/10.1136/jnnp.2003.026260

Witte, M. B., & Barbul, A. (1997). General principles of wound healing. *Surgical Clinics of North America, 77*(3), 509−528. https://doi.org/10.1016/S0039-6109(05)70566-1

Wu, J., Vallenius, T., Ovaska, K., Westermarck, J., Mäkelä, T. P., & Hautaniemi, S. (2009). Integrated network analysis platform for protein-protein interactions. *Nature Methods, 6*(1), 75−77. https://doi.org/10.1038/nmeth.1282

Xia, Y.-P., Zhao, Y., Marcus, J., Jimenez, P. A., Ruben, S. M., Moore, P. A., Khan, F., & Mustoe, T. A. (1999). Effects of keratinocyte growth factor-2 (KGF-2) on wound healing in an ischaemia-impaired rabbit ear model and on scar formation. *The Journal of Pathology, 188*(4), 431−438. https://doi.org/10.1002/(SICI)1096-9896(199908)188:4<431::AID-PATH362>3.0.CO;2-B

Xue, M., & Jackson, C. J. (2015). Extracellular matrix reorganization during wound healing and its impact on abnormal scarring. *Advances in Wound Care, 4*(3), 119−136. https://doi.org/10.1089/wound.2013.0485

Yin, X., Li, Q., McNutt, P. M., & Zhang, Y. (2022). Urine-derived stem cells for epithelial tissues reconstruction and wound healing. *Pharmaceutics, 14*(8), 1669. https://doi.org/10.3390/pharmaceutics14081669

Zhang, Z., Zhang, Y., Zhou, Z., Shi, H., Qiu, X., Xiong, J., & Chen, Y. (2017). BDNF regulates the expression and secretion of VEGF from osteoblasts via the TrkB/ERK1/2 signaling pathway during fracture healing. *Molecular Medicine Reports, 15*(3), 1362−1367. https://doi.org/10.3892/mmr.2017.6110

Zhao, R., Liang, H., Clarke, E., Jackson, C., & Xue, M. (2016). Inflammation in chronic wounds. *International Journal of Molecular Sciences, 17*(12), 2085. https://doi.org/10.3390/ijms17122085

Ziraldo, C., Mi, Q., An, G., & Vodovotz, Y. (2013). Computational modeling of inflammation and wound healing. *Advances in Wound Care, 2*(9), 527−537. https://doi.org/10.1089/wound.2012.0416

Author index

Index

Note: 'Page numbers followed by "f" indicate figures and "t" indicate tables'.

Printed in the United States
by Baker & Taylor Publisher Services